Mechanical Engineering Series

For further volumes:
http://www.springer.com/series/1161

The Mechanical Engineering Series features graduate texts and research monographs to address the need for information in contemporary mechanical engineering, including areas of concentration of applied mechanics, biomechanics, computational mechanics, dynamical systems and control, energetics, mechanics of materials, processing, production systems, thermal science, and tribology.

Giancarlo Genta · Lorenzo Morello
Francesco Cavallino · Luigi Filtri

The Motor Car

Past, Present and Future

Springer

Giancarlo Genta
Lorenzo Morello
Francesco Cavallino
Luigi Filtri
Department of Mechanical
and Aerospace Engineering
Politecnico di Torino
Torino
Italy

ISSN 0941-5122 ISSN 2192-063X (electronic)
ISBN 978-94-007-8551-9 ISBN 978-94-007-8552-6 (eBook)
DOI 10.1007/978-94-007-8552-6
Springer Dordrecht Heidelberg New York London

Library of Congress Control Number: 2013955051

Printed on acid-free paper

Springer is part of Springer Science+Business Media (www.springer.com)

Foreword

Motor vehicles are complex machines and are likely to become more and more complex in the future. The requirements their designers must satisfy are increasingly demanding and it is possible to state that designing a motor car is one of the most difficult tasks engineers must face. It is true that there are machines that are more complex, must operate in a more hostile environment or must satisfy more demanding requirements (just to mention two examples, a nuclear submarine or a space shuttle, but the list could be much longer), but what makes the design of a modern car so difficult is the conflicting nature of the design requirements and the complex nature of its development and production processes.

A car must supply the required performance in a scenario of changing and more sophisticated expectation from the customer, satisfying increasingly strict safety, environmental and energetic standards, at a tag price that is appealing to the customer, involving acceptable maintenance costs with consistent performance while being ready to perform its tasks for most of the time, remaining safe even if misused and not maintained properly (within reasonable limits). All this in a machine that can be built in large numbers, whose design can survive for a number of years to reasonable changes in the overall scenario in which cars are manufactured and in a market characterized by fierce competition.

In addition, the disciplines involved in designing a car include not only mechanical and thermal engineering, as obvious, but also styling, a kind of creative design, marketing, a social science studying the customer's needs, and ergonomics, dealing with a wide variety of user interfaces. On the purely manufacturing side, the large volume of numerous models requires a multiplicity of dedicated machines and tightly structured assembly processes, both of which continue to undergo intensive study by manufacturing engineers.

This situation has become increasingly difficult in the recent years, starting from the 1970s when new problems, regarding safety, environmental issues, fuel consumption and market fragmentation emerged, or at least became much more compelling, driving all manufacturers to introduce innovations that deeply changed both the product and the way it is designed and manufactured.

In spite of all the forecasts of an end of what has been called 'the motor car era', motor vehicles are still one of the typical features of our age and our society owes to their widespread use many of its characteristics. In particular, the widespread use and ownership of small passenger vehicles gives ordinary people a freedom of movement that was earlier unheard of, but also causes a number of problems, ranging from road accidents to pollution, from traffic jams with related economic losses to the need for large supplies of liquid, mainly fossil fuels.

The growing diffusion of cars in the twentieth century caused the growth of the mechanical industry to the point that it was a common opinion that the health of an industrial economy could be assessed from how well the automotive industry was faring. Although many voices denouncing the dangers connected with this situation and announcing the end of the ‘motor car era’ were raised, in particularly starting from the end of the 1960s, the situation has not changed, and it is easily predictable that individual mobility will be based on privately owned motor cars well into the twenty-first century and that the automotive industry will remain one of the foundations of the economy of developed countries.

The automotive industry proved to be able to innovate and to adapt itself to new needs and requirements in an ever-changing scenario and the motor car of the first decade of the twenty-first century is deeply different from the motor car of the 1960s and 1970s. The automotive industry is based on large-scale production, and as such it has an inertia that prevents changes to be sudden and swift, but this has not prevented changes to be pursued and implemented, sometimes even at an unpredictable rate.

These are the reasons why the authors feel that a book about basics of automobile design and production, as well as the automotive market, would still be interesting reading and a useful source of information for students and for the general public.

Self-propelled ground vehicles represented a fairly recent achievement in the history of technology; except for a few precursors that had little effect on technology or found practical applications, they did not really appear until quite late in the industrial revolution. It was only in the nineteenth century that working models of self-propelled vehicles could be built and they became truly practical only at the end of the century.

As usual with any technological development, the development of motor vehicles, particularly in its early phases, can be seen from two conflicting viewpoints: one that recognizes discontinuities and one that emphasizes the slow evolution of ideas and designs. In the first case a number of heroic figures of inventors and their revolutionary contributions to the technology are described. Each new step is presented as the result of an original idea, often defined in a patent, and of the work of an enterprising individual, often one who had to fight against influential people with opposing views.

The other approach stresses the fact that the development of new technologies proceeds usually in small steps, by accumulation of infinitesimal changes, and that in most cases new ideas and solutions are attempted several times, often with small differences, before being incorporated into the mainstream development of a product. The patents themselves, as the results of the work of inventors, deal mostly with small changes and usually the parts of the patents that find practical application are those dealing with small improvements of previous ideas.

In this view, it is possible to speak of evolution in the technological field; indeed since Charles Darwin published his book *On the Origins of Species by Means of Natural Selection* in 1859 a parallel is often drawn between biological evolution and the evolution of technology.

Some historians of technology, like George Basalla,[1] support an evolutionary theory of technological change based on the idea that this is more than just an analogy. Technological objects have no genes and cannot procreate, thus cannot transmit their genes to their offspring, but require human intervention in the design of new machines, carrying over from the old ones most of their features and introducing the few changes. In this sense, technological evolution is more like the artificial selection depended upon by farmers and stock-breeders than like natural selection. After all, the term natural selection used by Darwin was suggested by his observation of artificial selection.

One of the main strong points of this view is the wide diversification that occurs when a new technology is developing. This is also the case of automotive technology and it is amazing to observe the large variety of technical solutions used to accomplish an assigned function in early cars, in particular if compared with the almost total standardization in current mechanical products. This proliferation is often so large that many technical solutions that are thought to be recent inventions were in reality attempted in old products, only to be eventually discarded. This may have been due to several reasons, such as the lack of adequate materials, of analytical techniques allowing development to an operational stage, or of constructional techniques allowing implementation in a cost-effective and technologically satisfactory way. In some cases they were even abandoned just because another alternative appeared to be better without really detailed studies. When studying the evolution of motor vehicles in the last two centuries, we realize that almost all architectures presently in use for car components were already conceived during the first years of car history and then abandoned for problems met in their development.

This diversification followed by the selection of a few configurations is typical of biological evolution.

A point that introduces an essential difference between biological and technological evolution is the role that chance has in it. From the beginning of his studies, Darwin contended that evolution did not imply finality and that changes occur at random; causality enters the game only in the subsequent stage of selection. This appears to be quite different from what goes on in technological changes: the common opinion is that new artifacts are invented (or developed, to stress more continuity than revolutionary changes) under the pressure of needs, from the simplest biological ones such as food, shelter, and defense, to the more sophisticated needs arising in developed societies.

This opinion is however likely to be wrong, at least in most cases. Technology is mostly not developed as a direct response to human needs; the push as a direct

[1] G. Basalla, The evolution of Technology, Cambridge University Press, Cambridge, 1988.

response to human needs and the push toward changes are more linked to irrational factors than to a pondered evaluation of human needs and of the tools suited to satisfy them. There are a few well-known examples of it, but the best one is that of the laser, which was defined at the beginning as an 'invention without an application'. After lasers were developed a large number of applications were found and, if now the mentioned definition seems to be absurd, it was not at the time this invention was produced.

In the specific field of automotive vehicles, in the second half of the nineteenth century little need was felt for a self-propelled road transportation vehicle: most people knew how to ride horses and there were many stables in cities and along country roads inns with livery could be found almost everywhere. Upper class homes had spacious courtyards and stables for horses and carriages. Moreover, early cars were not up to the task of being usable and reliable means of transportation and offered a comfort, reliability, and safety well below the standards of horse-driven carriages. In this situation motor cars were big-boys' toys for a small number of wealthy sportsmen who were ready to face the discomfort and the dangers the new machine presented as a part of the excitement and the fun it offered. It was only later, and earlier in the United States than in Europe, that a more utilitarian use of motor vehicles started and that the very existence of the new product created a need for it. The trend of replacement of horse-driven vehicles by cars can be inferred from Fig. 1. In the United States until 1915 there were more horses than cars, and it was only after World War Two that the motor car became the predominant road transportation means.

Often the inventor of a technological project has an idea of its possible uses and applications having little to do with the actual applications that in time will prevail, and there are many cases in which the invention is produced just for the sake of it, with no idea of what applications may be.

This randomness makes it particularly difficult to predict future developments in any field of technology. If predicting the future is always a dangerous exercise, to do so in the technological field is even worse. An example, that can be amusing, is what is said in the mentioned book by G. Basalla[2] regarding personal computers. The author states that already in the mid-1980s it was clear that personal computers were a short-lived (and costly for the manufacturers) fad and that those who bought them, not knowing what to do with a computer and having no need for it, ended in using them just for videogames, an activity that soon lost all its allure. The conclusion (drawn in 1988!) was that personal computers had been an utter failure, that caused the bankrupt of their manufacturers.

Other examples are the many wrong forecasts of the past (supersonic air transportation, thinking machines, and Moon and Mars colonies within the year 2000, etc.) and the failure to predict technological products that quickly changed our everyday life (Internet, cell phones, etc.). This is not to say that the technologies that have not entered our lives will never do so: in particular we likely will

---

[2] G. Basalla, *op. cit.*, p. 260 in the Italian Edition.

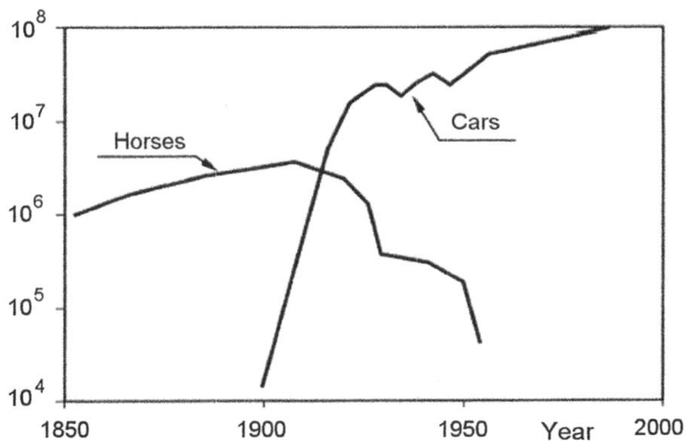

Fig. 1 Estimation of the number of horses used for transportation and cars in the United States from year 1850 to year 2000 (from: J.H. Ausubel, Cities and Their Vital Systems: Infrastructure—Past, Present and Future, Wiley, New York, 1988)

have in the future supersonic air transportation, Moon and Mars colonies and perhaps even thinking machines, although surely not in the forms and within the times of past predictions, but to stress how little reliability lies in all the technological forecasts we so often make.

The same can be said for the future of automotive technology: in the 1950s and 1960s flying cars were expected to materialize in a few decades and the application of gas turbines to cars and industrial vehicles seemed to be the future of road transportation. Later electric vehicles, drive-by-wire technologies, and many other applications of ICT seemed to be at hand, only to be seen as technological dreams in a shift toward a more indeterminate future.

But in spite of this difficulty in forecasting the paths that will be taken by the evolution of automotive vehicles in the future, we need to have visions and long-term projects to inspire research and the development of new ideas and prototypes. The last part of this book must be read in this context: it is not meant to make technological predictions that are likely to be wrong, but to discuss the trends of automotive research and innovation and to see the possible paths that may be taken to solve the many problems that are at present open or that we can expect for the future.

Turin, December 2013

Francesco Cavallino
Luigi Filtri
Giancarlo Genta
Lorenzo Morello

References

J.H. Ausubel, *Cities and their Vital Systems: Infrastructure—Past, Present and Future* (Wiley, New York, 1988)

G. Basalla, *The Evolution of Technology* (Cambridge University Press, Cambridge, 1988)

Acknowledgments

The authors are pleased to express their gratitude to all the persons who made this book possible, with their help, their encouragement, and their suggestions.

In particular, Paolo Scolari who had for many years been in charge as Design and Development Vice-president at Fiat Automobiles. When Fiat and the Politecnico di Torino decided to cooperate in founding a course in Automotive Engineering, Paolo Scolari participated in the design activities, giving a wide personal contribution. In particular, he insisted on including a course on the historical development of motor vehicles, a course that was later entitled "Motor vehicles and their evolution." It was a first-year course, having the aim of introducing freshmen to the world of the automotive industry, through lectures and visits to production plants and design and test facilities.

Paolo Scolari was in charge of teaching this course from its institution in 2001, till his untimely passing away in 2008. He collected much material and wrote notes for the students, which were distributed in paper form and then as a CD. The authors, who were his colleagues in their earlier jobs in the automotive industry, had the pleasure to work with him in the preparation of this course and in writing the notes which were the starting point for this book.

Nevio Di Giusto, the former Managing Director of Fiat Research Center. Since the beginning of his involvement in the Automotive Engineering course, he was particularly aware of the importance of this teaching in generating new human resources for the future of the automobile and granted the authors all his cooperation with suggestions and updated illustration material.

Donatella Biffignandi, in charge of the Information Center of the Automobile Museum of Torino, and Maurizio Torchio, in charge of the Fiat Historical Center. They supplied an essential support in the preparation and illustration of the historical section.

All those who granted their permission to reproduce figures and drawings. The authors made every effort to seek permission from the original copyright holders of the figures, and apologize if there are cases where they were not able to achieve this objective. This is particularly true for some figures that come from the mentioned notes prepared for the students and for those taken from the Web.

Contents

Acronyms

4WD	4-Wheels Drive
4WDS	4-Wheels Drive and Steering
ABS	Anti-lock Braking System (from German: Antiblockiersystem)
AC	Alternating Current
ACEA	European Automobile Manufacturers' Association
AD	Analog to Digital
AFC	Alkaline Fuel Cells
AFR	Air-to-Fuel Ratio
ALV	Autonomous Land Vehicles
ANFIA	Associazione Nazionale Filiera Industria Automobilistica
AR	Aspect Ratio (tires)
ARC	Active Roll Control
ASR	Anti Spin Regulator
ATV	All-Terrain Vehicle
BAS	Brake Assist System
BDC	Bottom Dead Center
BEP	Break-Even Point
BER	Block Exemption Regulation
BEV	Battery Electric Vehicle
BOOM	Binocular Omni-Orientation Monitor
BRIC	Brasil, Russia, India, China
CA	Crank Angle
CAC	Charge Air Cooler
CAD	Computer Aided Design, Computer Aided Drafting
CAFE	Corporate Average Fuel Economy
CAE	Computer Aided Engineering
CAM	Computer Aided Machining
CAN	Controlled Area Network
CAS	Computer Aided Styling
CFD	Computational Fluid Dynamics

CI	Compression Ignition
CNG	Compressed Natural Gas
COP	Conformity Of Production
CRP	Carbon Reinforced Plastics
CRS	Common Rail System
CVS	Constant Volume Sampling
CVT	Continuously Variable Transmission
CFV	Critical Flow Venturi
CVVT	Continuous Variable Valve Timing
DARPA	Defence Advanced Research Projects Agency
DC	Direct Current
DMFC	Direct Methanol Fuel Cells
DMU	Digital Mock-Up
DNPR	Diesel NO_x Particulate Reduction
DOC	Diesel Oxidation Catalyst
DOHC	Double Over-Head Camshaft
DPF	Diesel Particulate Filter
DSC	Dynamics Stability Control
DSRC	Dedicated Short Range Communications
EBD	Electronic Brake Distributor
ECD	Exhaust Closing Delay
ECU	Electronic Control Unit
EGO	Exhaust Gas Oxygen
EGR	Exhaust Gas Recirculation
EIVC	Early Intake Valve Closing
EMC	Electro Magnetic Compatibility
EOA	Exhaust Opening Advance
EOBD	European On-Board Diagnostic
EPA	Environmental Protection Agency
EPS	Electric Power Steering
ESP	Enhanced Stability Program
ESV	Experimental Safety Vehicle
EUDC	Extra-Urban Driving Cycle
EVO	Exhaust Valve Opening
FC	Fuel Consumption; Fuel Cell
FEM	Finite Element Method
FWD	Front-Wheels Drive
GDI	Gasoline Direct Injection
GPS	Global Positioning System
GRP	Gross National Product, Glass Reinforced Plastics

GVW	Gross Vehicle Weight
HC	(unburned) Hydro Carbons
HCCI	Homogeneous Compression Combustion Ignition
HMD	Head Mounted Display
ICD	Intake Closing Delay
ICE	Internal Combustion Engine
ICT	Information and Communication Technology
ID	Ignition Delay
IOA	Intake Opening Advance
IPCC	Intergovernmental Panel on Climate Change
IRS	Injection Rate Shaping
ISO	International Standards Organization
ITS	Intelligent Transportation Systems
KERS	Kinetic Energy Recovery Storage
LDWS	Lane Departure Warning System
LIVO	Late Intake Valve Opening
LNT	Lean NO_x Trap
LPG	Liquefied Petroleum Gas
LRV	Lunar Roving Vehicle
MBT	Minimum spark advance for the Best Torque
MCFC	Molten Carbonate Fuel Cells
MPI	Multi Point Injection
MPV	Multi Purpose Vehicle
MUV	Multi Utility Vehicle
NAFTA	North American Free Trade Area
NCAP	New Car Assessment Program
NEDC	New European Driving Cycle
NHTSA	National Highway Traffic Safety Agency
NMHC	Non Methane Hydrocarbons
NSC	NO_x Storage Catalyst
NVH	Noise, Vibration and Harshness
OBD	On-Board Diagnostic
OEM	Original Equipment Manufacturer
OES	Original Equipment Supplier
ODE	Ordinary Differential Equation
OHC	Over-Head Camshaft
OHV	Over-Head Valves
OICA	Organisation Internationale des Constructeurs d'Automobiles
PAFC	Phosphoric Acid Fuel Cells
PAH	Polynuclear Aromatic Hydrocarbons

PAV	Personal Air Vehicle
PEMFC	Proton Exchange Membrane Fuel Cells
PFI	Port Fuel Injection
PLM	Product Lifecycle Management
PM	Particulate Matter
PRP	Product Range Plan
RON	Research Octane Number
RWD	Rear-Wheels Drive
SA	Spark Advance
SAE	Society of Automotive Engineers
SAHR	Self-Aligning Head Restraint
SCR	Selective Catalytic Reduction
SHED	Sealed Housing for Evaporative Determination
SI	Spark Ignition, Sistème Internationale
SMPI	Sequential MultiPoint Injection
SOF	Soluble Organic Fraction
SOFC	Solid Oxide Fuel Cells
SUV	Sport Utility Vehicle
TCS	Traction Control System
TDC	Top Dead Center
THC	Total Hydrocarbons
UDC	Urban Driving Cycle
UHEGO	Universal Heated Exhaust Gas Oxygen
VDC	Vehicle Dynamic Control
VGT	Variable Geometry Turbocharger
VII	Vehicle Infrastructure Integration
VNT	Variable Nozzle Turbine
VOC	Volatile Organic Compounds, Voice of Customer
VR	Virtual Reality
VSC	Vehicle Stability Control
VTT	Variable Twin Turbocharger
VVA	Variable Valve Actuation
VVT	Variable Valve Timing
V2I	Vehicle to Infrastructure
V2V	Vehicle to Vehicle
WHO	World Health Organization
WOT	Wide Open Throttle
ZEV	Zero Emission Vehicle

Part I
Past

Chapter 1
Introduction to Part I

For hundreds of thousands of years human beings lived without using any particular means of transportation. When they had to move an object, they simply lifted and carried it, if they were strong enough. If the object was too heavy, they arranged to drag it. It is likely that occasionally branches or other round objects were slipped under the load to reduce friction, but no evidence of this practice remains.

With the Neolithic revolution the need for transportation greatly increased at the same time that the practice of taming animals opened new perspectives. The development of agriculture created the need of transporting seeds to the field and crops back to the homestead. The number of objects that were considered important and necessary for humans to carry with them increased as a result of the new needs of village life.

Sleighs were used in Northern Europe before 5000 B.C., and their use in other places at that time can be inferred. Sleighs and sledges can actually be used for transportation not only on snow and ice but also on grassland (American Indians used the *travois* well into the nineteenth century), deserts and sometimes even on rock.

It is impossible to state when a sledge was mounted for the first time on a pair of wheels or who instigated this technical revolution. Ancient wheels were made primarily of wood, so that little direct archeological evidence would remain.

The potter's wheel was introduced about 3500 B.C. to produce pots with axial symmetry. The use of the potter's wheel can be inferred from the marks left on pots made with it. The supporting wheel for vehicles is thought to have originated at about the same time.

The most ancient evidence of a wheeled vehicle is a pictogram on a tablet from the Inanna temple in Erech, Mesopotamia. This document dates back to slightly later than 3500 B.C., and includes a small sketch of a cart with four wheels, together with a sketch of a sledge (Fig. 1.1a).

The vehicle shown in Fig. 1.1b has two features typical of all vehicles for more than a 1000 years: The wheels are discs made from three planks of wood, and the animals are harnessed to a central shaft.

G. Genta et al., *The Motor Car*, Mechanical Engineering Series,
DOI: 10.1007/978-94-007-8552-6_1,

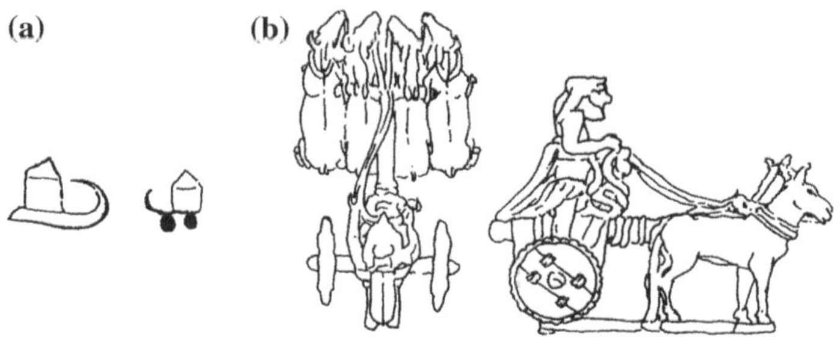

Fig. 1.1 **a** Pictogram on a tablet from the Inanna temple in Erech, Mesopotamia. The document dates back to slightly later than 3500 B.C., and includes a small sketch of a cart with four wheels, together with that of a sledge. **b** Copper model of a war chariot, driven by four onagers, found in the tomb of Tell Agrab, from the third millennium B.C. (from Genta and Morello 2009)

This uniformity of wheel types and driving systems, particularly when compared with the great variety of vehicle structures, has led to the opinion that the wheel was *invented*, or better, developed, in a specific place and then started a slow diffusion throughout the ancient world. In various places where the new vehicle was introduced, the local type of sleigh was adapted to it using the standard wheels and harness.

Where wheeled vehicles were first developed is not known, but it can be inferred that it was in Southern Mesopotamia, where the wheel was definitely used about 3500 B.C. The spread of the wheel was quite slow. Evidence of its use dates from 3000 B.C. in Elam and Assyria, 2500 B.C. in Central Asia and the Indus Valley, 2250 B.C. in Northern Mesopotamia, 2000 B.C. in Southern Russia and Crete, 1800 B.C. in Anatolia, 1600 B.C. in Egypt and Palestine, 1500 B.C. in Greece and Georgia, 1300 B.C. in China and about 1000 B.C. in Northern Italy. Some centuries later it reached Northern Europe.

It is impossible to know from ancient pictures whether the axle turned along with the wheels or was stationary. The fact that the central hole of the wheel disc was round has little meaning as a circular hole can also be explained by the ease of construction. It is likely that both solutions were used, as it is the case with people who use these primitive technologies today (Fig. 1.2).

It is, however, likely that the wheel did not derive from the roller: The types of wheels used would rule that out, and it is likely that, in the mind of the ancient wheel maker, the wheel and the roller had little in common.

As previously noted, only after animals were tamed could wheeled vehicles be propelled in a proper way. In Mesopotamia both transportation vehicles and war chariots were pulled by onagers. Oxen were doubtless used for transportation as well.

The spoked wheel, which appeared about 2000 B.C., accompanied the use of horses to drive war chariots. It is not known where horses were first tamed and used for that purpose, but the scarce archeological evidence indicates that it likely happened in north-east Persia, and that from that region the use of horses spread throughout the ancient world, from China to Egypt and Europe.

Fig. 1.2 Cart axle with two wheels dating back to the mid-twentieth century. Except for the iron tire, the structure with a disc made of three planks is the same as that of prehistorical wheels. The axle turns together with the wheels (picture taken in Urchisar, Turkey, in 2011)

Fig. 1.3 Egyptian war chariot found in a tomb near Tebe (fifteenth century B.C., from Genta and Morello 2009)

The structure of an Egyptian chariot of the fifteenth century B.C. is shown in Fig. 1.3. It represents without any doubt the best state of the art of its times, one that remained unchanged for centuries.

The progress from the Sumerian vehicle shown in Fig. 1.1b is great, and if the greater power of the two horses compared with that of the onagers is considered, it is easy to understand why some historians ascribed to the use of this weapon the expansion of the Hittites in Anatolia, the Achaei in Greece and the Hyksos, who in the eighteenth century B.C. invaded Egypt, teaching the new technology to the Egyptians.

Chariots became obsolete as military vehicles when the knowledge of riding became widespread. Donkeys were used as pack animals and for human transportation in the third millennium B.C. and horses were surely used in the same way, but

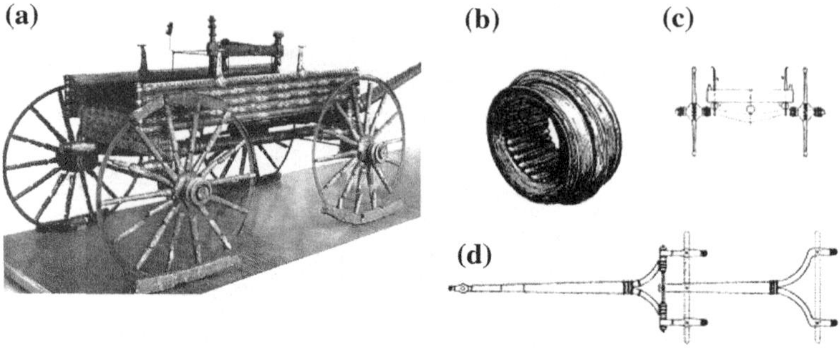

Fig. 1.4 Celtic carriage from the first century B.C. found near Dejbjerg: **a** picture, **b** wooden roller bearings in the wheel hub, **c** *cross section*, **d** *top view* (from Genta and Morello 2009)

only at the beginning of the first millennium B.C. the art of riding was developed enough and mounted warriors started to substitute chariots in the conflicts in the Middle East, North Africa and Europe. By the fourth century B.C. the armored cavalry introduced by Alexander the Great had a winning hand on most battlefields.

In Europe only the Celtic tribes continued to use war chariots, which were first carried north of the Alps by the Etruscans. Celtic wheelwrights learned the art of building wheeled vehicles and made significant progress.

The remains of the wagon found at Dejbjerg are shown in Fig. 1.4a. It is the first example of a wagon with steering on the front axle, but it can be considered as an articulated vehicle made by two chariots. It is, however, unlikely that this solution was actually used for transportation; it looks more an insulated example for ceremonial (burial) purposes. At any rate it incorporated other interesting features, such as the wooden roller bearings in the hubs.

When the military pressure was over, the progress of vehicles on wheels slowed down. Greeks and Romans used chariots only for ceremonies or races, while transportation on wheeled vehicles was hampered by the lack of suitable roads. The transportation vehicles of roman times, both in the form of two wheeled carts (the *birota*, with a payload of about 60 kg), or wagons (carrying up to 500 kg), were no more advanced than those of more ancient times, and above all wagons had neither steering nor suspensions. The Assyrian wagons of the third millennium B.C. remained practically unchanged for thousands of years.

The increase of commerce and the new needs of personal mobility which took place in the last centuries of the Middle Ages continued at a greater pace in the following period. A new confidence in progress and in the possibilities of technology led to try new experiments and to refine existing machines. It is believed that the steering of the front axle of four-wheeled wagons dates back to that period, even if some scholars think that this important innovation is older. The first coaches with the body suspended on belts or chains date from the beginning of the fifteenth century, but the innovation did not spread out fast. In 1665 steel springs were introduced for the suspension of vehicles, but only at the beginning of nineteenth century this practice became general and it was possible to suspend the heavier vehicles on steel springs.

At any rate, in spite of all the progress in the means of transportation, journeys on land were still made on foot or on horseback (more often on the back of donkeys or mules). Still in 1550, in Paris only five coaches were reported, including that used by the queen.

The widespread use of prime movers like the water wheel and the windmill lead some visionaries to forecast the construction of self-propelled wheeled vehicles, even if there was no practical means to implement this idea. Drawing of wind wagon, i.e. wagons carrying a windmill that operates the wheels through a number of gear wheels, were common in the 'theaters of machines' i.e. books containing the sketches of a number of actual and fantastic machines. Likely the first of such sketches was drawn by Guido da Vigevano in 1328; others can be found in the books by Roberto Valturio and by Mariano Jacopo, said Taccola, from the first half of the fifteenth century. These machines were never built, or at least, could not work as expected.

The use of sails could yield better results, and in 1599 Simon Stevin built a sail wagon for Maurice of Orange. It could carry 28 persons, and is said to have reached a speed of 12 km/h. It did not have however practical applications. Other attempts were made using clockwork mechanisms, and the German clock-maker Hans Hautsch built a wagon powered by steel springs that was said to be able to travel at about 1.5 km/h, but its range is not known.

In these applications the wheel had not only the usual task of supporting the vehicle, but also that of producing the traction needed for motion. A wagon with a fully developed steering and driving mechanism can be found in some drawings by Francesco di Giorgio Martini, who worked in the second half of the fifteenth century. The vehicle has four driving and steering wheels, powered by devices similar to the ones used in capstans, operated by humans or animals. It is then not an automotive vehicle, but incorporates devices that will be later essential to build self propelled vehicles.

In order to build a truly successful automotive vehicle it was necessary to wait until a viable thermal engine, with a power/mass ratio high enough to operate a vehicle able to transport the engine plus a certain payload, was available. This idea was around since at least the seventeenth century: For instance in 1681 the missionary F. Verbist reported that a steam vehicle (likely a reduced scale model) had been built in China and a sketch in a work of Isaac Newton shows a vehicle propelled by the reaction of a steam jet. It is, however, likely that the drawing is only a sketch to show the reaction principle and that Newton never thought seriously to that application.

In spite of these early ideas, it was only at the end of the eighteenth century that a suitable thermal engine was available, but a further century had to pass before the early attempts could produce vehicles that could find practical applications.

As stated in the preface, the early motor vehicles were built following a wide variety of technical solutions for almost all the functions that an automotive vehicle had to incorporate. About 30 years were needed for a few standard layouts to emerge and for the motor car to reach its maturity. For example, if we consider the function of transmitting the motion from the engine to the wheels, that will be described in details in a specific section, cars of the end of the nineteenth century used:

- Belt transmissions.
- Cylindrical gear gearboxes.
- Bevel gears gearboxes.
- Continuously variable transmissions based on friction wheels.

In the 1920s the variety of gearboxes virtually reduced to the cylindrical gears type, with dog clutches or sliding trains, all of them with a friction clutch as a start-up device; in the 1940s the first synchronizers appeared, developed by Mercedes, Porsche and Borg Warner.

Presently, differences in gearboxes are virtually inexistent, if we neglect differences due to the different powertrain lay-out. We can thus conclude that a manual gearbox now is a mature product.

The same cannot be said for automatic gearboxes, at least for Europe, where such a component is at its initial development stage. The attention of the market to purchase and running costs and the interest of customers in the fun of manually shifting gears worked against their diffusion till some years ago. The present situation of congested traffic is shifting major shares of manual transmissions in favour of automatic transmissions, significantly improved by the application of electronics. A different environment has shifted the balance between costs and benefits.

Present European automatic transmissions still show many different solutions:

- Planetary gears with multi-disc clutches of many different families.
- Conventional gears with a clutch for each gear.
- Conventional gears with a clutch for shaft (dual clutch gearbox).
- Continuously variable gearboxes with variable geometry.

In a near future some of them could prevail and bring the synchromesh manual gearbox to extinction. Many other similar examples will be shown in the next sections addressed to body system, chassis systems and powertrains.

At the end of the 1920s the general architecture of cars was standardized but the 1929 crisis marked a setback also for the automotive industry, particularly in the United States.

After World War Two the car industry restarted with a new pace, and the aim of a car for each family seemed reasonable, first in the United States and then in Europe. This goal was reached and, in the more developed countries, many families had soon more than one car. Technological development went on at a lower pace and by the middle of the 1960s the automotive industry seemed mature and little innovation seemed to be in sight.

As often happens, when a technology seems to be mature, new requirements and new problems forced the automotive industry to change deeply. The first problem that had to be tackled was safety. A new consciousness of the social cost of car accidents and new laws extending the responsibility of manufacturers forced to look at safety issues in a different way. New definitions of active and passive safety and new regulations forced deep changes in the design of all car subsystems, particularly of the chassis and the body, and new vehicles, defined Experimental Safety Vehicles (ESV), were presented at all car shows. While ESVs soon faded away, and rightly since they were true monsters, massive and fuel-hungry, the work done on safety issues lead in a few years to vehicles which were much safer than those built in the past.

Before the research work on safety could produce any effect, a new issue came out, and the focus shifted on the pollution caused by the use of motor vehicles, particularly in urban environment. New regulations on emissions, more and more restrictive, were, and still are being, enforced in all countries.

While the automotive industries were dealing with the two mentioned problems, safety and pollution, a third problem, namely the sudden increase of the cost of oil, came out. The new imperative was soon to decrease fuel consumption, an imperative often in contrast with the attempts to satisfy the other two mentioned requirements—it is enough to mention the high fuel consumption of the ESVs.

Under the pressing requirements caused by the changed scenario in which the automotive industry had to operate, a technology that was assumed to be mature had to innovate deeply. By the end of the millennium deep changes had reached the average customer, and the long wave of change is far from being over. For sure the new design methods allowed by the introduction of computers first in the design and then in the production of vehicles helped. Computer simulation and virtual prototyping allowed solution at the design stage of many problems that before could be tackled only by costly and tiresome experimentation and automation allowed to produce motor vehicles of increasing complexity, as required by the new needs, at costs still compatible with the possibilities of consumers.

Computers and generally information technology did not only change the way cars are designed and built, but entered the vehicles themselves. In present motor vehicles many functions that before were performed by mechanical systems or by the driver are now entrusted to microprocessor-based control system, and this has changed, and likely will change in the future, the way motor vehicles operate and are used.

A chronology of the most important technical events that affected car evolution is presented as a closure to this introduction.

3500 B.C.: Introduction of the wheel for ground vehicles.
2000 B.C.: Introduction of the spoked wheel.
1328: First drawing of a wind wagon by Guido da Vigevano.
1595: First vehicle suspended on springs, described by Fausto Venanzio.
1599: Simon Stevin builds a sail wagon.
1630: First description of tractive wheels (operated by humans on board) for a vehicle (they had already been used on war engines).
1673: First idea about an internal combustion engine operated by an explosive charge according to the Dutch physicist C. Huygens.
1690: D. Papin develops an idea about a steam engine suitable to move a motor carriage.
1769: N. J. Cugnot builds up a steam carriage with a pressure vessel and two cylinders acting on the front single wheel axle.

1802: R. Trevithick builds his first steam coach, that opens the era of the steam coaches in England.
1803: R. Trevithick invents a steam engine on rails; an horizontal cylinder moves the rear axle with a rod and crank mechanism.
1807: I. De Rivaz builds a vehicle with an atmospheric internal combustion engine, working on hydrogen and electrical spark plug.
1817: Ackermann, English agent of the German inventor G. Längensberger, files a patent about steering of four wheels vehicles in cinematically correct conditions.
1822 : The first commercial service of steam coaches (London–Birmingham) is started.
1823 : First electric car made by Jacobi.
1824 : S. Carnot publishes his treatise on thermal engines where he defines an operating cycle and the related thermal efficiency.
1828 : F. Pecqueur invents and files a patent on a differential to be applied on a steam carriage.
1839 : C. Goodyear invents the vulcanization process, to improve the mechanical properties of rubber.
1845 : R. W. Thompson files a patent about a tire, to be applied to a horse carriage to reduce the traction force and improve comfort.
1846 : Thomas Hancock builds a solid rubber tire to be added to the steel tires of the wheels of coaches.
1856 : E. Barsanti and F. Matteucci file a patent about the first combustion engine working automatically.
1860 : E. Lenoir introduces a double acting gas internal combustion engine, without compression; it is the first to be produced industrially.
1862 : A. Beau De Rochas gives a description of a four cycle internal combustion engine with compression.
1865 : The Red flag law requiring the presence of a man walking in front of the motor vehicles with a red flag and a speed limit of 6.4 km/h is passed in England. It ended the era of the steam coaches.
1867 : N. A. Otto and E. Langen start production of an atmospheric internal combustion engine.
1878 : C. Jeantaud, a French car manufacturer, develops a link mechanism to obtain Ackermann conditions in practice.
1880 : A. De Bollée builds the steam cars Obeissante, Mancelle and Rapide; the last one, a 6 places car, is capable of a speed of 60 km/h.
1884 : The patents of N. A. Otto are cancelled, because of prior claims by A. Beau De Rochas. From this date internal combustion engines are free of any licence burden.
1884 : G. Daimler develops the first four stroke single cylinder internal combustion engine for vehicles, featuring 0.5 HP (0.35 kW).
1886 : K. Benz develops the first car driven by an internal combustion single cylinder engine featuring 0.8 HP (0.6 kW).
1886 : First Daimler car.
1887 : R. A. Bosch files a patent about electrical magneto ignition.
1888 : J. B. Dunlop files a patent about bicycle tires, but the patent is invalidated in 1890 because of Thompson's priorities.
1890 : Panhard and Levassor presents the first car with front engine and rear driving axle with an engine built under Daimler's licence.
1891 : Peugeot introduces his car, with Panhard and Levassor engine. Panhard and Levassor and Peugeot are first in the world to produce cars with industrial processes.
1892 : R. Diesel files a patent about a constant pressure internal combustion cycle.
1896 : E. and A. Michelin apply a pneumatic tire to a car.
1899 : L. Renault develops the first direct drive gearbox, a shaft transmission with universal joint and, immediately after, the first sedan car.
1899 : The French government enforces the first Highway code with a maximum speed of 30 km/h in the country and 20 km/h in the town; the driving license is also introduced.
1905 : De Dion-Bouton develops the first single disc friction clutch.
1910 : Isotta-Fraschini introduces the first four-wheel braking system.

1912 : Cadillac introduces the first electric starter and the breaker ignition according to the ideas of C. Kettering.
1920 : Artz in Germany develops the first stamping press for steel sheets.
1921 : Duesemberg develops the first hydraulic braking system with Lockheed components.
1922 : Lancia develops the first unitized body.
1927 : Budd and Dodge develop the first electrical spot welding process.
1927 : Bosch develops the first high pressure diesel injection pump.
1928 : Packard applies the first synchromesh gearbox.
1933 : Lancia develops the first unitized body for a sedan car.
1936 : Mercedes introduces the first car with a prechamber diesel engine.
1939 : Packard applies the first air conditioning system.
1947 : Goodrich invents the tubeless tire.
1948 : Buick applies the first automatic transmission with planetary gears and torque converter.
1949 : Michelin introduces the first radial tire.
1952 : General Motors applies the first hydraulic power steering system.
1954 : Mercedes develops the first direct gasoline injection engine for a car.
1954 : Ford and Chevrolet apply the first pneumatic power brake system.
1955 : Citroën applies the first disk brake to a car.
1959 : Volvo applies the first three point safety belt.
1960 : The first emission regulations are presented by Senator E. Muskie in the USA.
1960 : Bosch develops the first oxygen sensor.
1960: Bosch introduces the first diesel rotary pump.
1962 : Ford introduces the first rack-and-pinion steering box.
1967 : Bosch introduces the first electronic gasoline injection system.
1973 : Engelhard develops the first three way catalist for exhaust gas aftertreatment.
1974 : Bosch introduces the first air flow meter for gasoline injection systems.
1978 : Bosch introduces the first ABS system.
1988 : FIAT applies the first direct injection diesel engine to a car.
1990 : Pioneeer introduces the first navigation system based on GPS technology system.
1995 : Bosch introduces the first vehicle dynamic control electronic system system.
1997 : FIAT introduces the first diesel common rail injection system in a car engine.

Chapter 2
Body and Car Architecture

The structure of early cars consisted in a chassis and a body.

The chassis was the main structural frame on which all mechanical components, such as the engine, transmission, suspensions and steering system, were mounted. It was stressed by relevant reaction forces due to all these concentrated loads. The body was the container defining space for passengers and luggage; in early cars it was mounted on the chassis like the other mechanical components while, in modern cars, it is integrated with the chassis, contributing in this way to structural strength and stiffness of the vehicle.

The word chassis, deriving from the French language, has in English a double meaning: it means either the assembly of the structural element of the vehicle or the assembly of the mentioned structural elements plus the mechanical parts that provide to the vehicle motion (suspensions, wheels, steering, brakes, etc.).

In its latter meaning, the chassis is therefore a vehicle subassembly that can be actually available at a given stage of the assembling process, as happened until about the end of the 1940s for most cars, or can exist only virtually, as is the case for more recent vehicles. Nowadays the chassis includes the same components as at the beginning of automotive technology, but it cannot be disassembled as a whole entity from the vehicle.

At the dawn of the motor car era, the innovators who dedicated their intellectual and financial efforts to the development of this new product, concentrated on the components that were more directly involved in the motion of the vehicle, such as engine, transmission, suspensions, steering and braking systems. They became therefore designers and manufacturers of chassis.

At that time body technology was not considered to be fundamental and was directly taken from horse driven carriages and coaches, without significant modifications; many coach builders became body builders, as the first manufacturers of car bodies were called. This situation determined a significant separation between these two industrial branches.

Car manufacturers, or—better—chassis manufacturers, applied mainly metallic materials and were equipped with machines for metal casting, stamping and machin-

G. Genta et al., *The Motor Car*, Mechanical Engineering Series,
DOI: 10.1007/978-94-007-8552-6_2,

ing; because of the precision required, they built parts according to drawings, with tools suitable for batch production.

Body manufacturers applied instead wooden structures or composite structures of steel and wood and were equipped with machinery for working wood or wrought iron components, with tools suitable for single specimens, often copying models, without the help of drawings. This tradition came from the fact that wood was much easier to work than metal and could generate curved shapes, as the current fashion required.

Varnishing, necessary for protecting surfaces from the environment and for giving an acceptable appearance, was another point in favor of wood, taking into consideration the existing oil varnishes. The complete separation of the chassis from the body assembling process was also justified by the need to prevent varnish from being damaged by mechanical components during the complex assembling process.

Several hundreds working hours were required for completing the painting of a car body, only to apply the many layers of varnish; to this figure the drying time of each layer must be added to obtain the total time required for the painting cycle.

The availability of a stand-alone chassis was therefore justified by the existing technology and industrial organization. The chassis manufacturer could thus ship a finished product to the final customer or to a body builder.

A chassis could, in fact, be driven, tested and sent elsewhere; in addition, it could demonstrate its performance to customers also without a body. The temporary test body, shown in Fig. 2.4, had precisely this purpose.

In many cases, the final customer bought a chassis, which was subsequently shipped to the body builder for further works, under a different contract. Nevertheless there were examples of body builders that bought chassis on their own, to sell complete cars, or of car manufacturers that had their own body production capability.

This situation was common until the beginning of World War One.

The needs of mass production encouraged many car manufacturers to develop their own body design and production capability, under pressure to reduce costs. This favored a smooth transition from wood to steel technology, with hybrid wood-steel structures.

The introduction of synthetic enamels, in addition, shortened the painting time by an order of magnitude and favored a deeper integration between chassis and body assembling.

This new organization was developed mainly in the United States and imitated by the major manufacturers in Europe, starting from the beginning of the 1920s. About 20 years later, the first unitized bodies started to be applied in Europe.

The evolution of chassis and body can be described by considering three partially overlapping periods, including respectively the separable chassis, the partially integrated chassis and the unitized bodies.

A following section will be addressed to describe the exterior shape evolution and its influence on aerodynamic performance. The last section will supply a summary about car architecture evolution.

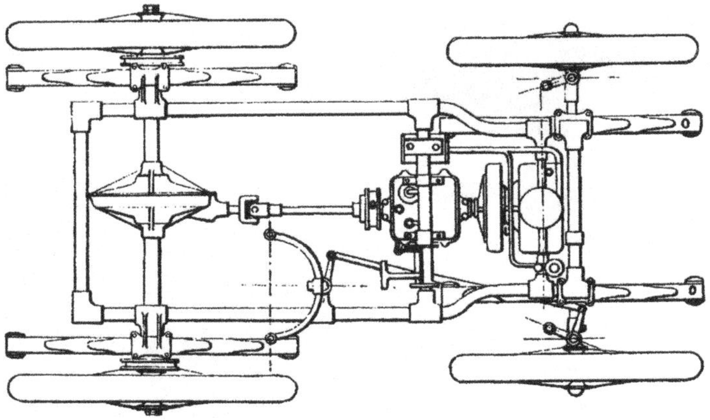

Fig. 2.1 The steel tubes chassis of the 1899 Renault Type A (Courtesy of Renault)

2.1 Separable Chassis

The chassis, in its structural meaning, was the bearing structure of a vehicle and had to support all mechanical components and the complete body that in this case was only a ballast without any particular structural function. In addition, the chassis had to offer an easy organization of the assembling process, being practically the mounting fixture for all mechanical components.

Many of the first car chassis were based on the existing technology developed for bicycles. The 1899 Renault shown in Fig. 2.1 offers an example of this solution. This kind of frame was made with pieces of seamless steel tubes that were cut and bent to their final shape; the different pieces were joined to iron shells with layers of brazing material and cooked in an oven; these iron joints are shown in the figure at the corners of the chassis and where cross beams are joined to the two longitudinal beams.

A second, completely different, solution was applied in the 1907 Sizaire and Naudin, shown in Fig. 2.2; this car is characterized by a chassis made with solid wooden beams, with iron reinforcements, although the body is covered by curved and bent steel panels. There is no apparent reason to explain this strange choice. We could argue that the manufacturer had a previous experience in wooden horse coaches chassis, while it was found easier to use steel panels for exteriors because they could avoid the use of wooden skeletons.

A third solution, used for instance by the Panhard & Levassor firm in its first cars, applied wooden beams covered and reinforced by a thin steel sheet layer nailed on them.

As soon as the weight of mechanical components increased for the application of the first four cylinder engines, the so called *grillage* structures were developed, as shown in Fig. 2.3, where an example of chassis of the beginning of the twentieth century is shown. The chassis was made by two longitudinal beams, made with

Fig. 2.2 The 1907 Sizaire and Naudin had a chassis made of wooden beams (National Automobile Museum of Torino)

Fig. 2.3 Typical grillage chassis of a car of the beginning of the twentieth century, made of stamped steel (from Morello et al. 2011)

stamped steel with a C cross section; the distance between the two longitudinal beams was reduced in the front part of the vehicle to make steering easier, while it was increased in the rear part to offer a more suitable interface with the body. These width variations were obtained either by bending longitudinal beams or by applying them according to a trapezoidal scheme.

The front and rear ends of the longitudinal beams were bent also in the vertical direction, to leave clearance for suspension motions and were tapered to take advantage of the decrease of bending moment in terms of weight reduction. These curvatures were increased in more recent cars, to lower the floor between the axles.

The two longitudinal beams were joined by bolts to a number of cross beams that built the grille structure; also cross beams were bent to allow space for the engine and gearbox. Cross beams were generally positioned where the most relevant loads, such as engine and gearbox, or the suspensions were applied.

Car bodies used with these chassis were quite similar in shape and technology to those of horse coaches. The body shapes in use were many, as shown in Fig. 2.4 where the shapes in use during the 1920s are listed; their names derived directly from the corresponding ones used for horse driven coaches.

Early cars were open, such as phaetons, spiders or torpedoes; only in 1899 Renault had the idea of introducing a covered car, a sedan with internal driving seat . Closed

Fig. 2.4 Typical car bodies from the beginning of the twentieth century: (1) Spider, (2) 4 seat coupè, (3) Double phaeton, (4) Coupè de ville, (5) Limousine, (6) Cabriolet; (7) 4 door saloon, (8) Test Body, (9) 2 door saloon, (10) Triple phaeton, (11) Race spider, (12) Brougham, (13) Spider with additional seat, (14) 2 seat coupè, (15) Milord, (16) Torpedo, (17) Limousine, (18) Cabriolet (from Morello et al. 2011)

Fig. 2.5 Coach during the fabrication process, before exterior panels were assembled (from Morello et al. 2011)

bodies represented 20 % of the market in 1920; this percentage grew up to 80 % in the following years maintaining a dominant position until now.

The body fabrication technology used on the early cars was the same as that used in coach building. Figure 2.5 shows a picture taken at the beginning of the twentieth century illustrating a coach during the fabrication process, before exterior panels were assembled. The complex skeleton is made by shaped beams, cut from solid wood boards and assembled using mortise and tenon joints; the picture shows the coach before assembling exterior panels.

These panels, like the interior ones, are made with thin plywood with pronounced curvature. The curvature of the panels is obtained by bending them in place after they had been weakened with a jet of steam; glue and nails allowed them to keep the final shape. This manufacturing technology allowed very rigid and light structures to be built with curved shapes complying with the current aesthetic rules.

The wooden skeleton had not only a structural function but it was the fixture used to model the skin to its final shape.

The drawings of these parts were quite basic, sometimes made directly on the wooden boards; if many equal parts had to be manufactured, the first of the batch was made with hard wood, such as mahogany, to be used afterwords as a template to draw the following samples.

Panels with double curvature could be made with planks nailed on formers and shaved to their final shape as in wooden boats. This kind of body style was named *bateau* after the French language; the phaeton in Fig. 2.4 can supply an example of this style. Each piece of the body had to be adapted in place to the neighboring elements; joints were hidden with plaster and varnish or with profiles, according to the different styles.

Fig. 2.6 The wooden beams of the skeleton were placed under the skin, where significant curvature changes were present and according to a scheme of ribs and formers (from Morello et al. 2011)

The wooden beams of the skeleton were placed under the skin where there were significant curvature changes, as shown in Fig. 2.6 and according to a scheme of ribs and formers, as in the same figure in the roof area. Ribs and formers were applied under the areas of higher curvature, such as the roof, the rear closure and the transition surface between the body sides and the hood. The roof cover was often made with a sheet of pegamoid, as it will be later explained for Weymann bodies.

The torsional stiffness of grillage chassis was quite low, roughly determined by the sum of the torsional stiffness of each of the two longitudinal beams, considering the limited contribution of cross beams. The torsional stiffness of each longitudinal beam was itself low, due to their open cross section. On the contrary, the torsional stiffness of a new wooden body was significantly higher, in consideration of the kind of joints used in carpentry, such as dovetail joints or mortise and tenon joints.

The significant deformations of the flexible chassis induced premature cracks and ruptures in the joints of the body, due to the higher stiffness of the latter. In fact, body and chassis are two flexible structures put in parallel subjected to equal displacements in their joints; the stiffer the element connected to the joints, the larger the stress.

Wood joints could not bear very high loads and increased their clearances: this produced annoying squeaks and rattles; in addition, a car parked on an uneven road could make door opening difficult.

Increasing the torsional stiffness of the chassis became an important issue for its effects on the body; the problem became more relevant with the increase of the speed cars could reach. The remedies to this problem followed two opposite strategies: To increase the body stiffness or to make its deformation less subject to squeaks and rattles.

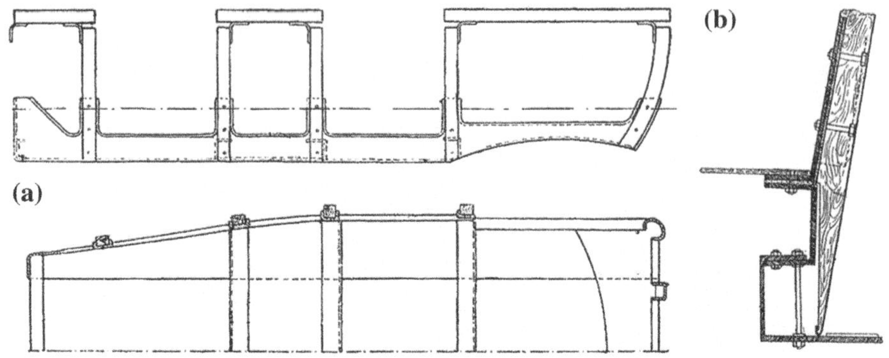

Fig. 2.7 Two opposite remedies to body squeaks and rattles: To make deformation less noisy **a** or to increase the body stiffness **b** (from Morello et al. 2011)

Figure 2.7 shows some details of these two strategies. In Fig. 2.7a the cross section of a body designed according to the first strategy is shown. There was a heavy auxiliary steel frame at the bottom of the wooden body; the frame had an L cross section and was joined to the wooden skeleton. The vertical wing of the L section was bolted to the struts of the skeleton, while the horizontal wing was interfaced with the chassis and supported the wooden boards of the floor. The wooden struts of the body skeleton reached the steel running board, covering the sides of the chassis.

While the body was made much stiffer than the original wooden structure, the joints with the chassis were made flexible with layers of rubber or fibers in between; with such a design the deformation of the flexible chassis was allowed by the joints without major involvement of the body.

An alternative strategy was developed by some body shops, among which Weymann, an airplane manufacturer , was the most famous; in fact this technology is usually associated to his name. He developed his body structure concept during the 1920s, deriving it from the aeronautical technology of that time used to make wings and fuselages.

The body skeleton shown in Fig. 2.7b was still wooden, with beams set at the rims of the body shape or according to a scheme with ribs and formers; the steel joining elements between beams were intentionally flexible and the wooden beams were mounted with significant gaps. With this architecture there was no danger of contacts between wooden parts and of consequent squeaks and rattles.

The body skin was made, as in airplanes, with a layer of cloth, covered with a waterproof varnish, having the appearance of leather and called pegamoid or leatherette. Hood covers and firewall were made of steel or aluminum, as in other bodies for reasons of resistance to heat. This kind of body manufacturing technology was particularly appreciated for sport cars because of the reduced weight.

The technologies explained above refer to the time period till the end of the 1920s; in the following decade intensive use of steel started to spread, making possible different design solutions.

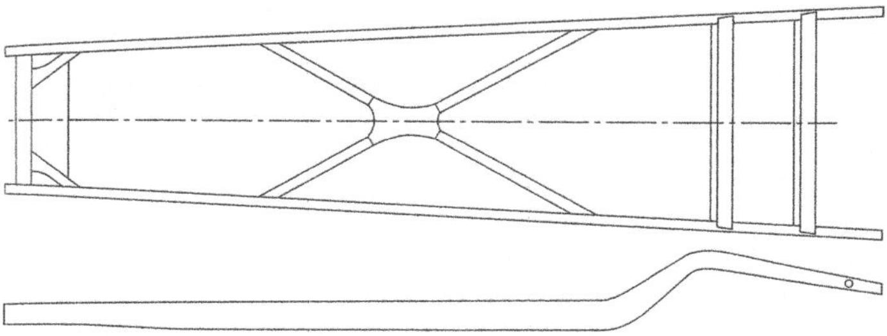

Fig. 2.8 Stamped steel chassis with an X cross beam (from Morello et al. 2011)

2.2 Partially Integrated Body and Chassis

In partially integrated architectures the body and the chassis are bolted together, with an increase in the bending and torsional stiffness of the car; the assembling cycle can remain the same as in previous case, with a complete body mounted on a complete chassis at the end of the assembling line.

Some of the chassis of the 1930s were developed according to a configuration different from grillage schemes; in particular the application of *X cross beams*, as shown in Fig. 2.8, introduced a significant improvement in the stiffness performance.

The X cross beam, that was usually riveted to the longitudinal beams, because of its shape, reacted to chassis torsion by bending; this is easily understood because the longitudinal beams are no longer parallel when twisted and this causes the X cross beams to bend. Since the flexural stiffness of cross beams is larger than their torsional stiffness, the chassis overall torsion stiffness is increased.

A further improvement to this architecture was given by a chassis with only one central longitudinal beam, having a closed cross section; this became feasible with the introduction of large stamping presses and of welding in the manufacturing shops. *Central beam* chassis have not only an advantage for the increased stiffness of closed cross section, but also for the possibility of reducing floor height with a consequent improvement in car weight and roll-over resistance. A chassis of this kind is shown in Fig. 2.9.

Steel stamping allowed any part of the chassis to be designed to withstand local stress, again with a better control of the weight; the only disadvantage was the impact on investment cost due to the larger amount of stamps and presses necessary to the manufacturing process.

Rear driven cars with the engine in the front had the transmission shaft installed inside of the central tubular beam. With older architectures, also cross beams had to be bent to clear the bulk of the transmission shaft.

This kind of stiffer chassis could now be bolted directly to the body, made mostly of steel, because the torsion stiffness of both parts was better balanced; the result so obtained was a good increase in the vehicle overall stiffness. Also large steel panels were now feasible thanks to the availability of large presses.

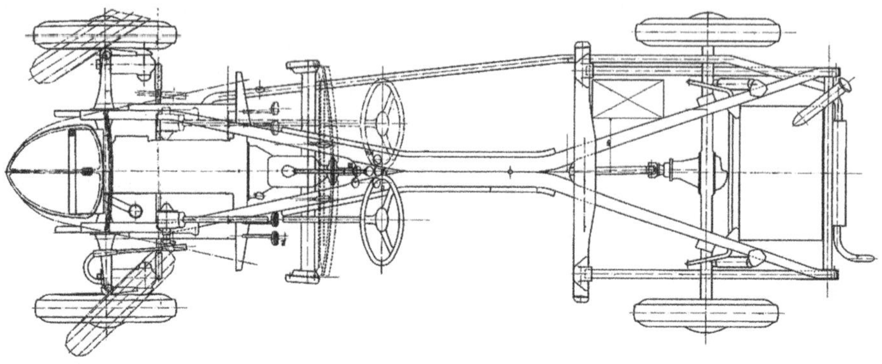

Fig. 2.9 Chassis with central tubular beam; all chassis members feature closed cross sections (Courtesy of FIAT Historical Archives)

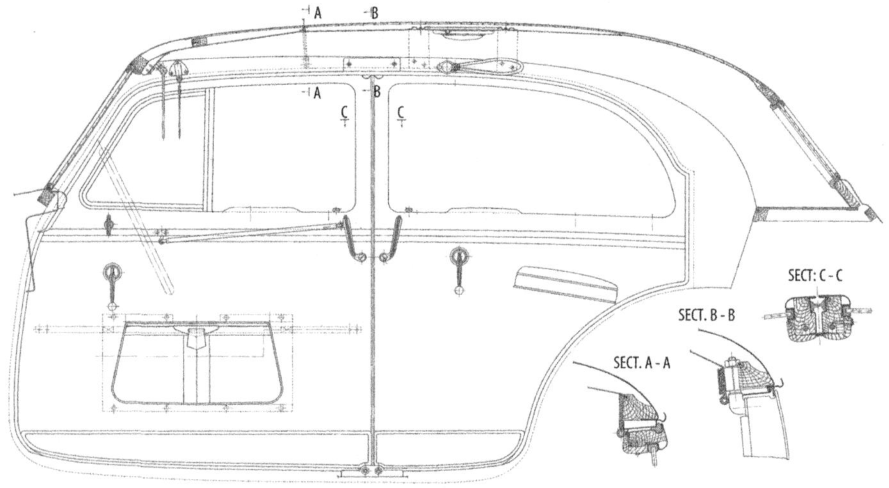

Fig. 2.10 Body structure made with wooden skeleton and steel panels. Notice the wooden elements in cross sections (from Morello et al. 2011)

Two alternatives were developed for partially integrated steel bodies. In the first the steel skin of the body was joined to a wooden skeleton with wood screws. The function of the wooden beams was double: to reduce the shape instability of the thin sheet panel and to offer an easy attachment for the internal trims and upholstery. The shape of the steel panels could be simple, being reduced to portions of the exterior shape, with small flanges at their contour to be bolted to the wooden skeleton. An example of this kind of body architecture is shown in Fig. 2.10a.

For aesthetic reasons, screws were set only where they could not be seen from the outside, i.e. at the external contour of doors and on door posts.

A second alternative, introduced in the same period of time, was characterized by a body skeleton made also of stamped steel sheets and more complicated flanges, at the contour of body panels. Also in this case, the body was bolted to the chassis.

The contemporary existence of these two architectures today might seem difficult to understand, but the explanation is again linked with cost and production volumes. The all-steel solution was typical of large production volumes, to absorb the impact of a larger expense in stamps and presses. At the same time, the low salaries of that time did not discourage the application of more complicated wood and steel structures, in case of small production volumes, because of the advantage of reducing investments.

The structural adequacy of the two alternatives was almost equivalent, with the advantage, in case of wood, of a better corrosion resistance; in fact, many 60 years old cars with a skeleton of ash-wood are still kept in good shape by collectors.

Till the end of the 1940s, luxury sedans, sport cars, station wagons and vans had this kind of body architecture; many fashionable station wagons of this time (called familiarly Woodies) had also part of the exterior panels made of natural wood.

This intermediate solution of partially integrated body and chassis may be justified by remembering that the manufacturing process was organized with chassis and body shops in different locations; a fact that delayed the introduction of unitized bodies. In addition, the non integrated solution was preferred because different bodies could be installed on the same chassis, with positive impact on investments, and wealthy customers could continue to obtain separate chassis for their custom-made bodies.

2.3 Unitized Bodies

The integration of body and chassis in a single unit was considered as the best chance to improve the structural performance, with reductions in weight and cost. In closed bodies, like those of sedans or limousines, the distance of the roof from the longitudinal beams allowed building of a very stiff structure, provided that these element were joined by other vertical structural elements (pillars) stiff enough and well fitted in the joining points. In addition, if the body skin panels were suitably stamped, they could further increase the stiffness of the skeleton by their resistance to shear.

Probably the first attempt to develop such an architecture was done with the 1922 Lancia Lambda; its structure without exterior skin, but complete with all mechanical components, is shown in Fig. 2.11. The body structure is made with steel panels about 2 mm thick, obtained by milling and bending; they are joined with a variety of joints, including rivets and continuous and spot welding.

The sides were made with reticular beams; many structural elements were set above the large sills (in fact, door dimensions are limited) and increase the overall structure stiffness, in comparison with traditional chassis.

The contribution to the overall stiffness of the sides is increased by very rigid cross diaphragms, corresponding to the dashboard, the seat back rests and the tail. The front suspension is connected to the body with a tubular portal structure, surrounding the

Fig. 2.11 The 1922 Lancia Lambda body structure integrating chassis function (National Automobile Museum of Torino)

Fig. 2.12 The body of the 1922 Lancia Lambda was very low over the ground, thanks to the integration of the chassis function (National Automobile Museum of Torino)

radiator, that contributes further to the stiffness. Welded panels covered the skeleton with further increase of bending and torsional stiffness even if the use of rolled and bent components prevented obtaining of very stiff closed cross sections.

Apart from the structure performance, the impact of this new architecture on the body volumes layout is immediately clear from Fig. 2.12: the body rides much lower

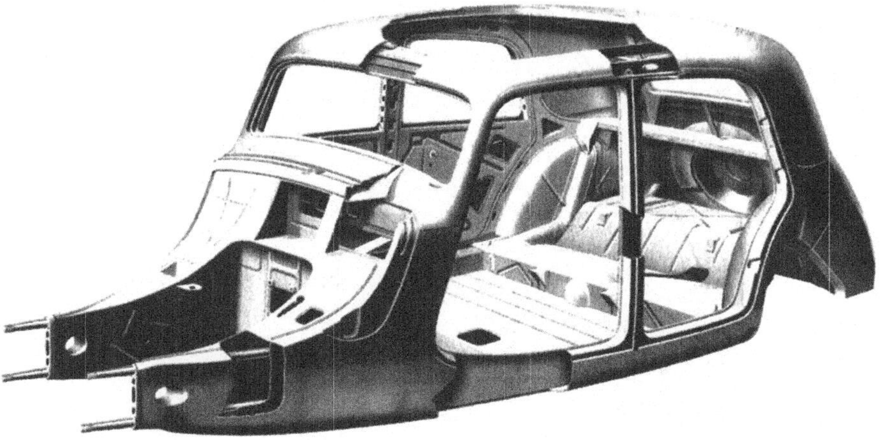

Fig. 2.13 Unitized body of the 1934 Citröen 11 CV (from Morello et al. 2011)

on the ground. The patents of Lancia included also the transmission tunnel that gave a significant contribution to height reduction, with positive impact in aerodynamic and handling performance.

It must be pointed out that the success of this car was affected by the fact that body builders were not equipped for this significant change in aesthetics; in the following cars, Lancia offered a specific traditional chassis for custom-made bodies, and a factory-made unitized body as an alternative.

The structure designed by Lancia at that time was suitable for torpedo bodies only; sedan bodies were obtained by adding a Weymann top cover (*ballon*), without contribution to stiffness. A further step was made about 15 years later with a more extensive application of spot welding and of deep stamped steel sheets.

One of the first examples of a very effective structural integration between body and chassis is offered by the 1934 Citröen 7, 15 and 11 CV, whose body is shown in Fig. 2.13. This car is also the first application on large production volumes of the front wheel drive scheme. All steel panels of the body have particular shapes, suitable to build up stiff closed sections by welding them along the flanges. The result was a structural cage, including sills, pillars and roof beams.

The sills integrate, in their front part, flanges to which the powertrain and the front axle were assembled. Also roof beams show a closed cross section, as it can be seen in Fig. 2.14; the aesthetic damages of spot welding on the roof panels were masked by filling with melted tin, that was then polished.

The chassis did not exist anymore as a physical element, nor it could be separated from the body, without destroying the welds. This completely changed the assembly cycle and body shop organization in comparison with the past. The unitized body was welded and then painted and only later the mechanical subsystems of the chassis could be mounted on the painted body. Low volume custom-made bodies became now very expensive, because the body shop could take care also of the complete car assembly.

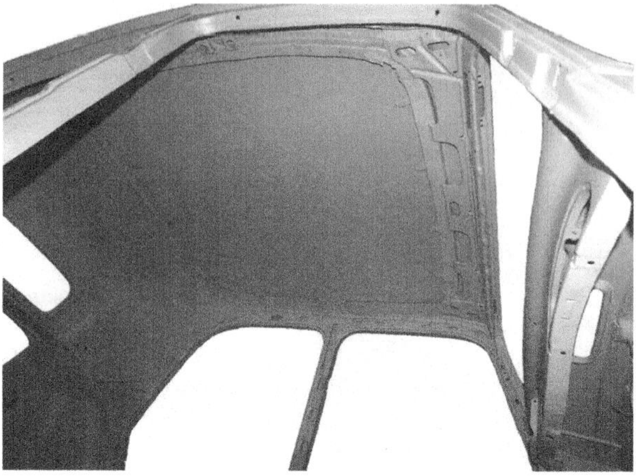

Fig. 2.14 Roof structure of the Citröen 11 CV unitized body (Picture taken at Rétromobile, Paris, 2009)

2.4 Aerodynamic Performance Evolution

It is incorrect to state that aerodynamic performance was completely neglected in the early cars, as their squared shape could suggest.

The control of the aerodynamic drag had been imported from shipbuilding industry; design criteria were however naive, for lack of experimental data or calculation procedures. Some cars show a particular attention to this problem; among them the Jamais Contente, an electric car with a torpedo like shape, the first to exceed a speed of 100 km/h in 1899 and the Alfa Romeo 40–60 HP, with the body designed by Ricotti, shown in Fig. 2.15.

Both bodies were inspired to the slender body shape, the shape sporting the minimum drag resistance in an uniform airflow. These shapes were premature with reference to the real customers' needs and not suited for building a vehicle; the advantages that could be obtained were practically useless (the Ricotti's car claimed an increase of top speed from 125 km/h, of the standard car, to 139 km/h).

Five sedans made by FIAT are shown in Fig. 2.16 to illustrate the aerodynamic evolution of car shapes. In the early period, up to 1915, the body shapes were squared, particularly in the front, where the hood joined the firewall below a flat windshield; an example being the 1908 FIAT 35–45 HP, shown in Fig. 2.16a. This shape was the direct consequence of the manufacturing technology used.

The 1919 FIAT 501 shown in Fig. 2.16b is representative of a new style, typical of the period between 1915 and 1930. Shapes are now more rounded and show the changes that can be obtained by the use of steel sheets instead of wood. The surfaces have simple curvature, feasible with rolling and bending presses; fenders have double curvature and are obtained by beating sheets by hand on open stamps.

Fig. 2.15 The streamlined body of the Alfa Romeo 40–60 HP designed by Ricotti (Alfa Romeo Museum at Arese, Italy)

There is a continuous profile between the hood cover, the cowl and the body sides; the radiator too is rounded at its edges.

A completely new style is shown by the 1934 FIAT Balilla Sport Berlinetta, (Fig. 2.16c); this is the first body developed according to systematic studies of the aerodynamic behavior, performed partly in an aeronautical wind tunnel, partly on the road directly. The windshield inclination was optimized on the road, with an increase of top speed of 5 % around 100 km/h. The hood profiles are completely aligned with those of the body sides; the windshield and the radiator are inclined. Each corner is rounded and the shape of the body is tapered at the front and tail end. The fenders tend to be integrated into the body sides.

Many studies of this kind were performed in Europe at this time; those conducted by Lay in the United States in 1933 are still interesting today, showing the effect on aerodynamic drag of many different body arrangements.

The leader of the revolution of the streamlined design was the famous 1934 Chrysler Air Flow, shown in Fig. 2.17, that however was not received by the public with enthusiasm.

The 1935 1500 (Fig. 2.16d) is the reinterpretation, given by FIAT's stylists, of this streamlined design and show further improvements of the shape of the Balilla, such as the headlights in the fenders and the rounded radiator; this is one of the first cars with an aerodynamic coefficient lower than 0.5.

The chart plotted in Fig. 2.18 shows the evolution of the aerodynamic drag coefficient as function of the decade of commercial launch, representing with bars the best and the worst value. Tests are recent and made in the same wind tunnel.

Fig. 2.16 Main milestones of the evolution of the car body shape, shown by some FIAT sedans (from Morello et al. 2011)

The 1950 FIAT 1400 (Fig. 2.16e) is one of the first examples in Europe of the ponton style, where body sides integrate the fenders completely in a single rounded surface; running boards have disappeared.

The oil crisis of the 1970s stressed further more the need of reducing the aerodynamic drag; the new improved aerodynamic performance is not only due to new body shapes, but also to a painstaking care to many small details of the outer surface and of the underbody.

2.5 Vehicle Architecture

Vehicle architecture may be defined as the lay-out of the main components, such as engine, gearbox, transmission, axles, etc. Also in early cars space was never enough and an unsuitable lay-out could easily sacrifice the volume available for the passenger compartment.

Fig. 2.17 The 1934 Chrysler Airflow (from a Chrysler's Catalog)

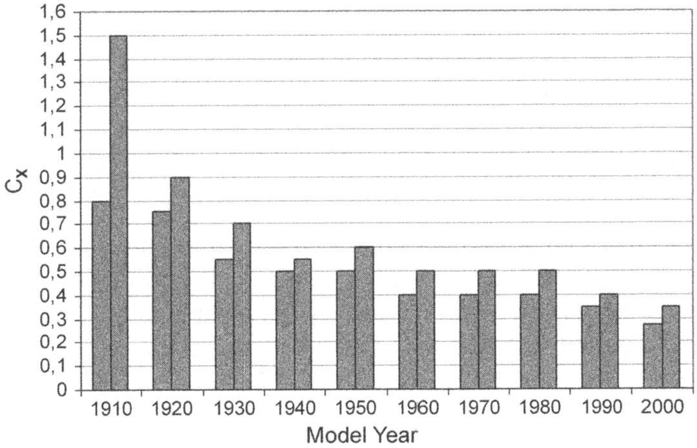

Fig. 2.18 Evolution of the aerodynamic drag coefficient as function of the decade of commercial launch. The best and the worst value for each decade are shown (from Morello et al. 2011)

By considering only the main features and neglecting for simplicity the shape of the body, the vehicle architecture can be identified as follows:

- The number of wheels;
- The position of the driving axle;
- The position of the engine;
- The kind of mechanical linkage between engine and wheels;
- The position of the passengers.

Fig. 2.19 The 1886 Benz Patent Motorwagen tricycle (Mercedes-Benz Museum, Stuttgart)

Apart from two wheel vehicles not considered in this book, the majority of ground passenger vehicles were and are provided with four wheels, organized in two axles, except for some of the earliest vehicles that were based on a three-wheels architecture.

The first motor car (meaning the first car manufactured for sale and not for scientific experiments), the Benz Patentwagen, was a tricycle indeed, as it was the first Italian car (Bernardi) and the first best seller on the French market (De Dion-Bouton). The tricycle option introduced some simplification in the steering mechanism, if the single wheel was placed at the front axle, as in the 1886 Benz Patent Motorwagen. In this way the steering motion of the wheel was cinematically correct by definition and the related control was simple, as can be seen in Fig. 2.19

If the single wheel was placed as a rear driving axle, the transmission could be simplified avoiding the differential, as in the case on the 1894 Bernardi, (Fig. 2.20).

In addition, a tricycle was a statically determined system in its interface with the ground: The forces acting between wheels and ground are univocally determined by the position of the centre of mass and these forces are, within certain limits, independent from the ground, that is never flat nor horizontal.

As a drawback, the effect of the centrifugal force in a curve or of the lateral slope of the ground could impair the roll-over stability of the vehicle. Mainly for this reason, four wheels vehicles were eventually preferred, despite their complication.

The application of four wheels made the vehicle statically undetermined, i.e. the forces on the ground cannot be computed without taking into account local deformations; moreover, a rigid link between wheels and vehicle could produce loss of contact between one wheel and the ground, if the latter is not perfectly flat.

Daimler, with his 1889 Stahlradwagen (Steel Wheel Car), shown in Fig. 2.21, tried to bypass this obstacle by mounting the front axle on a tilting beam, like a

Fig. 2.20 The 1894 Bernardi tricycle (National Automobile Museum of Torino)

Fig. 2.21 The Daimler 1889 Stahlradwagen (Deutsches Verkehrsmuseum in Munich)

balance, able to compensate for the effect of uneven ground. However, in this way the lateral stability was equivalent to that of a tricycle, having a front single wheel in the position of the hinge of such a balance.

Fig. 2.22 The 1769 Cugnot's Fardier (Conservatoire National des Arts et Métiers in Paris)

This is the reason why elastic suspensions were introduced, not only for reason of comfort, but also to keep all wheels in contact with the ground (see Appendix B.5).

Restricting to two axles vehicles, with a single driving axle, it is possible to identify the following architectures, which are listed roughly in chronological order:

- Early front-wheel drive vehicles with front engine;
- Early rear-wheel drive vehicles with rear engine;
- Early rear-wheel drive vehicles with front engine;
- Evolved rear-wheel drive vehicles with front engine;
- Evolved rear-wheel drive vehicles with rear engine;
- Evolved front-wheel drive vehicles with front engine.

Because horses were hitched in front of the carriage, it looked natural to many inventors to install the engine in the front part of the vehicle. The first motor vehicle, the 1769 Cugnot's Fardier (Truck) had a front driving wheel with the steam engine and the boiler rigidly mounted on the front steering wheel, as shown in Fig. 2.22.

Later, some motor-axle assemblies were developed, to replace directly the front axle of a coach; some of these motor-axles (with electric, steam or internal combustion engines) were still available at the end of the nineteenth century, to convert existing horse driven vehicles to the new technology. An example of this approach is shown in Fig. 2.23.

Locating a heavy mass, like an electric motor and its batteries, without any kind of suspension obviously caused problems of discomfort and severe limitations to the maximum attainable speed. When installing the engine on the suspended body and considering the problems linked with the transmission of the power to the driving wheels, it was more expedient to have the rear non-steering axle as the driving axle. At any rate, it was necessary to develop some mechanical transmission device allowing to connect two elements (engine and axle) that could move with respect to each other.

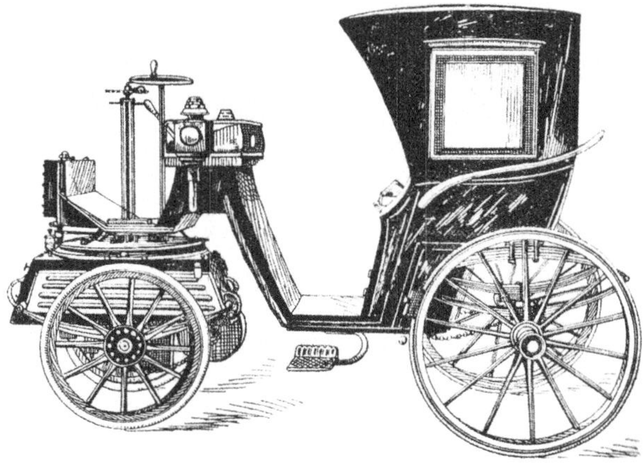

Fig. 2.23 An example of a horse driven coach transformed with an electric front axle (redrawn from Pedretti 1908)

It looked also easier to place the engine in the rear, not only because it was closer to the driving axle, but also to leave a better visibility to the driver. If the engine was placed in the back, seats were set in a *vis à vis* (face to face) arrangement; if it was in the middle, in the *dos à dos* (back to back) arrangement; both architectures were already applied to small coaches (*vis à vis* and dog-carts).

An architecture of this first kind is shown in Fig. 2.24, depicting the 1890 Benz Victoria, whose engine is installed in the back of the car and transfers its motion to an auxiliary shaft, in the middle of the vehicle, through a couple of leather belts. These belts acted as a two speed gearbox and also as a clutch. The auxiliary shaft contained the differential mechanism, needed to drive the rear wheels at different speeds.

The transmission between the auxiliary shaft and the wheels was made by two chains; since the chain could not change in length, the centre of the auxiliary shaft was coincident with the centre of rotation of the leaf spring, close to its front eye. Similar schemes were applied till the beginning of the twentieth century.

Thanks to the developments of Panhard & Levassor, a more rational solution was identified in 1894, as shown in Fig. 2.25. It consists in a front-mounted engine, that operates, through a clutch and a gearbox, an auxiliary shaft with the differential, followed by a chain transmission to the rear wheels.

A further improvement to this architecture was introduced by Renault in 1899 (Fig. 2.1) with his first car, the Model A; chain transmissions were replaced by a single shaft transmission with two universal joints, while the differential was installed in the axle. This kind of transmission was less vulnerable and easier to lubricate. The imitation of this improved solution by competitors was delayed by the patents that Renault filed to protect his invention.

These two architectures became rapidly common and the latter is still present in some contemporary cars; both approaches allowed placement of two or three rows of seats all fronting the road.

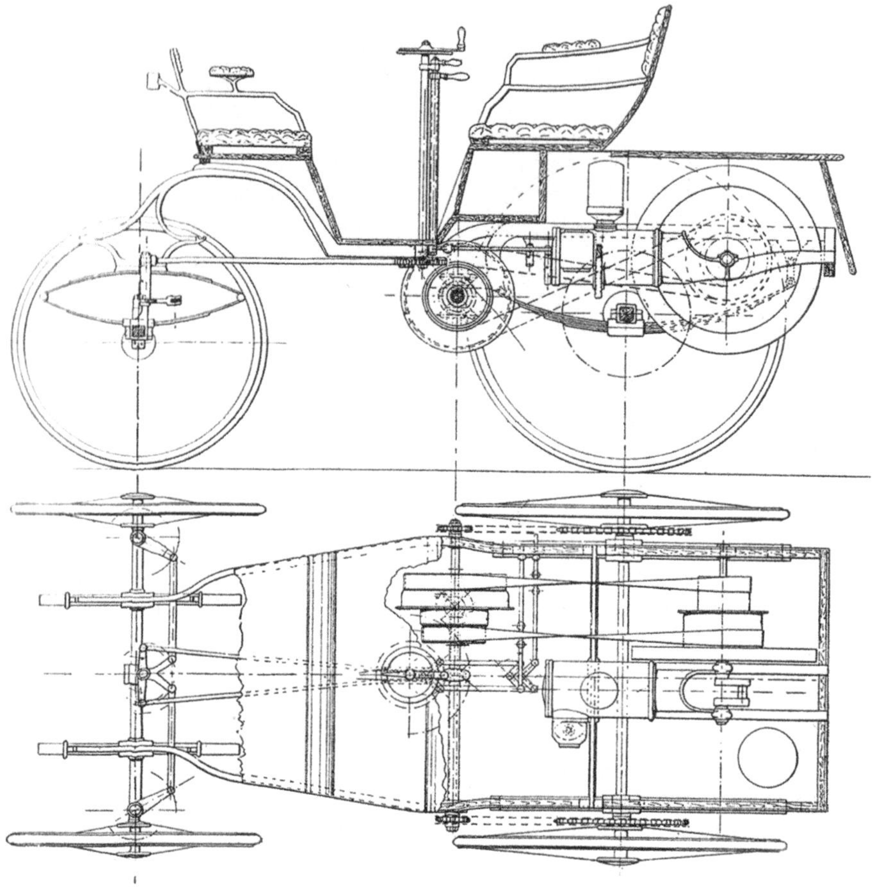

Fig. 2.24 Scheme of the chassis of the 1890 Benz Victoria (Redrawn from Baudry de Saunier 1900)

The transmission shaft has always been an obstacle to obtaining a comfortable passenger arrangement, particularly in small cars that suffered from a floor position that was too high and a small luggage compartment, because of the presence of the differential in the rear axle.

A good improvement to this situation was obviously to concentrate the powertrain on the driving axle, avoiding the transmission shaft; the first option was to develop rear wheel driven car with rear engine, because joints suitable for steering wheels were not economically available by the time when small cars became interesting. The first attempts were made in the 1930s by Mercedes and KdF (later Volkswagen).

Many small cars with this kind of architecture helped the European economical recovery after World War Two. A well known example is the 1953 FIAT 600,whose

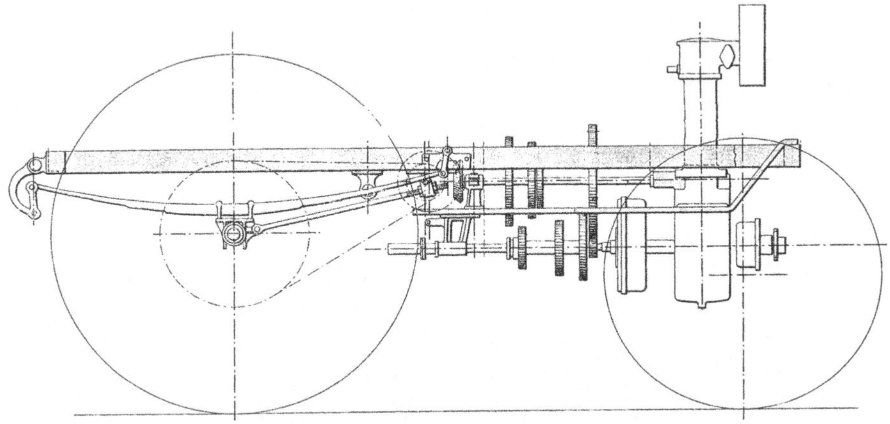

Fig. 2.25 Scheme of the chassis of the 1894 Panhard & Levassor (Redrawn from Baudry de Saunier 1900)

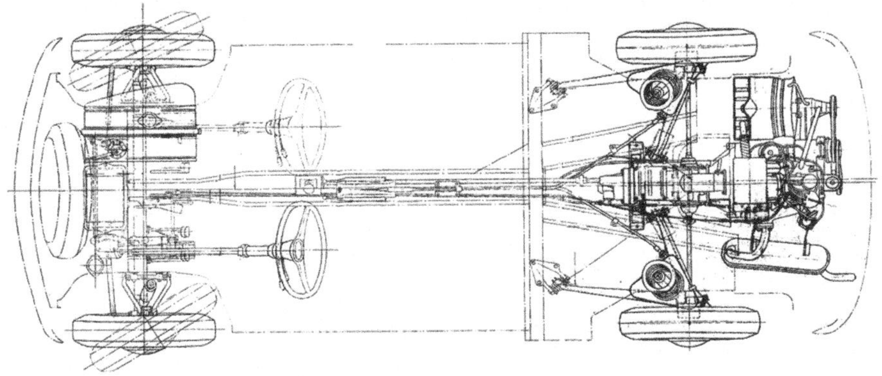

Fig. 2.26 Scheme of the 1953 FIAT 600 (Courtesy of FIAT Historical Archives)

chassis is shown in Fig. 2.26. In a length of about 3.2 m there was space for four passengers and their luggage. This architecture lasted about 40 years, from 1940 to 1980; but at present it constitutes an exception.

The solution that proved more expedient where the internal space is concerned, particularly for the luggage compartment, was obviously front wheel drive with front engine; this architecture was attempted several times without success, and found its conditions of economical feasibility only when some specific components were developed, namely:

- The constant speed ball joints (Ford 1936);
- The Mc Pherson front suspensions (Ford 1948);
- The powertrain with engine and gearbox aligned (Autobianchi 1964).

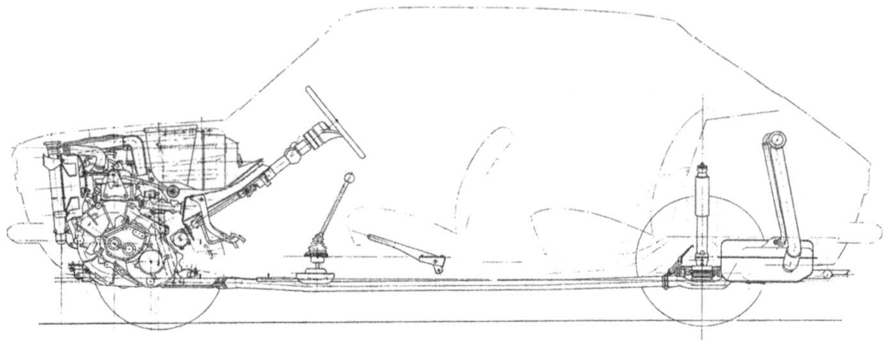

Fig. 2.27 Scheme of the 1971 FIAT 127 (Courtesy of FIAT Historical Archives)

As an example of this architecture, the chassis of the 1971 FIAT 127 is shown in Fig. 2.27; it allows a large passenger compartment with sufficient space for luggage, not too high on the ground. Most contemporary cars are based on this scheme.

Chapter 3
Chassis

This section deals with the evolution of the chassis subsystems in its broader meaning, and thus includes suspensions, steering system, wheels, tires, brakes and transmission systems. This order does not imply any priority, even if suspensions and steering system had a more significant influence on vehicle architecture and its evolution. The steering system is described together with suspensions because its configuration is strictly linked with design of the latter.

The primary function of the vehicle suspension system is to isolate in the best possible way the body, often referred to as the sprung mass, from the disturbances coming from traveling on the uneven surface of the road. To do this, the wheels and the related elements, the so called unsprung masses, are connected to the body with mechanical linkages that allow their relative motion, primarily in the vertical direction, while the forces acting along this direction are transmitted to the body through elastic and damping elements. However, the path of the wheel motion is neither exactly vertical with reference to the road nor occurs along the vertical axis of the car body, but can have components in the other two directions as well.

A second function for the suspensions is to guide the wheel, during its displacement (suspension deflection), so that undesired motions, in terms of lateral and longitudinal displacement and steering rotation, are prevented.

Four wheel vehicles are, as far as the exchange of forces with the ground is concerned, statically undetermined systems: A further task of the suspensions is thus distributing the forces exchanged with the ground and allowing all wheels to contact the ground even if the road is not flat.

Many of these considerations were not known at the beginning of the motor-car era and were developed quite recently, as compared with the hundred and twenty years of automobile evolution, mainly because of the increase in vehicle speed and improvements in road conditions.

G. Genta et al., *The Motor Car*, Mechanical Engineering Series,
DOI: 10.1007/978-94-007-8552-6_3, © Springer Science+Business Media Dordrecht 2014

Fig. 3.1 This coach, built around 1650, shows a suspension made by four leaf springs and leader belts (National Automobile Museum of Torino)

3.1 Solid Axle Mechanical Linkages

The importance of the suspensions in granting the required passenger comfort was already known in the fifteenth century. Coach bodies were suspended through a set of leaf springs attached to the chassis framework, a rigid structure bearing wheel hubs. The free end of the springs was connected to the body through leader belts; Fig. 3.1 offers an interesting example of this configuration in a coach dated about 1650.

The steel leaf springs were introduced at this time while, formerly, they were made of wood. There is no component explicitly dedicated to damping suspension oscillations in this coach but the internal friction of leaf springs and belts should have been high enough to allow the expected comfort level.

Secondary motions were not an issue because wheels were rigidly connected to the chassis frame; the suspension system was statically determined because of the clearances in the front steering axle.

Elliot, an English wheelwright, was credited with the invention of the single solid axle suspension, using semi-elliptical steel leaf springs; in this design the front axle can steer on a pivot in the centre of the axle. This suspension layout, used on coaches and carriages, was also applied in the nineteenth century for the early steam road vehicles. An example is shown in Fig. 3.2.

The weak point of this system derived from carriages was a reduced stability of the vehicle to rollover in a turn; when lateral centrifugal forces were applied, the rollover line was no more the line joining the two contact points with the ground of the wheels on the same side, but was shifted closer to the centre of the vehicle.

Längensberger envisaged in 1810 a new steering layout, where the front axle was always parallel to the rear axle but only the front wheel hubs were articulated to the axle using a *king pin*; the two stub axles were connected through *track-rod arms* and a *track-rod*, forming an articulated parallelogram. Using this device the vehicle rollover stability was unaffected in turns.

Fig. 3.2 The Bordino's steam coach was built in 1854. The suspension system is strictly derived from that of a horse carriage. The rear axle is shaped as a crankshaft, on which the connecting rods of the two cyclinders operate directly (National Automobile Museum of Torino)

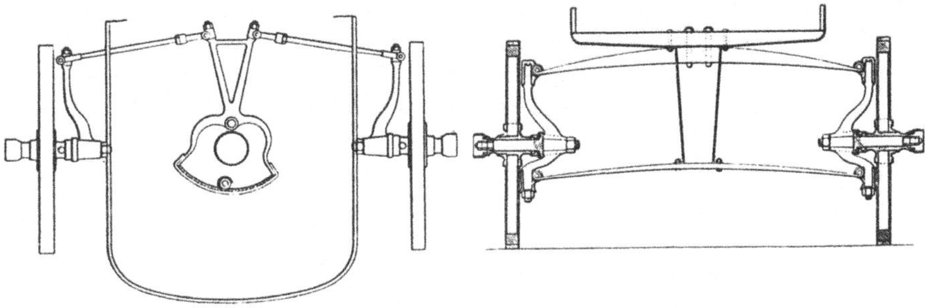

Fig. 3.3 Drawing of the front steering axle of the 1878 Bollée Mancelle; the independent double wishbone suspension with transverse double leaf springs is clearly shown (redrawn from Baudry de Saunier 1900)

A new patent was filed in 1818 in London in the name of Ackermann; this patent described the law the steering angles of the two wheels should follow to allow wheels to roll correctly, with their symmetry plane containing the local speed vector. This invention found no immediate practical application because there was no real need for carriages to have a steering system different from the turntable steering, and there was no idea about how to satisfy this law in a simple way.

Jeantaud, another wheelwright, proposed in 1878 a mechanism, perfected from the idea of Längensberger, in which the two track-rod arms were slightly inclined toward the centre of the vehicle so that their axes crossed close to the centre of the rear axle. This mechanism realized Ackermann's law with acceptable approximations and is still in use today on solid steering axles.

Independent of Jeantaud's idea, the solution presented by Bollée in his Mancelle, also in 1878, was not far away from the Ackermann's law. The interest of this idea lies mainly in the fact that this is probably one of the first steering systems specifically designed for an automobile. The scheme of this front axle is shown in Fig. 3.3: It is an independent suspension with transversal leaf springs, equivalent to the double wishbone mechanism.

Fig. 3.4 The first cars powered by internal combustion engines, the tricycle by Benz and the *Stahlradwagen* by Daimler, show no evidence of a great attention to the suspension system (from Genta and Morello 2009)

It would be wrong to assume that independent wheel suspensions were invented in the 1940s, at the end of an era that saw only solid axles with leaf springs; independent wheel suspensions are sporadically present even in the earliest cars. Nevertheless, the advantages obtained on the existing roads, at the speeds then achievable, can be assumed to have been negligible when compared to the design complications involved.

The first cars with an internal combustion engine, the tricycle built by Benz in 1886 and the *Stahlradwagen* by Daimler in 1889 (see Fig. 3.4) show no evidence of a great attention to the suspension system. The first is in fact a tricycle with a single non-suspended front steering wheel; the second is a four wheel vehicle with no suspensions, in which the front axle is balanced on a central horizontal pivot to make the vehicle statically determined. It can be assumed that the undoubted genius of these two precursors was focused on developing a reduced weight engine with a significant storage of energy on board, rather than on passenger comfort.

The most important feature of solid-axle suspensions with leaf springs is that the axle linkages are in this case integrated with the elastic element.

As far as suspensions are concerned, two typical problems of solid axles are:

- If the engine is part of the sprung mass (and this is usually the case), it is necessary to develop a mechanism to connect the gearbox output shaft with the wheels, two parts that have a variable relative position.
- Because the steering control (steering wheel or steering bar) must be within reach of the driver, a mechanism must be developed to connect the stub axles to the steering wheel without causing the steering angles to be affected by the bouncing motion of the suspension.

The solution of these two problems was a difficult task for the early car designers, particularly that related with the steering system; the multiplicity of solutions in the early cars is an evidence of the difficulty encountered in finding an adequate technical solution to the problem.

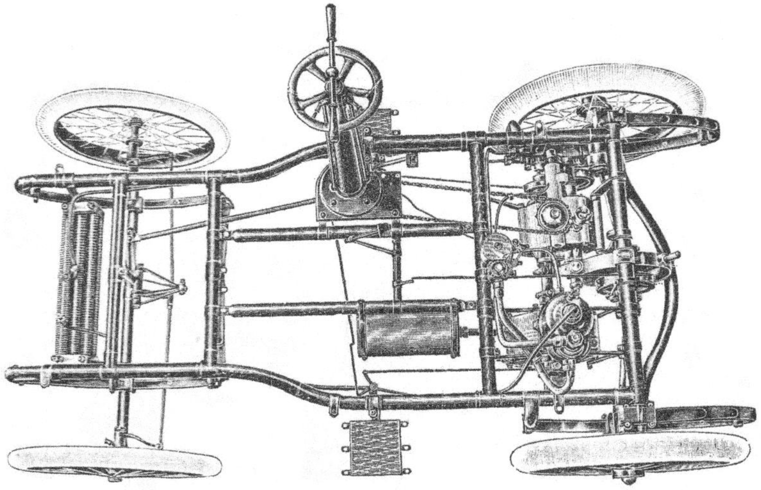

Fig. 3.5 A typical De Dion-Bouton chassis of the beginnng of the twentieth century shows four leaf spring suspensions and front wheels steering on king-pins (redrawn from Baudry de Saunier 1900)

On the Benz car the problem is bypassed, considering that there is no front suspension, while on the Daimler *Stahlradwagen* the steering control is fixed to the balancing axle close to the pivot point.

A few years later almost all suspensions were solid axles mounted on longitudinal leaf springs, with steering front wheels rotating on king-pins, as it can be seen in Fig. 3.5 for a De Dion-Bouton chassis. This suspension architecture was widely applied in the first twenty years of automotive history.

The Jeanteaud steering mechanism connected to the steering bar through a longitudinal linkage can be seen on the same chassis; it is also interesting to note that the linkage knuckle mounted on the sprung mass was positioned in the center of curvature of the trajectory of the axle in its bouncing motion, to avoid undesirable changes of direction while riding on uneven roads.

Leaf springs, as seen above, were already known in the seventeenth century; at the beginning of the twentieth century they were built and applied with a satisfactory level of technology. Leaf springs integrated in a single element the elastic and the structural functions; the elasto-kinematic performance of this component was fully adequate for the needs of the cars of that time.

The spring was made with leaves of different length, reduced in constant steps from the top to the bottom, as shown in Fig. 3.6; this assembly simulates with good approximation the performance of a constant stress structure, allowing the maximum value of deformation at a given stress level. This fact was clearly described in the first automotive engineering handbooks. This structure is quite flexible in the vertical plane but could be quite stiff in the horizontal plane.

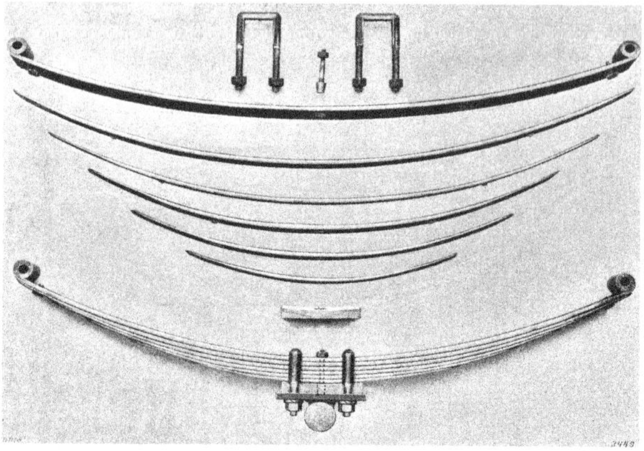

Fig. 3.6 A leaf spring is made by leaves with different lengths, reduced in constant steps from the *top* to the *bottom*; this assembly is similar in performance to a constant stress structure (redrawn from Baudry de Saunier 1900)

The spring was mounted to the chassis using eyes on the *mother* (the longest one) *leaf*, one end was fixed and the other could move thanks to a *swinging shackle*, or a sliding element; these devices allowed the spring to change its length because of the deflection. The fixed eye was on the same side of the fixed point of the transmission line and of the steering system, typically behind the front axle and forward of the rear axle.

This suspension architecture could have many different variants, according to the objectives of comfort, cost and weight of the unsprung mass. Leaf springs were classified after their shape, with reference to the ideal elliptical shape, made of two mirror-like leaf springs; the solutions shown in Fig. 3.5 were referred to as semi-elliptical for the front axle and 3/4 of ellipse for the rear.

The most commonly used solution featured two semi-elliptical springs for each axle; the 3/4 of ellipse solution (Fig. 3.7a) was sometimes used for axles showing wide load variations, where the side beam of the chassis structure was fit directly to a 1/4 ellipse spring. This solution allowed increased flexibility, with the drawback of an increased weight of the unsprung mass and a higher cost.

A 3 semi-elliptical spring solution was also used for the rear axle in heavy cars; the function is comparable to that of the 3/4 ellipse arrangement, where the two 1/4 ellipse springs are integrated in a single element, fit in the middle to the structure of the chassis. In this case, swinging shackles are made using a universal joint (Fig. 3.7c).

A particular cantilever spring, shown in Fig. 3.7b, was developed and distributed by Ford. A single semi-elliptical spring was fixed to a cross beam of the chassis structure, at its mid-span; it performed like two 1/4 ellipse cantilever springs, mounted on the axle; in this case, two swinging shackles had to be used. Undesired motion of the axle in the longitudinal direction was avoided through the use of two thrust beams.

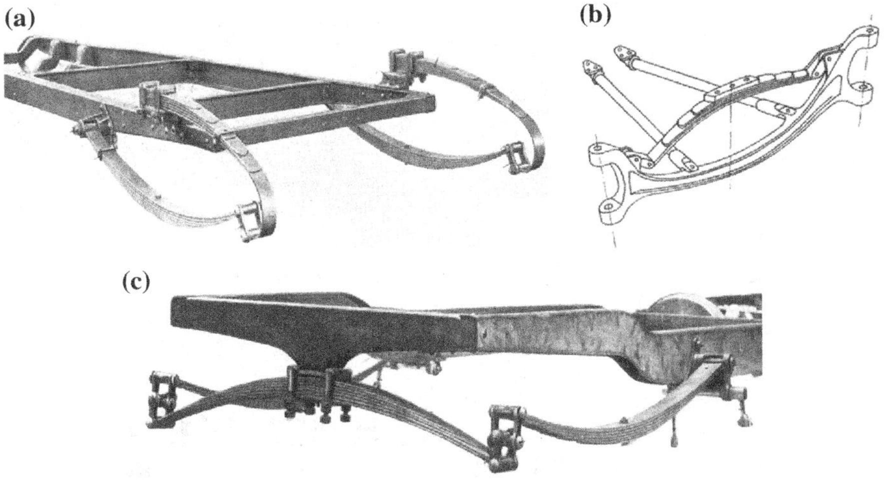

Fig. 3.7 Some application schemes for leaf springs. **a**: $\frac{3}{4}$ ellipse solution; **b** Ford scheme with a single spring for each axle; **c** Three leaf springs solution equivalent a $\frac{3}{4}$ ellipse spring (from Genta and Morello 2009)

A convenient suspension should not only have a compliant elastic member, but also a damping element; otherwise a permanent oscillation of the sprung mass could take place as a result of an external force input. This cannot actually occur, because of internal friction; nevertheless several oscillations could result.

With leaf springs this problem was not addressed for some time, because of the high value of internal friction due to the relative motion of the leaves; on the contrary, every effort was made to improve leaf lubrication to avoid annoying squeaks and permanent sticking because of oxidation. The first shock absorbers appeared in the 1910s, considered initially as aftermarket accessories to be installed by demanding costumers interested in sport driving.

Many solutions were considered, the most interesting ones being shown in Fig. 3.8. The first four are friction shock absorbers; (a) and (c) work on the extension stroke only, while (b) and (d) work in the compression stroke as well.

Many engineers thought that damping forces were undesirable in the compression stroke, because they increased the value of the forces applied to the sprung mass; others thought that reducing the danger of oscillation was the most important priority. Only later the importance of having a bilateral force, proportional to suspension compression or extension speed, was understood.

A friction shock absorber has some properties that work against comfort. In fact, a dry friction force allows suspension motion only above a threshold value of the force. It can happen that suspensions are completely blocked, when travelling on very smooth roads or on macadam pavement; in this case the only working elastic members are the tires. For this reason the shock absorber in Fig. 3.8d features an adjustment screw, suitable for changing the pressure force on the friction discs; this

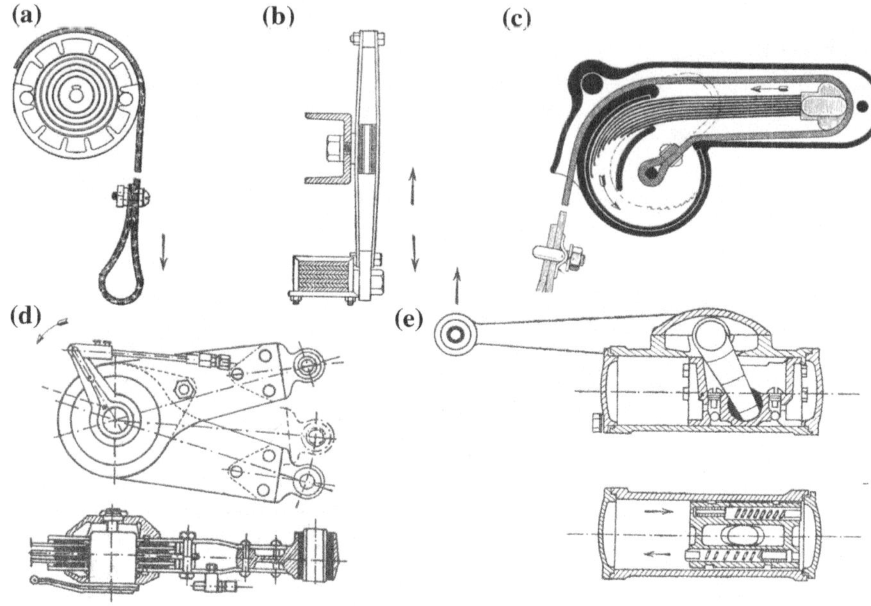

Fig. 3.8 Some types of shock absorbers. **a**, **b**, **c** and **d** work on mechanical friction; **d** can also be adjusted. **e** Refers to a hydraulic shock absorber (from Genta and Morello 2009)

screw can be adjusted by the driver from the inside of the car, opening the possibility of reducing damping on good road or at low speed and increasing it under opposite conditions: A first example of controlled suspensions.

The last sketch shows a hydraulic shock absorber, to be applied on the side beam of the chassis structure. This shock absorber works bilaterally with progressive forces, depending on the suspension extension or compression speed. Modern telescopic shock absorbers implement this principle in a simpler way; these were introduced at the end of the 1920s.

The success of leaf springs was also due to the reliability of the fabrication process for steel leaves. The microstructure of the material could be improved easily by plane forging; a failure of a leaf was never catastrophic, because leaves broke down one at a time and the consequent vehicle trim variation was easily detected. The necessary repair was simple and could be done by a competent smith. The success of coil springs was delayed by these facts: As a matter of fact, leaf springs were initially applied to independent suspensions as well.

With the increase of vehicle speed some critical aspects of the solid axle started to become evident: The most important problems were weight, secondary deformations of the axle and *shimmy*.[1]

[1] Shimmy is a general term indicating self-excited vibration of steering wheels or, better, of the whole steering system. It is applied to the vibration of the steering systems of cars and trucks, but

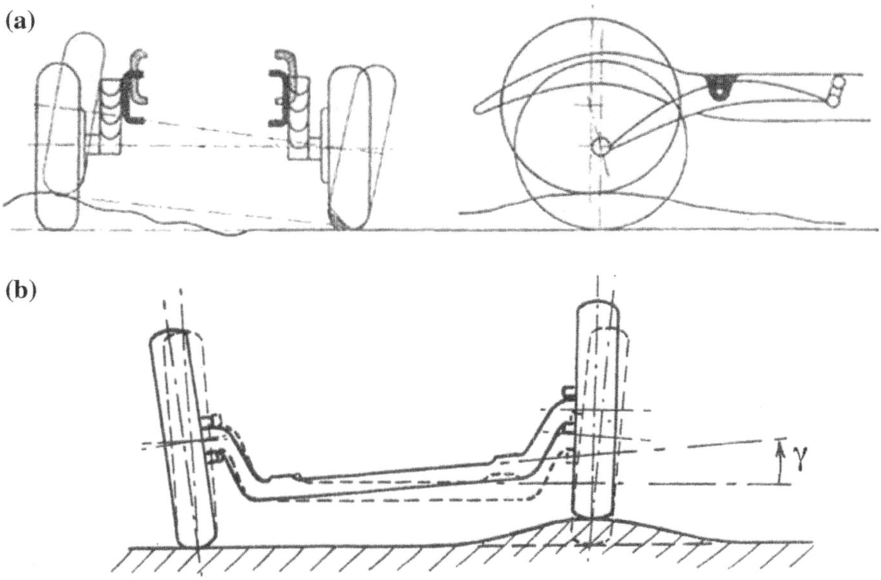

Fig. 3.9 The *upper scheme* illustrates axle steering of a Rolls-Royce suspension while negotiating asymmetric obstacles or in a turn. The *lower scheme* depicts the phenomenon of the *shimmy* (from Genta and Morello 2009)

The axle weight was remarkable; the amount of the unsprung mass as compared with the relatively high vertical flexibility of the tires could cause axle hopping on certain road, that was worrisome in turns and never suitable for comfort.

Secondary deformations of the axle were caused by braking (S-deformation) or by vehicle roll (different elongations of the two suspensions of the same axle). The first type of motion could cause annoying resonances of the entire driveline while starting up or braking on gravel or dirt roads.

The second one caused undesired steering of the axle. The roll motion of the body, due to the centrifugal force in a turn, causes a different longitudinal motion for the two points where the leaf springs are joined to the axle; in other words, the solid axle steers independently from the steering wheel while the vehicle is turning. The architecture dictated by the transmission installation (fixed eye in front and swing shackle in the rear side) causes the vehicle to oversteer or to increase the path curvature because of the deformation; this fact could cause instability or spinning out in curves that today appear undemanding.

The scheme in Fig. 3.9a illustrates this behavior for a Rolls-Royce rear suspension; the advantage is that while driving on an asymmetric obstacle a self aligning steering action can be provided. To avoid this inconvenience some solid axle suspensions

(Footnote 1 continued)
also of motorcycles (a more correct term is wobble), of airplane tailwheels and of any pivoting wheels, like those used in shopping carts. These phenomena are different, so that there is not a single explanation or mathematical model for wheel shimmy.

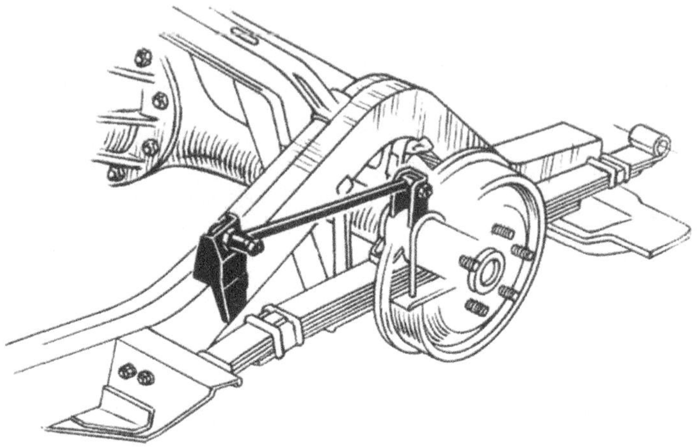

Fig. 3.10 Example of a rigid axle with leaf springs, showing additional linkages for reducing axle steering and S deformation (from Genta and Morello 2009)

were built with two longitudinal linkages that modified the kinematic behavior of the axle; an example is shown in Fig. 3.10, where two rods have the double function of avoiding the S-deformation and of giving the axle an understeer behavior.

Figure 3.9b explains the kind of *shimmy* that is typical in solid steering axles. Consider the situation where, riding over an asymmetric obstacle, the axle has a certain roll speed. Because the wheels, that in old cars had a significant moment of inertia, are rotating, a gyroscopic torque is applied to the steering mechanism and consequently the elasticity of the steering linkage (some elasticity was required to limit steering wheel vibration caused by the engine) could induce annoying oscillations. Because obstacles are primarily asymmetric, this phenomenon was well known; it was solved in steering solid axles by using additional shock absorbers, to be applied to the track rod.

A major advantage of the solid axle was highlighted only after independent suspensions became common: on solid axles the body roll does not affect the camber angle of the wheel, because the wheel is always perpendicular to the surface of the road.

3.2 Independent Suspension Mechanical Linkages

Solid axle suspensions with leaf springs dominated the market for a long time and they began to be replaced by independent suspension only in the 1930s; nevertheless many examples of independent suspensions are older, beginning with their first application, on the already mentioned Mancelle.

Fig. 3.11 The vertical telescopic independent suspension was probably introduced by Stephens in 1898 (from Genta and Morello 2009)

Steering suspensions are here described first. Many examples of suspensions in which the wheel hubs are guided by vertical linear guides can be found; this solution was perfected and produced by Lancia, but probably the first application of this kind appeared in the 1898 Stephens, shown in Fig. 3.11, where the front wheels are guided by a telescopic steering fork inspired by bicycles, while being supported by a transversal leaf spring attached to the body in its midpoint.

This architecture was later considered by Sizaire and Naudin and Decauville, whom many authors quote as precursors to Lancia. In any case we can doubt that these precursors had a clear view of the advantages they could claim.

It is interesting to note that many of the first advertisements for independent suspensions showed the wheel while driving over an asymmetric obstacle but never on a curved road. This suggests that the new architecture was developed to avoid the shimmy phenomenon, while no attention was paid to camber variations caused by the roll angle; it also explains why independent suspensions were first applied to steering axles. The problem of camber angle variations was studied and solved only in the 1940s, when the speed achieved in normal use of the vehicle increased significantly.

Other reasons in favor of a change were the advantages obtainable in terms of vehicle architecture. The front solid axle, with its relevant lateral width, could not be set too close to the ground and a certain clearance had to be provided over the axle to allow for the suspension compression stroke. As a consequence, the engine had to be positioned quite high over the ground and the front part of the body, the hood, was quite tall.

The independent wheel suspension made this layout no longer necessary; the engine could be moved down and forward, between the suspension arms, with advantages in terms of vehicle length and weight, a reduction on the height of the center of mass and consequent improvements in speed, because of the use of streamlined aerodynamic shapes.

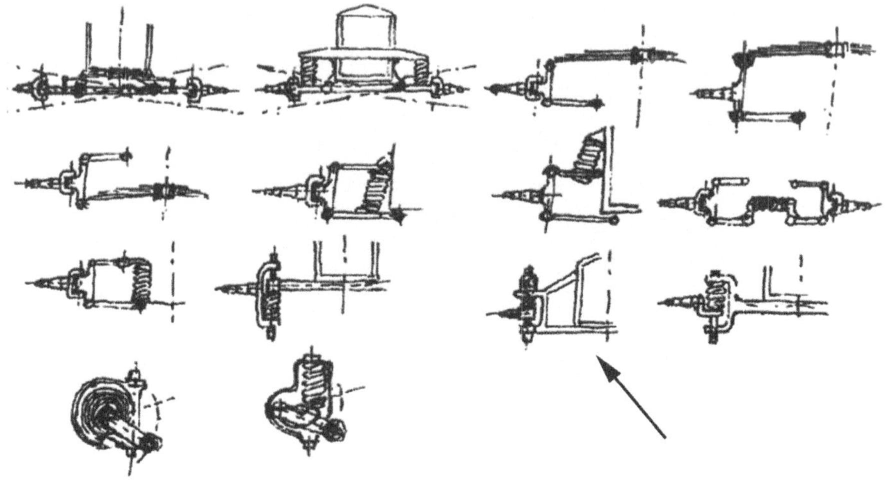

Fig. 3.12 This *table* shows the alternatives examined by Lancia while designing the front suspensions of the Lambda. The *arrow* points out the choosen solution (from Genta and Morello 2009)

Fig. 3.13 The *front view* of the Lancia Lambda shows the front suspension and the gate structure of connection to the body (National Automobile Museum of Torino)

Figure 3.12 shows a set of sketches drafted by Falchetto, the vehicle engineer who, under the guidance of Vincenzo Lancia, faced the problem of introducing an independent wheel suspension on the Lambda. The solutions considered on that occasion were almost all developed in the following years; the choice favoured the one marked with an arrow in the figure, possibly because of the problem of shimmy, but there is no record of the decision-making process.

Figure 3.13 shows a picture of the final vehicle launched in 1922; the most important feature of this design was the solution to the problem of efficient lubrication of the telescopic sliders by integrating the guide, the spring and the hydraulic shock

Fig. 3.14 An independent front suspension designed by Dubonnet and produced by Fiat, starting from 1935. A sealed cartridge (see *cross section*) includes both the spring and the damper (redrawn from Genta and Morello 2009)

absorber into a sealed element. The elastic element is now a coil spring; and this is likely the first application in which the car weight is entirely supported by a spring of this kind.

A different scheme of front independent wheel suspension is attributed to Dubonnet, an important independent car designer, and was exploited by many car manufacturers, among them Fiat, which applied this solution to production starting from 1935. It includes a sealed bearing element (*cartridge*) that integrates the helical spring, lubricated by the same oil used for the shock absorber, as shown in the detail on the right of Fig. 3.14; spring and damper work on a finger crossing through the cartridge. This finger is connected to one of the arms of a double trailing arms mechanism.

In Fig. 3.14b it can be seen that the cartridge is mounted on a non-suspended solid axle through a king-pin; in this case the entire suspension is steering together with the wheel and the steering mechanism is the same as for a solid steering axle, but the suspension motion of wheels is completely independent. The kinematic behavior is almost the same as for a vertical tube suspension, but the spring and shock absorber unit is easier to manufacture. The architecture is similar to that of a trailing arms suspension.

A suspension similar to the this scheme, based on a double trailing arm, is attributed to Porsche and is shown in Fig. 3.15; it was developed in 1931 and applied until the 1970s on the Volkswagen Beetle and other cars from this company. The two articulated parallelograms are mounted at the end of a double tube structure that also constitutes the front cross member of the platform.

The upper arms are flanged to the same torsion bar, contained in the upper tube; this bar limits the roll angle. The lower arms are flanged to two different torsion bars, each of them flanged at the other end to the tubular structure; these act as elastic

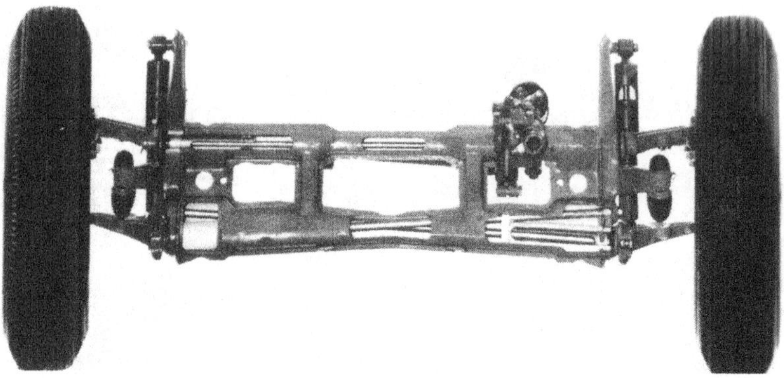

Fig. 3.15 The Porsche suspension was made with two trailing arms; it was developed in 1931 and produced by Volkswagen till the 1970s (from Genta and Morello 2009)

elements. The dimensions of this suspension are therefore constrained in the three directions.

While these suspensions were effective at reducing shimmy, they later demonstrated an undesired feature: They caused the wheel to have, during turns, a camber angle equal to the roll angle of the body. Since the roll angle, due to the centrifugal force, tilts the body to the outside of the curve, the camber angle of the wheels reduce the cornering stiffness of the tire: a better situation would be that the changes in the camber angle occur in the opposite direction, but the ideal behavior is to have no camber variation under any condition.

To improve this undesirable situation, double wishbone suspensions were developed with arms of unequal length; the shorter upper arm increases the camber angle when the suspension compresses and decreases it when the suspension is extended. The value of the camber angle is such as to compensate partially for the effect of the roll angle.

One of the first examples of this concept was introduced by Studebaker in 1939 and followed by many car manufacturers until the present. A drawing of this suspension is shown in Fig. 3.16. The elastic element is integrated with the lower arm and is still made by a transversal leaf spring; as a matter of fact many manufacturers were not keen to abandon leaf springs. The reasons were many, but the main one was reliability. The reliability of the production process was likewise important as were, last but not least, the existing investments for a high volume of production.

An interesting FIAT patent on this subject was applied to cars with rear engines, starting from 1955. A leaf spring was attached to the body through two symmetric bearing points: With a suitable distance between them it is possible to obtain different elastic characteristics on symmetric and asymmetric suspension strokes so that it is possible, with a single spring element, to obtain the functions of the main elastic element and of the anti-roll bar.

From Fig. 3.16 it is clear that the Jeantaud mechanism, that was present on previous steering systems, was no more used since the track rod can no longer be applied,

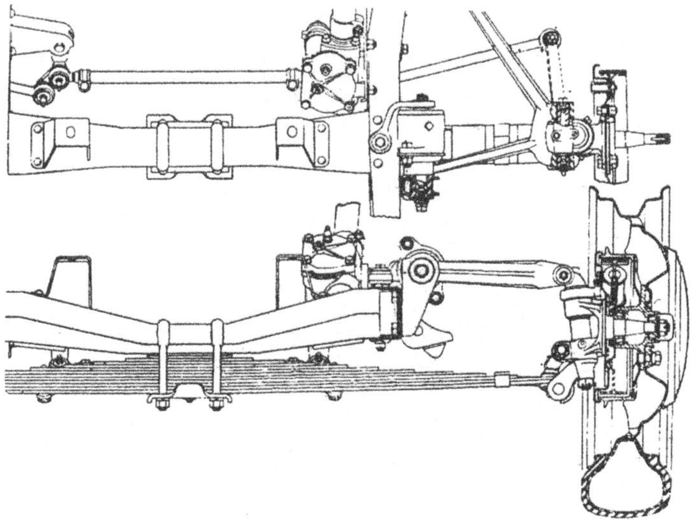

Fig. 3.16 Front suspension made by Studebaker in 1939. The double wishbone mechanism features an *upper arm* of reduced length. The *lower arm* integrates elastic function, being made with a leaf spring (from Genta and Morello 2009)

because the distance between the articulation points of the track rod arms changes with the suspension stroke. Before the rack steering systems were introduced starting from the 1960s, a small articulated parallelogram, whose rods were articulated to the body and rotated by the steering box, was employed. The two steering linkages were connected to suitable points on these rods.

This mechanism also approximates, with some error, the ideal Ackermann law, but the presence of ever increasing sideslip angles, due to increasing speed, made this error less important. It must be remembered that theoretical models for the behavior of tires when producing side forces were developed during these years and the concept of sideslip angle was introduced. Double wishbone suspensions with arms of different length spread quite rapidly in the following years and became almost universal in front axles during the 1960s.

Mc Pherson introduced the suspension layout that now is known under his name (Fig. 3.17) on French Ford cars starting from 1947. The kinematic equivalent of this suspension is a double wishbone layout where the upper arm has infinite length. An advantage of this suspension is that it simplifies the body structure thanks to the more rational distribution of the connection points.

The reduction of kinematic performance as compared with the double wishbone solution is not very relevant and this solution is also applied to contemporary sport cars.

Honda introduced in the 1980s a final innovation in the front suspension architecture. It conceived a swan-necked wheel strut (see Fig. 3.18), which allowed the double wishbone suspension to be installed also on front-wheels drive cars with transversal

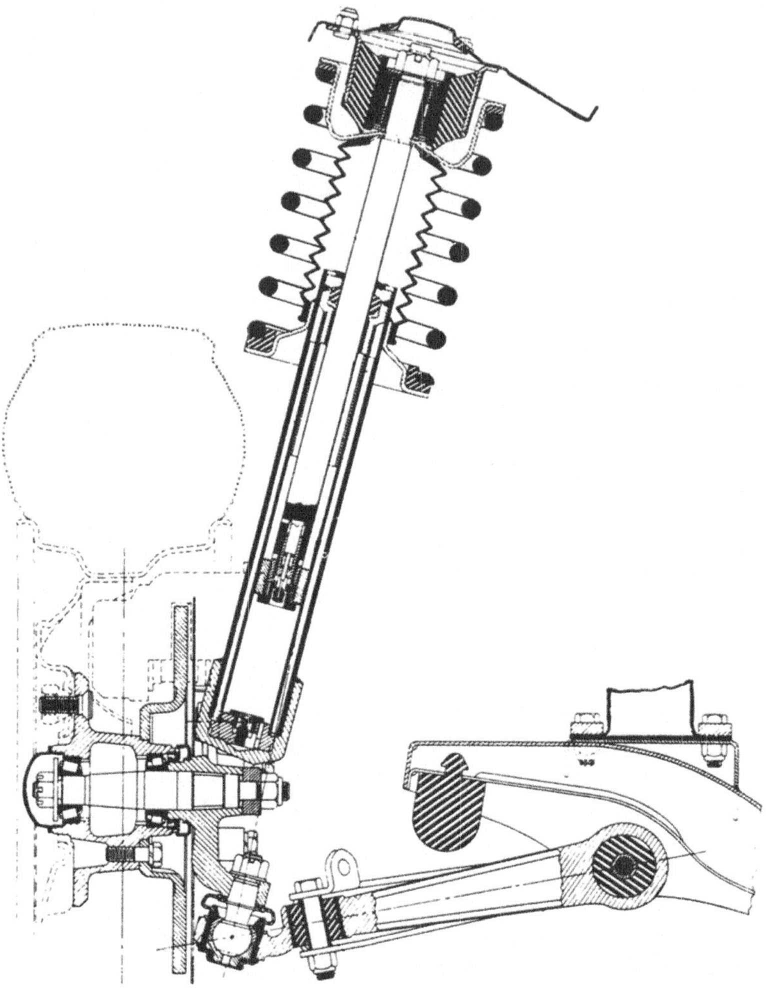

Fig. 3.17 Mc Pherson introduced on Ford cars this kind of suspension in 1947. The kinematic equivalent of this suspension is a double wishbone suspension where the upper arm has infinite length (from Genta and Morello 2009)

engine. The descendants of the Mc Pherson and Honda suspensions today share the market, with the introduction of many improved details that will be omitted here for the sake of simplicity.

Rear suspension history is much more complicated and the attempt to identify evolutionary trends risks oversimplifying the explanation.

It should be pointed out that solid axles, based on an elastic member more sophisticated than leaf springs and on additional linkages, had a long life both on front and rear driven cars. On front-wheels drive cars the rear solid axle is not very heavy

Fig. 3.18 Honda front suspension with high double wishbone; high refers to the position of the *upper arm* compared to the conventional case (from Genta and Morello 2009)

and its simplified layout, because of the absence of the differential and final drive, allowed the use of tubular structures that were quite light.

On rear-wheels drive cars, particularly in case of luxury and sport cars provided with solid axles, the weight was reduced by resorting to suspended differentials, while a good kinematic behavior was obtained with coil spring and more complicated linkages.

A particularly elegant example of solid rear axle is provided by Alfa Romeo.This scheme, shown in Fig. 3.19, was used on different cars starting in the 1970s and can be considered an improvement of the De Dion-Bouton suspension. The solid axle is made of a triangular structure, linked to a spherical joint in the front, which defines a precise position for the roll axis; the suspension stroke and the body roll do not affect axle steering thanks to the guidance given by a Watt mechanism in the back.

Eventually, the need to obtain a well shaped and roomy trunk imposed the development of rear suspensions different from the solid axle.

One of the first applications of an independent rear suspension with trailing arms appeared in the Lancia Aprilia in 1937.This suspension, shown in Fig. 3.20, is suitable for rear-wheels drive vehicles and consists of two longitudinal arms with the articulation point forward of the wheels; the elastic element is still a transversal leaf spring with the addition of a transversal torsion bar. Being an independent suspension, the differential and final drive is suspended on the body. This kind of suspension, in

Fig. 3.19 A particular example of solid rear axle was provided by Alfa Romeo on different cars from the 1970s. It can be considered an improvement of the De Dion-Bouton suspension (from Genta and Morello 2009)

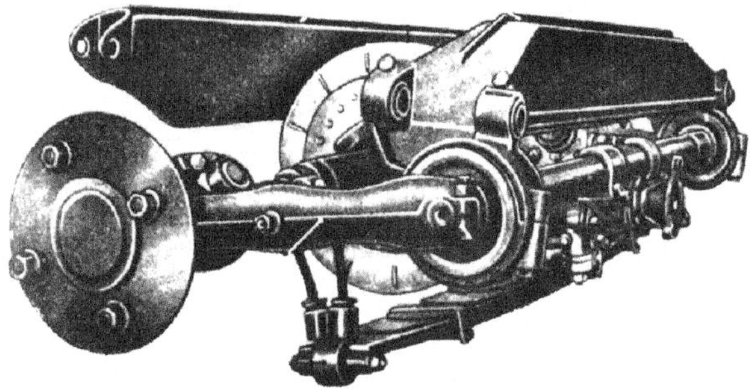

Fig. 3.20 One of the first applications of an independent rear suspension with trailing arms appeared in the Lancia Aprilia in 1937 (redrawn from Genta and Morello 2009)

connection with the application of coil springs, had considerable diffusion up to the present, mainly on cars with front wheel drive.

Such suspensions have the advantage of a reduced space used up by linkages and springs; the disadvantage is the poor kinematic performance and the weight of the trailing arms.

Rear-wheels drive cars with rear engine had a significant diffusion between the 1940s and 1960s; today they are no longer found, with the exception of some niche sport cars. The reason for the diffusion of this architecture was the excellent interior space in a small car, thanks to the absence of the propeller shaft, at a time when low cost and reliable components for front wheel drives were not yet available.

Solid axles could not be applied because of the short distance between the power train and the axle, thus the usual solution was the so-called semi-trailing arms suspension, because of its compatibility with inexpensive constant velocity joints.

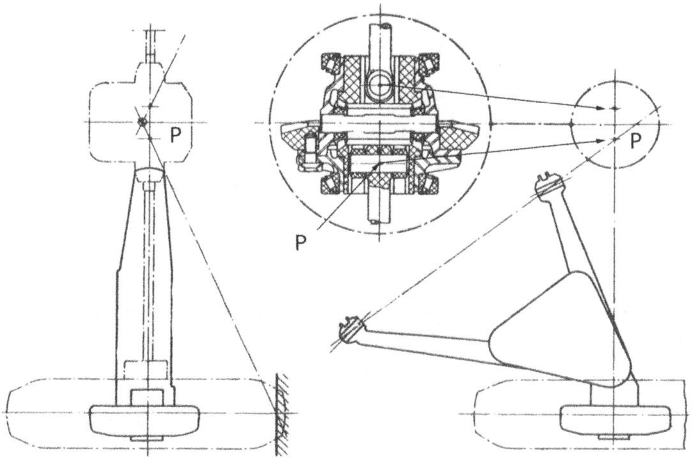

Fig. 3.21 Sketch of a semitrailing arms rear suspension: It shows how it is possible to locate the arm center and the differential center (see *points P*) close to each other; this condition reduces the joint lateral motions (redrawn from Genta and Morello 2009)

A sketch of one of these suspensions is shown in Fig. 3.21: It shows how it is possible to locate the arm center and the differential center (see points P) very close to each other, a condition needed to limit the lateral motions of joints.

The camber angle relative to the ground does not change because of the roll of the body, but varies remarkably because of payload variations and the look of the unloaded car is characterized by a noticeable positive camber (the wheel mean planes cross below the ground): In this condition handling is fairly bad. In addition, the track changes continuously on bumpy roads because of the suspension stroke, causing premature tire wear. This kind of suspension was also applied in many rear wheel driven cars with front engine because of its limited height, but today is no more used.

In 1969 Volkswagen conceived a new rear suspension quite suitable to front wheel driven cars: The so called semirigid axle or twist axle. In a short time it became one of the most common architectures. The original application, shown in Fig. 3.22, demonstrates quite clearly the advantages of this solutions: The tank and spare wheel find their place between the arms and the cross member, leaving the trunk bottom low, flat and wide.

From this point of view alone, trailing arms are slightly better. They do not allow any camber angle variation relative to the body during the suspension stroke; the sides of the wheel wells can be closer to the tire profile and leave more space for the trunk. Their elasto-kinematic behavior is however poorer, in terms of the cornering stiffness of the tires, as previously explained.

On twist axles the gate structure that bears the wheels is characterized by a cross beam made with a steel bar with open cross section, like a horizontal U, for example, open at the back. Using this feature the gate is flexible in torsion (differential

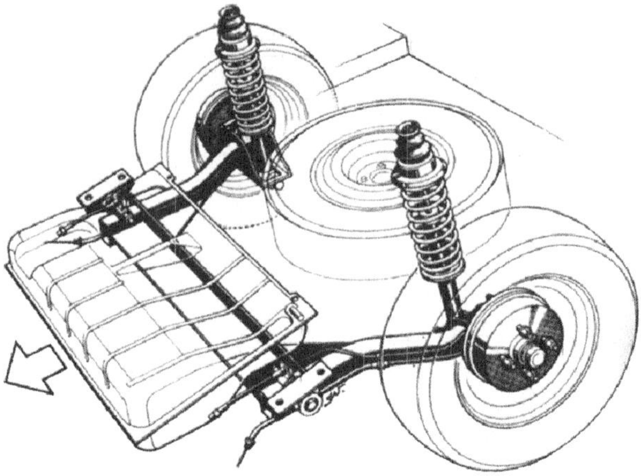

Fig. 3.22 In 1969 Volkswagen conceived a new rear suspension quite suitable to front-wheels drive cars, the so called twist axle. In a short time it became one of the most common architectures (from Genta and Morello 2009)

Fig. 3.23 Multilink rear suspensions were first mass produced by Mercedes in 1982 (from Genta and Morello 2009)

suspension strokes), but stiff in bending (side forces). The wheels are, therefore, effectively guided and, during body roll, they do not change very much their camber angle with respect to the ground.

This overview of car suspensions can be concluded by mentioning the multilink architecture, which represents the pinnacle of suspension evolution. It was introduced during the 1960s on race and sport cars, but was first mass produced by Mercedes in 1982 ; it is shown in Fig. 3.23.

If a suspension is considered solely from a geometric point of view, it can be stated that the suspension mechanism must leave the wheel a single degree of freedom, the stroke, not considering the wheel's rotation around its hub. If only kinematic linkages with spherical joints are considered, a maximum of five linkages can be used to reduce the six degrees of freedom of a free body in space to the single one. The multilink suspension features five linkages to obtain the maximum number of adjustable parameters, thus best approximating the ideal behavior.

With this arrangement camber variation with reference to the ground can be minimized. It is also possible to induce angle variations capable of optimizing the steering behavior and design the longitudinal flexibility at will. Optimum comfort can thus be obtained with no drawback on steering. The multilink suspension family is very large; it is now found on almost all rear wheel driven cars with front engines; many members of this family are also already present on large and medium front-wheels drive cars.

This section covers the last 60 years, but the evolution of independent wheel suspension has not stopped; although it is difficult to conceive new configurations of these mechanisms, many efforts are being made in the field of controls.

3.3 Wheels

Considering the important role wheels play on cars and the fact that wheels are the most ancient element used on ground vehicles, this historical outline will be extended to much more ancient times. As already stated, the oldest wheels date back to the middle of the third millennium B.C., and are solid discs made of three planks, like those of Figs. 1.1b and 1.2.

The need of building lighter wheels for war chariots probably led to the development of the spoked wheel, which is much more efficient for such use. The wheel with spokes was probably first used about 2000 B.C. and by 1600 B.C. had reached its fully developed form, particularly in Egypt. The central part of a wheel of that type (with 8 spokes) is shown in Fig. 3.24a. It is part of a chariot dating back to 1350 B.C., found in a tomb near Thebes. The spokes are fitted into the hub; the felloe is usually built in various parts but some examples of felloes in one piece, bent into a circular shape, were found as well.

A wheel which seems to be a stage in the evolution between the disc and the spoked wheel is shown in Fig. 3.24b. The provenience of this wheel leads one to believe that it was simply an attempt to copy a spoked wheel by a wheelwright familiar with disc wheels; but wheels of the same type are represented on more ancient Greek paintings.

In many cases, even in very ancient times, wheels used a hoop or tire or at least some device to strengthen the rim. Some disc wheels have a wooden rim in one or more pieces. Sometimes the rim of the wheel is inlaid with copper nails, to reduce wear or perhaps to hold a leather tire in position. Certainly many Egyptian war chariots had wheels covered with leather. In some pictures, even very ancient ones,

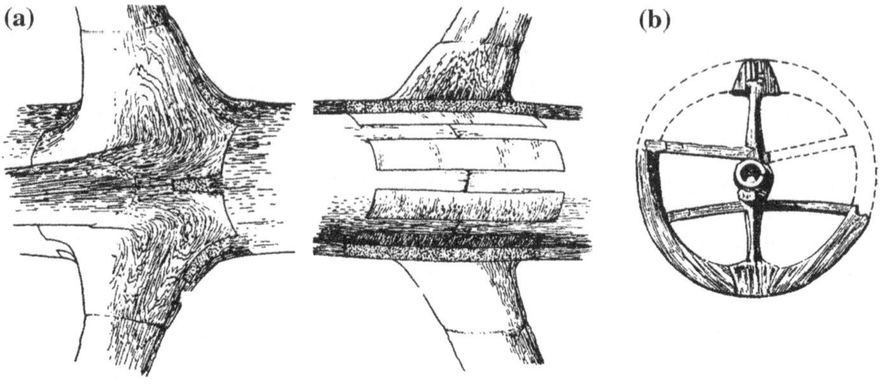

Fig. 3.24 **a** Cross section and view of the center part of a wheel of an Egyptian cart from 1350 B.C. found in a tomb near Tebe. **b** Wheel found in Mercurago, on the River Po flat, probably dating to about 1000 B.C (from Genta and Morello 2009)

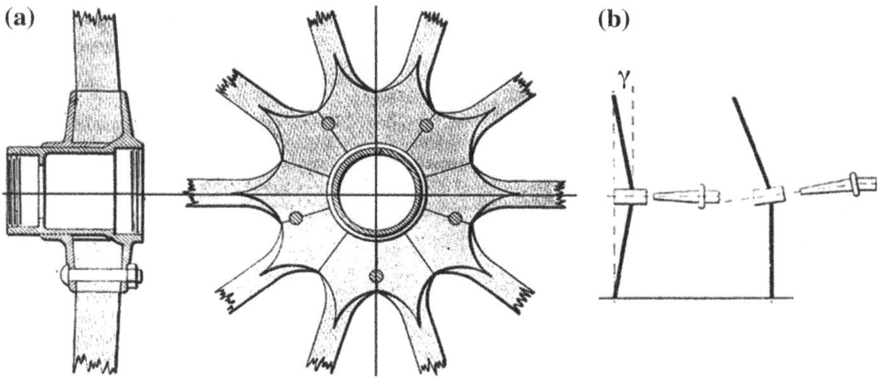

Fig. 3.25 So-called artillery wheels were commonly applied on carriages and on the first cars. Wheel spokes are slightly inclined, following a bell shape. This shape required the wheel to be mounted at a certain inclination γ, with respect to the vehicle, so that the spokes could work correctly (redrawn from Genta and Morello 2009)

something that looks like a metal tire can be seen. The evidence for such practice is however much more recent, dating back to about 1000 B.C. These metal tires were built in various parts, welded together and then shrink-fit to the wheel.

Ancient spoked wheels, like those of the wagon of Fig. 1.4, show many constructional details we can find on the wheels of more modern vehicles, like the so-called artillery wheels, that were commonly applied to heavy carriages and coaches at the end of nineteenth century and on many of the early cars; the hub of one of these wheel is shown in Fig. 3.25a.

The wheel is made by wooden spokes, whose base is shaped like a sector of the hub. They are clamped by two metal flanges, one of which is integral with the outer ring of the bearing; the rim is made of wooden felloes kept in place by a shrink-fit steel ring which forces the spokes into the hub. Spokes are inclined with respect to the hub according to a bevel structure.

This kind of design increases the lateral strength and stiffness of the wheel and gives it the necessary radial flexibility to allow for shrink-fitting the rim. It, however, required a suitable inclination of the wheel rotation axis so as to have the spokes working correctly under the vehicle weight; this angle γ was called *camber angle* and is shown in Fig. 3.25b. On metal wheels this kind of inclination was no longer necessary, but the term camber angle for the angle between the midplane of the wheel and a plane perpendicular to the ground remained.

On this kind of wheel a solid rubber rim could be mounted. Other applications showed a cycle type wheel with solid rubber tire, like those used on the cars by Daimler and Benz, shown in Fig. 3.4.

In Fig. 3.26 many of the fundamental evolutionary stages of the pneumatic wheel are shown. The wheel shown in Fig. 3.26a, in use until the 1910s, was similar to the artillery wheel; instead of a metallic tire, it had a channel ring designed to receive the pneumatic tire. These wheels could not be easily dismounted from the hub; initially, a punctured tire could be changed only directly on the lifted car, with enormous difficulty. The separable channel rim, shown in Fig. 3.26b, makes changing the tire easier, allowing to carry one or more spare rims with inflated tire on board.

Solid metal wheels first appeared in the 1920s and with them the mechanical strength was impressively increased. The new wheels, the so called Sankey wheels, were detachable from the hub through a bolted flange and changing a punctured wheel became much easier.

Spoke wheels were used until the 1950s because of their low weight, particularly on sport and luxury cars; the first type shown has a conventional flanged hub, while the second features a rapid dismounting Rudge Whitworth joint. This is characterized by a spline coupling, between wheel and hub, held in place by a large butterfly nut, a feature which became a style feature of fine cars.

In comparison with the early cars by Daimler and Benz, a different spoke arrangement was used in more recent models: Now two rows of spokes, with a different geometric configuration, are present. The first row, inside with reference to the car, has spokes with radial inclination only; these spokes transmit braking and driving torque and are inclined so that they can transmit forces only along their axes. The second, outer row has spokes arranged on a conical surface. They can thus transmit lateral forces applied to the wheel, together with the spokes of the first row.

In the same figure the Michelin wheel is shown as a last example. It includes a stamped steel disc welded to the channel rim; after a number of improvements, this solution is still in use.

3.4 Tires

Today a car without pneumatic tires is inconceivable. The term *tyre* or *tire* was used for the outer cover of the wheel made of leather, rubber or felt used on carts and coaches; they were applied particularly in urban roads to reduce the rolling

Fig. 3.26 Evolution of the car wheel: **a** and **b** are made of wood with metal reinforcements; **b** has a removable rim; **c** is the Sankey wheel, made entirely of steel and detachable from the hub; **d** and **e** have cycle type spokes; **e** has a rapid dismounting fit; **f** is the Michelin disc wheel scheme, still used on today cars (redrawn from Genta and Morello 2009)

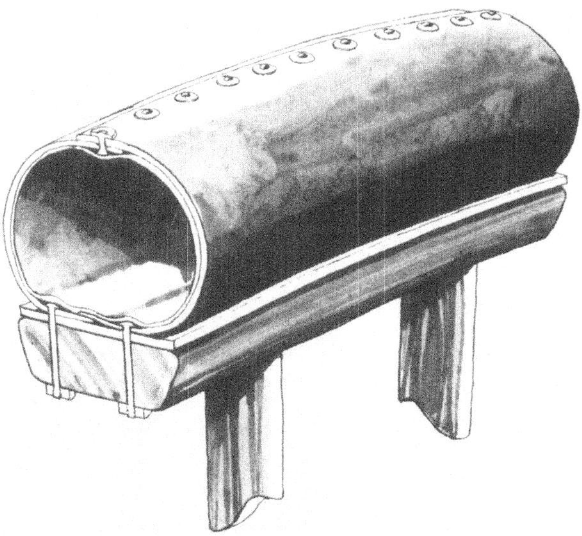

Fig. 3.27 The first prototype made by Thomson was built with a leather cover nailed on the rim of a wooden wheel (from Tompkins 1981)

noise and to improve the ride comfort. Pneumatic tires originated in Great Britain, more precisely in Scotland and the patent title, *pneumatic tyre* generated the current name of this component. The adjective pneumatic was added to point out the use of compressed air to enhance their elastic behavior.

The tire had two different inventors who acted separately without any knowledge of the work of the other. The first patent was filed in 1845 by Robert William Thomson, born in 1822 at Stonehouse in Scotland and was named *aerial wheel*. In the patent the tire is described as a hollow belt made with any material that should be tight to air or water as rubber or gutta-percha; this belt should be inflated with air in such a way as to present at any point of its periphery an air cushion as interface with the ground or rail where it rolls upon.

The first prototype made by the inventor according to his idea was built with a leather cover nailed on the rim of a wooden wheel, as shown in Fig. 3.27. A second tube, made with cloths soaked with rubber to be air tight, was placed inside the leather tube.

The 1849 Mechanics' Magazine of London, whose cover is shown in Fig. 3.28, reported the result of a pull test performed on a coach on macadam and gravel ground; the application of the pneumatic tire on the wheel reduced the traction effort from 45 to 28 lb in the first case, and from 120 to 38.5 lb in the second. The relevant reduction of tractive effort can be explained by the lower pressure exerted on the incoherent ground, with consequent reduction of the penetration of the wheel at the contact point. On incoherent soils the rolling resistance is primarily due to the plastic

Mechanics' Magazine,

MUSEUM, REGISTER, JOURNAL, AND GAZETTE.

No. 1233.] SATURDAY, MARCH 27. [Price 3d.

Edited by J. C. Robertson, 166, Fleet-street.

THOMSON'S PATENT AERIAL WHEELS.

Fig. 1.

Result of Experiments tried by Messrs. Whitehurst and Co., and the Patentee, for ascertaining the comparative Draught of R. W. Thomson's Patent Aerial Wheels and the common Wheels. Tried in Regent's Park, March 17, 1847.

Weight of Carriage, $10\frac{1}{2}$ cwts.	Common Wheels. Actual draught in pounds.	Patent Wheels. Actual draught in pounds.	Saving of draught by Patent Wheels,
Over a smooth, hard, macadamized level road	45	28	60 per cent.
Over new broken flints	120	$38\frac{1}{2}$	310 per cent.

Fig. 3.28 The cover of Mechanics' Magazine reporting the results of a pull test performed on a coach on macadam and gravel ground. The application of tires reduces the traction effort noticeably

displacement of the ground under the contact pressure; the contribution of the rubber hysteresis is in this case negligible.

The inventor died in 1873 without having the chance to make any business with his invention; the invention was forgotten for the small interest of the advantages for a coach in view of the negative impact on cost and service life.

The second patent on tires was filed in 1888 by John Boyd Dunlop, again a Scotsman, born in 1840 in Dreghorn. This time the invention had more chance to be exploited because it was proposed for a bicycle that could take more advantage from the use of tires, while posing smaller structural problems; the idea was transformed into a successful industrial product that made the inventor famous around the world with his own trade mark.

The first tire was made by wrapping up the wheel rim in a rubberized sailcloth bandage; this bandage crossed the wheel spokes and could not be disassembled from the wheel. Because of its appearance this application was named *mummy tyre*.

The invention had immediate success, owing to the reduced rolling resistance; in sport meetings amateur bicyclist with tires were reported to beat famous professional bicyclist with conventional hard rubber wheels. Also benefits on comfort were much appreciated.

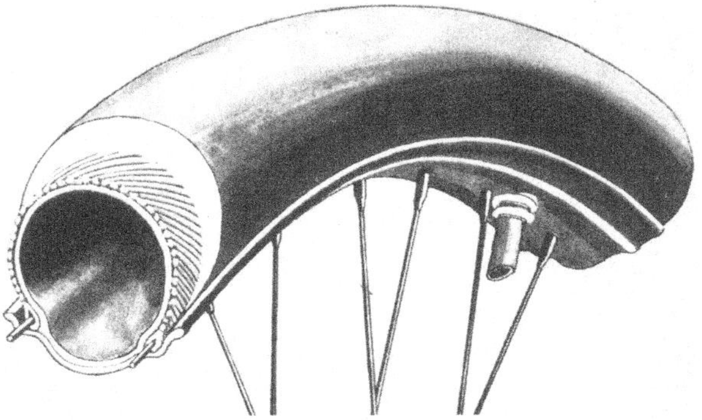

Fig. 3.29 The Welch's invention consisted in the separation of the tire from the tube; the tire received an open cross section kept in contact with the wheel rim by steel cables (from Tompkins 1981)

The Pneumatic Tyre Ltd. was set up in Dublin for the industrial production of tires. Many requests for production licences were presented by interested companies, but Dunlop liked better to fabricate tires an exclusive product, to take the maximum advantage.

This very binding situation stimulated other potential producers to find out different ways to produce tires and probably during their researches the 33 years old Thomson's patent was rediscovered; Dunlop's patents were invalidated after a legal suit. Dunlop's bona fides was straight and incontrovertible; his patent was the consequence of the lack of attention of the patent office due to the difficulty in consulting old documents and to the reduced notoriety of the first invention. The deletion of the patent coverage gave an advantage to the entire industrial system that could take immediate benefit of the new invention.

An unpleasant characteristic of the first Dunlop's tire was the difficulty in repairs and the impossibility to disassemble the tire from the wheel. Other inventors developed and improved the original idea.

The first improvement was developed by Charles Kinston Welch and was patented in 1890. His invention, described in Fig. 3.29, consisted in the separation of the tire from an inner tube for air containment, both made with rubberized cloths of different thickness; the tire received an open cross section, like a horse shoe; at its two edges steel cables kept the tire in contact with the wheel rim.

A second type of tire-rim interface was invented by William Erskine Bartlett, an American living in England; according to his proposal the steel cable could be avoided if the tire had an Ω cross section matching with the rim with the shape of the bead.

By installing this tire on a wheel with a suitable rim shape, as in Fig. 3.30, the air pressure itself keeps the tire in place because of the matching bead. Two different wheel and bead shape are shown in the figure.

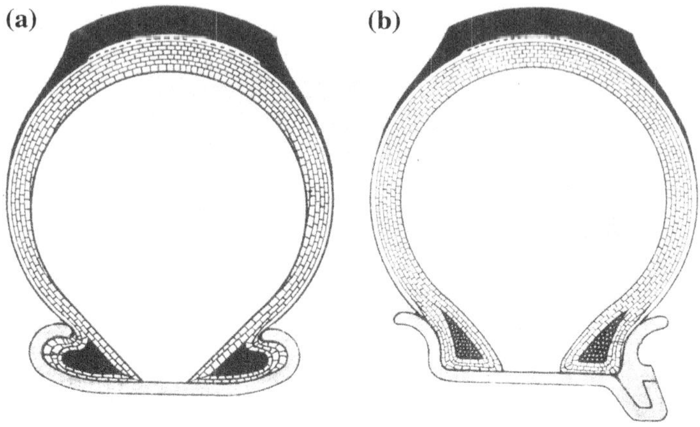

Fig. 3.30 According to the Bartlett's patent, the steel cable could be avoided with a suitable shape of the bead, matching with the wheel rim (redrawn from Tompkins 1981)

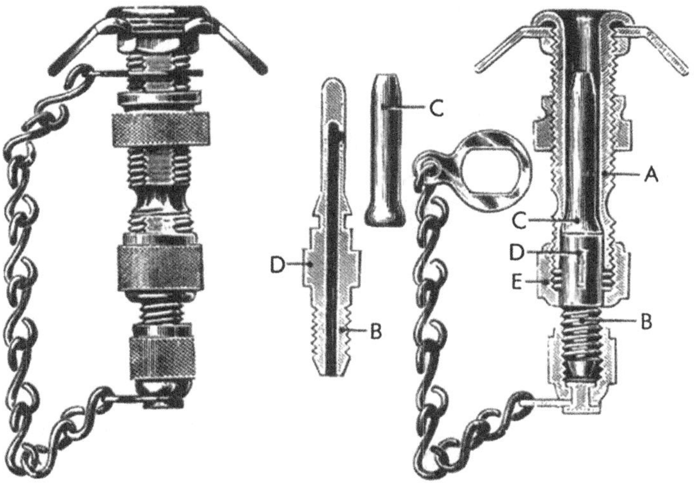

Fig. 3.31 A Wood's valve is characterized by a body D with two holes that intersect each other with a right angle. On this *cylindrical body* a rubber tube C is fitted, working as a check valve (redrawn from Tompkins 1981)

The inflation valve, patented by Charles Woods is shown in Fig. 3.31; it is characterized by a body D with two holes that intersect each other at a right angle. On this cylindrical body a rubber tube C is fitted; during inflation it enlarges because of the air pressure so that the hole opens, while during normal operation it remains tight.

Integrating the cylindrical body with a nipple E, it is possible to inflate the tire with a pump connected to the joint B; to deflate it is sufficient lo loosen the threaded nipple. This kind of valve is still in use in some English bicycle, while for common application a lighter and smaller poppet valve was later introduced.

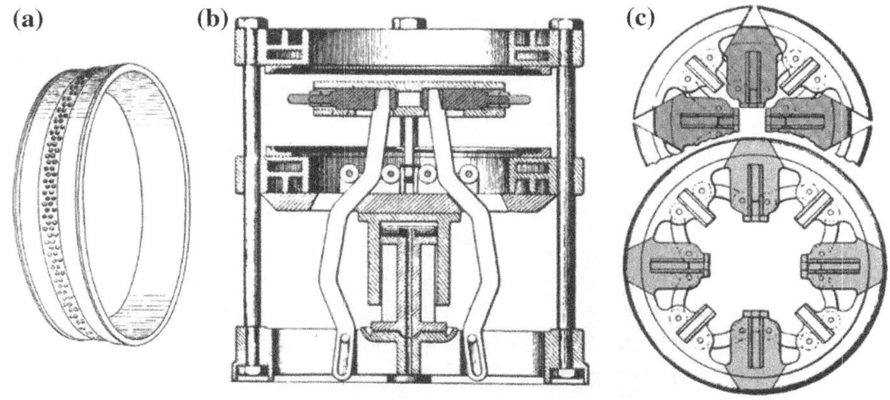

Fig. 3.32 **a**: Die for tire reverse stamping; **b** and **c**: Expandable die for tire direct stamping (redrawn from Tompkins 1981)

A major issue was the limited service life of tires; this drawback was not only due to the quality of the rubber available at that time but also to the weakness of the armature cloth.

The structure of a tire cloth may be thought as a number of layers of rubberized fabric, with cords running in the direction of the warp and in that of the weft. The structural behavior of a cloth, when withstanding a uniform load, as the inflating pressure, is similar to that of a shell made by its threads; the stress of the threads is increased where their curvature in increased because they cross each other. The strength of the fabric is thus increased when the number of crossing points is decreased but, on the other hand, the cloth can be manipulated with minimum damage only if the number of crossings is large enough.

John Fullerton Palmer obtained a patent in 1892 by proposing the use of a cloth made with warp only, with no weft at all; according to this patent the structure of the tire is obtained by bands (*cords*) made with parallel threads, kept together with a layer of rubber. The rubber maintained the threads in position till the complete hardening of the rubber. This idea is still applied in contemporary practice.

At the end of the nineteenth century stamping of tires was made on the reverse side. A steel die used to create the tire had a cylindrical shape with the drawing of the tread and trade marks on its outside surface, as it can be seen in Fig. 3.32a; the positioning grooves for cables and beads were located at its ends.

The rubber was spread on the die, then the cloth and the two rubberized cables were set in position; these cylinders were then cured in a heated autoclave for polymerization and vulcanization. At the end of this process the tire was reversed and the final shape was obtained by the first inflation; in this way, tires were stressed even before their first use.

Doughty designed and patented in 1896 the first press for stamping tires to their final shape. This press is sketched in Fig. 3.32b and c: The die featured three parts sliding on vertical columns, where the intermediate part, for obtaining the inside of the tire, was made with eight expandable elements, moved by suitable levers. The tread grooves and trade marks were printed on the upper and lower part.

The process of assembling cloths and rubberized cables was performed outside of the mould, according to the final tire shape; in this way the cloth could adapt to the die shape without stress, having its threads free to move in the liquid rubber. The final shape was more precise and independent of the inflating pressure.

At this point, the tires produced for bicycles were already quite similar to the present ones; however, when the first cars were produced, the same evolutionary process repeated again, apparently with no account for the experience gathered with bicycles. The reason was that tires were not suitable yet to the higher load of a car and to the higher driving and braking forces of the new application.

Another issue was that the way a bicycle tire operates, with almost only radial load, was quite different from that of the car, subject to significant loads perpendicular to the wheel plane. The most common solution was again using solid rubber tires, already abandoned by bicycles.

A significant contribution to the development of the pneumatic tire for cars was given by André and Edouard Michelin who applied their tires for the first time in the Paris—Brest race of 1891; the Michelin brothers had already an experience with bicycles tires. The expected advantages were again the same: a lower rolling resistance and a better comfort that could imply also lower stress for the chassis structure, particularly considering the poor effectiveness of suspensions, mainly at high frequencies. These advantages were however accompanied by frequent repairs of punctured tires.

Despite the time spent for 5 punctures and their repairs in the 1,200 km of the race, the car of the Michelin brothers arrived first, 8 h before the second.

But this was not enough for the diffusion of tires on cars. A second major issue was the monopoly of the Pneumatic Tire Ltd, renamed Dunlop Rubber Co. after 1900, on many important patents, particularly of the Welch's patent on the steel cable in the bead. The remaining manufacturers, among them Michelin, had to settle with the Bartlett's bead whose licence was offered. It is not surprising that the diffusion of tires on cars started only in 1904 when the Welch's patents expired.

In fact, the bead proposed by Bartlett (Fig. 3.30a) was not sufficient to keep the tire in position safely, against the forces applied in curves and many manufacturers tried to solve the problem by applying additional devices to improve the steadiness of the tire-rim connection.

An example is given in Fig. 3.33, showing the 1907 Itala 35–45 HP, winner of the famous rally from Peking to Paris. In this car the rim can be disassembled in thee parts, the wheel itself and two side rings kept in position by the screws shown in the picture; the tire can be disassembled without any deformation of the bead that can be, in this case, very massive and robust.

A different approach to the same problem is shown in Fig. 3.34; on the 1917 Panhard & Levassor the one-piece rim cannot be disassembled from the wheel but the tire can be changed operating as shown in the sketches on right part of the figure. A number of wedges with bolts and butterfly nuts kept in position the tire in the rim

Fig. 3.33 In this car the rim can be disassembled in thee parts; the tire can be removed without any deformation of the bead that can be, in this case, very massive and robust (National Automobile Museum of Torino)

Fig. 3.34 In the 1917 Panhard & Levassor the one piece rim cannot be disassembled from the wheel. The figure on the right shows how tire and tube can be changed

and the rim on the wheel; the rim profile had to be modified to retain the tire more efficiently.

Obtaining a sufficient grip for the tire on the road in longitudinal and transversal directions was more difficult in cars than in bicycles, owing to the weight of vehicles and to the road condition, often dusty or muddy. In many cases additional covers

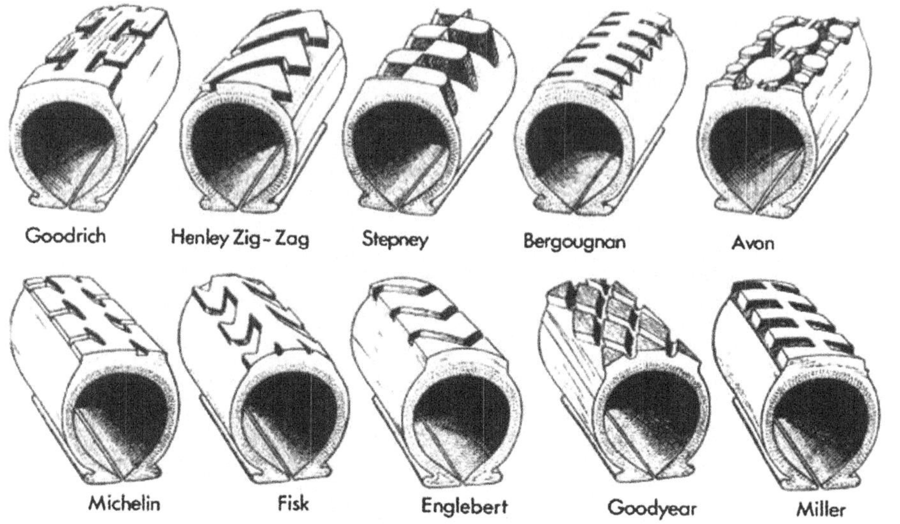

Fig. 3.35 State of the art of tread design in the 1920s (redrawn from Tompkins 1981)

with nails were mounted on the tire, kept in position by belts and buckles; nails had the function of increasing grip.

Some tire manufacturers included this kind of additional cover in the tire; others deformed the tread before of hardening with tubes in order to obtain a sort of groove on its periphery or added hard rubber spheres during hardening.

New developments were available for the tread when the stamping technologies were improved using direct stamping that made obtaining more elaborated shapes possible. The set of drawings in Fig. 3.35 show the state of the art in the 1920s. The most relevant target was obtaining grip improvements on incoherent ground with protruding knobs (*blocks*) that could exploit the shear resistance of the soil, without using add-on devices that were often lost on the road.

Each design, strictly empirical, was protected by patents and considered as a selling point; the improvement in grip had to by paid by a reduced comfort and an increased rolling noise; some manufacturer, like Rolls-Royce, used only peripheral grooves with reduced longitudinal grip performance.

On the other side the introduction of blocks on the periphery of treads proved to be suitable to avoid the premature appearance of cracks, as reported in solid rubber tires: In fact, blocks increased flexibility and reduced operating temperatures, by reducing heat generation in repeated bending and by increasing heat exchange with the environment.

A good compromise in reducing noise and increasing service life and grip was obtained again by Dunlop in 1927 with a particular design shown in Fig. 3.36.

The blocks were connected with rubber bridges to avoid lateral deflections that made edge contacts with the ground more likely: Edge contacts were partly responsible for noise and discomfort and for premature wear of the tread. A further refinement

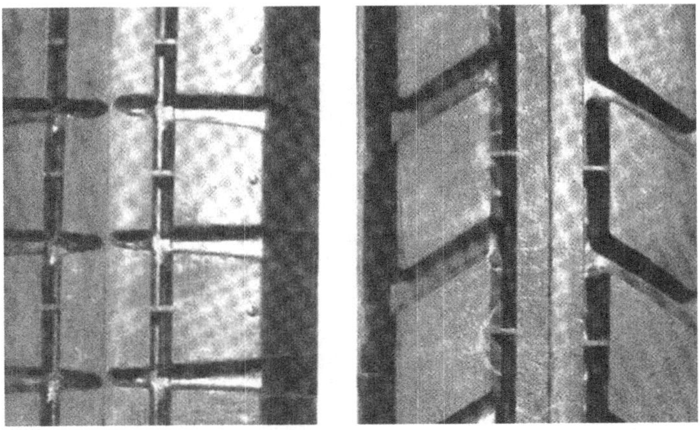

Fig. 3.36 Particular block design developed by Dunlop to limit edge contact of blocks with he ground, with benefits in comfort and service life

consisted in a central area of the tread without blocks. Due to the curved shape, only the central zone of the tire was in contact on a paved road, with a small reduction in grip and no effect on comfort; on the contrary, on muddy roads, a wider contact area made the blocks effective for grip improvement. This approach was imitated by competitors and is applied also in contemporary tires.

The improved fabrication process made it possible to build tires of larger size, particularly for what the width is concerned, leading to an increased comfort. This new breed of tires born in the 1930s, called *balloon* or low pressure tires, required a lower inflation pressure and allowed therefore an increase in vertical flexibility.

About at the same time the potential contribution of *blades* in improving grip was discovered; blades were made by parallel cuts on the surface of blocks; the increased local flexibility caused the blades to scratch the ground with their edge, with improvements in traction on wet or muddy grounds. Initially they were actually cut on the blocks after the vulcanization process, afterwards they were obtained by modifying the dies.

At the end of the 1950s *tubeless* tires were invented, with reduction in weight, stiffness and cost; this new design was made possible by treating the internal surface of the tire for tightness and by reducing the dimensional tolerance of the wheel rim. An additional improvement was the reduction of the deflation speed in case of puncturing, with benefits for safety.

In the same period the so called *safety shoulder* was introduced, again by Dunlop. It was an additional portion of tread added at the sides of the tire, as shown in Fig. 3.37. The function of this modified tread was reducing tire sensitivity to longitudinal small obstacles, such as streetcar rails or pavement joints. In the following developments the shoulder became a rounded area of the tread at the sides of tires.

After World War Two some french manufacturers, like Citroën and Panhard, introduced the first small cars with front wheel drive; they represented a new challenge

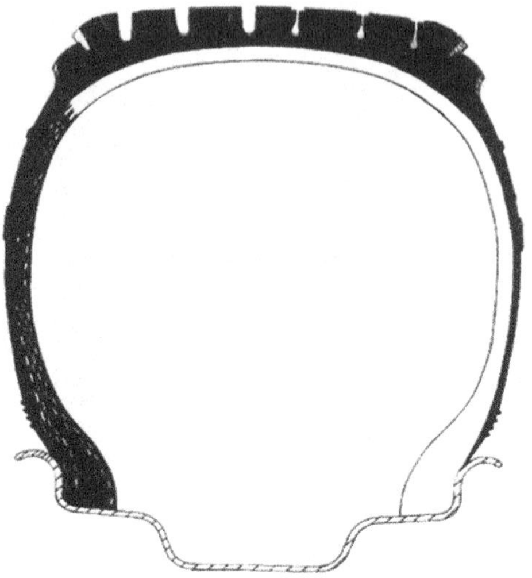

Fig. 3.37 The so called *safety shoulder* was an additional portion of tread on the sides of the tire (from Tompkins 1981)

for local tire manufacturers since front wheel tires were subjected to drive and steering forces at the same time.

The plies of the carcass of existing tires were made of cord tissue, i.e., cloth without weft. The threads of the warp were running in a helical direction; cord layers with symmetrical inclination were provided to grant the tire a sufficient stiffness to driving, braking and side forces. This structure was called *cross-ply* from the layout of cords.

This particular layout caused a significant torsional deformation of the tire contact area, as evidenced by the side slip angle of the wheel path; in a curve the combination of steering and traction forces produced a significant amount of heat that could damage the tire in a short time.

The idea behind the X tire, invented by Michelin to solve this problem, was quite simple: Being impossible to improve the rubber sensitivity to heat, the solution of the problem was to reduce the deformations of the contact area with a stiffer cord that in this case was made of steel. A new cord lay-out was developed for the purpose, providing a set of steel threads almost circumferential, the so called *belt*, and a set of almost radial cords of rayon at the sides of the tire; this new structure was quite stiff to side slip deformation and flexible to radial forces, with good result for comfort. A scheme of this structure is shown in Fig. 3.38.

The structure of this new tire, called *radial tire* from the cord layout, not only lead to a significant increase of the service life, but also to an increase of the cornering stiffness.

Owing to the increased cornering stiffness of this kind of tires, solid axle suspensions or independent suspensions not carefully controlled in their motion could cause an unpleasant steering behavior. As a consequence, this new generation of tires lead to the development of a new generation of front axle independent suspensions

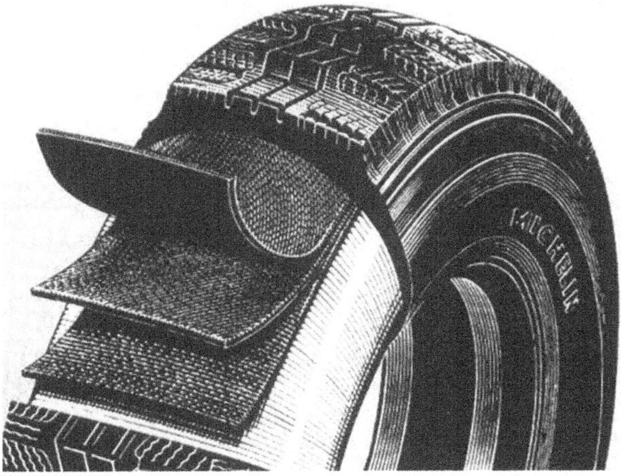

Fig. 3.38 Structure lay-out of the Michelin X belt or radial tire

with a more careful design. Radial tires were developed by other tire manufacturers with belts made with conventional material as well.

A last aspect of the tire evolution can be shown by their geometry, characterized by the rim diameter D, the tire width W, the rolling diameter D_r (tire diameters are traditionally measured in inches, tire widths in mm) and the aspect ratio AR, defined by the equation (in which consistent units are used):

$$AR = \frac{D_r - D}{W} .$$

Early tires were characterized by a large diameter and a small width; their aspect ratio was not standard but its value was about 100 %. The need to improve comfort, increase cornering stiffness and reduce car height caused a reduction in diameter while the width increased.

Figure 3.39 show the evolution of these dimensions for average family cars with a curb mass in the range of 1,000 kg; the rim diameter D decreases from 25″ in the 1910s to 13″ in the 1970s with a subsequent small increase; the tire width W doubled from the initial value of 90 mm to the value of 180 mm in recent cars.

The aspect ratio AR decreased consequently from an initial value of about 100 %, to 80 % in the 1940s to the present value of 50 %.

3.5 Brakes

The evolution of brakes is linked with the increases of both speed and traffic density, with the related need of stopping cars in shorter distances. It is reasonable to think that braking early cars in the traffic of their time was not a serious concern. The earliest

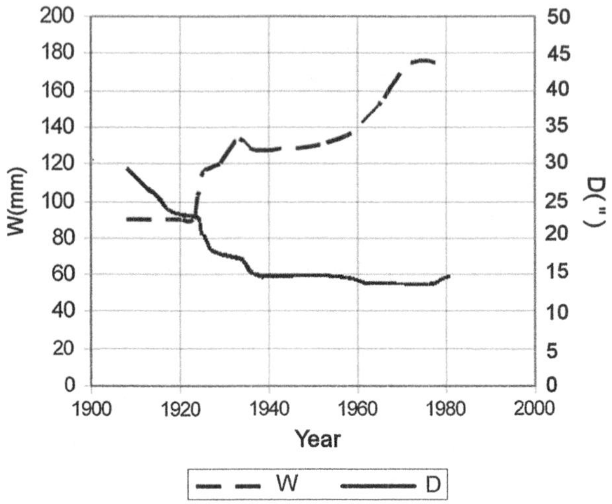

Fig. 3.39 Evolution over time of tire dimensions for cars with curb mass in the range of 1,000 kg (from Genta and Morello 2009)

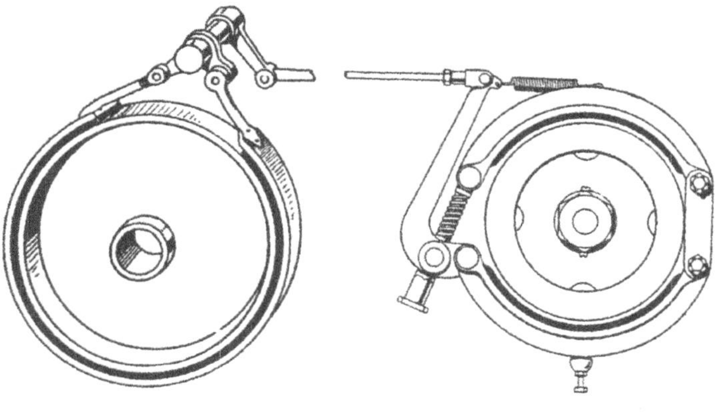

Fig. 3.40 Band brakes and external shoe brakes, to be applied on the transmission line before the differential and final drive (from Genta and Morello 2009)

cars featured brakes that worked on the external surface of the wheel with moving shoes, like those on horse drawn carriages and railway cars; this simple device was not suitable for wheels with rubber tires because of the potential wear.

For this reason external shoe brakes and band brakes working on the transmission line, like those shown in Fig. 3.40, were introduced. The choice was motivated by the need of exploiting the differential also for distributing uniformly the braking force on the wheels of the axle, a thing difficult to achieve with mechanical control linkage. Such brakes were also virtually insensitive to machining tolerances.

Fig. 3.41 Mechanical control of front wheel brakes; the contact point of the cam, actuating the expansion of shoes, is close to the king pin axis of the wheel (from Genta and Morello 2009)

Brake liners were missing, but some examples of *Ferodo*® liners, a synthetic material developed by Frood, were already present on shoe brakes in coaches; this practice spread on a large scale in the 1920s.

Drum brakes were developed later and were applied on rear wheels only; they were more difficult to machine but showed a superior ability of dissipating the heat generated over long braking distances. Band and shoe brakes were not initially abandoned, but continued to be applied on the transmission as supplementary or parking brakes. Disc brakes were introduced into mass production by Citroën in 1955.

Drum brakes were controlled by a pedal and were usually designed to stop the vehicle; the transmission brake, on the other hand, was controlled by a hand lever which could be locked in the desired position. It was usually designed for continuous braking on downhill slopes and for use as a parking brake.

The position and function of driver controls was informally standardized only in the 1930s and the solutions initially chosen could be much different; a system that had a certain diffusion, and which is seemingly unjustified today, provided for combined control of brake and clutch to avoid stalling the engine when stopping the car.

The use of brakes on the front axle began only in 1910, thanks to Isotta Fraschini, a luxury car manufacturer. This innovation did not regard the brakes themselves, which were similar to other rear brakes, nor the realization of their usefulness, but the control system for steering wheels, as shown in Fig. 3.41.

Brake controls were in fact mechanical, as already shown in Fig. 3.40. They were made by linkages and cables put in tension by a pedal; in drum brakes the internal shoes were expanded by a cam. What had to be avoided, at any rate, was that the braking force could be affected by wheel steering angles; the patent by Isotta Fraschini

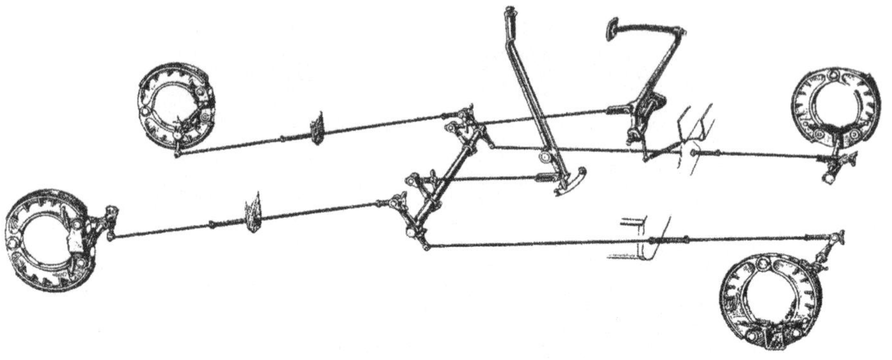

Fig. 3.42 Mechanical control for a four-wheels braking system; the articulation points between levers and tie rods had to be located at the centre of instantaneous rotation of the suspensions (from Genta and Morello 2009)

consisted in rounding the cam in such a way that contact points would always be concentric to the king-pin axis. Brake control was thus made insensitive to steering; this patent was followed by other mechanical solutions and four wheel braking was state of the art by the end of the 1920s.

What was said about steering motion can be also applied to the suspension motion of both axles. The distance between the articulation points of tie rods could not be allowed to change by more than a small value during the suspension stroke; if this happened the brake could be actuated by road bumps or the tie rods could be broken by excessive pull.

For this reason, as shown in Fig. 3.42, the articulations points between the levers and tie beams had to be located at the centres of instantaneous rotation of the axles, located near the fixed eyes of the leaf springs during the suspension stroke. Since the distribution of braking forces was influenced by the lengths of the rods, the latter had be adjustable in length, to compensate for the wear of the shoes.

All these problems were solved by the application of hydraulic controls, which spread during the 1930s. They were proposed by Lockheed and introduced on a car for the first time by Duesemberg in 1921.

The last component that contributed to the development of the braking system was the power brake, based on manifold vacuum, introduced in 1954 by Ford ; this device enabled also weak driver to brake strongly and immediately.

Before that, as early as 1910, many luxury and race cars were provided with a completely mechanical power brake, as shown in Fig. 3.43. The mechanical energy needed to operate the brakes was derived from the engine, transferred to the braking system through a clutch sliding in a controlled way, managed by the brake pedal. Car inertia also, through the propeller shaft, contributed to slowing the car.

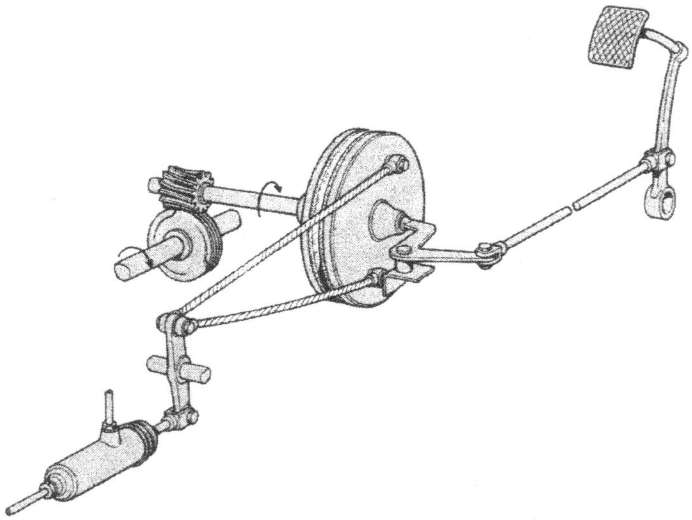

Fig. 3.43 Mechanical power brake introduced on luxury and racing cars; the needed energy derives from the propeller shaft and is transferred to the braking system by a clutch, operated by the brake pedal (from Genta and Morello 2009)

3.6 Wheel Transmission

The transmission of motion from engine to the wheel of the first Benz and Daimler cars was made by a leather belt where possible distance variations between the driving and the driven shafts were compensated for by a spring loaded moving tensioner. The leather belt had the advantage of integrating in a simple way the functions of the transmission with those of the gearbox.

The motion of the driven shaft was transferred to the wheels by a secondary chain drive, as shown in Fig. 3.44, for a slightly more recent Panhard & Levassor.

Car designers took care to place the center of the sprocket of the chain transmission at the same position as the rotation center of the sprung mass in the suspension motion; the application of this rule allowed the axle to move without changing the length of the chain. Similar chain transmission were applied also together with gearboxes.

A different transmission system was applied by Renault in 1899, as seen in Fig. 3.1; it included a simple transmission shaft with two universal joints, to allow for the suspension motion of the rear driving axle. This simpler and better system was applied by other car manufacturers with some delay, because of the patent protection in favor of Renault but, when these patents expired, it became a standard solution for rear-wheels drive cars (since about the 1920s).

In parallel with Renault, De Dion-Bouton applied a different system, to avoid chains and not to infringe the Renault's patents. As shown in Fig. 3.5, engine, gearbox and differential were integrated in a powertrain mounted on the chassis that operates

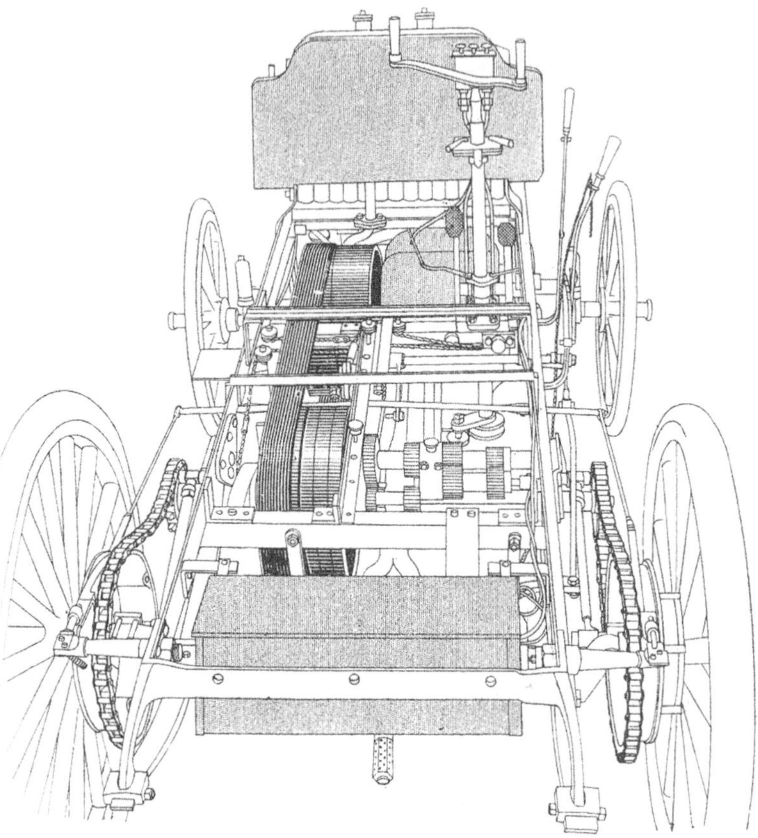

Fig. 3.44 Complete transmission of a Panhard & Levassor prototype, including a first stage with belt drive, operating as a clutch, a gearbox and a second stage with chain drive (redrawn from Baudry de Saunier 1900)

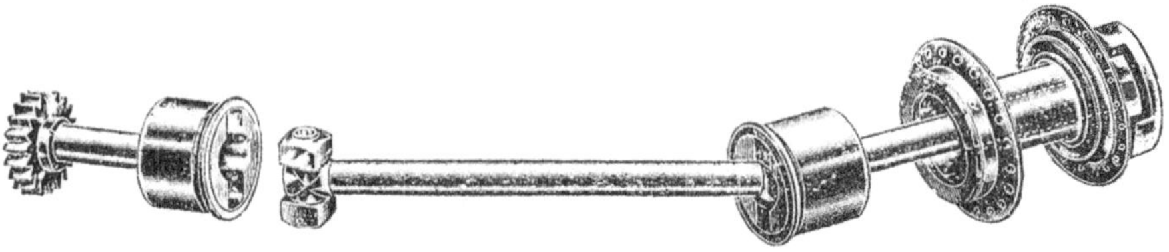

Fig. 3.45 Transmission axle shaft applied by De Dion-Bouton to drive rear wheels, featuring two bipod joints (redrawn from Baudry de Saunier 1900)

the wheels of a solid rear axle through two articulated shafts featuring bipod joints. One of these joints is shown in Fig. 3.45.

This solution, improved with the application of constant velocity joints, was applied also in more recent times (an example is shown in Fig. 3.19) because

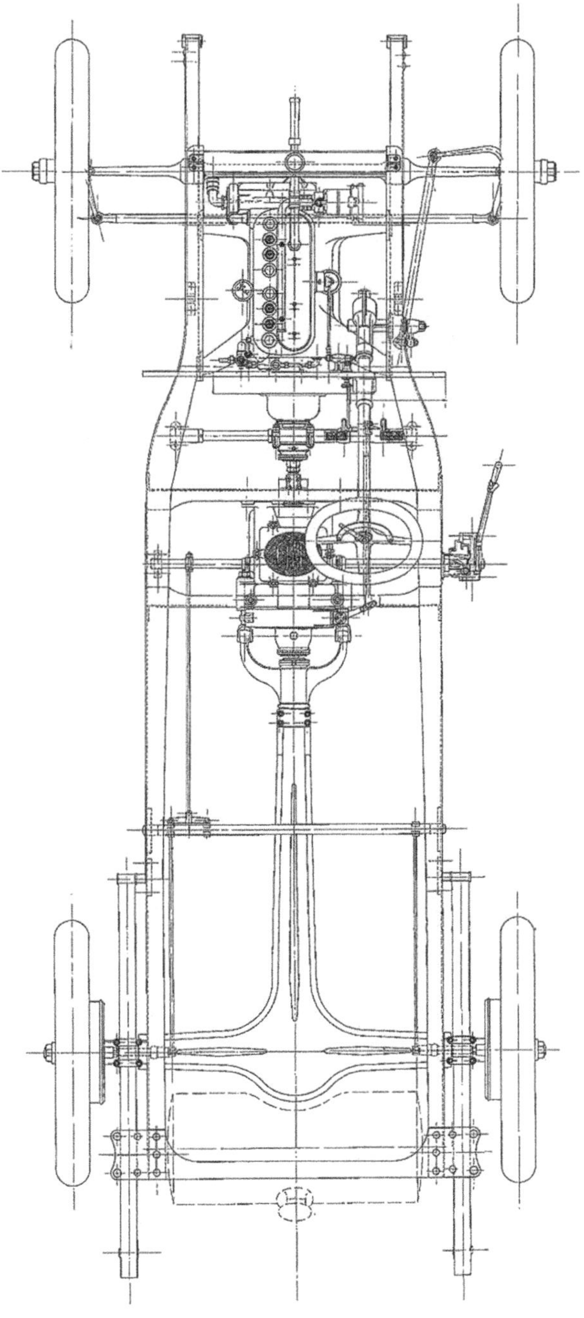

Fig. 3.46 FIAT chassis featuring a single universal joint shaft transmission (courtesy of FIAT Historical Archives)

it allowed reduction of unsprung masses in comparison with more conventional solutions.

At the beginning of the twentieth century shaft transmission were also applied with a single universal joint, again to avoid infringing Renault's patent, as shown in Fig. 3.46 for a FIAT chassis. With this solution, the speed of the wheels was not perfectly constant as with two joints, but there was the advantage of connecting the rear axle with a hinge to the chassis directly, reducing in this way any secondary motion due to driving or braking forces.

Front-wheels drive transmission could not use the De Dion-Bouton solution because of the larger working angles needed to steer the front wheels. Before constant speed joints were developed, the Tracta joint was applied. It included a very short transmission shaft with two universal joints, with their two forks integrated in a single piece.

Chapter 4
Powertrain

The powertrain includes the engine, in modern cars mostly a reciprocating internal combustion engine, and the gearbox, that converts the engine torque into the torque needed for vehicle traction. Some engines of this type that were developed before the invention of the automobile are first described: The De Rivaz, Barsanti, Lenoir and Otto were important milestones in the development of present spark-ignition engines, sometimes improperly called *Otto engines*. The description will be limited to this family of engines, neglecting, for sake of simplicity, compression-ignition engines, sometimes referred to as *Diesel engines*.

An historical perspective of car spark-ignition engine, developed from the end of the 19th till the end of the twentieth century, will then follow. Dedicated subsections are addressed to the technological evolution of the main auxiliary systems of the spark-ignition engines, such as those for air-fuel mixture preparation, lubrication, ignition and starting.

A major issue in spark-ignition engines and, more in general, in internal combustion engines is their torque characteristic, i.e. the plot of the torque versus the rotation speed.

More specifically, reciprocating internal combustion engines

- cannot supply any torque below a certain speed (*idling speed*);
- the maximum value of their torque (*wide open throttle torque*) is almost constant with speed, with the consequent impossibility to accelerate the vehicle at low speed with high power.

An ideal engine should deliver a torque even at zero speed, to start-up the vehicle and should be able to accelerate the vehicle at its rated power at any value of speed. These deficiencies compelled use of a suitable gearbox and a clutch. For this reason a section is dedicated to the evolution of the gearbox and its start-up device, in its two main variants of manual gearbox with friction clutch and automatic gearbox with torque converter.

In consideration of the complex nature of the transmission of the early cars, it is possible to argue that the weight advantage of the spark-ignition engine (including its tank) was partially jeopardized by those heavy gearboxes. In addition, the combined

G. Genta et al., *The Motor Car*, Mechanical Engineering Series,
DOI: 10.1007/978-94-007-8552-6_4,

operation of the clutch, accelerator and brake by the driver, and the starting of the engine by means of a crank were perceived as discouraging by many customers of the first automobiles, who felt they needed the assistance of a professional driver or, at least, mechanician.

The above problems stimulated many manufacturers to apply to their car either electric motors or steam engines that could be governed by a single and simple control. For this reason, an additional section will be dedicated to these two kinds of prime movers.

4.1 Combustion Engines Before the Automobile

The first thermal engine is ascribed to Heron who developed it about 130 B.C.; it was operated by the reaction torque due to two steam jets mounted on a rotating pivot.

Reciprocating steam engines, that later inspired internal combustion engines, were developed between the second half of the 18th and the beginning of the nineteenth century. One of these early steam engines, powering the first self-propelled vehicle developed by Nicholas Cugnot and tested in 1769, is shown in Fig. 2.22.

It is impossible to identify a single person as the inventor of the internal combustion engine; many different attempts were made and many different solutions were tried. One of the first ideas for such an engine can be attributed to the famous astronomer and mathematician Christiaan Huygens who designed a machine at the end of the seventeenth century to raise the water for the gardens of the castle of Versailles. His engine was inspired by a piece of artillery and, thus, its fuel was gun powder.

On the contrary, the priority for the first vehicle operated by an internal combustion engine is certain; this vehicle was invented, designed and tested in 1807 by Issac De Rivaz, after many fruitless attempts performed since 1780 to build a steam vehicle. A scaled down model of this vehicle is shown in Fig. 4.1. Its engine, forerunner of a generation of similar machines, is based on a piston, free to move in a vertical cylinder, together with a rod connected to a chain. The chain was wound on a pulley with horizontal axis: This chain-pulley device is equivalent to a rack-pinion mechanism.

The fuel for this engine was gas contained in a balloon; it could be lighting gas, obtained from the combustion of coal, or pure hydrogen, obtained in a laboratory by the reaction of sulfuric acid with metallic zinc. A certain quantity of gas was introduced into the combustion chamber at atmospheric pressure and ignited by an electric spark. The energy of combustion launched the piston upward; at the end of the piston stroke the combustion gases were exhausted through a valve, leaving the combustion products in the cylinder at atmospheric pressure; at this time the exhaust valve was closed.

The cooling down of the gas caused the cylinder pressure to decrease below its atmospheric value. The piston weight and the ambient pressure forced the piston to return to its initial bottom position slowly; the energy coming from the combustion was partly changed into useful work by the piston descent. A ratchet mechanism

Fig. 4.1 Scaled down model of the vehicle of Issac De Rivaz, tested in 1807 (from L. Morello, *Evoluzione tecnologica dei motori automobilistici ad accensione comandata*, ATA: Ingegneria dell'autoveicolo, 2004, N° 7/8 e 9/10)

connected the wheel with the chain moved by the piston; this mechanism moved the driving axle through a rope transmission.

This family of internal combustion engines, where the internal pressure becomes lower than the ambient pressure during the work phase, were called *atmospheric engines*. The useful work was produced by the atmospheric pressure working against the partial vacuum in the cylinder.

Timing the engine, such as opening or closing intake and exhaust valves or igniting, was accomplished manually, by the driver. De Rivaz described in his reports a vehicle of 900 kg of mass; the engine piston had a bore of 365 mm and a stroke of 1.5 m; the vehicle speed depended on the ability of the driver to operate engine controls quickly. During the official test, the vehicle drove for 9 m, at each engine stroke, accomplished in 5 s, corresponding to a speed of about 7 km/h.

A similar mechanical architecture is shown by the engine of Barsanti and Matteucci; however an important improvement was achieved by timing the engine automatically, as in modern engines. The fuel was still gas, therefore not very suitable to be carried on board a vehicle. A scale model of this engine is shown in Fig. 4.2.

Niccolò Barsanti (1808–1864) was a professor of Mechanics and Hydraulics at the Ximenian Institution in Florence; Felice Matteucci (1821–1887) was a brilliant mechanical engineer specialized in the design of canals and dams, to reclaim the marches in Tuscany. They started studying gas combustion in 1851 and in 1853 they filed officially a sealed document at the Academy of Georgofili. This document described in details the results that could be obtained with a gas combustion engine

Fig. 4.2 Scale model of the first atmospheric engine of Barsanti and Matteucci tested in 1854 (from L. Morello, *Evoluzione tecnologica dei motori automobilistici ad accensione comandata*, ATA: Ingegneria dell'autoveicolo, 2004, N° 7/8 e 9/10)

they invented; the contents of these documents were disclosed only in 1863, when they had demonstrated the performance of their engine with working prototypes.

The first prototype was working in 1854 and in the same year they obtained a patent in England. Later they patented the engine in different Countries. These patents did not regard the engine thermodynamics, already known from the papers of Carnot, but the mechanisms that enabled the engine to work automatically. The first engine delivered to a customer was built by Benini Works in Florence and installed to operate the railway machining shop in the same town.

This engine features two pistons; one of them is shown as detail *a* in Fig. 4.2 (the figure numbers refer here to the drawing, taken from the original patent, reported in Fig. 4.3). The piston rod is connected through a ratchet to a gear wheel *b*, operating the shaft *c*, that can be seen in Figs. 4.1 and 4.4. The piston is launched upwards by the combustion energy and stops as the combustion products are exhausted by the sleeve valve *f* in Fig. 4.3. From this time on, the return phase begins under the action of atmospheric pressure that is higher than the internal pressure. The descent motion causes shaft *c* to rotate; the speed of this shaft is made regular by a flywheel (see Fig. 4.2) and by a second piston with opposed phasing.

The same shaft moves, with a crank and a connecting rod, the yoke *d* in Fig. 4.3; its motion is phased with that of the piston and opens and closes the sleeve valve *f*; the function of the piston *e* is to improve the scavenging of the combustion chamber

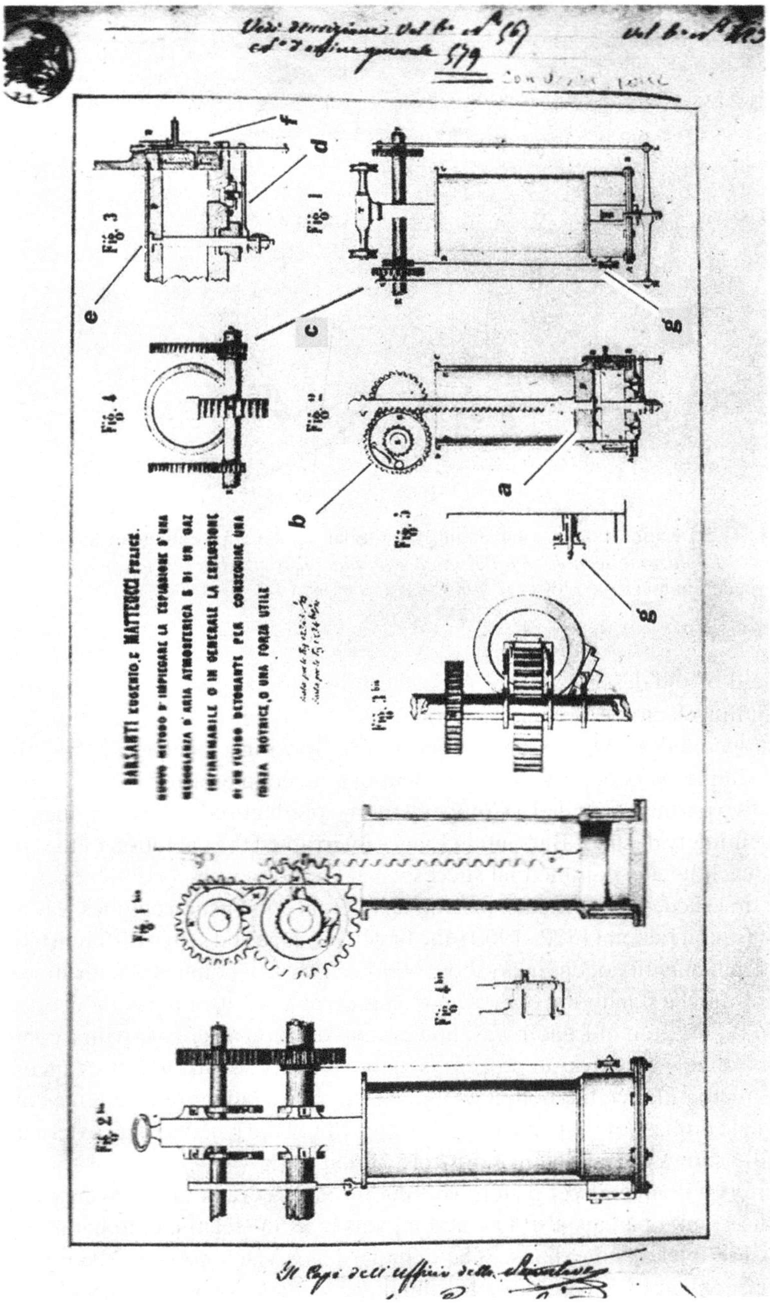

Fig. 4.3 The Barsanti and Matteucci engine features two pistons; one of them is shown as detail *a*. The piston rod is connected with a ratchet to a gear wheel *b* rotating together with shaft *c* (from L. Morello, *Evoluzione tecnologica dei motori automobilistici ad accensione comandata*, ATA: Ingegneria dell'autoveicolo, 2004, N° 7/8 e 9/10)

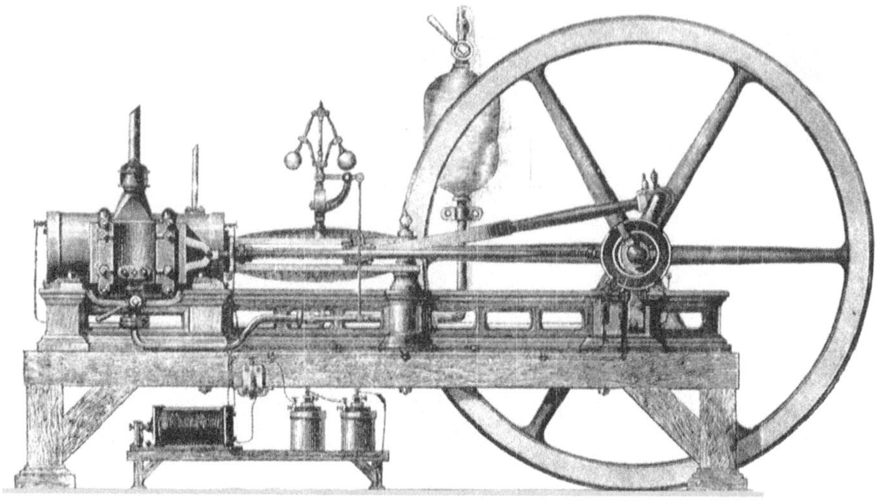

Fig. 4.4 The Lenoir's combustion engine is similar to a reciprocating steam engine (from L. Morello, *Evoluzione tecnologica dei motori automobilistici ad accensione comandata*, ATA: Ingegneria dell'autoveicolo, 2004, N° 7/8 e 9/10)

at the end of the descent stroke. An electric ignition is given by an electrode *g* in Fig. 4.5; this electrode is connected to a battery, that has a voltage multiplier with a contact operated by the motion of yoke *d*. The drawings, labelled as '*bis*' refer to a second simpler version of this engine, without the counterpiston *e*.

The two partners founded a Company for the production of these engines in 1860, but the untimely death of Barsanti in Lieges interrupted the operation of the company before reaching any commercial success.

The first successful company producing internal combustion engines was founded by Jean Joseph Lenoir (1822–1900), the first to produce and sell in different Countries a significant quantity of engines (about 600)backward, if compared with atmospheric engines from the standpoint of thermodynamics, because it works without a compression phase. The aim of Lenoir was, however, to develop an engine with a continuous and controllable combustion process, suitable to work as easily as a steam engine.

As a matter of fact, the Lenoir's engine is a reciprocating steam engine converted to internal combustion, capable of operating without the penalty of an external boiler. Figure 4.4 shows the engine in a drawing of that time.

There is a double effect piston working through a crank mechanism with crosshead; the crank operates also two sliding valves as in steam reciprocating engines, for mixture intake and exhaust. The cylinder has a water cooled jacket; two spark plugs are ignited by a battery with Rühmkorff voltage amplifier, located under the cylinder.

In actual operation, spark plugs controlled combustion at partial loads only, while at wide open throttle self-ignition prevailed, making the engine very noisy and rough, as customers claimed. Also the high gas consumption was considered as a weak point.

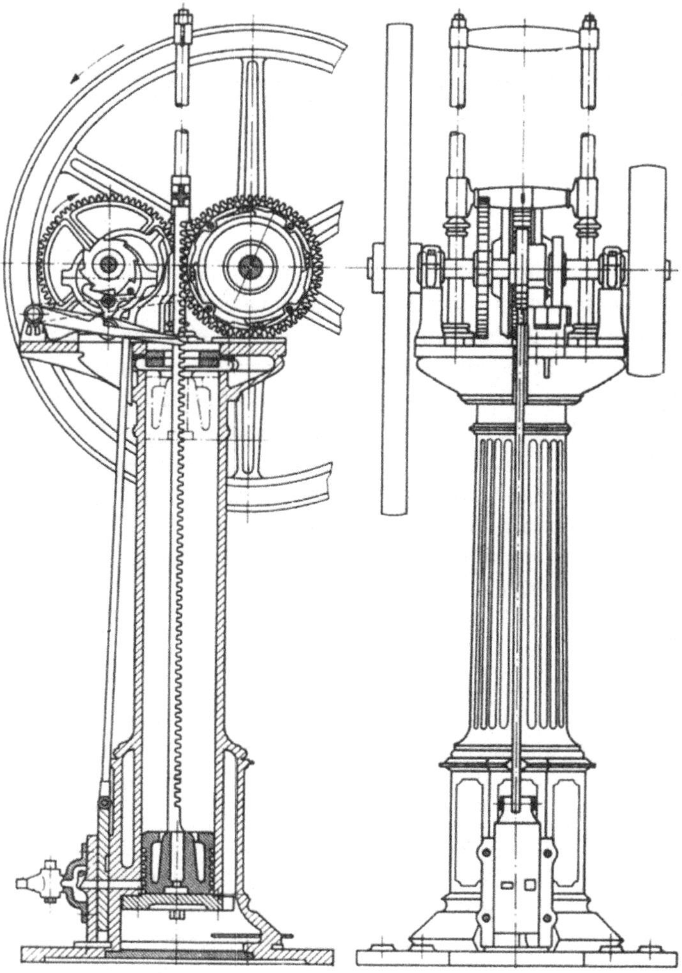

Fig. 4.5 An atmospheric engine by the Otto Co.; its strong point was the free wheel developed by Langen (from L. Morello, *Evoluzione tecnologica dei motori automobilistici ad accensione comandata*, ATA: Ingegneria dell'autoveicolo, 2004, N° 7/8 e 9/10)

There is no mention of any vehicle working with his engine, even if there is a sketch from Lenoir on this subject.

Nicolaus August Otto (1832–1891) did not have the chance to receive a technical education suited to his ambitions, but his job as a travelling salesman allowed him to study Lenoir engines and to develop an interest in the subject. He spent all savings he could make from his revenue to experiment with and improve Lenoir engines and to make their combustion more gradual and controlled. He tried to introduce a compression phase in their working cycle, but without success. As a consequence, he started to build atmospheric engines.

When he had his first prototype under test, he had the chance to meet Eugen Langen (1833–1895), a dynamic technician who had already reached success in the introduction of new inventions. In 1864 they established together the Otto Co., with the aim of producing and selling compression engines.

Commercial success came after the 1867 International Exhibition in Paris where the Otto engine was granted a golden medal, as prize for the internal combustion engine with the lowest fuel consumption. About 3,000 engines similar to that in Fig. 4.5 were produced till 1882.

This engine was quite similar to that of Barsanti and Matteucci but the patents of the two Italian inventors concerned the automatic timing of the engine only; the free wheel connection of the Otto engine, developed by Langen, was different and better if compared with the ratchet wheel of Barsanti and Matteucci. A law suit for infringement started by Matteucci against Otto was lost by the former.

Encouraged by the success of the atmospheric engines, Otto devoted himself to accomplish his former idea of introducing the compression phase into an engine similar to that of Lenoir; Carnot's theory suggested this solution to be the best, but in practice was not achieving good results yet, because of preignition and self-combustion, problems caused by hot spots in the combustion chamber that were difficult to eliminate with the available technologies.

After many tests Otto developed a four strokes engine, in which one stroke was dedicated to mixture compression; in addition, the mixture inside of the combustion chamber was stratified, being richer in the region around the plug and leaner at the combustion chamber walls and piston top. With this idea, hot spots were virtually no longer dangerous. In his patent for a four stroke engine, filed in 1876, three claims regarded mixture stratification, only one four strokes operation.

This new engine, shown in Fig. 4.6, was provided with a shaft parallel to the cylinder axis, operating a sleeve valve 1 and a poppet valve 2, the first for intake and ignition, the second for exhausting combustion gases. Three ducts were machined in the head: one for combustion gas, one for air, the last for ignition. A flame of an external burner was brought to the combustion chamber at the time of ignition.

During the intake stroke, air was introduced at first and reached the combustion chamber 3, followed by a rich mixture of air with gaseous fuel, which remained in the recessed portion 4; and then the ignition flame was transferred into the recess, to start combustion. In consideration of the low speed of the engine, there was practically no mixing of the rich mixture with the dilution air, before ignition, obtaining the desired stratification.

Otto Co. and its licensees built about 40,000 engines of this kind, from 1876 to 1889. The patent made the concept of four strokes operation an exclusive asset of Otto Co.; it was protected very carefully by stopping similar projects developing in Germany and other European countries; however, a paper was discovered in 1884, describing in details the advantages of an engine based upon a four strokes operation. The content of this chapter, published by Alphonse Beau De Rochas in 1862, remained almost unknown for years and was not protected by any patent. A competitor brought this chapter to court and obtained a ruling invalidating Otto's patents, whose contents became available to everybody.

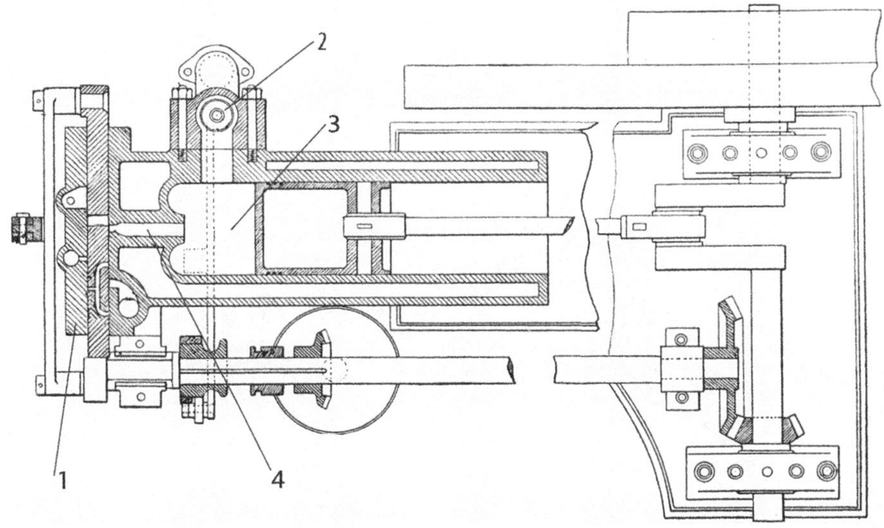

Fig. 4.6 The new Otto four strokes engine has a shaft parallel to the cylinder axis operating a sleeve valve *1* and a poppet valve *2*, the first for intake and ignition, the second for exhausting combustion gases (from L. Morello, *Evoluzione tecnologica dei motori automobilistici ad accensione comandata*, ATA: Ingegneria dell'autoveicolo, 2004, N° 7/8 e 9/10)

Neither Otto's reputation nor his business were affected, but this sentence should be considered as a major event in history of technology, because it opened the market to many new engines and prepared the birth of the automotive engine.

4.2 Automotive Internal Combustion Engines

4.2.1 Mechanical Architecture

The first engine specifically developed for vehicular applications was designed by Gottlieb Daimler (1834–1900) and Wilhelm Maybach (1846–1929), both former employees of Deutz Co., the new name of the Otto Co., we mentioned in the previous section.

The main objective of their studies for a self-propelled vehicle—cars, bicycles, boats and railcars were considered by these inventors—was weight reduction and power increase. This target implied a different mechanical design, allowing a higher rotation speed and the application of a new fuel, easier to be carried on board the vehicle, with a higher energy density than coal gas.

Their choice fell on *ligroin*, a very light liquid oil distillate, with a vapor pressure slightly higher than current gasoline. The ease of evaporation made *carburetion*, the

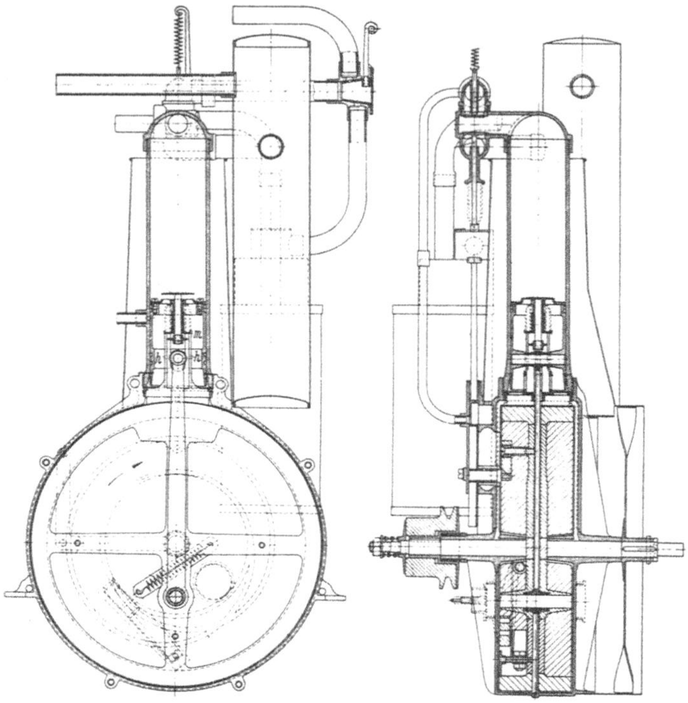

Fig. 4.7 In the first Daimler's engine the charge stratification is favoured by the long stroke and is obtained by a second intake valve on the top of the piston, opening at bottom dead center (from L. Morello, *Evoluzione tecnologica dei motori automobilistici ad accensione comandata*, ATA: Ingegneria dell'autoveicolo, 2004, N° 7/8 e 9/10)

process of mixing air with a liquid fuel, simpler, but posed major safety problems. The carburetor was invented and designed by Maybach, after studies performed at Deutz; this Company, however, was not interested in this invention, addressed to fuels that were more expensive and unnecessary for stationary engines, the main product of Deutz.

The first Daimler's engine started operating in 1883; its displacement was 212 cm^3, with a maximum power of 0.5 HP at 600 rpm. Its specific power of 2.35 HP/l was much higher than the value of 0.5 HP/l, reached by stationary Otto engines. These engines received from their particular shape the nickname *Standuhr*, grandfather's clocks in German. The cross section of this engine reported in Fig. 4.7 shows some unusual detail.

The cylinder was made with a thin steel tube, cooled by an air stream, moved by a fan rotating with the flywheel. The charge stratification was favoured by the long stroke (about 1.7 times the bore) and obtained by a second intake valve on the top of the piston, opening at bottom dead center, through which a quantity of dilution air was blown into the combustion chamber by the pressure built up in the crankcase.

Fig. 4.8 The head of the Bernardi engine, with one of the first jet carburetors and, beside the valves, the ignitor (Museum of the University of Padova)

The main intake valve was automatic. The exhaust valve was operated by a double action cam, made by a spiral groove cut in one of the very heavy cranks. The cranks work also as flywheels and contribute to reduce the dead space in the crankcase, to increase its compression ratio.

The ignition was obtained by a glow plug, without possibility of adjusting the spark advance. The plug was made by a tube crossing the cylinder, with one end in the combustion chamber and the other end facing the flame of a lamp.

This engine was demonstrated, as a vehicle propulsion system, for the first time in 1886 on a motorcycle; the first car, presented in 1889, had a two cylinder V engine, derived from the first, with a displacement of 565 cm^3. This kind of engine had a wide diffusion in cars by many licensees all around the world. In particular, it powered the cars built by Panhard and Levassor and Peugeot in France, who were, at the beginning of the twentieth century, the most important car manufacturers in the world.

About ten years later engines were developed with both intake and exhaust valves opened by cams, as we can see on the tricycle built in 1894 by Enrico Bernardi (1841–1919) in Italy. The engine head with one of the first jet carburetors and the ignitor, is shown in Fig. 4.8. This single cylinder engine, with a displacement of 624 cm^3 delivered 2.5 HP at 785 rpm (about 4 HP/l).

No regulation valve was applied to the carburetor except for air to fuel ratio adjustment in cold operating conditions. Therefore the engine worked at wide open throttle always; its speed was controlled by a Watt regulator, like in steam engines, closing the intake valve to regulate the speed. The engine was therefore regulated with an on-off operation around the desired value of speed.

A similar regulation system is shown by the Daimler Phoenix engine in Fig. 4.9, built under licence by Panhard & Levassor. The Watt regulator, with centrifugal masses *e*, acts in this case on the exhaust valves; they will not be opened beyond a regulated speed limit, stopping combustion.

The combustion chamber featured two lateral appendages where the valves were installed with their stem parallel to the cylinder axis: the automatic intake valve *a* is at right and the exhaust valve *b* at left; close to the intake valve, the glow plug *c* and its heating lamp *d* can be seen.

This third generation engine of Daimler, still inspired by the first, shows a water cooling system and a more conventional cam shaft; the specific performance was doubled as compared with the first.

The devices addressed to obtaining a stratified charge were abandoned by Daimler at this time, because of an improved control of the mixture ratio, obtained with a jet carburetor, and the elimination of hot spots with water cooling.

In the following years, at the beginning of the twentieth century, no significant improvement in specific performance was made; to increase the performance of luxury and sports cars almost all manufacturers introduced engines with larger cylinder displacement and a higher number of cylinders.

This approach to increased performance can be justified by noting that the cars of that time were an exclusive product for wealthy customers: For them the high performance reputation of a car was important in making the purchasing decision. A high power (the most powerful commercial cars of that time featured about 60 HP at 4 HP/l) could only be obtained with a large displacement, it being still impossible to increase engine speed and compression ratio.

An example of this trend is shown in Fig. 4.10. The structure of these engines, mainly with four cylinders, was very bulky, made of two blocks (three blocks for six cylinders) including cylinders and heads in a single cast piece; a four or six cylinders engine was composed by installing two or three blocks on a single crankcase.

The combustion chamber of these engines was wider than the cylinder cross section, containing valves on a lateral appendage (lateral or bilateral valves, when intake and exhaust valve are at opposite sides); both valves were operated by camshafts that, despite their higher complexity, allowed pumping losses reduction as compared with automatic valves.

Most of these engines featured automatic carburetors, where air to fuel ratio was almost constant with engine speed and load and where load regulation was obtained by a throttle valve; speed regulators were therefore gradually abandoned.

On top of each cylinder head a tap (*decompression tap*) made hand cranking of the engine easier. It allowed introduction of an additional quantity of gasoline in the cylinder closer to compression, to compensate for the condensation of fuel on

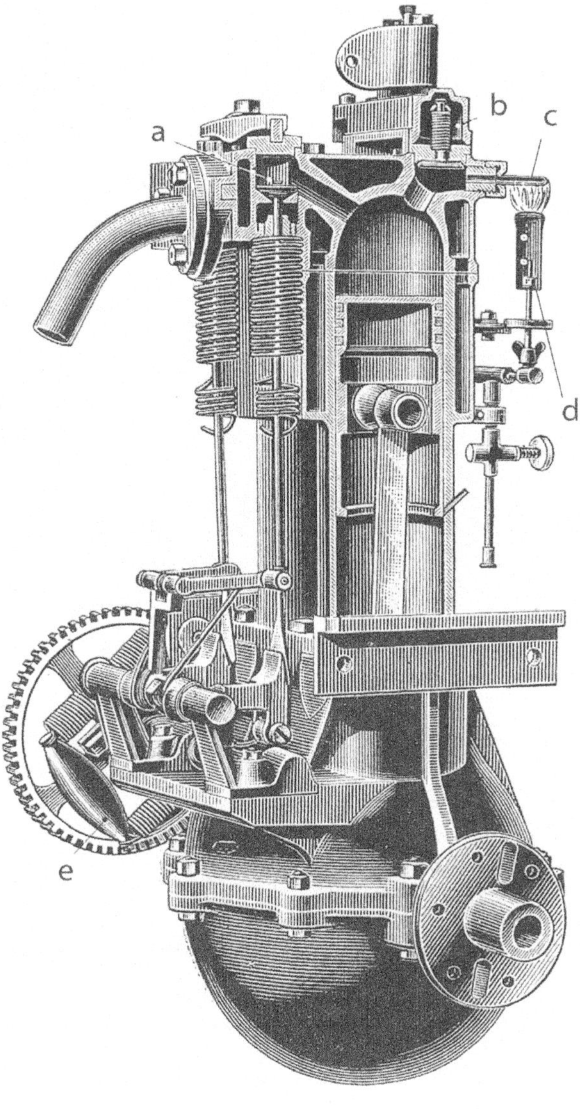

Fig. 4.9 Third generation engine by Daimler: the Phoenix. It was still inspired by the first, but features a water cooling system and a more conventional cam shaft (redrawn from Baudry de Saunier 1905)

cold manifolds and to reduce compression pressure in non-firing cylinders. The cross section of a two blocks engine with this architecture is shown in Fig. 4.11.

The layout with lateral valves on one side of the block became widely common and was applied for its simplicity till the 1940s. The long connecting road, used to reduce the lateral loads on the cylinder wall, was typical of the engines of this generation.

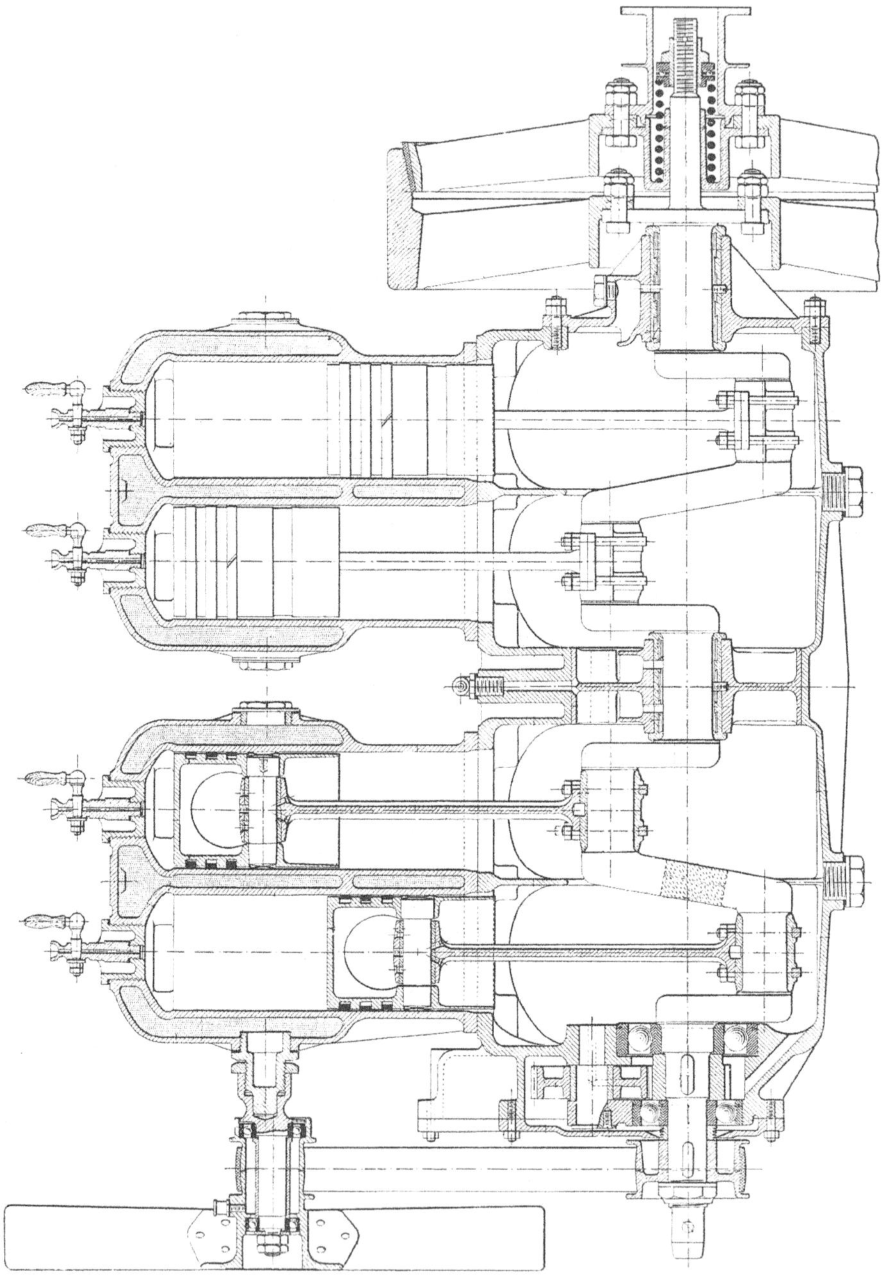

Fig. 4.10 The structure of four cylinders engines of the 1910s featured two blocks, including cylinders and head in a single cast piece (redrawn from L. Morello, *Evoluzione tecnologica dei motori automobilistici ad accensione comandata*, ATA: Ingegneria dell'autoveicolo, 2004, N° 7/8 e 9/10)

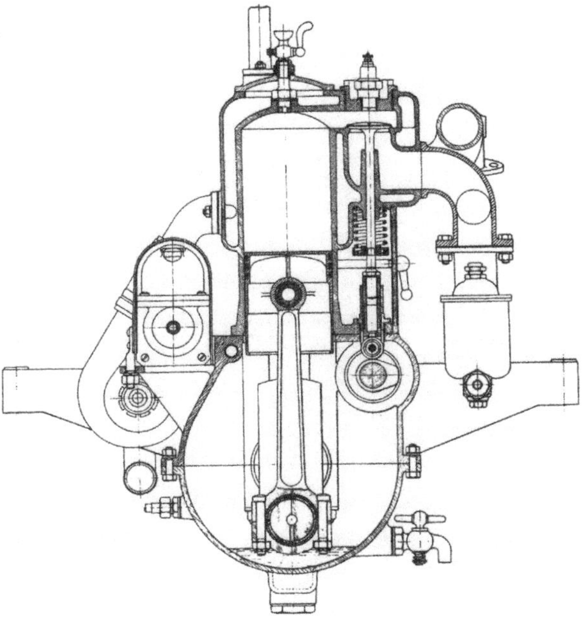

Fig. 4.11 Cross section of a two blocks four cylinder engine, the first produced by Alfa Romeo starting from 1910 (from L. Morello, *Evoluzione tecnologica dei motori automobilistici ad accensione comandata*, ATA: Ingegneria dell'autoveicolo, 2004, N° 7/8 e 9/10)

The valve seats were reached from the top for machining, because it was impossible to disassemble the head from cylinders, each valve chamber was closed by a removable bronze plug: One of these plugs could be drilled to obtain the spark plug hole. Glow plug ignition was abandoned in favour of magneto ignition.

The 1922 Lancia engine, shown in Fig. 4.12, is one of the first where performance was improved by increasing the engine speed. In these years, particularly in Europe, a trend toward decreasing displacements was encouraged by the high price of gasoline and to the taxes that almost every country applied on the basis of the engine displacement.

In this and other engines two typical features of modern engines appear: a structure including a detachable cylinder head and a cylinder block that includes the crankcase and the oil sump (*monoblock*) and overhead valves. The Lancia engine had other technical features that made it unique among competitors, such as cylinders in a narrow V arrangement and an overhead camshaft. The 1934 Bianchi engine, shown in Fig. 4.13, featuring a cam shaft in the block with push rods is probably a more appropriate example of the state of the art of these years.

Nevertheless, many engines of this time still had lateral valves, particularly in cheap high volume cars; the ratio between the length of the piston and connecting rod and the bore are closer to the current state of the art, because of improvements in bearing materials. Molded parts had also a reduced thickness.

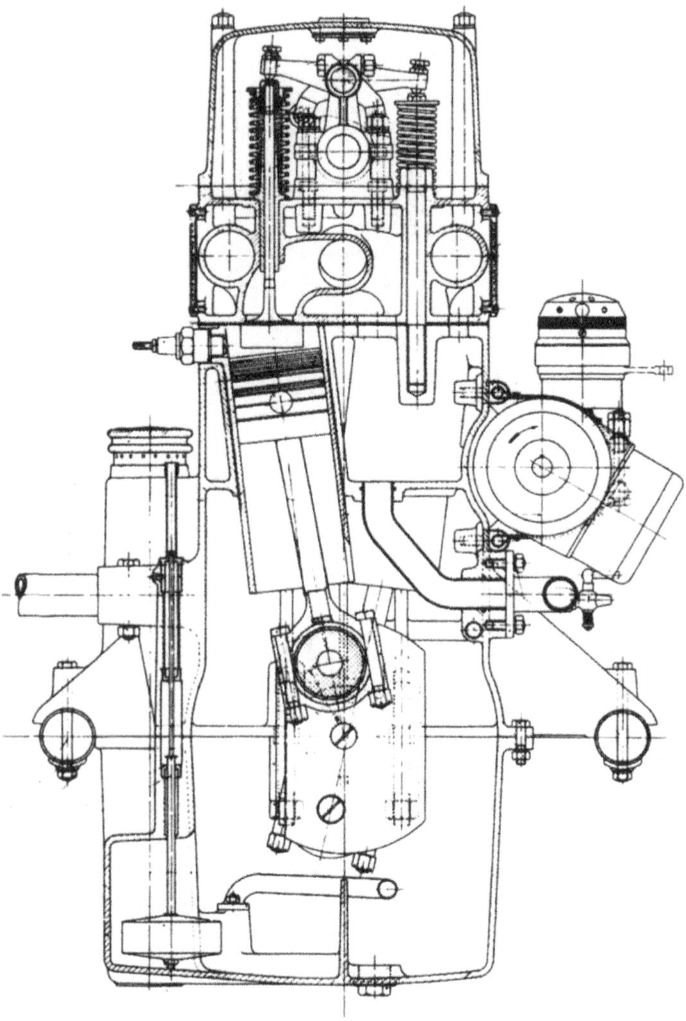

Fig. 4.12 A 1922 Lancia engine in which two modern features appear: detachable cylinder head and overhead valves (courtesy of FIAT Historical Archives)

Some engines built between 1910 and 1930 were very different from the dominant rules. Poppet valves, that had been widely applied since the beginning of the automotive industry, were a source of many problems. They implied fabrication problems, particularly in non-separable heads, and operation problems because of wear and dirtiness of their seats. This last issue caused the need for periodic cleaning and re-machining of the seat. A last reason for complaints by customers was their noisy clicking, so that some manufacturers tried to avoid all these negative points, at least for luxury cars.

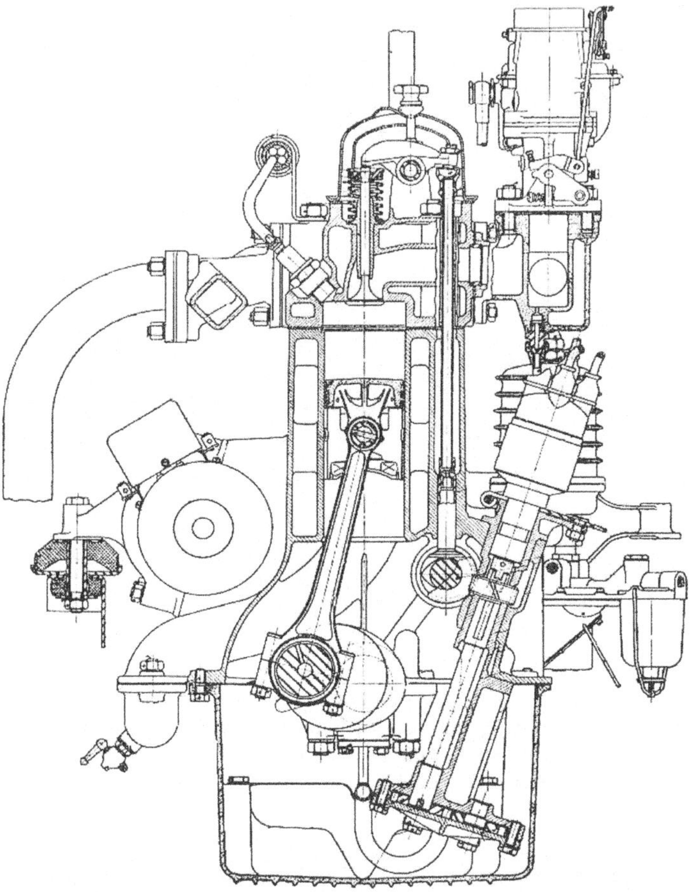

Fig. 4.13 1934 Bianchi engine, featuring a cam shaft in the block, with push rods (redrawn from L. Morello, *Evoluzione tecnologica dei motori automobilistici ad accensione comandata*, ATA: Ingegneria dell'autoveicolo, 2004, N° 7/8 e 9/10)

Many proposals of different valve systems were made; the Knight's patents, developed in the United States in the first years of the twentieth century, were bought and applied by some European car manufacturers, as Panhard and Levassor in France, Daimler in Germany, while Itala, in Italy, developed its own patent.

Figure 4.14 shows a cross section of the 1912 Panhard and Levassor '*Avalve*' engine, probably the first application of this kind of engine that remained in production till the 1930s. The meaning of *avalve* was *without valves*, but valves still existed in a different arrangement, made with sliding sleeves (*a* and *b* in the figure) around the piston. On these sleeves suitable openings were cut that constituted the intake and exhaust ports: the internal sleeve was also used as piston liner. Both sleeves were moved by a crank and rod mechanism *c*, turning at half of the speed of the engine.

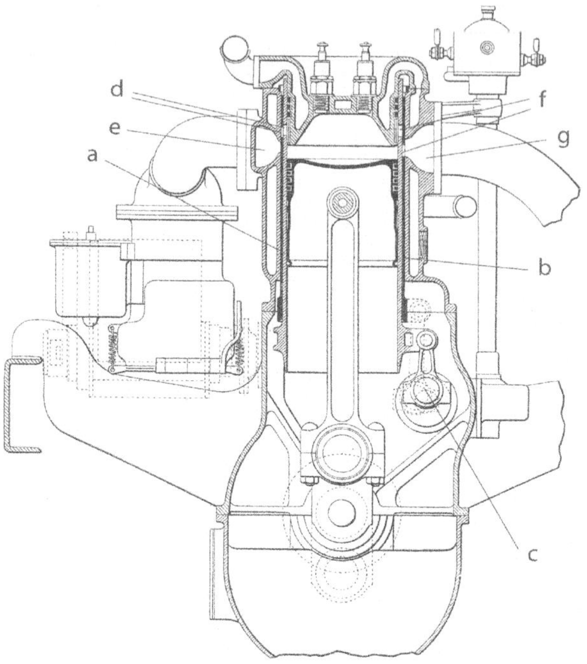

Fig. 4.14 Avalve engines had valves made with sliding sleeves, fit around the piston. Both sleeves *a* and *b* were moved by a crank and rod mechanism *c* turning at half of the speed of the engine (redrawn from L. Morello, *Evoluzione tecnologica dei motori automobilistici ad accensione comandata*, ATA: Ingegneria dell'autoveicolo, 2004, N° 7/8 e 9/10)

Sleeve cuts *d* opened the intake ports *e*, while sleeve cuts *f* opened the exhaust port *g*; segments were necessary also on the cylinder head to have appropriate tightness.

These engines were claimed to be so quiet as it was impossible to hear the engine idling; as a major drawback, a significant friction between sleeves and cylinder liner increased consumption and affected performance negatively.

The improvements in poppet valves materials, in machining tolerance and particularly in gasoline, that reduced tar deposits on seats, made the advantages of sleeve valves to disappear.

A new valve design was introduced in aircraft engines and transferred to race car engines and later to high performance commercial engines; these engines were provided with semi-spherical combustion chambers with inclined poppet valves. The shape of the combustion chamber, with its walls closer to the spark plug, reduced engine octane requirement and made larger poppet valves applicable with advantages in performance.

Examples of this new technology, widely used in present engines, were the Alfa Romeo engines after 1928; overhead valves were operated by a twin overhead camshaft making top speeds of 4,500 rpm, very high for that time, possible.

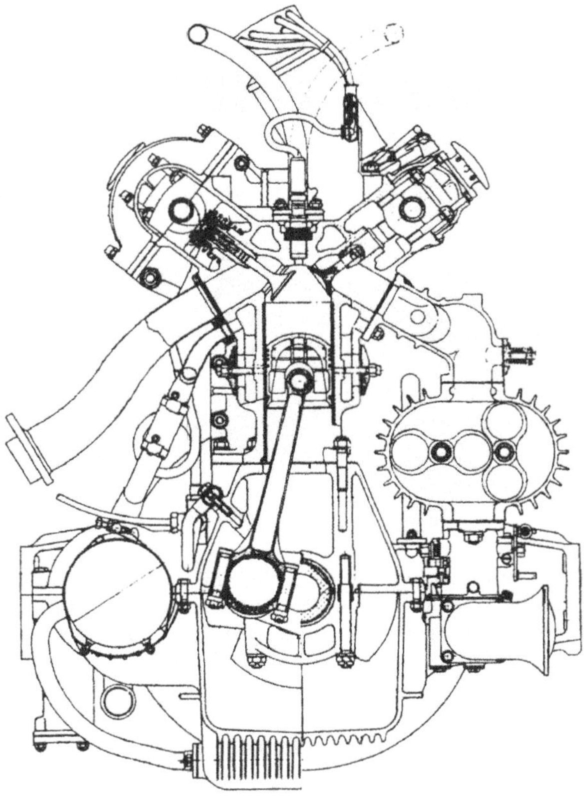

Fig. 4.15 A new design of 1937 with inclined overhead poppet valves was based on a semi-spherical combustion chamber (redrawn from L. Morello, *Evoluzione tecnologica dei motori automobilistici ad accensione comandata*, ATA: Ingegneria dell'autoveicolo, 2004, N° 7/8 e 9/10)

One of these engines is shown in Fig. 4.15; It is the 1937 8C 2900 B, an in-line eight-cylinder supercharged engine that reached about 62 HP/l, an impressive figure for that time. Most engines, however, featured parallel valves till the 1950s when many examples of semi-spherical combustion chambers were in production, both with overhead or push-rod cam shafts.

This short report on architectural evolution is concluded by the 1974 Volkswagen engine, shown in Fig. 4.16. It is an example of a new breed of cheap and rational engines, with parallel valves and overhead camshaft, operated by an elastomeric timing belt. Valves are aligned to simplify the head design, and a suitable shape of the combustion chamber is obtained by designing a cavity in the piston top (cup chamber) or around the valves (pent-roof chamber).

Fig. 4.16 The 1972 Volkswagen engine is an example of a new breed of rational engines with parallel valves and overhead camshaft, operated by an elastomeric belt (from L. Morello, *Evoluzione tecnologica dei motori automobilistici ad accensione comandata*, ATA: Ingegneria dell'autoveicolo, 2004, N° 7/8 e 9/10)

4.2.2 Structural Components Technology

A picture of the results obtained in about 100 years of engine evolution can be obtained by comparing the performance of the last engine described with that of the first Daimler engine:

- Specific power: from 2.3 to 60 HP/l (from 1.7 to 45 kW/l).
- Rotation speed: from 600 to 6,000 rpm.
- Compression ratio: from 2.5 to 9.
- Specific mass: from 180 to 1.35 kg/HP (240 to 1.8 kg/kW).

If the most important commercial car engines produced in Europe from the beginning to the 1980s are taken into account, the plot of the engine specific power versus the year of commercial launch shown in Fig. 4.17a can be obtained. This progress was allowed by the evolution of materials and production technologies and by new inventions.

An important issue in engine performance was the tightness of exhaust valves and their life since leaking valves caused loss of performance and started a process of quick deterioration of their seat. After sleeve valves, traditional in steam engines, were abandoned, poppet valves became a standard solution but they required a perfect match between the bevel surfaces of the valve and its seat. This result was initially obtained by adjusting each valve on its seat, because machining tolerances were too wide to provide tightness after assembly.

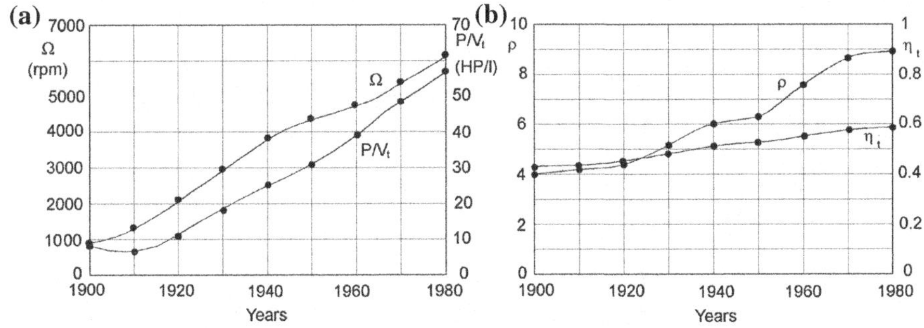

Fig. 4.17 Evolution of reciprocating spark ignition engines in the years 1900–1980. **a** Specific power P/V_t (in HP/l) and rotational speed Ω (in rpm); **b** Compression ratio ρ and thermodynamic efficiency η_t (from L. Morello, *Evoluzione tecnologica dei motori automobilistici ad accensione comandata*, ATA: Ingegneria dell'autoveicolo, 2004, N° 7/8 e 9/10)

The valve was temporarily assembled on the engine head and then rotated by a screwdriver, after having sprinkled the matching surfaces with abrasive dust; an easy access to the valve was thus required for this operation that had to be repeated from time to time as a current maintenance operation.

The seat and the stem guide had also to be machined and therefore the engine head had to offer a suitable access to machining mandrels. When the early car engines were manufactured, no existing material could offer a suitable sealing to combustion pressure between engine head and cylinder block and therefore these two parts were made in a single piece, where access holes were provided for drilling and machining valve seats. These holes were closed by plugs.

This feature is shown in Fig. 4.18a for an engine with bilateral valves and in Fig. 4.18b for an engine with one overhead and one lateral valve; in both cases, a hole in the centre of the head (decompression tap in a, intake valve in b) is provided, to allow access for the mandrel used to bore the cylinder surface. In both cases the valve size was reduced for machining, a thing that decreased the engine volumetric efficiency and the overall performance.

Another design detail, that was quite different from current practice, was the bearing length, that had to be as wide as the shaft diameter because of the low value of oil pressure, usually introduced in the bearing by dripping. The consequence of the long bearings can be seen also in the connecting rod head width and, more in general, in engine length and weight. Moreover, the low lubrication pressure caused also piston wear and, to provide a sufficient engine life, the connecting rod had to be long enough to reduce lateral thrust.

To better understand the difference between current and old engines, the connecting rods of two engines of about the same displacement but dating back to the 1920s and to the 1960s, are compared in Fig. 4.19. The bulk and weight of the blocks and the whole engines were obviously a consequence of these dimensions.

Moreover, casting technologies initially available affected the design practice. Casting processes were developed to give iron ingots a shape as close as possible to

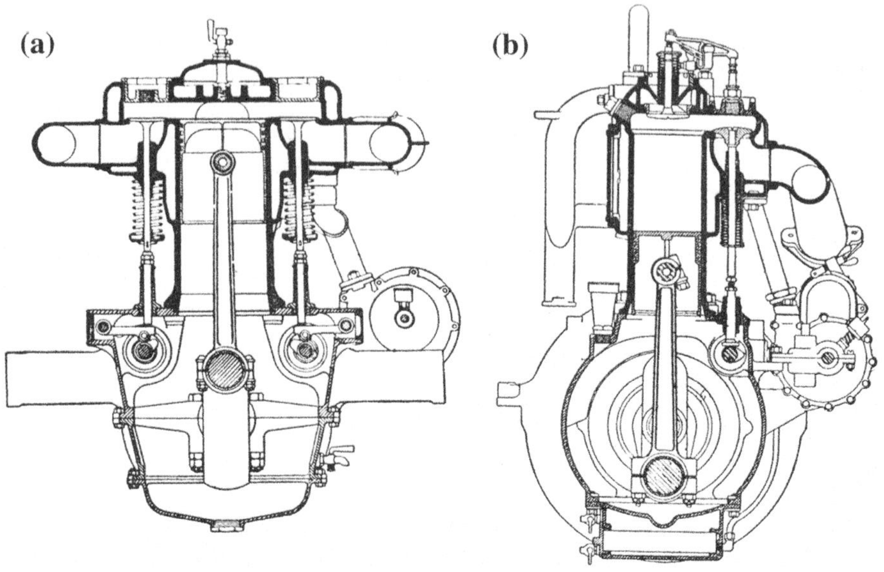

Fig. 4.18 Valve seat arrangement in an engine with bilateral valves (**a**) with one overhead and one lateral valve (**b**) (from L. Morello, *Evoluzione tecnologica dei motori automobilistici ad accensione comandata*, ATA: Ingegneria dell'autoveicolo, 2004, N° 7/8 e 9/10)

the final shape of the part to be produced. Casting took benefit from the relatively low melting temperature ($<1{,}100\,^{\circ}C$) of this material and similar processes were applied to produce castings made of aluminum, brass or bronze. The traditional method, thousands of years old, for casting metals used sand moulds.

Simply stated, in sand casting a pattern is placed in sand, to make an imprint, and the resulting cavity is filled with molten metal. Then the metal is allowed to cool, until it solidifies to the desired shape, the sand mould is broken away, and the casting is removed. The outer shape of the pattern must reproduce the shape of the desired casting, taking into account the volume reduction (*shrinkage*) of the metal during cool-down and must incorporate a gating system.

Even if the origins of sand casting date back to ancient times, this process is still the most widespread. Typical parts made by sand casting were and are engine blocks, cylinder heads, gearbox and differential housings.

Silica sands (SiO_2) are mostly used. Sand is a product of the disintegration of rocks over extremely long periods of time. They are inexpensive and suitable as mould material because of their resistance to high temperatures. Several factors are important in the selection of sand for moulds: sand having fine, round grains can be closely packed and form a smooth mould surface. Good permeability of moulds and cores allow gases and steam produced during casting to escape easily.

The mould should have good collapsibility to avoid defects, since castings shrink while cooling. The selection of the sand involves certain trade-offs with respect to

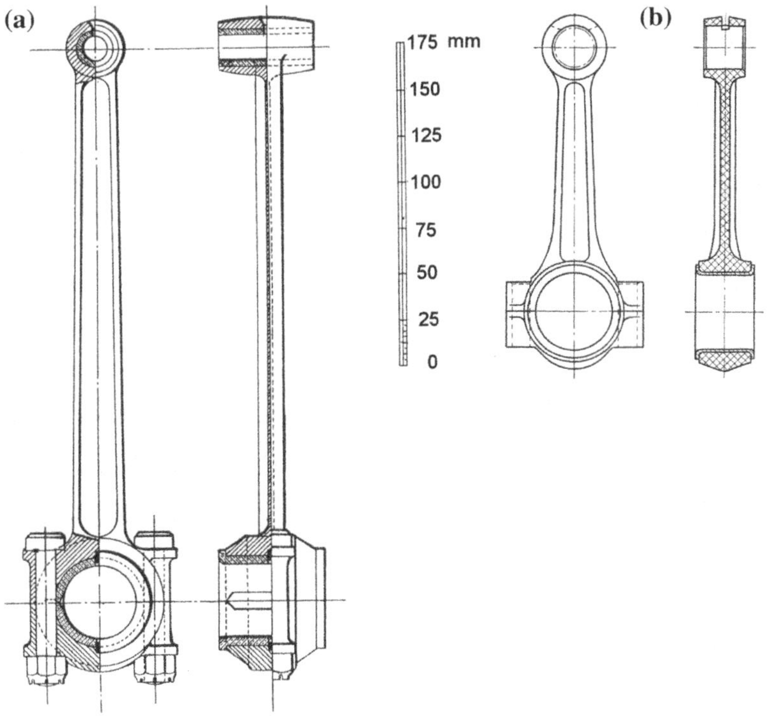

Fig. 4.19 Drawing of a connecting rod from the 1910s (**a**) and from the 1960s (**b**) for a cylinder with a displacement of about 500 cm^3 (from L. Morello, *Evoluzione tecnologica dei motori automobilistici ad accensione comandata*, ATA: Ingegneria dell'autoveicolo, 2004, N° 7/8 e 9/10)

properties. Sand is now conditioned before use and mills are used to uniformly mix sand with additives. Clay is often used in moulds as a cohesive agent to bond sand particles, increasing sand strength.

There are three basic types of sand moulds available now: green-sand, skin-dried and dry-sand. The most common mould material is green moulding sand and was the only one available to produce the early engines. The term *green* refers to the fact that the sand in the mould is wet while the metal is being poured into it: Green moulding sand is actually a mixture of sand, clay, and water; it is the least expensive and most easily used material for moulds.

In the skin-dried method the mould surfaces are dried, either by storing the mould in air or drying it with torches. Skin-dried moulds are generally used for large castings because of their higher strength.

Sand moulds dried in ovens are stronger than green-sand moulds and allow one to obtain better dimensional accuracy and surface finish. In modern sand casting, various organic and inorganic binders are blended into the sand to chemically bond the grains for greater strength. These moulds are dimensionally more accurate than green-sand moulds but are more expensive; dimensional accuracy allows one to obtain sharper edges and thinner walls.

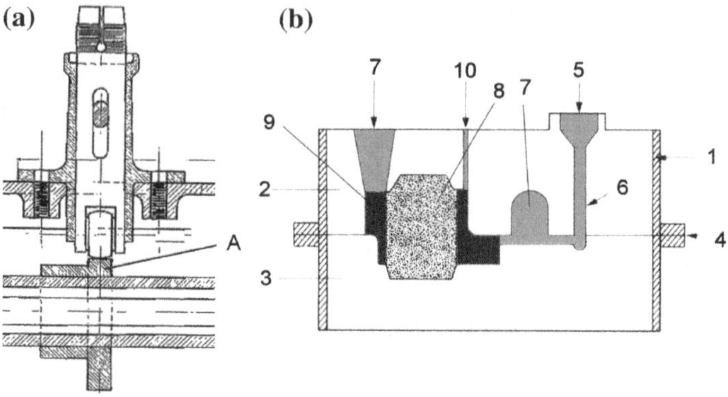

Fig. 4.20 The cam A in sketch **a** is taken as an example to describe a typical casting mould in **b**. The moulded piece *9*, i.e. the cam, needs a core *8*, to obtain the internal hollow. The mould is contained in a flask *1* and is divided in two parts, the cope *2* and the drag *3*, separated by a parting line *4*, positioned following the shape of the piece. The liquid metal is poured into the pouring cup *5*, through the sprue *6* and overflows through the risers *7*. The cooled metal contained in elements *5*, *6* and *7* must be removed from the casting

The most important parts of a mould are shown in Fig. 4.20, where the mould for one of the cams of the camshaft of an engine of the 1910s is sketched.

- The mould is supported by a flask 1. A two-pieces mould, like this, consists of a cope 2 on top and a drag 3 on the bottom. The seam between them is the parting line 4. The cope and the drag are filled with sand, represented in white in the sketch; the cam 9 to be cast is represented in black.
- The molten metal is poured into a pouring cup 5.
- The molten metal flows downward through a sprue 6.
- Risers 7 supply additional material to the casting as it shrinks during solidification. In the figure a blind riser and an open riser are represented.
- Cores 8 are inserts made from sand. They are placed in the mould to form hollow regions as the cam hub to be fit on the camshaft.
- Vents 10 are placed in moulds to carry off gases produced when the molten metal comes into contact with the sand.

After cooling, solid metal has filled also the areas 5, 6, 7 and 10, shown in grey. They must be cut away from the casting.

As a consequence of the brittleness of sand cores, the water jackets of the early engines had to be oversized; also fillet radii had to be conveniently wide to avoid cracks at the edges of the green sand contained in the top and in the drag.

A guideline drawing with suggestions for safe dimensions, from an engineering manual of the 1920s, is shown in Fig. 4.21 and gives an impression of that situation. As a result of these design rules, cylinders had to be grouped in blocks of two or three, so that four or six cylinder engines had an additional penalty in their length.

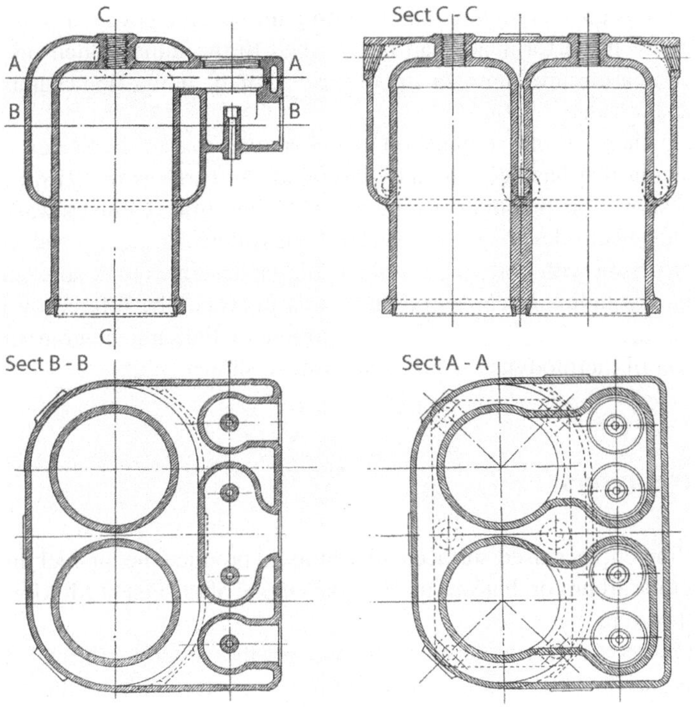

Fig. 4.21 Suggested proportions for an engine block with 1910s foundry technology (redrawn from L. Morello, *Evoluzione tecnologica dei motori automobilistici ad accensione comandata*, ATA: Ingegneria dell'autoveicolo, 2004, N° 7/8 e 9/10)

Later, the use of cured sand allowed reduction of casting thickness and obtaining of sharper edges, with a consequent reduction in overall dimensions.

A few figures highlight the difference in size between two engines, one from the 1910s, the second from the 1970s:

- Ratio between connecting rod length and stroke: from 3.5 to 1.4 at present;
- Ratio between engine height and stroke: unchanged at 6;
- Ratio between the distance between the cylinder centres and the bore: from 1.3 to 1.1 at present;
- Block walls thickness: from 8 to 4 mm at present.

The unchanged value of the ratio between the engine height and stroke refers to different engine layouts, from a lateral valve engine with camshaft in the block, to an overhead valve and camshaft engine.

While the ratio between the weight and the displacement of engines didn't change much, the reduction of the ratio between weight and rated power is clear. The impressive increase of engine speed, shown in Fig. 4.17a, can be ascribed to the reduction in weight of the reciprocating parts and the improvements in lubrication and wear resistance.

If the curve is compared with that regarding the specific power, a significant part of the increase in performance can be ascribed to the improvement in rotational speed. Other reasons are improvement of the design of combustion chambers and of gasoline octane value.

The wide shape of old combustion chambers was due to the design limitations on valve layout that has been already described. As early as in 1919, combustion studies had shown the potential advantages of combustion chambers walls closer to the spark plug electrodes. As soon as technologies allowed overhead valves of larger size in comparison with bore to be applied, engine designers took advantage of this opportunity. An indirect demonstration of this trend can be offered by Fig. 4.17b where the compression ratio evolution is reported; on the same diagram the ensuing improvement of thermodynamic efficiency is also shown.

4.2.3 Carburetors

Early gasoline engines used surface carburetors to produce the air-fuel mixture. An example is the carburetor that was used on the engine of the Patent Motorwagen built by Benz, shown in Fig. 4.22a.

In carburetors of this type, an air stream entering the carburetor through the suction port A, was enriched by gasoline vapors simply by contact with the fuel F contained in the tank; the mixture left the carburetor through the outlet C, where some device, like the shield D, was provided to avoid fuel droplets being swept away by the air stream. The float B allows measurement of the gasoline volume.

This was the reason why a very volatile fuel was preferred in the first engines; this system featured heavy drawbacks in safety, because any backfire from the engine was likely to ignite the whole tank content.

Also Daimler's carburetor, designed by Maybach, was a kind of surface carburetor where some improvements had been introduced: Here the mixture formation was obtained by bubbling air through gasoline, as shown in Fig. 4.22b. The surface of the gasoline in the tank was completely covered by a float, to keep circulation of air in the fuel constant in spite of the changing fuel level. Again, separators were provided to avoid that fuel droplets could be swept away by air stream. Some wire mesh was also used to avoid backfire from the engine reaching the fuel.

The first jet carburetors appeared at the beginning of the twentieth century; in this kind of carburetor the air and fuel mixture was obtained by spraying gasoline in the air stream flowing in a tube, called *barrel*. Air was sucked through an intake *a* and a regulation valve *b* (in this case a rotary sleeve valve) was introduced in the intake manifold, as it is shown in Fig. 4.23. The gasoline was sprayed through a nozzle *c* by the pressure gradient between the manifold and the ambient.

Since the gasoline was sent to the carburetor by gravity or by other means, a float valve *d* in the carburetor avoided fuel spill; the float was designed to bring gasoline to the mouth of the nozzle when vacuum was not applied.

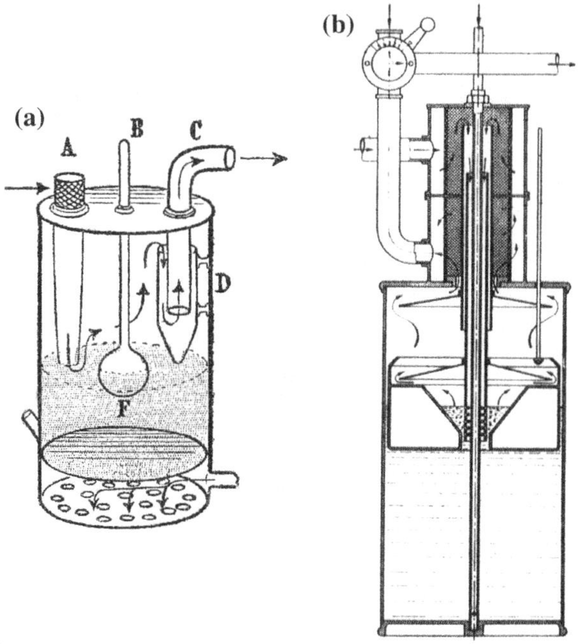

Fig. 4.22 Surface carburetors. **a** In Benz's surface carburetor a stream of air, entering through the suction port A, was enriched by gasoline vapors, simply by contact with the fuel F contained in the tank. **b** In Maybach's surface carburetor some improvement have been introduced; here mixture formation was obtained by bubbling air through gasoline (redrawn from L. Morello, *Evoluzione tecnologica dei motori automobilistici ad accensione comandata*, ATA: Ingegneria dell'autoveicolo, 2004, N° 7/8 e 9/10)

When it was impossible to install the fuel tank higher than the carburetor, a manual pump was used to pressurize the tank; sometimes the exhaust gas pressure was used for this purpose.

Carburetors featured usually two barrels, as in the example in the figure; the larger barrel was used for normal operation, while the smaller was dedicated to idle operation.

Since fuel rate depended upon manifold vacuum, high at low torque, low at wide open throttle, the mixture tended to be leaner at high load and richer at low load; to avoid this natural tendency it was necessary to adjust manually the fuel orifice within an acceptable range. This adjustment was avoided by modifying the larger carburetor barrel with the application of a convergent/divergent restrictor *a*, as shown in Fig. 4.24. This was introduced in the 1910s.

The restrictor, shaped using the geometry defined by Venturi in 1780, caused a local pressure drop that depended on air flow rate but not on downstream vacuum, regulating automatically the air to fuel ratio with sufficient precision.

In the following years other improvements were introduced in carburetor design, with the aim of keeping the air to fuel ratio constant in all operating conditions. In

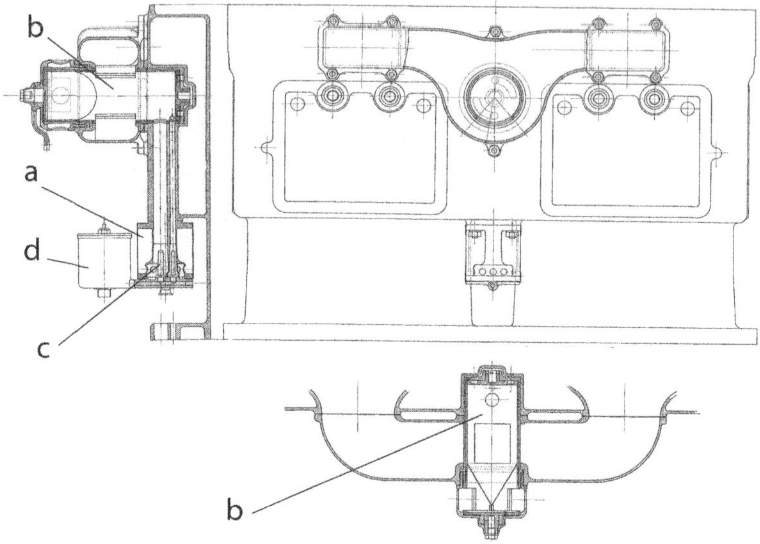

Fig. 4.23 In jet carburetors the air and fuel mixture is built up by spraying gasoline in the air stream. Air was sucked through an intake and a regulation valve close to the intake manifold (courtesy of FIAT)

particular the ratio tended to be leaner when the engine was cold, because of fuel vapor condensation on the walls of the intake manifold and ducts. In addition, fuel droplets in the air stream tended to aggregate in larger drops during sudden manifold pressure rise, as for instance in acceleration; larger drops burned only partially with an effect similar to leaning.

The first problem was solved by adding a second throttle valve upstream of the restrictor. This valve, called *choke* or *starter*, increased vacuum significantly causing an additional quantity of fuel to be sucked up by the engine. After the engine had started, the choke valve had to be opened gradually until a correct engine operation was obtained. The mixture leaning during sudden transients was avoided by installing a simple injection pump, operated by the throttle valve motion and sensitive to opening speed.

Since the application of a carburetor introduced an unavoidable additional pumping loss, in high performance engines multiple carburetors were installed, with a carburetor per cylinder in race engines. In some commercial engines with improved performance two (for four cylinder engines) or four (for eight cylinder engines) carburetors were integrated in a single body, the so-called multiple barrels carburetors.

Electronically controlled fuel injection systems that were introduced in the 1970s solved radically all the regulation problems that still affected mechanical carburetors.

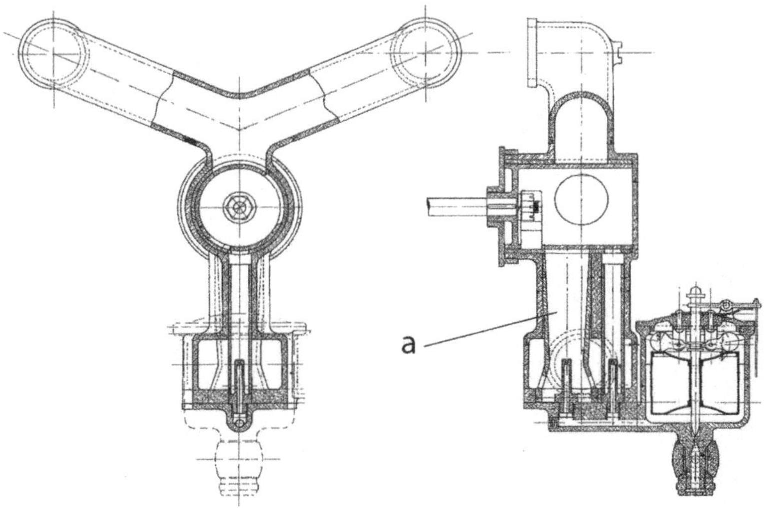

Fig. 4.24 The restrictor in the carburetor barrel, shaped following the geometry defined by Venturi in 1780, caused a local pressure drop depending on air flow rate only and not on downstream vacuum with an automatic regulation of the air to fuel ratio (courtesy of FIAT)

4.2.4 Lubrication

A basic problem in internal combustion engines has always been the lubrication of the many rotary (as crankshaft bearings) or sliding (as cylinder liners) couplings subject to wear. A solution to this problem, widely applied in the early engines, was to install oil dispensers close to the most critical points, such as connecting rod bearings, crank shaft bearings, cylinder liners, etc. These *oilers* were provided with a glass reservoir, to be easily inspected, containing a quantity of oil that was dripping into the coupling, as shown in Fig. 4.25a. Taps were provided to adjust oil leakage to the correct value and to avoid leakage when the engine was stopped.

A glass window (T in the figure) allowed the user to count the drops of oil released in a given time; user manuals reported the oil rate desired with reference to operation conditions; all oil was wasted after lubrication. This very simple system was improved by installing a single oil tank, usually on the dash board, with many outlets to the different lubricated points; each outlet was provided with a regulation tap and an inspection window.

It could be useful to remember that nobody drove alone without the assistance of a second driver or a mechanic, taking care of all auxiliary operations such as lubrication, fuel pumping, etc.

These central dripping lubrication systems were improved further, by introducing an automatic feed pump from an oil tank and by applying a suitable piston pump to each oiler, to adjust the oil rate to the optimum value automatically. Each pump was operated by the engine itself, from the cam shaft as in the example in Fig. 4.25b with

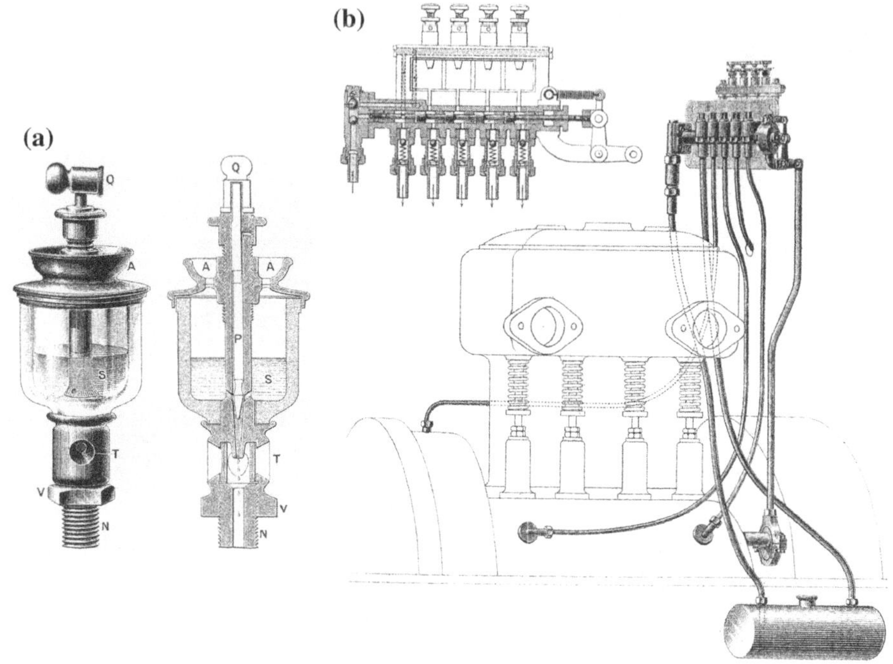

Fig. 4.25 **a** Oilers used in the early engines were made by a glass reservoir (S), to be easily inspected, containing oil dripping into the coupling. Taps (Q) were provided to adjust oil leakage to the correct value. **b** Drop lubrication systems were improved by introducing an automatic feed pump from an oil tank and by applying a piston pump of suitable size to each oiler (redrawn from L. Morello, *Evoluzione tecnologica dei motori automobilistici ad accensione comandata*, ATA: Ingegneria dell'autoveicolo, 2004, N° 7/8 e 9/10)

the double advantage of proportioning the lubrication rate to the engine speed and to interrupt lubrication when the engine stopped.

The next step, in the 1910s, was applying a closed lubrication circuit, where oil was reused after leaving each lubrication point; engines were closed by a lower cover, the *oil sump*, where the oil leaking from the engine was recovered and contained. The oil was splashed to each lubrication point by engine cranks diving into the oil sump; oil pockets were provided in the engine block, storing oil temporarily, to compensate for changes of oil level in the sump, due to accelerations (acceleration, braking, driving in a curve or on a slope). An example of this practice is shown in Fig. 4.26.

The final improvement, introduced in the 1930s, was to distribute oil in the closed circuit by suitable tubes provided in the engine block and head; the lubrication circuit was under pressure to increase oil lift in bearings and to allow oil distribution in any operating condition. Gear oil pumps have been used almost always for this purpose until now.

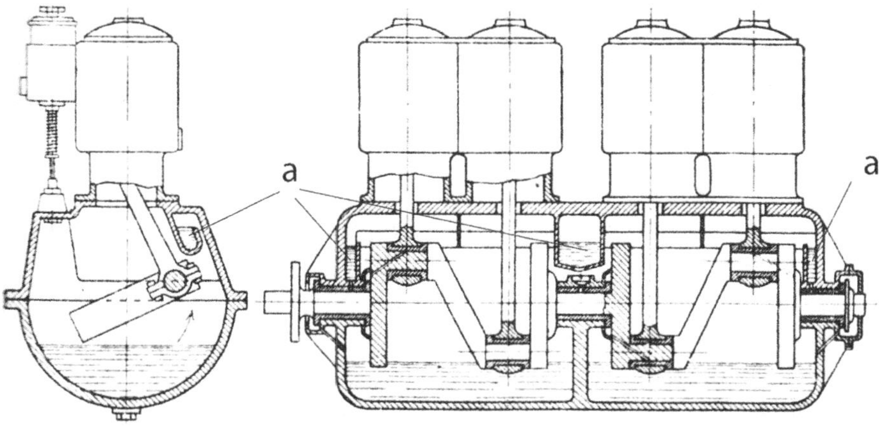

Fig. 4.26 In early closed circuit engines, oil was splashed to each lubrication point by the engine cranks diving into the oil sump; oil pockets *a* were provided in the engine block to store oil temporarily, to compensate for the change of oil level (from L. Morello, *Evoluzione tecnologica dei motori automobilistici ad accensione comandata*, ATA: Ingegneria dell'autoveicolo, 2004, N° 7/8 e 9/10)

4.2.5 *Ignition*

It is possible to classify the many different ignition systems developed for internal combustion engines in three families according to how combustion was started:

- electric spark;
- hot spot;
- continuous flame.

It would be wrong to think that present electric spark systems were the final choice after years of development of different solutions. As a matter of fact there was an immediate convergence of all engine inventors toward electricity: The 1807 De Rivaz engine had an electric ignition as well as the 1854 Barsanti and Matteucci engine. The basics of the electric spark had been studied by Alessandro Volta in his experiments on electric ignition of hydrogen mixtures, dating back to 1775.

The so-called Volta's gun demonstrated that an electric spark was suitable to ignite a hydrogen mass and produce a quantity of energy suitable to launch a bullet; the following development of De Rivaz and Barsanti demonstrated that with a suitable machine it was possible to control the release of this energy and produce usable work.

Nevertheless, different ignition systems were proposed and developed, because of the difficulty in producing and distributing electric energy on board of the early vehicles. Electric energy is in theory easily transportable and can be easily controlled by switches. This fact was particularly appreciated to ignite the combustion in connection with the crankshaft position and electric switches were suitable components to have engines working automatically.

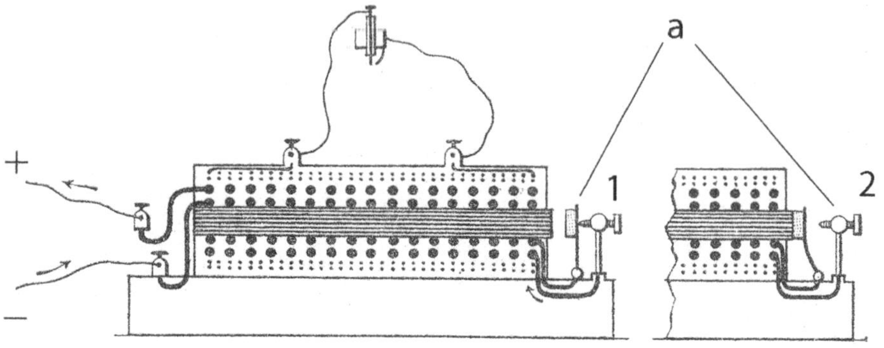

Fig. 4.27 In a Rühmkorff's coil the continuous electric current generated by a battery was conveyed into an electric coil wound around an iron core; the magnetic force in the iron core attracted the moving part *a* of a switch interrupting the electric current (redrawn from Baudry de Saunier 1905)

The first electric ignition systems used a device similar to the Rühmkorff's coil. According to its inventor, the continuous electric current generated by a battery was conveyed into an electric coil wound around an iron core (Fig. 4.27); the magnetic force, generated in the iron core, attracted the moving part of a switch *a* (1 switch closed, 2 switch opened), interrupting the electric current. Current interruptions released the switch, thereby feeding the coil again. The result was a chopped current into the coil.

A second coil with a larger number of turns, wound around the first, supplied a chopped current at higher voltage suitable to deliver sparks at an electrode. A typical Rühmkorff's coil, also called *vibrator* by engineers, because of its way of operating, is shown in Fig. 4.27.

Vibrators worked continuously generating a high voltage; the spark electrode was connected to the electric circuit when the intake phase of the air and hydrogen mixture was completed. However, this way of producing and conveying electric energy, developed in the 1850s, was more suitable to physics laboratories than to industrial machinery; this was the reason why the ignition system used in Otto atmospheric engines was a continuous flame, burning the same lighting gas used as engine fuel.

A valve system had to convey this continuous flame periodically, with a suitable timing, to the combustion chamber, avoiding exhaust gases blowing through the same duct extinguishing the flame. Such a system was considered to be unsuitable to a moving engine where air motion could easily affect its operation.

For this reason, Daimler, still thinking that electricity was unsuitable, decided to use a hot spot ignition. A lamp, fed with the same liquid fuel used for the engine, heated a metal pin mounted in a hole through the cylinder wall to glowing temperature. The pin had one end facing the flame, while the other was facing the combustion chamber. The lamp had to be used only for a while, after engine cranking, because the combustion heat itself could keep the pin in glowing conditions. The pin started

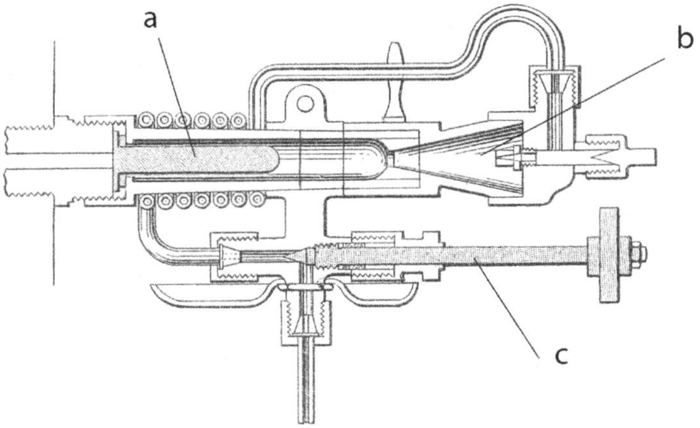

Fig. 4.28 A hot spot ignition system: the glow pin *a*; the small burner *b* and the manual valve *c* for fuel regulation are shown (redrawn from L. Morello, *Evoluzione tecnologica dei motori automobilistici ad accensione comandata*, ATA: Ingegneria dell'autoveicolo, 2004, N° 7/8 e 9/10)

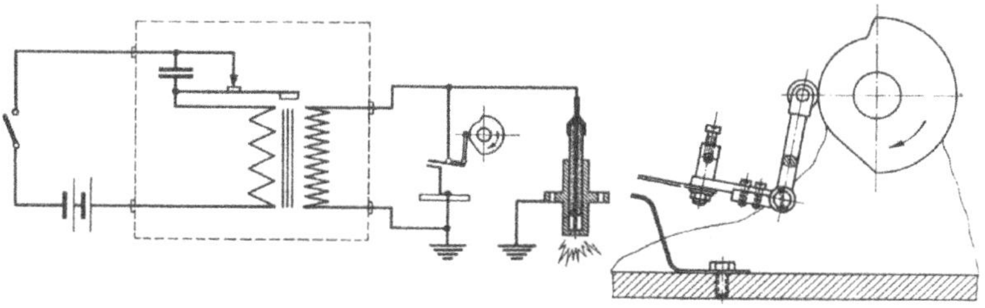

Fig. 4.29 Electric ignition system used in the 1886 Benz automobile engine (from L. Morello, *Evoluzione tecnologica dei motori automobilistici ad accensione comandata*, ATA: Ingegneria dell'autoveicolo, 2004, N° 7/8 e 9/10)

combustion as soon as compression had brought the mixture to the self combustion temperature.

A similar system is described in Fig. 4.28; the glow pin *a*; the small burner *b* and the manual valve *c* for fuel regulation are shown. The fuel is fed to the device by gravity. The advantage of this system was its simple layout and high reliability, the main disadvantage was the impossibility of adjusting the spark advance to a suitable value.

One of the first automotive engines made in 1886 by Benz used instead an electric system, similar in principle to that used by Barsanti and Matteucci; a scheme of the electric circuit and a detail of the rotary switch operated by the engine is shown in Fig. 4.29.

At the end of the 1910s the electric spark ignition was applied almost by all manufacturers; the differences between the different systems was how the electric energy was generated and managed.

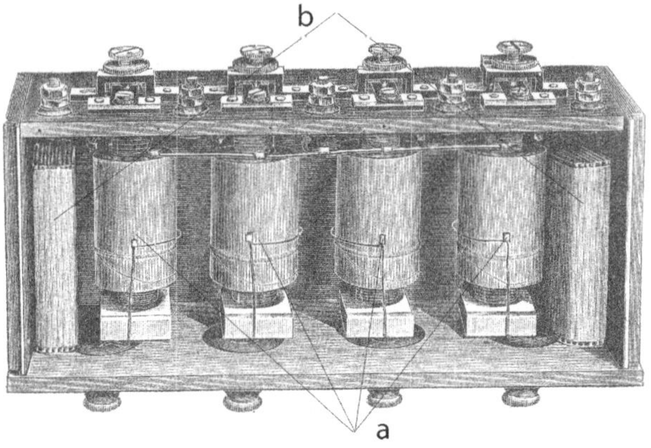

Fig. 4.30 Electric ignition system for a four cylinder engine, with four vibrators *a* and two batteries *b* (redrawn from Baudry de Saunier 1905)

Four different systems can be identified:

- vibrator ignition;
- low voltage magneto ignition;
- high voltage magneto ignition;
- breaker ignition.

The application of a manual advance adjustment was a feature common to all systems: The driver had to adjust the advance using certain empirical procedures, like setting a low advance during cranking, to avoid engine counter rotation; advancing the ignition after start to the highest possible value, to improve performance, but avoiding *knocking* (detonation).

In the first already described solution, electric energy came from lead-acid batteries that had to be recharged at home from time to time. Figure 4.30 shows an electric ignition system for a four cylinder engine. The fact that the source of energy was depletable was considered a major drawback; this system was abandoned for a time until suitable electric generators were available to recharge batteries during engine operation; but the relevant cost increase could be accepted only after other electric appliances had to be added on board of vehicles.

Magnetos were considered as simple electric generators able to supply the current necessary for ignition. The low voltage version of this electric machine was simply a set of coils put in rotation inside a continuous magnetic field; the current collected by sliding contacts was suitable, with simple transformations, to generate the spark. Figure 4.31 shows two different versions of a low voltage magneto: The magnetic field was generated by a set of horseshoe permanent magnets, for instance two (a), or a two poles rotor spinning inside the magnet (b) had a suitable number of coils that had to be insulated electrically.

Wire insulation, today easily made by immersion of the copper wire in a synthetic resin, was initially made by winding a silk thread around the copper; small displace-

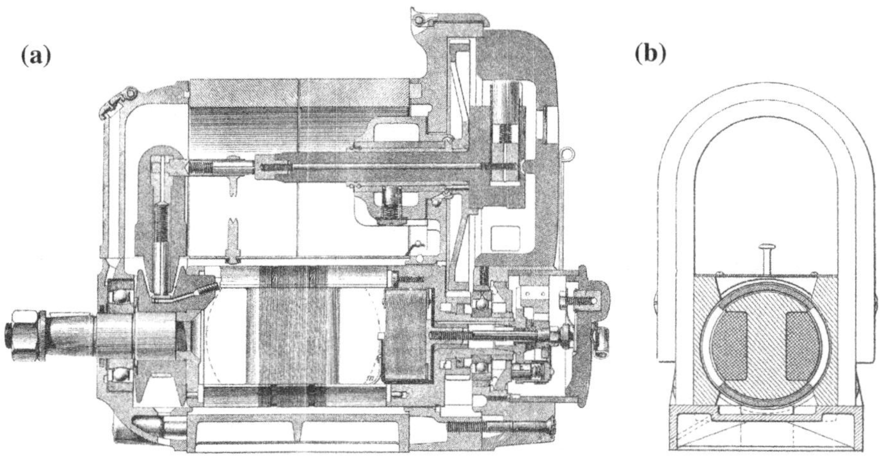

Fig. 4.31 Two versions of a low voltage magneto; a rotary coil version (**a**) and a fixed coil version (**b**) (from Baudry de Saunier 1905)

ments in the silk thread caused electric contacts with local short circuits reducing the output voltage. For this reason the voltage could not be high enough to make an electric spark with sufficient energy directly available.

Two sliding contacts were provided, one axial, at right, one radial at left; the radial contact connected with as many circuits as the cylinders in the engine, while the axial contact closed the electric circuit, usually using the steel mass of the engine and the chassis as return circuit. The multiple output of the magneto could be rotated for advance adjustment; the magneto was rotating at half the speed of the engine for correct timing.

To avoid one sliding contact, some magnetos had a fixed coil while a rotary iron element was provided to generate the necessary periodic magnetic field variation, as shown in the figure at right.

The voltage increase necessary for sparks was obtained by opening switches and by utilizing the voltage peak generated by the circuit inductance. These switches were directly installed inside the combustion chamber; their contacts were closed in short circuit for all the time but at the time of spark.

Figure 4.32 shows one of these switches; they were opened by additional cams on the camshaft with a profile suitable for very quick opening. Since magneto speed was not enough for firing during hand cranking, sometimes additional vibrators were provided that were switched off after the engine had started.

In high voltage magnetos an additional secondary coil was placed around the primary rotating coil, having the function to increase the voltage. High voltage magnetos were connected to conventional fixed electrode spark plugs, but this obvious solution was available only after copper insulation could be improved.

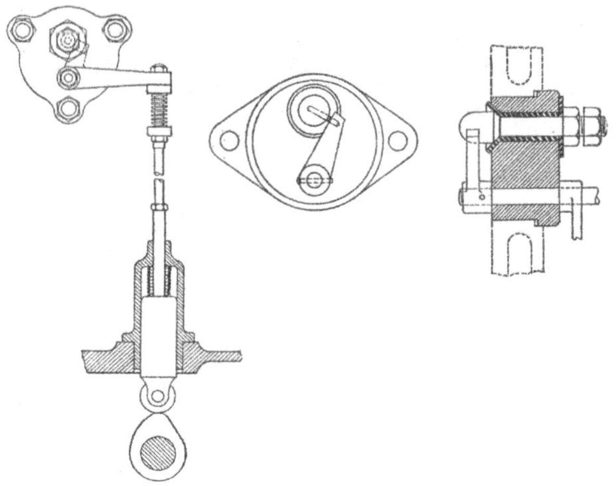

Fig. 4.32 Ignition switches were directly installed inside the combustion chamber; their contacts were closed in short circuit for all the time but at the time of spark (courtesy of FIAT Historical Archives)

Charles F. Kettering of DELCO patented in 1908 the first ignition system operated by a battery, recharged on board of the vehicle. Systems of this kind were in use till the 1970s when electronically controlled ignition systems were introduced.

The *breaker* (switch), controlling the current transient in the coil, is placed in the low voltage circuit, fed by the continuous current generated by the battery. The breaker is no longer a vibrator, but is operated by the engine with a cam, generating a contact at each second turn of the crankshaft. A condenser reduced the temporary peak of voltage, to avoid too strong sparks from damaging its electrodes. Spark plugs were connected to the high voltage circuit of the coil by a distributor, switching the contact on each cylinder.

A major improvement consisted in an automatic device controlling the spark advance as a function of engine speed and manifold pressure. This device could rotate the switch with a centrifugal regulator and a pneumatic actuator.

4.2.6 Starters

In the early electric ignition systems electricity was provided by dry cells; they produced electricity through an irreversible chemical reaction that made the cell unusable when discharged. An improved system included batteries to be recharged by connecting them to the home electric network. Lead-acid batteries, reusable many times, were invented by Plante in 1859. They were made by rolling up thin sheets of lead with separators made by a porous and insulating material. This roll was placed in an insulating jar, containing a diluted solution of sulfuric acid.

A battery like this had to be charged and discharged for many times before the porous spacers could contain the suitable mass of lead oxide, necessary to store the desired quantity of energy; this technique was expensive and inefficient.

Foure improved the lead battery design in 1881, by developing a battery where electrodes were made by lead grids with their mesh full with lead oxide paste. According to this design, flat electrodes were piled up with insulating separators in between, to obtain the desired voltage. The fabrication time of the battery was reduced because the lead oxide could be obtained by a separated process, not including electrolysis.

Each pair of plates could generate a voltage of about 2.25 V; more plates in a stack, connected in series, could deliver the desired voltage while more stacks in parallel could generate the necessary amount of current.

Almost immediately, batteries with a voltage of about 6, 12 and 24 V were standardized; 6 V batteries disappeared in the 1970s because their low voltage made high currents and consequently heavy copper wiring necessary, while 24 V were used only on large industrial vehicles. This kind of battery is still in use today, with some improvements, as service batteries of internal combustion engines for vehicles.

A contribution to lead acid batteries diffusion on cars was made in 1912 by the first application of an electric starter; also this invention was made by Kettering, under request of Leland, President of Cadillac at that time. Rumors say that Leland was impressed by the death of one of his friends, due to injuries he suffered while hand cranking his automobile.

Before the invention of electric starters engines were started manually by a hand crank connected to the engine crankshaft; the joint between the crank and the engine shaft was made by front teeth with asymmetric profile, designed to disengage as soon as the engine was started.

But the driver had to perform other complex maneuvers before engine cranking. The first step was mixture enrichment, if the engine was cold. This purpose was accomplished in earlier engines by introducing in the cylinders close to firing a suitable quantity of gasoline; each cylinder had a tap on its top with a little funnel for this purpose (see for instance Fig. 4.11). In more recent engines the enrichment function was accomplished by a choke valve in the carburetor.

The second step included manual spark advance adjustment; the purpose of this adjustment was to avoid slow combustion in the cold engine from taking pace too late, at open exhaust valves, and preventing the engine from starting backwards due to an early ignition.

At this time the engine could be cranked. Cranking required a strong effort, particularly in large displacement engines and many would-be drivers were discouraged by this operation. In addition, an unsuitable spark advance adjustment could start the engine backwards, sometimes causing severe injuries to the operator since the crank joint could not disengage in counter rotation.

Electric starters were therefore a welcome simplification that made larger distribution of the automobile possible, but the introduction of the starter motor required suitable batteries and current generators to be installed.

The system that Kettering developed was based upon the concept that a single electrical machine could either work as starter or as generator; this is in theory

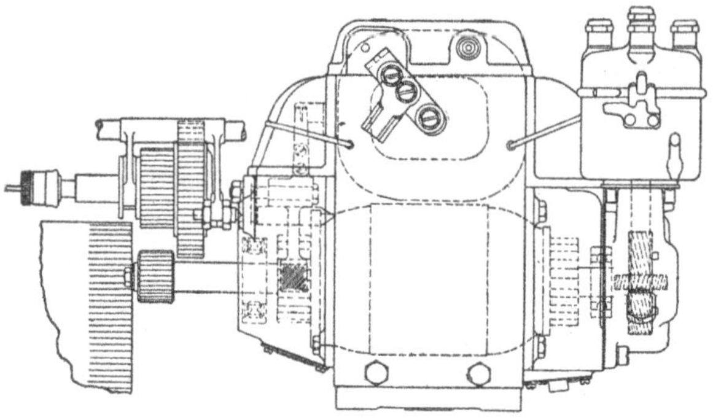

Fig. 4.33 Kettering's electrical machine includes a suitable gearbox that can engage with two different transmission ratios; the first, about 1:1, to generate current, the second 1:100, for engine starting (from L. Morello, *Evoluzione tecnologica dell'impianto elettrico*, ATA: Ingegneria dell'autoveicolo, 2008, N° 1/2)

possible because the two machines can be mechanically identical. In practice, engine starting requires high torque and, consequently, high currents at low speed, while current generation, in normal working conditions, requires a much lower current to recharge the battery, involving lower torque at higher engine speed.

This problem was solved by Kettering in a straightforward way, with a suitable gearbox that could engage the electrical machine with two different transmission ratios: About 1:1, to generate current and 1:100, for engine starting. A drawing of this machine, integrating also the distributor and the ignition breaker, is shown in Fig. 4.33. The electrical machine had two different windings and blade collectors, connected in series and in parallel, in consideration of the different current values. The mechanical engagement with the flywheel for engine starting was made by operating a dedicated pedal.

Later, generator and starter functions were obtained by separate electrical machines, since the added cost was paid off by avoiding a gearbox. However, the very large range of speed of the generator was difficult to be accepted. In fact, designing the generator to charge the battery at high engine speed, caused the battery to discharge at low speed, during city driving. On the contrary, designing the generator to charge the battery at low speed, caused too high a voltage at high speed: A voltage regulator was required.

A mechanical regulation was initially applied. Figure 4.34 shows a centrifugal regulator that disengages the generator when a speed threshold is overcome. In this system, the generator is able to recharge the battery at low engine speed, but it is stopped at high speed.

Later, electrical regulators were applied. Two alternatives were considered: Continuous and on-off regulators. An example of the first alternative was Kettering's centrifugal rheostat, controlling the generator excitation current to a suitable value, to avoid battery overcharging. The most common on-off solution included a set of three relays. A minimum voltage relay operating when there was a danger of

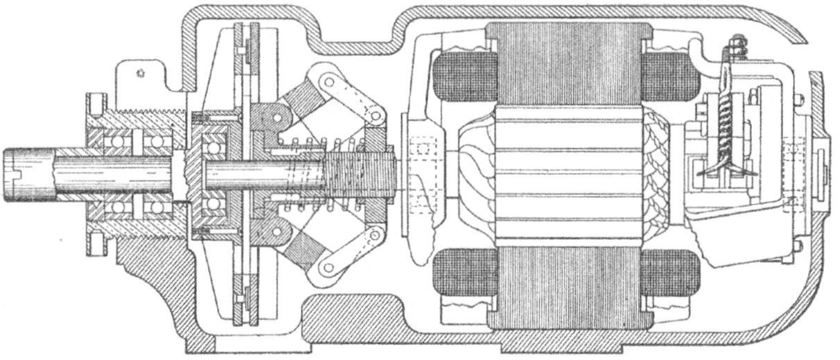

Fig. 4.34 Generator with mechanical regulator: The generator is disengaged when a speed threshold is overcome (from L. Morello, *Evoluzione tecnologica dell'impianto elettrico*, ATA: Ingegneria dell'autoveicolo, 2008, N° 1/2).

discharging the battery, because of too low generator voltage; a maximum current relay and a maximum voltage cutoff relay that interrupted battery recharge beyond a safety value.

The alternator, applied since the early 1970s, simplified the regulation being less sensitive to the speed; in this case a set of solid state diodes converted the alternate current to direct current.

4.3 Gearboxes

A gearbox with a clutch, or a start-up device of other kinds, are essential for adapting the transmission ratio between engine and wheels to the needs of the vehicle and the characteristics of the engine. This transmission ratio should be very high, ideally infinite, at vehicle start-up.

In the early cars, the gearbox was confused with a device for speed adjustment: The term of *speed* to name a given transmission ratio comes from this misinterpretation.

4.3.1 Manual Gearbox

Patent documents of 1784 give evidence that Watt foresaw the use of a constant mesh gear box, with dog clutches, to improve the traction performance of a steam engine; it is rather difficult to demonstrate how this idea might have influenced the design of cars, which did not follow this scheme indiscriminately.

The first commercialized internal combustion engine cars are, without doubt, a result of the efforts by Benz and Daimler in 1885 and 1886 , and the transmission problem was solved by using a scheme completely different from that proposed by Watt.

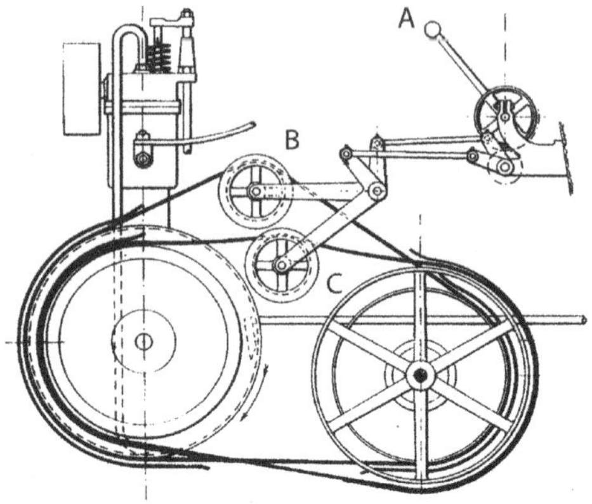

Fig. 4.35 Complete transmission of the 1885 Daimler car. Two transmission ratios are available, using a leather belt transmission. The slipping of the belts is exploited to start-up the car (redrawn from Genta and Morello 2009)

The complete transmission of the Daimler car is shown in Fig. 4.35. Two different transmission ratios are obtained using leather belt transmissions with pulleys of different diameter. The belts are always wound on their pulleys, but the motion is transmitted by only one of them, when one of the two tensioners *B* (for the high speed) and *C* (for the low speed), operated by the control stick *A*, engages the corresponding belt. Belt slip, occurring when the tensioner is not completely engaged, is exploited to start-up the vehicle from a standstill.

In this case the driveline design is simplified due to the lack of wheel suspensions. Many improvements to this layout were used in the following Benz car. The two speed transmission was inspired by the power transmission in contemporary workshops, where a single steam engine or water wheel operated a number of working machines through belts. This kind of transmission was probably invented by Anderson in 1849.

The two driven pulleys (in the top view in Fig. 4.36) are coupled with as many idling pulleys (at the outboard of the driven pulleys); the latter have a slightly smaller diameter than the driven pulleys and the active cylindrical surface of the driven pulleys is rounded to match the active surface of the idle ones. The two driving pulleys (in the back of the car, aligned with the engine crankshaft) have an adequate width to bear the belt on both driven and idling pulleys. Two tensioners can shift the leather belts from matching the driven pulleys to matching the idle ones.

The belts are crossed in order to increase the winding angle on the pulleys; belt tension is adjusted periodically, by changing the distance of the center lines. The rounding on the surface of the driven pulleys makes belt shifting easier; the lower diameter of the idling pulleys decreases the tension of the belts when they are not working, making the time between two subsequent tension adjustments longer.

The start-up function is again performed by exploiting belt slip; the suspension motion of the driving wheels is compensated for by two chain transmissions that connect the driven pulleys with the rear wheel hubs.

The literature on these and other cars of this time did not suggest a sequential use of gears to accelerate the car, but the car could be started in either one of the gears,

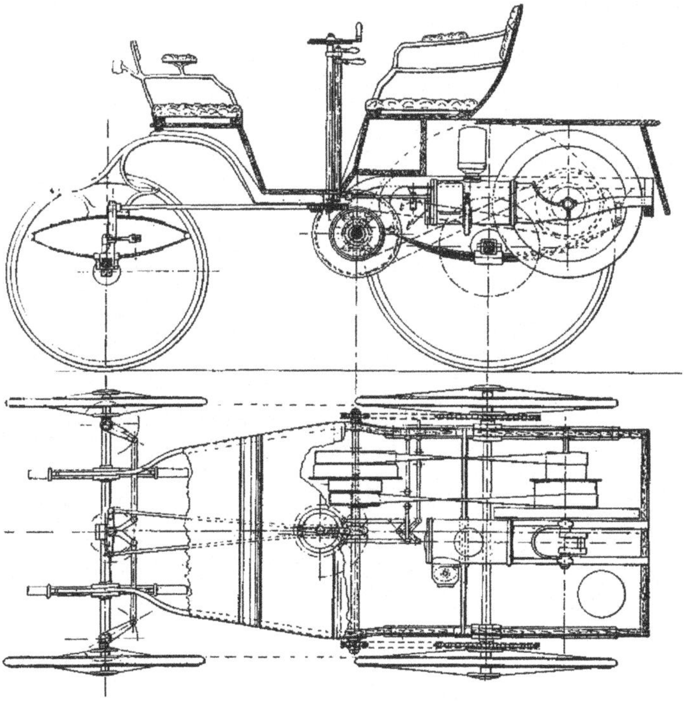

Fig. 4.36 Two speed transmission of the 1886 Benz car, of the leather belt type. Two tensioners can shift belts from a position engaged with a driving pulley to that engaged with an idling pulley (redrawn from Genta and Morello 2009)

and the choice depended more on the desired cruise speed than on the traction force needed for starting.

The idea introduced by Watt was then applied to the next generation of gearboxes. This is usually noted by saying that the first cars demonstrated how the technical skills of their inventors were polarized on engines, sometimes neglecting the other achievements that had been already obtained in the contemporary state of the art. This fact could be also explained by the difficulty of exchanging ideas within the technical communities of different countries.

A typical example, similar to contemporary layouts, is reported in Fig. 4.37, where a drawing of the FIAT 8/16 HP of 1902 is shown. This gearbox features three speeds and a reverse; in the oldest cars the reverse gear was missing.

The engine rotates the lower shaft, at the left, in the view showing a longitudinal cross section of the gearbox. This gearbox has a single reduction stage, where all driving gear wheels are aligned on the input shaft and all driven wheels on the output shaft. On the engine side the reverse gear is first, followed by first, second and third gears. An idle gear wheel, used to reverse the rotation speed of the output shaft is visible on the cross section.

Fig. 4.37 Single stage, three speed gearbox of the FIAT 8/16 HP of 1902. The internal shifting mechanism is made by a cylindrical rod, mounted inside the driven shaft (redrawn from Genta and Morello 2009)

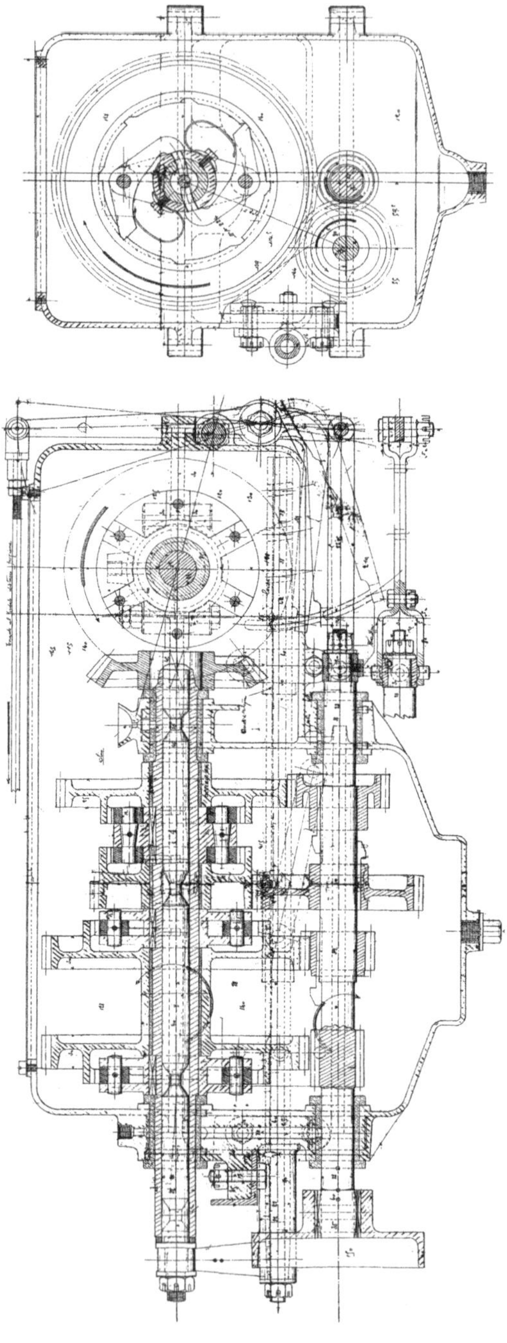

Driving wheels always mesh with driven wheels and must, therefore, be free to rotate on the output shaft. The output shaft operates a bevel gear, which moves two shafts through a differential. These are coupled through drive chains to the rear driving wheels, according to the scheme already seen in previous figures.

A tapered friction clutch, not represented in this figure, allows start-up of the vehicle and separates the gearbox from the engine during gear shifts (see for reference Fig. 4.45).

Three new mechanisms not yet introduced are here present: The reverse gear, the differential and the friction clutch. Their origin is for sure older than this car: The reverse gear was introduced by Selden in 1879, the differential by Pecquer in 1827 and the tapered friction clutch by Marcus in 1885. During the first years of the century these three mechanisms were integrated in a transmission suitable to automotive application.

The gear shifting mechanism is rather sophisticated, made by a cylindrical rod mounted inside a cavity in the driven shaft that can be moved axially. This rod shows some annular indentations, which in certain positions allow two ratchets to engage with the driven wheels; the detail of the reverse gears ratchets can be seen in the cross section.

When one of the grooves faces a pair of ratchets, two leaf springs provide their engagement with the ratchet gear; when the rod is shifted, ratchets retract, leaving the wheel again idle. The position of the grooves is operated by a sequential shift stick, where the positions of reverse, first, second and third follow each other.

This kind of gearbox architecture (*constant mesh gears*) was applied to many cars of that era, but was soon abandoned because of its complexity and consequent fragility. The next scheme is that of *sliding gear trains* (*trains balladeurs*, in French).

Sliding trains gearboxes can be exemplified by the FIAT 60 HP car of 1904, whose gearbox is shown in Fig. 4.38. This invention is older than its automotive applications since it was designed by Griffith in 1821 for workshop machinery.

It is not the first time that a better performing architecture (constant mesh) is abandoned in favour of a less evolved alternative (sliding trains), because a particular component (the dog clutch, the synchronizer, in this case) has not been developed yet. Sliding trains will soon be abandoned again, in favor of constant mesh, when the technological improvements made it possible.

The gearbox of Fig. 4.38 is still of the single stage type and has four speeds plus a reverse speed; gear wheels, grouped in two trains, from the first to the fourth, can slide on the upper driving shaft. The engine (not shown) is on the left, while on the right the bevel gears operating the pinion of the chain transmission through the differential can be seen.

The first and second speed wheels are located on the first train, while the third and the fourth speed wheels are on the second; these trains are mounted on a part of the shaft with a square cross section that allows them to turn with the shaft, being free to shift along it. Train sliding is accomplished by suitable sleeves that bring one wheel at a time to engage with its counterpart; the gears mesh only when their correspondent gear is selected.

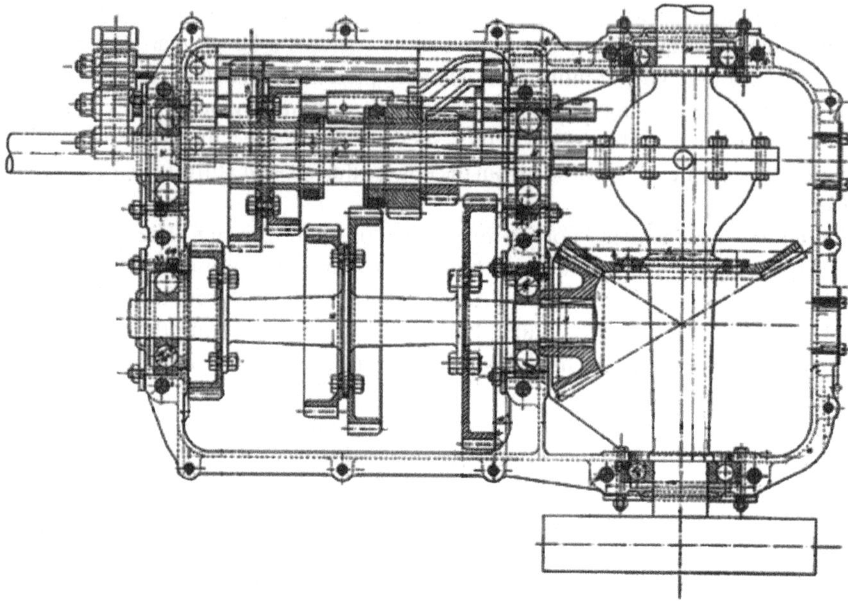

Fig. 4.38 Sliding train gearbox with four speeds of the FIAT 60 HP of 1904. The two gear trains are made by the driving wheels of the first and second speed and by those of the third and the fourth speed (redrawn from Genta and Morello 2009)

Reverse speed is obtained using a train of idler gears only, not represented in this figure, meshing with the first speed gears, when their train is in the idle position.

The sleeves are moved by forks fixed to sliding rods, partially visible under the gearbox shafts; there are two rods for moving the train for the first and second speeds and that for the third and fourth speed, while an additional rod is dedicated to reverse.

This kind of layout created the need for a particular kind of control stick, where gear shifting is no longer sequential, but is characterized by two separated motions: one across, to select the train to be engaged, the second, longitudinal, to engage a specific gear. This layout consolidated in common practice, surviving unchanged in our contemporary cars.

Sliding train gearboxes required a particular skill from the driver, who had to synchronize the wheels by means of the engine during idling, before engaging the next gear (double clutching); an imperfect maneuver was accompanied by the noise due to scratching of the gears getting in contact at an inappropriate speed.

Some manufacturers improved this architecture, actually returning to the Watt's idea of constant mesh gears; to show this new layout, the drawing of the gearbox of a small truck of the end of the 1910s is reported in Fig. 4.39.

In this gearbox three speeds are available together with a reverse speed; the gearbox is connected to the engine on the right of the figure, while the output flange is on the left. The flywheel on the output shaft is not part of the engine, but is the drum

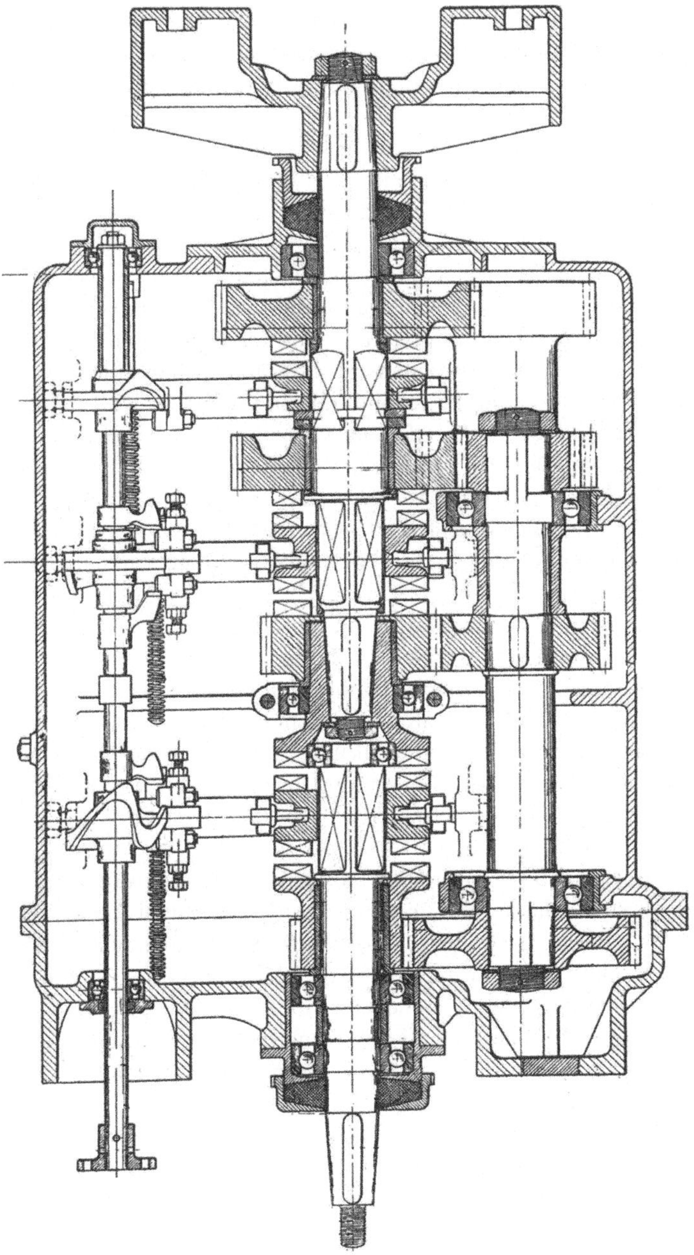

Fig. 4.39 Dog clutch gearbox of a truck built in the 910s. Sliding trains are now reduced to small front tooth gears working as clutches (redrawn from Genta and Morello 2009)

of the transmission brake. The input and output shafts are now coaxial; this layout is appropriate to the application of a universal joint transmission, that started to be applied in these years. Gear wheels are always meshing and carry dog clutches at their sides and the sliding trains are now reduced to the pairs of dog clutches.

The scratch problem is not solved, but possible damages are localized on an auxiliary component, the dog clutch, that can be sacrificed, with smaller impact on the operation of the gearbox. Dog clutch teeth can also be rounded, making shifting easier, without penalty for wheel size.

Input and output shafts are coaxial, the upper shaft made in two parts, free to have different rotation speeds. The lower sliding train can alternatively engage the two parts directly (third speed) or move a third shaft (*countershaft*), located at the right in the same figure. With this dog clutch only engaged it is possible to obtain the first and the second speeds, using the second train in the figure; with the last train it is possible to engage the reverse speed, connecting the countershaft to the output shaft and then to a pair of idlers (in practice a second countershaft), partially visible on the right of the figure.

Sliding train motions are operated by front cams, that allow in this case sequential control. A similar architecture, but for the control, is still present in front engine, rear wheel driven vehicles.

However, at that time these technical solutions were not consolidated and, as often occurs at the dawn of a new technology, deviations from what we consider is the main evolutionary path were many. After the engine, the gearbox was the field where early automotive engineers concentrated their innovative efforts.

A comprehensive classification of all the solutions that were attempted is well outside the scope of this section, even limiting the discussion to manual transmissions and only those that could be considered as the most original solutions are here considered.

Between 1924 and 1938 Frazer Nash built in England different cars, all with sport performance but affordable price in their category. Essential to these cars was the driveline, shown in Fig. 4.40. The gearbox is made by chain transmissions, three for forward speeds, one for reverse speed. Driving chain wheels (at right in the drawing) are operated by a bevel gear box, connected to the engine; this bevel gear box includes no differential gear. On the right half of this intermediate shaft the sprockets for the first and the reverse speed are located; on the left half shaft there are the sprockets for the second and the third speed.

Notice the idler reverse system with chain and gears. Driven wheels are directly mounted on the rear one-piece shaft; they can be moved along their shaft for chain alignment. A chain transmission was also used for compensating for suspension motion. The wheels were engaged by simple dog clutches.

The lack of a differential made the car difficult to drive, but very manageable for an expert driver on unpaved roads, common at those times. Admirers of these cars extolled the easy road holding, the excellent gearbox maneuverability and the ease with which broken parts could be replaced in the transmission.

A car with different technical details, also original and uncommon, is the 1907 Sizair Naudin, also known for featuring one of the first independent suspensions. Here

Fig. 4.40 Frazer Nash gearbox of 1924, made by chain transmissions. Three speeds and a reverse were available (from Genta and Morello 2009)

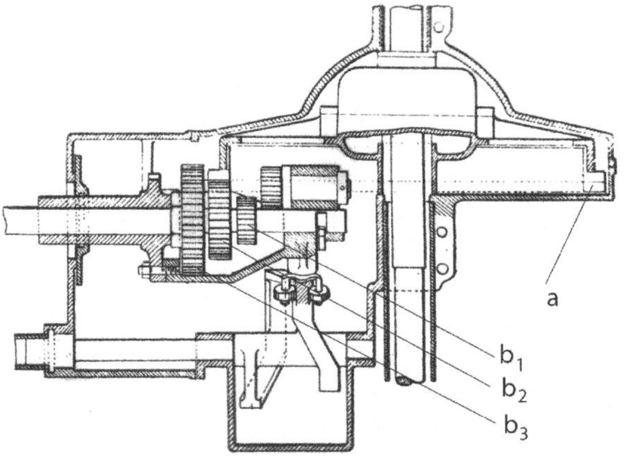

Fig. 4.41 Sizair Naudin gearbox of 1907. This design can be assimilated to a sliding train gearbox, with bevel gears (redrawn from Genta and Morello 2009)

the design goal should have been to contain gearbox cost, or at least the number of gear wheels, at that time quite expensive.

The gearbox architecture shown in Fig. 4.41 can be assimilated to a sliding train gearbox. There is a single driven wheel, the crown gear a. The only sliding train, made by spur gear wheels b_1, b_2 and b_3, is installed on a bearing swinging around the lower shaft in the drawing; the swinging motion of the bearing is necessary to mesh the driving wheel in use with the crown gear at different distances, depending on the driving wheel diameter.

The swinging motion of the driving shaft is compensated for by a universal joint transmission between engine and gearbox. The cam at the bottom of the drawing combines shift and swing motions and bears gearing forces. Bevel gears should have been used instead of the spur gears, for correct matching; teeth shape is, nevertheless, approximated by a cylindrical shape, with consequent contact errors. The reverse idler is also present on a dedicated shifting train.

A last example of amazing engineering ingenuity is given by the 1904 Turicum (a Swiss automotive trademark) shown in Fig. 4.42, with a drawing of the complete chassis: Here one of the earliest continuously variable transmissions used on automobiles can be seen.

The motion is transmitted to the rear axle by two friction wheels *a* and *b*, the first made of solid iron, the second with a rubber tread on its rim. The wheel *b* is connected to the shaft *c* through a spline and groove, that allows the wheel to be shifted along the shaft; the shaft *c* rests on a swinging bearing *d* and allows wheels *a* and *b* to be pressed against each other by the spring *e*.

By pulling the lever *f* it is possible to change the contact point between the two pulleys, between the center of the wheel *a* (infinite transmission ratio: transmission idling) and its rim (transmission ratio about 1:1). In the idle position, the friction is eliminated by unloading the spring *e*; the slip between the two wheels is used to start the vehicle up.

The proposals above were not imitated by other manufacturers and were probably abandoned by their own inventors. The evolution of manual gearboxes concentrated on perfecting a countershaft or double stage architecture, that became universal on all cars with front engine and rear drive.

An example of this evolution is offered by the gearbox of the FIAT Balilla of 1934 (four speed version, the first gearbox of this car in 1933 was a three speed design) shown in Fig. 4.43. This gearbox shows two different sections: The rear, for the first, second and reverse speeds, features a sliding train with cylindrical straight teeth. The front features helical gears (always meshing) and synchronizers.

This compromise was justified by the high cost of synchronizers, considered high technology components at that time. Synchronizers were limited to the more frequently used speeds, which could also benefit of helical gears, with gearing noise reduction.

Earlier engineering manuals suggested, as a good practice, to design the top ratio (in this case the final differential ratio) with values slightly higher then those ideally needed; this rule was addressed to limiting the number of gearshifts necessary to maintain the car at cruise speed, and demonstrated the difficulty for drivers to change speed with sliding train gearboxes. It is thus possible to assume that synchronizers brought benefits not only in shifting quality, but also in noise and fuel economy.

The gearbox of the FIAT 1400 of 1950, shown in Fig. 4.44, has, like many cars of this time, synchronizers on all speeds but the first, again for reasons of economy; the first is included in a sliding train mounted on the sleeve of the third and fourth speeds. The reverse speed is obtained with an idler, not shown in the picture, meshing with the wheels of the first gear when they are in neutral position.

During the following period, synchronizers were improved and made less expensive thanks to higher volume production; since the 1970s also economy cars had synchronizers on all speeds.

Fig. 4.42 Turicum chassis of 1904; A continuously variable transmission based upon two friction discs *a* and *b*, where the first is made of solid iron and the second has a rubber thread to improve contact friction, can be seen (redrawn from Genta and Morello 2009)

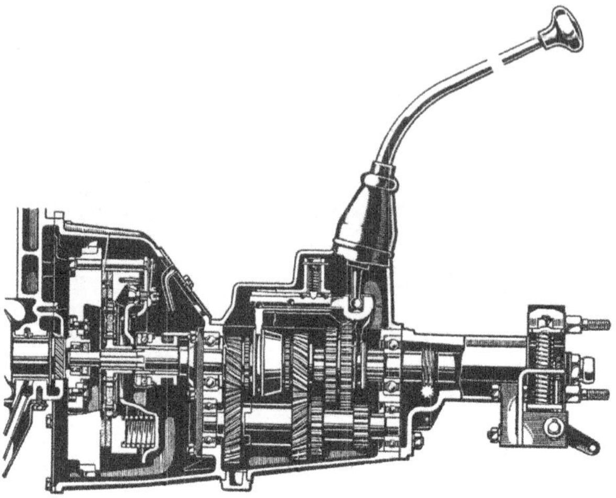

Fig. 4.43 Four speed gearbox of the FIAT Balilla of 1934. The drawing shows a longitudinal cross section; the rear of the gearbox has a sliding train for the first, second and reverse speeds; the front features synchro-mesh gears for the third and fourth speeds (from Genta Morello 2009)

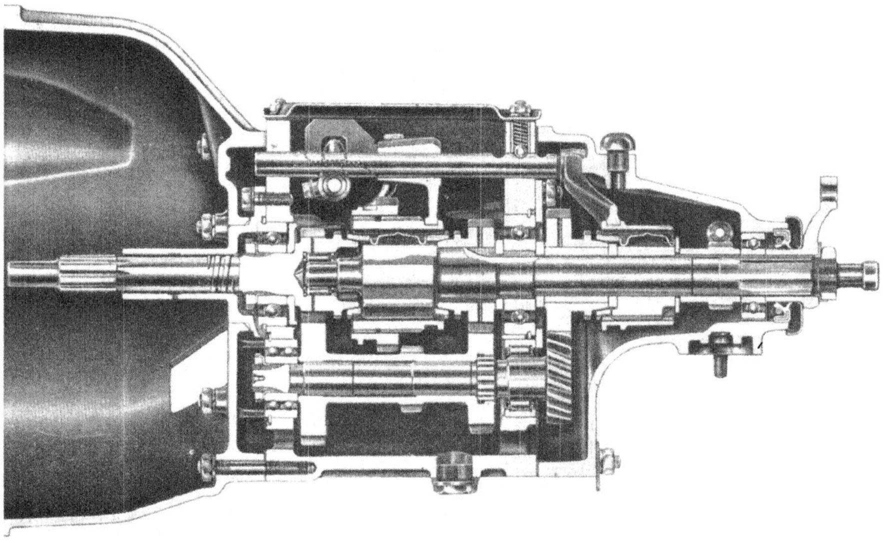

Fig. 4.44 Four speed gearbox of FIAT 1400 of 1950, with full synchronization, except for the first gear. The reverse speed is obtained through a sliding idler, not shown in this figure (from Genta and Morello 2009)

4.3.2 Friction Clutch

In the earliest cars, belt transmissions integrated the start-up function into the device used to change speed, but this was not the case in gearboxes, where a dedicated mech-

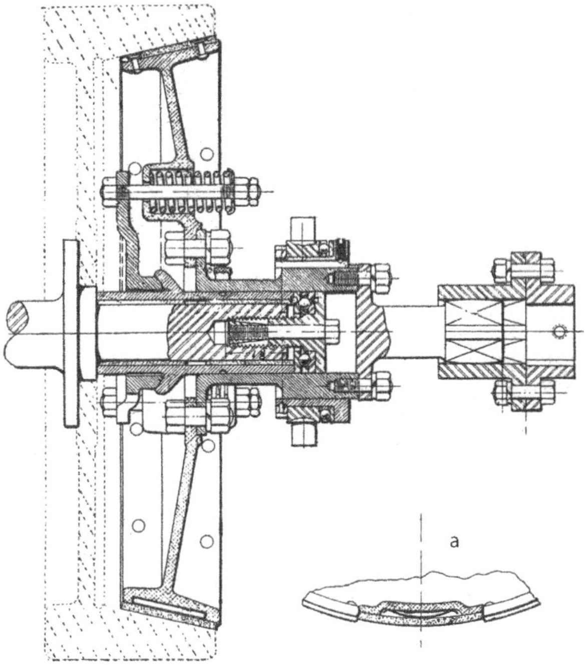

Fig. 4.45 Bevel clutch with leather linings. Clutch conicity (1:2) is limited by the available friction coefficient to avoid irreversible sticking (redrawn from Genta and Morello 2009)

anism was needed. This mechanism, a friction clutch, or, simply *clutch,* posed designers many problems related with its operating life. As already seen, bevel clutches were used since the beginning of the automobile era.

An example of bevel clutch of the first years of the past century is shown in Fig. 4.45. The friction surface was covered by a leather lining, riveted on a bevel pulley of cast iron. Although the famous synthetic material called Ferodo® had been invented by Frood in 1897, it became widely applied only in the 1920s. Leather had a friction coefficient similar to that of modern materials, but offered a limited performance in terms of heat dissipation and duration, requiring active surfaces having a large area. On the other hand, leather was, at those times, cheap and easily repaired or replaced.

In this application there was a single active surface shaped like a cone, a geometry that was chosen to limit the disengagement force on the pedal, which depended upon the cone diameter and engine torque. Due to the difficulty of integrating the leather lining into its support disc, the active surface was usually single rather than double, as it is in modern clutches.

With reference to the previous figure, the engine flywheel features a short shaft, bearing the reaction structure of the load springs working on the friction surfaces. Many coil springs (only one is shown in the cross section) pushed the tapered friction

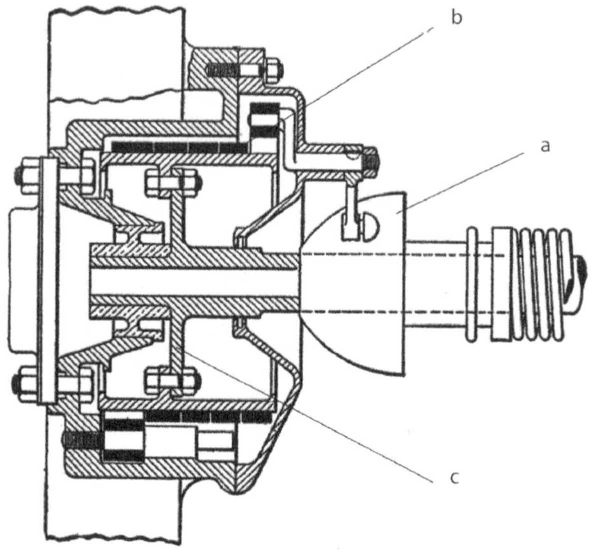

Fig. 4.46 Spiral spring clutch. The clutch pedal moves the ogive *a* that closes the spring *b* onto the input shaft *c*, creating a friction force; the same force increases the spring tensile force (redrawn from Genta and Morello 2009)

disc against the flywheel. The leather lining was riveted on this disc; very thin leaf springs (see detail *a*) were set between the lining and the disc to make the engagement more progressive.

With this kind of architecture, the gearbox input shaft must be able to slide on a square counterpart. Friction conicity (1:2 on this drawing) was limited by the friction coefficient between leather and iron, to prevent irreversible sticking of the clutch after engagement.

The large engine displacement of many cars and the limited size of the flywheel made many clutches too heavy to be operated; for this reason other mechanisms were also developed. The idea was to exploit the mechanical property of wound linings to reduce working forces. Here the friction force is itself used to increase contact pressure, as in leading shoes of drum brakes. This principle was applied through band clutches.

An application of this principle is shown in Fig. 4.46; a coil spring with rectangular cross section and with the coils quite close to each other (*b*) is installed in a cavity in the flywheel. One end of the spring is fixed directly to the flywheel, through the eye in the lower part of the figure; the other end is connected through a rocker arm.

If an ogival body *a* is moved closer to the rocker arm, the spring is twisted and its internal diameter is reduced. The gearbox input shaft *c* is surrounded by the spring with a small clearance so that when the ogive is advanced through the clutch pedal the spring closes on the shaft with a resulting friction torque.

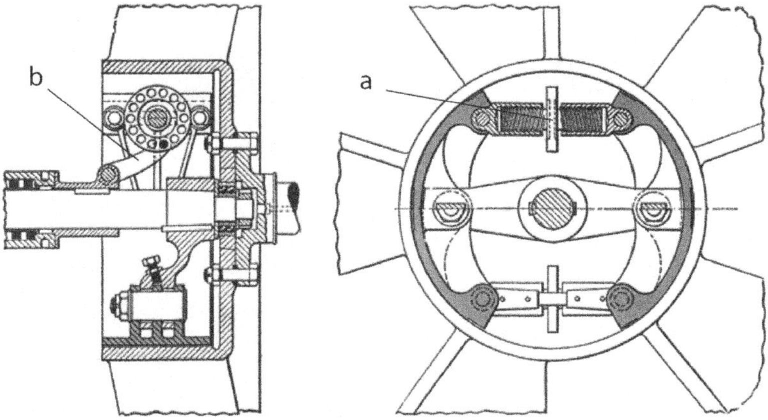

Fig. 4.47 Radial shoes clutch. The shoe displacement is caused by a screw mechanism *a*, operated by the clutch pedal through a crank *b* (from Genta and Morello 2009)

The friction tensile force along the spring coil increases the tangential force toward the spring eye, without increasing its reaction on the rocker arm; the resulting friction torque is an exponential function of the winding angle, which can be increased indefinitely. With a modest friction coefficient between metals it is possible to transmit the desired torque with a reasonable force on the pedal; the drawback is the roughness of the engagement maneuver, only partly eased by the spring elasticity.

A very different configuration based on the same principle is shown in Fig. 4.47; the torque is transmitted, in this case, by two shoes that expand in a drum, as in a drum brake. The shoe motion is created by a screw *a* that is moved by a crank and rod mechanism *b*; the disc shape of the crank is chosen to allow a simple play adjustment, to compensate for lining wear.

Faced with a difficult problem, inventors investigated many different solutions before consolidating and improving the best one; to solve these problems electric and hydrostatic transmissions were also investigated and applied.

The final solution was consolidated in the 1930s with the single disc clutch with synthetic friction linings; one example from this period is shown in Fig. 4.48. The friction surface is now flat and double, so that it is possible to transmit a double torque with the same force. The friction disc is mounted within two surfaces (the flywheel and the pressure disc) that are compressed by a number of coil springs against each other; a set of levers *a* on the pressure disc are used to release the clutch with the axial motion of a thrust bearing *b*.

This kind of clutch received its last improvement by the application of disc springs; these were introduced at the end of the 1970s and allowed many advantages, such as a further reduction of pedal force and a general simplification of the mechanism.

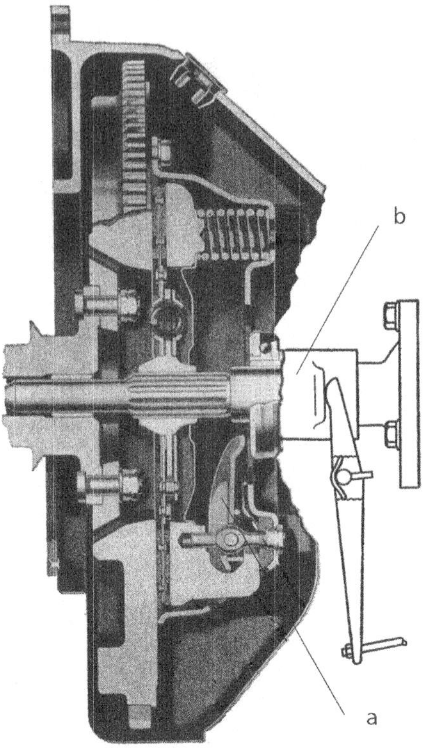

Fig. 4.48 Dry single disc clutch with coil pressure springs. A set of release levers *a* articulated on the pressure plate is used to disengage the clutch, through the displacement of the throwout bearing *b* (redrawn from Genta and Morello 2009)

4.3.3 Automatic Gearbox

Automatic gearboxes had their own history, one that received a crucial contribution from the American automotive industry. This is not to suggest that Europe failed to contribute to this development; many fundamental inventions in this area were actually developed on this continent, but nevertheless the European market, smaller and more fragmented, did not justify the mass production of this gearbox until recently.

The problems to be solved in developing an automatic gearbox included a different mechanism for engaging gears and starting the vehicle, easier to operate with the available automatic controls. These could be mechanical (exploiting centrifugal forces) or hydraulic (exploiting the pressure variation of the oil in a rotary pump).

Today this problem appears in a new context, because electronic microprocessors allow also an easy automatization of synchronizers and friction clutches, in use on manual gearboxes; many existing vehicles already substantiate this statement.

The first step was the development of gearboxes where speed shifts were possible without dangers to gear wheels and other parts that had to be synchronized. From this point of view the single stage, two speeds, manual gearbox of De Dion-Bouton, developed at the beginning of the past century and shown in Fig. 4.49 can be consid-

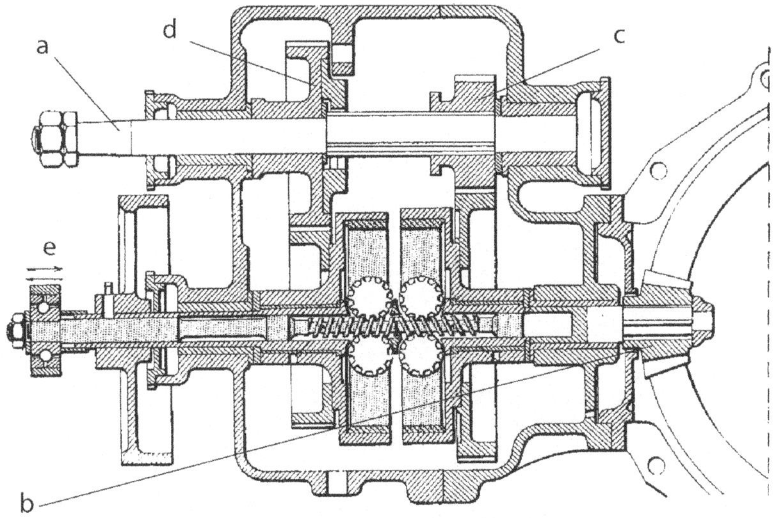

Fig. 4.49 The De Dion & Bouton gearbox can be considered as a precursor to power-shift gearboxes. By shifting a shaft with the bearing *e*, it is possible to engage one of the two shoe clutches available on each gear; a start-up clutch is not required (redrawn from Genta and Morello 2009)

ered as a precursor of automatic transmissions. In the figure the input shaft *a* and the output shaft *b*, which moves through a bevel gear the pinions of the chain drive are shown. The two driving gears *c* and *d* always mesh with the driven wheels idling on the output shaft. The engagement of the wheels is made by shoe clutches, similar to those already discussed in Fig. 4.47, controlled by screws.

By shifting the shaft with the thrust bearing *e*, it is possible to engage one clutch and to disengage the other one. As a consequence, a start up clutch needs not to be used during gear shifts. Although developed for manual gearboxes only, this kind of clutch is a relevant precursor of the powershift gearbox with band brakes and multi-disc clutches.

A second gearbox of historical relevance is that of the Ford Model T of 1908, the first car to be produced by the millions. This gearbox; based on planetary gears instead of ordinary gears, is shown in Fig. 4.50.

Planetary gears had not been invented by Ford, since they were already used in other applications. They may have been invented by Bodmer in 1834, although there is evidence that these mechanisms were already known to the ancient Greeks for devices used in performing astronomical computations. The three satellites v, r^1 and r^2 (the unusual position for the subscript, not to be confused with an exponent, is taken from an original figure), rotating on a single carrier, fixed to the engine flywheel are clearly shown in Fig. 4.50. They mesh with the corresponding sun gears s, s^1 and s^2. If the sun gear s is locked, by rotating the flywheel and the carrier a reduced output speed in the opposite direction is obtained at the sun gear s^2, fixed to the output shaft.

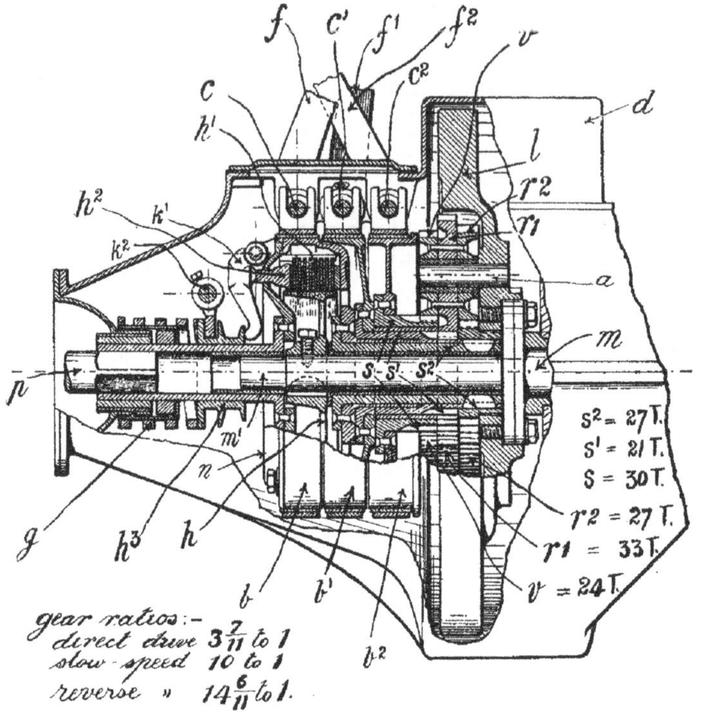

Fig. 4.50 Gearbox of Ford model T of 1908 based on planetary gears. One reverse gear, one reduced forward speed and a direct drive are available (from Genta and Morello 2009)

On the other hand, by keeping the sun s^1 locked, a reduced speed in the same direction, again at the sun s^2 ensues. A note in the figure reports the transmission ratios that were obtained, including the transmission ratio at the differential.

If the multi-disc clutch h^1 is engaged by shifting the sleeve h^3, it is possible to put the gearbox in direct drive by locking the hub h, rotating with the crankshaft, with the output shaft.

To obtain the different states of the gearbox, the sun gears rotate with the drums c, c^1 and c^2, which can gradually be stopped using the band brakes that are in turn controlled by front cams, moved by pedals. The lower part of these pedals is shown in the figure with the letters f, f^1 and f^2. The pedals have a spring system that locks them in either the released or depressed position: each pedal raises when another is depressed.

When engine and car are both stopped, the pedal f must be depressed so that the clutch h is engaged and the vehicle is in parking condition. By releasing the clutch h, through a lever, the engine is disengaged and can be cranked. The car is still braked. By depressing one of the pedals f^1 or f^2, the pedal f is raised, the car is left free to move and will start-up in low gear, either forwards or backwards: The vehicle can be put in reverse even when the vehicle is moving, and start-up on slopes is made easier.

The pedal f^1 can be released, as soon as the suitable speed is reached, by engaging the clutch h. The car is now in direct drive.

The gearbox is controlled by driver actions, but clutch management is performed automatically during gear shifts. The way to obtain a fully automatic gearbox from this situation was still long, but these achievements brought the final result closer. The configuration of this gearbox allows us to understand why planetary gears were preferred to ordinary ones in automatic transmissions: The main reason was the ease of integrating brakes and clutches.

A further step was made in 1928 by Wilson in England, who proposed a gearbox with two planetary gear trains in series, with the carrier of the first gear connected to the ring gear of the following one. With two planetary gears it is possible to obtain three speeds forward, one of them being a direct drive, and a reverse speed.

These gearboxes, similar in use to those of the Model T, were semi-automatic with manual preselection; according to this concept, a small lever close to the steering wheel was used to select in advance the next gear to be used. At this point no gearshift could yet start, but the brake mechanisms were arranged for the gearshift to be eventually made, as soon as the driver depressed a pedal specifically provided for this purpose. This last pedal was located where the clutch pedal normally is positioned.

With this mechanism the driver was helped in performing a coordinated maneuver of the gear stick and clutch; although the energy needed for gear shifting was still produced by driver's muscles through a pedal.

A particularly advanced semi-automatic gearbox was introduced by Cotal in 1934, in France. This gearbox, whose cross section is shown in Fig. 4.51, includes three planetary gears. In the figure the engine is on the left, the output shaft on the right. In this gearbox the brakes are substituted by toroidal electromagnets, used to lock some selected elements of the gear train. The first electromagnet puts the corresponding gear set into direct drive, by locking the sun and ring gears together; the second one is used to obtain a reduced speed, while the third, located at the right, to obtain a faster speed. The last electromagnet is used to set the gearbox in direct drive.

By energizing electromagnets in combination, two reduced speeds, a direct drive and an overdrive can be obtained. A small switch with five positions, located close to the steering wheel, allowed the four gear ratios to be obtained automatically, without the use of clutches, whose function was controlled by the timing of the electromagnets and the inertia of the parts to be accelerated or slowed down during shifts. The remaining fifth position of the switch was used for putting the gearbox into idle, with all electromagnet circuits open.

The first planetary gear a is operated, instead, manually, when the car is not moving and the transmission is in idle position; a control lever moves the carrier back and forth, to engage with the ring gear, obtaining a forward speed. If the carrier is locked a reverse gear is obtained. Vehicle motion can be obtained, after this manual shift, with the first gear, controlled by its electromagnet. The most relevant drawback of this gearbox were its heavy weight and large size.

Semi-automatic Wilson and Cotal gearboxes were used primarily by European manufacturers specializing in luxury cars; the World War Two crisis caused many of

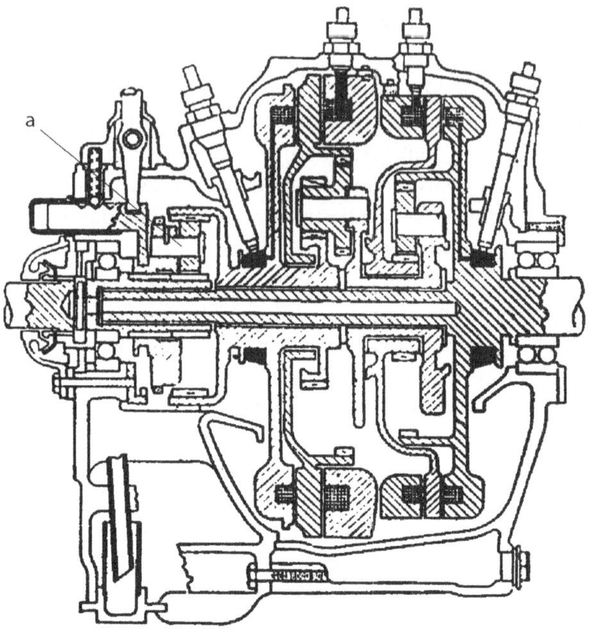

Fig. 4.51 Semi-automatic gearbox produced by Cotal starting from 1934. The different elements of planetary gears are braked by electromagnets. The reverse gear is engaged manually (from Genta and Morello 2009)

these manufacturers to disappear, and these transmissions disappeared together with them.

The final step toward modern automatic gearboxes was taken by using hydraulic torque converters. A torque converter had been introduced by the German shipbuilding industry, after the invention of Föttinger in 1905, well before their application in cars. He patented a torque transmission system connecting in the same hydraulic circuit a centrifugal pump and a turbine. With this device, the torque transmission is obtained by the momentum variation of the flow through the rotating blades, and is possible even when the pump, connected to the engine, is rotating and the turbine, connected with the vehicle wheels, is stationary.

The idea was developed further through the design of an integrated device of reduced size, that was almost interchangeable with the conventional friction clutch. In 1910, a patent for a hydraulic clutch, simplified by the elimination of the stator, was filed.

Again in Germany, in 1928, the research consortium Trilok developed a torque converter, that integrates the performance of the torque converter and the hydraulic clutch in a single machine. This was done by mounting the stator (that actually now is no more restrained from rotating) with a freewheel.

The first automotive automatic gearbox was produced by GM; called the Hydramatic, it has been in production since 1939: A cross section of this gearbox is shown in Fig. 4.52, where the hydraulic clutch *a*, followed by two planetary gear trains *b*, able to obtain three forward speeds and a reverse speed can be seen. The gears are engaged and disengaged by two band brakes *c* and two multi-disc wet clutches *d*.

Brakes and clutches are operated by oil pressure, generated by a gear pump, and modulated both by servo valves and a manual control located on the steering wheel. Gear shift automatization is based upon the comparison of the oil pressure generated by this first pump (dependent on engine speed) with the pressure generated by a second pump driven by the transmission output shaft (dependent on vehicle speed). The difference between these two pressures is used to move the gear shift servo valve. This valve is also made sensitive to the accelerator pedal position through a spring loaded mechanical link.

This system worked quite well on level road, with upshifting occurring at higher vehicle speeds when the accelerator pedal was more depressed; but on slopes or on winding roads the automatic control had to be corrected by the manual selector. The hydraulic clutch is always transmitting the engine torque and was used both for starting the vehicle and to damp out driveline torque vibrations.

The bulk of these gearboxes was absorbed by war production; their application to commercial cars began only in 1946 and was much appreciated by the public.

The Dynaflow gearbox, also from GM, has been produced since 1948. It introduced many improvements over the previous model. The planetary Wilson gear train, much simpler, allowed three forward and a reverse speed to be obtained, with two band brakes and a multi-disc clutch used in combination.

The most relevant step forward was the introduction of an improved torque converter, featuring a two stage reactor on freewheels; with this device it was possible to start-up the car with a torque transmission ratio greater than two (instead of one, by definition the ratio on the hydraulic clutch), allowing in the meantime the torque converter to work as a clutch, obtaining a better efficiency when input and output torques of the converter were equal.

This layout is still present in automatic gearboxes, even if the need for a higher number of transmission ratios justified the application of additional planetary gear trains.

The Gyromatic gearbox designed in 1949 by the Dodge Division of Chrysler is also worth mentioning for its original features. The clutch of this gearbox, that includes a hydraulic clutch and friction clutch operated by a pedal in series, is shown in Fig. 4.53. Some gear shifts always demanded operating a pedal clutch, but they were rare, thanks to a particular automatization device.

The twin friction and hydraulic clutches allow damping transmission vibration and a smooth start-up, even if the pedal is released without particular skill; in addition, the car can be kept stopped on a slope simply through the use of the accelerator pedal. The following start-up is considerably easier. Similar twin clutches were also applied in combination with conventional manual gearboxes on some European cars in the 1950s.

A clutch of this kind could be used in connection with conventional manual gearboxes where the clutch disengagement was made electrically, by simply touching the gear stick.

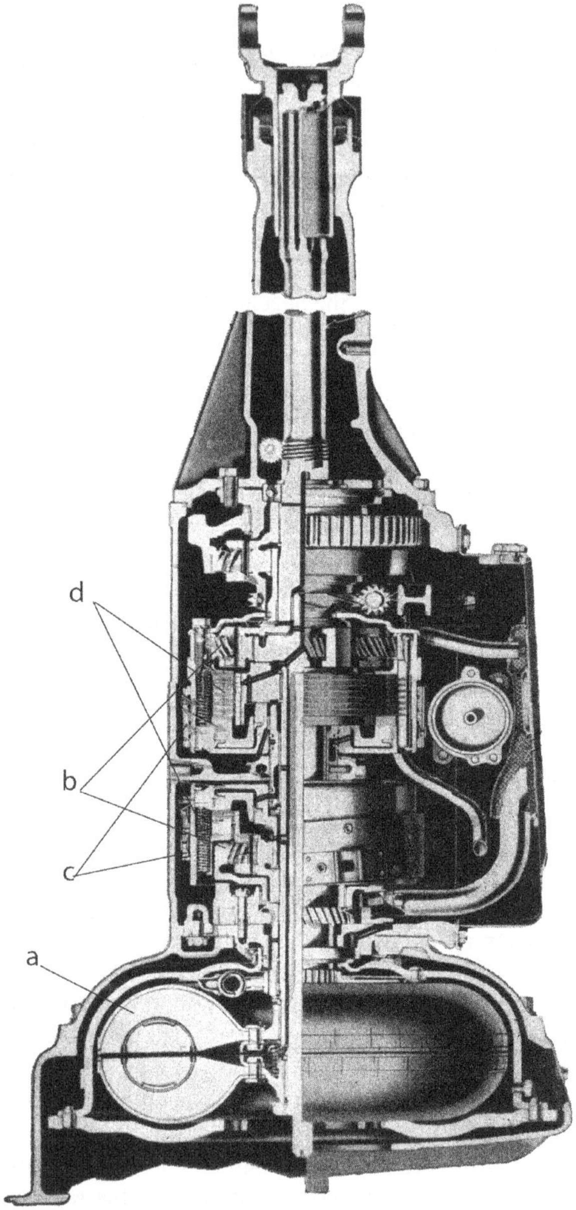

Fig. 4.52 GM Hydramatic transmission. The hydraulic clutch *a*, followed by three planetary gear trains *b*, able to obtain three forward and a reverse speed, can be seen (redrawn from Genta and Morello 2009)

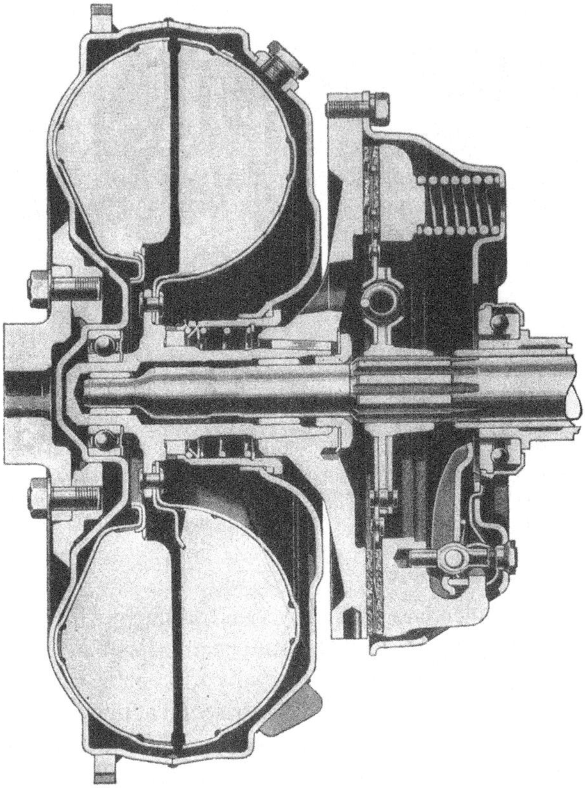

Fig. 4.53 Hydraulic clutch of the Gyromatic semi-automatic gearbox, produced since 1949 by Dodge. It is also provided with a conventional friction clutch with pedal control (redrawn from Genta and Morello 2009)

A particular European contribution to automatic gearbox development, the Variomatic transmission, was introduced by DAF Daffodil in 1950. This was probably the first reliable application of the continuously variable transmission to a car. This transmission, suitable for front engine, rear-wheels drive cars, is shown in Fig. 4.54. The engine drove two variable diameter (expandable) steel pulleys through a transmission shaft and a differential. These pulleys drove similar pulleys connected to the driving wheels through a special rubber belt.

The sides of the driven pulleys were compressed by coil springs that guaranteed the correct friction with a rubber belt; the sides of the driving pulleys were, instead, compressed by centrifugal masses and engine manifold pressure. Through this device the speed ratio variation took into account engine speed and required torque. A centrifugal friction clutch made car start-up completely automatic.

This transmission received no further application because of its strong impact on the vehicle architecture.

Fig. 4.54 Automatic Variomatic transmission of the DAF Daffodil of 1950. It is made of expandable pulleys and rubber belts, reinforced by cords (from Genta and Morello 2009)

The concept was completely reworked by Van Doorne (the DAF holding company), introducing a completely new design. This study had as its objective the development of a steel belt variable transmission of reduced size, capable of being interchanged with conventional manual gearboxes. An experimental application was made by Fiat and Ford and was followed later by mass production. This kind of automatic gearbox has now seen a number of applications on cars of different brands.

4.4 Alternative Powertrains

Propulsion systems different from internal combustion engines were considered at the beginning of automobile history; in particular steam engines have been on the market since the end of the eighteenth century, while electric cars since the 1880s, about ten years earlier than the first gasoline automobile. Both were mainly used in much heavier vehicles.

This situation was justified not only because internal combustion engines were still in their initial development stage, but also because there were technical reasons to believe that this kind of engine was unsuitable to vehicles.

Consider an engineer from the end of the nineteenth century, making a decision about the powertrain to be installed in the vehicle he is designing. At that time, internal combustion engines fuelled by gas with the related fuel tank had a mass and a bulk that were not smaller than that of alternative solutions, and the reduced clearances needed for their correct operation, that were affected by a wide range of operation temperatures, introduced severe design complications. These difficulties were mitigated only after carburetors were introduced by Daimler and Benz, allowing the use of liquid fuels, having a better energy density.

A second issue, difficult to be solved, was the need to apply a *torque converter* (gearbox and clutch) to adapt the engine torque to the traction requirements of the vehicle. From the first automobile magazines it is clear that the correct combined use of clutch, gearbox and accelerator controls was a major concern for early drivers and engine cranking practically made driving impossible to women. None of these problems were present in vehicles powered by electric motors or steam engines.

There was a further point in favour of electric motors: the maximum available power, particularly important in race applications, was higher.

Until the first years of the twentieth century, foundry and machining technologies allowed production of small displacement engines only, with no more than two cylinders. On the other side, considering feasible values for the top engine speed, a power of 20 HP, suitable to reach, with a heavier vehicle, a performance comparable with a horse driven carriage, could be obtained only with a displacement of at least 4 l. This goal could be achieved only with 4 or 6 cylinders engines, that were put in production between 1903 and 1905.

Electric or steam engines of that time could already reach this goal and they were consequently chosen for heavy and fast vehicles: trucks, buses, taxis and vehicles designed to achieve speed records were powered by engines of this kind. The first vehicle to reach a speed of 100 km/h, setting a new world record in 1899, was the electric vehicle Jamais Contente designed by Camille Jenatzy. This record remained unbeaten till 1902 when Leon Serpollet (driver and car manufacturer) reached 120 km/h with a vehicle powered by a steam engine. At the end of the same year, this barrier was broken by a vehicle powered by an internal combustion engine, but the second major milestone of 200 km/h was again reached by a Stanley Steamer in 1906.

This date marks perhaps the end of the golden age of steam engines, followed a few years later by the decline of electric motors.

4.4.1 Electric Cars

The fame of the *Jamais Contente* (in English, Never Satisfied) by Camille Jenatzy is linked with the early attempts to break the 100 km/h barrier with a road vehicle.

The history of electric vehicles started earlier, with Charles Jeantaud, one of the most important car manufacturers in France, operating between 1881 and 1906 in Paris. He inherited his shop from his father, a famous coach builder; and developed at least two important inventions: the electric car and the steering mechanism for solid axles, according to Ackermann's rule. There is a relationship between these two inventions.

The most important share of the Jeantaud's production were the so-called *fiacres,* the taxis for the city of Paris. A similar vehicle is shown in Fig. 4.55. The reasons for choosing this layout for one of the first self-propelled vehicles is unknown, but a

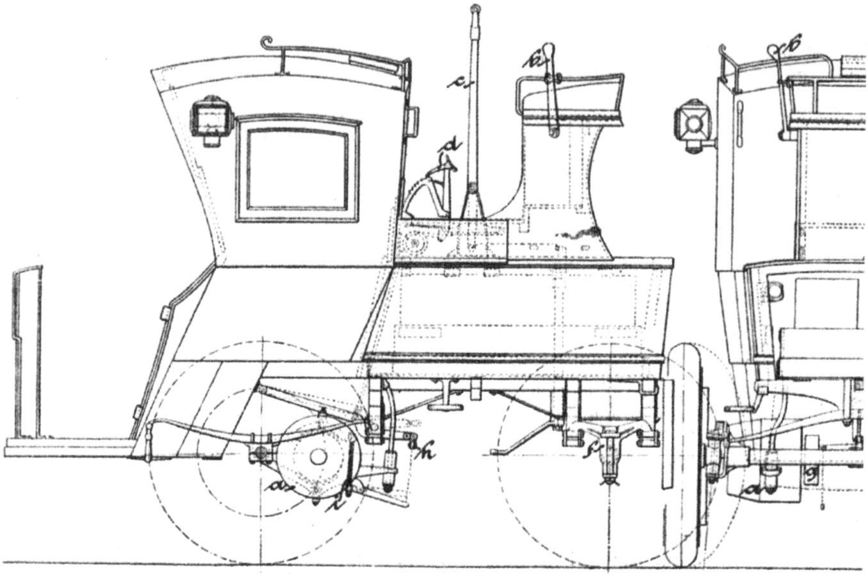

Fig. 4.55 One of the electric Fiacres of Paris (From Dinglers Politechnischer Journal 1898)

reasonable guess was that Jeantaud based his design on similar horse-driven vehicles. Horse-driven Fiacres had passenger seating in the front row, for a good visibility of the road while the driver was seated on top of the vehicle to effectively supervise the surrounding area.

The choice of front wheel drive was likely dictated by the analogy with horse driven carts: Like a horse was pulling the cart and steering the vehicle, the motor was expected to perform the same functions and thus placed on the front axle. Figure 4.56 shows how Jeantaud designed the front driving and steering axle, with a king-pin steering mechanism and a bevel gears transmission able to drive the steering wheels.

Jeantaud was one of the first to realize the potential of sport events to increase the visibility of his new invention and related sales. In 1898 an electric Jeantaud car, driven by Gaston De Chasseloup Laubat powered by a 35 HP electric motor reached a top speed of about 63 km/h, setting a world speed record for cars. The fact that the record was obtained by a car must be stressed, since an earlier record was reached by a bicycle, with 67 km/h, and rail vehicles had already broken the barrier of 100 km/h.

Camille Jenatzy was the ideal challenger for Chasseloup because he was both his rival in other sports and a competitor of Jeantaud, with his Compagnie Internationale des Transport de Paris, producing and managing a fleet of electric Fiacres. Jenatzy was also one of the first manufacturers of rubber products in France and, because of this, probably had the chance to meet the Michelin Brothers.

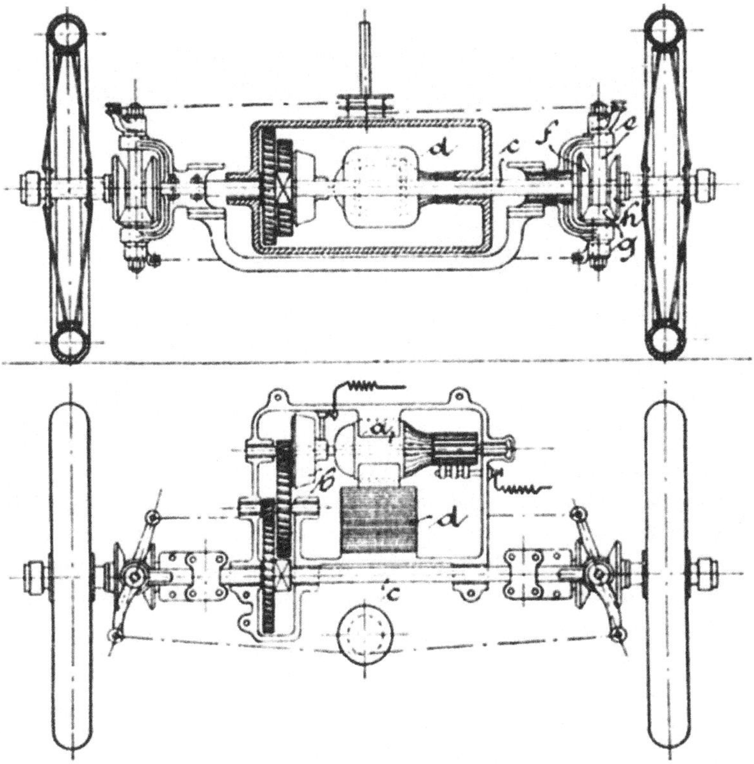

Fig. 4.56 Front and top view of thedriving and steering front axle of a Jeantaud Fiacre (From Dinglers Politechnischer Journal 1899)

To challenge Chasseloup he modified one of his electric vehicles, powered by a 68 HP electric motor and adapted a set of wheels to the new Michelin pneumatic tires. Road tests started in January 1899 and, after three different tests followed by design improvements, Chasseloup was still the winner with 92 km/h. Jenatzy did not give up, introducing in his car a streamlined body, made with light weight aluminum alloy and a larger set of lead-acid batteries.

In April 1899 the car, shown in Fig. 4.57, was timed at about 106 km/h. This record remained unbeaten for a while. The threshold of 100 km/h was considered to be critical, because of alleged breathing difficulties and because human reactions were thought to be too slow to operate the controls at this speed.

The Jamais Contente may be impressive to us for the naiveté of some details, but embodies some deeply innovative ideas. The body shell is in fact shaped as a slender body, with an elliptical cross section: It is probably the first car where a particular design effort was aimed to reduce aerodynamic drag. Its shape is very different from that of early automobiles, which maintained the basic body of a horse driven carriage with just some rough shield around the engine and the chassis subsystems added.

Fig. 4.57 A full scale replica of the1899 electric vehicle Jamais Contente, the first car to break the barrier of 100 km/h (picture of a replica, presented at Rétromobile, Paris, 2009)

The misunderstanding of some aerodynamic features, like the pointed nose, were common to the early airships and aircraft fuselages.

A second important point about the body is that it is made by curved light alloy panels joined by rivets, like airplane fuselages of much more modern times. This alloy was developed and patented for this application, with the commercial name of *partinium*, an aluminum alloy with copper and zinc. The chassis was, on the contrary, traditional, bearing some similarity to that of a production Fiacre. Anyway, it represents the state of the art of the time and was made with steel beams with iron suspension joints.

Suspensions were made by two rigid axles with semi elliptical leaf springs; rear springs were doubled to support the weight of the electric motors and batteries. The rear axle was obtained from a riveted steel plate structure including the two electric motors and the wheel hubs. The motors thus were located on the unsprung mass. Wheels were wooden artillery wheels, but had pneumatic tires. The contribution of tires to reduction of the rolling resistance on macadam roads was impressive and represented one of the most advanced characteristics of this car.

The energy storage system included 100 lead-acid batteries made by Fulmen, generating a voltage of about 200 V and the short circuit current reached 250 A. The rated power thus reached 50 kW (about 68 HP) without taking into account electrical and mechanical losses which were at any rate low due to the use of two electric motors with separate excitation, each driving one of the rear wheels directly, without gearbox and differential.

A pedal combination switch allowed to connect rotary and fixed windings in series and parallel with a total of six combinations of voltage-current characteristic, so that it was possible to accelerate the vehicle almost always at rated power without a gearbox. Many electric cars with similar technical features but lower performance were sold by different manufacturers in this period of time.

Electric cars were more popular in the United States than in Europe, because of specific factors, like a more intense use in urban environment and the existence of many women interested in driving a car: As already said, the mentioned complications of internal combustion engines discouraged women from driving. The most famous

Fig. 4.58 Pope Model 1901; note the electric motor integral with the rear axle (National Automobile Museum of Torino)

manufacturers were Pope, Studebaker, Detroit Electric, Baker, Rauch & Lang. The 1901 Pope, shown in Fig. 4.58, may be an example of a two seater cheap electric car.

Also in this car the electric motor was unsprung, directly acting on the rear axle, but through a couple of gears to reduce the size of the motor. Regulation was again performed using a combination switch.

An example of luxury city car may be the nice 1911 Rauch & Lang Brougham, shown in Fig. 4.59. It appears that this car inspired Walt Disney in designing the famous Grandma Duck's car. Its technology is the same as in previous models but its body suggests its popularity among women customers; front seats can turn to look ahead or behind, to favor conversation with the rear sitting driver and trimming is particularly refined and elegant.

The decline of electric cars started in 1912, at least in the United States, when electric starters began to be applied to internal combustion engines, solving one of the most important negative aspects of start-up. No relevant technical improvements able to make electric cars competitive with internal combustion engine cars, except in very particular situations, were introduced until the beginning of the twenty-first century, when Lithium batteries appeared on the market.

Fig. 4.59 The 1911 Rauch and Lang Brougham is an exampleof an electric luxury city car that may have inspired Walt Disney in designing the famous Grandma Duck's car (Auto Museum of Sacramento)

4.4.2 Steam Cars

Before discussing in details some important steam cars it must be stated that the most critical issue of this propulsion system has always been the time needed to pressurize the boiler, since all the water in the boiler had to be heated before having some steam available for propulsion. A second major issue was water consumption, because the expanded steam was directly exhausted to the environment. However, water consumption was not higher than that of cars powered by internal combustion engines, because in the early cars of the latter type the consumption of cooling water was higher than fuel consumption by a factor of ten.

The attempt to solve these two problems was the driving force behind the development of steam cars during the about 30 years in which they were built.

The steam car built by Enrico Pecori in 1891 (Fig. 4.60) is emblematic of other contemporary cars made by Serpollet, De Dion-Bouton and Olds; the choice of speaking of this car to describe the initial state of the art in this technology is motivated by the excellent conservation state of this old car.

The mechanical layout is based on a vertical boiler located in front of the driver's seat, as can be seen in the figure, so that the driver could feed the furnace and the rear driving axle could carry most of the load. The chassis was made according to bicycle technology of that time, with tubular beams and radial spoke wheels with solid rubber tires. Moreover, the car had no suspensions.

Fig. 4.60 In the 1891 Pecori steam car the position of the boiler allows the diver to feed the furnace (National Automobile Museum of Torino)

Controls were unusual, if compared with the current praxis: A handle at the right of the driver was turned to steer the front wheel, while a small crank in front of the driver could apply a band brake on the differential.

The Cornish boiler, like other boilers of this kind, was made by a vertical water tank surrounding the furnace which was fed through a front door and exhausted the smoke though a central chimney. This kind of boiler was not completely filled with water, because its upper part was used to store the steam under pressure. A solid fuel (wood or coal) was introduced through the furnace door as soon as the steam pressure was too low. Steam was stored in the boiler, so that the timing of fuel introduction was not critical. Boiler, chimney and cylinders were lined by wooden slats, to protect the driver from burns and to limit heat waste to the outside.

Steam pressure, proportional to the torque potentially available for traction, was monitored through a pressure gauge, while actual working pressure was regulated by turning the tap in front of the driver. Opening this tap gave the vehicle an immediate smooth start-up. The warm-up time of a boiler of this kind can be estimated in the range of half an hour.

The steam engine included two horizontal cylinders; one of them is shown in Fig. 4.61 with its sleeve valves on the side. The crankshaft was directly connected to the differential of the rear diving axle through a chain. Many other taps were used to discharge the water condensed in steam piping during stops, to avoid water entering the cylinders, causing severe damage.

Like electric cars, also steam cars were more popular on the American market: In 1902, for instance, the yearly car sale to this country (900 vehicles) comprised 50 % of steam cars, 25 % of electric cars and only 25 % of internal combustion engine cars.

Fig. 4.61 One of the two engine cylinders acting on a transversal crankshaft; this drives directly the differential through a chain (National Automobile Museum of Torino)

The problem of warm-up time was eventually solved by Serpollet in France and by Stanley in the United States, who developed the so-called fast boilers, containing a very small amount of water. The 1904 Stanley Spindle Seat shown in Fig. 4.62 may be used as evidence of the progress made in the 15 years during which steam car technology had constant improvements. The most important were the fast boiler and the distributor regulation shown in Fig. 4.63. The propulsion system was located under the driver's seat.

The boiler is the big vertical cylinder, in the background of Fig. 4.63. It carries many vertical steel tubes that reduce the quantity of water and increase the heat exchange coefficient between water and flames or hot fumes. The fuel was no longer wood or coal, but liquid petroleum distillates and the hot gases no longer heated the boiler by natural draft, but by a more efficient forced ventilation. The heat rate was regulated by the fuel pressure. In this way the boiler warm-up time was reduced to a few minutes.

A detail of a similar engine, featuring two double effect cylinders, is shown in Fig. 4.64. Being a two strokes cycle engine, it offers a driving smoothness that is similar to that of an eight cylinders conventional four strokes cycle engine. The double effect arrangement does not allow joining the connecting rod to the piston directly, as in internal combustion engines, but requires the use of a cross head mechanism.

The engine was directly connected to the rear differential, with no gearbox or clutch between them. The most important difference from the previous example is

Fig. 4.62 The 1904 Stanley Spindle Seat was one of the most popular steam engine cars built in the United States (National Automobile Museum of Torino)

Fig. 4.63 The most important developments in the 1904 Stanley are the fast boiler and the distributor regulation of the propulsion system (National Automobile Museum of Torino)

torque regulation, that was no longer achieved through a choke valve, but by changing the timing of the valves, with reduced energy losses. The intake and the exhaust had a common duct, as shown in Fig. 4.64. A sleeve valve connected these ducts with the

Fig. 4.64 A Stanley double acting two cylinders steam engine; the picture on the right shows the detail of the sleeve valves with adjustable timing (from www.stanleysteamers.com)

Fig. 4.65 Didactic mock-up of the Stephenson's mechanism used to adjust sleeve valves timing

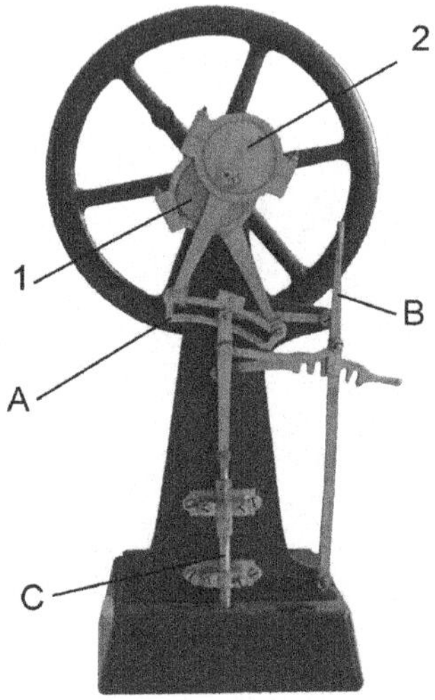

steam intake (in the center) or, through the exhaust, directly to the atmosphere and the engine torque was regulated by changing the intake retard angle and, symmetrically, the exhaust advance angle. This was performed by changing the sleeve valve stroke.

Figure 4.65 shows a didactic working model of the Stephenson's mechanism used to change the timing of a sleeve valve. The mechanism includes two eccentrics 1 and 2 connected to a link-block A, that moves the sleeve valve C.

A lever B can change the position of the fixed articulation of the link-block. Since the resulting motion of the sleeve valve is a combination of the motion of the eccentrics with opposite phases, any combination can be obtained between maximum direct and reverse torque, including zero torque, when the fixed articulation of the link-plate is coincident with the sleeve valve articulation.

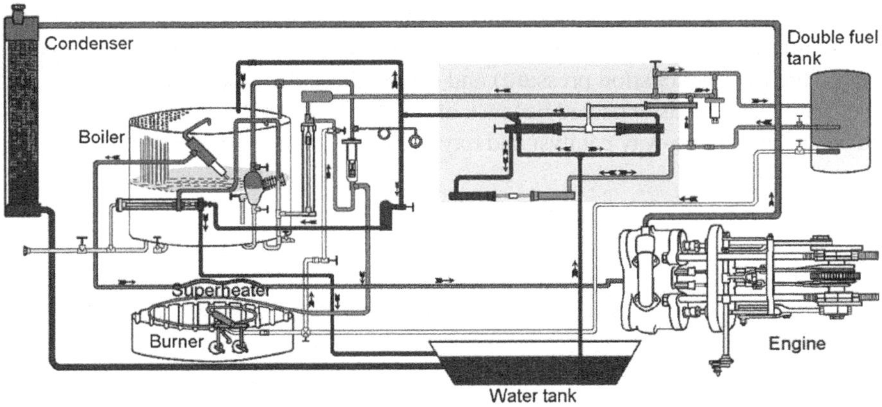

Fig. 4.66 Scheme of steam engine provided with a fast boiler, provided with an additional heater (superheater), close to the hottest area of the burner, and a condenser (redrawn from www.stanleysteamers.com)

A major drawback of this system was exhausting steam to the outside, causing high water consumption, residual heat loss and, not least, an unpleasant fog curtain following the car. All these problems were eliminated by application of a condenser. The scheme in Fig. 4.66 shows the high degree of complexity reached by the latest Stanley engines, at the end of the 1920s.

The fast boiler was provided with an additional heater (*superheater*) located close to the hottest area of the burner, that makes steam available at a pressure of about 40 bar. After expansion, the steam is conveyed to a condenser, similar to, and in the same position as, the cooling radiator of conventional cars. The steam is thus condensed to water (at low pressure) that is then transferred by a pump to the water tank so that water consumption could be limited to the water leaking through the seals.

A further development was related to warm-up speed. Since combustion speed was controlled by the fuel injection pressure, that was in turn generated by steam pressure, manual pumping was the only way to obtain the required pressure at start-up. This problem was solved by using a double fuel tank and by using gasoline for cold starting and then shifting to cheaper kerosene or similar distillates after warm-up. Operating in this way, the greater volatility of gasoline avoided the need of pressurizing the tank when the engine was cold. It is interesting to note that the two fuels were only separated by their density, being contained in the same tank.

As a last remark about steam engines, it may be interesting to recall that some car manufacturers worked on the development of new steam engines at the end of the 1970s. This was due to the fact that the State of California had determined that the main cause of the complex mechanism of atmospheric pollution in the Los Angeles area was the interaction between nitrogen oxides, produced by internal combustion engines, with other pollutants and that, as a consequence, a new law about emission limits was announced. Since at that time no way to reduce nitrogen oxides emission in internal combustion engines was known, a revival of steam engines seemed to be the best solution. Steam engines could allow both reduction of nitrogen oxides emission (due to the low combustion pressure) and use of alternative fuels, an important goal since an oil crisis was expected because of the depletion of American oil fields. The introduction of three-way catalyst and oxygen sensors made this revival useless only a few years later.

Chapter 5
The Technologies of Automobiles

A description of how the automobile evolved from its very beginning to our days would not be complete without a description of the evolution of the technologies aimed to design and build this product and of the evolution of the related market. If this is true for any industry, this link between product, production process and market is even closer in the case of the automotive industry.

As a first point it must be realized that no design or manufacturing technique was specifically developed for automobiles, with the exception of the technologies related to pneumatic tires, which were outlined in Sect. 3.4. The new industry, however, adapted and optimized for the new purpose many already existing technologies, laying in particular the stress on process organization as dictated by the high complexity of this product and the impressive extent of its mass production.

The development of a new product includes not only the design of the product itself, but also the design of production equipment and lay-out, where this second aspect has a particular relevance in the case of large scale production. These aspects are primarily supported by instruments and techniques supplied by the engineering disciplines.

No design process could, however, be started without the results of a third fundamental activity, supported by marketing disciplines, addressed to understand the market's needs and tastes.

The time scheme identified by Womack[1], who identified three different subsequent periods in automotive production, will be used. They are:

- the age of *craft production*, from the beginning, to 1920,
- the age of *mass production*, ending in 1980 and, finally,
- the age of *lean production* that is still going on.

It would be incorrect to think that these dates can be considered as milestones or that every manufacturer rigidly followed this scheme, or any other that could be introduced. On one side, each innovation was introduced gradually, at first by

[1] J. P. Womack, D. T. Jones, D. Roos, *The Machine That Changed The World*, Rowson Associates, New York, 1990.

G. Genta et al., *The Motor Car*, Mechanical Engineering Series,
DOI: 10.1007/978-94-007-8552-6_5, © Springer Science+Business Media Dordrecht 2014

innovators and then by others who copied and probably improved it; on the other side some manufacturers succeeded in delaying, or avoiding altogether, the introduction of these innovations because of the characteristics of their particular market.

5.1 Craft Production

The first of the mentioned periods spans from the invention of the automobile as an industrial product, about in 1895, to the 1920s when, particularly with the contributions of Ford and General Motors in the United States, new technologies were developed and exploited. It is difficult to identify a manufacturer that could be considered as emblematic of this age, because the relevant technologies were not peculiar to this industrial sector yet, but were simply imported from the industries manufacturing horse driven carriages and general purpose mechanisms.

However, French manufacturers and in particular Panhard and Levassor identified the most suitable technologies for this initial period. This company, that is almost unknown today outside of the milieu of fans of ancient cars, was the first in the world to overcome in 1902 the production barrier of 1,000 cars per year.

5.1.1 Market

Even if some pioneers of the automotive industry pursued without success the objective of producing comfortable, flexible and cheap transportation means, the majority of customers were looking for a sophisticated toy for wealthy adults, to satisfy their ambition to make a good impression in the society where they lived. This is testified by the consideration that most advertisements, aimed to promote the very idea of the automobile and to claim the superior quality of a brand, made a wide use of the results of sports events, mainly speed races and rallies.

As soon as technical developments made them possible, big luxury cars were built, with large displacement engines, featuring high performance in consideration of what was allowed by existing roads, regardless of their cost. Many manufacturers, which initially believed in small cars, had to convert their production to luxury products, to obtain an acceptable return for their investment.

Low price was not a major issue in determining the success of a car, considering that its costumers had revenues well above the average; even the cheapest car was sold at a price not smaller than 100 times the yearly per capita gross domestic product of a major European country.

Nobody asked a car to be reliable and unobjectionable, because everybody realized that it was already a miracle that it could move without any animal traction; it was taken for granted that being a car owner implied the availability of a chauffeur to take care of driving, maintenance and of the frequent repairs.

Relationships between customers and manufacturers were very close; customers became accustomed to meeting manufacturers to choose their car and to specify each product feature. The body was almost always built on order and sometimes also important chassis subsystems were built according to customers requests. Servicing of cars was done directly by the manufacturer's production shops, at least when the customer's chauffeur was not able to do it himself.

These characteristics explain why the first car manufacturers served a market localized around their main shop or had to establish subsidiaries to serve foreign markets.

The characteristic of addressing a limited market made of wealthy people who buy a new product mainly for recreational purposes has, however, been common to many technologies at their initial stage: the same was true of the machine tool industry that supplied lathes and other devices to noblemen who pursued the hobby of metalworking, and was at the base of attempts to develop the space industry through space tourism.

5.1.2 Production Process

The production processes identified at the end of the nineteenth century were based upon manufacturing machinery not dissimilar from the present one. As far as metal processing was concerned, iron production, steel refinement, iron foundaries, cheap cutting processes and heat treatments had already been invented and perfected in the previous century.

Nevertheless, powering and regulating low cost machine tools was still a critical issue, because electricity was not available in existing production plants and machines were operated by a complex systems of overhead belt transmissions that received their motion from a central prime mover, which was at the beginning usually a water mill but later was a steam engine and sometimes an internal combustion engine. The general look of a machining shop of that time is shown in Fig. 5.1.

Production machines were not particularly suitable or optimized for automotive production, but were the same as used in other mechanical industries. Many automotive parts were made of wood, particularly in the body, but sometimes also in the chassis, as was the usual practice for horse driven carriages and coaches. This was due to the higher cutting speed of wood and the fact that wood machinery had already reached a high level of effectiveness and could reproduce complex shapes from templates almost automatically.

The workers involved in the manufacturing process had to have a high level of skill and be able to build parts of cars starting from an outline drawing, deciding autonomously which kind of machinery had to be used. They worked in close contact with the car designer, sometimes one of the top managers of the company. Cutting tools were controlled manually and parts that had to be identical, built by the same worker on the same machine, seldom had identical dimensions.

Fig. 5.1 Before the widespread diffusion of electric power distribution, all machinery of production plants was moved by complex systems of overhead belt transmissions that received their motion from a central engine (Courtesy of FIAT Historical Archives)

An important detail of metal processing was that hardening heat treatments had to be made after cutting, because tools hard enough to work hardened steels had not yet been introduced. Heat treatments caused, as it is well known, permanent deformations, with consequent variations of the final shape of parts.

Because of these facts, an essential phase of every assembling process consisted in part adjustment, i.e. adapting the dimensions of matching parts, like a pin in its bearing, a valve stem and its guide or a mounting flange of a cylinder block and the corresponding flange on the crankcase. The need to adjust all fits by hand can be deduced from the layout shown in Fig. 5.2, where a number of benches equipped with vices are set close to each assembling post. Parts had to be hand scraped, verifying the result by a trial and error procedure, since the chip cutting process did not guarantee direct part matching with adequate clearances.

The so called *engineer's blue*, a blue varnish easy to be abraded, was used after those assembly attempts to show areas where some material had to be still removed. This manual adjustment operation was avoided by introducing finishing and grinding operations and by applying dimension tolerances.

Giving dimensions with tolerances was introduced by Loewe (a weapons manufacturer) in 1903 in Berlin in order to reduce the number of parts scrapped during the assembling process. Other German and foreign manufacturers gradually introduced this system in their plants.

Fig. 5.2 In the assembling shop of that time, a number of benches equipped with vices were set close to each assembling post. Because the chip cutting process did not guarantee an immediate part matching with adequate clearance, parts had to be hand scrapped, verifying the fit by a trial and error procedure (Courtesy of FIAT Historical Archives)

An important event outlining the application of these new concepts to automobiles was the Dewar Trophy established in England to award a prize to the automotive manufacturers who introduced major innovations, or new manufacturing techniques. Cadillac won this prize in 1908 thanks to the introduction of its standardization method, consisting in producing parts to be matched in series on the same machine; the expected result was full interchangeability.

A test was performed on the already famous Brooklands circuit, after an endurance test on three Cadillac Model A cars: The engines were disassembled and reassembled choosing the parts at random without any adjustment, then the reassembled cars were tested again without major changes in performance to confirm the success of this new production method.

Before applying this new practice, and owing to the need to adjust matching parts before assembly, a single worker would assemble one of the car's subsystems, such as the engine or transmission, or would integrate them into the chassis. The assembling operations had to be performed in a large room where all components to be assembled were stored around a number of cars that were being processed, as shown in Fig. 5.3. Many parts were adapted in place starting with raw materials, like cloth, leather, cables, pipes, etc. Pipes, for instance, were cut from a spool and bent by hand in their final position. Even two identical cars of the same model had no really identical or interchangeable components.

Because of this kind of organization, the cost of the product was insensitive to the volume of production and the high quantity of semi-finished parts inside the

Fig. 5.3 Assembly line where all parts had to be adapted in place starting from raw materials, as cloth, leather, cables, pipes, etc (Courtesy of FIAT Historical Archives)

plant contributed to increasing the drain on working capital, with consequences to financial cost. Finally, the need for skilled and well trained workers caused difficulties in expanding companies according to market demands.

5.1.3 Development Process

The most crucial issue in designing the early cars was to obtain a machine that actually worked correctly, certainly not an easy task because no consolidated experience or reference models were available. An indirect demonstration of this is the variety of technical solutions used for car subsystems and car architecture.

Conceptual design was therefore the most relevant part of the development process, upon which the attention of engineers was concentrated. The main tools used for this purpose were drawings, made to verify the layout and the motion of the parts. A design department of that time was a room where a small number of engineers worked on tables with pencils and rulers, as can be seen in Fig. 5.4. A number of wooden mock-ups whose aim was showing the operation of mechanisms completed the equipment.

Engineering calculations were not yet used and testing was mainly addressed to verifying correct operation rather than to assessing the duration or the reliability of components and assemblies. As can be deduced from the books of the early manufacturers, every car they produced, even the prototypes rudimentarily built, was sold to a customer.

Fig. 5.4 Design department in the craft production era; a few engineers worked on tables with pencils and rulers while some wooden mockups were used to show the operation of mechanisms (Courtesy of FIAT Historical Archives)

Drawings were rough, more or less simple sketches, with few prescriptions and the choice of the most suitable manufacturing cycle was left to the worker's interpretation, perhaps after discussions with the engineer. Field-testing was left to the customer who returned his car to the manufacturer only for the most difficult repairs; it was taken for granted that an object resulting from the most advanced technologies of that time could not be perfectly reliable.

Because many technical solutions were developed and applied for the first time, much effort was addressed to the preparation of patent applications; many parts or mechanisms that today are in common use, and whose patents may have long ago expired, were then proprietary applications. Some examples are the belt transmission and the radiator, patented by Daimler or the double universal joint transmission shaft patented by Renault. Much effort had also to be addressed to find or craft solutions that avoided infringing competitor's patents.

The organization was rather simple, and generally included a technical director and an accounting director under the owner's authority: The former had to deal with design, manufacturing, purchasing and customer service while the latter was in charge of bookkeeping and payments. Sales were made following direct contacts between the company owner and the potential customers, sometimes introduced by business matchmakers.

5.2 Mass Production

The second period spans from the 1920s to the 1980s in the United States, while it started significantly later in Europe. The first reason for this delay was that the basic principles of this technology were developed in the United States, where the market was richer and had a higher capacity to absorb a multiplicity of products. The main actor in this revolution was Ford, followed by General Motors, both of which introduced significant improvements. The second reason was that the diffusion of this technology in Europe was delayed due to years of severe drains on resources and the massive destruction caused by the two World Wars. Worldwide extension of mass production occurred only in the 1950s, after the reconstruction, even if these new technologies were well known, because Ford, in particular, never denied to his European competitors a visit to his innovative plants.

The assembly line is usually taken as a symbol of mass production, although it was not really the most important tool in this new era. An organization of the assembling process very close to the assembly line had already been introduced by Oldsmobile in Lansing (Michigan) when it was still an independent manufacturer. This solution was designed to start production after the fire that totally destroyed the Detroit plant in 1901. Everything had to be changed, since all production equipment, drawings and components were lost together with most prototypes.

The ideas on which the new plant was designed for resuming production were quite innovative. A shed was cleaned of the wreckage and the only remaining prototype was completely disassembled; its parts were ordered on tables with labels specifying their function. Then each worker took some of them and started visiting all shops in Detroit with the aim of finding companies able to reproduce the parts in sufficient quantity.

Many suppliers participated to this project, which can be considered as the first example of a completely outsourced automotive production. Some of these suppliers became later car manufacturers themselves, like Buick, Maxwell, Leland (the first kernel of GM) and Fisher, a famous body builder.

While a suitable assembly plant was still missing, the State Authority of Michigan offered for this purpose the premises of the Fair, thinking that it could be a good business. An innovative characteristic of this plant was that cars under production were moved on wheel carts along a line of assembling posts where workers stood accomplishing always the same set of tasks. This feature could have inspired Ford in designing his assembly line.

5.2.1 Market

The market increased its demand gradually with new customers that considered the automobile a transportation mean and a tool to increase productivity. This new trend started in the United States where the general layout of the cities favored a lifestyle with large distances between houses and the places where production activities were performed, either in the field of agriculture, industry or commerce. In addition, the cost of fuel, that in Europe was so high that it prevented daily travels by car, in the States was not an obstacle to the diffusion of the automobile.

Some American manufacturers like Oldsmobile and Ford stimulated this new need by introducing a new kind of car with reduced size and price; both manufacturers devoted themselves right from the beginning to develop cars for lower income people instead of luxury high performance vehicles. However, these new products alone could not have satisfied this new market if these precursors had not invented means to increase productivity and reduce costs with an increasing production volume. As seen above, one of the characteristics of craft production organization was on the contrary that production costs remained almost unchanged at increasing volume.

In addition, a larger diffusion of cars required a simpler and more reliable product or, if high reliability could not be achieved, it had to be easily repaired. For this reason a large effort was addressed to set up a sales organization able to make cars and spare parts available to customers in a short time; to achieve this result cars had to be standardized and to be based on interchangeable parts. A last important contribution to market development was offering financial services to customers so that they could purchase a car even if they didn't have the immediate availability of the corresponding capital.

This new way of producing cars allowed Oldsmobile to sell about 18,000 cars in 5 years, from 1902 to 1907 and Ford about 16,000, from 1903 to 1908, in an era when the highest production rates grew to the range of 1,000 units per year. The most astonishing result was the production of 15,000,000 Ford Model Ts in 15 years thanks to the new system, once it was fully implemented.

The price of these cars was as low as twice the per capita Gross National Product (GNP) of the countries where they were produced and at the close of this era, in all countries where mass production principles were applied, the price of economy cars further reduced to a value corresponding to the per capita GNP.

5.2.2 Production Process

The ultimate goal of the mass production principles was to produce cars at a reduced price, thanks to savings in all expense items, including materials, manpower usage, distribution cost and profit per unit. The lower profit per unit should have been largely repaid by the sale of a larger number of cars, but the growth in volumes could only be obtained if non-specialized workers could be hired for this additional production.

Most of these goals were reached by obtaining parts interchangeability. Only this condition could allow avoiding adjustments, simplifying and shortening the assembling cycle. Moreover, only interchangeable parts enabled customers to substitute defective parts themselves, making repairs easier and cheaper. A dimensional tolerances system was set up and guaranteed by gauges that were developed and adapted to each fundamental step of the manufacturing process; these gauges were designed for a single purpose and were frequently controlled. They were primarily Go/No Go gauges, applied at the end of each operation to check results; discarded parts were reworked or scrapped.

A relevant contribution to the reduction of scraps came from the development of harder steels suitable to cut metals even after hardening treatments; in this way it was possible to correct distortions by chip cutting operations.

Interchangeability was not the only progress obtained by the new production system; a further contribution came from the integration of parts formerly separated into a single part. The best example in this field is the integration of the two blocks of two cylinders each or the four separated cylinders into a single monobloc containing the four cylinders of 'four-in-line' engines. A second example was that of gearboxes, that were usually separated from the engine crankcase that were substituted by new gearboxes, directly flanged to the crankcase. The former required aligning their input shaft with the engine crankshaft, then applaying adaptors of different thickness between the mounting flanges of the gearbox and the chassis, adaptors that were no longer required in the latter case.

The wide application of these design principles and of interchangeability allowed Ford to reduce the assembly time of a complete car starting from its subsystems from 750 to about 100 min of work by a single man and that of a complete engine from 600 to 200 min, always referred to a single worker. In addition, the assembling cycle was divided in a sequence of simple operations assigned to different workers. At Ford plants, according to this scheme, the cycle time assigned to each worker was reduced from 8 h to about 2 min.

This reduction of cycle time of each worker was enabled by major changes in the assembly shop. In the former organization, cars undergoing work were kept at a fixed post, around which a few workers operated, taking the parts available in the shop and performing their assembly job. In the new organization, cars were always kept moving at a constant speed by a chain, from one assembling post to the following, where workers received the parts to be installed. Figure 5.5 shows the end of the floor conveyor line for assembling the chassis and of the trolley conveyor line for assembling the body; at the confluence of these lines these two systems were joined to build up a complete car.

A moving line with fixed working posts avoided any hindrance between workers for moving parts and simplified parts supply, because each part had to be moved to a single place along the line.

Job simplification shortened training time dramatically and allowed new huge plants to be built by hiring personnel from a field of green workers for whom this was their first experience in the automotive industry. Simplification also made workers easily interchangeable, allowing continuous production in three shifts per day and the substitution of workers who were not present.

Thanks to these innovations, the Ford River Rouge plant arrived at a production of about 7,000 cars per day. In this organization workers employed for part fabrication and assembly had no parallel or indirect activity to accomplish. Materials were supplied by different workers or by machines, and other workers provided cleaning and maintenance.

Fig. 5.5 End of the line for assembling the chassis and of the line for assembling the body. At the confluence of these lines the two subsystems were joined to build up a complete car (Courtesy of FIAT Historical Archives)

Different jobs of different complexity were thus established; a ladder of jobs of increasing complexity and salary allowed freedom for each worker to climb it from one position to another, after having indirectly learned the new job. Building and maintaining production equipment was the job at the top of the ladder.

The moving line system was also applied to parts manufacturing, such as steel sheet stamping, metal casting and chip cutting. Simple operations were assigned to each working post. They had to be performed on machines built for the purpose, which did not require either tool setting or control and the complete manufacturing cycle of each part was accomplished by transferring it from one post to the following, along a line of dedicated machines. Workers were assigned the simple job of loading and unloading a machine at the start and at the end of each operation. A line of huge presses that produced a complete body side starting from a steel sheet coil is shown in Fig. 5.6.

Sometimes a suitable line was dedicated to transferring semi-processed parts from one press to the next in that line. Temporary stores were provided between the manufacturing posts to avoid the propagation along the line of possible manufacturing stops in one of the posts. Small parts were stored on carousels. To speed up production many operations were matched, to be performed at the same time; in this case it was necessary to develop new production equipment not only dedicated to a kind of operation but, sometimes, dedicated to a single car model.

The machines that were built for drilling engine blocks are an example of these specialized tools. Typically, the monobloc of a four-cylinder engine with lateral

Fig. 5.6 A line of presses producing a complete body side starting from a steel sheet coil (Courtesy of FIAT Historical Archives)

valves had 10 holes for head bolts, eight holes for valve guides and a number of additional holes to install accessories or to drain lubrication oil and cooling water. In craft production, each one of the holes was drilled separately, probably in three steps, roughing, finishing and threading, by changing tools of the same multi-purpose machine. The block had to be moved 18 times or more on the same drilling machine.

With mass production, a line of only three different machines (for roughing, finishing and threading) each one drilling 18 or more holes contemporarily made the entire job. Each machine had a multi-mandrel head, with each drill-bit in the suitable position and a dedicated positioning fixture to place the block correctly. The worker's job consisted in placing the block on the fixture, starting and stopping drilling motors and unloading the block to the carousel leading to the next machine. Long and expensive operations were required to retool a machine of this kind for a different job and therefore it was practically dedicated to a single job for the entire life of a car model.

Other examples of this approach were the complicated positioning fixtures developed to shave many blocks at the same time on the same milling machine or the line to produce copes, drags and cores in casting process. In the foundry shop other lines were used for castings shakeout, de-gating and polishing.

Body and chassis welding was made in multiple spot-welding machines designed to produce a complex part in few operations; an example is the floor-welding machine for a unitized body, shown in Fig. 5.7.

This new machinery and these new construction techniques, that were typical of the automotive industry only, stimulated car manufacturers to establish new job positions for production engineers in their factories. These jobs were subdivided in

Fig. 5.7 Underfloor welding of a unitized body was made in multiple spot-welding machines designed to produce complex parts in few operations (Courtesy of FIAT Historical Archives)

two new specializations, aimed at studying production cycles and developing new equipment.

The effort aimed at increasing productivity urged car manufacturers to extend this new approach to the entire production chain, from mining raw materials, to delivering the car to final customers. Large companies like Ford managed mines, shipping lines for hauling raw materials, steelworks, rolling mills, glassworks, railway lines to exchange semi-finished parts between different production plants, trucks for delivering cars and, last but not least, training schools. This tendency to gigantism was depicted as shifting work *from invisible to visible hands.*

The best results could be obtained by standardization; again Ford supplied the best example of this approach with its Model T, almost unchanged for 18 years, with a single chassis and nine body variants. Referring to this car, Henry Ford claimed that "any customer can have a car painted any color that he wants so long as it is black."[2]

GM adopted Ford's production methods and organization but had to develop some peculiar element that contributed to further improvements. GM was born by merging many different manufacturers in different times and, owing to this, suffered from overlapping product ranges, internal competition and facilities multiplication. The cure that was applied was to organize the company as a corporation with operating divisions with a central financial control, each one behaving as independent profit center, with its own facilities to obtain this result.

[2] H. Ford, S. Crowther, *My Life and Work*, Doubleday, Page and Company, Garden City, New York, 1922. This is not only a joke: the only color that could dry at the speed of the assembly line was black; other colors had to be treated on separate lines.

There were divisions for car manufacturing, as Chevrolet, Oldsmobile, Pontiac, Buick and Cadillac and other for part manufacturing, as Delco for electric systems, Saginaw for mechanical systems and Rochester for carburetors. Each car division covered a different market segment, with the aim of accompanying the customer in his economic growth, offering cars characterized by the reliable simplicity of a Chevrolet, up to the opulence of a Cadillac. Each car was conceived with a kernel of standard parts invisible to the customer with a long lasting life and a visible external cover, proprietary of the brand. The latter was subject to frequent changes to stimulate the interest of the market. A large accessory range, conceived as add-on devices, could enrich the value of each model, without changes in structure or in production tools and organization. Standard parts were shared by different divisions.

Most major European manufacturers imitated this kind of production organization after World War Two, but the three largest manufacturers in the United States produced 75 % of the cars in the world in 1955.

5.2.3 Development Process

This new way of producing cars introduced new actors in the development process. In addition to product engineers, designing cars and related components, production engineers, with two different professional branches, entered into play to identify the optimum sequence of operations aimed to produce a given part or subsystem and to design production equipment.

But specifying a car according to the market request could not be a task assigned to product engineers only and for this purpose a new job was identified, suitable to marketing experts, to effectively interpret customers needs. Aesthetic aspect of a car had an important role in determining its success, for its influence in the purchasing decision making of customers and this caused a new job to be introduced, that of the stylist, a creative designer with knowledge of product manufacturing and ergonomics.

This variety of jobs involved in the process increased complexity and caused an undesired increase in bureaucracy because only a competent and ubiquitous boss could make a trade-off between the conflicting proposals prepared by specialists of such different disciplines.

Manual drafting, made on drafting machines was the common tool of most of these specialists. One of the huge drafting rooms in use by most of the car manufacturers of that time is shown in Fig. 5.8. Drawings became more detailed, containing all figures necessary to specify each part, including dimensions, tolerances, materials, heat treatment, joining techniques, etc. Each car required the preparation of several thousand drawings, not including those necessary for production equipment and plant lay out.

Huge archives stored the detailed drawings of parts and the assembly drawings, usually cross sections, with the purpose to verify geometrical compatibility of parts and to be used as starting point for designing production equipment and assembling cycles.

Fig. 5.8 One of the huge drafting rooms in use by most of the car manufacturers at the end of the 1950s (Courtesy of FIAT Historical Archives)

Calculations became part of the design process; they were addressed to verify structural adequacy of parts and to foresee material quantity, weight and cost. Also calculations departments were organized according to the principles of mass production as shown by the accounting room shown in Fig. 5.9.

An expert designer or accountant defined the correct procedure to perform a calculation. This procedure was then divided into simple operations, easily executed on calculators, usually electromechanical machines, capable of the four main basic operation and calculation steps were described in printed tables, the so-called calculation forms. This way of proceeding allowed freshly hired technicians to perform the most complicated calculations, without risk of mistakes.

Electronic computers were introduced into the design process at the end of this era, substituting calculation forms and allowing complex mathematical models, like those dealing with structural analysis or vehicle dynamics, to be implemented.

5.3 Lean Production

Lean production philosophy began to be applied by all world car manufacturers in the 1980s, but was developed earlier by Japanese manufacturers, particularly by Toyota. Actually the lean production approach is sometimes called *Toyota production system*.

Fig. 5.9 Accounting office organized according to the principles of mass production (Courtesy of FIAT Historical Archives)

Toyota started producing cars in 1936, but less then 4,000 automobiles had been produced from the beginning to the early years after World War Two, because of concentration on the war effort and low responsiveness of the Japanese market. Automotive production was restarted in the early 1950s with modern technologies under the stimulus of the Japanese government that envisaged in this industrial branch the capability of strengthening the economical system.

A journey by Toyota's top management to Ford's production plants, to learn about production technology and organization, made clear the huge amount of investments that was needed and how difficult it would have been to transfer this kind of organization to Japan. In addition, hiring a large amount of fresh manpower for completely new jobs would have been problematic in their country, taking also into account that any kind of migration to Japan was practically impossible. Toyota's top management considered a different approach and started designing a new production system for the purpose, which proved to be a sort of way in-between craft and mass production. About 30 years were needed to tune up this system that, however, proved to be winning over the previous one, not only as far as product costs and investments were concerned, but also for product quality and workers' motivation.

5.3.1 Market

In the 1980s the market in developed countries was close to saturation, and new car sales were made primarily to substitute existing obsolete cars. Cars, having become an easy asset, had lost their initial appeal of high technology product and were approached rationally. They were essential to modern lifestyle and this made customers more demanding for quality and reliability.

The disaffection with mass products brought the market to look for cars that were different from previous best-sellers, like sedans and station wagons, with a renewed interest for sport cars and the birth of SUVs (sport utility vehicles), MPVs (multipurpose vehicles), crossovers and other new product categories.

The first oil crisis in 1973 started a period of continuous increase in fuel price. In Europe and Japan there was a further increase in customer interest on fuel consumption figures, an interest that was already high because of heavy oil products taxation in these countries. In the United States, a kind of disaffection with large sedans started to spread. Since they were the only models produced by local manufacturers, this caused European and, particularly, Japanese import cars to enter that market.

Because of these facts, former high volume models decreased their sales and some of the fundaments of mass production started vanishing.

5.3.2 Production Process

An emblematic example of the new principles of lean production may be represented by Toyota's approach to the steel sheet-stamping process. The situation that Toyota's top management observed during their visit to American stamping shops was characterized by high output presses, able to produce more than 10 parts per minute, suitable for a daily volume of thousands of automobiles. They could have observed the same features had they visited high volume European car manufacturers.

A complete production cycle of a body part, including forming or drawing, coining, piercing and blanking was performed on a line including four or five presses. The installation of a stamp in a press was a complex operation consisting in moving a heavy and cumbersome stamp from the store to the press, the installation of the parts of the stamp in the press and the accurate adjustment of moving parts with reference to fixed ones, to avoid sheet sticking at stamp opening or defects in sheet surfaces. An operation of this kind required about one working day by a team of specialized workers and the consequent stopping of production. For these reasons, stamp changes were limited to routine maintenance, unless some defects appeared in the parts they produced.

The huge investments made necessary by this approach were unreasonable since Toyota could expect at that time to produce a few thousands of cars per year. The solution they conceived was to develop stamps that allowed a quick change without the need for adjustments. In addition, the stamping shop had to be designed to contain

Fig. 5.10 Temporary semi-finished part store near a press, managed according to the mass production approach (Courtesy of FIAT Historical Archives)

a set of different stamps, needed to produce the expected production mix, close to each press. Moreover, each press had to be served by a system of roller tables to move stamps easily into and out of the press. The target was reducing the amount of presses and producing small lots economically.

At the end of the 1950s it was possible to change a stamp by stopping the production for just 3 min. The typical production time between changes lasted about 3 h. A very important point was that the production workers performed the stamp change operation by themselves. Advantages were clearly a reduced investment and the ability to produce a high variety of parts at reasonable cost. Other advantages could be easily guessed by looking at Fig. 5.10 showing a temporary semi-finished part store near a press managed according to the previous approach.

Producing small lots could:

- Reduce working capital owing to fewer parts stored in the shop.
- Improve the quality, because defective part lots, if any, were smaller and defects corrected more frequently during stamp changes.

A key point for making this approach viable was convincing production workers to accept their role in performing stamp maintenance and parts quality control. To understand how this problem was approached we must go back to the end of the war when production was restarted. Market stagnation of those years made it impossible to maintain the previous work force and companies decided to lay-off about half of the workers; this decision was immediately contested with strikes and occupation of production plants. An agreement between workers and company, reached sometime later, accepted layoffs from one side but on the other side, some top managers had to resign, in recognition of their failure.

More important, a new contract was defined that provided workers:

- To be permanently hired.
- To have salaries strongly depending on seniority.
- To receive in their salary a part of the possible profit.

This kind of agreement was adopted also by other Japanese manufacturers and led to set up a particular liaison between companies and their workers.

On one side, workers had lost any interest in changing company, because of consequent penalties in salary; on the other, companies had to identify alternative activities to be performed, if production had to be decreased. The awareness of workers of sharing their future with their company made them ready to contribute also to quality improvement and waste reduction.

Organization strategies similar to what has been described for stamping presses were developed also for other departments. The term *lean production* was derived from the seriousness of its impact on investment funds, enabled by an assignment of jobs to workers broader than what was usual with mass production.

Under the earlier organization, a kind of mistrust of people led to transfer of more activities to machines in expectation of their infallibility; in the new one, the intelligence and motivation of people was considered a way to resort to a less intensive usage of expensive investments. Nevertheless, it must be pointed out that, particularly in Europe, mass production organization changed also because of the development of new machinery with higher production flexibility in comparison with traditional single use systems.

The body welding line shown in Fig. 5.11 may represent an example of this trend; if this picture is compared with that of the floor welding line in Fig. 5.7, it can be seen that single clamps moved by programmable robots replaced multiple spot welding clamps in a fixed position. With this new system welding more than one kind of body on the same line was possible as well as changing body style in a reasonable time by modifying operating programs only.

A similar approach for an engine assembly line, where programmable automatic transportation carts move the engines along the assembly line, is shown in Fig. 5.12. In this case, the assembly line features fixed posts, accomplishing broader operations than in a conventional line and the transportation system moves the engines to the posts dedicated to accomplish the various tasks, according to the planned product mix. A similar strategy is implemented in lean production but with much simpler hardware.

Former assembly lines never had to be stopped; if for any reason defective or incomplete assemblies were generated, they were transferred to the next station in any case and at the end of the line defective cars were identified and sent to a repair shop.

The elementary operations assigned to the workers in the lean production assembly line were in general as broad as to enable the worker to evaluate the result of his operation. Workers were organized in teams, accomplishing a complete assembly of

Fig. 5.11 This body welding line may represent a step in between mass and lean production; if this picture is compared with Fig. 5.7, it is possible to see that multiple spot welding clamps in a fixed position were substituted by single clamps moved by programmable robots (Courtesy of FIAT Historical Archives)

Fig. 5.12 Flexible engine assembly line; programmable automatic transportation carts move engines along the line (Courtesy of FIAT Historical Archives)

a functional subsystem; a team leader had the job of substituting temporarily missing team members and identifying possible problems. This team was also assigned indirect activities as cleaning, maintenance and quality control.

A basic difference was that if any member of the team identified defective assemblies, the line was immediately stopped, till the complete elimination of the defect, regarding the operation performed by this team or by any team upstream in the assembly line. When the error was corrected at the end of the line by a repair team, the result was at best the complete elimination of this defect, but never the elimination of the chance of repeating the same mistake in the future. Correcting the error by the same person who identified it could more easily lead to the elimination of its causes, to avoid this mistake being repeated.

One of the most important additional tasks assigned to the teams was defining new operating conditions that could improve quality and avoid the detected mistakes to be repeated. This activity was performed in meetings of the team at the end of the production shift that were called *quality circles*. During these meetings the problems met and temporarily solved during the working shift were analyzed and improvement presented and discussed.

A famous rule observed in this analysis was that of *five whys*; this rule asked that each mistake encountered had to be explained by at least five hierarchies of causes; root causes only, at the fifth level, the highest in the cause-effect chain, had to be corrected. Other subjects of discussion were improvements in the work environment or in the product quality and waste reduction.

This new way of managing work teams lead to institution of a continuous improvement process, the so-called *kaizen* in Japanese.

A last fundament characteristic of lean production could be identified in the relationship with suppliers. Car production volumes at the end of the 1980s could not be reached if car manufacturers alone had to perform all activities; something between 25 and 75 % of the value of a car was built by suppliers, with a relationship with the car manufacturer that could be defined as *unstable*. This instability derived from the facts that supply contracts were frequently signed and abandoned, or simply ignored in favor of offers of new lower prices, and that supplies were interrupted or increased according to market fluctuation, never accurately forecasted.

The desire of car manufacturers to have a constant rate of production led to consider the dealer more as a flywheel to dampen market fluctuation than as a means to understand customer's need; in these conditions small decreases in order quantities caused sudden production stops. Car manufacturers with a limited knowledge of their production systems designed what suppliers produced, with negative effect in cost. In addition, suppliers had to maintain important material stocks to avoid undesired production stops.

Again Toyota studied how to stabilize the relationship with its suppliers, with the objective of reducing cost and improving quality. A key point was to establish a durable relationship with the supplier, instead of stopping and restarting this relationship at each model change: Because it was essential to have always the assurance of a sufficient technical and economic performance, it was decided to call directly on the supplier's capital.

The object of supply was extended also to designing new components and to producing in accordance to the same rules of the parent-client company; the diffusion of lean production principles was enhanced by personnel exchanges at any level of responsibility. Also working capital reduction was pursued in the supplier's premises; a new method, the so-called *kanban*, that was very simple.

First of all, cars were built on order and according to the specification of an identified customer and, second, a reasonable stock to be maintained at the car manufacturer's premises was determined for each component. This stock determined the amount of containers that were needed along the assembling line. When a container was empty, it was sent back to the supplier with an implicit new shop order.

Also in this case, reducing stock gave benefits not only in terms of working capital but also of quality. An organization of this kind was also established between first level and second level suppliers, obtaining production on orders along all supply chain.

Lean production systems are still in use and are applied worldwide; a tangible consequence of this wise investment management is the economical production of many different models in small quantities, according to the requests of the present market.

5.3.3 Development Process

The organization model introduced by mass production allowed identifying a number of new disciplines that, in addition to automotive engineering, were needed to develop the specific products of this industry, such as marketing, design, and production engineering. An undesired result of this know-how specialization was that new disciplines caused new departments to be established in each company, with the result of complicating the decision making process.

The development of a new product involved thus a long and complex sequence of operations. First, car specification had to be developed and transferred from the marketing to the style department, where style models had to be developed. Then style models had to be transferred from style to product engineering department that developed a large number of drawings. This chain continued with the transfer of these drawings from product to production engineering department, in charge of designing production tools, calculating investments and ordering machinery. This introduced a lot of bureaucratic activity with negative consequences on development time and cost.

Still in the early 1950s about twelve months were sufficient in Europe to start producing a car from the day the initial decision was taken, while at the end of the 1980s this time was up to 4 or 5 years. It should also be taken into account that, in the same time, cars had become more complex, because of the increase in body and engine versions and functions and of local market requests. Further causes of this increased complexity were also the laws and regulations regarding safety, pollution and fuel consumption that were imposed by regulating bodies.

The gigantism of these organizations introduced a general decrease of the efficiency of the various companies due to a number of reasons among which the duplication of activities, and conflicts in decision-making were of paramount importance.

The basic concepts of lean production, brought to product development, suggested creating a multidisciplinary task force that could overcome the organizational barriers between different departments. In this case the new organization provided for summoning temporarily a mix of specialists from their departments and having them to work together for a common goal, under the supervision of a project manager, whose only goal was starting the production of a new car or car family. In this way, conflicts were overcome immediately by comparing disciplinary points of view and evaluating their impact on the final result.

Sharing the same office for a while gave these specialists working in interdisciplinary teams the chance of understanding that many documents were useless and that many activities were duplicated in each of the departments. Avoiding formal information transfers saved a lot of paperwork and time. This organization by development teams, instead of by disciplines, is often called *platform* organization, when addressed to development of a family of cars built on the same platform, the assembly of common components installed on the same chassis or underbody.

Old departments were not cancelled because they were considered essential for professional training and skill development; the new organization simply required a temporary transfer of resources to platform organizations.

A useful example of this change can be offered by the development of the car body, the subsystem of the automobile that was renewed more frequently. In the former organization, the style designer prepared a set of alternative pictorial sketches, describing the body shapes; after this creative activity a choice was made and tridimensional full-scale models were built for further decisions and refinements. Body engineers measured surface coordinates of the chosen model, so that the drawings of each steel sheet part could be prepared and the fit of the parts in assemblies could be verified. At this point some features of the style model were often deemed unfeasible and in this case a second cycle was started. After body drawings completion, including steel and plastic parts, they were transferred to stamps designers who started their activity, and they frequently discovered mistakes in the previous development process or at least parts that could not be produced in a cost effective way, so that the work had to be restarted from the beginning.

The contemporary operation of professional experts of different disciplines, the so-called *simultaneous engineering*, made possible to perform all the required verifications in parallel with the creative activity, saving time and avoiding corrections and repetitions; in addition, simultaneous engineering offered the chance to discover that many drawings performed for different reasons could be unique as, for example, those describing style surfaces, stamped parts or stamps.

The concept behind this approach was simple. The work of every actor in the development process consisted in setting up alternatives and then choosing the best ones with different tools, like clinic tests with potential customers, evaluation by mathematical models or testing prototypes. With the previous procedures, each piece of information obtained in a step of the process was passed to the following actor only when a choice was made according to the evaluations performed; on the contrary, when applying simultaneous engineering, each actor proceeded in his work starting with uncertain assumptions, to correct his work eventually when modifications were needed. The workload was virtually unchanged but development time was certainly shortened.

Consider as an example a part of the body, like a door. It contains elements that have a technical function only (like frames, sidebars, hinges, latches, etc.) and elements with also an aesthetic function (skins, glasses, trimming, etc.). Since the former can be designed without risk of modifications, their design can be defined even if the style is not frozen yet. In a similar way some parts of a stamp are independent from the body shape and their design can be started early in the design process. The risk of repeating some work because of modifications can thus be limited with a suitable organization.

A significant help in the implementation of this new approach came from Computer Aided Drafting (CAD) systems with parametric and associative features because they allow conceiving a drawing as an archetype that can later be modified in one of its parts, being sure that the rest is consistently and automatically remodeled. With these computer based tools, modifications are introduced easily, regardless of their extension.

The engineers of suppliers participate in simultaneous engineering teams and are invited to cooperate in the definition of the design. The power of existing Internet functions allows the team members working in different places to work simultaneously cooperating with each other.

Part II
Present

Chapter 6
Economic Figures

6.1 Manufacturers and Brands

In the last century, the automotive industry went through a number of deep transformations, developing from a group of minor mechanical shops, where the individual cars were crafted and repaired by hand, to the present-day situation. The areas where the starting process took place were concentrated in France, Germany, Italy and England; now a number of global-scale enterprises are organized following the integrated production principles.

The first revolution in motor car production organization was promoted by a few US manufacturers who, starting from the first decade of the twentieth century, promoted and organized the first mass-production methods and introduced a number of technological innovations.

The second one took place in Europe starting from the 1950s, when manufacturers were able to apply mass-production methods while, in addition, engineering a wide product differentiation in the attempt to both overcome the US production and, subsequently, to reduce fuel consumption by widening the available engine choices to improve overall fleet efficiency.

A third transformation stage took place at the end of the 1960s in Japan, where many innovations, both in production organization and in product differentiation, were introduced to improve quality and at the same time to lower costs. During this period, supra-national, international and national Authorities started introducing environment and safety Regulations and to update them to meet the new, stricter, expectations.

A new stage is now determined by Far-East Countries, mainly North Korea and, later on, China and India which, starting from the above mentioned achievements, are capable of further boosting mass production, both locally and for export.

The time frame in which the car industry operates is more a short-term time frame than a long-term one, in particular because this industry is fully market oriented, and then the technological evolution must always find its economical support from the market acceptance and the consequent selling volumes.

G. Genta et al., *The Motor Car*, Mechanical Engineering Series,
DOI: 10.1007/978-94-007-8552-6_6,

This section is aimed at outlining the worldwide situation of Manufacturers and Brands, and refers to those brands that operated in the 2010 scenario and thus determined the present-day market situation. The overall production capability, which is often addressed to both passenger cars and commercial vehicles, is mentioned in term of units sold. The overall relative weight of these two sectors is 75/25 %. When both productions are present, the technical and financial situations of the involved manufacturer are heavily integrated, considering that, for example, many powertrain solutions are in common among passenger cars, SUVs and light duty commercial vehicles. At the end, it is almost impossible to distinguish between these two activities and, as mentioned, the final ranking of the manufacturer is usually based on its overall production.

Mass-production, starting in the USA, determined the first move toward concentrating more manufacturers under the control of the strongest one. This latter, usually, gave the name to the new group and the previous ones became brands of the group. Since many decades ago, this process determined the concentration of American manufacturers into the, so called, Big Three, which were able to consolidate their activities with the following acquisitions:

- General Motors, during the 1920s and the 1930s, took over the original Buick and than Cadillac, Chevrolet, Pontiac, Oldsmobile and Hummer. This consolidation of local Companies was followed by further steps aimed to widen the geographical area of interest with the acquisition of European and Asian Companies: Opel in Germany, Vauxall in UK, Saab in Sweden, Isuzu and Daewoo in South Korea and Holden in Australia.
- Ford took over Lincon and Mercury in the USA, Aston Martin, Jaguar and Land Rover in UK, Volvo in Sweden, Mazda in Japan and set up Ford factories both in Germany and in UK.
- Chrysler took over Jeep and Dodge.

When one brand is no longer profitable, its facilities can be sold out or discontinued; this is the case of Volvo Car Division, sold to the Chinese Geely, of Jaguar and Land Rover, sold to Indian Tata, of Hummer, that is being sold to Tengzhong, or the case of Saab, that was discontinued in 2012.

In some cases, a famous Manufacturer's name, whose production had been discontinued in the past, was resumed, as a brand, by a manufacturer having the financial capabilities to use the historical brand to rev-up selling in a specific car segment. This occurred, for instance in Europe, where:

- Volkswagen: took over Audi, Skoda, Seat, Lamborghini, Bentley and resumed Bugatti;
- FIAT: took over Abarth, Alfa Romeo, Ferrari, Lancia, and resumed Maserati;
- BMW: took over Roll-Royce and resumed Mini;
- Daimler AG: based on Mercedes-Benz activities, took over Smart and resumed Maybach;
- Peugeot: took over Citroën and set up the new PSA (Peugeot Societè Anonime) group.

In some other cases, a new joint-venture was extablished. This is the case of:

- Renault who joined with Nissan including, as brands, Dacia from Eastern Europe and Samsung Motors from North Korea.
- FIAT who, after a previous step, took over the majority of Chrysler group.

Another group of manufacturers is active in the regions of the former Soviet Union. The main ones, which for different reasons did not take part in the above mentioned concentration process, are:

- AvtoVAZ, better known as Zhiguli for domestic market and Lada for export, was established in the late 1960s in collaboration with FIAT and now is owned by Renault at the 25 %;
- GAZ that, with Volga models, became also a symbol of higher status in the Soviet nomenklatura in competition with the more exclusive, but very limited, ZIL. GAZ production is now declining;
- UAZ producing off-road vehicles.

In Japan, some manufacturers followed the European experience, while others belong to bigger industrial conglomerates:

- Toyota, that took over Daihatsu;
- Suzuki, that took over Maruti from India;
- Mazda, that was taken over by Ford;
- Subaru is the automotive division of transportation conglomerate Fuji;
- Mitsubishi is the automotive division of Mitsubishi Heavy Industries;
- Honda is the automotive division of a big industrial conglomerate, involved in motorcycles and internal combustion engines production.

In South Korea, in addition to the already mentioned Isuzu and Daewoo (now Chevrolet):

- Hyundai, that took over Kia.

To complete the scenario, the move of Toyota, Nissan and Honda is interesting: they decided to launch new premium brands, respectively: Lexus, Infiniti and Acura.

In the initial stage, the number of manufacturers essentially corresponded to the number of marketed brands. From that period onward, the concentration trend determined a reduction of manufacturer's number leaving unchanged, or even with a small increase, of the brands, showing the importance of branding in sales policies. More recently, the concentration process developed further and took global proportions under the bias of a quest for competitiveness which expanded the market to world size, with the aim of expanding the production capability and optimizing the efficiency. Such consolidation brought about advantages to the entire sector, reducing the production capability excess and determining fewer but stronger competitors on the market. These initiatives offered manufacturers the opportunity of improving their position on the market, by increasing the company size and, at the same time, reducing costs, through economy of scale and operation synergies. The above mentioned situation occurred in Europe, North America and Japan, and substantially continues with the BRIC (Brazil, Russia, India, China) countries manufacturers and many joint-ventures with them.

Historically, the activities of European and USA automakers started at the end of the nineteenth century or in the first decades of the following one, while operations of Japanese manufacturers started between the 1920s and the 1940s. South Korean manufacturers started around the end of the 1960s. Chinese car industry has been mainly driven by the central Government's decisions, but a certain autonomy was subsequently given to local Regions (Anhui, Jiangxi, Hunan, Chongqing, Guangzhou, etc.), with the aim of boosting their industrialization, promoting, very often, their conversion from military supply to the private car and commercial vehicles business. Finally, in 1997 private capital was allowed to enter the market.

In order of units sold, based on 2010 production, the main Chinese automakers are:

- Shangai Automotive Industry Corporation (SAIC), one of the first to sign a joint-venture agreement with Volkswagen since the 1980s and with GM in 1998. In 2004 SAIC acquired a 49 % stake in Korean SsangYong and in 2007 took over Nanjin Automobile. Subsequently, it took over activities with Italian Iveco and British MG;
- Dongfeng Motor Company, know also as Second Automotive Work, started producing military trucks for the state owned FAW Group Corporation. From the 1990s it signed joint-ventures with PSA Group, Kia Motors, Honda and Nissan;
- First Automobile Works (FAW), established in 1953, initially produced large saloon sedans for Chinese state officials and trucks in cooperation with the former Soviet Union. From the 1990s it signed joint-ventures with VW and, in 2009, with GM;
- Beijing Automotive Ind. Co. (BAIC), which set up a joint-venture with Chrysler in 1983;
- Chana, also known as ChangAn Motor, initially producing machine guns, signed a manufacturing licence with Suzuki in 1984 and subsequently joint-ventures with Ford and PSA. In 2009 it acquired two smaller domestic automakers, Hafei Motor and Jiangxi Changhe.

In addition to the previous ones, many other automakers are operating and some of them are:

- Geely, founded in 1986 as a refrigerators-maker, in 1998 was transformed into a private automaker selling cheap products initially based on the Daihatsu Charade to Chinese consumers. In 2010 bought Volvo Cars from Ford and in 2013 also the small brand LTI producing the London city taxis formerly owned by Manganese Bronze Holding;
- Chery, whose production started in 1997 using a chassis licensed by VW Seat Toledo. Vehicle adapted to EU market are sold in Italy though DR motor company;
- BYD, established in 2003 as a part of BYD Co Ltd, a rechargeable battery maker. As a consequence, it is very active in electric and hybrid passenger cars;

- JAC, basically a truck maker that in 2007 gained government approval for passenger cars production;
- Brilliance, founded in 1992 and linked with BMW;
- Great Wall, that started in 2006 producing the Peri model, similar to FIAT Panda, and now mainly oriented to the production of Pick-ups and SUVs;
- Jiangxi Jiangling Automotive Co. or JMC, producing a compact car based on Ford Fiesta and, from 2005, a SUV;
- Hunan Jiangnan Automotive, a large scale military enterprise, more recently involved in the private car business;
- Chongqing Lifan, that started manufacturing automobiles only in 2005 with compact sedans, microvans, and hatchbacks, that now has a large share of export.
- Guangzhou Group (GAC), a subsidiary of Guangzhou Automotive Industry Group (GAIG), is currently also known for its joint-ventures with Honda and with FIAT.

The main automakers operating in India are:

- Tata, formerly known also as Telco, started being, and still is, very active in the commercial vehicles business. Recently entered the extra-small car segment, with the Tata Nano and in the premium area with Jaguar and Land Rover brands. Is now linked in joint-ventures with FIAT;
- Maruti started its production as a state-owned company with a car based on the Suzuki Alto. Now Suzuki has its complete control;
- Mahindra, quite active in the SUV segment, is now entering the market of passenger cars with a model based on the Renault Dacia body;
- Premier in 1999 discontinued the production of FIAT Padmini and recently resumed its activity in the light SUV area.
- Hindustan, traditionally linked with the British Morris, is now producing Opel models in joint-venture with GM;
- Bajaj Auto, division of Bajaj Group, a strong producer of scooters and motorbikes, is active in the low-cost car segment and announced a cooperation with Renault/Nissan and many other OEMs.

In Malaysia operates:

- Proton, heavily linked with Mitsubishi. In 1996 took over Lotus from UK.

The main 40 car manufacturers recorded in 2010 following the OICA/ANFIA data base, are shown in Table 6.1. As usual, this classification includes the total number of vehicles produced by each single manufacturer. Within this list, Isuzu, Volvo Trucks and MAN do not produce passenger cars but only light and heavy duty commercial vehicles; similarly BMW, Geeley, AutoVAZ, BYD and Hunan produce only passenger cars. In all other cases a combination of private cars and commercial vehicles is always present.

The breakdown between the different vehicle cathegories can be seen in Table 6.2.

As mentioned, many automotive groups form joint-ventures, and thus produced vehicles are counted into one group only: their national group.Therefore, quite often,

Table 6.1 Worldwide 2010 production of the first 40 manufacturers

	Make	Units	Make	Units	
1	Toyota	8,557,351	21	Chery	692,438
2	G.M.	8,476,192	22	Fuji	649,954
3	Volkswagen	7,341,065	23	Dongfeng	649,559
4	Hyundai	5,746,918	24	Beijing Aut.	615,725
5	Ford	4,988,031	25	AutoVAZ	545,767
6	Nissan	3,982,162	26	BYD	521,232
7	Honda	3,643,057	27	Isuzu	488,484
8	PSA	3,605,524	28	JAC Aut.	439,327
9	Suzuki	2,892,945	29	Brilliance	434,182
10	Renault	2,716,286	30	Great Wall	398,692
11	FIAT	2,410,021	31	SAIC	346,525
12	Daimler AG	1,940,465	32	Mahindra	292,149
13	Chrysler	1,578,488	33	Hafei Motor	215,558
14	BMW	1,481,253	34	Volvo Trucks	191,560
15	Mazda	1,307,540	35	Jiangxi Changhe	190,906
16	Mitsubishi	1,174,383	36	Jiangxi Jiangling	173,577
17	Chana Aut.	1,102,683	37	Proton	172,360
18	Tata	1,011,343	38	Hunang Jiangnan	135,648
19	FAW	896,060	39	MAN	127,729
20	Geely	802,319	40	Chongqing Lifan	126,402

Their total production is 73,079,860 units

Table 6.2 Breakdown of the different vehicle categories produced in the world

Vehicle type	Units produced	Share (%)
Passenger cars	58,470,700	75.1
Light commercial vehicles	14,730,700	18.9
Medium/Heavy comm. vehicles	4,257,100	5.5
Buses	392,000	0.5
Total	77,850,500	100

mainly for Chinese and Indian manufacturers, the figures quoted in the table may not equal their total world production because, for example, VW and PSA models locally assembled, are attributed to the national group. Viceversa, the bigger Chinese automakers previously mentioned have been ranked as the result of original and joint-venture brand units. As a consequence, the local production for SAIC, Dongfeng, FAW, Beijing and Chana are respectively 3.6, 2.8, 2.6, 2.5 and 2.4 million of vehicles. For the remaining manufacturers the difference between the two production sets is not so large.

No original manufacturer operates in the Mercosur countries: the units produced there are consequently all attributed to the national groups.

In addition to the above list, another 69 smaller car companies are officially registered; some were already mentioned and are mostly regional, or operating in niche markets. Their overall production sums up to roughly 4.8 million of vehicles.

Among these, it is possible to mention:

- Porsche in Germany, financially linked with VW group;
- Tofaş in Turkey, technically linked with FIAT group;
- Aston Martin, still almost independent;
- Mc Laren and Force India, today involved in Formula 1 racing activity.

At the end of 2012 the first three positions in Table 6.1 remained unchanged, with an increase in the number of vehicles sold. Toyota, for instance, reached 9.7 million units and Volkswagen reached 9.0 million. This increase was mainly due to the good performance of USA and Chinese markets.

6.2 Production and Car Density

Today, the mentioned Original Equipment Manufacturers (OEMs) organize their production facilities almost worldwide, the interested countries being:

- European Union (15 Countries): Austria, Belgium, Finland, France, Germany, Italy, Netherlands, Portugal, Spain, Sweden, UK;
- European Union (new members): Czech Republic, Hungary, Poland, Romania, Slovakia, Slovenia;
- Other Europe: CIS, Belarus, Ukraine, Uzbekistan, Turkey;
- NAFTA (North America Free Trade Area): Canada, Mexico, USA;
- Mercosur (Mercado Comun del Sur): Argentina, Brazil, Paraguay, Uruguay;
- Asia - Oceania: Australia, China, India, Indonesia, Iran, Japan, Malaysia, Pakistan, Philippines, South Korea, Taiwan, Thailand, Vietnam;
- Africa: Egypt, Morocco, South Africa.

Based on OICA (Organisation Internationale des Constructeurs d'Automobiles) /ANFIA (Associazione Nazionale Filiera Industria Automobilistica) data base, the production of passenger cars and light duty commercial vehicles, subdivided by geographical areas, is shown in Fig. 6.1 which underlines how the rapid growth between 2000 and 2010 came largely from the BRIC (Brazil, Russia, India, China) countries rather than from the traditional high production areas: North America, European Union and Japan.

The production volumes, the market interchanges and the economical situation of each area determine the number of cars present in that area and, as a consequence, the car density. These figures, referred to year 2009, are reported (in millions of car and number of cars per 1,000 inhabitants respectively) in Table 6.3. The car density has large variation from area to area, and a saturation of the car market is close to being reached, or has already been reached, where this parameter reaches the highest values.

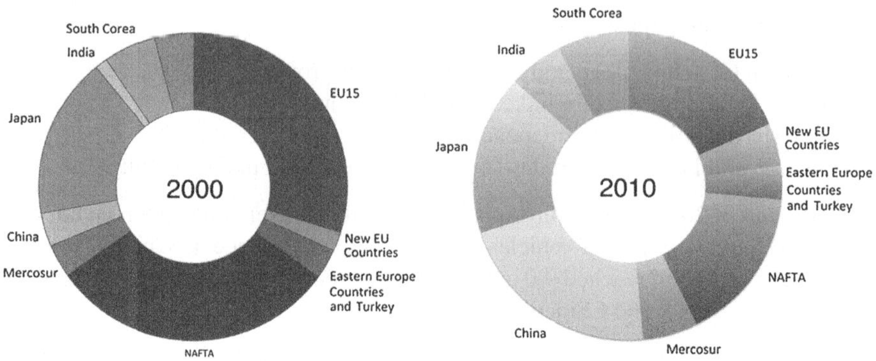

Fig. 6.1 Evolution of production breakdown in the first decade of the 2000s

Table 6.3 Worldwide car population (in millions of car) and car density (in number of cars per 1,000 inhabitants)

Area	Car number	Density	Area	Car number	Density
USA	178.8	576	Russia	35.7	250
Canada	19.9	584	Brazil	23.6	121
Australia	12.1	542	Turkey	7.1	98
EU 15	207	506	China	31.4	23
Japan	58	459	Africa	21.9	21
New EU countries	37	357	India	14	11
South Korea	13	270	Total	837	121

Figures referred to year 2009

6.3 Market Segmentation

It is well known that there are differences in the car types that must be distributed in the different geographical areas to meet a number of requirements such as:

- Income. The customer requirements vary with their levels of income. Consumers in high-income countries are willing to pay for more sophisticated vehicles, and even small vehicles sold in advanced-country markets have structural characteristics and add-on features frequently not found in developing countries.
- Standards and Regulations. Different regulations on matters such as safety, emission standards and recycling are stated in different countries, even though developing countries tend to raise their own standards. Compliance with such regulations has a big impact on technical content, price and performances.
- Driving conditions. In developing countries roads and fuels are frequently of poorer quality than in developed countries. This requires vehicles to be adapted to local conditions, particularly with regard to strengthening the body, suspension, steering and fuel system materials.

- Consumer preferences. These arise in response to the characteristic of particular societies. In the small vehicle segment it is usual the preference for 2-volume solution in high-income countries while the 3-volume solution is preferred in Brazil, China and India, recently entering into car market. Even more, in spite of the influence of NAFTA (North American Free Trade Area) big cars, Brazilian customers, since the beginning, were better oriented towards European cars. Another item is the preference for manual or automatic gearboxes according the different geographical areas.
- Taxation. Tax policies can have a significant impact on vehicle demand. This was seen clearly in Brazil in the 1990s where tax concessions on popular cars shifted demand toward smaller and cheaper vehicles. Similarly, in Thailand a favorable taxation for pick-ups created a market in which demand was overwhelmingly oriented toward light pick-up trucks. In 1996, passenger cars accounted for only 30 % of all vehicle sales in Thailand, compared with 76 % in neighboring Malaysia.

The above mentioned requirements tend to prevent the definition and marketing of a "world car", even if many attempts were performed. Nevertheless, many companies use common platforms for a variety of vehicles, extending platform commonalities across markets. This not only reduces design costs, but also increases the speed with which new models can be introduced in different markets and reduces exposure to the volatility of demand in any particular market.

Each market segment tends, although with some deviations, to include vehicle models having comparable technical contents, prices and final customers. In fact, as previously shown, cars and commercial vehicles are market oriented, adapting their technical content to the market needs. The starting point and the driving factor in the development of each model is the market demand, that may be broken down into a number of specific aspects, like first purchase or replacement, potential and expectations, quality and quantity, etc. This mix defines the primary demand, i.e. the quantity required by all potential buyers in a certain period, usually 1 year.

It is, however, clear that the technical contents and solutions used within each segment may be differentiated in a complex way. More specifically, each segment is identified within a portion of demand characterized by a maximum internal uniformity and a maximum external differentiation. The manufacturer's goal is to use a large number of equal components within each segment, but to select the remaining ones with different technical specifications or style, to attract different customers within the same segment. It is thus possible to define independent commercial planning schemes for each segment. The definition of the optimal segment composition is one of the primary efforts of any company and entails a deep awareness of the market demand related to the final product, as well as promotion policies, price policies and choice of selling channels. The evaluation of customer's reactions and expectations is usually known as Voice Of Customer (VOC).

The passenger car segmentation is quite different among the various countries. In USA and, quite similarly, in Canada the passenger car classes are defined based on the interior volume index in liters, as follows:

- minicompact (under 2,407 l);

- subcompact (range 2,407–2,831 l);
- compact (range 2,832–3,114 l);
- midsize (range 3,115–3,397 l);
- large (over 3,398 l).

A similar criteria is used to classify the station wagons.

Vehicle segments in Europe do not have formal characterization or regulations. Model segments tend to be based on comparison to well-known brand models mainly based on external or internal volumes and prevailing use, as follows:

- Segment A: Mini cars;
- Segment B: Small cars;
- Segment C: Medium cars;
- Segment D: Large cars;
- Segment E: Executive cars
- Segment F: Luxury cars;
- Segment S: Sports cars;
- Segment M: Multi purpose cars;
- Segment J: Sport Utility Vehicles (SUVs), including off-road vehicles;

Pick-ups are considered in a specific category.

In Asia the prevailing parameter is the length and, for instance, the segmentation according Indian regulation is:

- Segment A1: up to 3400 mm;
- Segment A2: Compact (from 3401 to 4000 mm);
- Segment A3: Midsize (from 4001 to 4500 mm);
- Segment A4: Executive (from 4501 to 4700 mm);
- Segment A5: Premium (from 4701 to 5000 mm);
- Segment A6: Luxury, (from above 5000 mm) ;
- Segment B1: small vans;
- Segment B2: MUVs (Multi Utility Vehicles) and MPVs (Multi Purpose Vehicles);
- Segment SUV: Sport Utility Vehicles.

In China the situation is quite similar, with five classes.

6.4 Suppliers of Parts and Components

Initially, car assemblers built in-house all the components they needed and were organized to perform the complete cycle: engineering, prototyping and production. Later on, the first suppliers, most of which are still on the market today, started their activity and, at the end of the nineteenth and in the beginning of the twentieth century, provided single parts, mainly tires and electrical components. After World War Two, for instance in the 1960s, an automaker, when designing a vehicle component (for example, a seat, the dashboard, etc.), made the detailed drawings of a number (20 or 30) of separate elements, found a supplier for each one of them and assembled the final component. Later a shift in design activities from assemblers to suppliers occurred, thus shifting toward a greater customizing, with suppliers selling their parts to many automakers, now identified as Original Equipment Manufacturers (OEMs), satisfying their needs and progressively moving independently.

Today the most qualified suppliers receive performance specifications about the interface with the remaining car components and their functional interactions from OEMs, thus being able to define the components using their know-how and their own technologies. The quality system certification, according ISO (International Standards Organization) standards, becomes more and more important due to the increasing importance of delivering certified parts and meeting the demands of just-in-time production systems. Complying with ISO standards facilitates the automakers' task to cope with worldwide production, which makes it easier to standardize platforms and models across their constituent divisions, to reduce development costs, obtain economies of scale and facilitate trade between regions. At present, there is a growing preference for using the same supplier in different locations.

A list of the first 40 suppliers, able to comply with the above mentioned requirements, is reported in Table 6.4; the data came from Automotive News publication and are based on the 2010 revenues.

The evolution taking place in the last decades followed the trends listed below:

- The in-house component activities of the major OEMs encouraged an increasingly extended know-how of suppliers up to the point in which they were in a position to offer their components to other companies. Some of the best examples may be mentioned: Delphi, (GM's component activities), Visteon (formerly belonging to Ford), Magneti Marelli (belonging to FIAT group);
- a wave of takeovers and mergers affected even the largest component suppliers. In areas such as seats and braking systems, only a few companies remain on the market;
- logistics, cost, engineering and sometimes protectionism make local production of many items a necessity. In these cases the supplier "follows" the OEM to new locations. Finally, an automaker, entering in a new emerging area, is likely to encourage a "follow sourcing" by the preferred suppliers. This policy can present some risks when, for example, the follow sourcing is eroded by the volatility of the demand in emerging markets. On the contrary, in most of the cases where the 'follow source' strategy was not used, the OEM chose another transnational firm to supply the relevant parts.
- the above mentioned developments show that global supply networks are becoming increasingly important in the auto industry. OEMs and suppliers develop parallel networks across the world.

This suppliers' network is identified as first-tier suppliers and they need individual components and basic products made by second- and third-tier suppliers.

The ex-factory price of a vehicle from the original equipment manufacturer is the price of the car leaving the factory. Out of this industrial price, approximately the

Table 6.4 Revenues (in million US Dollars) of the 40 worldwide major suppliers in 2010 (tire manufacturers, like Michelin, Bridgestone, Good year and Pirelli are not listed)

	Company	Revenue	Product		Company	Revenue	Product
1	Bosch	34,565	Elc, Elct	21	Valeo	7,952	Trans, Elct
2	Denso	32,850	Elc, Elct	22	Visteon	7,320	Elct
3	Continental	24,819	Brk, Tires	23	Autoliv	7,171	Safety
4	Aisin	24,613	Trans, Elct	24	Magneti Marelli	6,754	Elc, Elct, Mech
5	Magna	23,600	Engr	25	Mahle	6,628	Mech
6	Faurecia	18,220	Exht	26	Benteler	6,365	Exst, Mech
7	Johnson Control	16,600	Int	27	Dana	6,109	Trans
8	ZF	15,748	Trans	28	Toyoda	6,000	Rbr, Plst
9	LG Chem	15,500	Int, Plst	29	Cummins	5,846	Diesel engines
10	Hyundai Mobis	14,433	Int, Elct	30	DuPont	5,671	Rbr, Plst, Mtrs
11	TRW	14,400	Brk, Safety	31	BorgWarner	5,653	Trans, Mech
12	Delphi	13,817	Elct	32	Schaeffler	5,400	Trans, Mech
13	Yazaki	12,531	Elc	33	NTN	5,297	Trans
14	Lear	11,955	Seats	34	NSK	5,279	Mech
15	Sumitomo	11,228	Elc, Elct	35	Mitsubishi	5,265	Elct
16	BASF	10,400	Plst, Exst	36	Tenneco	4,768	Exht, Mech
17	Toyota Boshoku	10,400	Seats, Int	37	Behr	4,630	Clng, A/C
18	Calsonic	8,775	Elc, Int	38	Brose	4,609	Body comp.
19	JTEKT Corp	8,285	Trans	39	NHK	4,519	Mech
20	Hitachi	8,011	Elct	40	Koito Ltd.	4,390	Light

Elc electric components (starter motors, alternators, etc.), *Elct* electronic components (engine/vehicle control systems, thermal controls and car information systems), *Engr* engineering of a complete body and powertrain application, *Brk* mechanical brake components and systems electronically controlled, *Mech* mechanical components of engines and chassis (steering, suspensions, etc.), *Trans* gearboxes, clutches, joints, all-wheel-drive systems, axles, driveshafts, *Mtrs* advaced composite materials, specialty chemicals, advanced polymers, etc., *Int* internal components, door and instrument panels, floor consoles, seat systems, *Plst and Rbr*: paints and coatings, interior and exterior trims, fuel tanks and lines, sealing devices, *Clng, A/C* engine cooling and air conditioning systems, *Safety* airbags, seat belts, occupant-restrain systems, security systems, *Exht* manifolds and exhaust lines, catalytic converters, diesel after-treatments systems, *Body comp* window regulators, door modules, seat adjusters, closure system, power head restrains.

60 % is needed to provide the direct materials, either supplied by the manufacturer or by its suppliers. The material costs is approximately shared as:

- 20 % ferrous metals;
- 18 % plastics and rubber;
- 14 % non-ferrous metals (aluminium, copper, etc.) and glass;
- 8 % chemical products.

The remaining 40 % represents financial costs which include:

- cost of labour (salaries);
- amortization costs for the manufacturing equipment;
- financial burdens or expenditures, i.e. the money rate for bank loans;
- profits.

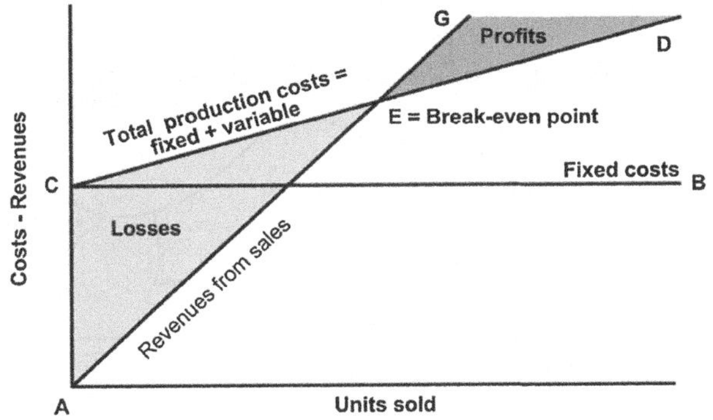

Fig. 6.2 Break-even point: Balance between costs and revenues

From the above, it is clear that the automotive industry boosts significant side-effects in the area where it operates, such as a supplies and spare part industry, distribution and marketing services, technical and administrative services (e.g. car registration, taxes, etc.), financial activities (credits, loans, etc.), public and private car fleets management, various types of insurances, road and parking infrastructures, fuel distribution network and finally raw materials and energy supply industries.

In the European Union, 1.2 million jobs are directly linked with the automotive industry; by adding the workforce employed by direct suppliers, this number rises to 1.9 million. In addition, the above mentioned side-effects induce further benefits promoting employment and creating new job opportunities. Therefore, according to European Commission estimates, each job in the automotive industry creates another 10 jobs.

The importance of the motor vehicle industry is confirmed when considering the vehicle segment taxation compared with the domestic fiscal income that, for Italy, was 16.6 % in 2010. Similar value can be found for the other main European countries.

6.5 Break-Even Point: Prices and Revenues

The price of a car is determined following various criteria. In the literature, it is said that the price must correspond to the customer's interest. In the motor vehicle sector, matters are simpler and more difficult at the same time: simpler, because in practice the price is determined by taking into account the direct competition and the need to cover costs; more difficult, because competition is fierce and widely motivated by the need of saturating production capabilities when considering the very high sales costs and the low profitability.

To survive, a company must follow at least one rule: to reach the Break-Even Point (BEP) shown in Fig. 6.2, that is specific for each version of a car model. The plot shows the relationship between total product costs and produced units in a simplified

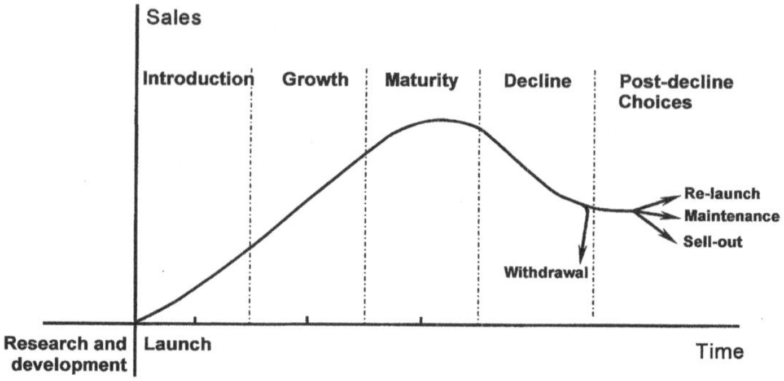

Fig. 6.3 Car lifecycle

form. The total economic burden consists of two types of fundamentally different costs:

- fixed costs (i.e. costs that are present also with zero production, such as development costs, investments for facilities and manufacturing equipment and their amortization, money rate for bank loans, energy costs, general expenditures, etc.)
- variable costs (i.e. costs proportional to production volumes, such as raw materials, components, salaries, etc.).

Profitability is obtained only after reaching the break-even point E at the intersection of the revenues from sales with the total costs. The break even point is thus defined as the point of balance between total production costs and sales revenues where the total revenue and total cost lines meet.

The launch of new and better products ensures survival and development of all companies. In the case of cars and commercial or industrial vehicles, this concept takes an even higher importance, because it involves maintaining a competitive presence and ensuring high economic profits able to support the huge investments needed to develop each new vehicle.

New products are therefore the result of major decisions and aim at well defined commercial, technological and financial targets. The need for renewal or restyling comes from the concept of lifecycle, i.e. the period of time during which a vehicle remains appealing to customers.

After the development phase and product launch, the car lifecycle is a consequence of the following main steps, shown in Fig. 6.3:

- Introduction, characterized by the commercial launch of a new model,
- growth, characterized by a rapid expansion of its demand,
- maturity, in which the model reaches the saturation of the interested market, and
- decline, during which the decrease of interest of the customers takes place.

During the following steps the break-even point must be re-defined and specific strategies, like restyling or face-lifting, are almost always chosen when approaching the declining step, to re-launch or maintain sales volumes at a profitable level.

A continuously updated planning approach is almost always chosen by companies before launching new products. New products are represented in the so-called Product Range Plan (PRP), in which the innovative contents are identified. The Product Range Plan is the result of the demand forecasts and evolution of expectations arising from the study of the various scenarios.

The innovation process is based on Product Range Plans of products, engines and other significant components. Each manufacturer strictly monitors the competitor's Product Range Plan, because it is necessary to reply to or keep ahead of the competitor's moves. During recent years, the model lifecycle became gradually shorter and new brands and models entered the market. This intensification of competition commits manufacturers to a spasmodic renewal and expansion of product ranges.

6.6 The Sales and Maintenance System

E-commerce and new European laws caused a substantial change in product distribution in the last decades. More specifically, in the automotive 'trade marketing' configuration, the manufacturer is still responsible for some aspects and possible actions such as product definition, price list, advertising and communications in general. In turn, dealers in their showrooms promote and sell vehicles and services to customers, including warranty and servicing, loans and more.

The first step is technically called sell-in and the second one is referred to as sell-out, to indicate sales to dealers and then to end customers, respectively. However marginally, manufacturers maintain a direct presence in the distribution structure by means of branches or sales centres.

In general, all manufacturers are working to increase the number and quality of their services. It may be said that a traditionally 'strict' customer culture, i.e. an approach only intended to the bare selling action, which tends to be less sensitive to market expectations and more focused on sales, is rapidly shifting into an attitude in which relationships are properly valued and customers considered as a premium.

While, normally, the manufacturer focuses in depth on the new vehicle market, the enormous importance of used car sales must not be underestimated. According to the European Automobile Manufacturers' Association (ACEA), the age of used cars is, in Europe, 8.2 years as an average and, during the lifecycle of a car, various changes of ownership may be expected.

The car maintenance is performed:

- by the official Original Equipment Manufacturer's Dealers, and
- by the Original Equipment Suppliers (OES) that operate independently from a single Original Equipment Manufacturer's organization.

The fundamental services include, as a first thing, repair services to cover production defects and to manage different types of warranties, defined to satisfy the customer's expectations with various contractual agreements, such as:

- the mandatory warranty applied to the new vehicle that, since 2001, was extended to 2 years by an EU Directive,
- the extended warranty, in which the manufacturer provides the additional service of replacing or repairing parts or systems that fail within the extended period and
- a warranty addressed to used or second-hand cars, considering that users, mainly the first, change their vehicle even if it is still in good conditions.

The Block Exemption Regulation (BER) Directive was published in October 2003 to enhance the competition between Original Equipment Manufacturer's Dealers and Original Equipment Suppliers. Prior to 2003, automobile owners in the EU region risked to nullify their vehicle warranty when the vehicles were serviced or repaired in workshops not belonging to the vehicle manufacturer or its dealers. This barrier was broken by the BER Directive, which granted vehicle owners the freedom of having their servicing and repairs done at their chosen workshops, with the final goal to reduce the money spent on servicing.

Despite the above consideration, the consolidated experience suggests that, up to 3 years, the service activity is performed by OEM's workshops and dealers. As an average the first change of ownership takes place after 3.5 years and between 4 and 7 years the OEM's workshop activity is remarkably reduced, down to 50 %, favouring independent shops. After 7 years the activity of the latter is definitely prevailing.

Maintenance activities were, in the past decades, mostly aimed to assuring the basic function of the car: its transportation capability. Since the 1990s this goal widened with the aim to assure the car's correct behavior, addressed to maintaining the level of safety and environmental control that are verified during the Type Approval process, as it will be better explained in Chap. 7.

Within 4 years from the first registration, and then every other year, the road-worthiness level of any vehicle must be verified by performing the suitable checks specifically focused on:

- safety components (brakes, tires, lights, steering, suspensions, wheel alignment, transmission, windscreen and wipers, chassis, seat belts and warning devices), and
- emission control system. This check is performed by measuring the idle and fast idle carbon monoxide level and air-to-fuel ratio for spark ignition engines and smoke for compression ignition engines. In addition the on-board diagnostic system, installed since 2001, alerts the driver of the likely exhaust emission exceedance, asking for an immediate repair action.

Unavoidably, the above mentioned activities boost a market for replacement parts cheaper than the original ones. Cost reduction in this area is achievable, considering that, in this case, innovation is no more required because the design is copied from existing components and reverse engineering is frequently used.

Chapter 7
Regulations

7.1 Technical Regulations

Motor vehicles must comply with the Technical Regulations applicable in the country they are offered for sale. In European Countries, compliance with the above Regulations, required to obtain European Certification, is verified through the *Type Approval* process performed by the manufacturer who must either test the car in front of the Government's Authority or perform the tests in a Government Laboratory. The aim of these tests is to verify that all regulations are duly complied with.

By applying the testing procedures required by the applicable regulations, the manufacturer must demonstrate that the car meets the standards and limitations stated by the regulating bodies. Almost worldwide, the relevant Authorities are the Ministry of Transportation and the Ministry of the Environment of the Country where the car is intended to be sold.

The main goals of the Technical Regulations are to:

1. Guarantee engine and vehicle performances, and in particular the fuel consumption declared by the manufacturer;
2. set limitations on the emission of pollutants and greenhouse gases through the exhaust;
3. set limitations to the external noise;
4. verify the efficiency of the systems needed to provide safe driving (e.g. brakes, steering, etc.);
5. protect car occupants and pedestrians in case of accidents;
6. protect electronic devices against external radiofrequencies;
7. verify the efficiency of some key components usually delivered by the supplier's network (e.g.: tires, lighting devices, towing hook, etc.);
8. provide for recycling the car body and related materials at the end of the vehicle's life.

The manufacturer has the responsibility and the freedom to select suitable technologies to meet the above mentioned requirements.

G. Genta et al., *The Motor Car*, Mechanical Engineering Series,

DOI: 10.1007/978-94-007-8552-6_7, © Springer Science+Business Media Dordrecht 2014

The Technical Regulations cover also the administrative, control and inspection procedures, as follows:

- Field of application, i.e. definition of vehicle types to which regulations address;
- application dates;
- procedures and administrative documentation needed to register and introduce new vehicles on the market for sale;
- requirements to satisfy the Conformity of Production (COP);
- manufacturer responsibility and penalties in case of non conformity;
- periodic inspections to verify the vehicle roadworthiness.

The definition of reliable testing and measurement procedures required a large and long-lasting study that started in the 1960s under the responsibility of technical organizations operating, at that time, in the traditional industrialized areas: USA, Europe and Japan. This process lead to the definition of the Code of Federal Register in the USA and of the Directives and Regulations in the European Union.

A number of Governments elsewhere in the world accepted the above mentioned legal documents and applied them during the Type Approval process. Documents related to administrative requirements and periodic inspections are, however, usually decided and issued by individual Governmental organizations and their representatives.

The Technical Regulations are therefore defined at the following levels:

- Supra-national, when a Technical Regulation is defined by an institutional organization which represents many Countries. It means that individual Countries grant to this organization the ability to define the Regulations, like in the case of the European Union (EU). Within this regulatory frame, the single Country's Authority is entitled to witness and validate the Type Approval process: the Countries belonging to the Union accept the Type Approval so obtained with a reciprocal acceptance of the test results.
- International, when many Countries, even if located in different geographical areas, decide to recognize and to accept the validity of a certain number of Technical Regulations. It is the case of many Mercosur countries accepting the USA regulations or many Asian countries accepting the European ones.
- National, when a national Authority is forced to define and apply a specific Regulation to solve and cover a special need in that Country. For example, the National Highway Traffic Safety Agency (NHTSA) of the USA defined Regulations to limit the crash consequences and some European Countries took actions to promote the use of alternative fuels, e.g. Compressed Natural Gas (CNG) and Liquefied Petroleum Gas (LPG).

Historically, the first legislative action was the definition of vehicle categories to state the field of application of the rules. This classification was always based on the vehicle mass and its capability to carry passengers or goods.

For the EU, the fields of application were defined by Dir. 70/126. The first two digits identify the year in which the Directive was published in the EU Official Journal and the subsequent digits identify the order number of the Directive within that year.

The mentioned Directive defines the following vehicle categories :

- M: Motor vehicles having at least four wheels and used for transportation of passengers;
- M1: vehicles with up to 9 seats, including the driver;
- M2: vehicles with more than 9 seats, including the driver, and maximum loaded mass equal to, or smaller than, 5 t;
- M3: vehicles with more than nine seats, including the driver, and maximum loaded mass larger than 5 t. For urban buses a further distinction, within the M2 and M3 categories, takes into account the number of seated and standing passengers.
- N: Motor vehicles having at least four wheels and used for carrying goods;
- N1: vehicles with a maximum loaded mass smaller than, or equal to, 3.5 t;
- N2: vehicles with a maximum loaded mass within 3.5 and 12 t;
- N3: vehicles with a maximum loaded mass above 12 t.

In the USA, the field of application for light duty vehicles includes two main categories:

- Light Duty Vehicles: passenger cars and their derivatives carrying up to 12 passengers;
- Light Duty Trucks: vehicles intended for good transportation or for passengers (more than 12) and having a maximum loaded mass larger or equal to 8,500 lbs (3,856 kg).

When defining vehicle's categories for legislative purposes, the following basic parameters are always considered:

- Unladen (or curb) mass : mass in running order, i.e. without driver, passengers or loading but with accessories and the fluid fill-up necessary for correct working (fuel, engine oil, engine coolant and additional fluids as required);
- payload : useful load the vehicle is able to carry;
- maximum loaded mass : curb mass + payload.

7.2 European Type Approval

The minimum requirements regarding the above mentioned eight points that must be satisfied to obtain the European Certification are defined by fifty specific Directives, that are grouped in the following five areas:

1. Environmental protection , including external noise, exhaust emissions, smoke (for diesel engines), fuel consumption, engine power and recycling;
2. Active Safety , including steering wheel, horn, windshield wiper, theft deterrent systems, visibility angles, tachometer, mirrors, tires, windshield, brakes, headlights, vehicle identification and fenders;

3. Passive Safety , including seat belt attachments, exterior protrusions, seat retainers, steering column, fuel tank, interior fittings, safety belts, door locks and hinges, headrests, safety windows, frontal impact, lateral impact and pedestrian protection;
4. Lighting components, including stop lamps, turn signal devices, licence plate lamps, bulbs and headlights, fog lamps, fog warning lamps, reverse lamps and parking lamps;
5. Additional prescriptions, including emergency hook, electromagnetic compatibility, radio frequency, license-plate installation, identification plates, passenger compartment heating, masses and dimensions, reverse gear and trailer towing devices.

When writing a new Directive, the European Commission applies the following sequence:

- Title: identifies the scope of the Directive and its field of application;
- Preamble: indicates the reasons of the new Directive or of the updating;
- List of Articles: specifies dates of implementation, Commission mandate for further decisions (e.g. limit to other pollutants, introduction of new measurement methods) and administrative frame to introduce fiscal incentives in interested Countries;
- Technical Annexes: describe the testing details to be applied during the measurement sequence.

When applying a Directive, two dates are the important ones:

- The first date: Type Approvals, for new models. Comes into force when the Manufacturer decides to certify a new engine or vehicle not yet in production applying the Type Approval process to meet new requirements (e.g.: stricter emission standards).
- The second date. Comes into force when all the manufacturer's engine/vehicle types, already in production, must meet the new requirements before being registered for circulation.

The two dates allow a certain time gap: 1 or 2 years for engine components, but definitely a longer time, up to 4 or 6 years, when the car chassis is concerned in the case of safety requirements.

7.3 Environmental Protection

The improved quality of life achieved during the last century is the result of an increased rate of energy production and of the massive use of natural resources. As a side effect, waste production and atmospheric pollution were increasing. As a countermeasure, many improved chemical processes have been defined and are in operation to clean up and recycle many kinds of wastes. In spite of that, the continuing population increase and the consequent increase in energy consumption and larger use of different types of machinery threatens the ability to control the above mentioned negative effects.

7.3.1 Pollution as a Worldwide Concern

Since the beginning of the industrial revolution, the combustion of coal with a high sulphur content caused the formation of smog , initially in the UK, where high levels of fog are present during a significant part of the year. The term *smog*, introduced in 1911, identifies the simultaneous presence of smoke and fog. The natural fog becomes rich in sulphur, released from coal combustion, with consequent human health risks.

The same term is now also used to identify the pollution determined by Ozone. In this case the mechanism is different: the Ozone formation takes place in different geographical regions when high solar radiation is present. In the early 1950s, the biogeochemist Haagen-Smith, working in California, suggested that the formation of urban ozone, Los Angeles smog, was determined by the action of sunlight on reactive hydrocarbons and nitrogen oxides released by oil refineries and automobiles. This ozone built-up takes place in the lower atmospheric layers and causes a reduction in lungs functionality and an increase of respiratory symptoms.

Later on, during the 1980s, the *ozone depletion* process was identified in an atmospheric region between 20 and 50 km of altitude, where two distinct but related phenomena, became evident. They are:

- A steady decline of about 4 % per decade in the total volume of ozone in Earth's stratosphere (the ozone layer), and
- a larger decrease in stratospheric ozone over Earth's polar regions. This latter phenomenon is referred to as the *ozone hole*.

The most important process causing the ozone depletion is the catalytic destruction of ozone by atomic halogens released into the stratosphere by man-made halocarbon refrigerants (CFCs, freons, halons) mainly used in refrigerators and air conditioning systems. The observed and projected decreases in ozone generated a worldwide concern leading to the Montreal Protocol that bans the production of CFCs, halons, and other ozone-depleting chemicals such as carbon tetrachloride and trichloroethane. The above mentioned decisions affected the car industry, that had to find solutions focused to replace the banned chemicals with new ones.

During the 1950s, the rise in atmospheric pollution due to four main sources: industrial processes, plants for energy generation based on oil combustion, households (heating, air conditioning, lighting) and traffic, was a matter of concern.

A decade later the energy consumption due to road traffic, including light and heavy duty vehicles, reached, in the European Union, 20 % of the total consumption, with a further increase in the following years. Identifying, measuring, controlling and reducing the pollutants emitted by internal combustion engines became very important goals.

A strong commitment to devote large economical resources to reduce pollutants released by vehicles started in the 1970s. It was also recognized that, up to a certain degree, the amount of pollution is a consequence of the quantity of fuel consumed, and thus much attention started to be devoted to fuel consumption reduction. The results of these efforts are impressive: compared with the end of the 1960s, present day cars produce less than 5 % of pollutants for spark-ignition engines and 10 % for compression-ignition engines. At the same time, the fuel consumption is reduced by fifty per cent, as an average.

Now the goal of the continuing efforts in this direction is reaching emission levels close to zero, with even lower fuel consumption.

During the 1980s and the 1990s much attention started to be devoted to the greenhouse effect. This effect is a consequence of the warming potential due to the increasing carbon dioxide (CO_2) concentration in the atmosphere. CO_2 release is the direct consequence of fossil fuel burning since, unavoidably, the carbon contained in the fuel is transformed in CO_2 during the combustion processes.

To reduce the overall greenhouse effect, when changing fuels or introducing new technologies, the amount of methane (CH_4) and nitrous oxide (N_2O) must always be monitored. Actually, even if their concentrations are extremely low, their greenhouse potential are much larger than that of carbon dioxide.

In addition to exhaust emissions, it is mandatory to reduce the amount of solid wastes and, in this direction, the only effective way is represented by the recycling processes. Recycling is an important issue when considering that cars destroyed every year in Europe add up to 12 millions.

The first emission Regulations and related limits have been issued by the State of California in 1967 under the thrust of the ozone formation during many summer sunny days in the Los Angeles area. In the same period, the USA Federal Government issued a strategic document, the Clean Air Act , defining technical regulations and emission limits in the Code of Federal Register.

In Europe, the first emission Directive was the Dir. 70/220 which took technical prescriptions from Regulation n° 15, issued in 1970 by the ECE-ONU organization, located in Geneva, Switzerland.

In the same years also Japan decided to apply regulations to control traffic pollution. A specific regulation, mainly based on the USA experience, was issued in 1972.

7.3.2 Regulated Pollutants

This definition is applied to the pollutants considered in emission regulations: Carbon Monoxide (CO), Hydrocarbons (HC), Oxides of Nitrogen (NO_x) and Particulate Matter (PM).

Carbon Monoxide is a gas that can enter the blood circulation, combining with hemoglobin thus slowing down the release of oxygen to blood and to other human tissues and organs. People already suffering from different kinds of circulation diseases are more affected by CO poisoning. In case of healthy people, CO acts like a poisonous gas, determining a loss of both visual perception and reactions. Prolonged exposition to high CO concentrations is lethal.

Hydrocarbons are present both in exhaust gases and vapors emitted by evaporation from fuel tanks. In this latter case hydrocarbons are identified as Volatile Organic Compounds (VOC) because they contain only the lighter fraction of hydrocarbons. The HC family includes also more dangerous species like benzene. When exposed to ultraviolet radiation contained in sun light, hydrocarbons react with NO_x acting as ozone promoter in the low atmospheric layer. Ozone, being a strong oxidant, can damage human tissues and specific cells.For this reason, this gas is particularly dangerous for young people whose natural protection system has not yet been established, and for older people whose system has already been damaged. The European emission limits Euro 5 and 6, applicable since the 2010s, introduce a distinction between Total HC (THC, equivalent to the traditional HC content) and Non Methane HC (NMHC) to recognize that Methane is not a pollutant. As already mentioned, its quantity is always measured considering its greenhouse potential.

Regarding gasoline, the following relation applies:

$$\mathrm{NMHC} = 0.9 \times \mathrm{THC}.$$

Oxides of nitrogen (NO_x) are many; the prevailing one being nitric oxide (NO), in a lesser extent nitrogen dioxide (NO_2) and traces of nitrous oxide (N_2O). The mentioned nitrogen oxides are the ones relatively stable and can be isolated and measured; not the same for some other oxides which decompose rapidly, such as: N_2O_3, N_2O_4 and N_2O_5.

In the common language, the term 'oxides of nitrogen' indicates the sum of NO and NO_2, the latter being the component of greatest interest. In fact the current scientific evidence links short-term NO_2 exposures to adverse respiratory effects including airway inflammation in healthy people and increased respiratory symptoms in people with asthma.

Nitrogen oxides NO_x, in combination with HCs, contribute to Ozone formation and with sulfate compounds contribute to rain acidification with negative effect mainly on water supply and vegetables, trees and crops, in lower or higher degree depending upon the quality of ground, if acidic or basic. Finally, in combination with other atmospheric pollutants, NO_x contributes to the formation of particulate, known as secondary particulate.

Particulates are fine particles present in exhaust gases, mainly in those from diesel engines, that can reach the lungs and, unfortunately, remain there. Metabolic processes have some difficulty in removing these undesired components, mainly heavy hydrocarbons and aerosols, linked to the individual particles. The scientific community is concerned that, during metabolic transformation, some health risk can take place because some of the above mentioned substances are characterized by carcinogenic risks and, again, young and old people and individuals already suffering from respiratory diseases, can be more affected. Particulate matter emitted by car exhausts can be classified following their aerodynamic diameter, ranging between 0.01 and 5 μm (10 $\div$ 5,000 nm), as follows:

- Fine particles ($PM_{2.5}$), with a diameter below 2.5 μm. They can enter into the upper respiratory system;
- ultra-fine particles, with a diameter below 0.1 μm. They can enter into the primary bronchi;
- nanoparticles, with a diameter below 0.05 μm. They can enter into the bronchioles and alveoli.

The smaller the diameter, the higher are the health risks, considering that smaller particles can more easily reach the smaller breathing lung ducts, bronchioles and alveoli, increasing the tendency to accumulation rather than being removed by the natural air movements within the lungs.

7.3.3 Unregulated Pollutants

Additional types of pollutants are present in exhaust gases, even if their concentrations are definitely lower than those of regulated pollutants. The main ones are lead, sulphur dioxide (SO_2), benzene, Polynuclear Aromatic Hydrocarbons (PAH) and aldehydes. Their control consists of introducing additional specifications in fuel composition, such as:

- Lead. It was formerly present in gasoline as tetraethyl lead and used to increase the Research Octane Number (RON), thus preventing knocking. Similarly to other heavy metals, its detrimental effects on human brain tissues were recognized at the end of the 1980s and therefore its content was progressively cut to almost zero in the EU from 0.15 to 0.013 g/l. Unleaded gasoline, required by three-way catalysts to meet Euro 1 standard , was thus introduced. In 2000 a further reduction to 0.005 g/l took place to improve catalyst durability. This is now the maximum lead content for the gasoline types distributed in the EU: 91, 95 and 98 RON.
- Sulphur dioxide (SO_2) is determined by the sulphur present in every type of fuel. Starting from the 1990s, with subsequent directives, the sulphur content was gradually reduced to reach the same level both in gasoline and in diesel fuel. The maximum sulphur content was reduced from 500 to 350, to 150, to 50 and finally to 10 ppm; the latter content, allowing qualification of the fuel as sulphur free, has been mandatory since January 2009.
- Benzene (C_6H_6) is the simplest of the ring-shaped aromatic hydrocarbons. It shows the best behavior against detonation thus allowing the octane number to raise. Unfortunately, it is recognized as being carcinogenic and thus new anti-knock hydrocarbons and components have been introduced to partially compensate for the benzene properties, since starting from year 2000 the benzene percentage in

gasoline was lowered from 5 to 1 %. As a consequence, the average RON in Europe was reduced from the previous 98 to the present-day 95 RON.

- Polynuclear Aromatic Hydrocarbons (PAH) are also considered as being carcinogenic; they are due to the transformation of benzene and other aromatic hydrocarbons during the combustion process. Again the most effective measure is to reduce their quantities in fuels and therefore, starting from year 2000, a specific standard states that aromatic hydrocarbons must not exceed 11 % of diesel fuel.
- Aldehydes. The possibility to obtain fuels from agriculture is becoming of increasing interest; the most common biofuels being alcohols, either ethylic or methylic. These fuels, despite some advantages, promote aldehydes formation. Their smell, in the exhaust, is typical and is due to a specific molecular bond: CHO. As known, in Brazil, ethylic alcohol is obtained as a by-product of sugar cane cultivation. In the interested countries, a formaldehyde standard has been introduced considering its carcinogenic risks.

7.3.4 Emission Regulations

Starting from the 1970s and continuing every 5–7 years, Europe, USA and Japan reduced emission standards in order to achieve in perspective a level close to zero. Modifications to Technical Regulations and measurement procedures are less frequent and typically take place every 10–5 years. Emission regulations for gasoline fueled cars require reduction of CO, HC and NO_x. For many years emissions standards where applied to CO and, up to Euro 2, to the sum of HC + NO_x with the aim to:

- Reduce the two gases responsible for Ozone formation and
- offer a certain degree of freedom, when tuning the engine for emission reduction, limiting the sum rather than each individual pollutant.

Since 1989, improvement of measurement procedures allowed measuring diesel PM.

With Directive 91/441-Euro 1, enforced in January 1993, the EU reached the goal to reduce by 90 % the vehicle emissions with respect to 1970, when the first emission Directive was published (70/220). Further emission reductions Euro 2 and 3 were introduced in 1997 and in 2001. The dates reported here and hereafter refer to vehicle registration.

Euro 4 standards, enforced in January 2006, are a further step toward the zero equivalent emission level (see Table 7.1).

The emissions must be measured while following a well-defined driving cycle, able to simulate real driving conditions. In addition to the mentioned limits, gasoline cars must comply with the evaporative emission standard, set to a level of 2 g/test.

Subsequent standards, Euro 5 and Euro 6, are so strict that are measured in mg/km (instead of g/km) and, in fact, are considered as almost zero equivalent. The following dates apply to all newly registered vehicles:

Table 7.1 Regulated pollutants standards (in g/km) as prescribed by Euro 4 regulations

	CO	HC	NO_x	PM
Gasoline cars	1.0	0.10	0.08	–
Diesel cars	0.50	–	0.25	0.025

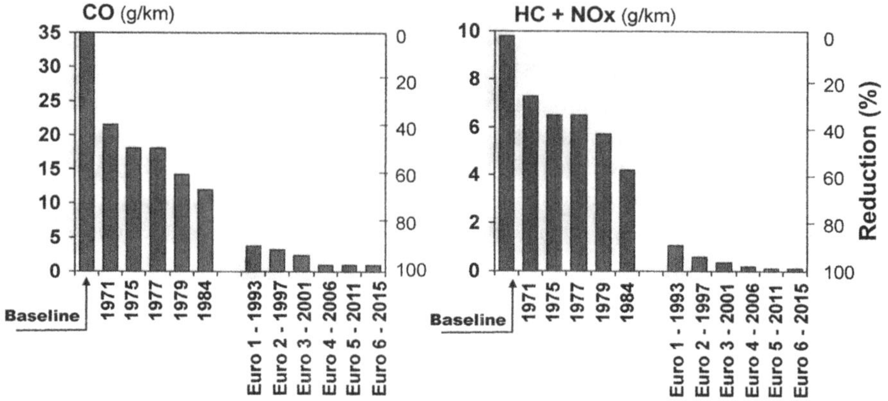

Fig. 7.1 Emission reduction through the years for spark ignition engines. The reduction obtained exceeds 95 % for all pollutants

Gasoline Euro 5, January 2011 and Gasoline Euro 6, September 2015

- CO: no variation, 1000 mg/km
- THC: no variation, 100 mg/km
- NMHC: new standard, 68 mg/km (average reduction by 25 %)
- NO_x: reduced from 80 to 60 mg/km (−25 %)
- PM measurement only for Gasoline Direct Injection (GDI) : first step, 5 mg/km, second step 4.5 mg/km.

For diesel engines, the situation is different and the almost zero level situation will be reached only at the Euro 6 stage. At that point the relevant emissions will be almost identical to that of gasoline engines and, for this reason, Euro 6 is considered close to a fuel neutral standard.

Diesel Euro 5, January 2011

- CO no variation 500 mg/km
- PM from 25 to 5 and finally to 4.5 mg/km (−82 %)
- NO_x from 250 to 180 mg/km (−28 %)

Diesel Euro 6, September 2015

- CO and PM: no variation
- NO_x from 180 to 80 mg/km (−55 %)

Figures 7.1 and 7.2 show the emission reduction requested by the EU from the beginning, in the 1970s, up to Euro 6 respectively for gasoline and diesel engines.

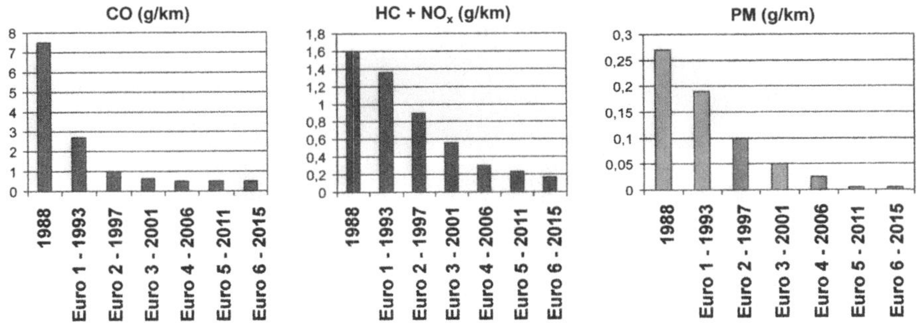

Fig. 7.2 Emission standard sequence through the years for diesel engines. The reduction obtained exceeds 90 % for all pollutants

For particles determination, in addition to the gravimetric value, the need is now accepted worldwide to provide a reliable measuring system capable of counting the number of particles emitted every kilometer and the new limit, applicable since Euro 6, has been set at 6×10^{11} particle/km for diesel engines and, some years later, also for direct injection gasoline engines.

The exhaust emission standards so far decided are set with the aim to meet the air quality standards through Europe established by the World Heath Organization (WHO) and measured as hourly or daily or yearly averages in $\mu g/m^3$. Considering that the air quality goals have been already obtained for many pollutants (e.g. CO, HC, SO_2, lead) applying the Euro 4 standards, the new standards are intended to reach a similar result also for PM and NO_x.

The effects of exhaust control on the air quality is shown in Fig. 7.3 in which the trend of CO atmospheric concentration in some European cities is reported. These results are even more striking when the traffic increase that occurred in the same period is considered. The period shown is the same as that of the Directive application and therefore the overall positive effect of the regulations is clear. This results convinced many Administrations of developing countries to enforce emission standards as an effective countermeasure to traffic pollution.

When deciding to enforce a new regulation, the technological and economical situations in emission control in that particular Country must be considered, and whether the proposed rules are based upon the European or the USA legislation. The new standards determine unavoidable additional costs and defines the need for suitable infrastructures. Other important factors like, for instance, the availability of fuels meeting environmental specifications, are important in reducing the content of unregulated pollutants like lead, sulphur, benzene and aromatics.

Considering that fuels meeting updated specifications are progressively available in an increasing number of Countries, regulations considering the use of subsequent catalyst generations and related engine technologies can be implemented more widely. Traditionally Canada, South America, Taiwan and North Korea recognize USA rules, while the rest of Asia and Africa are aligned with European Directives.

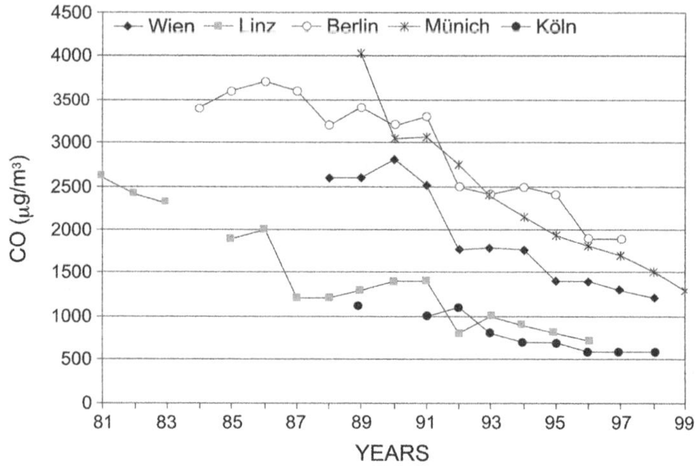

Fig. 7.3 Air quality improvements through the years in some European cities (courtesy of ACEA)

Japan is still following its own way, but, for the time being, no other Country accepts Japanese standards.

During the Type Approval driving cycle also CO_2 emissions are measured. By adding the carbon contained in CO and HC, it is possible to determine the total carbon quantity in exhaust gases, which obviously is the same as that contained in the fuel used during the cycle. This determination defines the carbon balance method, which allows determining the fuel consumption during the cycle. Considering that CO and HC quantities are now negligible, a good correlation between CO_2 emissions and fuel consumption exists.

The application of the carbon balance method is expedient, because it is performed during emission measurements, thus avoiding modifications to the fuel system which may affect the accuracy of the result.

The existing regulations do not prescribe a CO_2 maximum value like for noxious pollutants but, considering its greenhouse potential, vehicle Manufacturers agreed with EU Authorities on a strategy aimed at reducing, year by year, vehicle fuel consumption and, as a consequence, the CO_2 released in the atmosphere. The decisions taken on this matter are shown in the section on fuel consumption.

7.3.5 Exhaust Emissions Testing

Exhaust emissions are measured in laboratories allowing testing in which the vehicle is driven following a cycle whose profile simulates city driving conditions. Therefore, the climate control system must be able to maintain a constant level of temperature and humidity, within pre-determined levels, in spite of engine operation with its

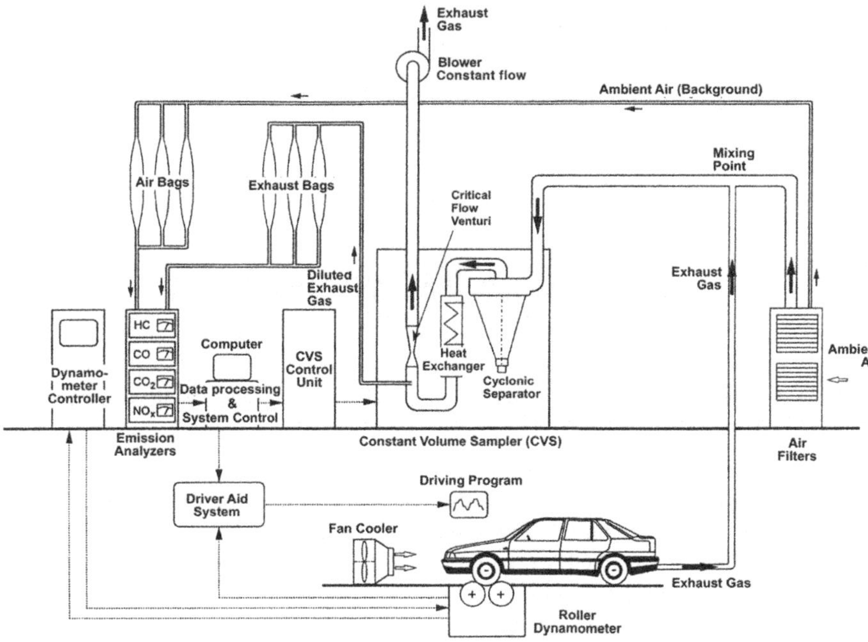

Fig. 7.4 Constant Volume Sampling system outlining the three main components for emissions measurement: dynamometer, critical flow venturi and analyzer assembly (courtesy of FIAT)

consequent heat release. The selected levels are within 20–30 °C and 30–70 % of relative humidity.

The key features of the measuring system are the following:

- Using a Constant Volume Sampling (CVS) device which provides a correct dilution and mixing of the exhaust gases with ambient air, thus reproducing the diffusion conditions taking place in the atmosphere. The dilution prevents the condensation of water contained in the exhaust gases.
- Simulating the resistance to motion experienced by the car both at constant speed and in transient conditions.

To achieve the above conditions, the laboratory must be equipped with the following facilities, shown in Fig. 7.4:

- A blower driven at a fixed speed, providing a constant flow of ambient air in which the exhaust gases are diluted;
- a chassis dynamometer equipped with one or two rollers, allowing the car to be driven by a professional driver according to the selected cycle;
- a cooling fan, placed in front of the vehicle radiator and delivering an air stream proportional to the vehicle speed;
- a system displaying the theoretical and actual speed traces on a monitor, thus allowing the driver to follow the test cycle selecting the relevant gears and operating

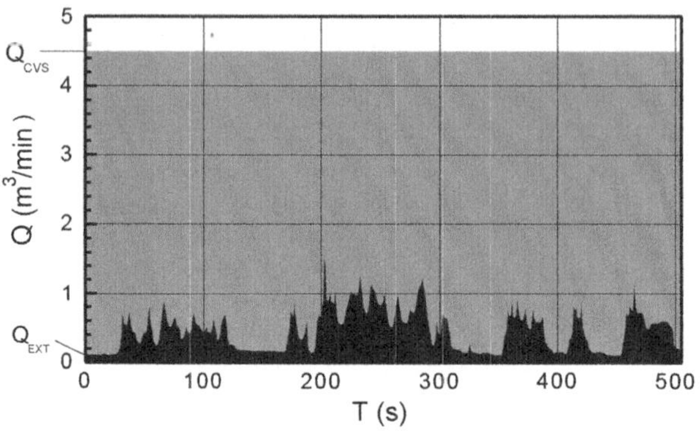

Fig. 7.5 Constant volume sampling system; variable dilution of exhaust gas flow (Q_{EXT}) with dilution air (Q_{CVS}) during the USA driving cycle (courtesy of FIAT)

the accelerator, brake and clutch pedals. The final comparison of the traces allows one to determine the accuracy with which the cycle was followed;

- a Critical Flow Venturi (CFV), to measure the volume of the mixture ambient air/exhaust gases determined by the blower during the cycle;
- a sampling point upstream the CFV and a sampling line to collect the diluted exhaust gases in suitable bags to measure the pollutant concentrations in the vehicle exhaust;
- a sampling point of ambient air to subtract the concentration of pollutants already contained in the ambient or dilution air, the so called background concentrations;
- an emission analyzer cabinet, to measure, in the respective bags, the average concentrations of polluants contained in the exhaust gases and in the ambient air;
- a computer which, based on the measured volumes and concentrations, calculates the emissions (in grams per kilometer) according to the Directives.

The dynamometer, in addition to the rollers, is equipped with an electric brake, able to simulate the external forces encountered during the vehicle movement: rolling resistance, aerodynamic drag and inertia forces.

For a given vehicle configuration the resistance to motion (the so called *road load*) is determined experimentally, as a function of vehicle speed, through a *coast down test* performed on a perfectly horizontal test track and in absence of wind. The coast down times are measured performing a sequence of free decelerations between two selected speeds. This sequence allows one to determine the rolling resistance and aerodynamic drag to be reproduced by the electric brake of the dynamometer, that can also simulate the car inertia, based on its curb mass value, incremented by the driver's mass (75 kg) and by an additional 25 kg mass.

As already mentioned, the blower provides a constant volume flow in spite of the transient conditions, from idle to acceleration phases and assures a dilution of the exhaust gases with ambient air with a ratio ranging from 1:10 to 1:15. A heat exchanger, when needed, stabilizes the mixture temperature. Figure 7.5 visualizes the constant ambient air flow and transient exhaust gas flow mixing together.

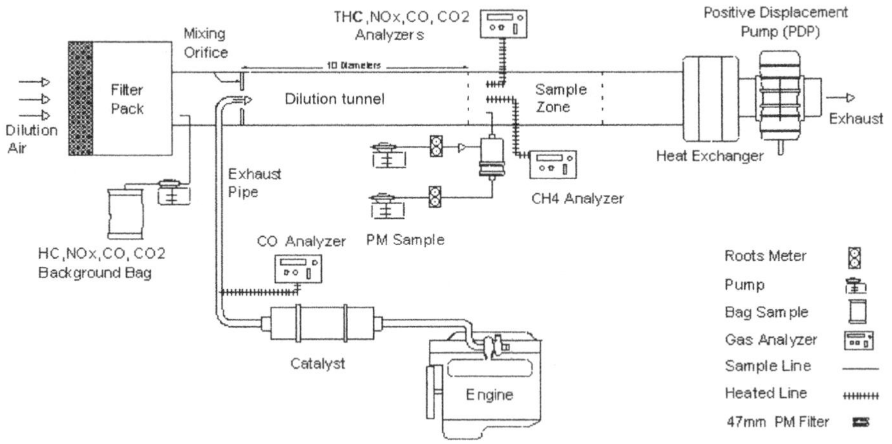

Fig. 7.6 Diesel exhaust dilution tunnel (courtesy of the South West Research Institute)

Emissions in g/km are determined through suitable computations after entering, as input data: the mixture volume, the concentrations of the diluted gas, the densities of individual pollutant gases and the distance run during the driving cycle.

Similarly to what is done for gasoline engines, CO, NO_x and CO_2 emitted by diesel engines are analyzed in the sampling bags, while specific devices are needed for:

- Hydrocarbons. Diesel fuel features a higher content of Carbon and its exhaust gases are emitted at a lower temperature, when compared with gasoline. This causes a further tendency to condensation along the walls of the sampling lines. To overcome this risk, the sampling lines are heated up to 190 °C.
- Particulate Matter. Since particulate emissions are not a gas, a homogeneous mixture with the dilution air must be achieved before performing their sampling. To achieve this result a dilution tunnel, as shown in Fig. 7.6, is added between the point at which the exhaust gases are mixed with dilution air and the sampling point; its diameter ranges between 200 and 300 mm and its length is, at least, ten times its diameter.

From the tunnel, the heated sampling lines reach, respectively, the analyzers and the filter used to collect the PM sample; the filter is weighed before and after the cycle.

To determine the particle number, as requested by Euro 6 standard, an additional device able to measure the number of dry particles, avoiding consideration of the aggregates of aerosols and water molecules, must be added. This quite complex system must be able to count the number of particles per kilometer in the size range of 0.023–2.5 μm (23– 2,500 nm).

Figure 7.7 shows an example of particle counting before and after a Diesel Particulate Filter.

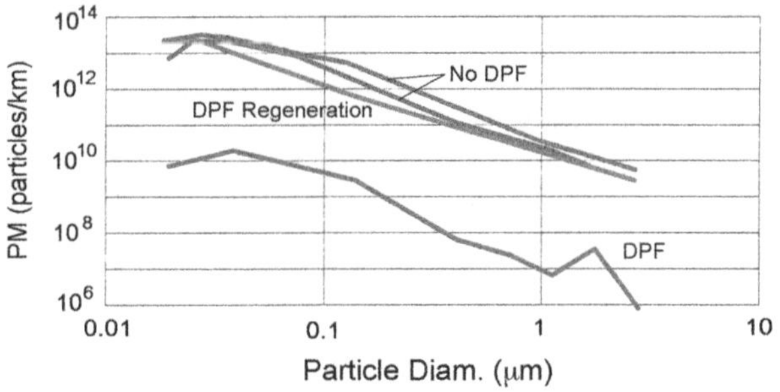

Fig. 7.7 Examples of particle number (PN) reduction dowstream of a Diesel Particulate Filter. In the PM_1 region the particle number decreases from 50 billions/km down to 10 millions/km, close or even lower than the background PN (courtesy of Stazione Sperimentale per i Combustibili, San Donato Milanese)

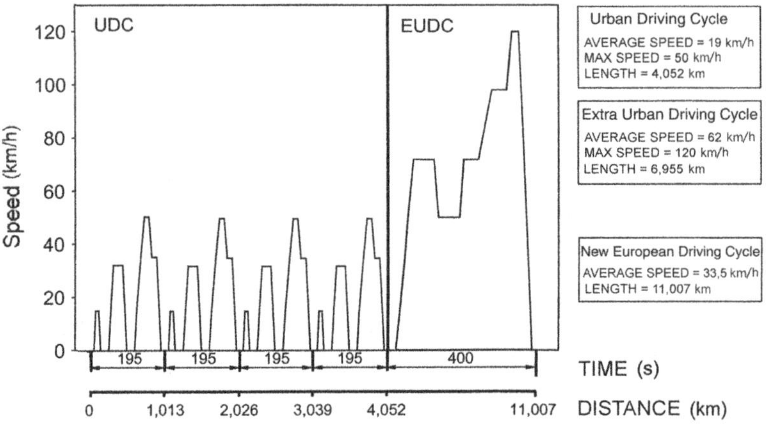

Fig. 7.8 Speed profile of the New European Driving Cycle, including urban and extraurban sections (courtesy of FIAT)

The two above mentioned methods are complex, require costly instrumentation and are not suitable to obtain data during transient conditions. Considering that dry particles, or soot, constitute the carbonaceous fraction able to absorb visible light, an opacimeter measuring the light absorption due to soot is used and the opacity is meadured in Smoke Units (SU) .

7.3.6 Driving Cycle

As shown in Fig. 7.8, the driving cycle is a sequence of idle periods, accelerations, constant speed running and decelerations, that simulates congested traffic conditions specific to the area in which the emission regulation is applied. For this reasons, different cycles are still used in Europe, USA and Japan.

The European cycle is known as the New European Driving Cycle (NEDC), has an average speed of 33.5 km/h, a total length of 11,007 km and requires starting the engine in laboratory conditions. This cycle includes the following sections:

- Urban Driving Cycle (UDC) obtained with four sub-cycles to be repeated for a total of four kilometers: average speed 19 km/h and maximum speed 50 km/h;
- Extra Urban Driving Cycle (EUDC) section with a length of seven kilometers: average speed 62 km/h and maximum speed 120 km/h.

Some international organizations undertook the task of defining a new profile of an emission cycle that is accepted worldwide. This goal has already been achieved for heavy duty commercial vehicles. If this task is not achieved, the European Commission is ready to define an updated version of the NEDC cycle for the Euro 7 step which likely will be enforced about year 2020.

7.3.7 Emissions and Durability

Emission regulations prescribe that emission standards are met for the useful life of the car, when used in normal conditions: a durability verification is thus required. Durability is a matter of continuing interest considering that emissions tend to worsen for the following two main reasons:

- Increase of baseline engine emissions, due to engine wear and, sometimes, to Customer's lack of maintenance;
- deterioration of the main emission components, namely the lambda sensor and the catalyst.

In the late 1960s, the USA Environmental Protection Agency (EPA) defined a specific cycle, 80,000 km long, aimed at verifying the emission deterioration to be completed before starting the type-approval process of any model. As requested by Regulations, the goal is to verify the emission standard compliance with the car both new and aged. This requirement was followed by a number of Countries, Europe included. Year by year this request is losing interest because a single type of cycle cannot represent all combinations of Customer uses.

To update the durability verification, the Regulations now include the possibility of measuring the emissions of aged Customer's cars randomly selected within 80,000 km. If the average of one or more pollutant exceeds the standard, the Authority has the right to oblige the Manufacturer to take suitable corrective actions, at its expense. This sequence is known as a recall campaign, which means calling back all cars and restoring them in the manufacturer's authorized maintenance shops. These additional requirements impose on manufacturers the obligation to design new cars with sufficient safety margins that take into account deterioration so that old cars

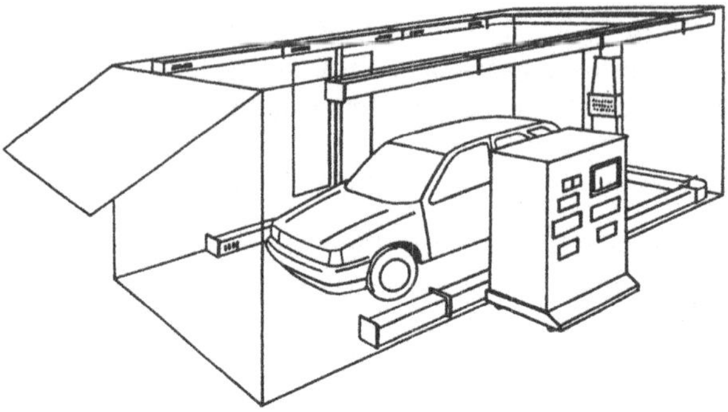

Fig. 7.9 Scheme of the Sealed Housing for Evaporative Determination (SHED) (Courtesy of FIAT)

can pass durability controls. The initial 80,000 km requirement was then extended to 100,000 km with Euro 4 standards, and to 160,000 km to comply with Euro 5 and 6 regulations.

7.3.8 Evaporative Emissions

Gasoline contains many light and, as a consequence, volatile hydrocarbons which are usually identified as Volatile Organic Compounds (VOC).

The air in the free fuel tank volume is saturated with gasoline vapors and the fuel tank must be vented to the atmosphere to avoid abnormal pressure increases or decreases. Usually the first takes place when the tank is exposed to a temperature rise and the volume of the liquid gasoline marginally increases while some of it evaporates. The second occurs when fuel is consumed in a low temperature environment. In the first case the vapors must be vented to the atmosphere through a suitable bleed, thus polluting the environment. It is therefore necessary to control this fuel tank bleed, connecting it to a filter containing active carbon to control the evaporative emissions. The active carbon absorbs the light saturated vapors avoiding their release into ambient air. The active carbon can be regenerated by a stream of fresh air induced through the carbon volume when the engine is running; in this desorption phase the engine can burn this excess of vapors.

During the 1970s, the EPA defined a measurement device and the relevant procedure to guarantee the required measurement accuracy for this type of test. This device is a cabin, of rectangular shape, that can be sealed: it is usually referred to as Sealed Housing for Evaporative Determination (SHED) (Fig. 7.9). The internal cabin surface is impermeable to hydrocarbon vapors and the roof surface is usually flexible to assure that the pressure inside the cabin remains at the atmospheric level, in spite of the temperature increase when a car with a hot engine is introduced. A leak proof gate allows the car to enter at the beginning of the test and to exit at the end.

An internal recirculation system and a heat exchanger assure that temperatures and volatile organic compounds concentrations remain homogeneous within the cabin during the test. These concentrations are measured, before and after the test, using an HC analyzer placed outside the cabin.

The SHED monitors two different conditions during which the gasoline vapor losses take place: Hot Soak and Diurnal Losses.

The test sequence includes the following steps:

- Saturation of the carbon canister with fuel vapors or with butane up to the point at which two grams of gasoline vapors, or butane, escape from the filter thus reaching the breakthrough condition;
- preconditioning of vehicle and canister running an Urban Driving Cycle cycle followed by two Extra Urban Driving Cycle and vehicle soaking in a constant temperature area (20–30 °C);
- a NEDC cycle is run, followed by one Urban Driving Cycle to reproduce the conditions during which the air regenerates the active carbon;
- the car is introduced into the SHED and left for 1 h, hot soak phase; at the end, the HC concentration is measured. This first phase simulates the car in parking conditions with a hot engine, which takes place after a driving period;
- the car is soaked again in the Laboratory area (20 − 30 °C) for a period of 6 − 36 h;
- the car is introduced again into the SHED and the diurnal loss phase which lasts 24 h is started. During this phase the temperature is varied according to a predefined sequence. It means that, starting from 20 °C, the car is heated up to 35 °C within 8 h and then cooled again to 20 °C during the remaining 16 h. At the end, the HC concentration is measured. This second phase simulates the natural day-night temperature variation with the car in normal parking conditions.

Having measured the HC concentrations at the end of the two phases and the SHED interior volume, it is possible to determine the evaporation, in g/test, of hot soak and diurnal loss phases. Their sum determines the final HC value to be compared with Euro 5 and 6 standards (2 g/test).

7.4 Fuel Consumption

The oil crisis of the early 1970s raised, for the first time, much concern about the dependence of road transportation on fossil fuels, a non-renewable source, unevenly distributed in the word. Starting from that time, fuel economy became an issue and the need for significantly improving it was recognized.

Following the above considerations, in 1975 the American Environmental Protection Agency (EPA) introduced the first limitations dealing with maximum fuel consumption values for new cars . The Corporate Average Fuel Economy (CAFE), which defines a standard for the weighted average fuel economy (in miles per gallon) for new cars, was introduced. It was to be applied yearly to the new cars sold by each individual Manufacturer. In case a Manufacturer is not able, with its fleet, to meet the standard, a Gas Guzzler Tax is applied; it is proportional to the difference between the Manufacturer's actual value and the CAFE value. This Tax must be paid for each car produced and thus affects the final Customer's price.

CAFE and Gas Guzzler Tax still represent the basic incentives for the heavy changes introduced to vehicle power trains and their architecture.

In the same period, in Europe, the Authorities accepted voluntary commitments focused to reducing European car fleet fuel consumption. The results obtained have been interesting and, in the period from 1978 to 1987, a 20 % average reduction was achieved.

The introduction of demanding regulations in terms of emissions and safety and the lower Octane Number of the unleaded gasoline, from 98 to 95 RON, needed to introduce three-way catalysts, slowed this trend during the period from 1987 to 1994.

Approaching the year 2000, increasing recognition of the greenhouse effect gave new strength to the drive toward reducing fossil fuel combustion. Among the other green-house gases, CO_2, released from fossil fuel combustion, was considered as the most important. Along this line, the European Union decided to

- take, as reference, the CO_2 weighted average in 1995, namely 186 g/km;
- set in 1998 a Voluntary Commitment with the ACEA (Association des Constructeurs Europeens d'Automobiles) , to reduce the average value to 140 g/km in 2008 thus calling for a 24 % reduction compared to 1995 and a 1.8 % yearly reduction;
- monitor the actual CO_2 emissions value in 2005 which was equal to 160 g/km, thus showing a 13 % reduction and a 1.3 % yearly reduction.

Considering the above mentioned pace, it was recognized that it was impossible to meet the Voluntary Commitment. In fact only Fiat, PSA and Renault, producing many low segment cars, could meet the 2008 Commitment.

A new goal was thus defined in 2007. This time, instead of being voluntary, it was mandatory thus implying fiscal penalties above 130 g/km to be achieved in 2015 with a 30 % reduction, compared to 1995, and a 1.5 % yearly reduction. This goal is applied to the average vehicle mass of the European car fleet (M_o) defined, in 2009, as 1372 kg. It means that, when the Manufacturer's average mass (M) is above this value, the related standard is higher and the opposite takes place when the average mass is lower according to the following relationship

$$\text{mass}\,(CO_2) = 130 + a\,(M - M_o) \tag{7.1}$$

in g/km, where $a = 0.0457$ is a coefficient defining the slope of the line setting the standard. In fact for M = 1,000 or 2,000 kg, the CO_2 limits become respectively 113 and 159 g/km. The straight line shown in Fig. 7.10 shows how CO_2 values correlate to car mass in 2010 year.

The final 2015 goal is 120 g/km. It must be obtained through an integrated approach: 130 g/km obtained improving car technologies and the remaining reduction obtained through the following measures: use of bio-fuels, introduction of a tire pressure monitoring system, fuel-efficient tires, improving the air conditioning effi-

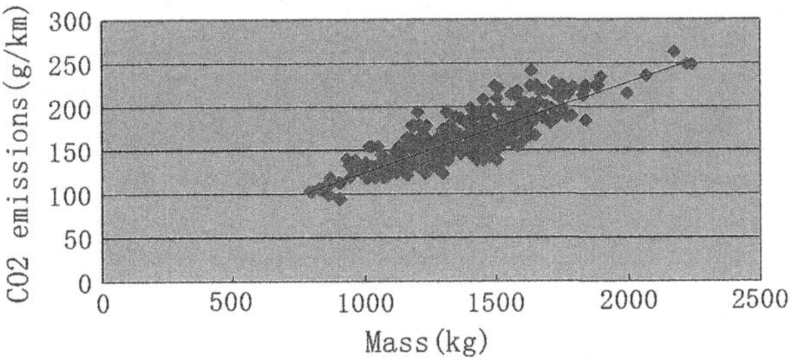

Fig. 7.10 Spark ignition engine: effect of car mass on individual CO_2 emissions value measured according to the NEDC driving cycle (year 2010)

Table 7.2 Correlation between fuel type, fuel consumption and CO_2 release

CO_2 produced per unit volume and mass of fuel			
Fuel type	Density (kg/l)	CO_2 (kg/l of fuel)	CO_2 (kg/kg of fuel)
Gasoline	0.750	2.33	3.10
Diesel fuel	0.835	2.64	3.16
Equivalence between fuel consumption and CO_2			
Gasoline		1 l/100 km emits 23.3 g/km of CO_2	
Diesel fuel		1 l/100 km emits 26.4 g/km of CO_2	

ciency, introducing gear shift indicators and implementing fuel efficient technologies in light-commercial vehicles.

The second goal has been set to 95 g/km to be reached in year 2020 and a third must be defined and applied in 2025.

The mass is heavily affected by the market tendencies and by safety regulations (restraint systems, air bags, etc.): these two effects cause an estimated 1.5 % yearly mass increase. It is recognized that the engine technology evolution should have been able to achieve, in the first 10 years, a 21 % CO_2 reduction instead of the above mentioned 13 %, if the car mass could be kept constant.

To further attract the Customer's attention toward this issue, the European Commission decided to publish a guide listing the fuel consumption and the CO_2 emissions values when running the NEDC cycle for exhaust emission determination. It is estimated that attracting more attention on this issue and considering that many mass-related passive safety components have already been installed, the yearly weight increase, in the next years, should decrease to 0.8 %.

The carbon balance method, applied running the NEDC cycle, allows determination of a direct correlation between fuel used, fuel consumption in l/100km and CO_2 release in g/km, as shown in Table 7.2.

In Europe the transportation sector contributed with a 26 % to the CO_2 release in recent years. The different shares are reported in Table 7.3 , referred to the first decade of the years 2000s; within the road transportation sector, the passenger car share is estimated at about 12 %.

Table 7.3 CO_2 emissions in the EU (courtesy of ACEA)

Source	Share (%)
energy plants	39
households	19
industry	16
passenger cars	12
other transp. means	14

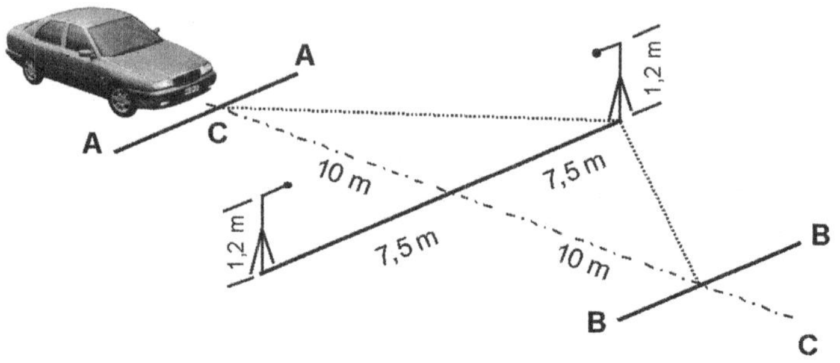

Fig. 7.11 Exterior noise measurement. The car performs a full load acceleration in second and third gear passing by the phonometers (courtesy of FIAT)

For the years to come, the European Commission estimates that the transportation share will most likely tend to increase, because in this sector fossil fuels are, in practice, the only possible energy source since all likely alternatives still show many difficulties in their practical application.

7.5 Vehicle Exterior Noise

Recent studies estimate that traffic noise releases most of the acoustic energy in urban areas, about 80 % .

Today the external noise of vehicles is measured using a test procedure which includes two phonometers placed perpendicularly with respect to the vehicle path, at a distance of 7.5 m, as shown in Fig. 7.11. Regulations state that the acoustic level released when the vehicle is driven at full acceleration on a length of 20 m must be measured. When considering a five speed vehicle, the test is repeated using the 2° and 3° gear. The final result is calculated as the average of the two sound levels.

Directive 92/97 (Fig. 7.12) has been enforced in Europe since October 1995 for new type approvals and October 1996 for all vehicles. It requires a reduction in car noise level to 74 dBA.

This reduction was important and required adequate technological efforts: the limit value is linked to the reduction of the acoustic energy, which is not proportional to the noise. In fact, a 3 dBA reduction, starting from 77 dBA, means a 50 % reduction

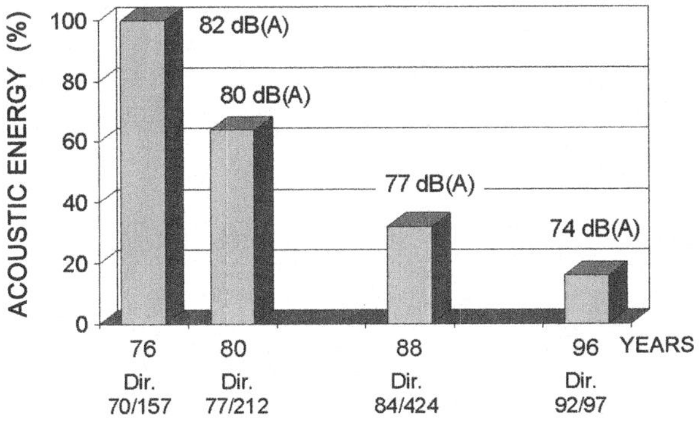

Fig. 7.12 Sequence of external noise limits according to EU Directives. The reduction from 77 to 74 dBA corresponds to a 50 % reduction in acoustic energy (courtesy of FIAT)

of acoustic energy. It is possible to translate the Directive noise limits into acoustic energy; in so doing, from 1976 till now, the energy has been reduced by 85 %. This means that eight cars meeting the 92/97 Directive emit an acoustic energy equal to that emitted by a single 1976 car.

The technical solutions applied to reduce the sound energy level, unfortunately, are not directly linked to the effective noise reduction experienced in daily traffic. This is due to traffic increase and to the limited degree of correlation between the testing procedure and the actual traffic noise generation. The above mentioned noise limitation caused a reduction of noise released by engine and mechanical components leaving, in practice, unchanged the noise determined by the contact of the tires on the road surface, which represents an important source.

A modification of Directive 92/97 is in progress with the aim of:

- Dividing vehicles in categories to ask for noise reductions achievable with technical solutions more focused on specific vehicle categories (e.g. passenger cars, SUVs, light duty trucks, etc.), and
- implementing new test procedures to evaluate the tire contribution, performing, for instance, specific tests at constant speed (70–90 km/h).

7.6 Vehicle End of Life

In recent years a growing importance has been ascribed to solid waste treatment and elimination. In spite of the fact that waste produced by the car scrapping process are only 0.5 % of the total waste produced, the problem is that the number of cars scrapped yearly in Europe sum up to 12 million.

As an average, 75 % of the car mass is made up of metallic components which have been subject to recycling for many years. The remaining 25 % are non metallic

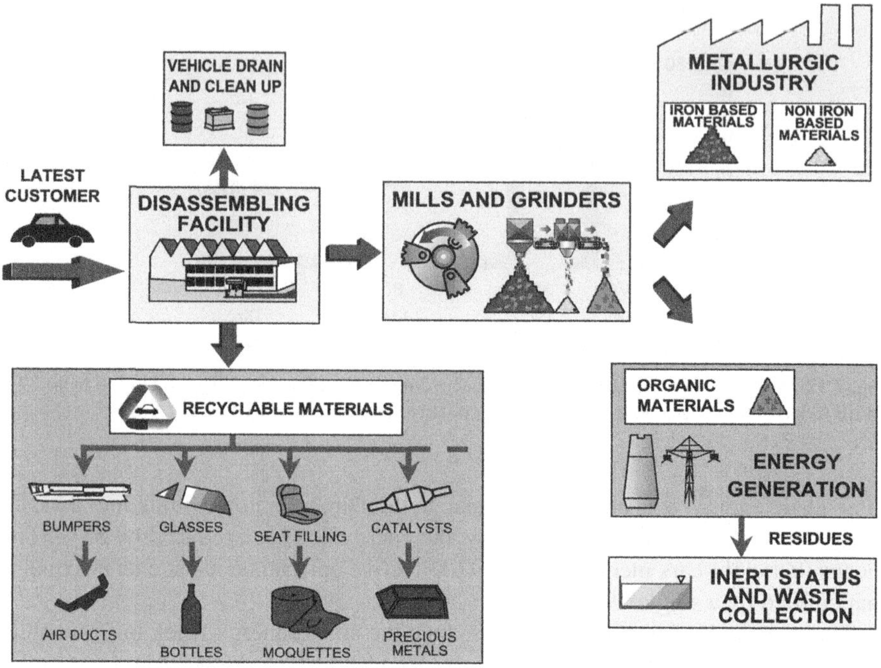

Fig. 7.13 End of life scrapping and material recycling and related organization (courtesy of FIAT)

parts usually sent to the waste collection system. It is now necessary to avoid that these are simply scrapped, rather than being re-utilized.

With this goal, at the end of the 1990s, the European Commission approved a Directive which defines responsibilities and objectives related to the quantity to be recycled. The correct scrapping of cars, at the end of their life, is a shared responsibility of all operators dealing with this activity and the objectives of recycling are the following:

- From 2006/2007: re-utilisation of 85 % in mass of the car as a minimum; 80 % with mechanical recycling, and 5 % maximum with energetic use;
- from 2015: re-utilisation of 95 % in mass of the car as a minimum; 85 % with mechanical recycling and 10 % maximum with energetic use.

The first step is to disassemble and dismantle the car, thanks to a net of operators able to disassemble the components which can be recycled by other Companies specialized in well identified materials (i.e. glasses) and interested in their transformation (Fig. 7.13). As a next step, cars are sent to special mills or grinders to separate and collect metals. Than the organic residue (plastics, elastomers, adhesives, noise absorber, etc.), up to now not collected, is burned, using suitable combustion processes, with energy recovery, exploiting the high heat value of this kind of materials.

The final car recycling process determines the lifetime cycle of the product managed with a global process which starts from the initial vehicle design, already meeting some specifications in view of the final recycle process. Materials re-utilisation allows proper conclusion of the lifetime period and allows opening of a new one, producing new components suitable for new cars. Primary material consumption is therefore minimized and also environmental problems are more easily solved.

7.7 Car Safety

It is well known that an accident is the result of a number of interrelated factors, which must all be considered to improve global safety.

Statistics show that:

- The driver (the human factor) is responsible for most (80–90 %) accidents;
- road configuration (the infrastructure) is in second place with a share of 7–10 % and
- mechanical deficiencies, mechanical wear or poor maintenance, are responsible for the remaining 3–5 %.

In many cases, driver's errors should be minimized with a better configuration of the road profile or its characteristics (e.g. effective separation of traffic flows, suitable structures along the road border to reduce crash intensity, signals, visibility, etc.). Another positive effect can be achieved by using active safety devices and preventive measures as it will be further discussed.

Only the aspects linked with the vehicle and the related safety features and Regulations will be discussed here.

The safety of a vehicle is usually subdivided into active and passive safety .

7.7.1 Active Safety

Active safety includes all those design specifications, equipment and systems installed on the vehicle with the goal of avoiding conditions which can cause or make easier the occurrence of an accident. It can be further sub-divided into driving safety and preventive safety.

Driving Safety

A first contribution to safety is applying a suitable body and chassis engineering, like body stiffness , suspension geometry, steering and brakes design. Further devices are today able to improve road-holding performance, like power-steering , power-brake , Anti-lock Braking System (ABS) , Dynamics Stability Control (DSC) , proximity radar detection systems , etc., and to assist the driver, like the navigation and communication systems.

Furthermore, many items are intended to help the driver to have a clear picture of the situation, like the window design, visibility angles, avoidance of improper reflections, defrosting capability, head lamps, turn signal devices, indicator lights and to keep the vehicle in suitable conditions and in a good state of maintenance and reliability, like an adequate presentation of the owner's manual.
Preventive Safety.

Items considered in this category are those able to reduce the circumstances in which an accident may occur. They deal not only with impacts with other vehicles, but also are aimed to preventing minor inconveniences suffered by occupants, due to lack of attention or due to a not sufficient car knowledge. This category includes the following devices:

- Warning lamps to alert if safety belts are not locked;
- external mirrors and devices suitable to improve visibility during manoeuvres;
- items able to improve the driver's comfort and driving concentration; within this category the air conditioning system is one of the key features;
- systems to avoid door opening from the inside, mainly due to an accidental actuation by a child. The rear doors are fitted with a safety latch having two positions selectable only when the door is open; when the latch is in its locked position, the door may only be opened from the outside;
- safety devices and operation strategies applied to power the windows, the sunroof and the folding top actuations. As it is easily understood, it is important that these movable parts stop operating when an obstacle interferes with the movement itself (i.e. a child interfering with a power window).

7.7.2 Passive Safety

Passive safety includes all those design specifications, equipment and systems installed with the goal of reducing the consequences of an accident after it has taken place. Passive safety can be further subdivided into interior safety and exterior safety.
Interior Safety.

The aim of interior safety is to provide the driver and passengers with an adequate protection and surviving space after the accident has taken place. A suitable structural behavior of the car body is that of offering a non deformable survival structure surrounded by external components provided with controlled deformation and energy absorbtion characteristics. The geometrical intrusion toward the occupants of rigid parts, like the engine, must be avoided. The engine hood must not intrude the survival cell through the windscreen. The steering column and the pedal support structure must be provided with a controlled deformation and a limited intrusion. Similarly, minor components, control levers and electrical actuators must be located in places where they cannot be dangerous to the car's occupants. As an alternative, and often in addition, it is possible to design them so that they can collapse above a certain load, as usually done for the inner rear view mirror. Finally, occupants must be protected

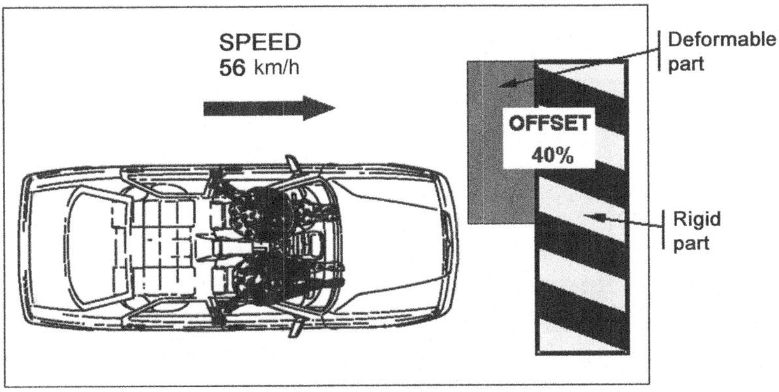

Fig. 7.14 Frontal impact: experimental layout required by the EU Directive 96/79 (courtesy of FIAT)

with suitable shields against high temperature parts. Electric conductors and gasoline pipes must be protected so that they cannot cause fires following a collision.

Having achieved the controlled geometry described above, suitable restraint devices must be installed: safety belts and related pre-tensioners, air bags, head restraints, hinges with improved resistance, effective seat fixing locks and devices to retain the luggage in the trunk.

Exterior Safety

Exterior safety refers to the car's exterior shape whose design should not be aggressive but should be able to limit the damages caused to other vehicles against which the relevant vehicle might collide. Special provisions are now required to limit the damages caused to pedestrians against which the vehicle may collide.

7.7.3 Crash Tests

The European Directives, dealing with passive safety, include two types of tests :

- Frontal impact against a deformable fixed barrier, with a 40 % offset, at a speed of 56 km/h (Fig. 7.14). The amount of offset is defined to expose to the crash only a part of the vehicle structure obtaining, in this way, a more severe test. The goal is to make the test itself more representative of the real accidents taking place on the roads between two cars running in opposite directions.
- Side impact of a deformable barrier against a non-moving vehicle which represents a typical impact in a downtown city intersection (Fig. 7.15).

A geometrical evaluation of the crashed structure takes place after the crash, and the steering wheel intrusion into the passenger compartment is measured. This has been one of the first crash criteria and is still used today.

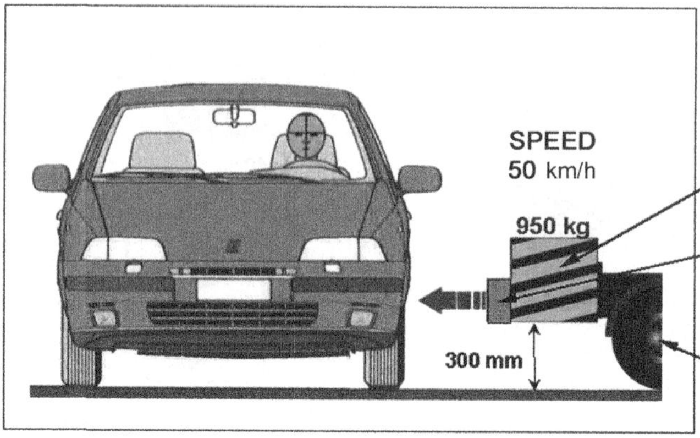

Fig. 7.15 Side impact: experimental layout requested by the EU Directive 96/27 (courtesy of FIAT)

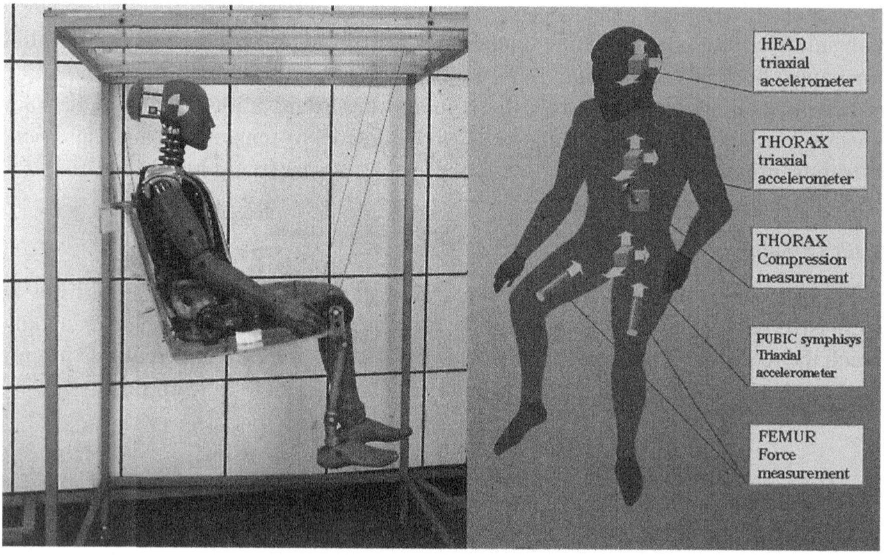

Fig. 7.16 Dummy configuration for bio-mechanical crash criteria

During the 1980s, crash tests started being evaluated by using dummies (Fig. 7.16) and bio-mechanical criteria. Dummies are measuring devices having a human shape and containing sophisticated electronic instrumentation able to record the type and severity of impacts suffered by the simulated human body. Dummies reproduce different kinds of human bodies: adults, both man and woman, and children.

Based on bio-mechanical criteria, a vehicle complies with the Directive if the impact severity recorded by instruments does not exceed the bio-mechanical limits applicable to different human segments (head, neck, thorax, pelvis, legs). As an

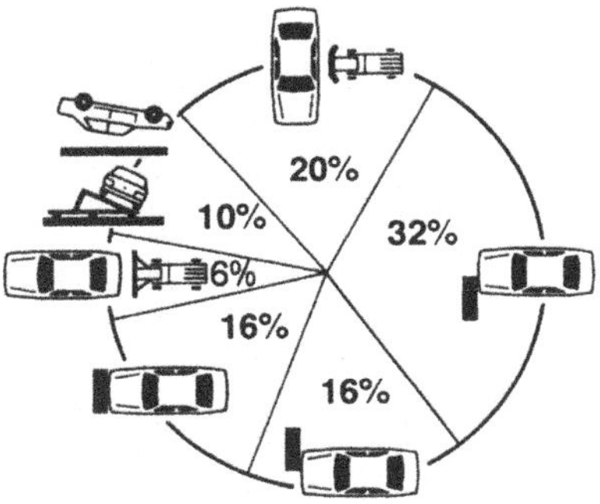

Fig. 7.17 Various crash types and their related probability (courtesy Bosch)

example, for a frontal impact, some bio-mechanical criteria, at 0 ms, are listed below with their indicative values, which are continuously revised:

- Head acceleration < 80 g;
- neck traction stress < 3.3 kN;
- neck shear stress < 3.1 kN;
- neck bending moment < 57 Nm;
- thorax compression criterion < 50 mm;
- femur compression stress < 9.07 kN;
- tibia compression stress < 8 kN;
- movement of sliding knee joints < 15 mm.

Similarly, bio-mechanical parameters are measured also in side impacts.

In addition to the above mentioned legal crash tests every manufacturer is paying much attention to achieving additional improvements, beyond legal requirements; an additional number of laboratory tests has been defined to evaluate many different crash situations according to their probability of occurrence. These tests are applied to vehicle prototypes during their development phase, but also at the start and during production, to verify that initial requirements are still met.

The most used ones are shown in Fig. 7.17:

- Offset frontal impact (right and left side), at a speed higher than the one required by regulation to allow a safety margin;
- side impact against a pole or a car to verify body deformation in case of high and localized load;
- central frontal impact, at medium speed to verify activation of passenger restrain devices;
- vehicle rollover to verify body deformations and stresses to passengers. It allows to verify, in addition, the possibility of providing a quick and effective first aid to passengers after a collision, and to assure the integrity of the fuel tank and fuel system;
- rear impacts at medium and low speeds (4–8 km/h) to verify bumper integrity when this small impact takes place.

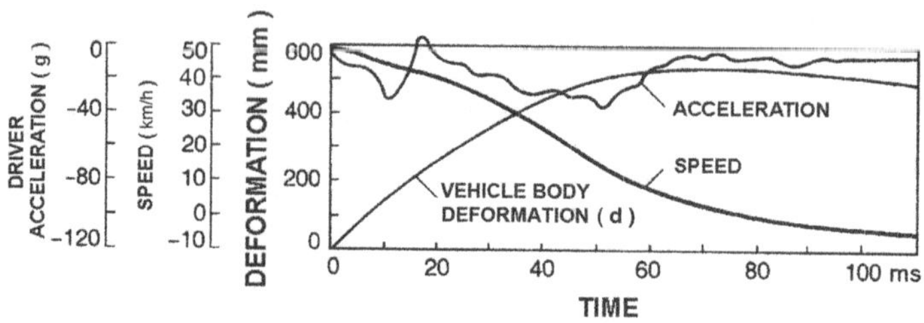

Fig. 7.18 Basic parameters entering the front impact severity evaluation (courtesy Bosch)

Consider now a frontal impact at 50 km/h. Figure 7.18 shows a typical record of the acceleration, deformation and speed of the point to which the safety belts are fixed. This point represents the conditions experienced by the driver.

Starting from the instant in which the vehicle enters into contact with the obstacle, the impact lasts, as an average, for 100 ms; the driver's seat reaches an acceleration peak close to -40 g while, in the same conditions, the front section of the vehicle body records typically an acceleration of -120 g; the body deformation, measured at the end of the test, is 500 mm. During the collision the vehicle absorbs a certain quantity of the initial kinetic energy; the remaining one is dissipated against the obstacle and as elastic return of the crashed parts.

For a suitable vehicle body engineering, the kinetic energy must be absorbed, in sequence, as deformation energy of the bumper, the vehicle front part, and, only in severe cases, the forward section of the passenger compartment. The deformable elements, with their higher deformations, provide the crash absorption capability and are therefore referred to as sacrifice elements. On the contrary, elements showing lower deformation offer the survival space within the passenger compartment, which, by definition, must undergo limited deformation. To sum up, a vehicle subjected to a frontal crash can be modeled, schematically, as a two mass system and two deformable elements of anelastic type in which the active force is constant and independent from the deformation.

One of the major problems when designing parts having a programmed deformation is represented by the behavior of the mechanical parts during the crash. This is mainly true for the engine which, being extremely rigid, prevents the formation of the desirable deformation or, even worse, can intrude, dangerously, into the passenger compartment. In the case of compact vehicles the solution is usually to install the engine in such a way that it slides below the body during a frontal impact, preventing additional damages.

7.7.4 Pedestrian Protection

Safety regulations dealing with pedestrian impacts are the most recent. Their first consequence is the need for a design of the exterior parts of car body and accessories

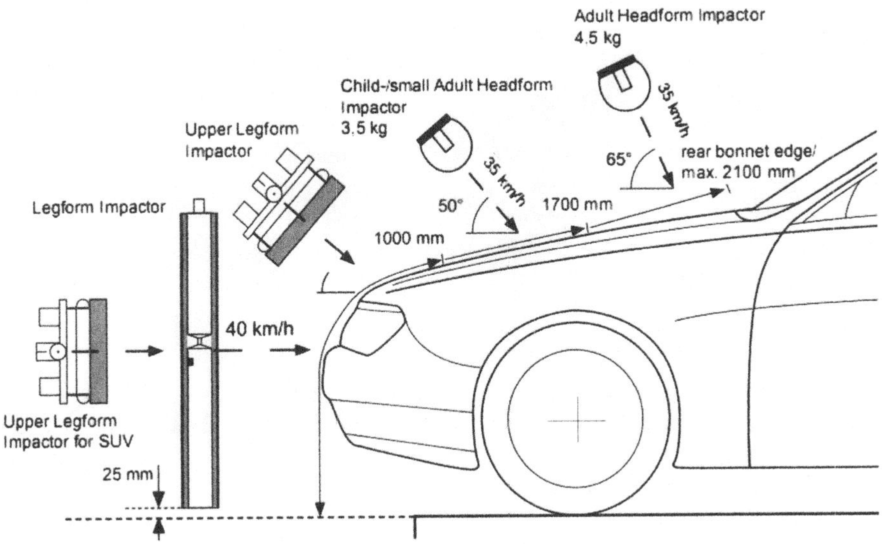

Fig. 7.19 Pedestrian protection: simulation of leg and head impact requested by the EU Directive 2003/102 (from http://www.carhs.de/en/training/order.php)

(side mirrors, handles, etc.) that avoids sharp angles and protruding parts that were so common in the bumper designs of the 1950s.

More recently, to limit the consequences of impacts suffered by pedestrians, the vehicle front elements which, during the impact, interfere with many pedestrian body segments, as shown in Fig. 7.19, had to be re-engineered. Within this test, some specific mechanical components simulate the body segments with special regard to injuries likely to take place to the legs and the head according to the pedestrian average size (adult or child).

A satisfactory result is achieved by designing bumpers with a continuous and smooth shape made with suitable plastic material having internal stiffening ribs able to adequately dissipate the impact energy. Adequate materials, able to absorb the impact energy, like structural polyurethane foam, must be inserted between rigid components, like the engine, the front cross member, etc. and the exterior surfaces. Suitable space must also be provided to absorb the impact energy with constant deceleration.

7.7.5 EURO-NCAP Program

A more severe impact evaluation has been performed according to the Euro NCAP (New Car Assessment Program) defined in 1996. This program is sponsored by the Consumer Associations, European Commission and some member States.

For example, the frontal impact against a deformable barrier is performed at 64 km/h. The final evaluation of crash consequences on the bio-mechanical body segments of the dummies is expressed with a qualitative evaluation (good, adequate, marginal, weak, not sufficient).

The overall evaluation is expressed with stars, from 1 to 5, and includes a judgment from technical experts who examine the vehicle's overall behavior during the crash (vehicle body deformation, air bag activation, etc.). These initiatives activate the Manufacturer's technical efforts in the direction of continuing improvements.

In the period 2008–2012, Euro NCAP included the evaluation of additional features, such as:

- Active systems: ABS, TCS, ESP, etc.;
- emergency brake assistance during pedestrian impacts;
- Self Aligning Head Restraint (SAHR) to prevent the whiplash which determines the neck shear stress;
- other safety devices, if significant.

Chapter 8
Body

The *car body* is the external shell of the vehicle designed to accommodate and protect the people on board and their luggage from the external environment (unfavorable climate, noise, etc.) and the dangers motion may involve. To perform this task, the body must be designed to comply with a number of specifications, and in particular the car body must:

- Withstand the external loads, mainly due to road surface discontinuities and air resistance, and the internal loads due to mechanical components attached to the body itself and needed for proper vehicle operations (engine, gearbox, fuel tank, etc.). These loads must not cause stresses or strains that can produce failures or trigger fatigue phenomena detrimental to vehicle durability.
- Withstand the above mentioned loads, with structural deformations remaining within the elastic range and not exceeding allowable limitations.
- Protect occupants in case of crash, of predetermined intensity, against other vehicles or fixed obstacles.

The components and subsystems constituting the vehicle body can be classified using the following definitions:

- *Body structure*: is the structural part of the body, mainly made of steel box-type sections and pillars, capable of contributing to the final mechanical strength and stiffness. Often, today, its configuration is a space frame made of pressed steel thin-wall elements.
- *Body shell*: is the structural panelwork, mainly steel panels, capable of contributing to the final mechanical strength and stiffness. It constitutes the external skin of the vehicle and accommodates the mechanical elements and the payload.
- *Body frame*: is the structural lower part of the body (underbody), to which the main mechanical components are attached.
- *Movable parts*: they are steel components which can be removed without destroying welded joints. They are connected to the body with hinges; i.e.: doors, engine hood or bonnet, luggage trunk or boot etc. In their closed position, they are capable of contributing to the final strength and stiffness.

G. Genta et al., *The Motor Car*, Mechanical Engineering Series,
DOI: 10.1007/978-94-007-8552-6_8,

- *Aesthetic parts*: they are mostly minor components now usually made of plastic materials (i.e.: aesthetic profiles and spoilers). The shape of the steel panels has, for sure, also an important aesthetic function.
- *Accessories*: they are devices capable of providing specific functions such as door locks, lateral window mechanisms, air conditioning, etc.

All above mentioned definitions have a certain degree of flexibility as to their form and function, and can vary with different manufacturers. This classification however is useful to give an idea of the complexity of the car body.

8.1 Road Loads

Road surfaces induce random and asymmetrical loads on the body. Each wheel follows a road profile with irregularities characterized by different distributions and geometry at different times and this determines, for each wheel, a specific position and, as a consequence, a specific load acting on the body as a result of suspension spring flexibility. Due to unavoidable obstacle asymmetry, these loads cause torsional stressing of the body. In addition, the masses located on the body are not symmetrical with respect to the geometric symmetry plane (xz plane), causing unsymmetrical inertia loads on the structure.

It is therefore of paramount importance to evaluate how the body reacts to forces determining torsional and bending deformations. In addition, considering the body as a reticular structure, deformations determined by torsion will cause variations of angles in the grid of the reticular structure, or, in others words, variations in its dimensions, (for instance, those defining the deformation of the door ring are shown in Fig. 8.1b) and, as a final result, variations of areas/volumes included within the structural segments.

To define the loads acting on the vehicle body, the payload must be specified. Once the payload mass and position are known, since the vehicle's unloaded mass is also known, the load applied to each axle can be computed. In addition, the most frequently used accessories must be considered.

During the vehicle type approval, the following weights must be declared: maximum loaded weight (or Gross Vehicle Weight GVW), maximum weight on each axis, weight in running order (with all fluids filled up: curb weight) and the number of seats. In spite of the fact that European rules contemplate the use of SI, the vehicle weight is still often expressed in kg instead of being expressed in N. More correctly, the mass of the payload and of the vehicle in the various configurations should be specified.

The loads and the static load distribution of the Alfa Romeo 166 2.4 JTD is shown as an example in Table 8.1 for different load conditions. Payload increase (number of passengers) determines a higher share for the rear axle.

When determining the static load distribution it must be considered that the vehicle can rest on only three wheels, like when a wheel is placed on a kerb high enough

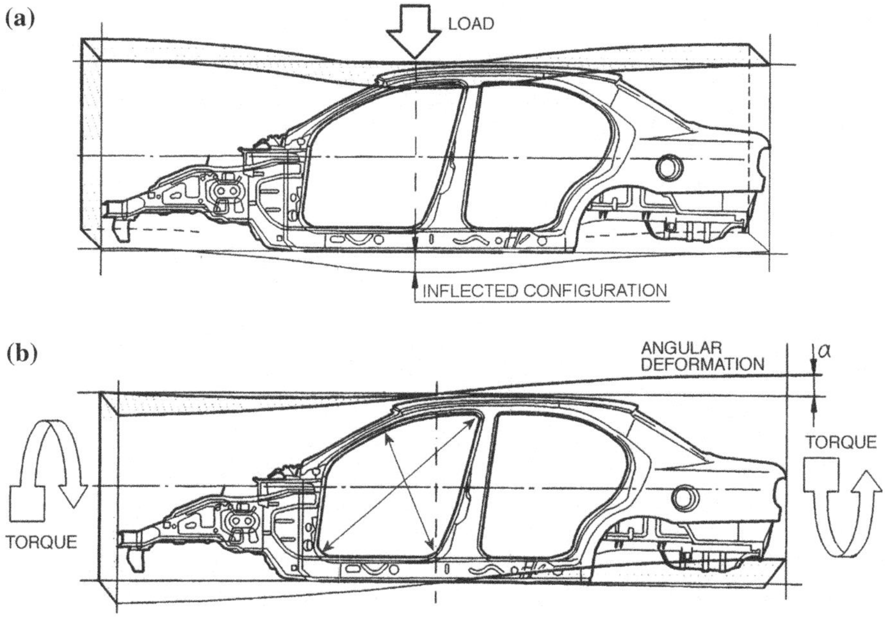

Fig. 8.1 Scheme for determination of the bending (**a**) and the torsional (**b**) stiffness. The *arrows* in (**b**) show the measurements defining the deformation of the door ring

Table 8.1 Loads and the static load distribution of the Alfa Romeo 166 2.4 JTD

	Mass		Axle weight		Distribution	
	Payload kg	Total kg	Front N	Rear N	Front %	Rear %
Empty	0	1,490	8,731	5,886	59.7	40.3
Driver	75	1,565	9,074	6,278	59.1	40.9
2 Passengers	150	1,640	9,418	6,671	58.5	41.5
3 Passengers	225	1,715	9,575	7,250	56.9	43.1
4 Passengers	300	1,790	9,732	7,828	55.4	44.6
5 Passengers	375	1,865	9,908	8,388	54.2	45.8
Maximum load	510	2,000	9,908	9,712	50.5	49.5

Number of seats: 5; curb mass without driver: 1490 kg; maximum payload capacity: 510 kg; maximum loaded mass: 2000 kg. These values are unique for each vehicle/engine combination and do not consider the accessory combination actually present

to cause one of the wheels to lose contact with the road. In this case all the axis load is transferred to a single wheel. Dynamic loads may be even larger than static ones. Centrifugal forces applied in the vehicle center of mass, for instance, can reach a value up to 100 % of the weight of the sprung mass (sprung weight). Similarly, dynamic forces must be considered when a road bump is encountered. For example, when the wheel overtakes a bump 100 mm high at 50 km/h, the load on that wheel may have a 5-fold increase.

Table 8.2 Values of the bending and torsional stiffness of some passenger cars

Model	Stiffness	
	Bending (kN/mm)	Torsional (kNm/rad)
Alfa Romeo 166	6.73–7.14	759–1,340
Fiat Nuova Punto 3 doors	8.90	720
Fiat Nuova Punto 5 doors	7.70	665
Fiat Marea sedan	6.10–6.20	550–930
Fiat Marea station wagon	5.29	700

8.2 Body Stiffness

The torsional and bending stiffness is computed using mathematical models and then measured experimentally using a suitable rig, that applies known loads to the body at the points where the suspensions are connected, and at the center of mass. For measuring the bending stiffness, four loads are applied at the ends of the two axles while the body is constrained at the center of mass; for measuring the torsional stiffness one of the axles is constrained while a pure torsional load, i.e. two equal and opposite forces, is applied at the ends of the other axle.

It is a common practice to measure the deformations at the points where the suspensions are connected with the body to compute the stiffness as the ratio between the force (or the torque) and the deformation. Other deformations are also important, like the changes of the diagonals of the door, hood and trunk openings. These data are useful to understand the potential problems related to the working of these parts and to the noise in actual use.

The *bending stiffness* (see Fig. 8.1a), is defined as the ratio between the load applied to the center of mass and the vertical deformation measured at the same point. The comparison between the results obtained without and with movable parts is significant. For the torsional stiffness the difference between the two values is mostly due to the contributions of the windshield and the rear window glued to their subframes using structural adhesive. It is also interesting to note the difference between the 3 and 5 door versions, which shed some light on the contribution to the stiffness of the rear part of the body sides that is not affected by the openings of the rear doors.

The *torsional stiffness* (Fig. 8.1b) is defined as the ratio between the torque and the rotation angle of the section in which the load is applied with respect to the constrained section. Some reference values for the bending and torsional stiffness are reported in Table 8.2. Other differences are related to the different vehicle body shape; the best values being those related to sedans having the back of the rear seat fixed (not foldable). In this case the torsional stiffness contribution of the separation panel between passenger and luggage compartment and that of the cross-member below the rear window are important.

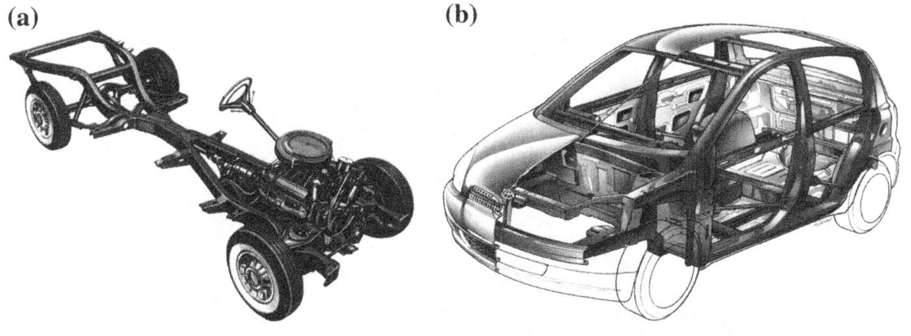

Fig. 8.2 The two more relevant technologies in car body structure construction; **a** the bearing chassis; **b** the unibody

8.3 Body Structure

As seen in Part I, the bodies of modern road vehicles are built following two different structural types, (Fig. 8.2): the *bearing chassis* and the *unitized body*, shortly unibody. If the body is joined to the bearing chassis without rigid mounts, the final torsional stiffness of the vehicle is usually poor, unless the chassis is provided with heavy stiffening elements. In the latter case, however, there are severe difficulties in obtaining the required compliance of the longitudinal members, which is important to insure the prescribed crash safety. For this reason the bearing chassis, after an initial success when the majority of cars were built accordingly with this configuration, was later abandoned in favour of the unibody approach. Solutions based on the bearing chassis are on the contrary widely used for industrial vehicles.

Crashworthiness, weight and a better stiffness/mass ratio are the actual advantages of the unitized body, Fig. 8.2b); for these reasons this configuration can be considered the present-day state of the art of car body technology.

For unutized body structure the definition *body frame* (see Fig. 8.3) is used to identify a semi-finished assembly that includes the floors, with related cross-members, the front rear longitudinal members and the struts.

Also in this case, the same body frame can be shared by many body configurations. However, in this case it is not correct to speak of a bearing chassis, since pillars and roof cross members are essential to obtain the required stiffness and strength of the unitized body. The body frame subassembly can be delivered to external coach builders to be completed, provided that they are able to complete the body frame with a body that not only has the required shape, but also is capable of achieving the final stiffness required according to the original project.

The design of a body frame has, normally, a longer lifespan than that of other parts that provide an aesthetic function. In other words, often cars of subsequent generations are based on a common body frame. Key sub-groups of the body frame are the central and the rear floor; they are designed as separate units to produce, along

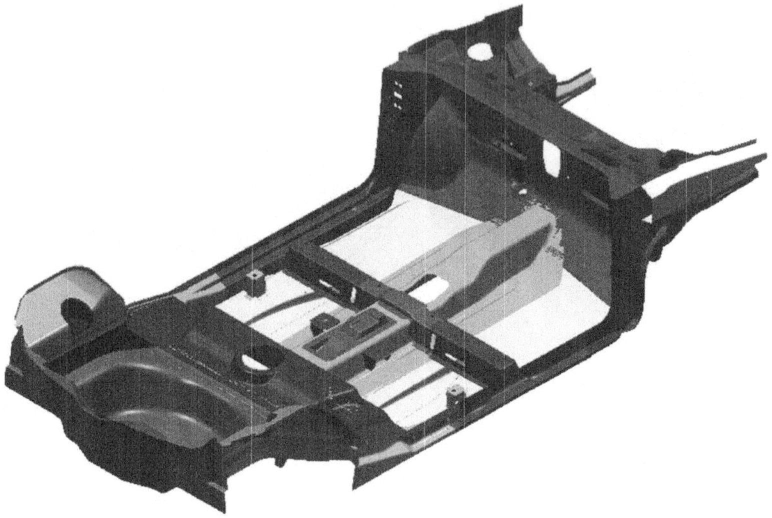

Fig. 8.3 A typical body frame for a unibody of present day configuration (courtesy of Fiat)

the same production line, bodies having different wheelbase or having different rear overhang (i.e.: 2 volumes, 3 volumes and station wagon versions of a given model).

8.3.1 The Unibody

The unibody is a shell of steel box-type sections and pillars welded together that constitute a reticular structure (dark grey elements in Fig. 8.2b) to which the steel panels, shown in light gray in the same figure, are welded. The windscreen and the external panels contribute also to the structural function, but only for what shear deformations are concerned.

In spite of the fact that the unibody constitutes a unique element and none of its parts can be mechanically disassembled, specific denominations are used for the various structural elements to easily distinguish their functions and to better characterize the various components before assembly.

Referring to Fig. 8.4, which represents a small European sedan car (1990–2000), the main components of the unibody can be easily identified

- Elements 1, 12, 6 are the windscreen pillar or A pillar, the central pillar or B pillar and the rear pillar or C pillar, respectively.
- Element 15 is the front bumper cross-member and elements 4 and 9 are the cross-members for central and rear parts of the floor. Usually an additional cross-member, not visible in this drawing, is welded to the front part of the floor.

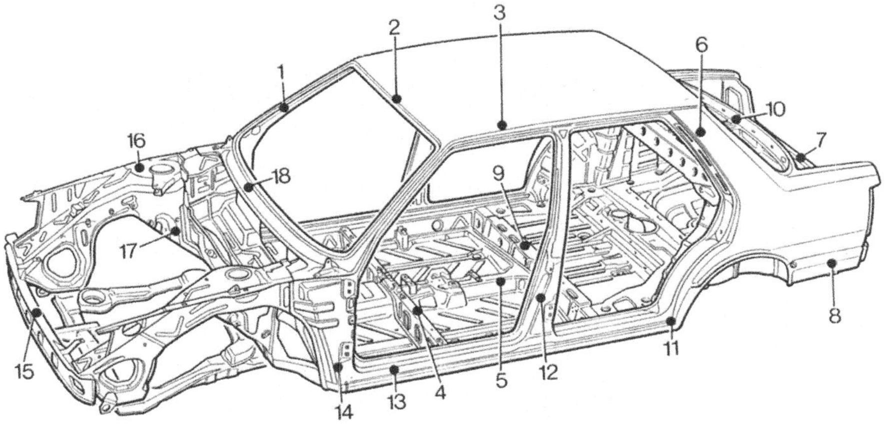

Fig. 8.4 Unibody components (courtesy of Fiat)

- Element 13 is the lateral side member; and element 3 is the roof small-section longitudinal member. The subassembly 3, 13, 14, 12, 11, 6, is often called the door ring.
- Elements 18, 10 and 7 are the cross-member below the windscreen, the cross-member below the rear window and the cross-member of the rear bumper.
- Element 16 is the front strut.

The elements that define the interior space are the roof, the front and rear segments of the floor, the firewall and the rear fenders. The front fenders are not shown in the figure because they are usually assembled with screws so that they can be disassembled. In specific cases this solution can be used also for the rear fenders. This solution has proved to be expedient to facilitate repairs of fenders and mudguards after a crash.

Thin steel panels, welded together, form hollow structures capable of providing a remarkably good structural behavior. Their effectiveness is due to their closed cross-section which offers a good stiffness in both bending and torsional loading. The cross section of some structural members is sketched in Fig. 8.5. The lateral side member (element 13 in the figure) is assembled to the floor using three different steel panels welded together. A second element is welded to the floor and this assembly constitutes a second lateral side member with a smaller section. Similarly, the central pillar (element 12) is assembled with three different panels welded together.

The protection of the occupants in the event of a crash is one of the most important structural functions of the unibody. To understand the severity of this condition it is enough to consider that the kinetic energy

$$E = \frac{1}{2} m V^2$$

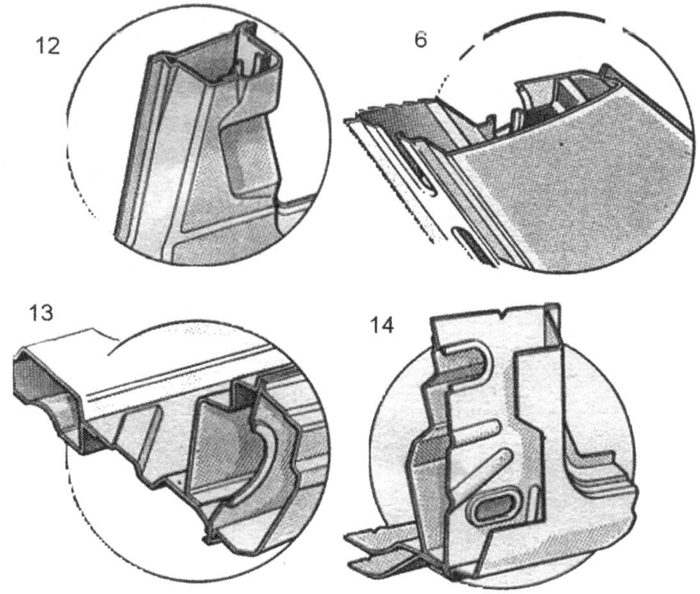

Fig. 8.5 Cross section of some structural members: **14** front pillar; **12** central pillar; **6** rear pillar; **13** lateral side member

of a vehicle with a mass of 1000 kg at 60 km/h is 140 kNm, corresponding to the energy the vehicle acquires when falling from a height of 14 m. This energy must be dissipated with no injury to the occupants. Since the impact energy increases with the square of the speed, it is impossible to build a car safe in the event of a crash against a fixed barrier like the one shown in Fig. 8.6, at any speed. Crashworthiness of cars is strictly regulated by standards to which all manufacturers must comply.

This protective action is assured by building the body in two conceptually distinct parts: the front part that can be sacrificed and the central part that constitute a *survival structure* acting directly as a passenger protection: All important safety components like seats, safety belts, air bags, etc. are directly mounted on the latter. The front part, being crushed during the crash, absorbs and dissipates a significant part of the vehicle kinetic energy, allowing the following part of the body to react to the shock forces without permanent deformations. In this way the vital space, available for the occupants, remains substantially unchanged during the impact.

The restraint components (safety belts, airbags) avoid impacts of the occupants against the passenger compartment structures. During the crash the passenger compartment is subjected to a deceleration which can reach values up to $1000\,\mathrm{m/s^2} \approx 100\,\mathrm{g}$ and the restraint components must be accordingly designed.

Fig. 8.6 The result of a crash test against a fixed barrier at 60 km/h performed on a Fiat Uno (courtesy of Fiat)

8.3.2 Unibody Components

Since different solutions for body construction are implemented by different car makers and in different models of the same maker, in the present section a short description of the unibody components of a specific car is presented. The vehicle taken as a reference is a typical Fiat mid-size car (2 volumes/five doors/hatchback). The aim of this section is to illustrate both the simplicity and the complexity of the unibody assembly, always remembering the design goals of body engineering: low weight, high stiffness, low production costs, good crash performances. The sequence of the unibody assembly, starting from the initial lower body frame, designed to support all other components while satisfying the crash-worthiness requirements, is summarized in Fig. 8.7. The side members, the windscreen ring and the roof components are then added; the body is then completed with the moveable parts (doors, fenders, bonnet and rear hatch). After the painting process the body is finally ready for the interior trimming .

The unibody fabrication scheme can thus be detailed as follows. The complete body frame is obtained by assembling the following subassemblies: front frame, floor, right/left side members, rear panel, headlight crossbar. The subassemblies are loaded into a positioning fixture either by hand or mechanically; spot welding is performed in the same way. The body frame is completed on an automated robot welding pallet line: geometry of the assembly is inspected in an optoelectronic station on the completion line.

Side panels are assembled on a separate line. Each side panel is obtained by assembling the following subassemblies: outer skin, upper strut insert, rear light bottom, rear side lower insert, fuel filler, bottom, windscreen pillar reinforcement, central pillar reinforcement, interior side frame, front pillar reinforcement. Also in

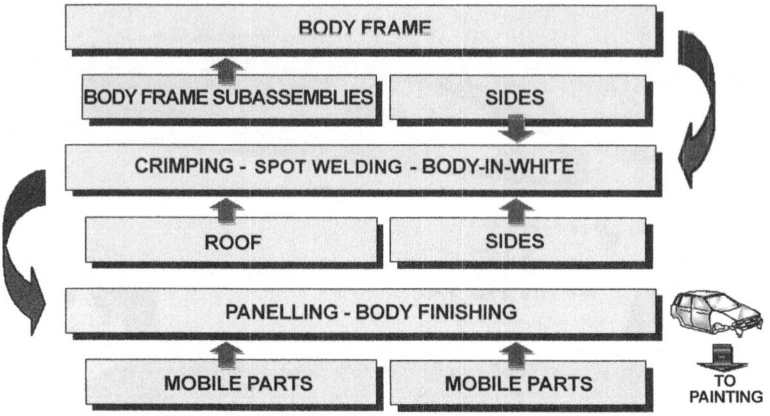

Fig. 8.7 Body frame fabrication scheme

this case, assembly is completed on a robot welding motorized pallet line and off-line spot inspection of the assembly geometry is performed by an automatic inspection machine.

The process then goes on with the assembly of the final body (Fig. 8.8) by tabbing, tacking and completing the body consisting of chassis, side panels and roof ribs. The body is completed by fitting the roof in the second step. This allows welding robots to work inside the body until the very end without hindrance.

Body Frame and its Sub-groups

Following the mentioned organization, the components lasting more than one generation are central and rear floor, instrument panel, *firewall*, front struts (two symmetrical) and rear struts (two symmetrical): They are the bearing structure of the already mentioned *platform*. The different components are welded to each other. Many other small components welded to a number of sub-groups cannot be shown in Fig. 8.9; they are mostly fasteners, fixing devices and reinforcement brackets.

The front strut has the structural function of supporting the front suspension, and a part of the powertrain group. In addition it cooperates, with its deformation, to dissipate a significant part of the kinetic energy in case of a high velocity front impact. A suitable front strut deformation assures a limited deformation of the passenger compartment. The strut and the front junction form a fundamental part of the structural front element. A reinforcement structure and two brackets constituting the connection area for the cross member and longitudinal rails, are welded to this element. The closed structure of the front struts, which complete the longitudinal members with the bracket for the suspension turret or dome (only partially represented), are connected to the previous elements. The suspension dome is, in addition, connected with the instrument panel to improve the connection with the different elements. The

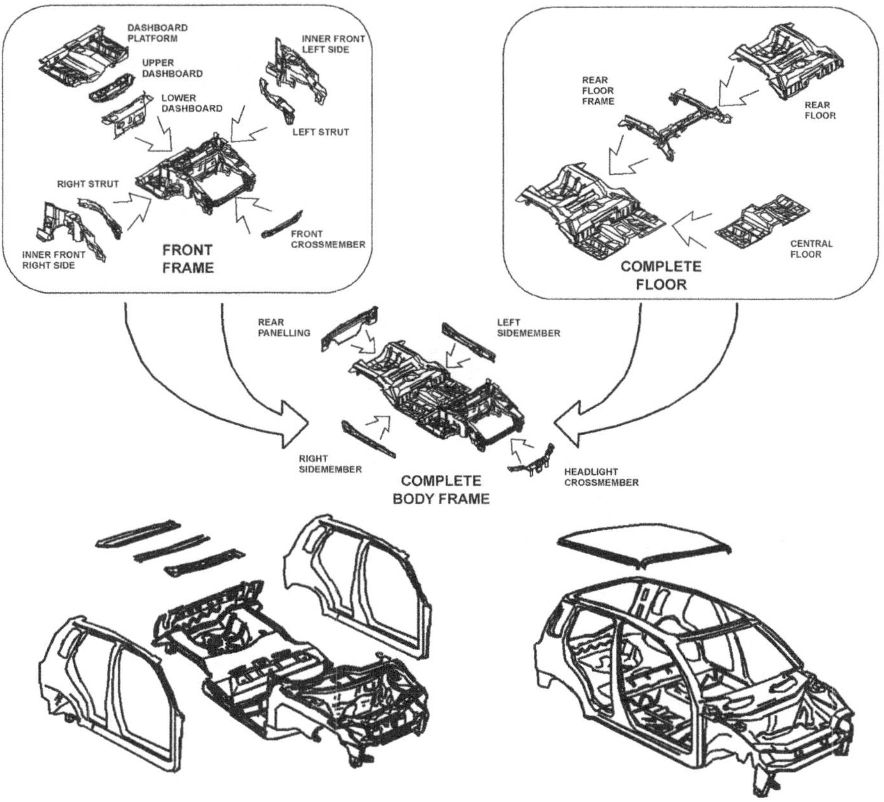

Fig. 8.8 From the body frame to the final unibody assembly (courtesy of Fiat)

two parts of the front pillar, contributing to the dome assembly, that contains and supports the spring—shock absorber assembly of the Mc Pherson suspension, are welded on the internal side.

The various body parts are here listed, in the order they are assembled along the assembling and welding line. First of all, the central floor which consists of the following parts:

- Two lateral floor parts are connected to a central part forming the tunnel. This has a stiffening function in addition to containing and protecting the exhaust pipe from collisions with the ground. Splitting the floor into three parts is preferable since it would be hardly feasible to make the floor in one single part and, in addition, longitudinal boxes can be formed in the joint areas, which are particularly useful in distributing the loads onto the rear structure in the event of a collision.
- Two transversal members are joined to the central part to form the central cross-piece. These parts will later be welded to the sides panels to form a solid structural assembly, capable of supporting the seats and absorbing side collisions.
- Local reinforcements are used to secure the seats and the handbrake.

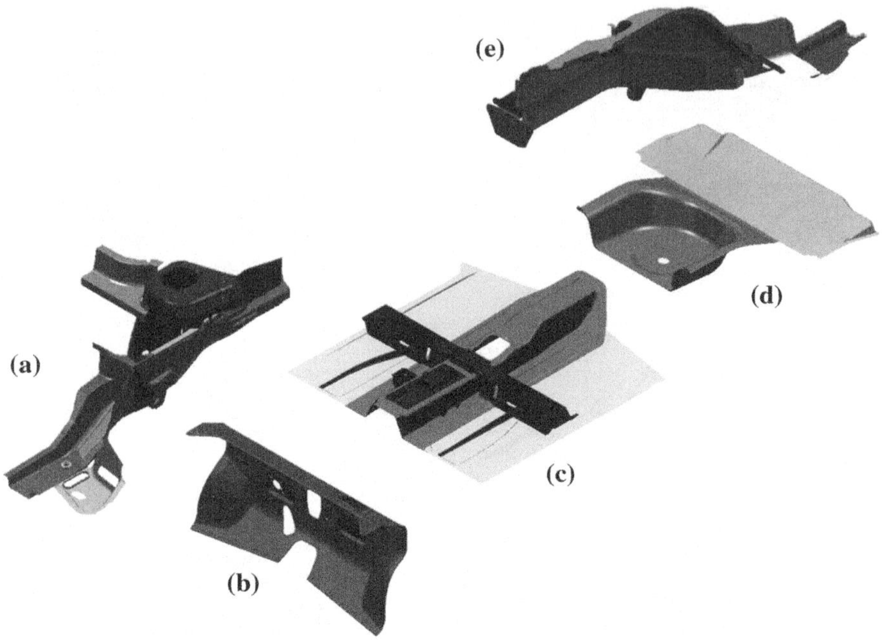

Fig. 8.9 The main sub-groups of the body frame: **a** front strut; **b** dashboard panel or firewall; **c** central floor; **d** rear floor; **e** rear sidemember

The rear side member is a relatively complicated part because it must be well connected to the wheelhouse to form a closed section. The functions it performs include supporting the rear suspension anchoring points (a torque axle in this case). It is an expendable part capable of absorbing the energy in the event of a rear collision.

The complete rear side member consists, in this case, of two parts joined together and welded to the rear floor. The side support is used to connect it to the two side panels, ensuring structural continuity. The wheelhouse is welded to the side and has the 3-fold function of containing the rear wheel and its movements, supporting the upper connection point of the shock absorber and acting as further connection element with the rear end of the side panel.

The central body is joined by the two struts (one on each side) previously described and by the dashboard (or firewall) whose function is to separate the engine from the passenger compartment. The front and rear parts are added to form two more crosspieces together with the floors. Moreover, two pressed parts are welded under the floor to form two additional central side members.

At this point, the structure is completed by two end side members, on the front and rear ends, by three crosspieces under the floor and by four side members, two central and two on the sides of the body.

The body frame is a complex assembly and the utmost care is needed, during its assembly, to obtain the required dimensional tolerance because many points will be used as geometric references for assembling other chassis parts, such as the suspensions and the engine mounts.

The body frame is therefore generally assembled in a fixture allowing positioning of all the main structural components precisely and forming of a reference frame. The welding fixture is usually a multiple jig, i.e. a press with electrodes for making most of the weld spots in a single step. Therefore, the welding electrodes are fitted in the right position on a mobile elements which closes on the fixture instead of being fitted on welding guns. The closing panels are then fitted on a second fixture. These consist of the dashboard, the central floor (in three parts) and the rear floor. The latter consists of two parts and also forms the spare wheel housing. Intermediate side member ribs (external part visible from the outside of the door ring) and two further front side members arranged on top of the ones already described, are added. These are used to support the front mud guards and also provide a structural contribution with the side members in front crashes. A further part connects the dashboard to the side members forming the inner, lower part of the box structure that will eventually form the front pillar.

Finally, a crosspiece closes the spare wheel compartment and forms the inner part of the bumper crosspiece. The body frame shown in Fig. 8.3 is so completed. Its torsional stiffness amounts to no more than 20 % of the total torsional stiffness of the complete body.

Sides, Windscreen Ring and Roof Components

The sides normally consist of a single, external part and various internal parts forming the boxed structure of the three pillars (Fig. 8.10). The inner wall of the lower part of the front pillar and the extended side member are already included in the body frame.

The roof consists of a closing panel which contributes also to forming the upper side member boxed structure (not shown in the figure). The windscreen ring (four parts) is the structural frame of the windscreen and joins the body frame to the two sides and the roof. The windscreen is glued to the body (this solution is now universally accepted) and provides a substantial structural contribution.

The aesthetic parts of the body-in-white are shown in Fig. 8.10 where, for clarity, only the right hand side has been represented. As previously mentioned, they normally include four major components: two symmetric external sides, the roof, the windscreen ring and the rear closing. Parts for separating the boot from the passenger compartment, i.e. rear window shelf and boot bulkhead, in addition to a number of reinforcements, may be fitted in three-box models.

The term *body-in-white* is related to the production process. As mentioned, the body-in-white process includes all the operations related to positioning and welding pressed parts and subassemblies. The term *panelled body* refers to the assembly and the squaring up of mobile parts before painting. This operation is called panelling. A complete body-in-white is shown in Fig. 8.11.

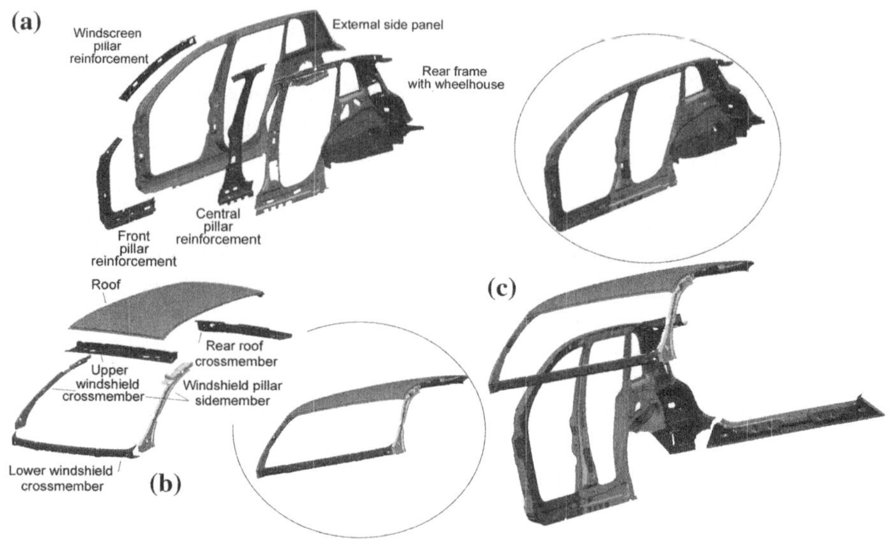

Fig. 8.10 Aesthetic parts of the body: **a** Side ring; **b** windscreen ring and roof; **c** aesthetic parts assembled (only the *right side* is shown) (courtesy of Fiat)

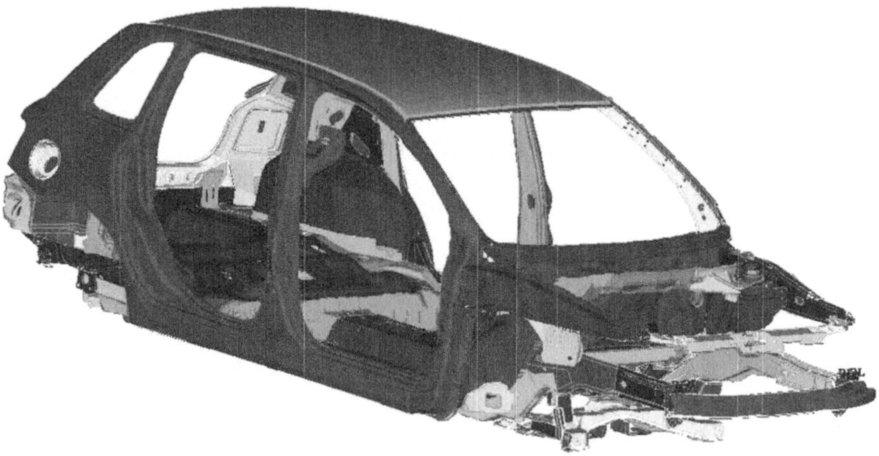

Fig. 8.11 A complete body-in-*white* (courtesy of Fiat)

Moveable Body Parts

The body is then mated to its moveable parts: the front and rear *doors*, the fenders. the bonnet and the tailgate. Some of them are simple parts, such as the mudguard bolted or welded to the body. Others, like the doors, are more complex, connected to the body with hinges and require a more in-depth analysis.

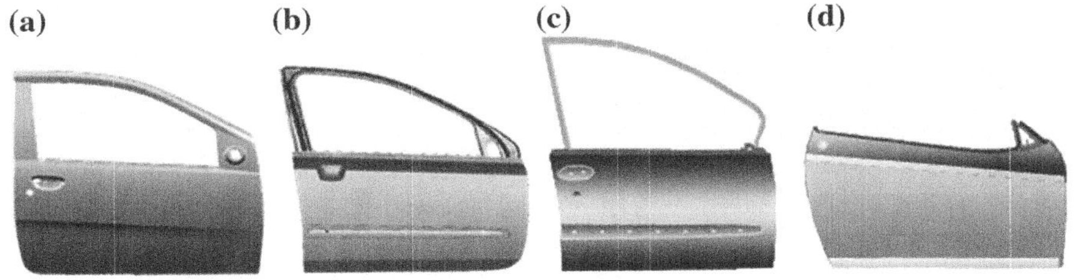

Fig. 8.12 Door layout. **a** Integral solution; **b** and **c** hybrid solutions; **d** solution for convertible cars (courtesy of Fiat)

Door layouts can be subdivided in four basic door types, as shown in Fig. 8.12: the integral solution, where the external paneling (or external skin), obtained from a pressed sheet of steel, is made in one single piece; the so called hybrid solution, with two main realizations: the first, such as the Fiat Stilo case, where the door seals are retained by a particular steel pressed frame and the second where the door seals are retained by a frame that is obtained by extrusion and bending; the fourth solution with no door seals is mainly applied to convertible cars. This classification is just a sketchy one, including the most common types.

The door consists of an internal part (frame) and an external one (skin), the latter being tacked and glued to the frame instead of being welded, to prevent aesthetic problems. It contains also additional mechanical and electric components. The front door frame has a wide opening, to reduce the structural weight and to make it possible to fit the many internal components, such as: locks and commands, window winders, power windows, speakers, etc.

The internal parts are additional lower reinforcements of the window frame: they are needed to secure the hinges and the side intrusion bar for protecting occupants from side crashes. The window seals are normally added only after painting (Fig. 8.13).

The doors are connected by means of hinges which, according to different design and technological solutions, may be either bolted to the body and to the mobile part, welded to the two parts or fastened with a combination of the two: the welded solution is preferable because it does not need further adjustments.

While during panelling the hinges are fitted on their interfaces, doors can be removed for painting and for repairs by removing the pin. The weight of the open door and a person possibly leaning on it can apply considerable stress to the hinge attachments which must therefore be appropriately designed using reinforcements in the points in which they are joined to the body.

Similarly, the tailgate is fitted on hinges to which similar considerations apply: additionally, removable elements, consisting of the tailgate support attachments in open position, are added; in many modern tailgates, the external panelling is incomplete because the rear window forms a structural non-framed closing element.

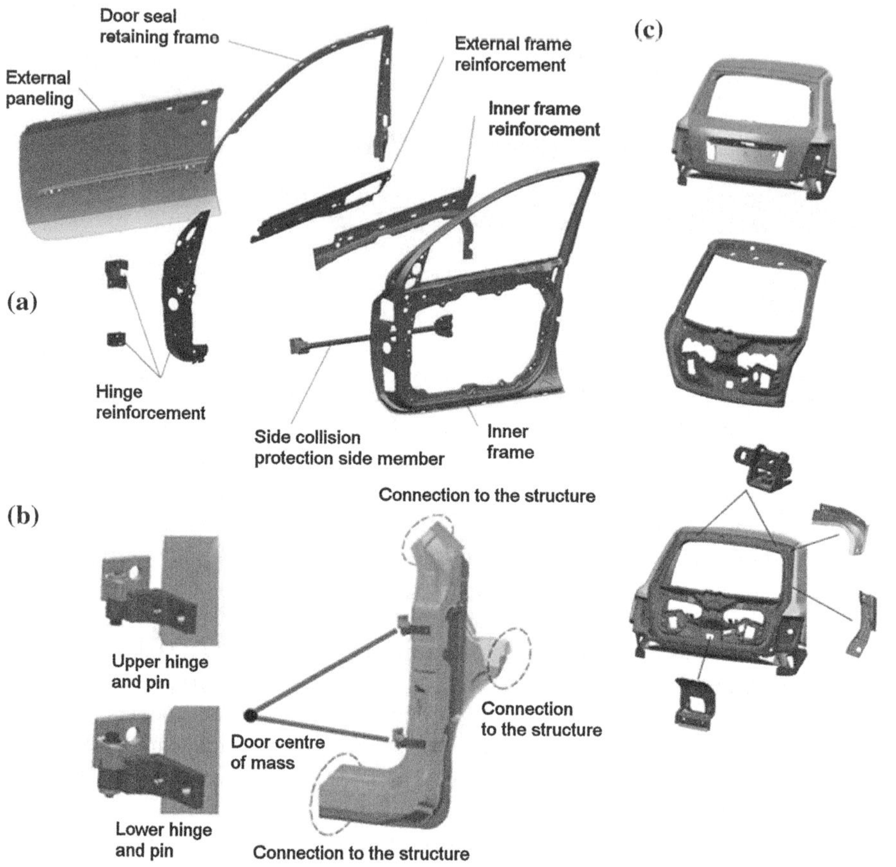

Fig. 8.13 The main moveable parts: a door with its hinges and the tailgate (courtesy of Fiat)

8.4 The Assembly Line

After panelling (assembly of mobile parts), the body assembly process proceeds with a number of crucial steps, such as painting and rust protection, and the body reaches the assembly line where the floor and roof are fitted.

The doors are normally removed from the body; after painting these are completed on a specific line which joins the main line for body assembly after the engine and suspensions have been fitted.

The scheme of Fig. 8.14 illustrates the operations performed on the assembly line and the major systems that are fitted in the car: mechanical systems and finishing systems. The main ones will be examined in the following sections.

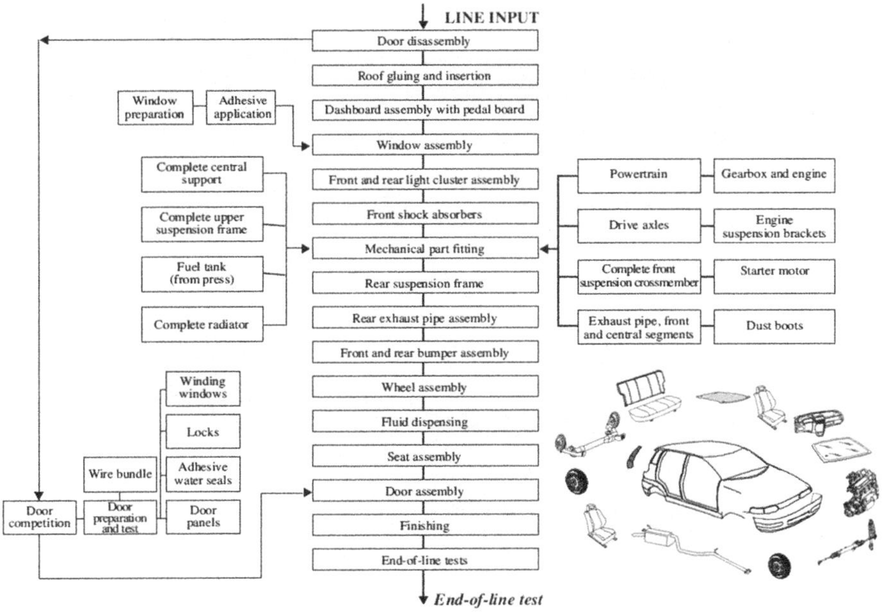

Fig. 8.14 A scheme of the assembly line

8.4.1 Dashboard

The *dashboard* is made of two parts: the insert and the outer structure or external skin. The insert, usually made of plastic material (ABS or similar), supports an outer structure forming a shell. The outer part of the dashboard also performs aesthetic functions: For this reason, its mould is especially designed to confer a better appearance to the component. This surface treatment is called *embossing*.

For foam-filled dashboards, a laminate shell (one side of which is embossed) is introduced together with the previously structured support into another mould. The plastic foam is introduced in the space between the two parts which may be more or less soft according to their density.

The final result is a dashboard like the one shown in Fig. 8.15. This component is then completed by fitting additional components, such as ventilation vents, on-board instruments, air channels, object compartments and flaps.

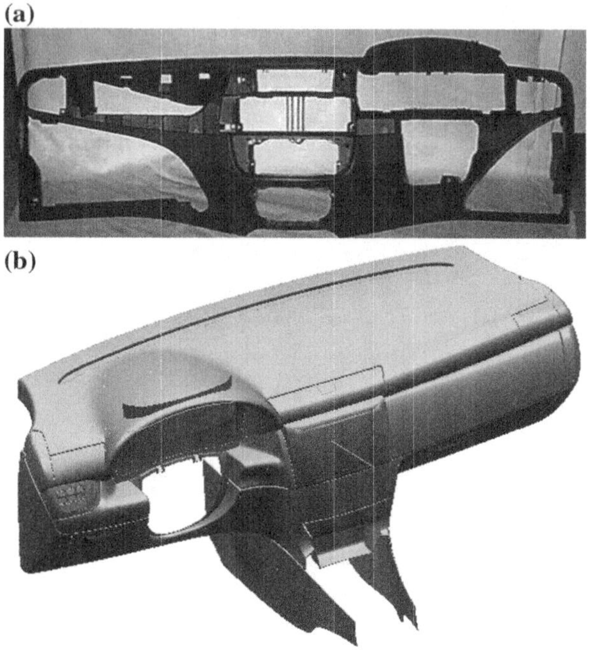

Fig. 8.15 Dashboard. **a** Insert and **b** complete dashboard

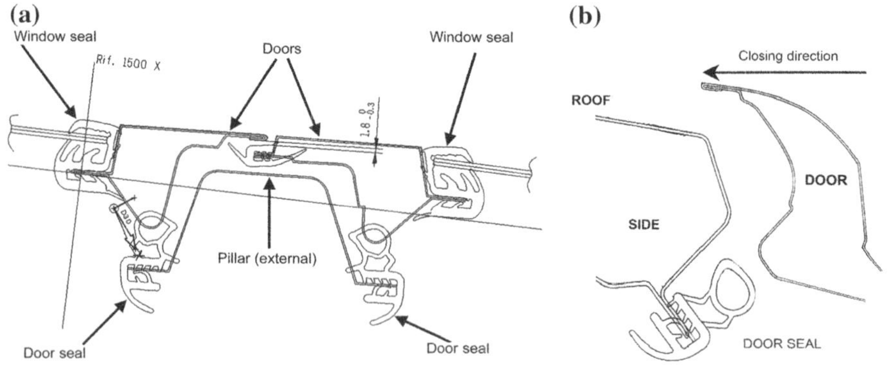

Fig. 8.16 The seals: **a** section on the B pillar; **b** section on the roof (seals are usually drawn according to their non-deformed shape, they have begùfore being assembled)

8.4.2 Door Seals

The *seals* are important trimming components. Two typical cross sections are shown in Fig. 8.16: the first is applied to the central pillar (external part only) and the second to the roof.

In the figure the wind-noise seal is represented in its undeformed configuration, as it appears when the doors are open. The wind-noise seal presses onto the pillar welding seams. It supports the pillar upholstery and makes the joint look trimmer. The wind-noise seal between the front and the rear doors seals the space between the two.

The window scrapers, normally provided with a structural metal core embedded in the rubber, are also pressed and conceal the inner reinforcement welding tab of the window frame. The functions of the seals include:

- preventing wind noise entering through the connection between fixed and moveable parts;
- isolating the passenger compartment from external elements, such as water, dust and noise;
- absorbing the energy impressed by the door when it closes;
- forming a water drip drain on the roof.

The windscreen seal is glued by means of an adhesive placed along the edge between body and windscreen. In the case, the function is mainly cosmetic.

8.4.3 Outer and Inner Additional Parts

The outer part of the vehicle is completed, after assembling the doors, with the assembly of the bumpers and the front and rear light clusters.

The central unit is fitted, along with the pillar upholstery and various interior components designed to improve the comfort and the functionality of the vehicle. Finally, the seats are fitted.

Fig. 8.17a shows a door from the inside after removing the cosmetic inner panel; the *window winder* mechanism is shown in detail. It consists of a parallelogram carrying the window attachment. The sector gear may be operated by either an electrical motor with worm screw or by means of a handle.

The *windscreen mechanism* must be designed to provide a wiping area as large as possible. For this purpose, the two blades can have different dimensions, and must be adapted to the shape of the windscreen, as shown in Fig. 8.17c. This task may also be performed by a single blade, sometimes capable of a motion more complicated than a simple rotation.

The most frequent case is shown in Fig. 8.17b. In this case, an electrical geared motor drives a crank performing a circular motion. The crank in turn drives the two blades in a reciprocating angular motion to by means of two connection rods. Different angular amplitudes can be obtained by using control levers of different lengths for the two blades.

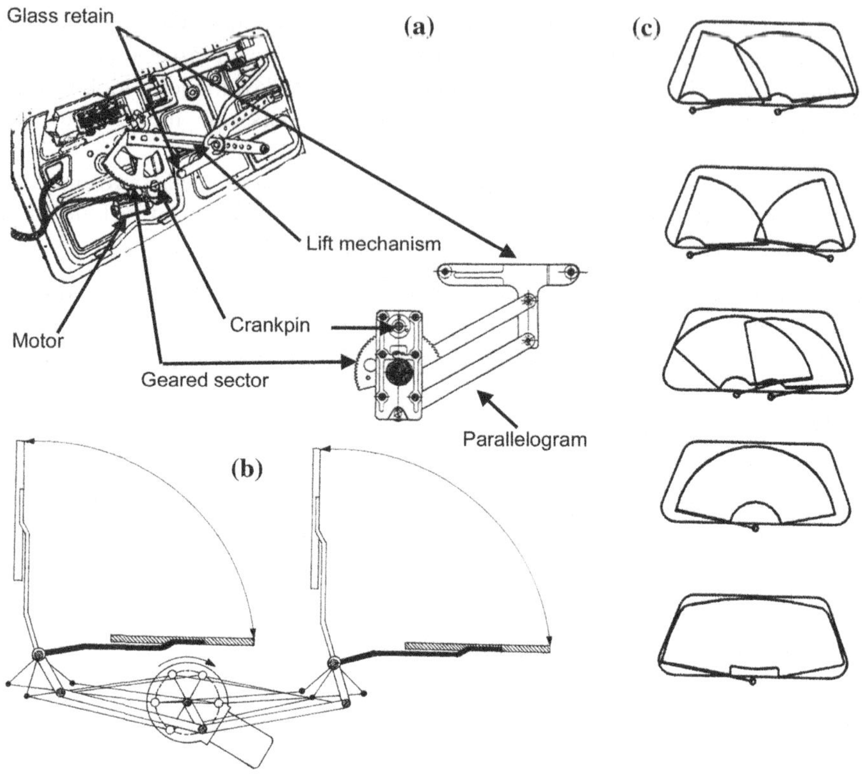

Fig. 8.17 **a** Inner view of the door with its windows winder. **b** Windscreen wiper mechanism. **c** Different wiper configurations

8.4.4 Headlights

The *headlights* were once made using a parabolic reflecting surface (rotational paraboloid) as shown in Fig. 8.18a in which a vertical section of the parabolic surface is shown, with lights reflecting from it. The property of the parabolic mirror is that any light ray emitted in its focus F is reflected in a direction parallel to the axis of symmetry of the parabola. The dimension "f", shown in the figure, is normally known as the 'focal distance'.

The headlight must generate two different types of light beams:

- High or main beam: nearly parallel to the ground, used for the maximum illumination when the driver is not crossing other vehicles. In this case the bulb filament is in the focus of the parabola;
- Low or dimmed beam: required by law when crossing other vehicles. It is angled slightly downwards, therefore providing less depth. It is obtained by means of a

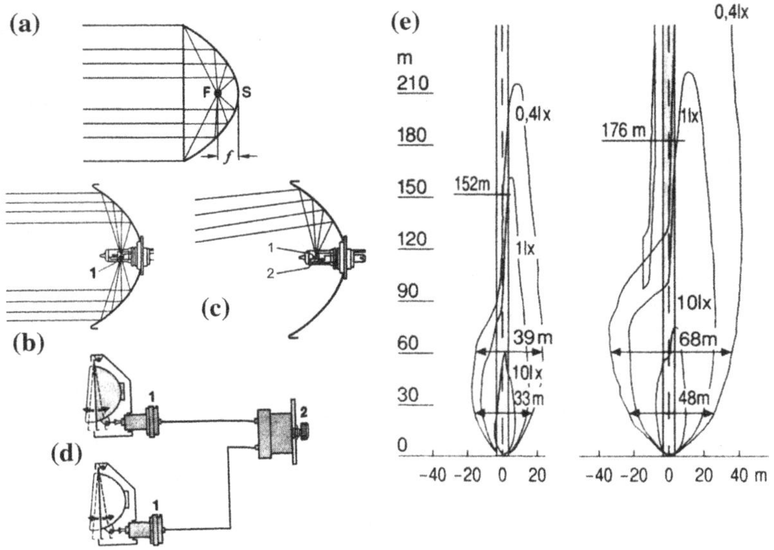

Fig. 8.18 Headlights. **a** Focal distance; **b** main beam; **c** dimmed beam; **d** adjustable beam height device; **e** light distribution map

second filament of the bulb in a slightly advanced position, with respect to the focus, and shielded downwards.

Recently, variously shaped headlights have been adopted in addition to parabolic mirror circular headlights. In this case, the operating principle is identical, but the shape of the reflecting mirror is considerably more complex, although still made of parabolic segments.

Finally, lenses with appropriate prisms are used to obtain asymmetric light beams. When 'clear', asymmetry is obtained only by means of the shape of the reflector. Light beam asymmetry accounts for the different lighting needs on the two sides of the road. The distribution of the light on the ground for two different lamp types is shown in Fig. 8.18e: it is easy to see how these differ from the elliptic shape which would be generated by a parabolic mirror.

As shown in the section on loads induced by the road, the different load conditions cause a different load share on the two axles. Loads affect both the vehicle's height from the ground and its pitch trim. For this reason, the headlights are fitted on hinges allowing the driver to adjust the beam height from inside the passenger compartment. Automatic devices that adjust the beam direction following the vehicle trim may also be used.

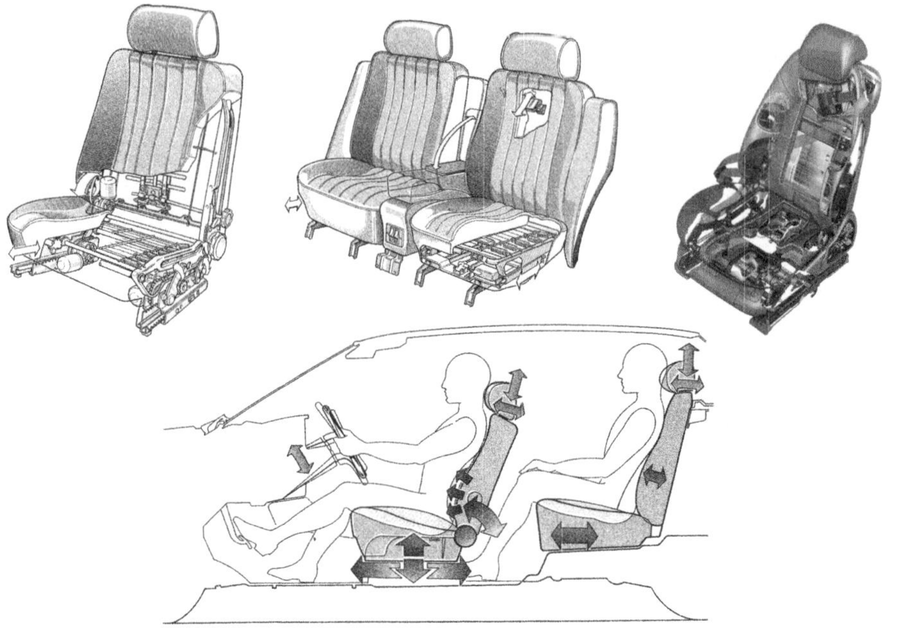

Fig. 8.19 Seats: structure of the front seats and adjustments of both the front and rear seats

8.4.5 Seats

The front *seats*, and sometimes also the rear seats, of a top range model can be moved by means of electrical mechanisms, either longitudinally or vertically. In any case, all seats can be adjusted in height by means of a specific lever system. The electric motor for adjusting the reclining backrest position and the lumbar support adjustment mechanism can be seen in Fig. 8.19. The back seats of the vehicle shown are also adjustable.

The seat of many recent top-range models has become increasingly more complex following the introduction of lumbar attachment features and even air cooling systems.

A further function of the seat is shown in Fig. 8.19: the seat belt retractor is arranged on the side of the seat, and in this case the seat structure must be designed also to withstand the loads of the seat belt. The advantage is that a three-point belt may be made available also in vehicles without the central pillar (convertibles) or for the middle seats in minivans fitted with the removable seat arrangements.

In the figure the padding is shown before being upholstered. It is supported by a system of springs, while in other cases the foam may be placed directly on a stiff backing (metal or plastic).

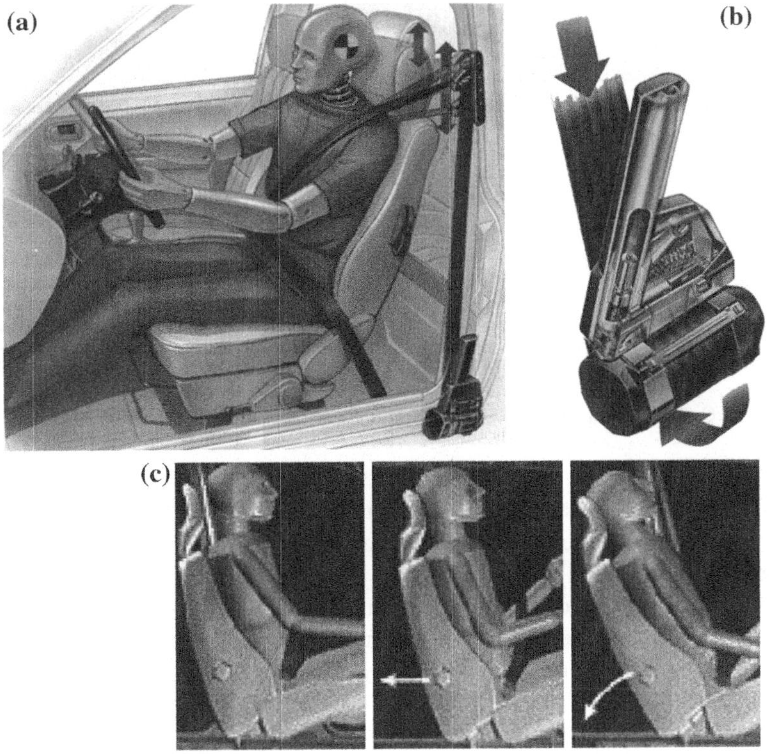

Fig. 8.20 Seat belts. **a** Three point seat belt with pretensioner; **b** pretensioner (detail); **c** anti-whiplash head restraints

8.4.6 Passenger Compartment Safety Devices

Each element of the passenger compartment must be carefully examined considering potential risks for passenger safety. The shapes and materials of the passenger compartment must be chosen so as to minimize danger in the event of an accident, and the materials must be fire-proof.

Occupants must be restrained to the seats in the event of a collision to prevent them from being hurled against dangerous parts of the vehicle or even out of the vehicle. Hooks must be arranged in the boot for correctly restraining the load which could become a dangerous projectile in a crash. It is currently assumed that the best restraint solution for occupants is the three-point *seat belt* with pretensioner, as shown in Fig. 8.20.

The *pretensioner* essentially reduces the seat belt operating times and allows to exploit the airbag effect to the most because, in a crash, the pretensioner keeps the seat belt tightened and reduces movements, holding the occupant in the most correct position. There are several different types of pretensioners, i.e. mechanical, pyrotechnic,

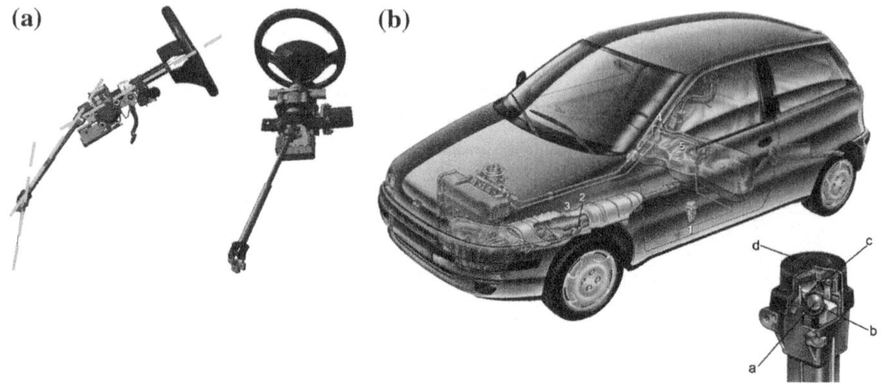

Fig. 8.21 **a** Collapsable steering column; **b** Fire prevention system (courtesy of Fiat)

etc. There are also adjustable seat belts with electronically controlled pyrotechnic pretensioners and front seat belts with special load limiter which attenuates its retaining force. A pyrotechnic pretensioner with a device for preventing the pyrotechnic load from being deployed when the belt is not worn is shown in Fig. 8.20b.

The head restraints (see Fig. 8.20c) are fundamental in preventing whiplash neck injuries following rear-end collisions. They must always be adjusted according to the occupant's height.

The steering column can become dangerous for the driver in the event of a crash of a certain severity: For this reason, the column is split into various, non-aligned segments that are joined together by universal joints, as shown in Fig. 8.21. In this way, the steering column collapses during a crash, avoiding that the steering wheel collides with the driver's thorax. Laws specify by how much a steering wheel may intrude following a collision against a barrier.

The risk of fire is an extremely severe danger in an accident: Devices are provided to block the flow of fuel to the engine when a certain deceleration threshold is reached. A typical Fire Prevention System of a recent car is shown in Fig. 8.21b.

Inertia switches that operate following a collision in any direction (also lateral) are used to stop the fuel flow. The switch consists of a steel ball with calibrated mass (a) kept by a magnet in a conical cavity (b). When the impact force exceeds the predetermined deceleration threshold (generally 8 g, corresponding to a crash at approximately 25 km/h), the ball moves up within the cavity and activates an electrical spring contact (c) which opens, interrupting the electrical supply to the fuel pump. After the collision, the contact may be restored simply by pressing a button on the inertia switch (d).

Lower and upper heat shields around the catalyst prevent fire risk to surrounding elements in case of catalyst overheating.

Airbags (Fig. 8.22) are inflatable devices drployed by the combustion gases of a pyrotechnic charge which are folded away in various areas of the passenger compartment (steering wheel, dashboard, seats, pillars, roof): They inflate in a predetermined

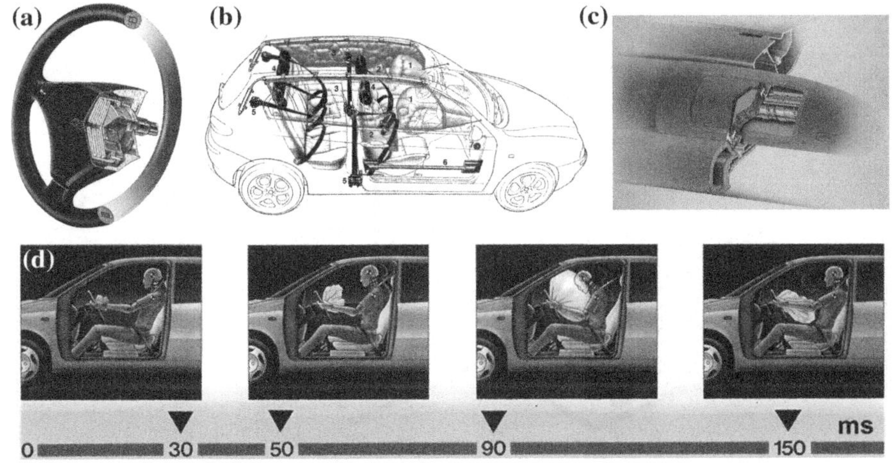

Fig. 8.22 Airbags (courtesy of Fiat)

way following a collision to attenuate the impact of occupants against stiff parts inside the passenger compartment. The steering wheel airbag and the dashboard airbag are shown in Fig. 8.22a and c, respectively.

The airbag deployment sequence in time (expressed in milliseconds) is shown from the impact starting point. Deployment must be synchronized with the occupant's movements subsequent to deceleration.

Airbags are designed to protect certain parts of the occupant's body (face, shoulders, knees). The front airbag improves head and chest protection in front crashes. When a crash exceeding a minimum predetermined acceleration threshold occurs, specific sensors activate the airbag propellant charge and the device is inflated in the matter of milliseconds. It is important that occupants wear seat belts and that their position is that specified at the design stage. Closer distances from the airbag may make its inflation dangerous. Side bags further reduce the risk of injury to the chest in side collisions, while window bags are useful in reducing injuries to the head in side collisions against high fixed obstacles and in roll-overs.

8.4.7 Climate Control System

In addition to containing many of the climate control system components, the dashboard is also used to distribute air flows within the passenger compartment.

Since the comfort of occupants is not only determined by temperature, but also by the local heat flow (as a first approximation, this can be represented as the product of the heat drop, the average air speed and the flow area), wide flows on large surfaces at low speed with low heat drops should be obtained. This is not always easy to obtain particularly if the air vents are small. Vents must also be designed to direct the flow of air onto the windows for de-misting.

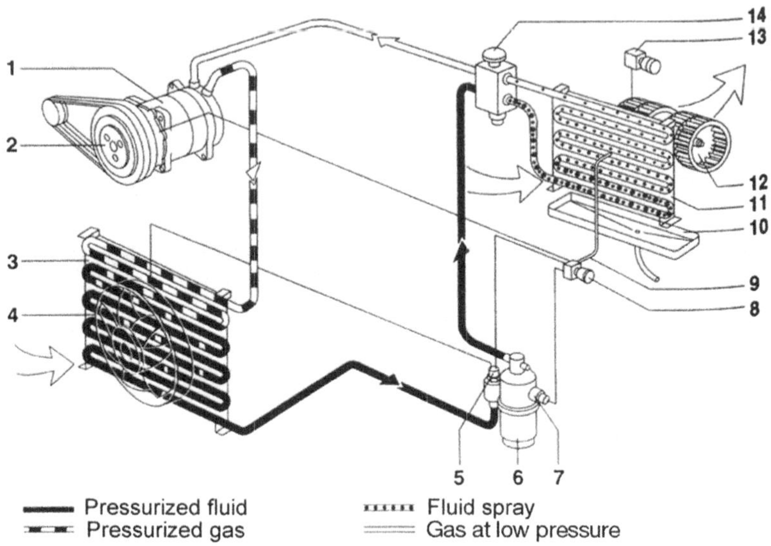

Fig. 8.23 Air conditioning system

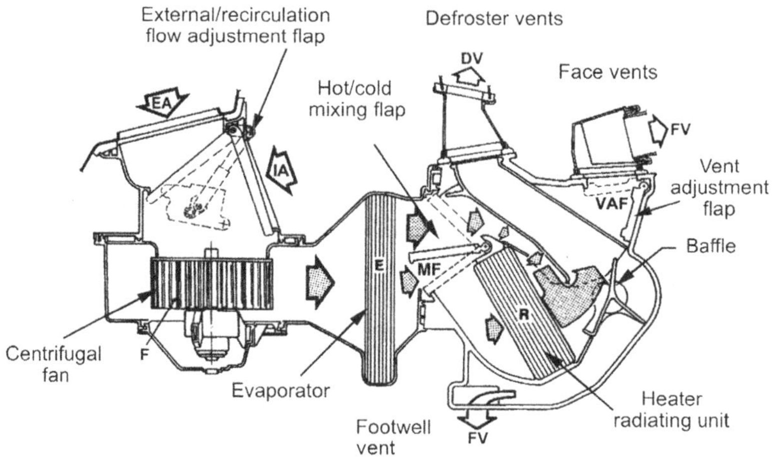

Fig. 8.24 Air distribution

The heating system is connected with the engine cooling system. Hot air is taken in by a speed adjustable cooling fan placed under the dashboard and connected to a heat exchanger. The temperature is adjusted by mixing the air flowing through the heat exchanger with outside air. When a temperature lower than the outside temperature is required in the passenger compartment, an *air conditioning* system is used (Fig. 8.23). A compressor (1), operated by the engine, compresses a gas acting as a cooling fluid: During compression, heat generation takes place. The cooling fluid is then pushed by the compressor through the condenser (3) where it is cooled, reaching the liquid state. Here the fluid releases the thermal energy received, partially in the compressor and partially in the evaporator (11).

The fluid, in liquid state, crosses the expansion valve (14), is atomized and conveyed to the evaporator where the evaporation process is completed. This process extracts heat from the external air flow directed into the passenger compartment and cools it. The humidity contained in the external air is extracted from the cooled air in the form of condensation water. An on/off device operated by an electrical clutch and a switch prevent freezing of humidity contained in the air on the evaporator fins.

A climate control system with adjustable centrifugal fan, evaporator and heater, is shown in Fig. 8.24. The hot and cold air mixing flap (or vane) and the front flap for taking in outside air or for recirculating the air from inside the passenger compartment can be seen. Other two vanes are used to distribute the conditioned air toward the occupant's face, toward the windscreen or into the footwell, respectively by means of a vent adjustment flap and a baffle.

Chapter 9
Chassis

As discussed in detail in Chap. 2, before the introduction of the unitized body, the term *chassis* was used to indicate a self-propelled system, sometimes capable of being driven, consisting of a load-bearing structure (a frame made of sidemembers and crosspieces) carrying the engine, the suspensions, the wheels, the driveline connecting the engine to the wheels, the braking system, the steering system, the controls and the engine accessories.

This system was taken to the final assembly line where the body and the remaining bodywork parts were fitted.

With the introduction of the structural body, the self-standing structural element (i.e. the frame made of sidemembers and crosspieces) did not exist any more, but the word chassis was, and still is, used to refer to these components as a whole. The chassis, according to commonly agreed definitions, thus consists of the following main systems, as can be seen in Fig. 9.1

- Powertrain (engine and driveline from engine to wheels),
- Powertrain mounts,
- Wheels and suspensions,
- Steering system and
- Braking system.

These are the systems which mostly contribute to giving each motor vehicle its character and peculiarities by determining its general performance and road behavior.

The chassis is therefore generally associated to the motor vehicle performance in terms of driving, handling, road holding and braking. The chassis functions, except for those of the powertrain described in another chapter, will be covered in the following sections.

G. Genta et al., *The Motor Car*, Mechanical Engineering Series,
DOI: 10.1007/978-94-007-8552-6_9,

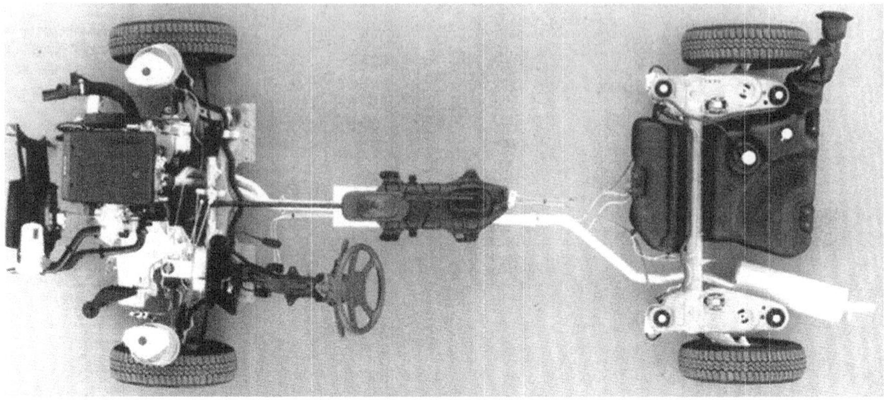

Fig. 9.1 Present-day car (transversal engine with front wheel drive): typical chassis configuration (courtesy of Fiat)

9.1 Tires

All forces the vehicle exchanges with the ground involve the wheel-tire assembly. These forces, which are used to drive the vehicle, are generated in the contact area between the tire and the ground, and are transferred to the suspension through the tire-wheel rim contact.

Almost all vehicles are supplied with pneumatic tires, inflated with pressurized air or in some case with a pressurized gas (mostly nitrogen). The air provides the tire structural properties and its ability of withstanding considerable forces while being an intrinsically elastic and flexible component.

The tire needs to maintain contact with the ground also in the presence of obstacles and must provide good grip despite being a lightweight and deformable structure. These features (deformability, lightweight, grip, ability of transmitting high forces) are essential to ensure good road holding and are assured by the composite structure of the tire, consisting of a braided nylon or steel wire carcass which is flexible in radial direction while being rigid in the longitudinal one, surrounded by a highly deformable rubber compound for high grip on the ground.

Basically the grip is the ability to produce forces parallel to the ground (longitudinal and transversal) in presence of perpendicular forces.

9.1.1 Wheel-Tire Assembly

Schematic drawings of a cross-ply (conventional) and a radial (or belted) tire are shown in Fig. 9.2; their main dimensions are also quoted.

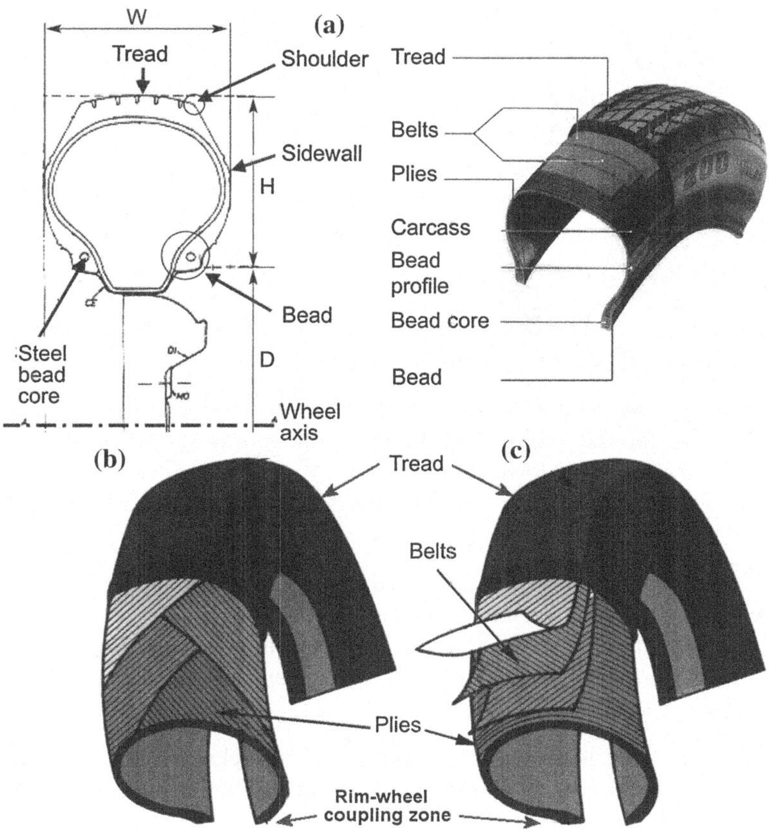

Fig. 9.2 **a** Cross section showing the main dimensions and the structure of a radial tire **b** and **c** plies orientation in a cross-ply and a radial tire, respectively

The main requirements of a rim-disk assembly are:

- Light construction: the wheel is one of the heaviest parts of the unsprung mass which, as it will be shown later, must be as light as possible.
- Strength and stiffness: the wheel must have a suitable mechanical resistance to withstand also accidental loads while the vehicle is travelling. It must also be adequately stiff to minimize the lateral flexibility under the loads due to the centrifugal acceleration when cornering.
- Tire anchoring to the wheel: the rim profile must allow easy removal and fitting of the tire, while providing a firm mechanical fit of the tire for transmitting motive and braking torques. It must also provide a suitable contact with the rim to retain the pressurized air.
- Wheel anchoring to the hub: the wheel must be secured to its hub in total safety while allowing for easy removal and fitting of the wheel when a replacement is required.

The main types of tires used in the recent past were *cross-ply* (Fig. 9.2b) tires. Nowadays tires are mostly of the *radial* type (Fig. 9.2c). They differ from cross-ply tires because:

- The plies have threads arranged in the radial direction in the sidewall.
- The plies in the sidewalls are fewer.
- They have more crosswise plies (crossed threads) in the belts below the tread.
- The sidewalls (rubber plus plies) are thinner.

The radial tire structure allow to obtain the following results:

- High tread stiffness in radial, longitudinal and transversal directions and
- low stiffness of the sidewalls to improve riding comfort.

They have a further advantage: the deformations involve a smaller volume of rubber (because the sidewalls are very thin) and the rolling resistance is, as it will be shown later, consequently lower.

Apart from the type of structure, tires may also be subdivided following the way air is contained as:

- *Tube tires*, in which the air is contained in an additional tube provided with the valve and arranged between rim and tire, and
- *tubeless tires*, in which there is no tube because the tire itself is airtight and the bead is appropriately designed to prevent air leakages in the zone in which it is in contact with the rim.

Tubeless tires are today used on virtually all vehicles because, in addition to eliminating a component, they do not deflate suddenly when punctured, increasing safety.

The identification symbols printed on the tire sidewall comply with international standards and state the geometrical dimensions and the maximum allowed speed. Following the European Directive 92/23, the first number indicates the width (in millimeters) of the carcass (W in Fig. 9.2a); the second number the height/width ratio (H/W in the same figure), then a letter indicates the type of structure, a third number indicates the rim diameter in inches (D), a fourth number expresses the load index and finally a letter indicates the maximum speed. Load and speed codes can be found in specific tables.

As an example, 165/70 R 14 81 H indicates a tire with a width $W = 165$ mm, a height $H = 115.5$ mm (ratio $H/W = 70\,\%$), a radial structure, fitting on a on a rim with a diameter $D = 14$" (355.6 mm), with load index 81 (corresponding to 4530 N) suitable for speeds up to 210 km/h (code H).

The rolling radius R can be determined from these specifications, taking into account a deformation of about 3 % of the radius, due to the load. In the example above, the rolling radius is thus $R_0 = 284.5$ mm.

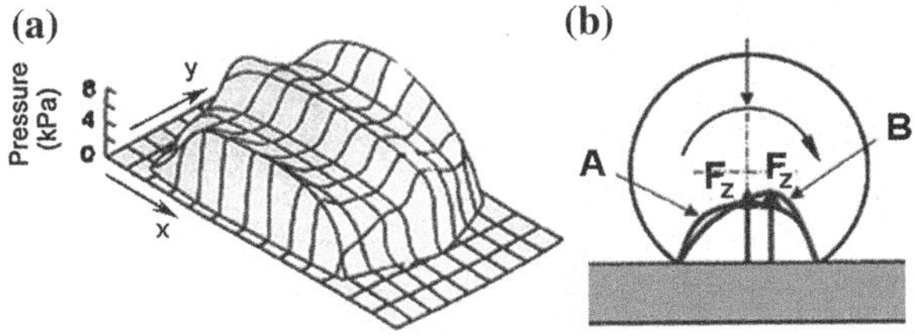

Fig. 9.3 **a** Pressure distribution on a non-rolling tire. **b** Pressure distribution and resultant of the normal force F_z in a non-rolling (A) and a rolling (B) tire (from Genta and Morello 2009)

9.1.2 Rolling Resistance

The term *rolling resistance* is used for the horizontal force which opposes the rolling of a tire when this is subjected to a vertical load. Consider the pressure distribution on the contact print (Fig. 9.3). If the wheel is not rolling, the pressure distribution is symmetric with respect to the center of the contact area and its resultant F_z goes through the center of the wheel (diagram A). When the wheel rolls, on the contrary, the pressure distribution is asymmetric and the resultant does not go through the center of the wheel (diagram B); this creates a torque which opposes tire rolling. The resistance is thus proportional to the vertical load. The asymmetry is caused by hysteresis: in other words, the work required to deform the tread making it adhere to the ground, is not totally returned when the tire takes again its natural shape, after its contact with the ground.

Rolling resistance is also proportional to the longitudinal size of the contact area; since the size of the contact patch is proportional to the load at fixed inflation pressure, this explains why an increase of pressure can decrease the resistance.

The ratio between the rolling resistance F_x and the vertical load is the *rolling coefficient*

$$f = \frac{F_x}{F_z}. \tag{9.1}$$

For radial tires rolling on good concrete or tarmac roads, the rolling coefficient takes values in the range from 0.008 to 0.01. Sometimes these values are expressed as $8 \div 10$ kg/t. The rolling coefficient for a tire is plotted as a function of the speed in Fig. 9.4. The curves, plotted for several values of the inflation pressure p, show that the rolling coefficient is very low at low speed, to increase slowly until a certain critical speed is reached. After that the increase becomes quick. This surge is caused by standing vibration waves that propagate in the whole structure.

The rolling resistance decreases with increasing inflation pressure and is more or less proportional to the vertical load.

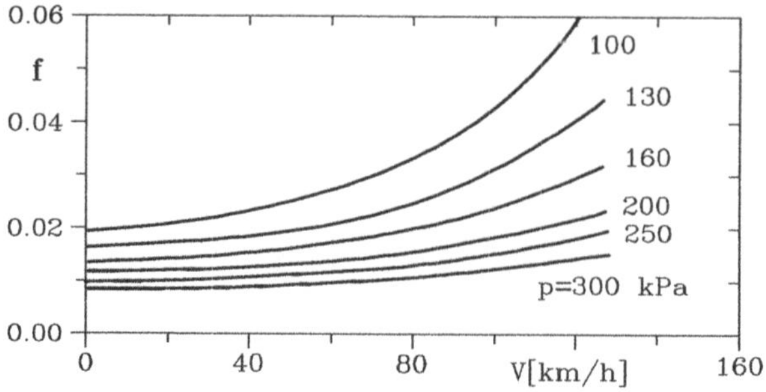

Fig. 9.4 Rolling coefficient f as a function of the speed V for several values of the inflation pressure (from Genta and Morello 2009)

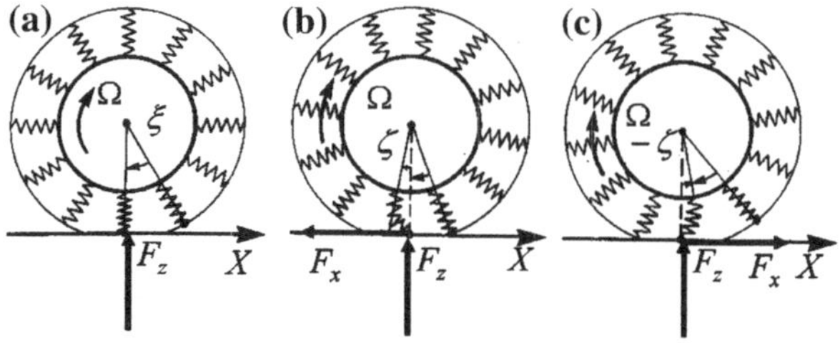

Fig. 9.5 **a** "Brush" simplified model of a tire; the tire is reduced to a discrete number of springs, featuring vertical, lateral and longitudinal flexibility and bearing a specimen of tire rubber on their tip. Static representation of radial and circumferential deformations of a tire, with simultaneous application of a vertical and braking **b** or driving **c** force (from Genta and Morello 2009)

9.1.3 Longitudinal Force

To predict, although only in a qualitative way, what will occur when applying force to a tire, a simple model, shown in Fig. 9.5a, can be used.

The tire is reduced to a discrete number of springs, featuring vertical, lateral and longitudinal flexibility, fitted on a rigid rim with a uniform angular displacement ζ and carrying a specimen of tire rubber on their tip. Assume, for the sake of simplicity, that the number of springs is such that there is a position where only one of them contacts the ground.

The spring tips are free to move, according to applied forces, with no connection to nearby springs. All simplifications can be removed without affecting the nature of the results we are going to study. We will call this extremely simplified model the *brush model*.

Under the application of the vertical force, a contacting spring will assume a certain radial deformation, as in an actual tire; if a suitable damping coefficient is associated to the spring, the model is also able to simulate rolling resistance proportional to the mechanical work dissipated by any contacting spring as it leaves the contact patch because of rolling. In addition, the rolling radius will take a value smaller than its unloaded value, because of the applied vertical force.

Assume that the tire undergoes a longitudinal force F_x in the $-X$ driving direction (Fig. 9.5 b) or the X braking direction (Fig. 9.5 c); according to the proposed model, an angular deformation ζ, proportional to the applied force, will be applied to the contacting spring and the tire.

It is easy to understand that if the tire has a rolling speed Ω, each contacting spring will return to a straight position when leaving the contact patch, while a new spring will take the angular displacement ζ. To the initial tire speed Ω a new deformation speed will be added, equal to the ratio between the angular deformation ζ and the time ξ/Ω needed to put the following spring in contact with the ground.

Under the action of a driving force the wheel should roll faster than when rolling free of force; on the other hand, if the wheel is braking, it should roll slower than when rolling free.

A *tire longitudinal slip*, usually indicated as σ can be defined as

$$\sigma = \frac{V - \Omega R}{V} \tag{9.2}$$

and is often expressed as a percentage of the speed V. If the brush model were accurate, these speed variations, and consequently the longitudinal slip, should be proportional to the applied longitudinal force.

If the total longitudinal force exceeds a certain limit, some elements, at first those located in the rear part of the contact print, exceed the grip limit and start slipping. If this limit value of the force is exceeded, the total force further decreases up to the point when complete slip is reached, and the force reduces to the friction force occurring between the rubber constituting the tire and the road in total slip conditions. The same occurs when exerting driving forces, but now the force and the torque have opposite directions.

The forces for accelerating or braking the car are therefore due to the difference between the peripheral speed of the tire ΩR and the forward speed of the car V, i.e. the speed of the wheel hub. This difference is due to the tire deformation, resulting from a partial or total slipping.

The longitudinal force generated during braking is plotted in Fig. 9.6 as a function of the longitudinal slip for various road conditions. From the figure it follows that:

- The longitudinal force is proportional to the longitudinal slip if the latter is small.
- In this small slip zone, the braking force is independent from road conditions for a certain tire.
- The braking force reaches a maximum value at slip values equal to $12 \div 20\,\%$ (tread elements are working without significant slip).
- The maximum grip force is equal or slightly larger than the vertical force on good, dry roads, but may be much larger in case of racing tires.
- Beyond the point of maximum force, the force decreases and continues to do so until the wheel locks (100 % slip).

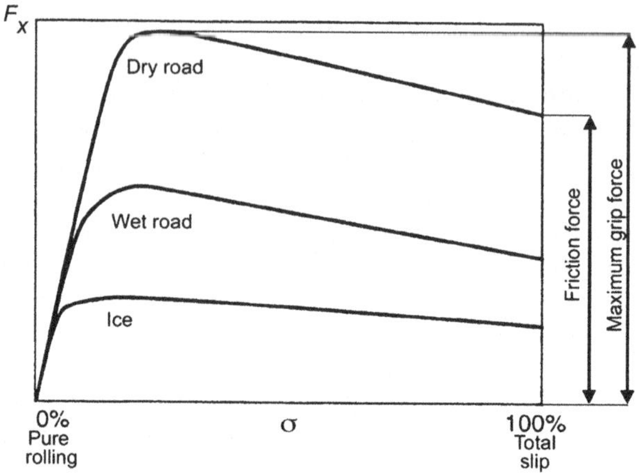

Fig. 9.6 Longitudinal (*braking*) force F_x as a function of the *longitudinal slip* σ (from Genta and Morello 2009)

- The maximum value of friction force strongly depends on the road conditions. On wet roads, the maximum value may be as low as $25 \div 50\,\%$ of the vertical load, and this value reduces to $10 \div 15\,\%$ on icy roads.

The *longitudinal traction coefficient* is defined as the ratio between the longitudinal force and the vertical force:

$$\mu_x = \frac{F_x}{F_z} \, . \tag{9.3}$$

9.1.4 Lateral Force

Consider a tire that is subjected to a vertical and a lateral force due, for example, to a centrifugal force F_y applied to the vehicle in a turn. In this case, the spring in contact with the ground are subjected to a lateral displacement in the Y direction. This phenomenon is shown in Fig. 9.7.

If the wheel is rolling, each spring will contact the ground as shown in the upper part of this figure, here seen relative to the side view on a certain point of the X axis.

The initial contact point will receive a lateral displacement Δy, shown in the lower part of the same figure; this displacement again is proportional to the lateral force. When the next spring contacts the ground it will take the same lateral displacement, which will be added to the wheel hub. As a result, the wheel hub will receive a lateral motion whose speed will be given by the ratio between this displacement and the time needed to place the next spring in contact with the ground, that is ξ / Ω.

If the model is accurate, because deformation is proportional to the force, we would expect the wheel hub to deviate in its path by an angle α, proportional to the

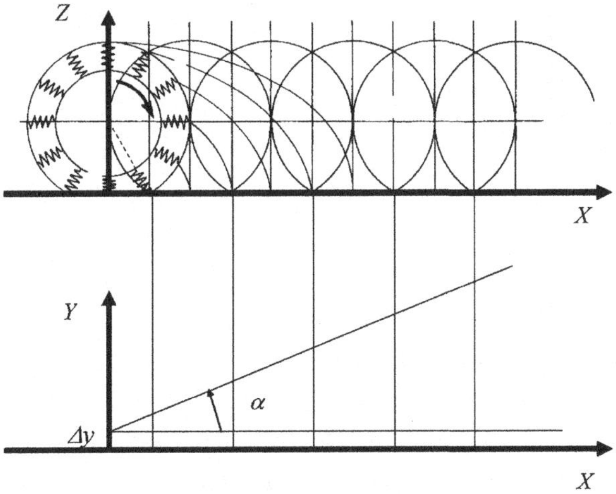

Fig. 9.7 Brush model; graphic representation of a rolling tire with simultaneous application of a vertical and a lateral force (from Genta and Morello 2009)

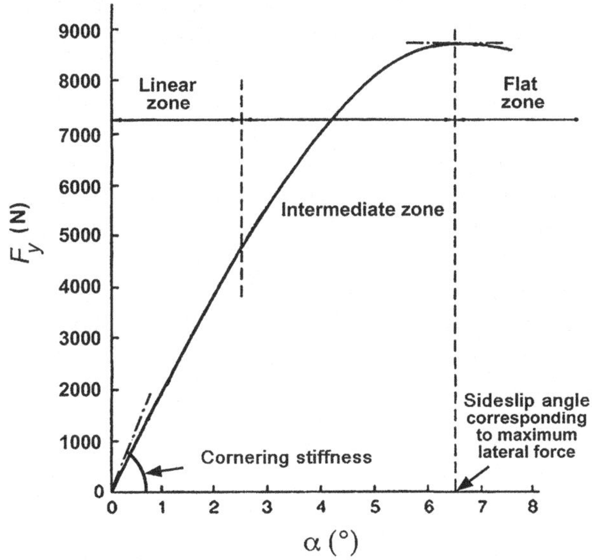

Fig. 9.8 Lateral force F_y as a function of the *sideslip angle* α

lateral force; this deviation is called *(tire) sideslip angle*, not to be confused with the vehicle sideslip angle or attitude angle.

In practice, the tire behaves, not very differently from the simplified model here described. The lateral force F_y produced by an actual tire is plotted in Fig. 9.8 as a function of the sideslip angle α. Initially, as stated above, the force is proportional to the angle. This linear part is followed by a zone where the increase of the force

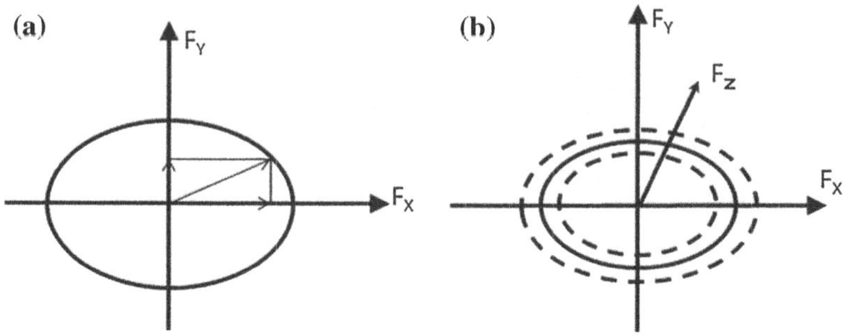

Fig. 9.9 Tire traction ellipse, **a** and the effect of the vertical force on the traction ellipse **b** (from Genta and Morello 2009)

is less than linear, because a loss of grip is experienced. The force does not increase any more and remains constant or, sometimes, even decreases. The plot in the figure was obtained for a vertical load of 8,200 N.

It the small-angles zones, where the side force is proportional to the sideslip angle, it is possible to define a proportionality coefficient defined *cornering stiffness*. Symbol C is commonly used for it.

If the equatorial plane of the wheel is not vertical (wheel camber inclination), the vertical force can be decomposed into a force perpendicular to the ground and a force parallel to it (the *camber force*). The latter causes a side deformation and a lateral motion even if no cornering forces are applied. The camber force is proportional to both the camber angle and the vertical force. When travelling on a straight road, a cambered wheel has a higher tire wear, mostly concentrated on the interested side of the tread, than a wheel travelling with no camber.

The *lateral traction coefficient* is defined as the ratio between the lateral force and the vertical force:

$$\mu_y = \frac{F_y}{F_z} \, . \tag{9.4}$$

9.1.5 Interaction Between Lateral and Longitudinal Forces

When a tire produces both longitudinal and transversal forces at the same time, it must be taken into account that the grip in one direction decreases the available grip in the other. The plot in Fig. 9.9a shows that the maximum lateral force F_y is obtained when the tire is not called to develop longitudinal forces ($F_x = 0$); similarly, when the longitudinal force is maximum, the transversal force is null ($F_y = 0$). The envelope of all forces which can be developed on the ground plane by the tire forms a polar diagram which is normally approximated by an ellipse called *tire traction ellipse*.

The dimensions of the traction ellipse is influenced by the vertical force acting on the tire; the available grip increases proportionally to the load.

Within the ellipse, i.e. far from the grip limit, the considerations on side and longitudinal slip which define the elastic behavior of the tire are applicable.

9.2 Suspensions

The basic function of the suspension system is to provide a flexible support so that the occupants can comfortably travel being isolated from the vibrations caused by road irregularities. It should be noted that riding comfort is an objective of paramount importance in automotive design. An additional important requirement of the suspension system is stabilizing the vehicle in all driving conditions, i.e. during changes of direction, braking, acceleration, etc.

These two goals (*comfort* and *handling*) are normally at odds because while a soft suspension is ideal for comfort, a stiff suspension is better for handling. Obviously, a successful suspension design will be capable of reaching the best trade-off between the two.

A last requirement is keeping the wheels in the correct position with respect to the ground at any position of the suspension. The ability of allowing only minor angular movements of the wheel with a large vertical travel is particularly important; in fact, steering and camber angle variations generate lateral forces which may cause deviations from the required path and negative effects on safety.

A suspension is characterized by many parameters (spring stiffness, damping of the shock absorbers, suspension linkage design, position of the attachment points, etc.) which, when designed correctly, allow us to obtain more than satisfactory results for any specific vehicle. It must however be said that the *universal suspension*, fit for all purposes, still remains to be found. This is the reason why there is such a large number of different suspension designs: each of them is generally a good response to the needs of a specific vehicle for a specific predicted use. The requirements of a small city car (in terms of simple design, light weight and low cost) are obviously quite different from those of a large sedan or a sport car. In the suspension design, perhaps even more than in the design of the other components of the chassis, it is thus essential to identify the objectives to be reached from the very beginning of the design task and then choose the best possible solutions accordingly.

The suspension system connects the wheel-tire assembly to the body frame. While the tire is responsible for exchanging vertical, lateral and longitudinal forces with the ground, in addition to filtering the vibration caused by road roughness, the suspension is responsible for determining the tire position with respect to the body and to the ground in the various conditions of use of the vehicle, while filtering the vibration at lower frequencies.

It must allow the wheel to shake, i.e. to move vertically with respect to the vehicle, but must also ensure tire grip on the ground even when crossing bumps, and filter the energy that the tire alone cannot damp. This function is implemented by arranging

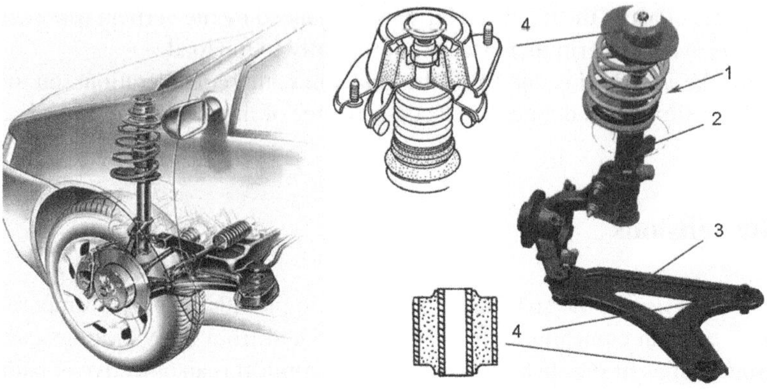

Fig. 9.10 The suspension system; *1* elastic elements (*springs*); *2* damping elements (*shock absorbers*); *3* rigid guiding elements (*rods, wishbones and struts*); *4* flexible damping elements (*pads and rubber mounts*) (Courtesy of Fiat)

elastic elements (the springs, (1) in Fig. 9.10) and dissipating elements (the shock absorbers, (2) in the same figure) between the wheels and the body.

The suspensions ensure, by means of linkages, the correct position of the tire with respect to the ground, which is essential for achieving the desired handling characteristics, while at the same time allowing the required freedom of motion to the wheels (vertical shaking, steering, minor elastic movements under load).

All the forces that the wheel exchanges with the ground load the suspension system: these include the forces exerted to accelerate, brake and steer the vehicle and the forces caused by collisions against obstacles and road bumps. The suspension linkages must therefore be able to withstand the forces applied to the wheel and, since the road irregularities excite vibrational motions, dissipating the vibration energy so that a smooth ride can be achieved.

To perform these two fundamental tasks, i.e. guiding the tire (*handling*) and absorbing shocks from the road (*comfort*), the linkages essentially consist of rigid guiding elements, such as rods, wishbones and struts (3) and of flexible elements, such pads and rubber mounts (4).

In terms of comfort, the suspension must allow a wide vertical stroke of the wheel (approximately 200 mm in total) and filter out the energy that the tire alone cannot dissipate.

The largest vertical movements are absorbed by springs and pads, which prevent the suspension from knocking against the body frame.

The stiffness of a suspension is governed by the sensitivity limits of the human body, both towards the lower end of the range (low frequency body motions, lower than 1 Hz, cause travel sickness) and the higher end of the range (high acceleration at high frequency). Minor movements, caused by bumps, are absorbed by the rubber bushings which are inserted in the suspension linkage joints.

The correct design of these rubber elements is particularly important to reach the targets in terms of comfort, requiring a high compliance, and handling, requiring limited wheel angles variations.

Another important parameter for determining the vehicle comfort is the mass of the wheels and of the other mobile suspension elements (*unsprung mass*) that must be a small fraction of the vehicle mass fraction supported by the suspensions (*sprung mass*).

If the unsprung masses are large, their large linear momentums cause large forces to be transmitted from the wheels to the body, which in turn produce large vertical acceleration of the latter, ultimately affecting comfort. This justifies, for example, the use of light (aluminium) or ultralight (magnesium) alloys for the wheel rims.

As a first approximation, when the vehicle crosses a road bump, the vertical momentum of the vehicle body can be assumed to equate the vertical momentum of the unsprung mass:

$$mv = MV, \tag{9.5}$$

where m is the unsprung mass, M the sprung mass, v the vertical velocity transmitted to the unsprung components following an impact against a road bump, V the vertical velocity transmitted to the body.

The body vertical speed can be obtained from the equation above:

$$V = v\frac{m}{M}; \tag{9.6}$$

since V must be as small as possible to increase comfort, therefore the ratio between unsprung and sprung mass must be reduced to a minimum.

In the following sections, before examining the most common designs of the suspension system and their main components, some attention will be devoted to the main parameters governing handling and comfort and to how these parameters can be correctly defined.

9.2.1 Wheel Characteristic Angles

The wheel position on the ground is of paramount importance for stability and handling, both on straight running and when cornering. It is determined by the *characteristic angles* of the suspension geometry (Fig. 9.11).

These angles are: the *toe-in angle* or *toe-out angle*, the *camber angle* , the *king-pin inclination* and the *caster angle* . Their values are important because they define the tire-ground contact geometry. A correct contact geometry allows one to obtain good stability, when driving both on a straight line and cornering, and acceptable tire wear (e.g. a duration of 20.000–40.000 km according to the driving style attitude).

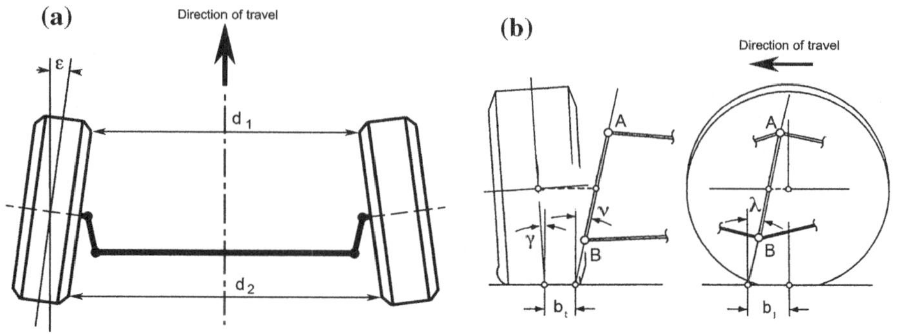

Fig. 9.11 Characteristic angles. **a** ε: toe-in angle, also measured by the value $d_2 - d_1$ in mm; **b** (*projection on the transversal plane*) γ: camber angle; ν: king-pin inclination; b_t: king-pin offset; (*projection on the longitudinal plane*) λ: caster angle; b_l: trail, lead

Toe-In or Toe-Out

The toe-in angle is defined as the angle γ between the longitudinal vehicle axis and the equatorial plane of the wheel; it is called toe-in when the wheels are angled as in Fig. 9.11a, toe-out in the opposite way. It is also measured in millimeters as the difference between dimensions d_2 and d_1: it varies approximately from $2 \div 3$ mm (toe-in) for rear wheel drive vehicles to $0 \div 2$ mm (toe-out) for front-wheels drive vehicles. The optimum value for this angle is zero, when wheels are rolling, but a non-zero value at standstill may compensate for the deformation due to the driving torque, taking place when the vehicle is moving. The aim is to assure an optimal tire print and a low tire wear.

Camber Angle

The camber angle γ is the angle between the equatorial plane of the wheel and the vertical plane parallel to the longitudinal axis of the vehicle, as Fig. 9.11b shows. It is measured on the projection of the suspension on a plane transversal to the car body. It is positive when the top of the wheel is slanted outwards with respect to the vehicle.

It should be noted that the camber angle is defined with reference to the vehicle, but what is relevant to the tire behavior, as it was explained in the previous section, is the camber angle referred to the ground; the two angles coincide when the vehicle is stationary or driving along a straight path but are different while cornering because of the roll of the body.

Because a camber angle different than zero produces lateral forces and increases tire wear, when cornering, the camber angles of the wheels of the same axle should be equal, with opposite sign, to the roll angle of the vehicle body, to maintain the wheel perfectly vertical on the ground. This result can be obtained if the wheel camber (referred to the body) is decreasing when the suspension is compressing (as do outer suspensions in a curve) and is increasing when the suspension is extending (as do inner suspensions in a curve). The wheels are perfectly vertical on a bending road if camber variations are equal to the roll angle; such a suspension is said to be able to *recover camber*.

It is obvious that such a suspension could generate undesired camber angles on a straight path, i.e. when the vehicle is braking or loads are changed; the design of camber angles and camber recovery is therefore a matter of compromise between straight path and cornering behavior.

King-Pin Inclination

The king-pin inclination ν is also determined projecting the suspension on a plane transversal to the car body, as in Fig. 9.11b. It is the angle between the steering axis AB of the wheel (or *kingpin axis*) and the vertical longitudinal plane. This angle promotes the *self-alignment* or *reversibility* of the steering system because, thanks to the king-pin inclination, the vehicle is slightly raised when steering.

This inclination, that is usually in a range between 5 and 10°, determines the value of the *king-pin offset* , or transversal base b_t, which has a direct effect on the steering torque, that has a minimum for $b_t = 0$. A positive value promotes a better driver's feeling, when cornering. In some cases, a slightly negative offset has been used to improve braking stability because, in these conditions, the toe-in is increased.

Caster

The caster angle λ is the angle between the kingpin axis AB and the vertical plane perpendicular to the vehicle symmetry plane. It is determined by projecting the suspension on the longitudinal plane. With the car moving, this angle can further increase the *self-alignment* or *reversibility* of the steering system when travelling, thanks to the offset of the lateral forces applied to the wheel. A positive caster angle (from 0.5 to 5°) is generally used on rear wheel drive vehicles, while a negative or slightly positive caster angle (from −0.5 to +1.5°) is used on front wheel drive vehicles. The caster angle determines the value of the longitudinal base b_l, which can be positive, *trail* , or negative, *lead*. In front wheel drive vehicles, the driving torque compensates, at the end, the geometrical static conditions, apparently unfavorable.

9.2.2 Suspension Kinematics

The main characteristic angles and their current values have been previously defined. It is important to remember that these values, defined by engineers during design setup, are related to a certain load condition of the vehicle. Since the suspension system elastically links the wheel to the body and grants a certain freedom to the wheel itself (i.e. vertical movements, steering rotations), the position of the wheel in space varies while travelling. All the characteristic parameters defined above (angles and offsets) will also vary as a consequence.

The control of these variations (due to kinematics and elastic effects) is a peculiar task of suspension design.

For this reason, the kinematic and elastic characteristics related to the degrees of freedom of a suspension will be examined in greater detail, with considerations and comments, and will be shown how such characteristics can affect the wheel positions and the related vehicle dynamics.

The six *degrees of freedom* that the wheel has as a free body in space will be considered: three displacements and three rotations about axes X (direction of travel), Y (transversal with respect to direction of travel) and Z (vertical). Displacements will be dealt with first. The three rotations of the wheel around the same axes will then be described.

Wheel Displacements

Displacement along axis X: the suspension must allow small wheel movements along this direction for reasons of comfort; they are allowed by the the geometry of linkages and by the elasticity of their bushings, in the order of a few millimeters. Indeed, the suspension absorbs road roughness better according to how well the wheel complies in front of an obstacle. The suspension must retract the wheel after hitting a bump.

Displacement along axis Y: the suspension must prevent major movements in this direction; excessive displacements would lead to unnecessary waste of grip and tire wear.

Displacement along axis Z: the wheel must be allowed to have large displacements along this direction; typical compression and rebound movements with respect to the reference trim are in the order of 100 mm in each direction. Indeed, in the vertical direction, the vehicle must be able to react to variations of mass (e.g. from one occupant to full load the change of mass is approximately 400 kg) and to the very high loads which may occur when negotiating obstacles. In order to elastically absorb such variations of load in the vertical direction, an elastic element (suspension spring) is provided. Bump stops, made using progressively stiffer elastic elements, are also applied to limit the vertical suspension motion as comfortably as possible.

Wheel Rotations

Rotation about axis X: this rotation changes the wheel camber angle. The suspension must keep this angle unchanged with respect to the ground; to do this it must allow minor angular variations, both kinematic and elastic (in the order of up to 3°) with the objective of compensating for the effects of body rolling. Roll causes loss of camber on the outer wheel if this is rigidly linked with the body. The suspension linkage is generally designed to provide negative camber angles during compression. This design allows recovering the loss of angle caused by rolling at least partially and holds the tire close to the vertical position with respect to the ground. Not all suspensions are however capable of *camber recovery*. This is an important parameter to be taken into account when choosing and developing a suspension system.

Rotation about axis *Y:* the suspension must leave this movement absolutely free. The wheel is for this reason connected to the suspension through ball or roller bearings.

Rotation about axis *Z:* a front suspension (steering axle) must allow this degree of freedom to the wheel with angular variations in the order of at least 35°. The angular position of the wheel is controlled, in this case, by the steering linkages governed by the steering wheel.

In the case of a vehicle with four steering wheels, the considerations made for the front suspension will apply also to the rear suspension, although the wheel angles in this case will be much more limited. In rear suspension the *toe-in variation* must be prevented, because they will cause self-steering, or either controlled by flexible bushings, to obtain the required under or oversteering effects.

The design of a steering suspensions is complicated by the fact that all vertical movements of the wheel may cause also variations of its angular position which are added to the steering angle imposed by the driver. In a properly performing suspension geometric or elastic displacements of the wheel, in a given direction, do not cause undue displacements or rotations in the other directions. For example, the linkage may allow the wheel to retract, following a collision with an obstacle, but should not allow movements in other directions, i.e. around *Z* axis which cause undue self-steering.

It is easy to guess that a more complex linkage design is more suitable to obtain these functions easily. For example, a suspension can be structured into several rods and the stiffness of the respective bushings can be calibrated to generate elastic deformation, so that during an emergency braking manoeuvre on a corner, steering angles can be slightly decreased to improve stability.

9.2.3 Suspension Components

The components of a suspension, shown in Fig. 9.10, can be split into three main categories according to the function they perform.

- *Structural components*: they ensure the position of the wheel with respect to the body providing only stroke movements and controlling the required characteristic parameters of the wheel (angles, offset, etc.) in the various conditions of use. They include kinematic and anchoring components. Kinematic components are wishbones, rods and struts; they are generally made of steel, cast iron or aluminium and must be designed to be safe in all conditions of use. Anchoring components ensure wheel control under load and provide damping of minor road bumps and vibrations to obtain the required level of comfort. They generally consist of rubber elements carefully designed for their damping characteristics.

- *Elastic components*: these components ensure riding comfort and allow major suspension movements. They elastically store the kinetic energy generated by the suspension stroke. They are arranged between the structural components and the body and may consist of helical springs, torsion bars or leaf springs. It is interesting to observe that, while helical springs and torsion bars perform their elastic function by being able of withstanding loads in one direction only, leaf springs are able of withstanding loads in several directions and can also act as a structural element in a simplified suspension design.
- *Damping components*: they cooperate with the elastic components and are used to dampen the oscillations of the suspension springs. Without these components, each oscillation impressed to the suspension would continue indefinitely in time. The hydraulic shock absorber is widely used: the energy dissipation is obtained by forcing oil through appropriately calibrated passages. Shock absorbers may be appropriately modified to integrate some structural functions: This is the case of the Mc Pherson suspension where they are able of withstanding lateral loads.

Also all components connected to the wheels need to be as light as possible. Indeed, the unsprung mass is determined by the mass of the wheels, tires and brakes as well as by the moveable part of the suspension.

9.2.4 Suspension Types

For the sake of simplicity, suspension designs can be classified into three major categories:

1. Independent suspensions.
2. Semi-dependent suspensions.
3. Dependent suspensions (solid axles).

In independent suspensions, a wheel may shake without affecting the other wheel of the same axle. Conversely, in a solid axle suspension, the two wheels are part of the same solid body. In a semi-dependent suspension, the two wheels are part of the same body which is not rigid but provided with a certain flexibility.

Not all suspension types can be used for both front and rear axles. The ease with which the different designs can satisfy the specific needs of the two axles has determined a natural selection.

In front axles there are two major constraints, the first being the bulk of the engine (either transversal or longitudinal) which requires a suspension design which does not intrude within the engine compartment. The second is the presence of steering wheels which may also be the driving ones. These constraints have been appropriately overcome only by a few types of independent suspensions currently used on virtually all vehicles:

- Mc Pherson suspension.
- Double wishbone suspension.

A general overview of rear axle suspensions shows that there is no dominant solution, but a balance between advantages and disadvantages of each of them leads to a correct choice. Rear axle constraints are in fact less binding than for the front axle. The most common types are the following:

- Solid axle suspension: limited to low-to-medium range models and commercial vehicles.
- Trailing arms suspension: it is an independent suspension applied to low-to-medium range models.
- Twist axle suspension: it is a semi-independent applied to low-to-medium range models.
- Mc Pherson suspension: it is an independent suspension applied to medium-to-high range models.
- Double-wishbone suspension: it is an independent suspension increasingly common in medium-to-high range models.
- Multilink suspension: it is an independent suspension becoming increasingly applied to premium medium-to-high range models.

The features of the most common suspensions will be described below.

Independent Suspensions

As seen above, the independent suspension architecture is applied both to front and rear axles. It is characterized by the fact that the two wheels are completely disconnected one with respect to each other. Reference is therefore commonly made to a reciprocally symmetric right and left suspension.

They are the best in terms of both comfort and handling because the movements of one wheel, induced by road roughness and forces, do not affect the other which holds its position.

Independent suspensions may be classified roughly according to the number of linking elements (generally rods and arms) which connect the wheel to the body. Logically, fewer linking elements in a suspension will correspond to higher functional integration and lower complexity (and consequently lower cost) but also lower functional performance ratings.

The following designs belong to the independent wheel suspension category and are ranked by increasing number of links in their mechanism:

- Trailing arm suspension (one link).
- Mc Pherson suspension (two links).
- Double wishbone suspension (three links).
- Multilink suspension (four or five links).

The features of these suspension designs are illustrated in greater detail on the following pages.

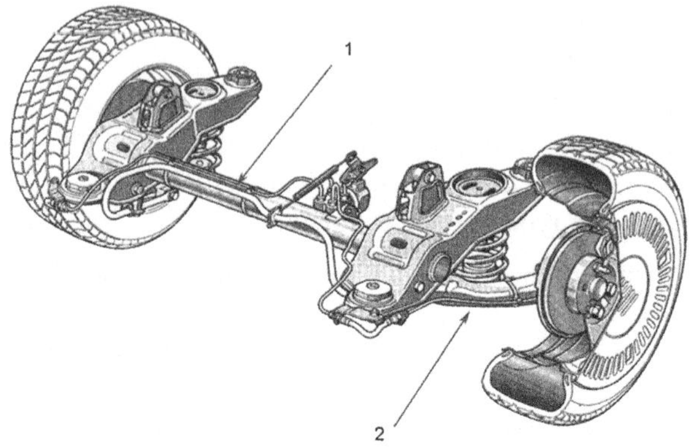

Fig. 9.12 Trailing arm rear suspension for a rear idle axle. *1* Cross member; *2* trailing arm (courtesy of Fiat)

Trailing arm suspension

The trailing arm suspension (see Fig. 9.12) is presently used for rear suspensions only, but some examples of front steering suspensions were applied in the past; in this case the arms were set in the opposite directions and called leading arms.

The wheel hub is connect to an oscillating arm (2) rotating about a transversal axis through a roller bearing; bearings are usually integral with a cross member (1). The wheel thus simply follows the arm without the possibility of changing its angles with reference to the body: The wheels remain parallel to the body and therefore, during rolling, the camber angle with respect to the ground will change accordingly.

A variant of this suspension design is obtained by slanting the axis of rotation of the arms on a horizontal plane. This is called a semi-trailing arm suspension. Better performance in terms of toe-in/out and camber are obtained during shaking, but such parameters vary also according to vehicle load and trim.

This suspension features the following advantages:

- Internal friction and, consequently, hysteresis are very low with roller bearing versions.
- No toe-in or toe-out variations with suspension stroke.
- Suspension intrusion into the baggage compartment is minimal.
- High simplicity and, therefore, low production cost.
- Ease of assembly.
- Suitable for both driving and idle axles.
- Reduced value for the unsprung mass.

The following disadvantages should be taken into account:

- Transversal deformations of trailing arms caused by cornering forces have an oversteering effect.
- No camber recovery.

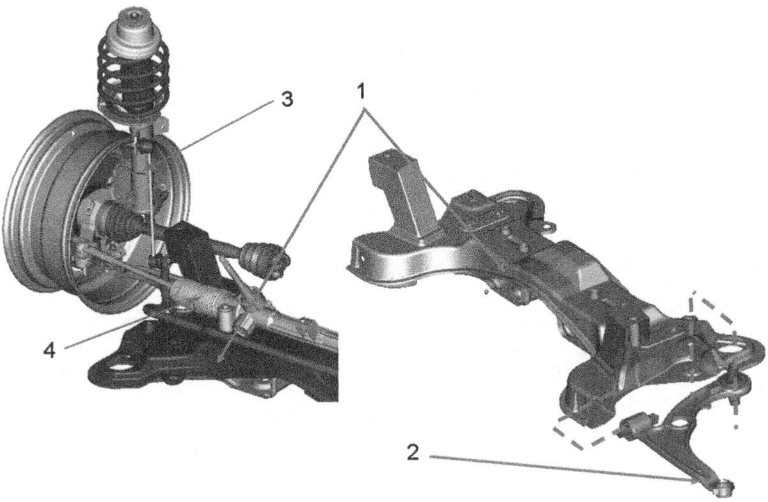

Fig. 9.13 Mc Pherson suspension for a front driving axle. *1* Cross member; *2* lower arm; *3* strut; *4* steering rack (from Genta and Morello 2009)

- Low longitudinal flexibility because of the rigidity of highly loaded bearings.
- There are no independent parameters to be tuned for improving the elasto-kinematic behavior.
- High vibration transmittance from the wheel because of bearing stiffness, again due to the value of acting loads.

This solution is used predominantly in low-end market segments.

Mc Pherson suspension

The Mc Pherson suspension shown in Fig. 9.13 is used worldwide in front axles, both in the driving and idle version; it represents a good example of how functions can be integrated. As for other suspensions, it is assembled on a structural cross member (1). It consists of a deformable triangle made with a lower arm (2) and a strut (3); the strut is a telescopic structural element integrating spring and shock absorber; the strut can rotate around its axis, providing steering function.

The Mc Pherson shock absorber in the strut differs from a traditional component only in terms of its structure: the operating principle is exactly the same. An advantage of a Mc Pherson shock absorber is the integration of several functions with consequent weight reduction, lower costs and a more compact size in relation to the engine compartment. A disadvantage is that, during its operation, it is always subject to lateral loads and the sliding stroke is not always free, presenting higher friction and hysteresis with respect to a traditional shock absorber.

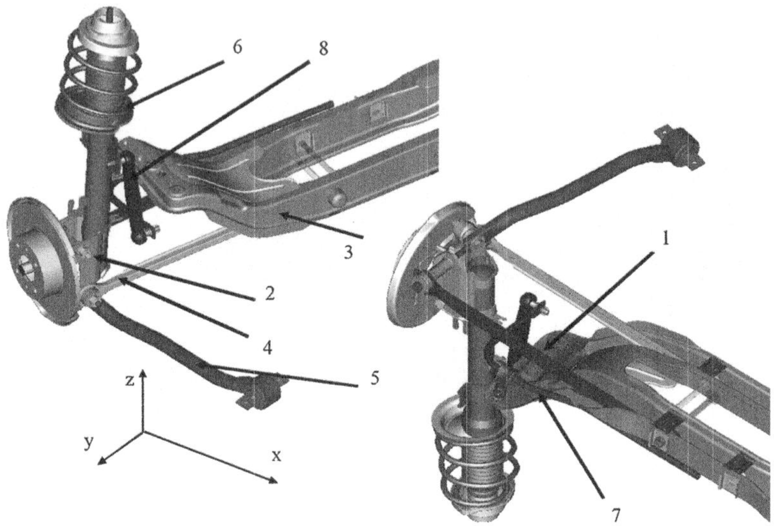

Fig. 9.14 Mc Pherson suspension for a rear idle axle. *1* and *4* transversal arms; *2* strut; *3* cross-member; *5* longitudinal arm; *6* spring and shock absorber; *7* and *8* roll stabilizer (from Genta and Morello 2009)

To compensate, at least, for the part of the lateral load transferred to the shock absorber by the vehicle weight which causes bending, the axis of the suspension springs are inclined with respect to the shock absorber axis.

The steering angle is applied by the steering rack (4) through two steering arms. It should be noted that the Mc Pherson suspension strut is intrinsically free to rotate therefore allowing steering of the wheel; in case of rear non-steering axles, instead of the lower arm, a three link system is usually applied to avoid, or control, wheel steering.

In the example shown in Fig. 9.14, the function of the lower arm is performed by a longitudinal arm (5) and two transversal arms (1) and (4); these two arms across avoid wheel steering, while the longitudinal arm allows mounting a very flexible bushing to grant a longitudinal compliance for improving comfort.

This suspension features the following advantages:

- Design simplicity and reduced cost.
- Because of the relevant separation of body joints, forces exerted on the body are low in comparison, for example, to a low double wishbone suspension.
- Higher suspension stroke than in other suspensions (a double wishbone one for example, because of the limitation on upper arm length).
- Limited transversal size, due to the absence of the upper arm; this is expedient in transversal engine installation.
- Possibility of designing with larger longitudinal flexibility, without greatly affecting the caster angle.

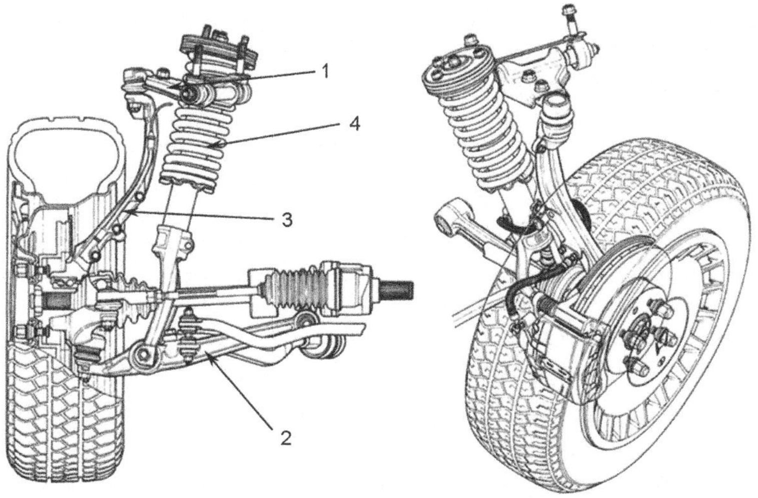

Fig. 9.15 Front double-wishbone suspension for a driving axle. *1* Upper arm; *2* lower arm; *3* strut; *4* spring and shock absorber assembly (from Genta and Morello 2009)

- The ratio between suspension and shock absorber stroke is near to one. Shock absorbers therefore work well with limited loads, low oil heating and valve wear.

The following disadvantages should also be mentioned:

- Lower performance in camber recovery in comparison with a double wishbone suspension.
- Suspension characteristic geometry causes a position for the upper pivot interface with the body, usually called *dome*, which is usually far removed from the stiffest structures of the body. This causes significant problems with suppression of vibrations and noise from the road.
- Shock absorber piston rod deformation can increase friction and hysteresis.
- Notable height for the upper pivot, so that the spring and shock absorber are set over the wheel; this could worsen the vehicle's aerodynamic shape and sporty body style.

For these reasons, Mc Pherson suspensions are found on nearly all low-to-medium range models and are replaced by more sophisticated designs in higher range models. Front and rear Mc Pherson suspensions account for nearly 50 % of all the suspension systems currently manufactured.

Double wishbone suspension

An example of *double-wishbone suspension* for a front driving axle is shown in Fig. 9.15.

The wheel is guided through a strut (3) by an upper (1) and a lower (2) arm, called *wishbones* for their triangular shape; the shock absorber and spring assembly (4) is not a structural component as in a Mc Pherson suspension. The difference in length of the two wishbones allows designing a suitable camber recovery, because the wheel is decreasing it camber angle, with reference to the car body, when suspension is compressing. For this reason it is also called SLA (Short and Long Arm) suspension.

There are two variants of a double wishbone suspension: low wishbone (mainly used in rear-wheel drive models with longitudinal mounted engine) and high wishbone (mainly used in front wheel-drive models with transversal engine); the version in the figure is a high wishbone version, where the upper arm is over the wheel; in low wishbone suspension the upper arm is behind the wheel.

The double wishbone suspension features the following advantages:

- Optimum design of elasto-kinematic parameters, particularly as far as camber recovery is concerned.
- Shock absorbers have no structural function; comfort can be improved, because of hysteresis reduction.
- Possibility of lowering the hood profile, particularly for the low wishbone version.

There are also the following disadvantages:

- Production costs are higher, because of the increased number of parts; an arm and its bushings are added, as compared with the Mc Pherson suspension.
- Additional parts for upper suspension attachment.
- The space taken by the upper arm is not small. Transversal engines demand the high version: The reduced length of the upper arm of this version compromises the possibility of reaching a long suspension stroke.
- The increased number of joints and bearings can affect wheel angles because of permanent deformations in the bushing rubber, with negative consequences on tire wear.
- The high value of braking loads can have a negative influence on longitudinal flexibility.

Double-wishbone suspensions (particularly the high wishbone arrangement) have become increasingly popular over the past years following the demand for better handling and comfort features and are commonly fitted on high and medium range vehicles instead of Mc Pherson suspensions.

Multi-link suspension

The most sophisticated independent suspension is the multi-link suspension (see Fig. 9.16) only fitted on rear axles. The wheel is linked to the body by means of five independent linkages thus providing a full control of all wheel angles and displacement.

This suspension could also be considered as a double wishbone suspension with an upper (2) and lower (3) cross arms, providing the desired camber recovery. Since

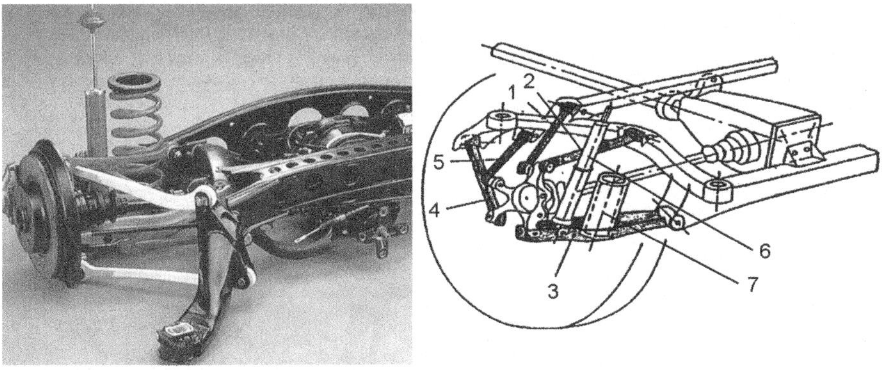

Fig. 9.16 Multi-link suspension for a rear driving axle. *1* Upper trailing arm; *2* upper cross arm; *3* lower cross arm; *4* lower trailing arm; *5* false steering arm; *6* shock absorber; *7* spring (from Genta and Morello 2009)

two cross arms do not grant the necessary longitudinal stability as two wishbones, two trailing arms are added, the upper (1) and the lower (4) ones; they provide longitudinal stability and their bushing can be designed for a sufficient longitudinal compliance for a better comfort, without affecting toe angles significantly. A fifth linkage (5), or false steering linkage, is used to control the toe angle.

The difference from a wishbone suspension is limited; nevertheless elasto-kinematic behavior will be different considering the different role played by the increased number of bushings.

When adopted to a rear driving axle, toe angle variations depend on the traction force, in addition to the cornering and braking forces; this fact must be taken into account while designing rubber bushings.

In rear wheel driven cars the wheels can increase their sideslip angle as a function of the traction force, owing to the interaction between longitudinal and lateral forces, thus affecting steering angles. This phenomenon is called *torque steering*.

To avoid a path change, the front wheel steering angle should be decreased, but an elastic bushing in the false steering linkage, suitably compliant to traction force can reduce this undesired phaenomenon.

In summary the design freedom allowed by this suspension is notably increased, particularly for a driving axle.

The following advantages of this suspension can be mentioned:

- Controlling toe angle variations as functions of cornering and braking forces.
- Camber recovery.
- Wheelbase increase at compression stroke.
- Reduced torque steering effect on rear wheel driven cars.

There are also the following disadvantages:

- High mechanical complexity.
- High production cost.

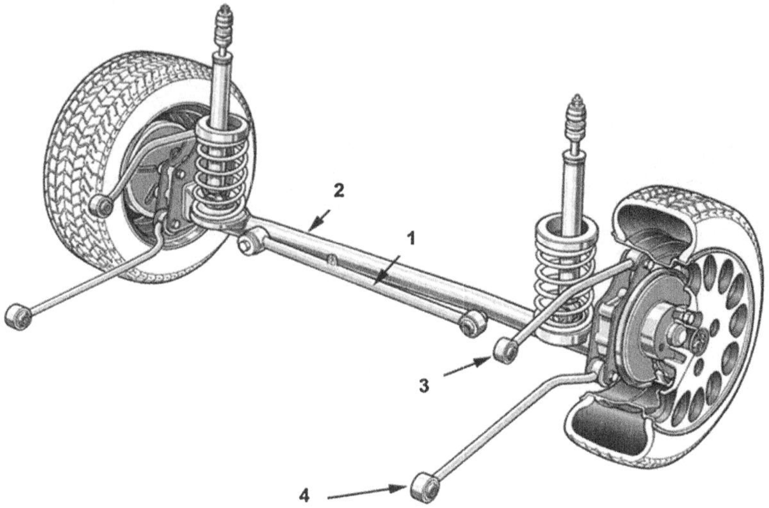

Fig. 9.17 Rigid rear suspension with Panhard bar; *1* Panhard bar; *2* solid axle *3/4* longitudinal rods (courtesy of Fiat)

- High volume and weight.
- High sensitivity to variation in the elastic behavior of bushings.

This suspension is the best choice for rear axles of premium cars of all dimensions, for both front and rear wheel drive.

Semidependent and Dependent Suspensions

Semidependent and dependent suspensions are only applied to rear axles. Dependent suspension, or solid axles, were applied in the past also to front axles but remain now on large industrial vehicles only.

They are characterized by the fact that the two wheels are connected to each other. Reference is therefore commonly made to a complete axle. The difference between semidependent and dependent suspension is only due to the stiffness of the mechanical structure connecting the wheels.

Solid axle suspension

A *solid axle suspension* suitable for rear axles of front wheel drive vehicles is shown in Fig. 9.17. A similar solution can be imagined for rear wheel drive vehicles.

This is the oldest and simplest of all suspension designs. Despite its age, it is the only design able of keeping the wheels constantly perpendicular to the ground.

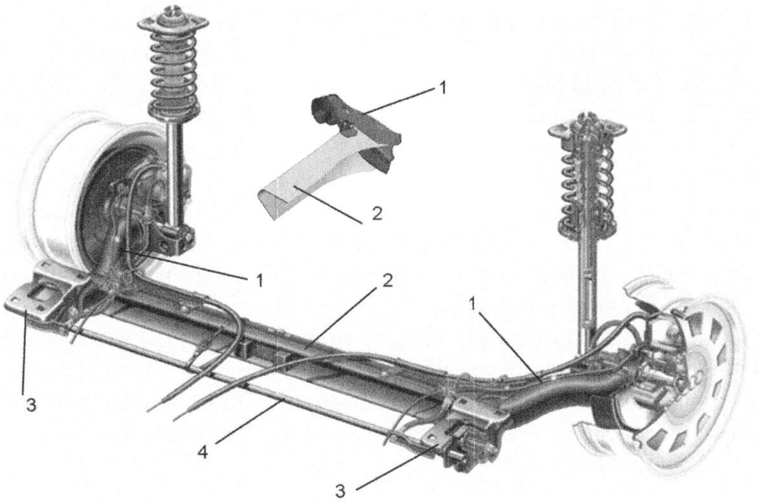

Fig. 9.18 Twist axle suspension. *1* Trailing arm; *2* cross member; *3* bearing; *4* antiroll stabilizer (from Genta and Morello 2009)

The axle rigidly joins the wheels and is connected to the body by arms of various type, position and size; in this case there are two pairs of longitudinal links (3) and (4), reacting to braking forces and one transversal link, a Panhard bar, reacting to cornering forces.

In older suspensions these links were substituted by two leaf springs, capable of exerting a limited kinematic control of the axle.

The obvious advantages of this solution are that the wheels are always perpendicular to the ground and the limited costs; the disadvantages are to be found in the high unsprung mass and in the limited control of toe angles.

This suspension is gradually abandoned in favour of higher performance types, namely twist axle suspension for the lower range and Mc Pherson, double-wishbone or multi-link for medium-to-higher range models.

Twist axle suspension

The only known application of this type is the *twist axle (or twist beam) suspension*; one example of this design is shown in Fig. 9.18.

In this suspension, the wheels are rigidly connected to two trailing arms (1) joined by a crossmember (2); the two arms are articulated to the car body through suitable bearings (3). A particular feature of the crossmember is high bending stiffness combined with a fairly low torsional stiffness; these apparently conflicting properties are achieved by designing the cross member with an open cross section (V or U shaped), like in the detail shown in Fig. 9.18. The high bending stiffness forces the arms to remain parallel also when the wheel strokes on each side are different; the low torsional stiffness allows this stroke difference with low material stresses.

A stabilizer bar (4) is arranged if the anti-roll effect of the crossmember is not sufficient. The shock absorber and spring assembly is arranged in vertical position next to the wheel. This configuration allows sufficient camber recovery and high stabilizing effect; the behavior of this suspension is a way in-between a solid axle and a trailing arm suspension.

These features, combined with light weight, low cost and sufficient space left for fuel tank, exhaust pipe and spare wheel make this the suspension design of choice for low and medium range vehicles.

Appreciated advantages of this suspension are the simple construction and the low number of parts, with consequent low weight and costs. Its disadvantages are the limitations in handling and comfort optimization. For example, it is difficult to control toe angle variations under side load and to obtain bearings with sufficient longitudinal compliance.

9.2.5 Anti-Roll Bars

The inertia forces applied to the centre of mass of the vehicle supported by elastic suspensions cause different kinds of motions of the body, displacements and rotations alike. In particular, transversal forces cause the *roll rotation* of the body around its longitudinal axis The sprung mass rolls about a virtual axis called *roll axis* whose position can be calculated by taking into account the suspension design and tire flexibility.

A certain amount of roll is advisable because it gives the driver a feeling of the centrifugal forces and the related danger when tackling corners at high speed. In other words, it provides a useful safety feedback to the driver. However, excessive roll makes the ride uncomfortable or can affect vehicle stability. The suspension system (linkages, elastic elements, damping elements) must therefore be carefully designed to take into account this target.

Anti-roll bars or *roll stabilizers* are used to limit roll within acceptable comfort and safety conditions. An anti-roll bar elastically connects the two wheels of the same axle together and reacts only to motions of each wheel with respect to the other, as shown in Fig. 9.19. The main springs (1) react to both jounce and roll (equal or differential strokes of the suspensions), while the anti-roll bar reacts to differential strokes only, when the sprung mass rotates by a roll angle θ or manages obstacles of different height.

The use of an anti-roll bar should allow in principle the application of softer vehicle suspension springs and consequently improve riding comfort. Indeed, when cornering, the anti-roll bar reacts to part of the roll load by decreasing the load on the main springs.

The bar therefore helps suspension springs in all circumstances when roll occurs and also when a wheel runs over an asymmetric bump. It should not be forgotten that,

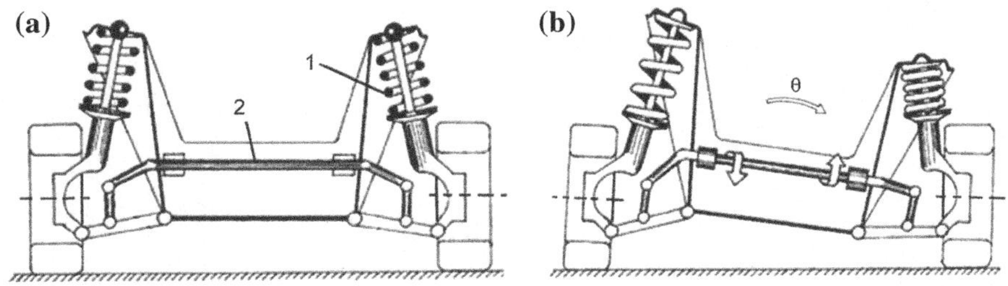

Fig. 9.19 Mc Pherson indipendent suspension with anti-roll bar. **a** No roll condition: indipendent suspension with 1 suspension spring; indipendent suspension with 2 anti-roll bar. **b** Condition with roll angle θ

since most of the bumps are asymmetric, roll-bars have a certain negative impact on comfort: again another parameter which is a matter of trade-off between handling and comfort.

9.2.6 Shock Absorbers

The sprung mass of a vehicle, together with the suspension springs connecting it to the ground, is an elastic system which can be roughly modeled as a mass-spring system. When the spring receives external energy, e.g. when tackling an obstacle, since the system is undamped, the mass would continue to oscillate for an infinity of time at its resonance frequency.

In actual systems, the friction in the bearings of the suspension arms and the internal damping of elastic materials will prevent this condition from happening. However, after encountering an obstacle, free oscillations of the suspension will occur for a certain time and this is not acceptable, either for comfort or handling.

The *shock absorber* is a component introduced to preventing oscillations, since it can dissipate the energy stored in the suspension system. This device feels the relative motion between body and wheel and generates forces which oppose the relative motion or better, the relative velocity. For this reason, one end is connected to the body and the other to a component integral with the wheel.

A scheme of a conventional double tube shock absorber is shown in Fig. 9.20. Oil fills the volumes A and C inside the shock absorber, except for the part over the volume C (compensation volume) which is filled with compressed nitrogen. Here, when the piston (6) penetrates the internal tube (2), the oil flows from the central body, pushed by the higher volume occupied by the piston and rod. The opposite occurs in extension when the oil flows back to the central body. The compressed gas reduces the natural tendency of oil to emulsify and the related hydraulic noise. An outer third tube (7) protects the inner tubes in which the piston runs.

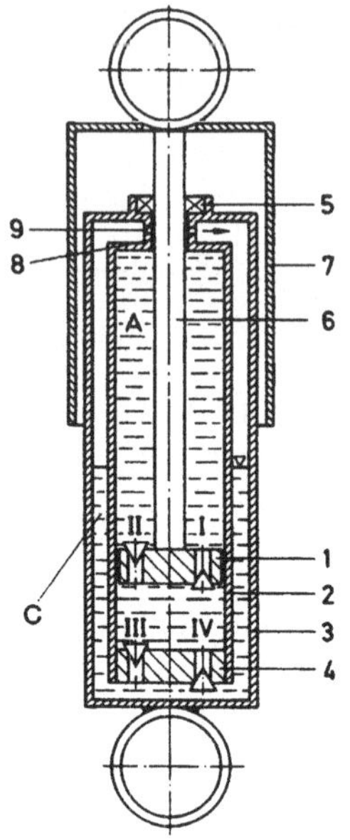

Fig. 9.20 Double tube shock absorber; *1* piston; *2* internal cylinder; *3* external cylinder; *4* bottom valve assembly; *5* seal; *6* rod; *7* protective tube; *8* rod guide; *9* leakage return holes; **a** Working chamber; **c** Compensating chamber; I, II, III, IV: valves (from Genta and Morello 2009)

The operating principle of a hydraulic shock absorber is based on the resistance opposed by oil forced through holes which are regulated by valves opening according to the direction of motion of the piston. The release valve II will open during compression and the release valve I during extension. The resistance opposed to the piston motion depends on its speed, the oil viscosity and therefore on its temperature.

The valves I and II can be designed to offer different resistance to the piston in the compression and extension stroke. The release valves III and IV do not play a significant role in the damping force but allow the main chamber A to be always full.

The force may take different values according to the piston speed (and therefore of the wheel vertical speed with respect to the body) and the variation laws can be designed by chosing the size of the valves and the stiffness of the springs of the release valves. Such laws define the damping force as a function of the piston speed in either extension (rebound) or compression.

These plots may be adjusted according to the internal parameters of the shock absorber, such as the size and shape of the oil passage holes, the calibration of the valve springs and the number of valves. The characteristic laws may therefore be customized for the various vehicle models, according to the type of suspension and to the required comfort and handling objectives.

Two characteristic diagrams are shown in Fig. 9.21 for a soft calibration (dash-and-dotted line) and for a hard calibration (solid line). The latter can be used to limit body movements in a sports car.

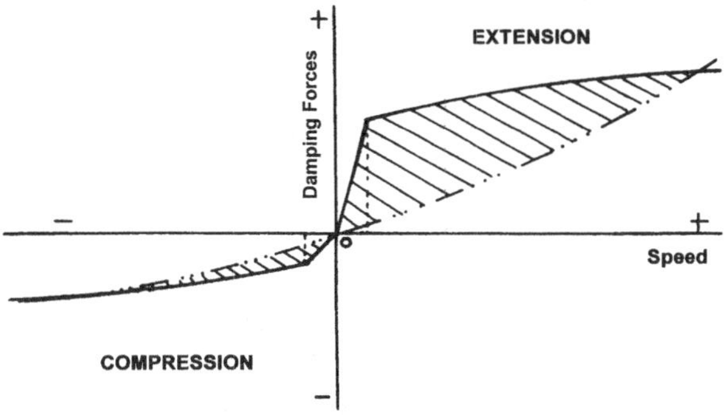

Fig. 9.21 Damping force versus piston speed (*extension and compression*); the diagram shows an example of what can be obtained from spings design for a soft calibration (*dash-and-dotted line*) and hard calibration (*solid line*)

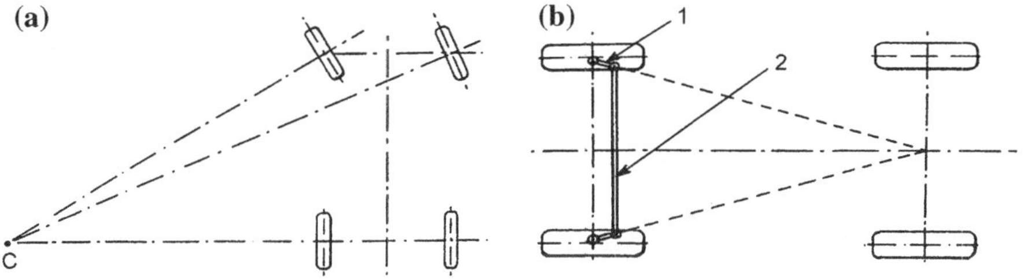

Fig. 9.22 Correct steering conditions. **a** Geometric relationship between wheel position and center of curvature of the path. **b** Scheme of the Jeantaud's steering mechanism: *1* steering lever; *2* steering coupling rod

9.3 Steering System

A vehicle can be driven along a curved path by steering the front wheels. If a four wheel vehicle is driven along a circular path with its center at point C, the lines perpendicular to the equatorial plane of each wheel must cross at this point, as it is shown in Fig. 9.22a.

If this geometric condition is not met, sideslip angles arise even at minimum speed and with negligible cornering forces. Therefore, to avoid sideslip when cornering, the inner wheel must have a higher steering angle than the outer wheel. These no-slip steering angles are called Ackermann steering angles, from the name of who filed the first patent on this subject.

A mechanism, satisfying this condition only approximately, was developed by Jeantaud for steering solid axles and is still in use on modern vehicles with this kind of suspension, such as trucks and off-roads. According to the Jeantaud's mechanism, the steering levers axes must ideally cross on the rear wheel rotation axis (in the assumption that they are not steering), as in Fig. 9.22b.

The parameters characterizing vehicle steering are heaviness, step response, precision, lane keeping and return.

- *Heaviness* is a parameter measured by the torque to be applied to the steering wheel. It depends on front suspension geometry, tires (dimensions, pressure, etc.), steering transmission ratio, and, obviously, on the effect of the power steering system when applied. An ideal steering wheel should be light when the vehicle is still or maneuvering at low speed but react with a torque proportional to the tire lateral grip at higher speed. For this reason, it would be wrong to reduce the torque on the steering wheel as much as possible because this could severely impair the driver's sensitivity in high speed manoeuvres.
- *Step response* indicates the vehicle's cornering response to sudden steering inputs. The time elapsed between the steering peak angle and the maximum lateral acceleration is measured when applying a step input to the steering wheel: the shorter the time, the more responsive the steering system. There must be a balance between obtaining fast cornering and maintaining sufficient control.
- *Precision* depends on how the vehicle can be driven on a constant radius curve without need of corrections, also when road bumps are met.
- *Lane keeping* is the capability of the steering system to hold a straight path without frequent corrections, also with road bumps. Precision and lane keeping are related to suspension geometry (suitable toe angle variations may contrast the lateral thrust impressed by bumps), to friction in the steering mechanism and to plays in the steering linkages.
- *Return* is the capability of the wheels to return spontaneously to the straight position while the vehicle is moving; it depends on the steering transmission ratio, friction and characteristic angles of the front suspension.

The transmission of motion between the steering wheel and the front wheels is made by two different kinds of mechanism: screw-and-sector and rack-and-pinion steering boxes.

9.3.1 Screw-and-Sector Steering Box

Screw-and-sector steering boxes are used both for independent suspensions and solid axles, as shown in Fig. 9.23. In the first case, the steering box (1) is moving an articulated parallelogram with two articulations fixed to the body (in black in Fig. 9.23a) which steer the front wheel steering levers through two steering rods; in the second case, the steering levers are connected by a coupling rod (3) and one of them is moved by a single steering rod (see Fig. 9.23b). The rack-and-pinion steering box, characterized by a higher mechanical efficiency, is currently preferred for independent suspensions.

The steering box could include a simple screw and a gear wheel sector moving the steering arm, as was often used in the past; this solution would imply high internal friction with the consequence that this kind of transmission is all but irreversible,

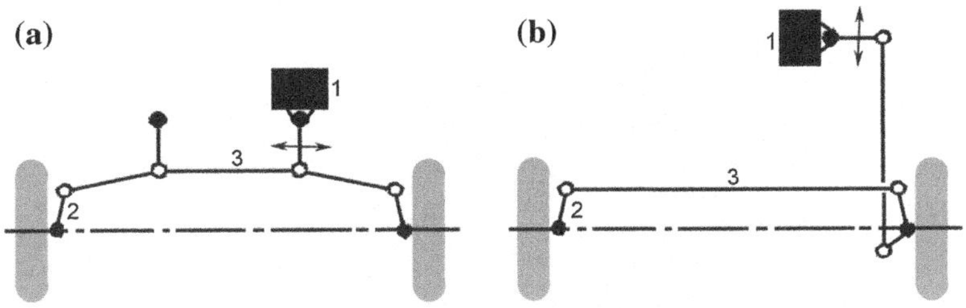

Fig. 9.23 Screw-and-sector steering mechanisms for **a** independent suspensions and **b** solid axles. *1* Steering box; *2* steering lever; *3* coupling rod

which means that the torque can be transmitted from the screw to the sector and not vice versa. This fact would imply lack of steering wheel self-alignment and lack of feeling for the existing friction between tires and ground.

Reversible kinematic pairs, working according to the same principle and using a multiple threads screw, can be produced, but with much lower transmission ratios. But in this case, the advantage of reducing the steering wheel torque almost disappears.

To counteract the irreversibility, a solution has been developed where the screw, operated by the steering column, is of globoid type and instead of a gear wheel sector there is a threaded roller, as shown in Fig. 9.24. The globoid screw has a small diamter in the center, to increase the number of working threads. The threaded roller, whose profile matches the screw, works as a rack, capable of motion according to the screw rotation, with a good reduction of mechanical friction. The circumferential motion of the roller is transmitted to the spline shaft, through a fork; the output shaft is connected to the steering arm.

9.3.2 Rack-and-Pinion Steering Box

A rack-and-pinion steering mechanism is shown in Fig. 9.25; it consists of a pinion 1, moved by the steering wheel and fitted on ball bearings (capable of withstanding the axial thrust), meshing with the rack 2; the rack, moving axially, is supported by two low-friction bushings, one under the pinion and one on the side. Side rods 3 connect the rack and the steering levers through ball joints on each end.

The steering rods must be carefully placed and sized according to the front suspension design to obtain suitable toe angle variations as a consequence of the suspension stroke.

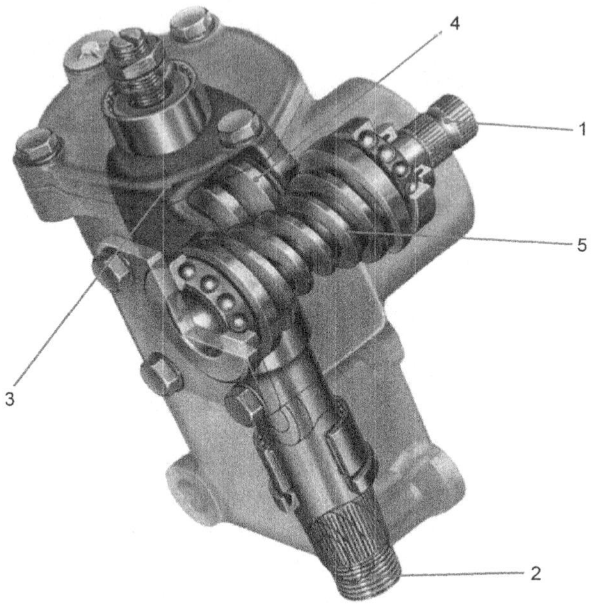

Fig. 9.24 Screw and sector steering box, to decrease mechanical friction a globoid screw is used. *1* Steering column; *2* spline shaft; *3* fork; *4* threaded roller; *5* globoid screw (from Genta and Morello 2009)

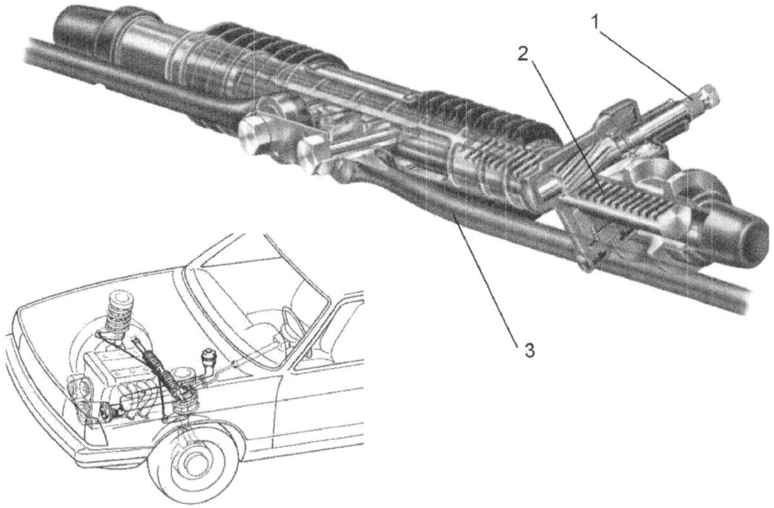

Fig. 9.25 Rack and pinion steering box with central spheric heads. *1* Pinion; *2* rack *3* steering rods (from Genta and Morello 2009)

9.3.3 Power Steering System

The constant increase in passive safety and comfort requirements and in interior room led to a general mass increase in all cars, in spite of the efforts devoted to applying light-weight materials and weight conscious design. This mass increase increased the forces between tires and ground and, therefore, the torque to be applied to the steering box.

The reduction of steering wheel torque to levels that are ergonomically acceptable was obtained by applying power assistance to the existing mechanical systems; power steering has spread from heavy luxury cars and industrial vehicles to virtually all cars.

The most widely used power assistance is hydraulic, sometimes integrated by electrohydraulic devices, to adjust the effect to vehicle speed; such hydraulic systems can be applied almost indifferently to rack and pinion or to screw and sector steering boxes. Electrical assistance is beginning to be applied on small and medium size cars and on electric and hybrid vehicles.

Hydraulic Power Steering

These systems take advantage of the hydraulic pressure of a flow of oil, generated by a pump driven by the engine that must always work. For this reason, if the internal combustion engine must be stopped from time to time, as in hybrid vehicles and when using start and stop fuel consumption reduction devices, electric power steering systems are preferred. The same obviously happens in electric vehicles.

The system includes an oil reservoir, a pump, normally of the blade type, driven by the accessory belt of the engine, and a modified steering box; a system of this kind is applied to both screw-and-sector and rack-and-pinion steering boxes. A heat exchanger is also included in the hydraulic circuit for cooling: Inadequate cooling can allow the oil to approach the boiling point, with lack of efficiency and danger of cavitation. Because of the choking valves and other circuit losses, this system has some damping properties, with benefits on steering wheel return and vibrations.

This system is relatively simple, but has a negative impact on fuel consumption; as a matter of fact, the fuel consumption increase is of about 2 or 3 % in the average driving cycle. To limit this negative effect, some systems have been developed in which the oil pump is driven by an electric motor and feeds a pressure accumulator. In this case the electric pump works only when the oil pressure in the accumulator drops below a certain threshold value, with potential reduction in energy consumption. This system also has the advantage of allowing a more flexible pump installation in the engine compartments of small cars with space problems but it had a shorter life and was abandoned in favour of simpler electric systems.

As shown in Fig. 9.26a, a double effect hydraulic cylinder is integrated into the steering box, in this case of the rack-and-pinion type, to move the rack. M and R are the input and output nipples of the oil flow and chambers S and D on the power cylinder assist left and right steering.

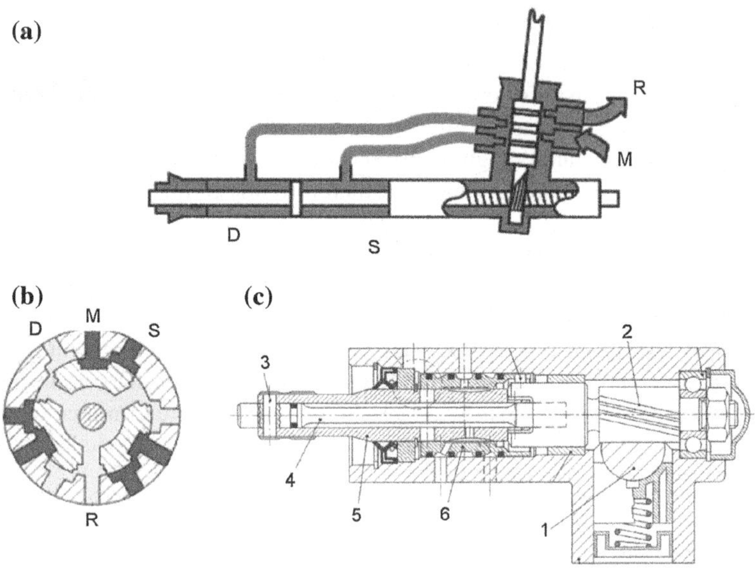

Fig. 9.26 Sketch of a hydraulically assisted rack and pinion steering box **a** detail of the hydraulic control valve **b, c** (from Genta and Morello 2009)

An enlargement of the distributor and control valve which pressurizes chambers S and D of the power cylinder is shown in Fig. 9.26b and c; the pressure force is added to that applied to the rack by the steering wheel.

The double effect power piston can have one of the sides at the intake pressure or both at the exhaust pressure, according to the angular position of the distributor and control valve. This valve is made up of two coaxial cylinders, with a relative rotation motion, featuring a number of windows, which can put the S and D chambers into communication with ports M and R, as shown in the cross section on the left.

In the position shown in the figure, the oil pressure is sent to the S chamber, while the R chamber is at the exhaust pressure. The higher the pressure, the higher the clockwise rotation angle of the inside cylinder of the valve, until a maximum opening of the window is reached; by counterclockwise rotation the pressure will be reduced until the window closes. If the counterclockwise rotation of the cylinder is continued, the same functions are performed for the other face of the piston

The inside cylinder 5, as can be seen in the right cross section, is connected through a pin 3 to one end of a torsion spring 4, loaded by the torque applied to the pinion 1 by the steering column. In the same way the external cylinder 6 is connected to the other end of the torsion spring.

The torsional stiffness of the spring 4 must be determined in such a way that the windows are completely open at the maximum steering wheel torque. At a lower torque, the spring deformation will be proportionally reduced; the steering wheel

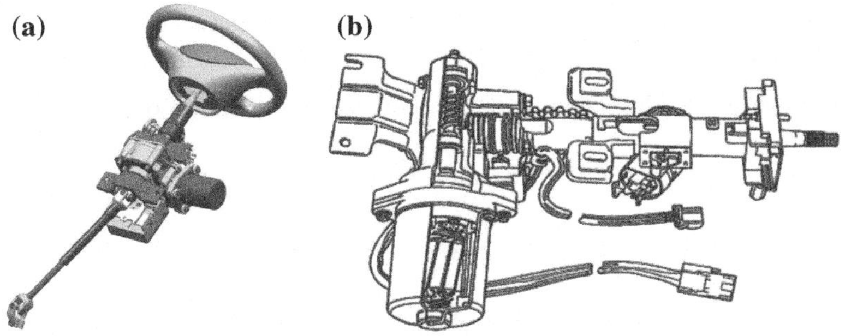

Fig. 9.27 Electric power steering system (EPS) showing an electric motor working on the steering column. **a** A view of the assembly and **b** a detail of the motor with reduction gearbox (courtesy of Fiat)

torque, even if reduced, will reproduce the aligning torque variation, thus giving a correct feeling for the dynamic behavior of the car.

This trailing mechanism reduces the assistance torque to zero as soon as the steering wheel torque is set to zero. The operation is the same when the steering wheel moves the wheels or when the wheels apply torque to the steering wheel, during steering wheel return or while driving over an asymmetric obstacle.

Pressure pulses or variations can generate vibrations and noise; therefore it is a good practice to apply rubber bushings to the steering box mounts and damping joints to the tubes under pressure.

Electric Power Assistance

Electric Power Steering (EPS) systems are now quite widespread. They are applied to rack and pinion steering boxes and operate through an electric motor adding torque to the steering column, or to the pinion in the steering box directly. Figure 9.27 shows the assembly including a modified steering column and the detail of the electric motor with the reduction gearbox.

The electric motor is driven by an electronic controller, which has the function of generating an assistance torque proportional to the steering torque. The control system includes sensors able to measuring the steering torque, vehicle speed, steering wheel speed and angle.

These data are used to calculate the optimum torque to be applied by the motor; this value is a function of the vehicle speed, to improve the driver awareness of small steering torque values. The assistance is increased at large steering angles, typical of parking maneuvers and narrow turns, where actuation speed is more important then sensitivity.

This optimum torque is applied in the correct direction, as given by steering angle acquisition. The motor is a simple direct current unit with permanent magnets, driving a screw gearbox.

This system has a number of advantages when compared with a conventional hydraulic unit:

- Oil circulation is eliminated, with consequent simplification of the engine compartment lay-out.
- Fuel consumption is reduced, because torque control is not based upon wasted power.
- Active safety is increased, because assistance is also available when the engine stalls or is switched off.
- There is the possibility to design a more sophisticated torque control than in hydraulic systems.

At present, a disadvantage is a limited available torque, due to the working voltage, which is standardized for cars. This limits the application of such systems to vehicles with a total mass of about 1,500 kg; the availability of higher voltages in electric and hybrid cars make this system suitable also to larger cars.

With increasing production volumes new dedicated configurations have been developed, in which the electric motor works directly on the pinion or the rack. This last application, in which the electric motor works on the rack through a recirculating ball drive, is more critical for what the design of the electric motor, working with larger torques, is concerned.

9.4 Braking System

The braking system has two different functions: To slow-down or stop the vehicle in a reasonably short space and to keep the vehicle still when it is parked. According to these two needs, a vehicle is equipped with two different systems: a service and a park braking system.

9.4.1 Service Brake System

Reducing the speed of a vehicle requires the dissipation of the vehicle's kinetic energy. This energy is converted into heat by the friction between two surfaces. One of these surfaces is fixed to the suspension (braking pad) while the other is rotating with the wheel (drum or disc); they are pressed together usually by an hydraulic pressure. When braking, the friction surfaces reach temperatures in the order of several hundred degrees.

A braking system consists essentially of four groups of components shown in Fig. 9.28:

- Control components: pedal, fluid pump and reservoir, booster.

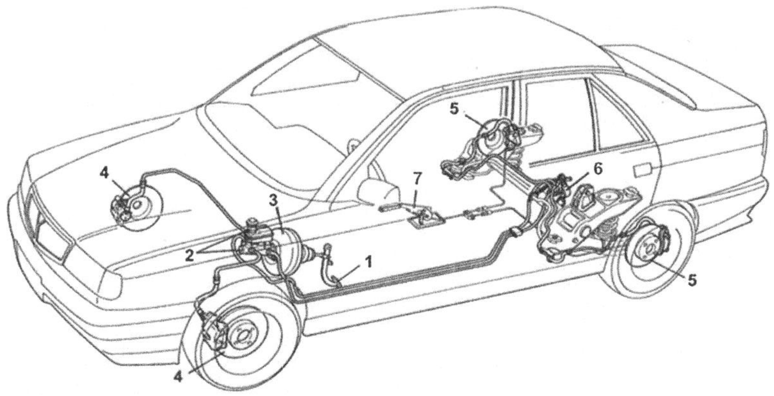

Fig. 9.28 Main components of an hydraulic service braking system: *1* pedal; *2* fluid pump and reservoir; *3* booster; *4/5* front/rear brakes; *6* pressure regulator. The park braking function is made by the hand brake *8* (courtesy of Fiat)

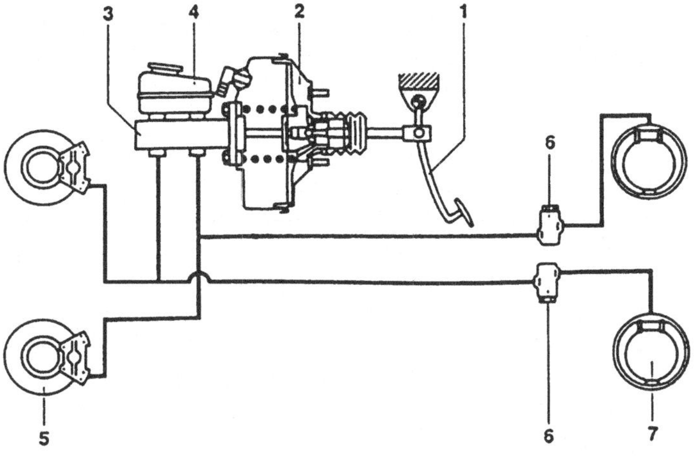

Fig. 9.29 Crossed circuit braking system; *1* brake pedal; *2* vacuum brake booster; *3* double circuit pump; *4* brake fluid reservoir; *5/7* front/rear brake; *6* pressure regulator

- Transmission components: a circuit made with rigid and flexible pipes, to allow wheel suspensions motion.
- Actuation components: front and rear brakes.
- Regulation components: pressure regulator valves.

The brake hydraulic circuit normally consists of two independent sections, as shown in Fig. 9.29; in this example, one operates the front left and the rear right brake, the other the front right and the rear left brake. This ensures an adequate braking capacity in the front axle (which has more load while braking) also if one of the two lines breaks.

Table 9.1 Braking circuits allowed by the regulations of the European Union; circuits 1 and 2 describe the still working brakes, when each of the flexible tubes, at a time, is broken

Symbol	Circuit 1	Circuit 2
TT	Front axle	Rear axle
K	Front r. and rear l. wheels	Front l. and rear r. wheels
HT	All wheels	Front axle
LL	Front axle and rear r. wheel	Front axle and rear l. wheel
HH	All wheels	All wheels
X	Front l. and rear r. wheel	Front r. and rear l. wheel

Braking circuits can be organized also in different ways; they are described in Table 9.1. Circuits 1 and 2 represent the possible configurations of the emergency circuit when one pipe is out of order. The choice among these schemes must take into account the vehicle load breakdown on the axles.

The scheme TT, for example, is suitable for a car whose center of mass is located approximately at the middle of the wheelbase, as can happen in a front engine, rear drive two seater. Schemes K or X (corresponding to Fig. 9.29) are, instead, suitable to a front wheel driven car, where it is necessary that at least one of the front wheels can brake, to satisfy the requested performance when loaded with the driver only.

The following schemes, such as HH, have the advantage of granting the emergency braking system a complete, or almost complete, braking circuit at a slightly increased cost; this allows the torque applied to the steering wheel owing to the unbalance of braking forces to be kept to a minimum.

The safety margin in *fading* conditions must be factored into this choice. Fading is here defined as a loss of braking efficiency caused by a partial evaporation of the braking fluid, as a result of high temperatures. When the fluid evaporates, it increases its compressibility dramatically and loses its ability to transmit pressure by compression. This phenomenon occurs in parts of the circuit close to a heat source and, therefore, close to a wheel. Because of this, circuits HT, LL, HH are more vulnerable when the fluid is evaporating at the front wheel.

The brakes are designed according to the full load mass of the vehicle and account for the load transfer which occurs during braking. Indeed, while braking, the rear axle becomes lighter and the rear wheels can easily skid: To avoid this, the hydraulic pressure in the rear circuit (and consequently the braking torque) must be limited. A *pressure regulator* (6 in Fig. 9.28) is fitted in the hydraulic circuit upstream of the rear cylinders to limit the pressure to a suitable value, as a consequnece of changes in the suspension position.

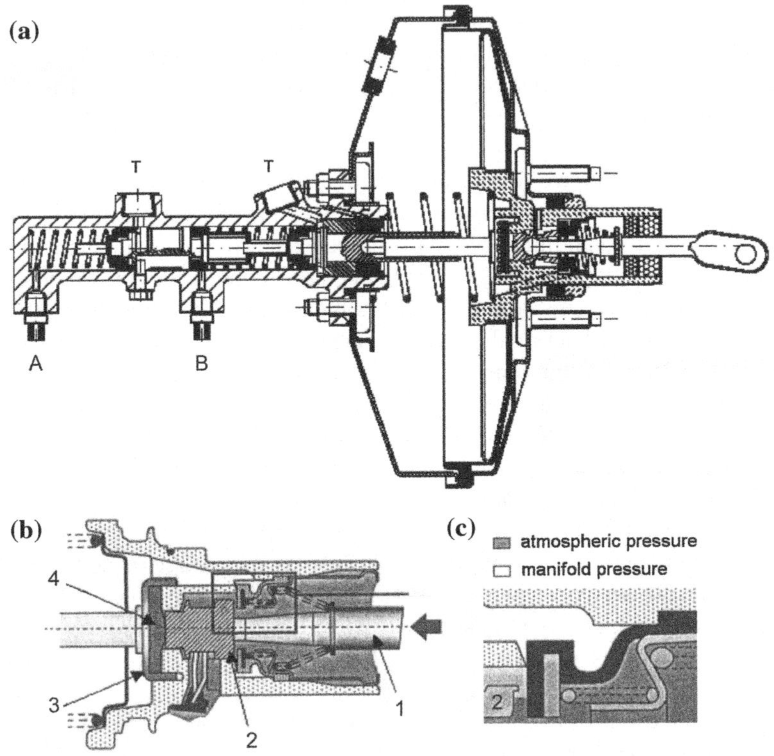

Fig. 9.30 **a** Cross section of the brake pump and booster. **b** and **c** Enlarged detail of the valve between the vacuum chamber and the ambient air, used to modulate the power assistance (from Genta and Morello 2009)

9.4.2 Park Brake System

The park system (7 in Fig. 9.28) is operated by a hand lever; the braking force is transmitted by means of mechanical linkages or flexible cables. Possible variants are related to the position of this lever (on the tunnel, on the dashboard, on the left of the driver's seat) or to the application of a pedal instead of the hand lever. In many recent cars, the park brake can be operated by an electric motor.

9.4.3 Brake Pump and Booster

The pedal operates the pump piston, through the push rod on the right of Fig. 9.30. Between this push rod and the pump the power brake is set, as will be explained later.

The *brake pump* is made by two pistons in series in the same cylinder; the openings T are connected to the oil tank, while the openings A and B to the two braking circuits. This particular pump feeds two completely separated circuits (service and emergency circuits), each of which can be connected to the brakes, according to the schemes in Table 9.1.

In rest condition, the two pistons are kept to the right by the coil springs shown in the figure; the two piston chambers are connected to the tank at atmospheric pressure; in this way the additional fluid needed to compensate for a clearance increase due to lining wear can be supplied. As soon as the pedal is depressed, the two holes T are closed and the pressure inside the circuits is increased, proportionally to the pedal force; this pressure will act on the pistons working on the pads or shoes.

This arrangement of the pump guarantees the operation of one circuit when the other has failed (spilling as a consequence of a pipe rupture); in fact, if one of the circuits should spill, the two independent pistons will contact each other and the pressure would increase in the still working circuit. The increase in pedal stroke warns the driver about the failure.

The power assistance system, traditionally called *booster*, is placed between the brake pump and the pedal and amplifies the force applied on the pedal by the driver, exploiting the difference of pressure between two chambers, one connected with ambient air and one with the intake manifold, in throttled gasoline engines. When the manifold vacuum is not sufficient, as in diesel or direct injection gasoline engines, a vacuum pump is driven by the engine.

With reference to Fig. 9.30, already used to explain the pump, the size of the actuator can be noted. The low value of the difference in pressure between the manifold and the ambient air causes the actuator to be large.

The actuator includes a cylinder and a piston made of steel sheet, with a suitable membrane seal; a front (at left) and a rear chamber are found in the actuator. The former is always in communication with the intake manifold, downstream of the throttle valve or with the vacuum pump.

Three different situations may be identified:

- Rest position, with pedal completely released,
- pedal depressed, and
- pedal released.

When the pedal is released or the pedal stroke is zero (as in Fig. 9.30), the two chambers are set at the same pressure p_v. This pressure is equal to that of the vacuum source. Because there is no pressure difference between the two faces of the power brake membrane, no assistance is present.

In the same figure, a detail of the shaft of the actuator's piston is shown in b. On this shaft there is a valve which puts the two chambers of the power brake in communication; in this figure the valve is drafted when the pedal is depressed at zero force, or when any play is set to zero.

Assume now that a pressure is applied to the pedal. After a while, the valve will cut the communication between the two chambers. In the detail of Fig. 9.30b, a plunger 1 closes the communication between the two chambers by a rubber surface. After a

Table 9.2 Description of the states of a vacuum power brake, as functions of the pedal force F and of its derivative over time $\frac{dF}{dt}$

F	Outside duct	Inside duct	Rear pressure	Assistance
$= 0$	Open	Closed	p_v	No
> 0	Closed	Open	$p_v < p \leqslant p_0$	Yes
> 0	Closed	Closed	$p_v < p \leqslant p_0$	Yes
> 0	Open	Closed	$p_v < p < p_0$	No

short while, the rubber element 4, compressed by the braking force, is deformed so that a passage 2 between the rear chamber and the outside pressure p_o is opened.

The pressure difference between the two chambers produces the assistance force. The opening of the passage is a function of the deformation of the rubber element 4, sensitive to the load applied by the pedal. This passage closes as soon as the driver reaches the desired load on the pedal; the pressure in the rear chamber will therefore be proportional to the rubber element deformation and thus to the applied load measured by the deformation. When the pedal is released, the communication with the outside closes and that between the two chambers reopens. Both chambers are thus at the same pressure, and no force is applied to the braking pump.

In summary, the possible states of the power brake are shown in Table 9.2, as functions of the load F applied to the pedal and of its time derivative.

9.4.4 Disc Brakes

Disc brakes (see Fig. 9.31) are universally used at the front axle and their application at the rear axle is constantly increasing (over 50 % of all rear brakes). They consist of a rotating disk, integral with the wheel hub, and of pads 5 which may be pressed against the surfaces of the disc by one or two hydraulic pistons that are integrated in a caliper 1.

A *floating caliper* disc brake has a single cylinder 4, working on the inside pad 5, while the outside pad 6 is pressed on the disc by the caliper body 1, able to slide on its mounts 7 along a direction perpendicular to the disc surface. This solution has the advantage of reducing cost and brake bulk, allowing the king-pin offset to be easily reduced.

Hydraulic cylinders are operated by the fluid pressure applied by the master cylinder; as soon as the pedal is released, the piston of the master cylinder returns to its initial position by means of a return spring, and the hydraulic pressure is set to zero.

The brake piston is pulled away from the disc by the elastic force, due to the lateral deformation of the sealing ring, which is mounted on the piston with a certain radial load; the shape of the groove (see the enlarged detail of the sealing ring in rest position) allows piston motion without sliding on the sealing.

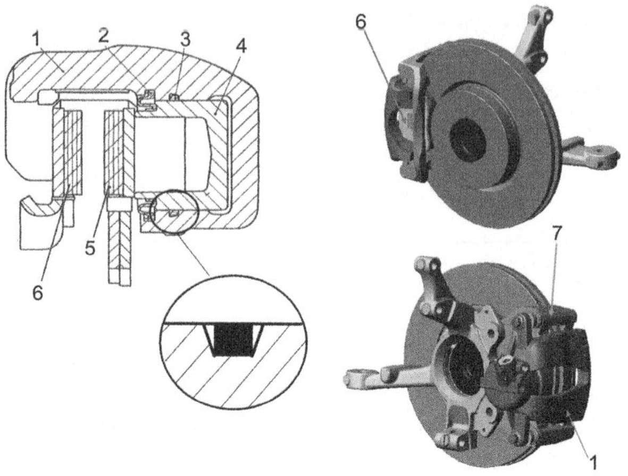

Fig. 9.31 Floating caliper disc with a single hydraulic cylinder. An enlarged cross section of the cylinder is shown with the detail of the sealing ring (from Genta and Morello 2009)

The movement of a floating caliper may be impaired or blocked by mud sediments or by corrosion. For this reason protection bellows and special coatings are applied to the slide pins. Impaired sliding of the caliper can cause different wear on internal and external pads; in some cases the caliper can be blocked by asymmetric braking, affecting the vehicle path.

The brake disc, generally made of cast iron, may be either full (the disc crown does not allow passages for the air) or ventilated (the crown is thicker and radial channels for air passage are provided inside).

In heavy and high performance cars a *fixed caliper* arrangement can be used, instead of the floating caliper; in this case the caliper cannot slide and each pad is pressed by its own hydraulic piston.

9.4.5 Drum Brakes

A drum brake, shown in Fig. 9.32, consists of a drum 1 bolted to the wheel hub 2 and a brake plate 3 that carries two brake shoes 4, covered by linings 5 that are pressed inside the drum by a hydraulic cylinder 6.

The disadvantages of this brake in comparison to the disc brake are a more difficult dissipation of the heat generated during braking and the consequent drum deformation; in addition, the longer approach stroke of the linings to the drum causes a longer idle stroke of the pedal. The only advantage is essentially a lower cost.

For this reason, the use of drum brakes is decreasing. They are only applied to rear axles of small front wheel drive cars, where the braking torque is limited.

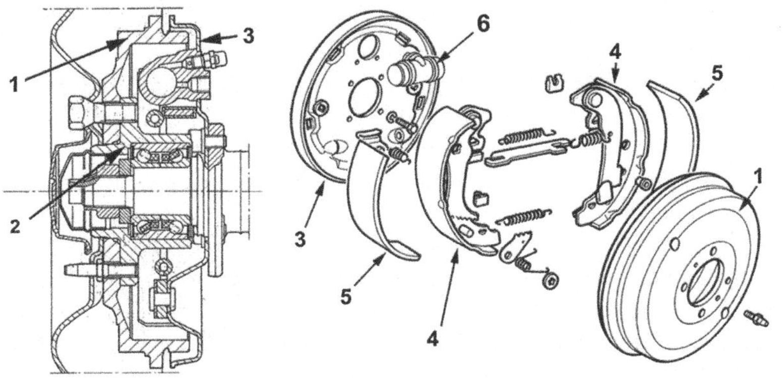

Fig. 9.32 Drum brake components: *1* drum; *2* wheel hub; *3* brake plate; *4* brake shoes; *5* brake liners; *6* hydraulic cylinder (from Genta and Morello 2009)

9.4.6 Brake Control Systems

Longitudinal forces generated by the braking system can affect not only the longitudinal, but also the lateral dynamics. Figure 9.9 shows how the maximum lateral force F_y is affected by the longitudinal force, decreasing when $F_x > 0$, to reduce to zero when the latter reaches its maximum.

Non-symmetric braking torques applied to the wheels can produce moments around a vertical axis, that can affect the vehicle path. Brakes can thus be used not only suitable for controlling the vehicle speed, but also for controlling its path and the traction of the wheels. The main control systems that are used to modulate the braking action are described in this section, using the trademarks and acronyms introduced by Bosch, since they have entered the common technical language.

ABS System

The objective of ABS (*Anti-lock Braking System*) systems is to avoid excessive longitudinal slip that can impair braking action and lateral dynamics control. This objective must be reached for every possible value of the traction coefficient (in principle, different for each wheel), of road slope and vehicle load.

In the meantime, the control system must minimize any disturbance resulting from sudden variation of the traction coefficient (such as driving on puddles, pits, manhole covers, etc.), suspension vibrations or tolerances in the fabrication or assembly process. In addition, braking capability should not be affected by transmission inertia (the vehicle inertia at the driving wheels is affected by the gear ratio and by the engagement or disengagement of the clutch). Finally, yaw torque disturbances caused by different traction coefficients of the wheels (the so-called *μ-split* conditions), are to be reduced, as are vibrations on the brake pedal caused by braking torque regulation.

Input parameters of this control system are wheel speeds, measured by sensors on the wheel hubs, while the controlled parameter is the braking pressure at each brake. ABS controls are unable to set a given value for the braking pressure, since the friction coefficient is unknown, but they increase or decrease it with respect to the current value.

While simpler systems have been built in the past, the four channels system, now universally applied, will be described; this system is characterized by a speed sensor and an independent pressure control for each wheel.

To achieve this function, in addition to a master pump with power system, as used in traditional systems, an ABS system must include:

- A brake actuator for each wheel (disc or drum) as in traditional systems; disc brakes are preferred to drum brakes, because they are simpler to regulate, being free of potential self-locking.
- A speed sensor for each wheel.
- A hydraulic pressure modulator, including an electronic control unit, regulation valves and a recirculation pump.

The control strategy of the ABS system is described first.

A typical plot of the basic parameters for a wheel are plotted versus time in Fig. 9.33. The figure refers to a progressive braking on a friction surface, with the intervention of the ABS system.

The first plot shows the time historyies of

- the vehicle speed V; this value is not measured by the control system, but is shown on the graph for reference.
- The reference vehicle speed V_{rif}; this parameter is estimated by the control system using algorithms that elaborate the speeds of the wheels.
- The wheel peripheral speed V_r; this is the product of the wheel speed Ω_r, measured by sensors, and the rolling radius R_0.
- The limit speed V_l; this is the threshold of the wheel peripheral speed, below which slip is too high to guarantee the control targets.

The second plot shows the peripheral acceleration of the wheel a_r, while the third shows the effective braking pressure on the actuator p_f as controlled by the electronic unit.

The electronic unit calculates the acceleration a_r, differentiating the speed V_r measured by the wheel sensors; particular techniques are applied to filter the signals from magnetic pick ups, thus suppressing noise.

Some characteristic threshold values are identified in the acceleration diagram; they are:

- Threshold a (value <0); this value is certainly higher than the maximum vehicle acceptable deceleration. It shows a marked difference between the actual braking moment and the maximum stable braking moment, due to excessive braking pressure or a sudden transient to the unstable area.

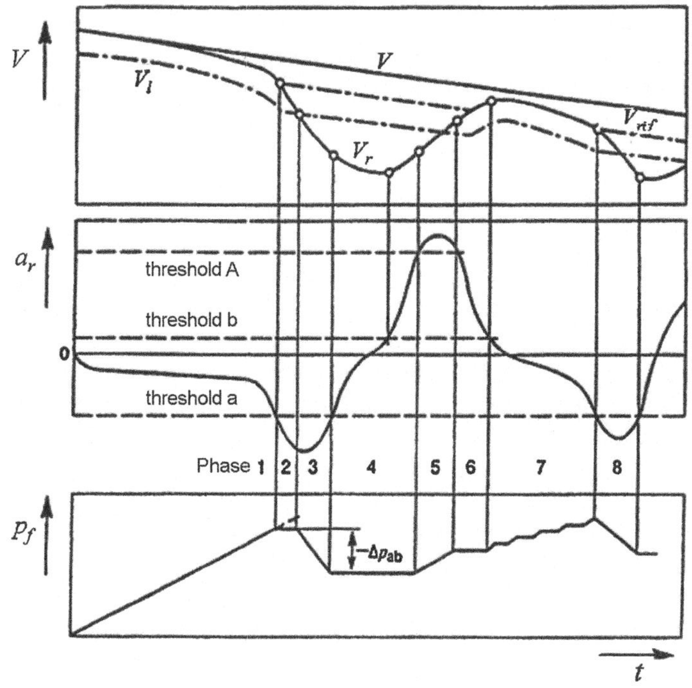

Fig. 9.33 Time histories of the speed V, the acceleration a_r and the braking pressure p_f in a braked wheel controlled by an ABS, in the different phases of the action of the control system

- Threshold A (value >0); this value shows a braking torque remarkably lower than the maximum stable braking torque; the corresponding slip has a small value, well below the maximum.
- Threshold b (value >0 but close to zero); this value underlines the return of the traction curve to a value close to its maximum.

Seven characteristic phases can be identified in a braking action occurring under the control of an ABS system.

Phase 1

Time $t = 0$ in the plots corresponds to the start of braking. Throughout phase 1 there is no intervention of the control system. In this phase the wheel speed V_r is greater than the threshold V_l and the peripheral acceleration a_r is negative and is close to the vehicle deceleration. The value of the latter is lower than the threshold a, that indicates the transition to the unstable field.

As a consequence, in a progressive braking, the piston pressure increases continuously, depending on driver actions and circuit delay. The acceleration will be maintained almost constant because the slip increase is such as to keep the friction torque increasing along with the braking torque. Before reaching the threshold a, the acceleration will show a sudden change, underlining that the boundary with the unstable area is approching.

Phase 2

As soon as the the acceleration gets larger than the threshold a, phase 2 begins, where braking pressure is kept constant. The presure must be larger than the threshold since a sudden high speed braking, characterized by a high braking torque gradient, in combination with a tire characteristic curve not very steep in its stable part, implies a slow increase of the friction torque. Therefore it is possible to reach the threshold a in stable condition and unnecessary to reduce the pressure in those conditions.

If the beginning of phase 2 represents the transition from stable to unstable working of the wheel, the value of the slip that has been reached is the optimum value, to be held constant during braking to produce good performance.

In this phase the wheel deceleration keeps decreasing. This explains why the braking torque is constant while the slip increases, with reduction of the friction torque (the slip is higher than the value corresponding to the maximum of the $F_x(\sigma)$ curve).

Phase 3

Below the value V_l the slip is very high close to the wheel lock; therefore crossing the corresponding threshold for the wheel speed V_r implies entrance into phase 3, characterized by decreasing pressure. The consequent decrease of the braking torque makes an inversion of the acceleration diagram possible, which even below the threshold a begins to increase, while the wheel speed decreases, but with a reduced slope.

In phase 3 the wheel is still in the unstable area and the slip is still increasing. The friction torque decreases more slowly than the braking torque. For this reason the acceleration of the wheel begins to increase.

Phase 4

As soon as the acceleration gets beyond the threshold a, phase 4, characterized by constant pressure, starts. While in phase 1 crossing the threshold a means entrance into the unstable area, in this case crossing means returning to the stable area, because in the meantime the braking torque has stabilized at a lower value, corresponding to Δp_{ab}, in the third diagram.

This allows the acceleration to continue increasing until it is positive. As a consequence, the wheel speed V_r decreases in the first part of this phase to a minimum and then starts increasing again. This increase is due to the fact that, while the braking torque is kept constant, the friction torque grows until it is larger than the braking torque, owing to the decrease of the slip.

In the middle of this phase the deceleration again gets over the threshold b. Should this not happen, it would be necessary to proceed to a further decrease of braking pressure. This could happen because the pressure decrease in phase 3 has not been sufficient to allow the friction torque to become higher than the braking torque.

Phase 5

As soon as the acceleration becomes larger than the threshold A, phase 5, characterized by a pressure increase, begins. In the first part of this phase, the longitudinal slip decrease causes an increase of the friction torque and a consequent increase of acceleration, until this increase changes the trend of the acceleration.

During this phase, the wheel speed continues to increase, because the acceleration is always positive.

Phase 6

Phase 6 begins as the acceleration crosses, with a negative slope, the threshold A, where the pressure is kept constant. In this phase the wheel speed continues increasing and the slip decreases, but because the slip is again in the stable region, the friction torque also decreases, together with the acceleration, until this crosses, always with negative slope, the threshold b.

Phase 7

When this occurs, phase 7, where the pressure increases in steps to avoid a steep increment of the braking torque, begins. This increase allows a better exploitation of the tires and decreases the influence of wheel inertia. At the beginning of this phase, the wheel speed continues increasing and the slip further decreases with the decrease of friction torque. The acceleration continues decreasing to zero, returning then to negative values.

This zero value corresponds to both torques being equal. As soon as the acceleration is negative, the slip again begins to increase toward the optimum value. These will be reached when the acceleration decreases below the threshold a, starting a new cycle, as described starting from phase 3.

When ABS is used, the pressure in the actuators cannot exceed the pressure of the master pump. In other words, the braking torque cannot override the action of the driver in control.

EBD System

The acronym EBD derives from *Electronic Brake Distributor*.

It is known that during braking, the inertia force, applied to the center of mass, produces a vertical load transfer, increasing the load on the front wheels and decreasing that on the rear wheels. If a braking moment simply proportional to the vertical static load were applied, the rear tires would loose their traction first, compromising the vehicle path stability.

To avoid this outcome, a distributor valve has been introduced on braking circuits without ABS control. As described in a previous section, this valve is addressed to limiting the rear wheel braking pressure. The same function can be performed by an ABS system too, and this added feature is called EBD.

For a given vehicle, it is possible to calculate ideal pressure distributions that guarantees the maximum braking force on both axles. This curve must be modified for each possible load combination on the vehicle. For instance, the full load curve allows a higher rear braking than the empty condition, because load variations primarily affect the load on rear wheels.

The EBD control tries to copy the ideal curve in a better way than mechanical brake distributors. To achieve this, the vehicle mass is estimated by comparison of the master pump pressure with the vehicle deceleration; the value of the mass can also be used to estimate mass breakdown.

If the control action on the brake pedal is minimal, the distributor follows the full load curve; additional slips of the rear axle are contained. At higher master pump pressures, the EBD control diversifies the pressure using the ABS valves. Such a strategy implies a precise calculation of the slip, which depends upon vehicle speed, which, however, cannot be estimated precisely from wheel speeds. A control system maintaining the difference between rear and front wheel slip is simpler and similarly effective.

The components needed by an EBD control system are the same as those seen for an ABS system; only the software is affected, because specific algorithms are added for this function.

An EBD system, like a mechanical distributor, is subjected to regulations on minimum deceleration in case of failure; the system must incorporate a recovery mode including a safe reduction of braking forces in the event of failure.

VDC System

VDC (*Vehicle Dynamics Control*) systems are also known through different commercial names, such as ESP (*Electronic Stability Program*) or VSC (*Vehicle Stability Control*).

Such systems are always aimed to obtaining a stable and predictable vehicle dynamic behavior and avoiding tire operation beyond their stable limit. The actuators of this system are again the wheel brakes, used so as to obtain a suitable moment around the vertical axis of the vehicle.

The components necessary to a VDC control include, in addition to those of an ABS system:

- A steering wheel angle sensor,
- lateral acceleration sensors and
- a connection to the engine electronic control unit.

Steering wheel angle and lateral acceleration sensors are used to identify the primary parameters describing vehicle lateral dynamics. The electric pump and valves are used for this system to generate pressure pulses for single brakes, independent of the driver's will.

When the vehicle is driven on a curve and the cornering force of one of the tires is less than that needed locally for equilibrium with the centrifugal force, the vehicle deviates from the previous path. If this deficiency is localized in the front axle, the phenomenon is similar to *understeering*[1] and the vehicle describes a path with a larger radius than in normal conditions; if it is localized in the rear axle, the phenomenon is similar to *oversteering* and the vehicle describes a path with a lower radius. The VDC control system tries to recover the initial path, with a correcting vertical torque compatible with tire limits.

To do this, the electronic control unit calculates the expected cornering acceleration on the maneuver from the steering angle and the vehicle speed. If there are differences from the measured acceleration, four braking torques are calculated that are useful for correcting this situation or at least reducing vehicle speed, if friction limits are surpassed.

A reduction in engine torque contributes also to this speed reduction; this is the purpose of the communication channel between the engine control system and the VDC control. The engine torque is reduced by reducing the spark advance, the injected fuel quantity or the throttle valve position in engines driven by wire.

These interventions must not induce slips that bring the tire into the region of side slip or longitudinal slip instability. Two control loops are required: One computes the braking forces needed to stabilization; an innermost loop controls the wheel slip so as to obtain the desired longitudinal and lateral forces.

A mathematical model must be suitable for predicting vehicle motion consequent to load variation, at different vehicle loads, road conditions and tire wear.

ASR System

This acronym derives from *Anti-spin Regulator*.

Critical driving situations can be due not only to braking or cornering. Accelerations and driving forces may also cause vehicle instability. ASR systems correct for these situations by avoiding too large a slip on the driving wheels.

---

[1] A vehicle is understeering if its path, at a given speed, has a lower curvature as that obtained at low speed, with the same steering angle; the opposite definition applies to oversteering.

This result is achieved either by temporarily reducing the torque applied by the engine to the driving wheels or by applying a braking torque from the control system able of generating braking torques autonomously. The ASR, as stated earlier, is thus an additional function of ABS systems, implying communication with the engine electronic control unit.

The ASR control unit is integrated with the ABS control and shares with it a number of components, such as wheel speed sensors and pressure control valves. If the throttle valve and the brakes are available as actuators, the best performance can be obtained; because most engine controls are now driven by wire, this option is no longer difficult to be implemented.

BAS System

BAS systems (*Brake Assist System*) have the task of applying a constant full braking pressure, independent of the driver's will, in an attempt to decrease the stopping distance as much as possible, in emergency situations. Such control systems are useful in panic braking by non-expert drivers; moreover expert drivers not used to ABS systems may instinctively decrease the force on the brake pedal, trying to avoid slips that the control system will in any case avoid. For impaired drivers, the pedal force may be inadequate to exploit the full braking capacity. A panic situation is usually detected when the braking pressure time gradient exceeds a threshold value.

Two alternative strategies may be applied:

- When a panic braking situation is detected, the system generates the maximum braking force independently of the driver's will, or
- when a panic situation is detected, a supplementary braking force is applied, modulated by the driver's action on the brake control.

In the second strategy, the deceleration is modulated by the force and speed of the driver's input. A wider field of intervention is available than in the first strategy.

The components needed to implement a BAS system are the same as for the ABS system, once the hydraulic circuit has been modified with a suitable storage capacity. The pressure modulator maintains a certain quantity of high pressure oil in a reservoir. The BAS system can apply a larger pressure than that determined by the master pump.

Brake Control Hardware Components

Wheel speed sensors are gear wheels with magnetic pick-ups. The gear wheel RF in Fig. 9.34a has a rectangular teeth profile. It is made from a ferromagnetic material and faces a magnetic pick-up; wheel rotation changes the magnetic flux captured by the pick-up.

Pick-ups are made by a magnetic core 1 and a soft iron core 2. The magnetic field created by the magnetic core is periodically altered by the rotation of the wheel

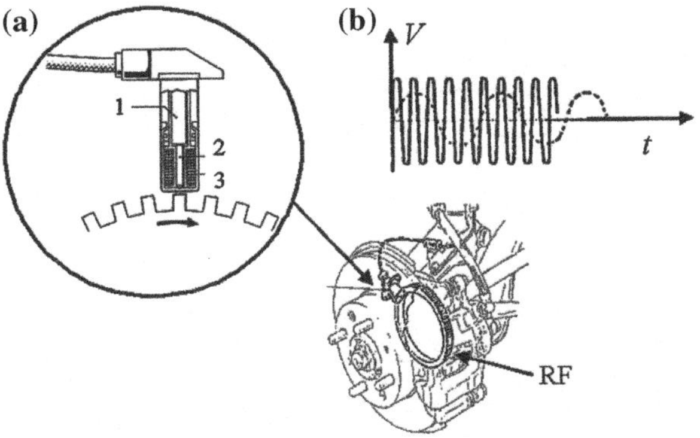

Fig. 9.34 Installation scheme for a wheel speed sensor; RF is the gear wheel. The enlargement shows the detail of the magnetic pick-up

and its value is measured by the coil 3, which supplies a voltage whose period and amplitude are determined by rotational speed.

The plot in Fig. 9.34b shows the voltage as a function of time, for a low (dotted line) and a high speed (solid line). This signal is amplified and squared by an electronic circuit; the peak count supplies the requested speed value. In current wheel speed sensors, electronic amplifier and signal conditioner are integrated into the magnetic pick-up, which can be interfaced directly with the control microprocessor.

The working principle of the accelerometer is simple and consists in measuring the deformations of a small structure subjected to the vehicle inertia forces. This measurement can be made by piezoelectric transducers, where an elastic deformation is converted into a voltage variations.

The steering wheel angle is measured by a potentiometer located on the steering column or by an optical encoder. The requested precision is in the range of a tenth of a degree. The encoder is still made by a wheel with teeth or a sequence of transparent and non transparent sectors, interrupting periodically the light beam received by a photo diode; the voltage pulse is measured and counted by an electronic circuit.

At the state of the art all these signals, after being processed by an AD (Analog to Digital) converter, are made available to the communication network of the vehicle, which makes them available to the brake control unit and to other services, as for instance, an electric power steering system.

The functional scheme of a hydraulic modulator for an electronic brake control system is shown in Fig. 9.35. Two independent hydraulic circuits can be identified, in this case, according to the X scheme, required to satisfy regulations on emergency circuits.

Six normally open electrovalves (ENO1 to ENO6) are present in the circuit together with six normally closed electrovalves (ENC1 to ENC6). Normally open means that the hydraulic circuit is open when the coil is switched-off. The circuit also includes two pressure accumulators, along with two electric pumps used to speed up pressure decrease by ABS and VDC modulation and to create pressure during VDC

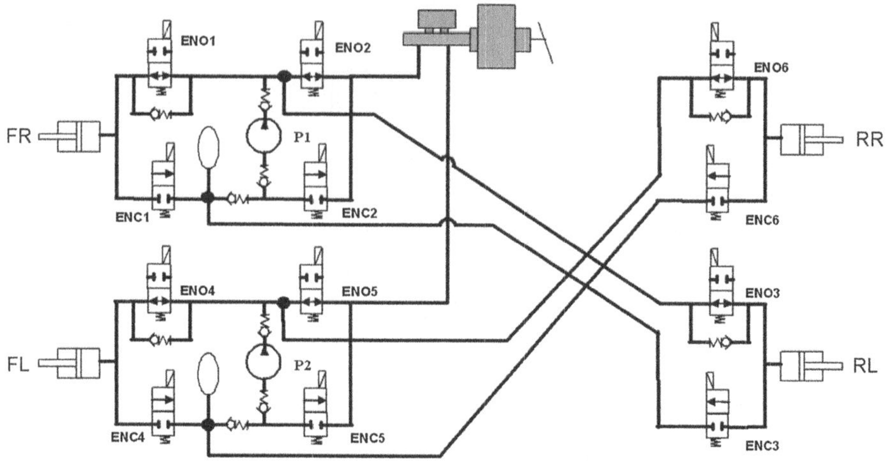

Fig. 9.35 Functional scheme of a hydraulic brake modulator for a four wheel vehicle. ENO valves are normally open; ENC are normally closed (from Genta and Morello 2009)

and ASR modulation. These pumps are provided with two radial pistons moved by an electric motor through a cam.

Referring to the front right wheel (FR) only, the default configuration, when no intervention of the control system is required and all electrovalves are switched off, is represented in the figure. There is a direct connection between the master pump and the brake actuator FR. This condition also allows conventional operation of the hydraulic circuit in case of control system failure.

When it is necessary to maintain a constant pressure in the brake actuator, independent of the driver's will, the electrovalve ENO1 is switched on and isolates the actuator from the rest of the circuit. If the pressure must be decreased, the electrovalve ENC1 is also switched on to allow some of the oil in the actuator cylinder to return into the master cylinder atmospheric pressure reservoir, with the help of the electric pump, if needed.

To increase the pressure independently of the pressure in the master pump (VDC and ASR regulation), the electrovalve ENC2 is switched on putting the pump P1 in communication with the reservoir on the master pump; in the meantime the electrovalve ENO2 is switched on to avoid pressurized oil returning into the reservoir.

9.5 Powertrain Mounts

During its operation, the engine transmits to the vehicle a set of continuously variable forces having different intensity and frequency. The motion of the reciprocating masses, i.e. the masses of the piston and of those parts linked to its motion, produces inertia forces, usually subdivided in primary and secondary inertia forces. The primary forces have a period equal to the time needed by the engine to perform one revolution (frequency Ω). They are generated by the motion of the projection of the crank pin along the stroke axis, i.e. the motion that the piston would have if the connecting rod had an infinite length.

The secondary forces have a period equal to the time needed for one stroke (frequency $2\,\Omega$). They are due to the fact that the motion of the piston is non harmonic, owing to the finite length of the connecting rod, and thus, once the inertia forces are developed in Fourier series, they have components with frequency $2, 4, 6\,\Omega, \ldots$

In current cars, the engine and the transmission are joined in a single unit, the powertrain, reacting as an almost solid body to external forces that can be classified according to three categories:

- Inertia forces generated by the motion of engine reciprocating masses. They depend on the number and arrangement of cylinders and by the application of crankshaft counterweights and balancing shafts (if present);
- reaction forces to the output torque. They also depend on the number of cylinders and are never constant, particularly at idling, because of the difference in the indicated pressure of each cylinder, due to occasionally incomplete combustions. These reaction forces are amplified by the transmission ratio of the gearbox, by about $4 \div 5$ times in conventional transmissions, or $16 \div 20$ times in units integrating also the differential and the final drive, as in front wheels drive cars or rear-engine, rear wheels drive cars, and
- inertia forces generated by the powertrain mass, because of the car motion (longitudinal, transversal and vertical accelerations).

The powertrain must therefore be connected to the body by means of elastic and, possibly, damping elements, able of reducing the negative effects of such forces on comfort. The design of such suspensions is complex because of their conflicting nature: the forces of the first type require a very soft suspension, but such a low stiffness could generate unpleasant powertrain vertical motions, coupled with the front suspension shake, or rotations because of torque variations. These motion may impair car driveability, owing to their inertial effects. In addition, an optimum solution does not exist but depends on the engine cylinders arrangement and on the body structure; moreover the structure response to the acceleration applied by the engine depends also on the local stiffness where the engine mounts are placed.

Only a few remarks about the powertrain mounts in four cylinder in-line engines will follow, since this subject is outside of the scope of this book. Even considering only powertrains of this type, the arrangement of the mounts changes according to the type of the transmission of the vehicle, whether the engine is located in longitudinal direction and the car has rear wheels drive or the engine is transversal and the driving wheels are the front ones (Fig. 9.36a and b).

In the first case, the front powertrain mounts are placed aside of the engine, in a position coincident with the crossings of the longitudinal beams under the floor with the front crossmember offering also support for the suspension mounts; a rear mount, with reduced load, is placed over a rear crossmember; because the output torque is multiplied by the gearbox alone, the offset between right and left front

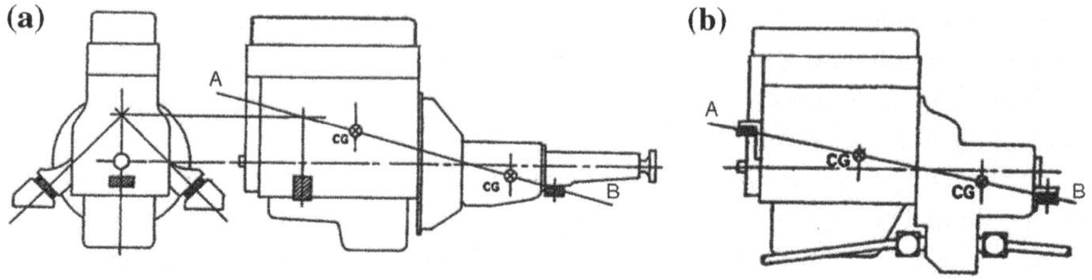

Fig. 9.36 The suspension centers of powertrain mounts are aligned on its principal inertia axis AB, almost coincident with the line through the centers of mass CG of the engine-gearbox unit. **a** Front-engine, rear–wheels-drive powertrain. **b** Front-engine, front–wheels-drive powertrain

mounts is sufficient to obtain a good compromise between vertical flexibility and torque motion. It could be demonstrated that the optimal solution for this suspension would be placing its centers along the principal inertia axis AB of the powertrain. The front center is obtained by joining the shake axis of the two front mounts, while the rear center is coincident with the rear single mount. The principal inertia axis is almost coincident with the line joining the centers of mass of the engine and the gearbox, shown in Fig. 9.36a.

Also in the second case the rule about the principal inertia axis can be applied, as shown in Fig. 9.36b; because of the transversal position of the powertrain, the two mounts must be placed in front of the engine and in the rear of the gearbox, the only positions close to sufficiently stiff locations in the car body, corresponding to the longitudinal front beams. Because such a suspension would be unstable against torque motion, a connecting horizontal rod (not shown in the figure), with elastic bushings, is added at one of the sides of the engine, to react to torque, leaving free any vertical motion. The vertical position of this rod depends again upon the body structure; if a crossmember is already provided for the front wheels suspension mounts, this will be certainly the most suitable place to provide the fixed mount for this rod.

Obviously the overall objective is to minimize the forces transmitted to the body and to obtain this result the elastic suspension should be provided with sufficient damping. The rubber itself, used in all powertrain mounts, has an intrinsic damping property due to the internal displacement that occurs during its deformation. For simpler applications the internal damping is enough to obtain a satisfactory result; this is the case of mounts loaded by loads that are not vey high like the mounts on the gearbox side or those of torque reacting rods, as shown in Fig. 9.37a and b. For heavier masses and loads, as for powertrain mounts on the engine side, an added damping function is recommended, particularly if outstanding comfort is included in the objectives of the vehicle.

An example of this kind of mount is shown in Fig. 9.37c; it is made with a rubber spring 1 that acts as a piston on a number of chambers filled with oil, whose arrangement is quite similar to that of a suspension shock absorber. The upper chamber 2 is forcing the oil to flow, back and for through the calibrated orifices 6, to the

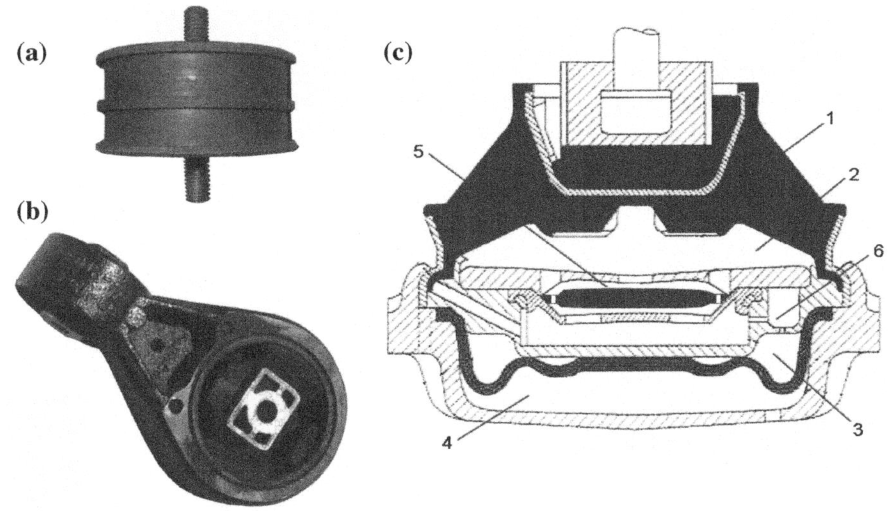

Fig. 9.37 Rubber engine mounts. **a** Rubber mount for *low loads*; **b** elastic rod for torque reaction; **c** rubber mount for *high loads* with hydraulic damper

lower chamber 3. The oil volume changes in the lower chamber are compensated by a membrane exposed to the ambient pressure at its lower side 4. The rubber disc valve 5 closes the passage to the orifices if the displacement is too high, limiting the damping function to small high frequency motions.

Chapter 10
Engine

Most road vehicles are powered by reciprocating internal combustion engines of either the *spark ignition* (SI or Otto-cycle) or the *compression ignition* (CI or Diesel-cycle) type. Moreover, almost all engines powering motor cars work following a four-stroke cycle sequence. Only these types of engines, and in particular

- spark ignition, indirect injection gasoline engines,
- spark ignition direct injection gasoline engines and
- compression ignition direct injection engines.

will be dealt with in the present section; leaving out carburetor gasoline engines or pre-chamber diesel engines. For them, and other types of internal combustion engines, that were sometimes used in the past, see Part I, while solutions based on electric or hybrid drivelines will be dealt with in Chap. 11 and in Part III.

10.1 Basic Engine Requirements

An engine must convert the chemical energy of fuel into the mechanical energy needed to operate the vehicle while being as user friendly as possible. Vehicle driving must be easy and safe in every driving condition, and thus the engine must be characterized by quick start-up and warm-up, followed by smooth driving with consistent performance, in any climate and situation from extreme cold to hot weather, under the sun or rain, from sea level to the highest elevations where roads exist. The comfort requirements dictate not only limiting disturbances, recently defined as Noise, Vibration and Harshness (NVH) but, even more, offering a pleasant acoustics.

Engine mass represents a significant part of the total vehicle mass and, as a consequence, its reduction allows improving performances like fuel consumption, acceleration and handling. Compact engines allow designers to refine aerodynamics, to increase the space available for occupants and luggage with reasonable external dimensions and, finally, to maximize the vehicle's safety.

G. Genta et al., *The Motor Car*, Mechanical Engineering Series,
DOI: 10.1007/978-94-007-8552-6_10,

An engine must offer the above mentioned characteristics with a suitable reliability, meaning the probability of efficient operation during the assigned mission, and durability, meaning resistance to wear over the prescribed mileage.

Starting from the late 1960s, in the USA, and from the 1970s, in Europe, the concern caused by environmental problems due to the increasing concentration of motor vehicles in densely populated urban areas became a technology-forcing issue and increasingly restrictive regulations were addressed to lowering the limits of exhaust emissions of carbon monoxide (CO), hydrocarbons (HC), nitrogen oxides (NO_x) and Particulate Matter (PM). This process is still continuing with the goal of achieving zero equivalent emission targets during the next decade. Reducing fossil fuel consumption and, later, the emission of green-house gases, mainly carbon dioxide (CO_2), became a further goal. Increased vehicle concentration causes also the need to reduce external noise and further restrictions will be unavoidably applied.

Engine production cost must be limited: similarly to many other products, the engine is facing the challenges of the world's open markets. Therefore the winning solution is also invariably the least expensive, provided all the above-mentioned requirements are satisfied. In fact, when choosing a new vehicle, usually to replace the previous one, the decision of the final Customer is based on a trade-off between driving pleasure, purchase, operating and maintenance costs and comfort.

The concept of driving pleasure is today more and more shifting from all-out performance (maximum speed, high speed acceleration, etc.) strictly linked with the engine power, to the engine capability of assuring both comfortable and smooth driving in congested traffic situations and an attractive driving feeling during acceleration and take-off. The latter can be achieved only if the engine is able to offer a good torque at low speed.

This was clear even in the past, when this result was achieved, in practice, only by increasing the engine displacement, with unavoidable increases in cost and weight. More recently it became possible to obtain an equivalent result by reducing the engine displacement and its weight while maintaining the same power levels, by increasing the engine power density. This goal was achieved by applying the latest generation of pressure charging techniques and is usually referred to as engine *downsizing*. This trend is completed by introducing lightweight materials and more sophisticated manufacturing techniques (e.g. thin wall shell castings) that, progressively, became applicable at reasonable costs.

The downsized configuration offers an innovative engine torque versus speed characteristic in which the so-called low-end torque is improved. The engine is thus able to produce a torque value close to its maximum already at low speed, typically 1,500–2,000 rpm. This allows us to improve driveability and a *fun-to-drive* feeling by improving engine flexibility and, at the same time, to reduce fuel consumption. This latter achievement can be further exploited because the low-end torque allows using longer transmission ratios, thus promoting the so called *downspeeding*.

Injection (either direct or indirect) spark ignition engines and direct injection compression ignition engines can satisfy the above mentioned challenging requirements, remembering that

- spark ignition engines are better for low emissions, good acoustic comfort and limited cost, while
- compression ignition engines are better for fuel economy and are fun-to-drive.

The challenge for the near future is reducing the drawbacks of both, while maintaining and enhancing their advantages at the same time meeting emission and noise regulations that will become more and more strict.

This picture is more related to the European situation, where the presence of the two types of engines is more balanced, with huge investments of car makers and Authorities in the development of both. In the United States the attention of manufacturers has been more focused on gasoline engines, emphasizing emission control and a fun-to-drive feeling, with lower efforts toward fuel economy, at least up to now.

Examining the ACEA (European Automobile Manufacturers' Association) fuel consumption commitment, the fierce competition between gasoline and compression ignition engines starting from 1995 is clear, with the goal of reducing CO_2 emissions below 130 and 95 g/km respectively in 2015 and 2020. These figures refer to the NEDC cycle, which includes transients and a cold start at 20 °C.

10.2 Engine Operation Principles

Internal combustion reciprocating engines include a component, the piston, which reciprocates between the Top Dead Center (TDC) and the Bottom Dead Center (BDC). The dead centers are so named because the piston speed reduces to zero when the motion is reversed at these points, as shown in Fig. 10.1a.

The piston slides along the cylinder and is connected to the crankshaft by the connecting rod. Four-stroke cycle engines perform one complete working cycle every two crankshaft revolutions (crank angle of 720°), during which the four strokes, usually referred to as *intake*, *compression*, *expansion* and *exhaust*, take place.

The main dimensions of the engine are defined as follows:

- Bore B, internal diameter of the cylinder (d);
- stroke S, distance covered by the piston from TDC to BDC;
- crank radius R_c, linked to the stroke by the formula $S = 2R_c$;
- cylinder cross-section area $A_c = \pi d^2/4$;
- unit displacement V, cylinder volume displaced or swept by a single stroke of the piston. It is also referred to as the swept volume, $V = A_c S$;
- total displacement V_t, unit displacement by the number of cylinders, referred to also as engine capacity;
- combustion Chamber Volume V_c, volume included between the piston top surface (at TDC) and the cylinder head;
- compression ratio ρ, defined as

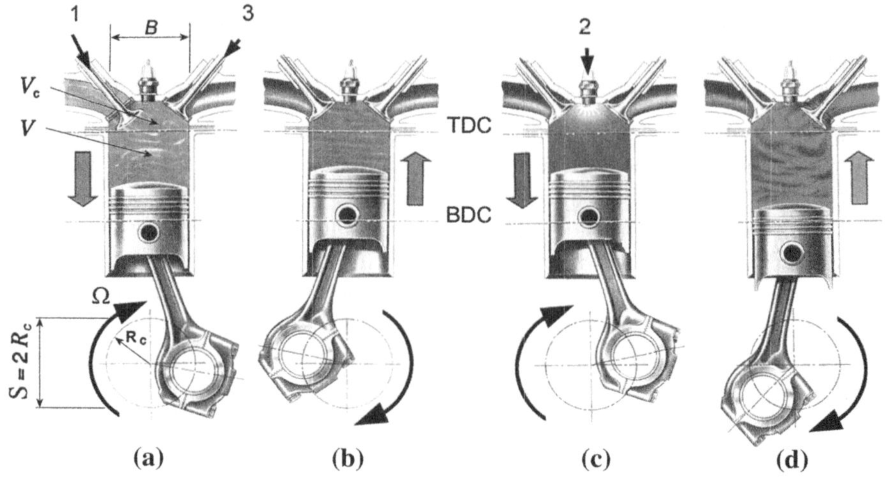

Fig. 10.1 Operation sequence of a four stroke spark ignition engine: **a** intake; **b** compression; **c** expansion; **d** exhaust. In **a** the swept volume (V), combustion chamber volume (V_c), bore (B) and stroke (S), are shown (courtesy of FIAT)

$$\rho = \frac{V + V_c}{V_c} \,. \tag{10.1}$$

For spark ignition engines the compression ratio varies from 10:1 to 12:1, whereas for compression ignition it varies from 16:1 to 18:1;

- shaft rotational speed or engine speed Ω: crankshaft angular speed, measured in revolutions per minute (rpm) or in rad/s;
- average piston speed u, defined as $u = 2S/\Omega$. It is used to correlate many engine parameters, such as fluid-dynamic losses during fluid replacement into the cylinder, inertia forces loading many mechanical components and energy losses during cooling; some typical values are shown in Table 10.2.

10.2.1 Four-Stroke Spark Ignition Engines

The sequence of the four phases (or strokes) in a four-stroke cycle engine operating following the Otto cycle is sketched in Fig. 10.1. As already mentioned, they are referred to as intake, compression, expansion and exhaust.

Intake

This phase takes place when the piston moves from the TDC to the BDC. The intake valve (1) opens before TDC, so that it is sufficiently open when the piston starts its downward stroke; it closes after the BDC, in order to take full advantage of the inertia of the incoming mixture flow even when the piston is already in its upward stroke. The piston downward stroke determines a negative pressure (relative to the atmospheric pressure), inside the cylinder that draws a fresh charge in, through the intake valve. The fresh charge consists of an air/fuel mixture, referred to as pre-mixed or homogeneous charge. Spark ignition engines are thus known as naturally aspirated or, simply, aspirated engines. The power is controlled by opening and closing the throttle valve positioned in the intake manifold which controls the amount of fresh charge entering into the cylinder.

Compression

As the piston moves from BDC to TDC, owing to the intake valve closure, the air/fuel mixture is compressed within the combustion chamber volume. During this phase, the mixture undergoes a considerable increase in temperature due to compression and, in part, to the heat released by the cylinder walls. The temperature at the end of compression reaches from 400 to 500 °C.

Expansion

A few crankshaft degrees before the TDC, the air/fuel mixture is ignited by the spark which takes place between the spark plug (2) electrodes. The term *spark ignition* or *positive ignition* characterizing this type of engines comes from this way of igniting the mixture. During combustion, a temperature peak of about 2,300 °C is reached, which causes a remarkable heat release toward the cylinder walls. At the combustion end, a pressure peak (typically 100 bar) is reached which acts on the piston pushing it toward the BDC. Expansion is the only active phase of the whole thermodynamic cycle.

Exhaust

Before BDC, the exhaust valve (3) opens to start the spontaneous removal of burnt gases from the cylinder, since their pressure is still significantly higher than atmospheric pressure. During the subsequent upward piston stroke, the exhaust phase is completed. The exhaust valve closes after the TDC to take full advantage of the inertia of the exhaust gas flow. The pressure quickly drops to the atmospheric value.

10.2.2 Four Stroke Compression Ignition Engines

Similarly to spark ignition engines, four-stroke compression ignition engines make a complete working cycle every two crankshaft revolutions (720°). The main differences between direct injection turbocharged compression ignition engines, today commonly used, and spark ignition engines are as follows:

Intake

Through the intake duct, air is directly admitted to the intake valve. The valve opens at TDC or just slightly before it, because the higher pressure, due to turbocharging, helps in allowing air in. Since diesel fuel is directly injected into the cylinder, the power is controlled by varying the quantity of fuel injected and no throttle valve is present. The pressure decrease in the intake duct is thus limited and substantially determined by friction in the duct and pressure drop in the air filter.

Compression

During the piston travel from BDC to TDC air is compressed in the combustion chamber reaching pressure and temperature values much higher than those typical of

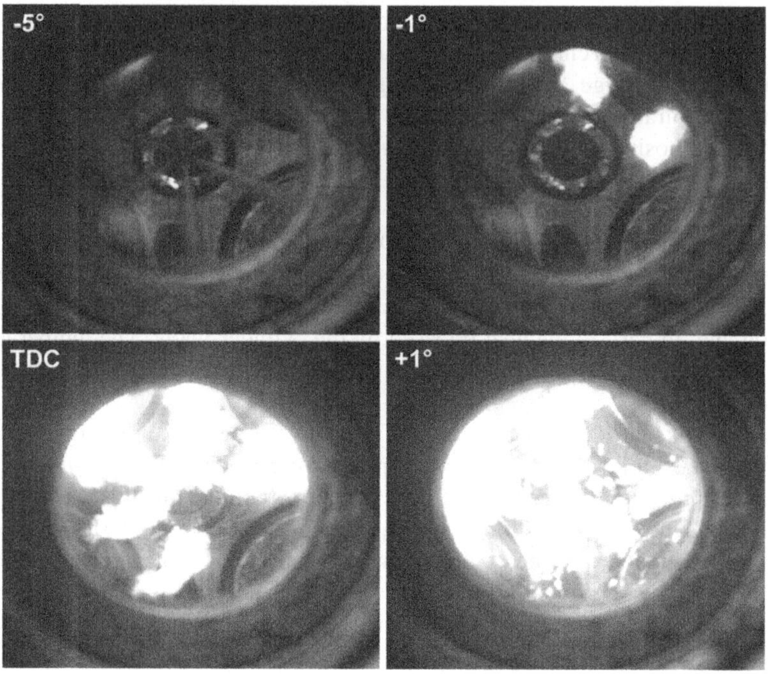

Fig. 10.2 Compression ignition and combustion in a Diesel engine at various values of the crank angle. Spray of the main injection and start of ignition are shown before the TDC. Combustion progresses immediately after TDC (courtesy Istituto Motori CNR, Naples)

spark ignition engines, due to the higher compression ratio. Temperatures as high as 750–800 °C are reached at the end of compression, close to the TDC, in present-day engines.

Expansion

The ignition mechanism is different: Contrary to what occurs in the spark ignition engines, the air/fuel mixture ignition takes place spontaneously due to the high temperature and pressure, and the term *compression-ignition engines* comes from this way of igniting the mixture. The injector sprays the fuel into the combustion chamber just a few crankshaft degrees before TDC; the spontaneous ignition of the initial part of the mixture spray and, consequently, of the bulk of the injected fuel generates a sudden rise of the pressure inside the cylinder, pushing the piston down toward the BDC.

The pictures in Fig. 10.2 show the sequence of the injection and spontaneous ignition immediately before and after the TDC ($-5°$; $-1°$; TDC; $+1°$). The pressure peak (typically 150 bar) takes place during combustion and the temperature reaches values higher than those typical of spark ignition engines. The air/fuel charge is obviously not homogeneous, since the injector sprays only fuel and only air enters through the intake valve.

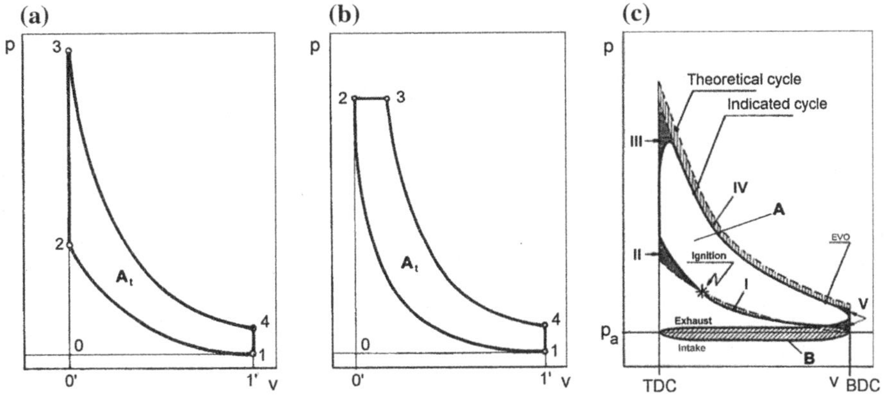

Fig. 10.3 Theoretical cycles: **a** Otto-cycle: constant volume combustion; **b** Diesel-cycle: constant pressure combustion; **c** Indicated cycle for a spark ignition engine. The areas from II to V, cause a loss of efficiency with respect to the theoretical cycle and reduce the useful area from A_t to A. The Exhaust Valve Opening (EVO) point is shown (from Giacosa 2000)

Exhaust

Similarly to spark ignition engines, at the end of an expansion downward stroke, the exhaust valve opens before the BDC to promote spontaneous gas removal and the phase continues as the piston moves toward the TDC. For reasons that will be better explained in Sect. 10.2.3, the temperature of the exhaust gases is always lower in compression ignition engines than in spark ignition engines.

10.2.3 Theoretical Cycles and Thermal Efficiency

The theoretical thermodynamic cycle is usually represented in a pressure-volume (p-v) diagram to highlight the different phases. The theoretical four-stroke spark ignition cycle is characterized by a constant volume combustion, whereas the theoretical compression ignition cycle is characterized by a constant pressure one. The theoretical cycles for a spark ignition and a compression ignition engine are plotted in Fig. 10.3a, b. They yield the maximum engine efficiency theoretically achievable, since they represent the best way in which the chemical energy of the fuel can be transformed into useful mechanical work.

The theoretical work, produced by the fluid when performing the theoretical cycles, is represented, on the p-v diagram, by the area A_t included within the curves (1-2-3-4-1). Area $0'$-3-4-$1'$ (or $0'$-2-3-4-$1'$ in case of the compression ignition engine) represents the expansion work W_1 (positive), while area $0'$-2-1-$1'$ represents the compression work W_2 (negative). In both cases their difference $W_1 - W_2$ represents the theoretical work.

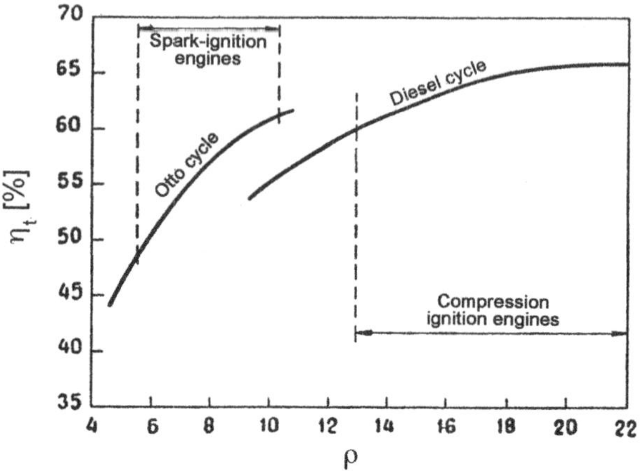

Fig. 10.4 Theoretical efficiency η_t versus the compression ratio ρ for Otto and Diesel engines (from Giacosa 2000)

The theoretical (or thermal or thermodynamic) efficiency

$$\eta_t = \frac{W_1 - W_2}{W_1} \tag{10.2}$$

is plotted in Fig. 10.4 as a function of the compression ratio ρ. The advantage of compression ignition engines is clear and is based on the higher values of the compression ratio ρ needed for a satisfactory combustion. For spark ignition engines the compression ratio is limited by detonation and causes only limited gain in efficiency when increasing above 12:1. In actual spark ignition and compression ignition engines used on passenger cars the combustion takes place partially at constant volume and partially at constant pressure and therefore the actual cycle lies in between the two shown. The compression ratio used in present spark ignition engines falls in the range between 10:1 and 12:1 and it is reasonable to assume a value for the theoretical efficiency close to 0.62. Compression ignition engines achieve a better value thanks to their higher compression ratio, spanning between 16:1 and 18:1.

10.2.4 Indicated Cycles

The actual p-v curves can be recorded during actual engine operation, thus obtaining the so-called indicated cycle (Fig. 10.3c).

The differences between theoretical and indicated cycles can be summarized as follows:

- Reduction of useful area from A_t to A due to the rounded corners which replace the sharp corners of the theoretical cycle and to further reductions during the compression and expansion phases;
- introduction of a second area B due to the work spent for fluid replacement.

The indicated cycle diagram is important because its area A quantifies the work actually transferred from the fluid to the piston while area B quantifies the negative work spent by the engine during the intake and exhaust phases. As explained above, the cycle area A, gone through in clockwise direction, produces a positive work, whereas area B, gone through in counterclockwise direction, represents a negative work. The reasons for the reduction of the former area from A_t to A, are listed below.

Heat losses

The heat losses are not accounted for in the theoretical cycle, while in the actual cycle they are significant. Since the cylinder is cooled so that the temperature of the cylinder walls does not exceed 200–240 °C in spite of the much higher combustion temperature of the gases, during combustion there is a certain heat transfer from the fluid to the cylinder walls. During compression, when the mixture is colder than the walls, the latter warm-up the mixture but, as the temperature increases, there is an instant after which the mixture starts to lose heat toward the walls. The nominal curve and the actual curve will therefore cross each other, while being almost superimposed. A loss of useful work thus occurs, corresponding to the algebraic sum of a small area (I) and a larger area (II) (Fig. 10.3c). A greater loss is determined by heat losses taking place during the expansion (IV).

Non instantaneous combustion

In the theoretical cycle the combustion is assumed to occur at constant volume and therefore is instantaneous. On the contrary, in the actual cycle the combustion takes place over a certain time. Should ignition take place at the TDC, the combustion would continue while the piston starts moving downward and the pressure value would be less than that theoretically predicted, with a loss of useful work. It is therefore expedient to anticipate the ignition, so that the combustion mainly takes place when the piston is closer to the TDC. This produces a rounding off of the theoretical heat introduction line and therefore a loss of useful work in the areas (II) and (III), but less than it would be without advancing the ignition.

Exhaust valve opening advance

In the theoretical cycle it is assumed that the exhaust valve opens at the BDC. In the indicated cycle, the exhaust phase extends for a longer time because the exhaust valve has to open in advance to allow part of the burnt gas to exit from the cylinder before the piston reaches the BDC, so that the pressure decreases to roughly the external value when the actual exhaust phase starts. This point is identified as Exhaust Valve Opening (EVO) and determines a loss of useful work in area (V). This loss is, however, less than the loss that would occur without an early opening of the exhaust valve.

Pumping work

The indicated cycle has another important difference when compared to the theoretical cycle: During the intake stroke the pressure is lower than the atmospheric pressure and during exhaust it is higher. Therefore the negative area B is equivalent to the work W_p lost during the two 'pumping' phases. This area increases proportionally to the throttle valve closing when the charge introduced and consequently the power must be reduced.

The above mentioned losses are applicable to spark ignition engines, but a similar situation holds also for compression ignition engines. All taken into account, the indicated Diesel cycle does not appear much different from the indicated Otto one.

In the case of a spark ignition cycle engine, it is reasonable to reduce the efficiency η_t by 20 % thus obtaining a value for the indicated efficiency η_i equal to about 0.50.

Area A is the indicated work W_i, while area B refers to the pumping work W_p. Not all the indicated work is actually available, since a part of it, the mechanical work W_r, is used to overcome the friction of the mechanical components and to drive the ancillary devices (water pump, oil pump, alternator, etc.). The effective work available at the crankshaft, is thus

$$W_e = W_i - W_r. \tag{10.3}$$

10.2.5 Mean Effective Pressure

The ratio of the indicated work W_i to the engine unit displacement V is a pressure, whose value is known as *indicated mean effective pressure*

$$p_{ime} = \frac{W_i}{V}. \tag{10.4}$$

The indicated mean effective pressure can be considered as a constant pressure which, if it were to act on the piston for the entire stroke, would generate the indicated work W_i.

The ratio of the effective work W_e to engine unit displacement is the *mean effective pressure* p_{me}.[1] It represents the constant pressure that, if it were to act alone on the piston for an entire stroke, would generate the effective work. In other words, it represents the work available at the crankshaft per unit displacement and per each cycle.

The *engine indicated power* P_i is obtained multiplying the indicated work per unit displacement W_i by the number of cylinders n:

$$P_i = \frac{nW_i\Omega}{60\varepsilon} = \frac{p_{ime}V_t\Omega}{60\varepsilon}, \tag{10.5}$$

[1] Instead of symbols p_{ime} and p_{me}, often the achronyms *imep* and *mep* are used for the indicated mean effective pressure and the mean effective pressure.

where $V_t = nV$ is the total engine capacity, Ω is the engine speed expressed in rpm (hence the constant 60 in the formula) and ε is the number of revolutions the shaft must complete to produce a full cycle; $\varepsilon = 2$ in four-stroke engines, $\varepsilon = 1$ in two-stroke engines.

As already mentioned, the indicated power is not totally available at the engine shaft on which the nominal power, or effective power, P_e is measured by running the engine on a dynamometer. It is expressed as:

$$P_e = \frac{nW_e\Omega}{60\varepsilon} = \frac{p_{me}V_t\Omega}{60\varepsilon}. \tag{10.6}$$

Considering the relationship between power and torque T

$$P_e = \frac{2\pi\Omega T}{60}, \tag{10.7}$$

(where, again, Ω is expressed in rpm) in the case of four-stroke engine ($\varepsilon = 2$), the mean effective pressure is linked to the torque by the equation

$$p_{me} = \frac{2\pi\varepsilon T}{V_t}. \tag{10.8}$$

The mean effective pressure is proportional to the engine torque, or in simpler words, to the engine load. In naturally aspirated four-stroke passenger car engines, the mean effective pressure reaches, at full load, the following values:

- $p_{me} = 10 - 12\,\text{bar} \approx 1.0 - 1.2\,\text{MPa}$ in spark ignition cycle engines, and
- $p_{me} = 7 - 9\,\text{bar} \approx 0.7 - 0.9\,\text{MPa}$ in compression ignition engines.

In case of turbocharged engines these values can significantly increase, up to doubling.

10.2.6 Mechanical Efficiency

As mentioned above, the indicated power is not fully available at the engine shaft and experience has shown that the power spent to overcome engine resistances and to drive the accessories is proportional in part to engine speed and in part to the square of the speed. It is therefore expedient to define a mechanical efficiency η_m as the ratio of the effective power P_e to the indicated power P_i:

$$\eta_m = \frac{P_e}{P_i} = \frac{p_{me}}{p_{ime}}. \tag{10.9}$$

The mean effective pressure p_{me} can thus be defined as the product of the indicated mean effective pressure p_{ime} by the engine mechanical efficiency. The effective power and the indicated power are shown in Fig. 10.5 as functions of the speed.

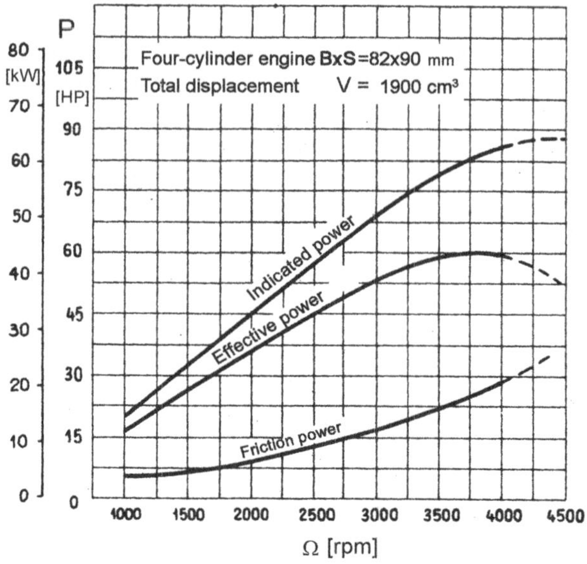

Fig. 10.5 Power dissipated by friction in a 4 cylinder, four-strokes-cycle engine, plotted as a function of the speed together with the effective and the indicated power

The values of the efficiency η_m are: $\eta_m = 0.89$ at 1,000 rpm and $\eta_m = 0.72$ at 3,750 rpm. The mechanical efficiency decreases when engine speed increases and an average value $\eta_m = 0.80$ can be assumed.

In the case of spark ignition-cycle engines, at the speed of nominal power, the friction losses are due, approximately, to the piston, connecting rod and crankshaft bearings for 55 %, to the valve train for 20 %, to the oil pump for a 15 % and to the coolant pump and auxiliaries for the remaining 10 %. A frictional mean effective pressure p_{fme} corresponding to the overall mechanical losses can be defined. At 1,000 and 3,750 rpm typical values for a small-medium size engine can be respectively 0.6 and 1.2 bar. Reducing the mechanical friction of the mentioned components allows us to decrease the p_{ime} and, consequently, the fuel consumption. In case of compression ignition engines the friction losses due to the piston and crankshaft bearings increase, considering that the higher combustion pressure calls for a stronger structure of the basic mechanical components and this is the reason why recently a tendency toward lower compression ratios in compression ignition engines emerged: at high values of the compression ratio, the gain in thermodynamic efficiency is often more than balanced by higher mechanical losses.

10.2.7 Indicated Cycles at Full and Part Load

Two indicated cycles of a spark ignition engine are shown in Fig. 10.6. The first is obtained at Wide Open Throttle (WOT), with the engine achieving its maximum

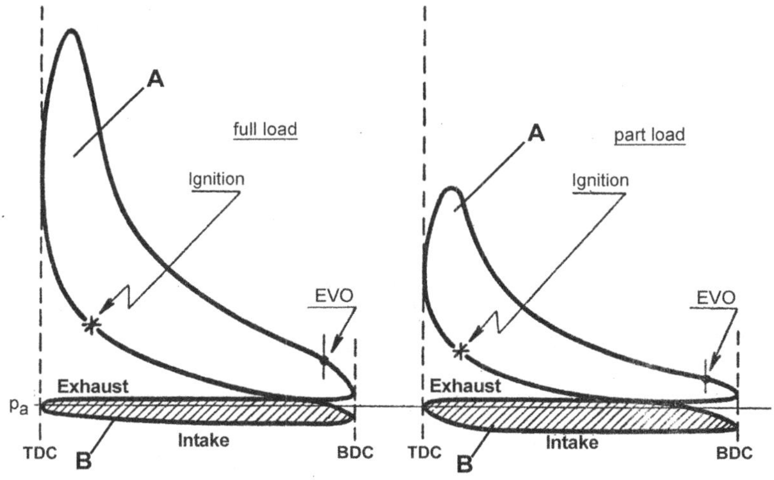

Fig. 10.6 Indicated cycles for a spark ignition engine at full and part load. The diagrams outline the relative weight of useful and pumping works when switching from full to part load (from Giacosa 2000)

performances, while the other corresponds to part load, with the throttle partially closed.

While the area A, representing the useful work, strongly decreases when closing the throttle (after all this is the reason for which the throttle is closed), area B undertakes a complete different change. While at full load the resistance to the air flow is minimal and the pressure in the cylinder when the intake valve opens is close to the atmospheric one p_a, when the throttle is partially closed the increase of the resistance to the air flow causes a pressure drop in the intake manifold downstream to the throttle. In the former case the work lost by pumping is minimum and the negative area is small, while in the latter a non-negligible amount of work is lost and the negative area B is larger. At the end the pumping work mostly identifies the work the piston performs against low pressure in the intake manifold.

In compression ignition engines this does not occur, because, as already mentioned, no throttle is used to decrease the power output and the pressure in the cylinder is close to the atmospheric pressure even in part load conditions: compression ignition engines have a significantly higher efficiency, particularly at part load.

When the engine is running at idle, with the throttle closed, area A is just a little larger than area B, the difference between the two areas yielding the work needed to overcome the mechanical resistances of the engine running at idle.

The exhaust temperature depends much on the temperature peak reached at the end of combustion, that depends on the engine load. For comparable levels of the specific power (40–45 kW/l) the exhaust temperatures are

- 550–850 °C, for spark ignition cycle engines at low/full load, and
- 250–650 °C for compression ignition engines at low/full load.

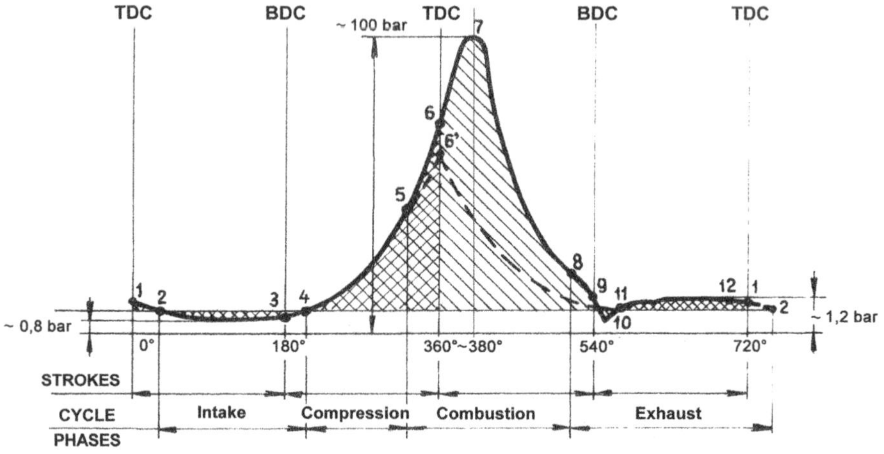

Fig. 10.7 Pressure versus crank angle in an Otto-cycle engine. The maximum pressure takes place at the end of combustion about 15–20° after the TDC (from Giacosa 2000)

The lower exhaust temperature in the latter are due to their better efficiency and to its excess of air, mainly in the low load conditions. In the last decades, the need to improve the after treatment system increased the importance of the above mentioned values.

10.2.8 Indicated Pressure Diagrams

Another way to represent the cycle is by plotting the pressure in the cylinder versus the crank angles, thus showing the pressure trend during the crankshaft rotation in the full 720° cycle. Figure 10.7, referred to a naturally aspirated spark ignition engine, shows that the actual cycle phases, with the corresponding gas physical and chemical state modification, do not take place within the geometrically defined dead centers but with some advance or delay. The corresponding points are well defined by the crankshaft angular position measured in degree. With reference to the geometrical phases, it follows:

Intake

At the start of intake, step 1–2, the cylinder pressure is slightly higher than the atmospheric pressure because the exhaust phase has not yet finished. From point 2 onward, the piston, approaching the BDC, draws in the air or the air/fuel mixture through the open intake valve. Due to friction resistance of the mixture along the duct, the pressure is lower than atmospheric pressure, and the intake vacuum proportionally increases with the inlet flow velocity. When at point 3 the piston starts its compression stroke to TDC, there is still low pressure inside the cylinder and therefore, in spite of the opposite motion of the piston, the mixture continues entering the cylinder until, at point 4, the internal pressure becomes equal to the atmospheric one. This is the correct point at which to close the intake valve, which varies depending on the intake duct length. In fact it is wise to take advantage from the inertia of the gases in the intake manifold closing the intake valve with a suitable delay. From point 4, the compression phase starts.

Compression

The compression of air/fuel mixture takes place when the piston moves toward the TDC. Considering that a certain time is required to complete the combustion, the ignition takes place at point 5, well before TDC to provide the best efficiency of the useful phase. Point 6′ represents the maximum value of the pressure without ignition (the dashed line corresponds to a cycle without ignition).

Combustion and expansion

With ignition at point 5, combustion starts before the end of the compression stroke; this generates a quick temperature and pressure rise with the maximum at point 7, taking place 15–20° after the TDC.

Combustion ends when the piston has covered the first part of the downward stroke and, after combustion, expansion takes place. The volume increases and the pressure drops rapidly, partly also due to heat transfer from the burned gases to the cylinder walls. The expansion should be extended as long as possible so as to take full advantage of the gas pressure, therefore until the BDC. But, to promote burnt gases exhaust, this phase is interrupted when the exhaust valve opens with a certain advance before the BDC (point 8), thus determining a sharp drop of gas pressure.

Exhaust

Gases, which at the opening of the exhaust valve are at a pressure substantially higher than atmospheric pressure, are discharged abruptly. This first part of the exhaust phase, that takes place almost at constant volume, is referred to as exhaust *blowdown*. The residual pressure, often exceeding 5 bar, lowers quickly and at point 9, when the exhaust stroke starts, the pressure exceeds marginally the atmospheric one, with a tendency to lower further during the first part of this stroke. Sometimes, depending on the exhaust duct length, the inertia of the gas column and the residual pressure, it may happen that a short period of negative pressure takes place, as at point 10 in the figure. At point 11 the second part of the exhaust phase starts, due to the piston moving toward the TDC. This takes place at a pressure slightly higher than atmospheric pressure, the so-called exhaust *backpressure*, due to fluid resistance and friction experienced by the exhaust gases passing through the valves and the exhaust ducts.

The piston, however, cannot completely discharge the burnt gases and a residual charge remains within the volume of the combustion chamber. As already stated, in point 1, at the end of the exhaust stroke, the pressure still has a value slightly higher than atmospheric pressure and, therefore the exhaust phase extends up to point 2. In the meanwhile, from point 1 to 2, the intake valve is again open, so that at point 2 it is already partially opened providing a suitable way in for the new intake phase.

The time history of the pressure described above is of paramount importance. In steady-state condition this sequence must repeat with the lowest possible cycle-to-cycle variation. The same time history must characterize all cylinders, with low cylinder-to-cylinder variations. Only in this way can a suitable combustion stability and even running be achieved, granting a satisfactory driving experience.

10.2.9 Valve Timing Distribution Diagram and Volumetric Efficiency

The valve timing characterizes the final engine set-up, for instance whether the engine is tuned for the lowest fuel consumption or the best performances and defines the volumetric efficiency η_v. The latter is the ratio between the mixture mass actually introduced, or trapped, in the cylinder or effective mass m_e, and the mass m_t which theoretically could be contained in the cylinder:

$$\eta_v = \frac{m_e}{m_t}. \tag{10.10}$$

Considering that, usually, valve timing does not change with engine speed, the best value of η_v is obtained for a well defined engine speed thus characterizing the final engine configuration. A good, but achievable, value of η_v which can be obtained in present-day engines, using a variable geometry intake manifold, is 0.90.

As mentioned, in ideal cycles the opening and closing of valves are assumed to take place instantaneously at dead centres. In actual operation, for the already mentioned reasons and due to the need that the valve speed and acceleration do not exceed given maximum values, the opening and closing of the valves take a certain time. The valve opening and closing, i.e. the valve *timing*, can be represented in a polar diagram, as in Fig. 10.8a, that shows the angular positions of the crankshaft at which the valves start to open and to close. The following four angles can thus be defined:

- A, Intake Opening Advance (IOA),
- B, Intake Closing Delay (ICD),
- D, Exhaust Opening Advance (EOA),
- E, Exhaust Closing Delay (ECD).

Angle C represents spark advance and the intermediate circular sectors identify respectively compression and expansion phases. Figure 10.8b shows the same points and time spans in a linear diagram and outlines the valve *overlap*.

The Exhaust Opening Advance, in relation to BDC, lowers the pressure of the burnt gases, before starting the exhaust stroke at a value close to atmospheric pressure, therefore reducing useful cycle work. This reduction is accepted and the optimal advance values is the result of a compromise, so that the work lost during the expansion phase is less than that saved at the beginning of the exhaust phase, by reducing the exhaust backpressure.

The Intake Closing Delay, in relation to the BDC, is recommended when considering gradual closing of the valve and the advantage in properly using the kinetic

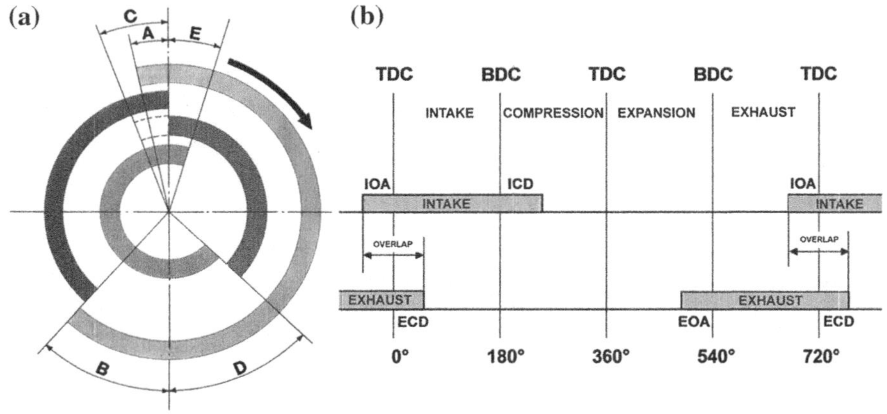

Fig. 10.8 Valve timing diagram: polar (**a**) and linear (**b**) representation. *A* Intake Opening Advance (*IOA*); *B* Intake Closing Delay (*ICD*); *C* Spark advance; *D* Exhaust Opening Advance (*EOA*); *E* Exhaust Closing Delay (*ECD*) (courtesy of FIAT)

energy of the intake mixture flow, through the valve, to promote inlet of the fresh mixture into the cylinder. In fact, the fresh mixture charge during the intake stroke reaches a velocity that, due to inertia, remains still high when the piston slows down, approaching BDC. Therefore, for a certain time, the fresh charge continues to enter into the cylinder even when the piston has already started the compression stroke. A large intake closing delay improves cylinder filling at high speeds but determines a greater reverse flow, or backflow, of the fresh charge toward the intake duct at low speeds.

The value of the volumetric efficiency η_v is plotted as a function of engine speed in Fig. 10.9a for three different values of the intake closing delay: it is clear that an increase of the ICD causes shifts of the maximum value of η_v to higher speeds.

The valve overlap is the sum of the intake opening advance and the exhaust closing delay, in relation to the TDC.

The intake opening advance is a consequence of the valve opening speed, because it is necessary that the lift action starts in advance with respect to the TDC to provide a sufficient valve opening at the beginning of the intake stroke. Similarly, considering the exhaust valve closing speed, it is advisable to provide a certain exhaust valve opening at the TDC to maintain the exhaust gas stream flowing through the exhaust pipe, thanks to its inertia.

A small overlap favours a better volumetric efficiency η_v at low engine speed, when the inertia effects are negligible, and thus improves running stability at idle and at low engine loads, which is extremely important to allow comfortable driving in congested traffic conditions. This setting is always beneficial to lower fuel consumption, considering that also the mechanical efficiency η_m improves at low engine speeds. This setting worsens NO_x emissions, but is beneficial to reduce HC emissions.

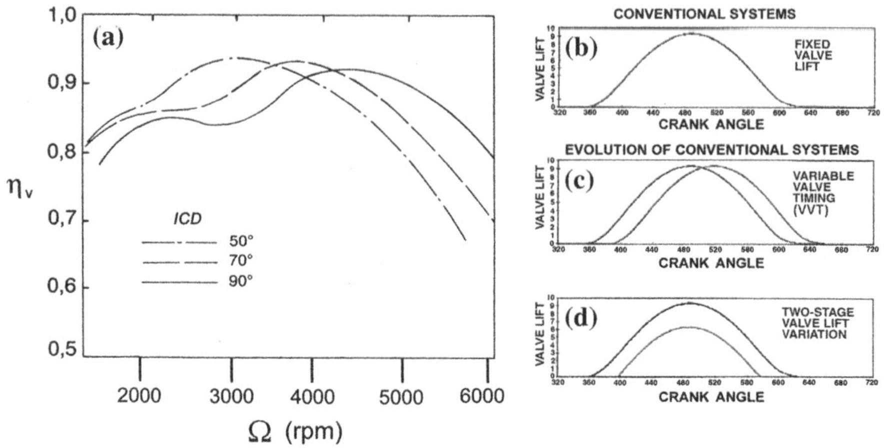

Fig. 10.9 Volumetric efficiency η_v versus engine speed for different values of the intake closing delay (*ICD*) **a**. Intake valve opening sequences: **b** conventional valve lift, **c** VVT system to modify valve timing and overlap, **d** two-stage lift to optimize the mechanical efficiency

A large overlap favours a better volumetric efficiency at high engine speeds, where its effect is quite similar to the intake closing delay increase. At low speeds a wide overlap must be avoided because in so doing part of the fresh charge is pushed back in the intake manifold, determining the already mentioned back-flow effect. This effect is also promoted by the intake vacuum which is stronger than the negligible inertia effect (at low speed) of the exhaust gases, causing an increase of the residual charge which worsens HC emissions but is beneficial to reduce NO_x. A large overlap promotes the best volumetric efficiency at high engine speed, therefore improving performance at the cost of an increase of fuel consumption due to a lowering of mechanical efficiency.

The cam profile with its timing and overlap is thus critical in characterizing the final engine configuration.

10.2.10 Variable Valve Timing

The term Variable Valve Timing (VVT) usually defines systems capable to modifying the phase angles of the camshafts, leaving unchanged both the duration of valve opening and their lift, thus providing two different overlap values as in Fig. 10.9c. The two settings are designed to optimize the volumetric efficiency for a wider range of engine speeds, thus optimizing the torque values within the same speed range. This system thus can also improve city driving with an engine tuned for high performance, offering a good driving feeling in any traffic situation. It also promotes

Table 10.1 Valve timing, in degrees, of different engines

Engine	IOA	ICD	EOA	ECD	Overlap
Fiat Fire 1.1 l	7	37	37	7	14
Alfa Romeo 2.0 l TS	−3/22	51/26	47	4	1/26
Alfa Romeo racing engine	62	65	71	58	120
Fiat 1.9 l Diesel	0	32	32	0	0

a lower emission level, thus allowing emission standards to be met in low load conditions, like the New European Driving Cycle (NEDC).

Variable valve timing devices normally operate only the intake camshaft, since intake timing variations are more effective in determining the previously mentioned effects. Timing variations are controlled by the engine management system, typically by delaying the opening of the intake valve at low speed, and advancing it at high speed.

This concept was first implemented by Alfa Romeo, in 1982. An updated application on the Twin Spark engine at the end of the 1990s is shown in Table 10.1. In this case the overlap increases from 1° to 26° from low to high speed, thus obtaining a good volumetric efficiency in a wide speed range. In fact, to increase η_v at high speed, the overlap effect is even larger than that of an increase of the intake closing delay.

In racing engines, both the overlap and the intake closing delay contribute to reach top performances.

The above mentioned situation is totally different in the case of compression ignition engines, since the turbocharger and absence of the throttle substantially improve the filling of the cylinder and the engine volumetric efficiency in a wide speed range.

The valve timings of four engines, typical of the mid-1990s, are compared in Table 10.1. They are:

- Fire 1.1 l. This engine, designed for city cars, features a fixed timing designed as an acceptable compromise to obtain a good torque at low speed, good fuel economy, limited cost, limited emissions and a specific power close to 45 kW/l (60 HP/l). A number of low cost engines for low segment cars still have a similar setting.
- Alfa Twin Spark 2.0 l. This engine, designed for high performing sedan cars, features the variable valve timing system which provides two overlap values specifically intended to obtain a comfortable city driving with a reasonable fuel consumption but also good performances with a specific power close to 60 kW/l (80 HP/l). When compared with the previous value, these two overlap values are almost symmetrical in order to promote both city and high performance driving conditions. The low overlap and a stable combustion are prerequisites to lowering the idle speed down to 700 rpm and even more, with the aim of increasing comfort and reducing fuel consumption.
- Alfa racing car engine 2.0 l. This engine, designed for a race car, provides a remarkable value of the specific power, close to 75 kW/l (100 HP/l), obtained without

turbocharging. In this case fuel consumption, comfort and emission control are not priorities.

- Fiat turbo diesel 1.9 l. This engine, designed for a medium sedan car, offers an excellent fuel economy with a specific power close to 52 kW/l (70 HP/l). Compression ignition engines maintain the basic setting without variable valve timing, considering that in compression ignition engines the priority is given to control the fuel delivery rather than the air delivery.

Figure 10.9b shows a fixed intake valve opening sequence, while Fig. 10.9c shows the sequence applied by the VVT system. This apparently limited shift, as described, can deeply modify the engine response. Considering normal production engines listed in the table, the opening angle of the intake valve is equal to about 224° and 228° for gasoline engines and 212° for the compression ignition engine.

10.2.11 Continuous Variable Valve Timing

The variable valve timing solution was widely applied by many Manufacturers on a number of engines and, at the end of 1990s, evolved into the Continuous Variable Valve Timing (CVVT) systems. In the last two decades, intake and exhaust valve phase variation proved to be one of the most efficient methods of improving volumetric efficiency over the whole engine map. The complete system should include phase variators on both intake and exhaust valves but, to decrease costs, often only one phase variator is used on the intake valve.

The continuous variable valve timing mechanism shown in Fig. 10.10 features a stator (1), driven by the chain operated by the crankshaft sprocket, and a rotor (2) connected with the camshaft and provided with blades (3); between the stator and the rotor blades a variable space is provided. The spring (4) maintains a suitable pressure between stator and rotor allowing the possibility to obtain an angular movement between the two components thanks to engine oil pressure. The movement is limited by the rotating limiting plate (5). Specific solenoid valves, managed by a control unit that processes the speed and load engine signals, send oil from one side or the other of the blades, so as to change the angular position between stator and rotor. When the camshaft reaches the pre-set position, the solenoïd valve stops the passage of the oil. The angular position setting is continuous and, for example, a 25° maximum range of the camshaft timing variation (50° as seen from the crankshaft) in both directions may be used.

More recently, in some advanced engine configurations, continuous variable valve timing allows activation of the scavenging function which increases the overlap angle at low speed. This function proved to be effective to better exploit turbo management and to provide better thermodynamic efficiency in turbo gasoline direct injection engines.

Fig. 10.10 Continuously variable valve timing device. It allows a continuous modification of the valve timing and overlap to optimize the volumetric efficiency in every point of the engine map. Timing modification is actuated by varying the oil volume between the rotor blades and the stator vanes (courtesy of FIAT)

10.2.12 Variable Valve Actuation

An interesting development toward the Variable Valve Actuation, was the solution developed by Honda with the VTEC system at the end of the1980s. This systems features the possibility of selecting a two stage valve lift variation, Fig. 10.9d, optimized for part and full load operations again with the aim of achieving low-speed stability with high-speed performances. A certain fuel consumption saving is obtained by reducing the mechanical work, owing to the reduced valve lift, thus lowering the frictional mean effective pressure. Beginning in year 2000, Honda updated its system introducing the i-VTEC version which combines the VTEC mechanism with variable cam phasing. To quote anther example, well known in Europe, in the same period Porsche launched the VarioCam Plus device and other manufacturers, e.g. Mitsubishi and Nissan, performed similarly.

The present-day variable valve actuation systems allow simultaneous variation of the valve timing and their lift. To obtain an efficient engine operation over its entire operating range, the valve variation must be fast enough to follow the speed and load combinations in the whole engine map. The possibility to vary the valve lift, allows elimination of the throttle valve, so reducing pumping losses during part load operations. In this way the size of area B in Fig. 10.7, responsible for pumping losses, is reduced and the engine volumetric efficiency improves.

Similarly to the continuous variable valve timing systems, also variable valve actuation devices are mostly applied to the intake valve. The control of the exhaust basically allows modification of the residual gas quantity and therefore the NO_x content.

These systems can be subdivided in two main families:

- Systems with no camshaft in which the valves are actuated by electromagnetic or electrohydraulic devices and
- systems using a camshaft to drive the tappets and actuating the timing and lift variation with mechanical or electrohydraulic devices.

The first had been actively studied considering their potential to offer full flexibility in management of intake and exhaust valve lift events. They still show interesting potential but, for the time being, suffer from intrinsic problems due to power consumption, safety and NVH levels. In addition, the usual 12 V supply is no longer sufficient to drive the valves electromagnetically, and an higher voltage supply (e.g. 42 V) is needed unless unreasonable values of current are accepted. A further difficulty is obtaining gradual valve closing avoiding mechanical shocks with consequent damage to the valve heads. The above mentioned situation prevented, up to now, starting of the production of systems belonging to the former family.

The systems presently used are therefore of the second type: a mechanical or hydro-pneumatic device, able to modify the valve lift between a minimum and maximum value, is interposed between the camshaft and the valve tappets. It is set so that the minimum value of the valve lift allows delivering to the engine only the small amount of air needed for idling and the maximum value corresponds to that available with the traditional system to obtain the required performance.

In both systems the main problem requiring a solution is their dynamic performance considering that at 3,000 rpm the crankshaft rotates 50 times per second so that the valve timing events must occur at precise times and must immediately follow engine load variations.

Example of a Mechanical Device

The Valvetronic® mechanical variable valve actuation system, introduced by BMW in 2001 is shown in Fig. 10.11. This system controls the intake valve lift while the timing is adjusted maintaining the previous continuous variable valve timing system, called Vanos.

The goal of Valvetronic is to control the power output through a variable valve lift, linked to the position of accelerator pedal instead of using the traditional throttle. Discarding the latter makes the system particularly noticeable at low loads during which the throttle is almost fully closed. Here a fuel saving up to 20 % can be obtained. This potential decreases proportionally with higher loads but average savings close to 10 % are achieved.

In this application the valve is actuated using the rocker arm through roller bearings (3) to reduce the friction. The arm is placed between the hydraulic tappet (7) and the valve stem and actuates the valve opening. The arm is usually directly driven by the cam; in this case it is driven by a floating intermediate lever (5) which pushes the arm through a roller bearing. The lever in its middle part is actuated by the inlet camshaft (1) with the interposition of a second roller and, in its upper part by an eccentric shaft (2) connected to a high-resolution rotation angle sensor. When the engine

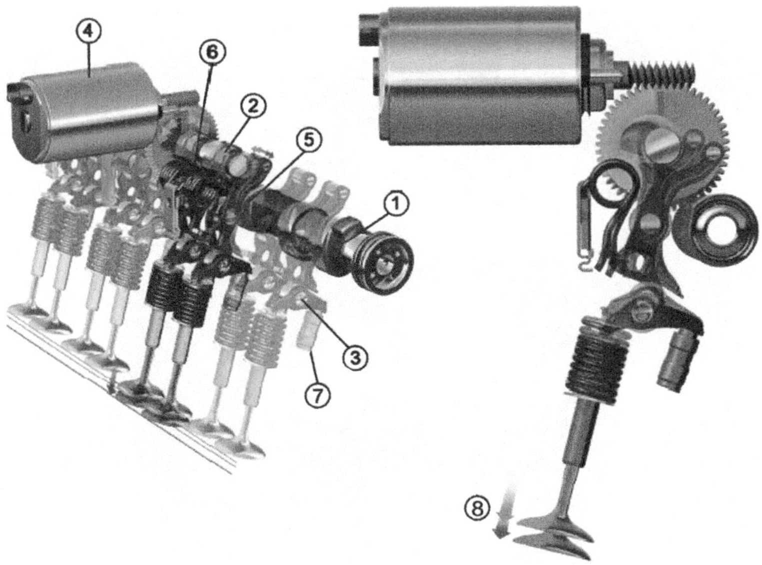

Fig. 10.11 Schematic drawing of the Valvetronic mechanical variable valve actuation system

management control unit registers a change in load demand a controlled signal causes the electric motor (4) to rotate and the connected worm-screw produces an eccentric shaft oscillation. The lever, in its lower part, is shaped according to a working contour which finally determines the valve lift (8). The camshaft thus provides the opening-closing law but the lift is determined by the shaft oscillation through the lever. This latter is not mechanically connected, but finds its position among the camshaft, the eccentric shaft and the rocker arm; its position is maintained by the return spring (6) which keeps the lever constantly in contact with the other mechanical components.

The engine power is thus controlled by varying the valve lift thanks to operation of the above mentioned components. The system allows a valve lift variation from a fraction of mm, to the typical value close to 10 mm, in a step-less way. At a speed of 2,000 rpm and a p_{me} of 2 bar, in the area of NEDC cycle load, the valve lift amounts to approximately 1.2 mm with a valve lift duration of about 120°.

The second generation of the Valvetronic system, introduced in 2004, was developed with the aim of further improving fuel economy and engine performance. This was achieved by a further pumping loss reduction and an increased turbulence level able to improve combustion stability and exhaust gas recirculation tolerance. The system has been recently extended to the PSA/BMW joint engine family with continuous updating.

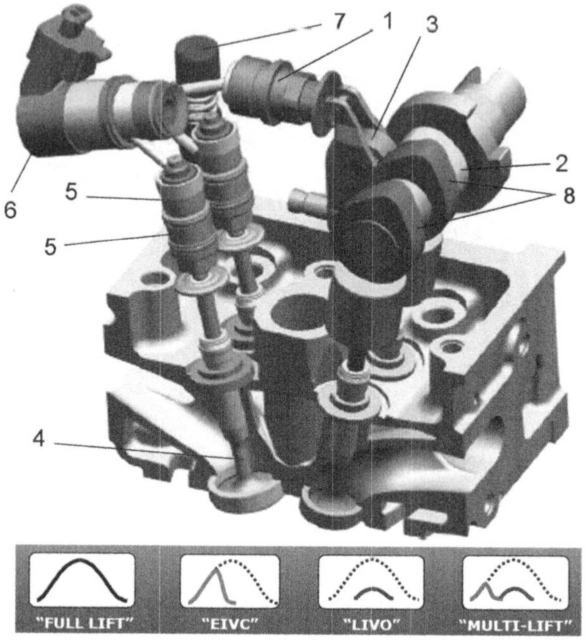

Fig. 10.12 Electro-hydraulic Variable Valve Actuation. This device actuates the valve lift variation by modifying the oil volume trapped in the high pressure chamber, thus providing the valve lift and timing combinations shown (courtesy of FIAT)

Example of a Electrohydraulic Device

The MultiAir® electrohydraulic VVA system, introduced by Fiat in 2009, is shown in Fig. 10.12. This system controls the intake valve lift being also able to modify its timing. It has been engineered with a mechanical layout which allows one to 'simply' add the system on a normal production engine thus reducing the investment costs in the production facilities.

The original double camshaft has been replaced with a single camshaft: this latter activates both exhaust and intake valves but, adding a rocker arm, their relative position and the pent-roof combustion chamber configuration remains unchanged.

The MultiAir system is able to perform the intake valve actuation independently from the engine speed. This is possible due to de-coupling between the mechanical cam and the intake valve movement by the interposition of oil held in a chamber that is pressurized closing the normally open on/off high response solenoid valve.

The operating principle of the system, applied to the intake valve, is the following: a piston (1), moved by the intake cam (2) and its rocker arm with roller bearing (3), operates the intake valves (4) through the hydraulic pressure chambers (5), which are controlled by the solenoid valve (6).

This latter is managed by the engine control unit, and when it is activated the closed volume of oil in hydraulic chambers behaves like a solid body and provides to the intake valves the lift schedule of the intake cam. When the solenoid valve is deactivated the hydraulic chambers are opened toward the low pressure circuit thus allowing the intake valves to close following the ‘lost motion’ principle. It means that the final part of the closing stroke is controlled not only through the spring but also following the action of a dedicated hydraulic damping unit, able to ensure a soft and regular closing phase in any engine operating condition.

As the solenoid valve piston moves, the oil present in the hydraulic chamber flows toward the oil accumulator (7), whose function is to hold the oil and keep a constant pressure in the circuit by the action of a pre-loaded spring. In so doing the oil accumulator, in the low pressure part of the circuit, helps in refilling the high pressure chamber for the following stroke thus minimizing energy losses.

It is therefore possible to distinguish the two circuits: a low pressure one, upstream of the solenoid valve and the second, at high pressure, downstream of the same valve.

The oil used in the hydraulic system is the same oil used for the engine lubrication, therefore special attention must be devoted to keep the oil properties, mainly its viscosity, as constant as possible otherwise the intake actuation could undergo undesired variations. The exhaust valves are operated by their usual mechanical cams (8).

The same figure shows the most interesting valve movement, thus making possible:

- Full lift mode: it is the position necessary to obtain full power. This actuation is performed by closing the solenoid valve in order to exploit the cam profile actuation. A 5–10 % of maximum power and torque increase can be obtained with a specifically suited cam profile.
- Early Intake Valve Closing (EIVC): this mode operates in part load conditions, in fact the intake valves close midway during the intake stroke and earlier than in the full-lift. During its initial opening it delivers the charge needed to the engine and significantly reduces pumping losses.
- Multi-lift: this mode has been proved successful in low part-load conditions; it favours a good combustion stability with reduced pumping losses.
- Late Intake Valve Closing (LIVO): this mode, applicable during idle and very low loads, is obtained with an open-close-open solenoid valve sequence which provides the proper air motion suitable to improve combustion stability.

In certain part-load conditions the pumping work can be reduced up to 40 %, as shown in Fig. 10.13. The EIVC mode supplies the charge during the valve opening period at a negative pressure that remains quite close to the atmospheric level (0.9 bar) which substantially differs from the negative pressure value traditionally present when operating the throttle valve (0.3 bar). As shown in the figure, during the intake stroke, the intake valve closes earlier and precisely when the desired mass of mixture has entered the combustion chamber. At the end, in part load conditions, the fuel saving and valve lift values are comparable with the ones already mentioned for the mechanical valve lift variation system.

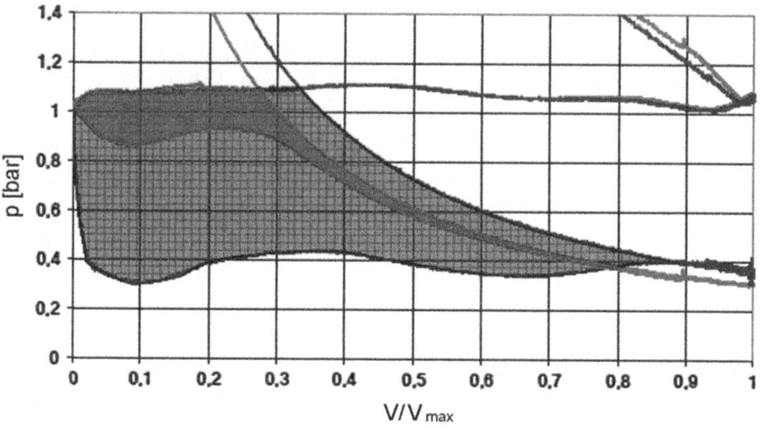

Fig. 10.13 Pumping loss reduction by 40 %, achievable with the Early Intake Valve Closing (EIVC) strategy (courtesy of FIAT)

After the initial application, FIAT is extending this configuration to the subsequent engines introduced in production both naturally aspirated and turbocharged including the 'extremely downsized' two-cylinder engine.

The electrohydraulic solution shown, activates the two intake valve lift simultaneously, i.e. tandem actuation, therefore some benefits related to intake charge motion can't be fully exploited. This option can be easily introduced adding a fully independent system for each valve.

Comparing the two systems, as described, the electro-pneumatic one offers the potential to differentiate the valve lift among the different cylinders. This can be a future interesting possibility when looking for a further fuel consumption reduction. In fact, with this goal, the cylinder de-activation becomes an interesting option for multi-cylinder engines.

When tuning these expensive and fairly complicated systems the manufacturer tries the best calibration to reduce emissions, basically looking for an effective catalyst light-off[2] and a suitable internal Exhaust Gas Recirculation (EGR, see below). Mainly for safety reasons, the traditional throttle body is usually maintained.

Quite often these devices retain the throttle valve that is kept in its fully open position and it is actuated only to overcome emergency situations or extreme ambient conditions.

[2] The term *light-off temperature* (T50), i.e. the temperature at which a catalyst begins to convert 50 % of the emissions into carbon dioxide, water and nitrogen. Its value spans between 200 and 250 °C.

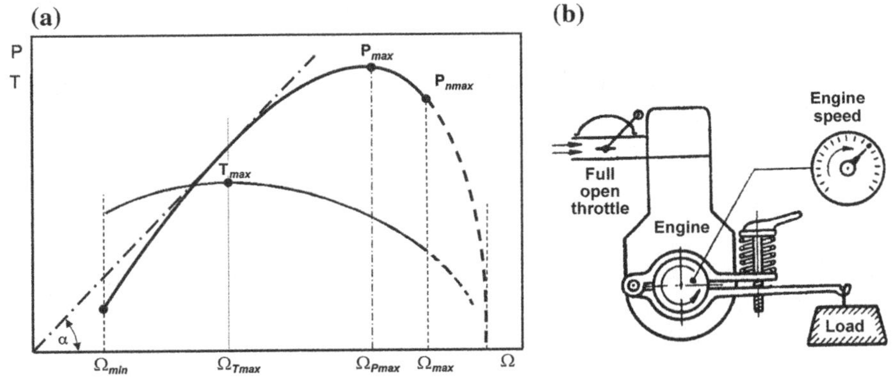

Fig. 10.14 **a** Power and Torque plotted versus the engine speed for a naturally aspirated spark ignition engine. The maximum torque occurs where a straight-line throught the origin is tangent to the power curve. **b** Mechanical engine dynamometer to determine the engine power in full load and part load conditions measuring the torque and the engine speed (from Giacosa 2000)

10.2.13 Torque and Power

The engine performance, in terms of torque and power, of an aspirated gasoline 4-cylinders engine is shown in Fig. 10.14a. The torque is reported as a function of the engine speed from the minimum speed $\Omega_{\min}$ (700–800 rpm), to the speed $\Omega_{P\ \max}$ at which the maximum power is obtained (5,500–7,000 rpm) and beyond, up to the maximum operating speed conditions. The corresponding values for a compression ignition are: 800–900 rpm and 4,000–4,200 rpm, respectively.

The current trend is to lower the idle speed to reduce the fuel consumption and to increase the maximum speed of compression ignition engines to improve performance.

Performance of both spark ignition and compression ignition engines is measured in full load conditions, testing the engine on a dynamometer. Spark ignition engines deliver their full power operating in Wide Open Throttle (WOT) conditions, while compression ignition engines when operating with the injection pump at its maximum delivery.

The scheme of a simple mechanical dynamometer assembly is shown in Fig. 10.14b. It allows measuring both the torque, by measuring the load applied to the arm of the brake, and the engine speed.

As already mentioned, the engine torque is proportional to the mean effective pressure. For this reason, the curve of the engine torque is, often, expressed in terms of p_{me} curve. The product of the torque by the engine speed, yields the available power, provided that both are expressed in consistent units. Often the speed is expressed in rpm, instead of rad/s, so that a coefficient $2\pi/60$ is included in the equation. The power should be expressed in W or its multiple, kW, and the old, non consistent, unit horsepower (HP, 1 kW = 1.36 HP) should be avoided.

Power initially grows almost linearly with speed, reaches its maximum value P_{max} for a certain engine speed $\Omega_{P\,max}$ and then, since the friction power increases approximately with the square of engine speed, decreases until the curve is interrupted at maximum engine speed Ω_{max}. The power $P_{n\,max}$, corresponding to Ω_{max}, is always lower than P_{max}. To avoid mechanical damages, a speed limiter, used in both spark ignition and compression ignition engines, cuts fuel delivery when reaching maximum speed. The power corresponding to the engine minimum speed Ω_{min} is the minimum power, always measured at full load. For lower speeds, engine running becomes irregular and unstable due to variability of the delivered torque between one useful cycle and the following one.

The straight line through the origin, tangent to the power curve meets this latter at the speed of maximum torque. The maximum torque and angle α in Fig. 10.14 are linked by the obvious relationship

$$T_{max} = \tan(\alpha), \tag{10.11}$$

which holds only if the same scales and consistent units (W and rad/s) are used for the axes.

10.2.14 Overall Efficiency and Fuel Consumption

A value of 50 % was previously mentioned as the maximum indicated efficiency for the Otto cycle. This value is further reduced in actual operating conditions and, typically, a spark ignition engine is capable of transforming into useful work a maximum of 34–36 % of the fuel chemical energy. The remaining 64 % is wasted and transferred as heat to the exhaust gases (about 26 %), to the cooling system (water, oil and air, about 18 %) and through convection and radiation (about 20 %).

The overall efficiency η_o of the engine can be obtained by multiplying the already mentioned indicated, volumetric and mechanical efficiencies: the mentioned value of 36 % is obtained as

$$\eta_o = \eta_i \eta_m \eta_v = 0.50 \times 0.8 \times 0.90 \simeq 0.36.$$

A similar evaluation holds for compression ignition engines, whose efficiency is slightly higher, close to 40 %.

Experimentally, the overall efficiency is determined by measuring the specific fuel consumption, i.e. the quantity of fuel consumed in one hour, divided by the power generated. Its experimental determination is carried out at full load and at the intermediate loads running the engine on the dynamometer and measuring the time needed to burn a known mass of fuel. It is usually expressed in the non consistent unit g/kWh, although the consistent SI unit should be kg/J = s^2/m^2.

Performing these measurements and joining the iso-consumption points on the plane in which the torque is plotted against the speed, the fuel consumption map is

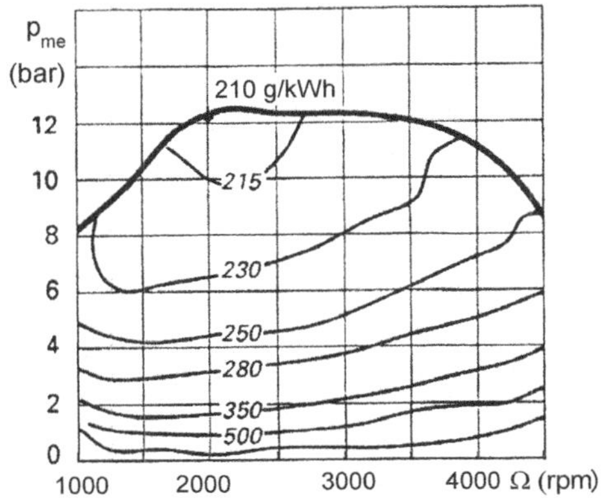

Fig. 10.15 Specific fuel consumption map of a turbocharged compression ignition engine in full and part load conditions (boost pressure 1 bar) (courtesy of FIAT)

obtained. Instead of the torque-speed plot, a plot of the mean effective pressure p_{me}, which is proportional to the torque, is commonly used and the final result is shown in Fig. 10.15.

The fuel consumption map shows:

- The maximum value of the p_{me} as a function of the speed (bold line curve) in the figure, and
- the specific fuel consumption for every load condition (pairs of values of p_{me} and speed).

The map for a turbocharged compression ignition engine plotted in Fig. 10.15 shows a minimum value of the specific fuel consumption equal to 210 g/kWh (~58 g/MJ). Considering that the chemical energy of Diesel fuel is 43.3 MJ/kg the value of the maximum overall efficiency η_o is 0.396.

In case of a gasoline engine, the minimum consumption point could be close to 240 g/kWh and therefore is 10–20 % higher, as an average. In part load conditions, particularly for gasoline engines, a significant fuel consumption increase is recorded, as already described. Usually no value is reported in the zone of the plot close to $p_{me} = 0$ and for negative values (i.e. when the engine is used as a brake, with the accelerator pedal fully released), since in these conditions the fuel consumption tends to ∞.

Steady state fuel consumption, however, does not give a complete information regarding the fuel consumption in actual operation, since transients are frequent in actual driving conditions. To overcome this limitation without carrying out practical tests of uncertain repeatability, specific test cycles have been defined and standardized. The European NEDC has been described in detail in Chap. 7.

Table 10.2 Values of geometrical parameters and related engine performances for some 2010 year Otto-cycle engines

#	V_t (cm^3)	n	V (cm^3)	B (mm)	S (mm)	ρ	F	T (Nm)	P (kW)	SP (kW/l)	Ω_{max} (rpm)	u (m/s)
1	875	L2	437	80	86	10.0	t	145	62	71	5,500	16
2	997	L3	332	73	79	11.0	na	90	48	48	6,000	16
3	1,086	L4	271	67	77	10.1	na	97	49	45	5,500	14
4	1,368	L4	342	72	84	10.8	na	130	79	57	6,500	18
5	1,742	L4	436	83	80	9.5	t	340	172	99	5,000	13
6	1,995	L4	499	84	90	12.0	na	210	124	62	6,700	20
7	6,262	V12	522	94	75	12.3	na	683	482	77	8,200	21

10.2.15 Numerical Examples

The values of the main geometrical parameters and related performance, namely the total displacement (V_t), number of cylinders (n) and configuration (in line, L or V), unit displacement (V), bore (B), stroke (S,), compression ratio (ρ), type of feed (F; naturally aspirated, na, or turbocharged, t), torque (T), power (P), specific power (SP), speed at maximum power ($\Omega_{P\ max}$), piston speed (u), for some year 2010 spark ignition engines are reported in Table 10.2.

From the table the more frequent ranges of the same parameters can be deduced:

- Unit displacement: 300–500 cm^3,
- Compression ratio ranging from 10 (indirect injection engines) to 12:1 (direct injection) with lower values for turbo versions,
- Specific power 45–75 kW/l for naturally aspirated and 75–100 kW/l for turbo versions,
- Specific torque ranges from 100 Nm/l for naturally aspirated engines to 200 Nm/l for turbo engines,
- Piston speed: 15–20 m/s.

10.3 Mechanical Engine Layout and Components

Four-stroke engines can be quite different from each other in their overall configuration. In the automotive industry, the cylinder number may vary from 2 to 12 (single cylinders engines, although much used in other fields, are not used on motor cars) and the most usual configurations are:

- In-line cylinders. All cylinders are lined along the crankcase, allowing a simple mechanical layout and improved accessibility. The number of cylinders today is limited to six, to avoid an excessive engine length and the four cylinder in-line is the most used configuration, even if engines with 2, 3, 5 and 6 cylinders are used in the automotive market.
- Horizontally opposed cylinders with flat or boxer configuration. The cylinders are mounted on the crankcase in two opposite banks that are staggered one with

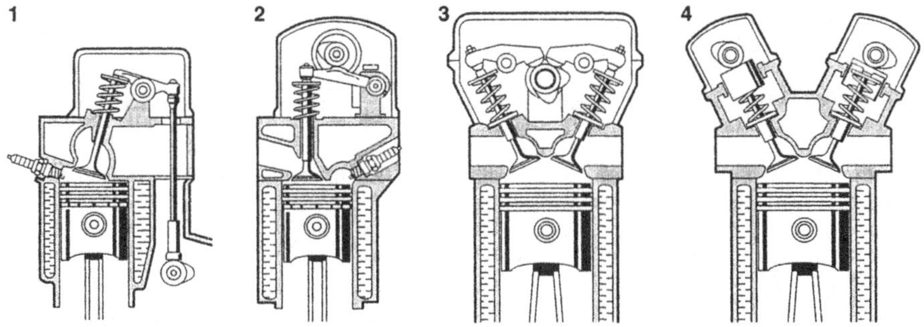

Fig. 10.16 Valve configurations and camshaft drive and positions: *1* Over-head valves (OHV); *2* Over-head valves and Single Overhead Camshaft (1OHC); *3* Over-head valves and Single Overhead Camshaft with double rocker arms; *4* Over-head valves and Double Overhead Camshaft (DOHC) (courtesy of Bosch)

respect to the other. The main advantages include an inherently good balance of the reciprocating parts, a low center of mass, which contributes to car stability, and a shorter engine structure. Typically they are produced with 2-4-6 cylinders mainly on front-wheels drive car and, less frequently, on rear-wheels drive with rear engine.

- V-configuration cylinders. The cylinders are mounted on the crankcase in two banks contained in two concurrent semi-planes which meets along the crankshaft axis and again staggered, allowing to obtain a more compact and lighter engine. In particular, the overall length of the engine can be appreciably reduced, so that both the structure and the crankshaft can be stiffer. Typically they are built with 6, 8, 10 and 12 cylinders.

Cylinder numbering usually starts from the front of the engine, where the camshaft drive mechanism is located.

With reference to cylinder head and valve arrangement, the more used solutions to operate the poppet-type valves are shown in Fig. 10.16:

- OHV, Over-Head Valves, with camshaft in the crankcase, push-rod and rocker arm assembly, peculiar of low segment engines in the fifties and sixties;
- OHV and 1OHC, Single Over-Head Camshaft, with single rocker arm and parallel valve configuration peculiar of low segment engines in the seventies and still used today;
- OHV and 1OHC, with double rocker arms and valve configuration peculiar of medium high segment engines in the 1980s and 1990s;
- OHV and DOHC, Double Over-Head Camshaft and valve configuration with direct valve actuation today widely used in medium and high segment engines.

The mechanical components that operate the valve from the cam must provide low friction and long durability. Those shown in Fig. 10.16 are: (1) pushrod and rocker arm, (2) pivoting follower, (3) directly actuated rocker arms and (4) directly actuated bucket-type tappets.

In spark ignition-cycle engines the fuel can be injected either in the intake manifold or in the combustion chamber:

- Multi Point Injection (MPI), indirect injection or Port Fuel Injection (PFI) in which the fuel is injected upstream the intake valve, in front of the intake port;
- Multi point direct injection or Gasoline Direct Injection (GDI) in which fuel is injected into the combustion chamber.

In compression ignition engines at present the injection always occurs directly in the combustion chamber.

The transversal and longitudinal sections of an Alfa Romeo gasoline engine featuring V-6 cylinders, 3.0 l displacement, with double overhead camshaft, direct valve actuation and multi-point, port fuel injection are shown in Fig. 10.17.

Figure 10.17 shows some already mentioned components typical of four-stroke-cycle engines: the intake and exhaust valves and ducts, the intake and exhaust ports into the cylinder head (a) and (b), the intake and exhaust manifold (c) and (d), the fuel rail to feed the injectors (e), the spark plugs with high voltage coil (f) and the pent-roof combustion chambers (g).

In the figures the camshafts (1), the left-hand cylinder head (2), the right-hand cylinder head (3), the crankcase (4), the cylinder wet-liners (5) with water cooling jacket in their upper part, the stud-bolts to fasten the head to the crankcase (6), the connecting rods (7), the crankshaft (8), the main bearing caps (9), the oil sump (10), the flywheel (11), the water (12) and the oil pumps (13) are also represented.

Many of the above components are mostly in common, or similar, to those used in compression ignition engines.

As mentioned, V-engines may have 6, 8, 10 and 12 cylinders and, to have a regular firing sequence and an optimal balancing, the angle between the two banks should be equal to $720°/n$ being n the number of cylinders. The V angle is thus 120°, 90°, 72°, 60° for V6, V8, V10 and V12 engines respectively. However, this rule is often disregarded when designing V engines and, in the 6 cylinder case, a narrower angle can be used, 60° as in the Alfa Romeo shown in Fig. 10.17, or 90°, to reduce the engine cross-section. In the case of 12 cylinder engines, a wider angle, can be used to lower the position of the center of mass.

From time to time the so-called narrow V configuration is used. The first successful example was the Lancia four cylinders, mentioned in Part I and, more recently, the engines with a V of 15° used in some models of the VW-Audi group. The W configuration is used by the latter group and also by Bugatti, i.e. a 16 cylinders engine with 90° opening between the three concurrent banks.

10.3.1 Crankcase

The engine crankcase acts as a support for all other engine components, allowing the rotation of the crankshaft through suitable bearings and the reciprocating movement of the pistons, using suitable liner surfaces along the cylinder walls.

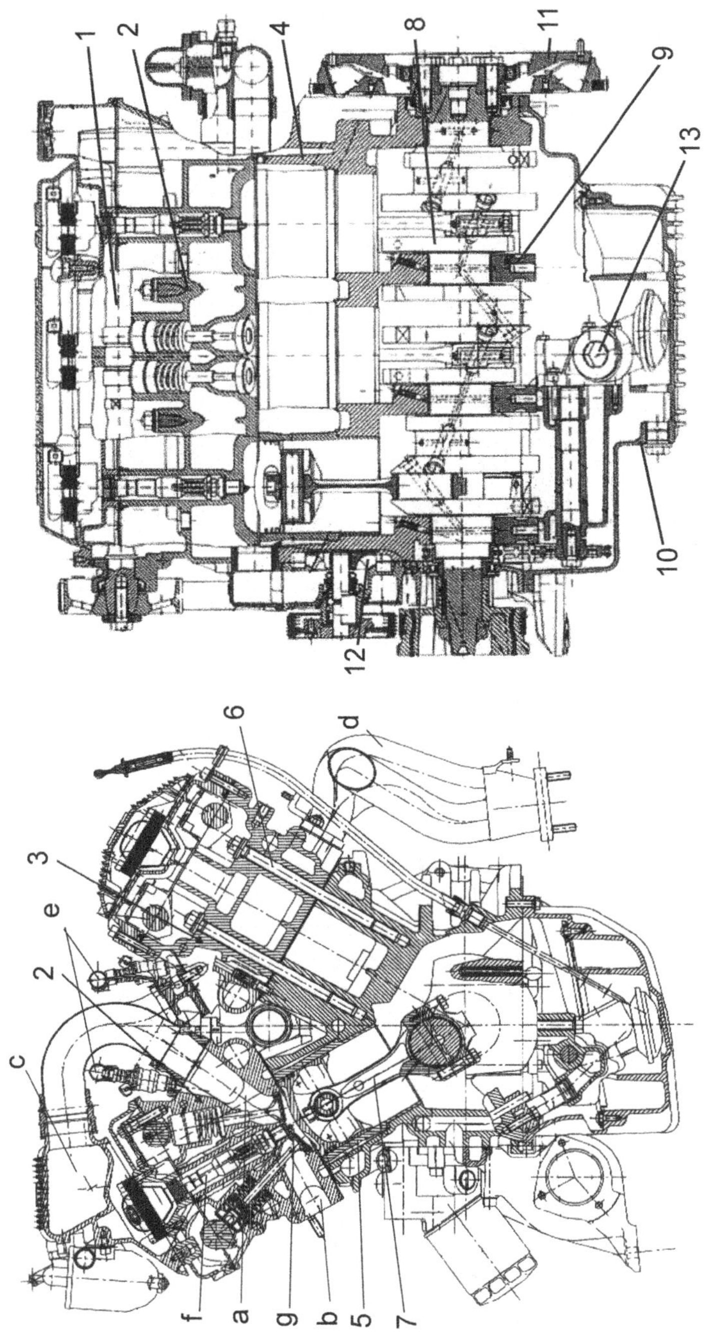

Fig. 10.17 Cross section of an Alfa Romeo six-cylinders engine, V-configuration (courtesy of FIAT)

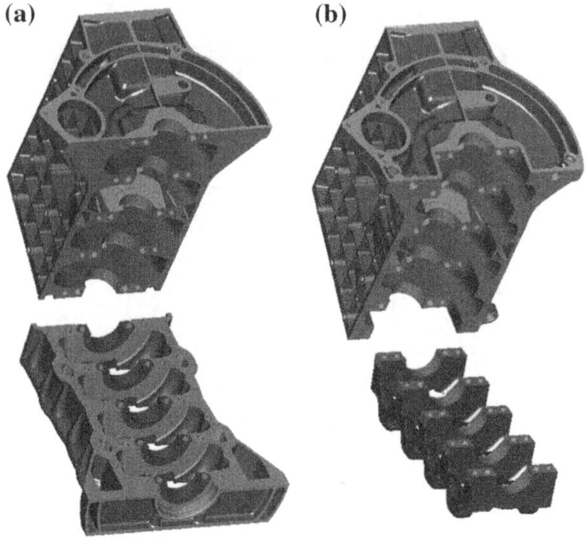

Fig. 10.18 Crankcase configurations. **a** Normal skirt; **b** Extended skirt. In **a** bed-plate bearing supports are used, while in **b** the supports are independent

The crankcase is closed in its upper part by the cylinder head and in its lower part by the oil sump. In addition, it contains the cooling fluid and the lubricant oil. This important component must offer adequate strength and stiffness, since it must withstand the mechanical and thermal stresses with their wide variations from idle to maximum power and in different climatic conditions, from extreme cold to extreme hot environment. The crankcase carries also the mechanical elements supporting the engine on the vehicle frame, all ancillary equipment on the engine and provides the gearbox-clutch connections. Considering that the engine mass is, as an average, 10 % of the vehicle mass, the material selection, namely either cast iron or aluminium alloy, is still a matter of debate. As an example, the latter allows a mass saving of about 20 kg for a 2 l engine with a mass of 130 kg. For lower displacements, about 1.0–1.2 l, this gain is limited to a few kilograms. When cast iron is used, the choice usually falls on grey cast iron. It is the preferred choice in case of compression ignition engines.

Figure 10.18 shows the two most common configurations for four cylinder in-line engines: normal skirt (a) and extended skirt (b). In the first case, common in cast iron construction, the lower part of the crankcase reaches down only to the crankshaft axis. In the second case, widely used for aluminium crankcases, the skirt extends well below the engine crankshaft. This, more costly, configuration usually provides for a higher crankcase stiffness.

In both cases, the lower part of the crankshaft bearings can have the bed-plate (a) or independent caps (b) configuration, as shown in the same figure. The first integrates, in a single rigid aluminium unit, the lower part of the bearings, thus providing additional stiffness. This solution is widely used in high performance engines, with aluminium crankcase. In the second, more traditional solution, the independent caps are connected, with suitable bolts, to the upper transversal ribs. It is widely used for cast iron crankcases.

Antifriction bearings are located between the crankcase and the crankshaft, providing suitable geometrical tolerances to assure the lubricating film formation in every running conditions. The lower parts of the bearing supports and the related connection bolts are heavily loaded by the connecting rod big-end trust.

Since the start and stop operation has been introduced, a special attention has been devoted to the bearing antifriction material, considering that the number of start-ups, traditionally in the range of 25,000, may reach 1 million.

The water jacket surrounds each cylinder. The crankcase has also passages to provide the oil return from the cylinder head to the oil sump and the holes in which the stud-bolts connecting the head to the crankcase are located. With cast iron solutions integral cylinder liners (or sleeves) are used: they are obtained by mechanical machining the cast-iron block. This solution is quite expedient in small displacement engines due to its simplicity.

When aluminium construction is chosen, the liner becomes a detachable dry-liner, basically made with cast iron, providing an anti-wear surface. The liners can be cold driven into the cylinder block with a press.

A significant problem with this type of dry-liners has been the circularity tolerance of the coupling: very often, due to the non perfect regularity, there could be microscopic discontinuities between the liners and the crankcase, that greatly impair the transmission of heat to the coolant. To overcome this problem, non detachable dry liners integrally cast with the aluminium cylinder block are now widely used. The thermal expansion of the different materials used for the liners and the crankcase must be at any rate considered.

Increasing the specific power output, the wet-liner solution is preferred or becomes mandatory to provide anti-wear resistance and a suitable cooling, as shown in the Alfa Romeo engine of Fig. 10.17.

In its upper part, the crankcase is connected with the cylinder head and this interface is referred to as 'the deck'. Depending on the final configuration, this interface may be of the 'open deck' or the 'closed deck' type: The former is more practical for small to medium engine configuration while the second is more expedient for more powerful engines.

10.3.2 Piston

The tasks carried out by a piston are:

- Transferring the driving force, developed by the combustion gases, to the connecting rod,
- guiding the connecting rod small end, and
- preventing combustion chamber gases from bleeding into the oil sump.

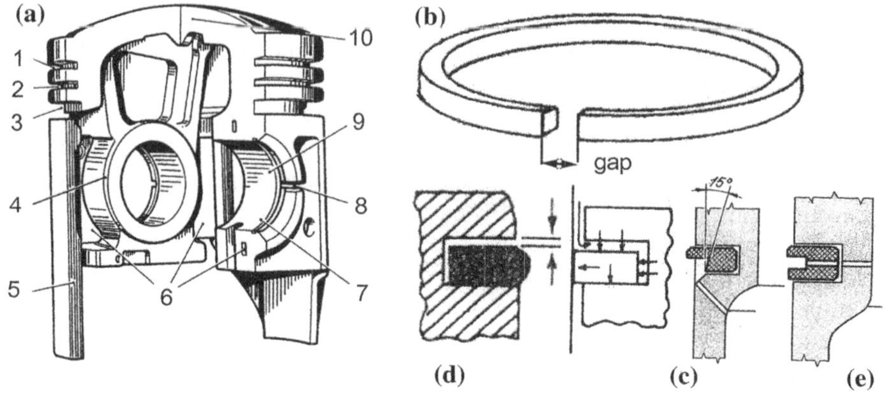

Fig. 10.19 **a** Piston. *1*, *2* and *3*: grooves for the elastic rings and (*4*) hole for the wrist pin installation. **b–e** Piston ring in their grooves. **c** and **d** Radial and axial clearances to provide free ring movements. **e** Configuration of the oil scraper ring, with the dicharge holes in the piston

The piston has, therefore, to withstand the load caused by the high pressures, the high combustion temperatures and the wear caused by friction against the cylinder walls; in automotive application the piston is made in a single piece, as shown in Fig. 10.19a.

The piston crown (10) is exposed to the gas pressure and may be either flat or suitably shaped, following the design of the combustion chamber. Below the crown, the ring belt accommodates the ring package in which three grooves are machined; the first and the second for the compression rings, (1) and (2), and the third for the oil scraper ring (3). A suitable clearance between the ring and its groove is of paramount importance to allow the ring movement, thus assuring their contact with the cylinder surface. The term crevices is commonly used for the clearances between the ring, the piston and the cylinder, that is significant when evaluating the exhaust emission formation.

The connection between the piston and the connecting rod is obtained by means of the gudgeon (or wrist) pin; the related hub (4) and seat (9) provide the connection between the piston and the connecting rod small-end. The snap ring groove (7) and the snap ring extraction groove (8) are the mechanical details allowing the wrist pin installation and removal. Recessed flats are provided at the wrist pin ends.

Below the ring belt, the skirt (5) is the main part of the piston and provides the cylindrical guiding surface offering the adequate bedding area to the cylinder wall to minimize the contact pressure and to provide a suitable heat dissipation. During engine operation, the skirt is not in direct contact with the cylinder wall, being separated by an oil film.

Cast iron pistons were used in the low-speed engines of early design. In present-day engines the pistons are always cast from aluminium alloy, combining a light weight with high thermal conductivity. They have a moderate silicon content to better retain their mechanical strength at operating temperatures within the range

280–350 °C, which in compression ignition engines reaches 360–420 °C around the bowl rim.

The piston crown, exposed to the gas temperatures, has a diameter smaller than the cylinder liner, to avoid hot seizing between piston and cylinder; the first and second compression rings thus must act as combustion gas seals. Along the skirts, the piston diameter is larger and the clearance between piston and cylinder is low, to avoid knocking. The skirt configuration underwent significant modifications in time, as can be seen by comparing Fig. 10.19, a piston dating back to the 1970s, with Fig. 10.52. The lower thermal expansion of cast iron pistons with respect to aluminium pistons must be considered: the steel plates (6) shown in the former figure, ensure the required dimensional stability with temperature variations. In more modern solutions the dimensional stability was achieved by using a more stable aluminium alloy, together with suitable design modifications and tolerances. As already explained, the piston ring package has the important task of separating the combustion chamber from the crankcase. To achieve this result each ring must be designed in a suitable way. Even when the ring provides a perfect gas seal, a certain gas bleed takes place through the ring joint gap (Fig. 10.19b) whose size spans from 0.2 to 0.3 mm when the piston is fitted into the cylinder.

The rings usually behave as follows:

- The first ring, top compression ring, has a plain rectangular cross section and is located in the upper part of the piston crown. Figure 10.19c shows how the combustion chamber pressure activates the ring sealing. Its purpose is, of course, to retain the gas under pressure preventing it from bleeding along the cylinder liner. To allow for the free radial movement of the ring, the axial clearance in its groove is approximately 0.035 mm (Fig. 10.19d).
- The second compression ring is not much different from the first and, often, it has a hook-like undercut section. Its shape contributes to compression control, providing also an initial oil scraping action (Fig. 10.19c). In the solution shown the oil returns back into the sump entering the sub-horizontal holes drilled in the piston wall.
- The third ring, or oil scraper ring, takes care of the bulk of the oil control. During the piston downward movement, it causes the oil to return into the sump and prevents that the oil film, adhering to the cylinder liner, enters into the combustion chamber taking part to the combustion. It takes the form of an outward-facing channel section from which the collected oil escapes through slots machined radially to the back of the ring. The oil then returns to the sump through holes drilled through the piston wall at the back of the groove.

The geometry now described applies to modern engines in which the higher mean cylinder pressures favours the sealing action; in early engines it was usual to fit more than three rings, up to five.

Traditionally compression rings and slotted oil control rings has been produced from cast iron, because this material provides a satisfactory wear-resistant surface and retains its elasticity at high temperatures. Today many gasoline engines use composite oil control rings including thin steel rails with chromium plated working edges, whilst compression ignition engines have a spring-loaded type ring in which a cast iron slotted ring is expanded by a stainless steel backing spring.

Beginning in the late 1980s, it was recognized that moving the piston rings closer to the piston crown was beneficial for HC emission reduction. The combustion flame quenches in the dead space above the top ring that, together with the crevice volume, becomes a source of unburned HC both in gasoline and compression ignition engines. In spite of the ring package, part of the oil reaches the combustion chamber and part of the gases penetrate into the crankcase with negative consequences for emissions.

In recent years, the downsized configurations lead to reach a specific power often exceeding 90 kW/l, a peak cylinder pressure in turbocharged compression ignition engines reaching 200 bar and a temperature of 400 °C. A stronger piston design requiring the use of steel is predicted for the future and has already been introduced in commercial vehicles. To avoid a weight increase, the higher density of steel is compensated for by smaller overall dimensions of the piston. Aluminium alloy pistons remain the most common solution in light-duty vehicles and significant improvements aimed at increasing their mechanical strength in spite of a continuous weight reduction were applied. To maintain acceptable temperatures in the piston crown zone, the oil cooling, described in the previous section, reaches the circumferential gallery shown in Fig. 10.52. This solution is widely applied in turbocharged spark ignition and compression ignition engines.

10.3.3 Connecting-Rod

The connecting rod (Fig. 10.20) is the mechanical component that connects the piston to the crankpin, converting the motion from reciprocating to rotary. The rod transfers the forces generated in the combustion chamber, through the piston, to the crankshaft producing the driving torque. This component is subjected to a combination of axial and bending stresses, the former arising from reciprocating inertia forces and cylinder gas pressure, and the latter from the centrifugal effect.

The connecting rod consists of the following parts:

- The connecting rod shank (5) has an "I" cross-section for maximum stiffness and minimum weight. It transfers the load from the piston (10) to the crankshaft. To reduce stress concentrations, the big-end arch (6) and the small-end eye (7) are connected with smooth fillets to the shank;
- the connecting rod small-end (7) where the wrist pin (8) is kept in its position thanks to the bushing (1) and snap-ring (9). The small-end is always smaller than the big-end, considering the reduced reciprocating and rotating inertia forces exerted between the connecting rod and the piston. A matter of concern is the small-end bearing lubrication, considering that this bearing has an oscillating rather than a rotary motion;
- the big-end (6) connects the rod to the crankshaft journal by means of its bearings (4) held in position by the cap (3) and its bolts (2). The connecting rod is under

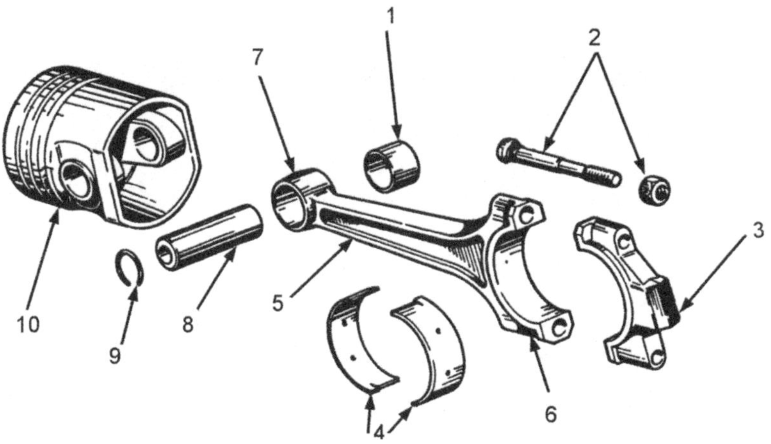

Fig. 10.20 Connecting rod and piston assembly (courtesy of FIAT)

heavy stress from the reciprocating load applied by the piston, actually being stretched and compressed during every rotation, with loads increasing with the square of the engine speed. Considering the above, great care must be devoted to applying the specified closing torque to the big end bolts.

The failure of a connecting rod always causes a catastrophic engine failure. Frequently the broken rod is forced through the crankcase skirt thereby destroying the engine. Failure can be the result of incorrect lubrication, fatigue stressing close to a physical defect in the rod, or incorrect bolt tightening. Such failures are quite rare on production cars considering the appropriate safety factors and systematic quality controls on each component and in their assembly.

New engines have connecting rods made using powder metallurgy, which allows more precise control of size and weight with less machining and less excess mass to be machined off for balancing. The cap is then separated from the rod by a fracturing process, which results in an uneven mating surface due to the grain of the powdered metal. This ensures that, upon reassembly, the cap is perfectly positioned with respect to the rod, compared to the minor misalignments which can occur if the mating surfaces are both flat.

A major source of engine wear is the sideways force exerted on the piston through the connecting rod by the crankshaft, which typically wears the cylinder into an oval cross-section rather than circular, making it impossible for piston rings to correctly seal against the cylinder walls. Geometrically, a longer connecting rod reduces this force, and therefore leads to a longer engine life but this causes the engine to be heavier and bulkier. In general, long connecting rods were typical of early engine designs.

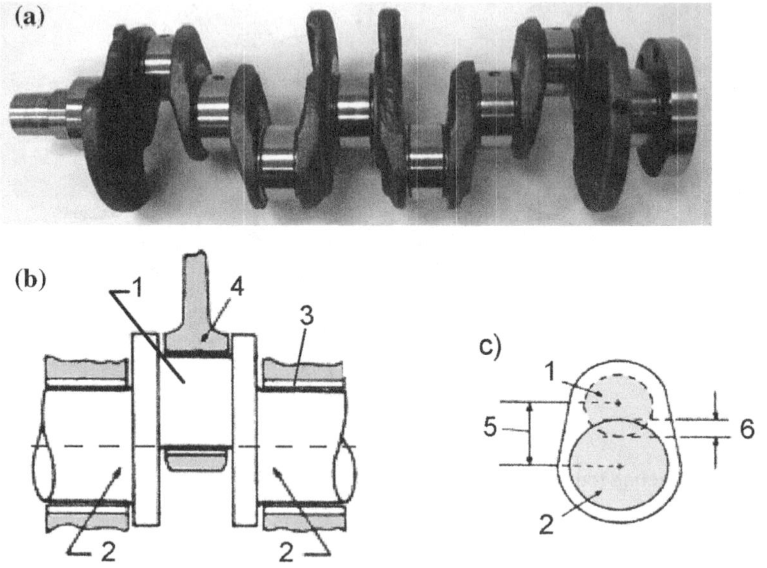

Fig. 10.21 Engine crankshaft of a four-cylinders in-line engine (**a**); webs connecting the main bearings with the crank pins (**b**); overlap between main journal and crankpin (**c**)

10.3.4 Crankshaft

Geometry

The crankshaft (Fig. 10.21a) is one of the most important engine component, since it coverts the reciprocating motion into a rotary motion. It carries the crankpins (1), to which the big ends of the connecting rods (4) are coupled, and the journals (2) of the main bearings (3). Each pin is connected to the nearby journals through two webs, as shown in Fig. 10.21b.

The crank radius (or throw) (5) is the distance between the axis of the journals and that of the crankpins and is equal to one half of the engine stroke as shown in Fig. 10.21c. The distance between the two subsequent crankpin center-lines is always somewhat larger than the cylinder bore, owing to the cylinder wall thickness. Based upon the stroke value, the overlap (6) is determined and it is straightforward to recognize that its value greatly contributes to the crankshaft stiffness. This is the usual case with normal production short-stroke engines as shown in Table 10.2, in which the stroke value is not much larger than the bore. Only in the case of the fastest engines, the last example in the table, the stroke is lower than the bore to keep the piston speed within acceptable levels.

This situation is much different from what was common in early engines, in which the stroke was definitely longer than the bore, thus causing the overlap to be either null or negative. The current tendency is to locate only one crank between two main journals: The advantages of this solution are an increased shaft bending stiffness, a better load distribution on the bearings and finally a better lubrication. Therefore, the crankshaft of a four-cylinders in-line engine is provided with five bearings (Fig. 10.21a), and the main bearings, with their thin friction layer, are similar to those of the connecting rod big-end.

The increased number of bearings allows use of a lighter shaft construction without risks of bending, even if it causes an increase of the costs for insuring an adequate alignment. In this respect, the V-cylinder configuration allows fitting of two cranks and the related webs between two main journals, thus allowing a shorter engine length, while keeping a good shaft stiffness (Fig. 10.17): a V-6 engine requires just four main bearings.

Another important task of the crankshaft is the distribution of the pressurized oil, through an appropriate hole system, in the main journals and crankpins, to lubricate the bearings (Fig. 10.17 and 10.67).

The crankshaft described above is made in one-piece obtained with a forging process and this is the most common solution, even if, from time to time, a built-up construction has been used to allow using rolling elements bearings. The most common production process implies the use of spheroidal graphite cast iron, for its better ductility and strength.

Centrifugal forces act on each crankpin during crankshaft rotation, determining shaft deformations and increasing the load on the bearings. Counterbalance masses are always applied to the webs, either integrally cast or separately attached, to limit these loads. The balance masses are slightly overdimensioned, so that holes can be drilled to make them lighter during the final balancing operation.

To achieve an acceptable engine balancing a suitable crankshaft design must be chosen; it features an uniform sequence of the useful phases and an overall symmetry with respect to the crankcase transversal plane (at least when the cylinder number is even).

The first rule allows one to obtain suitable ratios between the peak and the average torques, according to the values listed in Table 10.3, in which the effect of multi-cylinder configurations is shown. The second rule allows to achieve a better although not complete, balance of inertia forces.

Motion irregularities

As already stated, four-stroke engines complete one operating cycle every two crankshaft revolutions but only the expansion phase supplies useful energy to the shaft, while the other three phases (intake, compression and exhaust) absorb a part of this energy to overcome the pumping and friction losses. In addition to these resistances, the inertia loads, determined by the reciprocating motion of the pistons, imply additional motion irregularities.

The instant crankshaft angular speed varies considerably between a minimum and a maximum value, increasing during expansion to decrease during the passive phases, as shown in Fig. 10.22.

The purpose of the flywheel, installed at the end of the crankshaft, as can be seen in Fig. 10.17, is to store the kinetic energy during the expansion phase, and then to give it back to the crankshaft, to allow it to overcome the resistances to motion

Table 10.3 Ratio between peak torque and average torque R as a function of the number of cylinders n

n	1	2	3	4	5	6	8
A_{bp}	720°	360°	240°	180°	144°	120°	90°
R	8	4	2.8	2	1.7	1.4	1.1

A_{bp} is the angle between ignition pulses

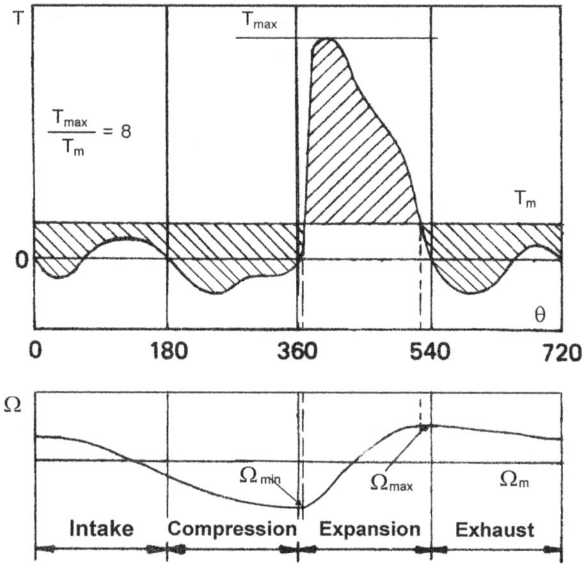

Fig. 10.22 Torque and speed as a function of the crank angle for a single-cylinder engine. Instant and average values along a full cycle (720°)

during the subsequent phases. Its final goal is thus reducing the variations of the instant engine speed.

The larger is the moment of inertia of the flywheel, the lower are the speed variations; however, it is not desirable to design a flywheel that has a too large inertia, as this would cause a slower engine response when a rapid speed change is requested and, in addition, its larger weight causes an additional load on the crankshaft bearing on the flywheel side. The effect of the flywheel in smoothing the rotation of the engine mainly applies when idling with the clutch open; under all other conditions, the contribution of the flywheel is negligible when compared with the vehicle inertia.

The torque diagram delivered by a single-cylinder engine running at idle is shown in Fig. 10.22, where the ratio between the torque peak value T_{max} and average torque T_{min} is equal to 8:1. The hatched area above the average value T_m is proportional to the energy stored by the flywheel and the area below the average line is the energy that is returned by the flywheel to the crankshaft during the passive phases.

In the single-cylinder configuration, widely used in motorcycle applications up to the 1950s, the flywheel dimensions were so large that, sometimes, this component was located outside the crankcase.

In addition, today the rim of the flywheel is used for other important functions, both in gasoline and compression ignition engines, providing:

- The crown gear, meshing with the starter motor pinion, to allow engine starting,
- the timing marks to check camshaft and ignition or injection settings relative to crankshaft angular position and
- the instant speed sensor today mandatory for misfire detection as a part of the on-board diagnostic system.

Engine crankshaft rotation and the reciprocating motion of the piston, connecting rod and related masses cause periodic load variations resulting in crankcase deformations transmitted to the engine supports as vibrations, whose intensity varies with the engine speed. The reduction or, even better, the full compensation of these vibrations has an important effect in achieving a better vibro-acoustic environment in the passenger compartment. It is possible to achieve a correct engine overall balancing by:

- Increasing the number of cylinders, and selecting a suitable arrangement,
- adequately designing the crankshaft and balancing it correctly and
- using specific balancing devices, as it will be further explained.

Engine balancing

As already stated, the need of reducing the torque irregularities without resorting to a heavy and bulky flywheel were solved by using multi-cylinder engines. Subsequent firing phases are thus closer in time and, as a consequence, also the torque and speed fluctuations become smaller. Comparing a single-cylinder with a two-cylinders engine, the time between ignition pulses is halved from 720 to 360° and, the ratio between the torque peak and its average value is halved too, from 8:1 to 4:1.

The addition of a third cylinder lowers the interval between two ignition phases to 240°. The torque peak is further reduced to 2.8 times the average value. The trend of these parameters with increasing number of cylinders is reported in Table 10.3. The table holds provided that the engine crankshaft has been designed according to the two mentioned rules.

At any rate, the piston itself undergoes two types of acceleration during its reciprocating motion, that are periodic in time. If the connecting rod had an infinite length, the motion of the piston would be harmonic with frequency Ω, i.e. with a period equal to the time required to complete two strokes. The reciprocating masses, i.e. the masses following the piston in its motion, would thus be affected by the corresponding inertia forces, that are usually referred to as primary (or first-order) inertia forces.

The length of the connecting rod is not infinite, and the motion of the piston is periodic but not harmonic. Since it is anyway periodic, it can be developed in Fourier series and the inertia forces due to the second harmonics of the acceleration, whose frequency is equal to $2\,\Omega$ are said to be secondary (or second-order) inertia forces.

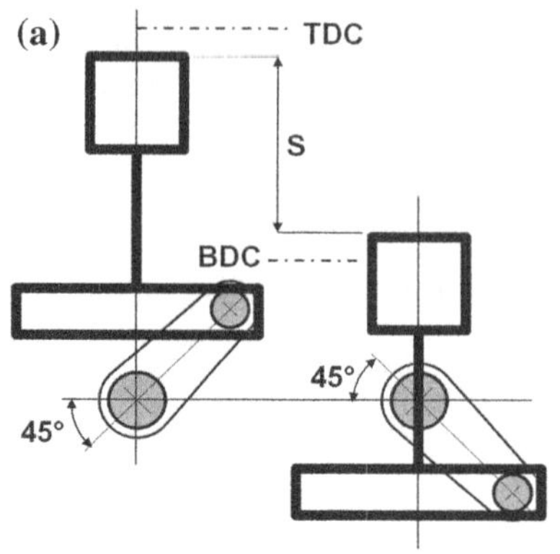

Fig. 10.23 Primary and secondary inertia forces due to the reciprocating masses. The slotted *bar* mechanism simulates an infinitely long connecting rod (**a**) while the actual geometry of the connecting *rod* is shown in (**b**). With the same +45° −45° crankshaft rotation the travel changes from s to s'

The latter result from the fact that the overall centre of mass of the pistons, instead of remaining stationary as it would happen if the connecting rods were infinitely long, has a small harmonic motion with frequency $2\,\Omega$. The other harmonic components are usually neglected.

The difference between a slotted bar mechanism, simulating an infinite length connection rod and a connecting rod-crank mechanism is shown in Fig. 10.23[3]: the piston travel is different in the two cases: s and s'. The first mechanical solution is unpractical and with the usual solution secondary forces must be accounted for. Balancing the inertia forces is an important stage in the design and development of an engine.

First-order inertia forces can be balanced by means of counterweights located in the crankshaft, usually as integral parts of the webs.

Second-order inertia forces can be balanced, particularly in four-cylinders engines, by the adding appropriate balancing shafts, geared so that they rotate in opposite directions at a speed twice the crankshaft speed, as shown in Fig. 10.24. When the pistons are at their top and bottom dead centres, the counterweights are always at their lowest position. The figure shows a solution used by FIAT in the mid 1990s, while this configuration was introduced by Mitsubishi in the mid 1970s.

More recently, the use of the counter-rotating shaft that, unavoidably, increase the fuel consumption by 1–2 %, has been made unnecessary by using a suitable bore/stroke ratio and more advanced engine supports.

Torsional vibration of the crankshaft

The crankshaft normally undergoes a permanent state of vibration induced by the strong torque pulses, transmitted by the connecting rod. As already mentioned, each power stroke accelerates the engine and tends to slightly twist the shaft; when

[3] From M. J. Nunney, *Automotive Technology*, 3rd ed., Butterworth, Heinemann, 1998.

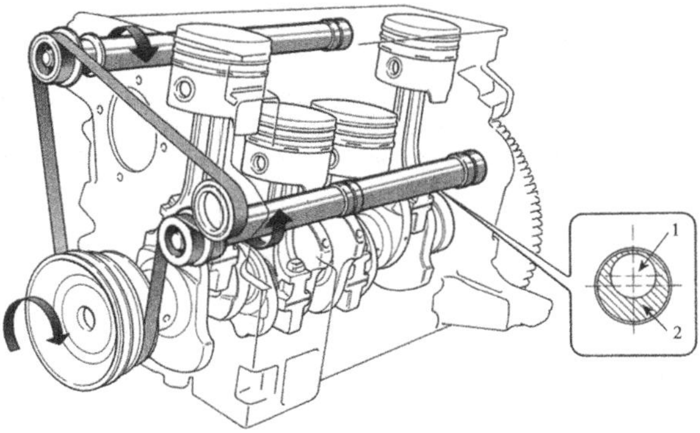

Fig. 10.24 Balancing shafts applied to a four-cylinders in-line engine. *1*: Rotation center of the shaft on its bearings; *2*: Counterwights applied to the shaft (courtesy of FIAT)

the power stroke ends the crankshaft untwists. At the end the crankshaft behaves like an elastic body having its natural frequency that depends on its geometry and on the material characteristics; long and slender journals connected to large crank masses, that are capable of damping the first-order alternating forces, lead to a low natural frequency. The variable forces due to the working gases and the inertias that are exerted on the crankshaft cause the latter to vibrate. When a vibration frequency coincides with one of the natural frequencies of the crankshaft a resonance occurs and the amplitude of the resulting torsional vibration may increase to the point of overstressing the shaft. Rough operation, noise and sometimes a fatigue failure may thus occur.

During World War I many systematic and catastrophic crankshaft failures of aircraft reciprocating engines occurred due to strong torsional vibration of the crankshaft. They were due to a torsional resonance, often lying within the engine working range. This example shows how important is also the rotational inertia of the masses connected to the crankshaft that, in the automotive case, depends on the configuration of the gearbox and of the whole driveline.

Since the engine torque has many harmonics, there are several torsional resonances, and thus several critical speeds.[4] The torsional vibration of the crankshaft usually causes the front end of the shaft to vibrate in counter-phase with respect to the flywheel, with a much larger amplitude. This vibration is superimposed on the rotation of the crankshaft. In present-day engines, the high crankshaft stiffness causes the natural frequency to be high, and only high order harmonics of the engine torque may cause resonances.

To prevent shaft damage, many engines are provided with a torsional vibration damper located on the front section of the shaft, where the vibration amplitude is larger. The dampers used on car engines, and particularly on compression ignition engines, are of the elastomeric type since the element that dissipates the vibration energy is a rubber element. The type shown in Fig. 10.25 uses a rubber ring (3) that

[4] In this case a critical speed is defined as a speed at which a torsional resonance takes place.

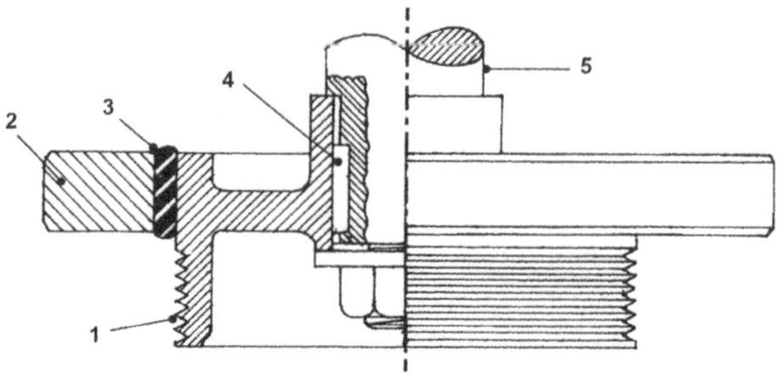

Fig. 10.25 Torsional vibration damper located at the front-end of the crankshaft

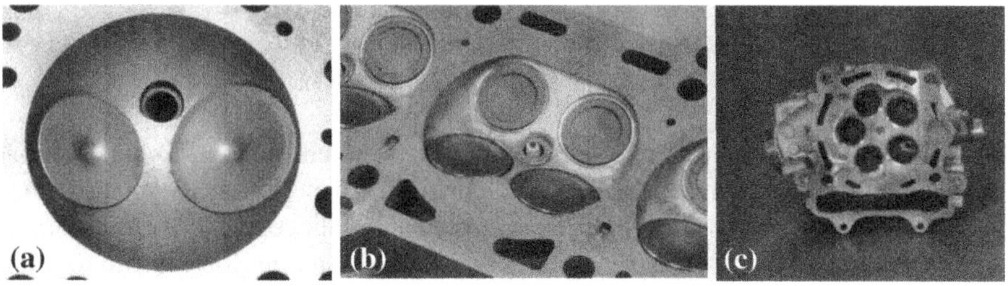

Fig. 10.26 Combustion chambers for cross-flow cylinder heads: **a** hemispherical two valves; **b** pent-roof with four valves; **c** with five valves

is compressed between an external cast iron ring (2) and an internal hub (1), that acts also as a pulley operating the belt that drives the accessories, applied to the front-end of the crankshaft (5) with a key (4). When the shaft does not vibrate, the ring supported by the rubber belt turns together with the crankshaft and the hub. If the crankshaft accelerates, for instance in an anti-clockwise rotation, the inertia of the ring mass tends to oppose this acceleration causing the rubber belt to be loaded by shear stresses. The vibration of the shaft thus causes the rubber element to be loaded by cyclic stresses that, owing to the internal damping of the material, produces a dissipation of the vibration energy that is converted into heat.

10.3.5 Cylinder Head

In gasoline engines, the lower deck of the cylinder head contains the cylinder combustion chamber which can take a number of different configurations. Referring to present-day configurations, three possible way in which the valves and the spark plug are located are shown in Fig. 10.26, namely: (a) hemispherical chamber with two valves, (b) pent-roof shaped chamber with four valves and (c) chamber with five valves. In all cases the intake valves is characterized by a larger diameter.

The common aluminium alloy construction, normally cast using a lost foam process, allows one to save weight and to obtain complicated shapes. It is therefore possible to provide a better heat distribution, to maintain a more uniform temperature and to obtain an easier cooling, pre-requisites to increasing the compression ratio as it will be explained below. These properties explain why aluminium is the preferred solution and, frequently, this head type is used on a cast iron crankcase.

The overall intake area increases when increasing the number of valves, thus promoting a higher volumetric efficiency. Switching from a two to a four valve configuration, the intake area increases by approximately 30 %. The better volumetric efficiency determines a higher value of the mean effective pressure: Based on a survey performed at the end of the 1990s, when the four valves layout was introduced, on roughly 400 engines, an increase from 10.2 to 11.3 bar (∼11 %) of the mean effective pressure was recorded. Solution (b), also shown in Fig. 10.17, features the valves arranged on the sides of a wedge and this configuration still proves its effectiveness thanks to the intake area increase and a lower surface-to-volume ratio. Moreover, the spark plug central location provides an equal flame propagation distance to reach the most remote points of the combustion chamber.

The five-valves solution, with three intake valves, applied during the 1990s on some high performance engines, did not achieve substantially better results because, in spite of a further intake area increase, it was penalized by a complex mechanical design. It was gradually abandoned. The figure shows the head of a single-cylinder motorcycle engine.

When the complexity of the internal geometry does not allow an integral casting without the risk of producing a high percentage of faulty parts, an overhead is used that houses the camshaft's bearings and the tappet sliding seats.

To take full advantage from multi-valve solutions, it is of paramount importance that the intake duct geometry, carved into the head, allows an unobstructed air movement entering into the cylinder during the intake phase. Each engine type requires a distinctive intake air flow to reach its best performance.

The mechanical configuration described above, in addition to improvements in the ignition system, allowed an increase in the compression ratio by about 10 %, typically from 9.5:1 to 10.5:1, using 95 Research Octane Number (RON) gasoline, determining a 2.5–3 % increase of the thermal efficiency.

The combustion chamber used on compression ignition engines is completely different.

The cylinder head is connected to the upper deck of the crankcase by cap-screws or by stud-bolts and, among the threaded engine connections, the cylinder head fastening to the crankcase is by far the most important. The connection of the cylinder head to the crankcase compresses the cylinder head while the screws or studs are stressed by a tensile force which tends to pull out and strain the metal around the threatened region on top of the crankcase.

The position of the holes with respect to the bore is critical: if they are too close, they can cause distortions in the upper part of the cylinder, endangering its circular section. On the other hand, if the holes are too far apart, the matching surfaces will tend to open during engine operation. This reduces the pressure on the gasket, thus reducing its sealing action against the gas in the combustion chamber, with the risk of spoiling the gasket itself. The upper plate of the crankcase has two rows of holes on the opposite sides of the cylinders, whose geometry must be properly selected to avoid bore distortions when the cylinder head is assembled. Reducing this distortion to a minimum so that the cylinder remains circular reduces the resistance to motion of the piston.

Although the matching surfaces of the cylinder block and the head are precisely machined to guarantee the best possible planarity, some irregularities are anyway present. The main ones are referred to as:

- Waving: the surface profiles show wide dips and ridges determined by deflections and vibration of the spindle of the milling machine.
- Roughness: overlays the previous profile and consists of a large number of tiny dips and ridges, determined by the grinding speed and by the size of the abrasive grains selected when running the final precision grinding process.

Therefore, if the surfaces were to be directly coupled, contact would take place between the irregular ridges of the metal surfaces. The function of the cylinder head gasket is to provide the required plasticity and elasticity to seal the fluids and the gases under all operating conditions, by means of the re-distribution of the tightening load over a wide surface and in an even manner. Copper satisfies the above requirements and therefore a solid sheet of copper has been frequently used in the past years. An integral sealing steel wire, around the circumference of the cylinder, is placed to provide a long lasting sealing device against the combustion gases, thus protecting the copper.

In addition, a wide variability of sealing pressure is required in different locations due to the different fluids to be contained: combustion chamber gases, oil and cooling fluids. Sealing against the combustion gases calls for higher specific pressure when compared with that required to seal against oil and water. In addition, the gasket chemical resistance must be obtained using suitable materials to withstand corrosion from fuel and combustion products, cooling fluids, engine oil and their additives. Last but not least, the gasket is exposed to widely fluctuating temperature and pressure, including those produced during detonation.

To fulfil the above mentioned requirement the gasket is, almost always, assembled using different materials (Fig. 10.27a). They are, basically:

- Rolled steel rings to withstand high temperatures, and
- armored resin with synthetic aramid fibres (Kevlar) to provide a good flexibility.

A completely metallic gasket is a valid alternative (Fig. 10.27b), since it takes the shape of an embossed steel sheet being used in heavy duty applications because it better retains the cylinder head torque tightening.

The most likely problems due to a failure of the gasket are compression losses, causing a reduction of the power output, or the exhaust gases being forced into the cooling system, with subsequent engine overheating.

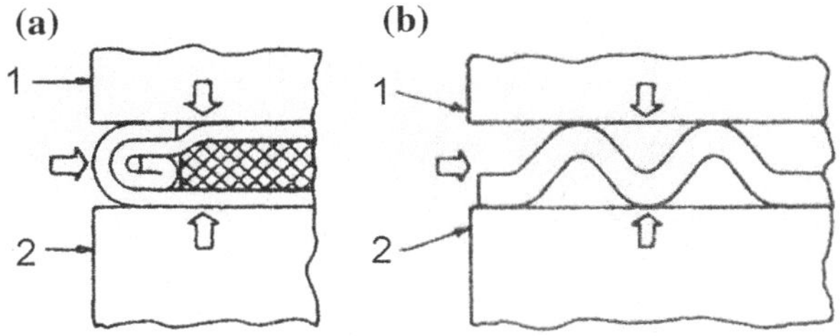

Fig. 10.27 Cylinder gasket between cylinder head (*1*) and crankcase (*2*): **a** rolled steel and armored resin, **b** completely metallic gasket

10.3.6 Camshaft Drive

While in four-stroke engines a complete thermodynamic cycle takes place every two crankshaft rotations, a complete valve opening and closing cycle, for both intake and exhaust, must take place during a single camshaft rotation. The camshaft must thus rotate at a speed that is half that of the crankshaft, while being strictly in phase with the rotation of the latter. The camshaft must be operated by a transmission with a gear ratio 2:1, able to keep the valve timing within strict tolerances. According to the different engine configurations, the camshaft positioning is the one of those shown in Fig. 10.16. The camshaft is operated by the crankshaft through:

- A gear train, or
- a chain and sprockets system, or
- a toothed elastomeric timing belt.

The gear train solution was used in early engines. It is the most expensive one and is still used in high speed racing engines, owing to the high engine speed (16–18,000 rpm), and on heavy-duty engines used on commercial vehicles, to assure a maintenance-free operation for the vehicle life. To limit the noise, helical gears are used but, frequently, the gear drive produces a high-pitched rattle not appropriate in a passenger car.

Chain drives for the camshaft (Fig. 10.28a) were, by far, the preferred solution since the early attempts at building four-stroke engines and today is competing with the timing belt configuration. The chain (1) is driven by the crankshaft sprocket (4) and drives the camshaft sprocket (5). The automatic hydraulic tensioner (2) is mandatory to compensate for the elongation of the chain, due to the thermal expansion of the cylinder head and crankcase, and by the wear of the sprocket teeth. The tensioner is applied on the driven side of the chain using a mobile runner (3) that is in contact with the chain thanks to the oil pressure thrust. In so doing, the optimal chain tension is maintained also thanks to the fixed chain guide (6) located on the driving side. The configuration shown in the figure provides also a separate chain (7) to drive the oil pump (8). Like gear trains, chains require a good lubrication to achieve a maintenance-free operation.

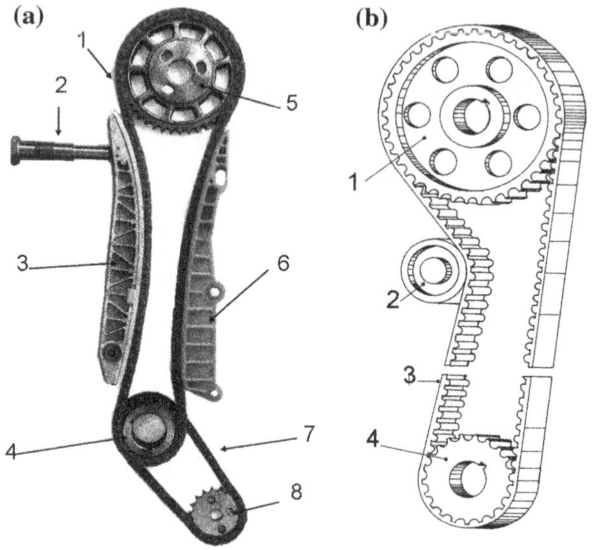

Fig. 10.28 Valve drive: chain and toothed timing belt solutions (from P. Scolari, *Motor vehicles and their evolution*, 10th edition, 2010)

Timing belt drives (Fig. 10.28b) were developed in the 1960s, and are still widely applied. A suitable phasing of the camshaft is granted, like in the case of chain drives, with the advantages linked with flexible belts. On the inner side, the timing belt (3) features neoprene teeth, shaped to mesh with the teeth of the crankshaft (4) and camshaft (1) pulleys. The selection of the mating materials is dictated by the need to provide a suitable grip. On the outer side of the belt an adjustable belt tensioner (2) is located. A self-adjusting hydraulic or mechanical tensioning device maintains the required belt tension, like in chain systems. The belt is reinforced by high tensile strength fibres (e.g. fiberglass or kevlar) to withstand the tensile load preventing the belt elongation since a high longitudinal stiffness of the belt is a prerequisite for obtaining a reliable valve timing. Correct belt tension is critical: if the belt is too loose it may whip, if it is too tight it can whine and put excess strain on the bearings.

Belt drives offer a number of advantages over chain drives: they are less expensive, quieter, more efficient and have no need for lubrication; they require a cheaper housing to protect the system against dust, mud and snow. On the other side, belt drives are more sensitive to high temperatures and fatigue, due to the repeated bending cycles. As a consequence, the belt needs to be replaced every 120,000–160,000 km or on a time base, e.g. 5 years; it is common to replace the belt tensioner each time the belt is replaced. Continuous efforts are underway to increase the belt duration and an innovative solution, in production since 2010, features an innovative belt, lubricated by the engine oil, which can last up to 240,000 km or 15 years, which is the duration usually considered for the engine.

A failure of the timing belt or chain results, almost always, in costly and often even fatal engine damage, since, due to the geometry of the valves and the piston and the high compression ratio, the mechanical impact between valves and pistons is unavoidable.

10.3.7 Direct Valve Actuation and Hydraulic Tappets

The poppet-type valves used in internal combustion engines are made of two parts: the head, which closes the duct opening, and the stem, that guides the valve movement. The seal is obtained on a conical surface, at the outer edge of the valve head. The valves must withstand strong mechanical stresses, caused by the high frequency operation, with the consequent impacts on the seats, and by the high temperatures which can cause deformations.

The diameter of the intake valve is always larger than that of the exhaust valve, since it is easier to evacuate the hot gases, thank to the exhaust pressure, than to intake the fresh gases that are subjected to a much lower pressure difference. Moreover, valve cooling improves using a smaller head diameter, longer guide and larger stem diameter, considering that the heat dispersion takes place through the contact between metal surfaces at the head−seat and stem−guide interfaces. The intake valve is, in addition, efficiently cooled by the air stream while the exhaust valve is always in contact with hot gases. Both fluidic losses and cooling requirements call thus for a smaller exhaust valve and larger intake valve diameter.

The solution used in the sliding followers, or direct drive configuration, applied to a wedge chamber is shown in Fig. 10.29: The two parallel valves are operated by a single overhead camshaft. This configuration features intake and exhaust ducts connected to the same side of the cylinder head and was widely used up to the 1980s. The valve actuation mechanism features the following components: valve (stem and head) (1), springs (2), split keepers (3), washers (4), oil seal gaskets (5), bucket-type tappets (6), calibrated shims (7), valve seat (8), duct and related port (9), cam (10), valve guide (11) and spark plug (12). In spite of the evolution of the combustion chamber shape, the mechanical components now described were maintained for many decades and the more significant improvement was the introduction of hydraulic tappets.

The valves are built using a high alloy steel and often special coatings are applied on the peripheral zone of the head that comes into contact with the valve seats.

The direct valve actuation features bucket-type tappets, with calibrated shims needed to provide the correct clearance. The thickness of the calibrated shim is chosen so that, with the valve closed, a small lash (0.4–0.6 mm) is present between tappet and cam. The deformations, induced by the normal engine heat release, cause an increase of the stem length up to the point in which the lash is zeroed. At this point, the lack of lash prevents the valve from closing completely, causing it to leak compression and to overheat because the metal-to-metal contact does no more take place thus impairing the correct valve cooling. The mandatory countermeasure is the

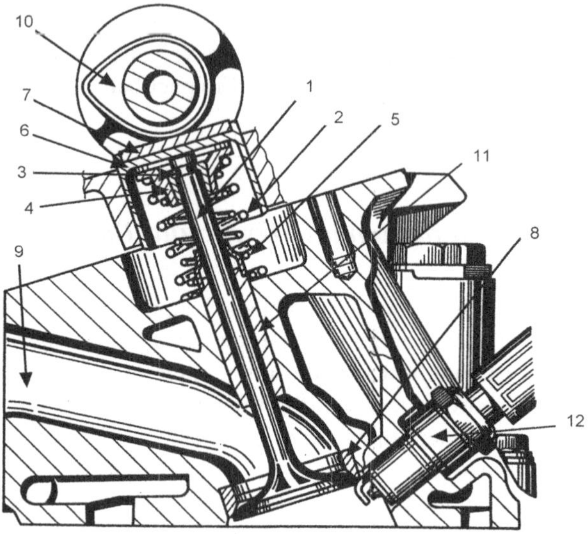

Fig. 10.29 Direct drive valve configuration, in which a single camshaft operates directly the exhaust and intake valves through bucket-type tappets (courtesy of FIAT)

periodic control of the valve lash. In case the latter becomes too small, it is necessary to restore it to the correct value, replacing the existing shim with a new one with the correct thickness.

To eliminate the valve overheating hazard, and the consequent periodical lash adjustment, hydraulic tappets were introduced in the 1980s and 1990s, with the added advantage of reducing operating noise that, unavoidably, takes place when the cam meets the tappet to start the valve opening. A hydraulic tappet (1) is shown in Fig. 10.30. It provides some internal oil passages and an internal piston (2) which allows the sliding of a plunger (3). The latter is the mechanical component acting on the valve stem. It is easy to recognize the other components already described.

During the valve opening phase (Fig. 10.30a) the cam acts on the tappet, and, as a consequence, on the piston, thus pressurizing the oil in the plunger internal chamber (6) owing to the closing of the ball valve (4) which acts as check valve under the trust of its spring. With this configuration the plunger provides the valve opening despite the minimum leakage between the piston and the plunger.

When the valve closes (Fig. 10.30b) the hydraulic tappet allows the tappet to be always in contact with the cam, thus zeroing the valve lash and the noise, but allowing a correct valve closing under the trust of its springs. In this phase, the tappet position allows the engine oil pressure to act from the lateral feed channel to the internal part of the tappet and finally to the check valve (4) which de-pressurizes the internal chamber (6). In this situation the return spring (5) provides to keep the plunger and the piston in a correct relative position thus allowing the tappet to maintain its contact with the cam.

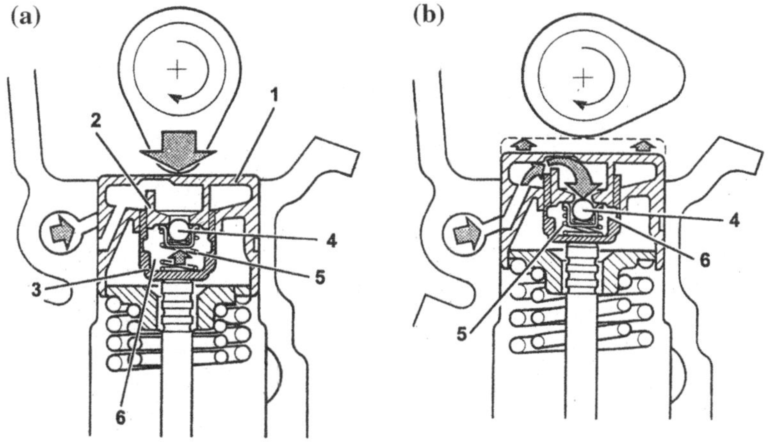

Fig. 10.30 Hydraulic tappets used to decrease noise and maintenance operations: **a** valve in open position; **b** valve in closed position (courtesy of FIAT)

During the valve closing period, since the trust on the tappet is limited to the engine oil pressure and not determined by mechanical forcing between components, as would happen in the case of mechanical tappets, the risk that the valve remains open is avoided.

10.4 Air/Fuel Intake Subsystems

The volume of air introduced into the engine determines the volumetric efficiency. Since the power is determined by the mass of air introduced, the intake system must, in addition, provide that this air volume has a high density, so that to maximize the mass of fuel and the chemical energy supplied to the engine.

10.4.1 Intake Manifold

A single pipe is the only component needed to transfer the air from the air filter to the inlet port on the cylinder head in single-cylinder engines; along this duct a fuel injector provides the requested gasoline quantity. On the contrary, an intake manifold of complex shape is needed in multi-cylinder engines, as shown in Fig. 10.17, where the plenum chamber and the ducts are clearly visible. The chamber volume is, approximately, somewhat larger than the cylinder displacement, by a factor 1.3 to 1.5, and provides an air volume from which the air enters the ducts with quite a low speed ($\leqslant 10\,m/s$). A circular cross section, characterized by the least resistance to

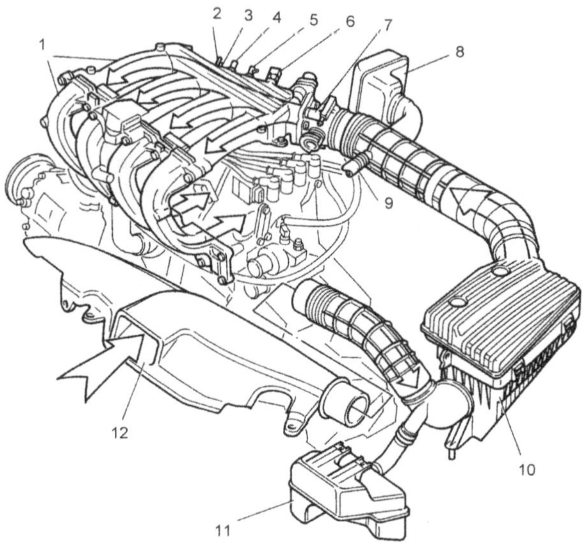

Fig. 10.31 Intake manifold complete with the external air inlet, air filter and throttle body (courtesy of FIAT)

air flow, is usually chosen for the duct; even if in case of a four-valve configuration the final duct section takes an oval or also a semi-rectangular shape, to feed the two intake valves. To provide the correct mixing between air and fuel the air velocity along the duct must increase, and typical values range from 70 to 100 m/s. The cross-section area is kept to a minimum, keeping however in mind that it must not impair the volumetric efficiency which, at the end, is the key parameter for a well designed intake manifold. The following conditions must be met:

- Optimizing the engine volumetric efficiency, i.e. providing the best possible charging of each cylinder, and
- providing an even distribution of the air quantity and of the air/fuel mixture to the individual cylinders.

As a first approximation, the achievable engine torque is proportional to the fresh mixture content in the cylinder charge: this means that the maximum torque can be increased by increasing the absolute pressure of the air entering in the cylinder and thus reducing to a minimum the friction losses through the ducts. The second pre-requisite for a correctly designed intake manifold is even distribution of the air flow to each cylinder. The fuel quantity supply and, finally, the correct air to fuel ratio (AFR) is provided by the injection system.

The layout of the intake system of an in-line, four cylinder, double over-head camshaft, multipoint port fuel injection engine with a displacement of 1.6 l is shown in Fig. 10.31. It includes the intake manifold (1) with its plenum chamber, downstream of the throttle body (7), from which the four branch pipes sweep down to the intake ports. The intake ducts are quite long to improve the volumetric efficiency in a range of low/medium engine loads, as further explained. The air intake (12) is placed outside the engine compartment to promote a good fresh air intake. The air duct layout is designed to locate the air ducts, upstream of the throttle body, in the coldest zone available underhood. The goal is to avoid heating the intake air before it reaches the plenum chamber. The air filter capacity (10) is designed to fit an air filter cartridge that can last about 40,000 km before being replaced. The connection to the blow-by vapors (9) is located downstream of the air filter, within an area of low negative pressure. Along the main air duct, upstream and downstream of the air filter, two Helmholtz resonators (11) and (8) are located. They act as noise dampers, able to reduce the so called induction pipe roar.

The relative pressure in the plenum chamber is always negative and its value is directly linked to the engine load, being proportional to the throttle valve opening. This partial vacuum is used to operate different functions and the intake plenum is connected with systems and components which need this source of negative pressure, namely the:

- Vacuum assisted brake booster (4),
- the fuel pressure regulator (5) which needs the vacuum signal to properly correct the fuel flow delivered by the fuel pump,
- the evaporative emission control system (2) and
- the exhaust gas recirculation system, when present.

The negative pressure value is proportional to the engine load and is measured by the absolute pressure sensor whose signal (3) is delivered to the engine Electronic Control Unit (ECU) to define the injector opening time thus providing suitable air to fuel ratio, as it will be better explained in the following sections. This latter ratio is further corrected by the signal of the air temperature sensor (6), which supplies the value of the air density.

In addition, this volume remaining at a negative pressure, has been used, since the early stage, to promote a fast evaporation of the fuel delivered to enhance the formation of the pre-mixed or homogeneous charge.

In the previously described throttle-less variable valve actuation system this vacuum source is no longer available. In this case a new mechanism is present, highly beneficial to the mixture formation and proper cylinder filling. In fact the reduced valve lift causes an increase of inflow speed from somewhere in the region of 50 m/s, part load, to more than 300 m/s. In engine configurations of this kind, as well as in compression ignition engines, the vacuum source must be provided by a specific vacuum pump, usually driven by the camshaft.

In the past, up to the 1990s, the intake manifold was an aluminium casting. The present day more complex geometries, and mainly the variable geometry intake manifolds, call for the use of polyamide (nylon), obtained with the injection molding technique, instead of aluminium. In spite of the higher cost of the basic material, it is possible to achieve an interesting cost reduction thanks to the possibility of easily integrating the connections for the above mentioned sensors and the actuators belonging to the engine management system. In addition, the smooth duct surfaces

(a) (b)

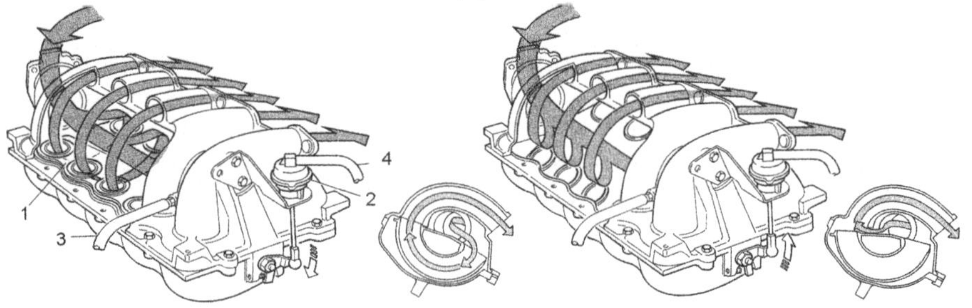

Fig. 10.32 Variable geometry intake system: **a** low speed, high torque and **b** high speed, high power conditions (courtesy of FIAT)

lower the wall losses, thus favouring a better air charge and improving volumetric efficiency.

10.4.2 Variable Geometry Intake Manifold

Suitable intake manifold and exhaust system geometries are important in optimizing the intake and exhaust processes. When the intake valve closes, a return pressure wave propagates backwards in the manifold. It then meets the quiescent air volume of the intake plenum chamber and is reflected back again, so that it returns in the direction of the intake valve at a speed equal to the speed of sound. The time T taken by this pressure wave for its travel from the inlet valve to the plenum chamber and back again is twice the duct length L divided by the sound velocity c:

$$T = 2L/c \,. \tag{10.12}$$

The resulting pressure fluctuations at the intake valve inlet section can be used to increase the fresh gas charge, thus improving engine torque. The above relation allows determination of the length of the ducts that produces a supercharging effect at the various engine speeds, known as ram-tube supercharging.

As a consequence, the length between the intake valve and the plenum chamber is inversely proportional to the engine speed so that an optimal supercharging effect can be obtained. Variable geometry intake manifolds were thus designed, in which this length can be modified, taking at least two values.

A variable geometry intake for a 2.5 l five cylinder in-line engine is shown in Fig. 10.32. This kind of variable geometry intake manifold features two fixed halves and an internal semi-circular movable element (1), operated by a pneumatic actuator (2) that can take two positions, thus providing two different lengths of the intake ducts. The actuator is driven by the intake manifold negative pressure (3) through the hose (4) and its open/closed position is controlled via the engine ECU which, based on engine speed and load signals, can decide the best actuator positioning. The actuator is in its closed position up to a certain engine speed, so that the air stream must follow the longer intake ducts (Fig. 10.32a). When the speed increases, the actuator opens and the shorter air column promotes a better volumetric efficiency (Fig. 10.32b). The shorter air duct has a larger diameter to easily allow the entering of a larger air volume, increasing the torque and power values at high speed.

10.4.3 Turbocharger

It is possible to classify both spark ignition and compression ignition engines according to their intake strategy, namely natural aspiration or forced induction. As already mentioned, the useful work, and therefore the mean effective pressure, depends on the amount of chemical energy available in the fuel. But the amount of fuel burned during combustion is heavily dependent on the mass of air entering into the cylinder, needed for providing a correct combustion. In previous sections it was shown how to optimize the mass of air for naturally aspirated engines. The air mass is determined by its volume and density, the latter depending on atmospheric pressure and pressure losses taking place in the intake plenum chamber and the ducts. In addition, the throttle limits the quantity of air in reduced load conditions, thus delivering the requested reduced power.

The pressure of the air filling the cylinders can be increased to further increase the mean effective pressure developed by a given engine. If air is forced to enter the cylinder at a pressure substantially above atmospheric pressure, thus increasing its density, more fuel can be burned per cycle even if the mean gas velocity through the intake ports remains unchanged for a given engine speed. The air volume still depends only on cylinder displacement but, since the cylinder ultimately receives a charge with a greater density, it is possible to achieve an increase in torque and power for a given engine displacement.

Forced induction technology was developed many years ago mainly for piston aero engines, to restore at high altitude, where the density of intake air is much lower, induction conditions similar to those occurring on the ground. During the 1970s this technology started to be applied to vehicle engines with a geometry close to the present-day one and a typical value of the specific power thus achieved was in the range of 75 kW/l. At that time, SAAB was recognized as capable of providing a suitable application in normal production engines.

A system with a turbo compressor driven by the exhaust gases is shown in Fig. 10.33. In this configuration, a radial-flow centrifugal compressor (2) forces the compressed air into the intake duct (4). The mechanical energy, needed to drive the compressor, is taken from the exhaust gases which, during their expansion, flow inward through the radial-flow centripetal turbine (1), thus lowering their temperature and pressure. In automotive applications, the turbine and compressor diameters are in the range of 30–60 mm, and therefore their speed is in the range of 150, 000–

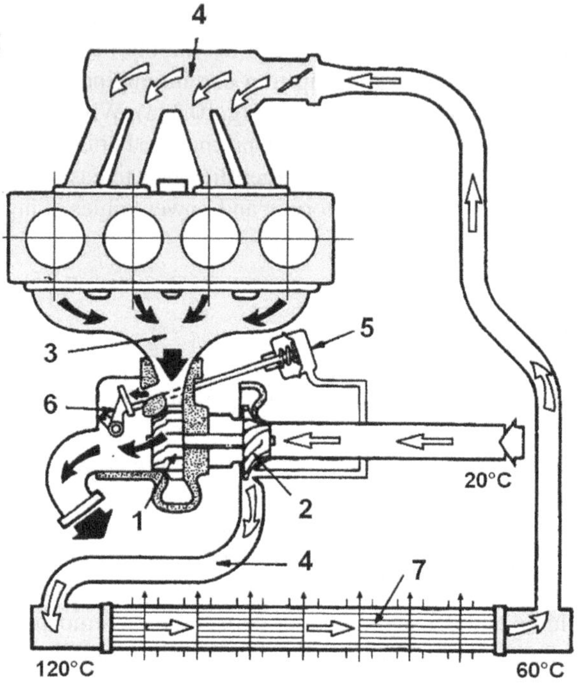

Fig. 10.33 Turbocharging system: turbine with wastegate and compressor. The intercooler lowers the air temperature, providing the required value of the air density (from P. Scolari, *Motor vehicles and their evolution*, 10th edition, 2010)

230, 000 rpm to provide a boost pressure in the range of 1.0–1.2 bar. In so doing the residual expansion energy of the exhaust gases, otherwise wasted, is recovered (Fig. 10.3a, point 4 of the theoretical cycle) and the exhaust gas temperature lowers by 100–130 °C. The losses, due to back-pressure increase in the exhaust system (3), are more than offset by the higher intake pressure, and the final result is an increase of the specific power. The wide speed range of gasoline engines calls for a governing mechanism able to maintain boost pressure at a relatively constant level through the engine operating range. The widespread solution is the wastegate valve (6), which controls the flow on the exhaust side, allowing a certain quantity of exhaust gases to by-pass the turbine. In so doing, this device regulates the rotational speed of the turbine and thus the output of the compressor. It is usual practice to limit the charge degree close to its maximum level for an engine speed close to 50 % of the maximum engine speed. The valve is automatically controlled by an actuator (5) following the intake duct pressure. Since the pressure increase causes an increase of the intake air temperature, with a consequent decrease of its density, the turbocharging system is usually provided with an air-to-air radiator, the Charge Air Cooler (CAC), or intercooler (7), to increase the density of the compressed air by decreasing its temperature, using external air as a coolant. Typical values of the air temperature are reported on the same figure.

During the 1990s, concerns about emissions and fuel economy caused this solution to be considered almost impractical and no longer interesting for the following reasons:

- High fuel consumption: to prevent detonation the compression ratio was no more than 8:1, for the same reason, the spark advance was retarded to avoid detonation, and the mixture setting was quite rich. Mixture enrichment was used not only to prevent detonation, but also to limit the temperature rise of the mechanical components considering the available materials: the extra fuel evaporation causes an effective cooling. These provisions cause an increase of fuel consumption.
- The turbo-lag effect, i.e. the time needed to obtain the required power in response to a throttle change. In fact, when a performance increase is requested, the turbine and compressor, connected to the same shaft, must be accelerated to increase pressure and density and this requires a certain time. This, linked to low compression ratio, results in a poor pick-up behavior and, at the end, at low load, the turbo version was behaving as a 'poor naturally aspirated engine'.
- Emission standards: along the exhaust, the turbine housing causes a poorer light-off performance of three-way catalyst due to the heat sink of the turbine housing, thus delaying the catalyst's capability to quickly reach its best efficiency.

In the first years of this century, with increased pressure to reduce fuel consumption, the turbocharging solution was reconsidered. The turbine diameter was reduced to reduce turbo-lag owing to lower turbocompressor inertia; this result can be improved by further reducing the rotor mass using a specific and costly technology to produce titanium impellers. A quicker response can be obtained by improving the fluid-dynamic performance using the twin scroll configuration shown in Fig. 10.34, which splits in two the inlet case of the turbine. It exploits the pulsation sequence determined by keeping separated the internal and external cylinders exhaust streams (1).

A further solution seems now close to being applicable: Electricity-assisted turbochargers, featuring the integration of an electric motor into the rotor of the turbocharger, as a means of augmenting the boost at low engine speeds. A 1.5 kW electric motor can allow the turborotor to spin up to 70,000 rpm.

The above mentioned solutions are not so much intended to increase maximum power but are mainly addressed to increasing low-end torque with the goal of optimizing the torque at low and medium engine speed. A good low-end torque and the reduced turbo-lag effect promote an attractive fun-to-drive feeling. The use of reduced engine displacement allows one to obtain a fuel consumption reduction in the range of 10–15 %.

This represents a significant step toward an effective engine downsizing, allowing the same performance of a larger engine using a smaller one, reducing friction losses in the crankshaft drive and piston assembly and so improving mechanical efficiency, and finally reducing fuel consumption. When the fun-to-drive goal is not the priority,

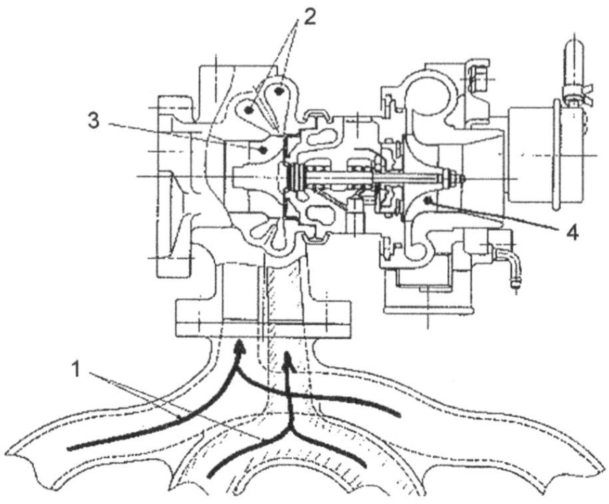

Fig. 10.34 Twin-scroll solution introduced to reduce the turbo-lag using cylinder pulsation sequence

it becomes possible to use longer transmission ratios thus obtaining the so called engine downspeeding, with a further fuel consumption reduction.

The difficulties in emission control were solved by installing the catalyst closer to the turbine and refining the air to fuel ratio control during engine warm-up.

In a gasoline direct injection engine configurations, which almost always include variable valve timing device, it becomes possible to further reduce the turbo-lag effect by activating the scavenging function. The wide overlap allows one to directly supply an additional amount of fresh air to the turbine, thus keeping a high turbine speed. A further evaluation of this function will be explained when describing gasoline direct injection systems.

A high exhaust temperature is still critical for cooling and lubricating the turbo-compressor, particularly with start-and-stop configurations in which the temperature peak at each stop can determine oil cracking with consequent loss of lubrication. Water cooling in the shaft area is thus now widely applied.

10.4.4 Volumetric Mechanical Supercharger

The basic design of the volumetric mechanical supercharger was defined and patented by the Roots brothers, around 1865 in the USA, as an air pump for industrial applications. Today it is used on gasoline engines and on large displacement two-stroke compression ignition engines for marine applications.

When applied on vehicle engines, a volumetric compressor is driven by the engine crankshaft through a belt. A cross section of a Roots type blower with two lobes rotors, where the second rotor is driven by the first through a pair of gear wheels with 1:1 ratio

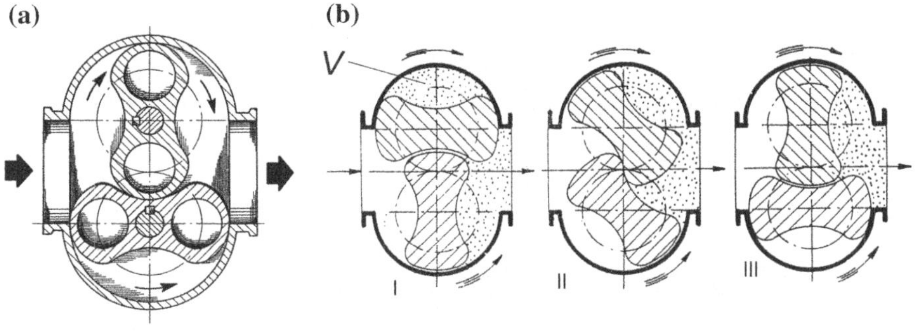

Fig. 10.35 **a** Roots volumetric compressor. **b** A volume of air V (*dotted area*) is trapped and forced into the intake manifold (courtesy of FIAT)

located outside the compressor housing, is shown in Fig. 10.35. The gear transmission is an essential feature, because the rotating lobes must not be allowed to make contact either with each other, or with their housing but the volumetric efficiency depends on maintaining the smallest practicable working clearances between both the lobes and the housing. Similarly to the turbocharger, it provides a pressure and density increase in the intake manifold thanks to the sequence shown in the same figure. The volume of air V trapped in the vane (step I), is forced into the intake manifold in the subsequent steps (II) and (III).

When compared with turbochargers, volumetric compressors do not allow one to reach high supercharging levels but, on the contrary, can follow more promptly the accelerator demand, thus mitigating the turbo-lag effect. When the highest possible torque is required in the complete engine speed range, a volumetric compressor can be used together with the turbocharger and this solution was applied on some Lancia rally cars and, more recently, on a VW 1.4 l engine.

The mechanical drive reduces, by 10–15 %, the engine power since it uses engine power to compress the air into the cylinders. The mechanical energy requirement increases fuel consumption due to the mechanical losses that are present even when the pressure increase is not needed (part load, idle, decelerations). Therefore it becomes almost mandatory to limit the actual supercharging only to conditions, like acceleration or high load operation, when it is really required. By so doing it is possible to limit the fuel consumption increase, with a turbo-lag reduction.

For an overall evaluation of the two systems, it can be mentioned that, out of the roughly 450 European engine configurations available in 2010, 25 % were supercharged, but only 2.5 % had a volumetric compressor. The Kompressor version used by Mercedes-Benz on 1.6 and 1.8 l engines remains the most common engines using the latter solution.

However, recent engineering developments by Eaton resulted in a more efficient type. In 2011 Nissan, on its 1.2 l, 3 cylinders, gasoline direct injection engine, introduced this type of supercharger with the aim of optimizing the low-end torque and improving fuel economy, with respect to the aspirated engine version. To achieve this goal a clutch limits the charger use only when needed, the fluid-dynamic efficiency

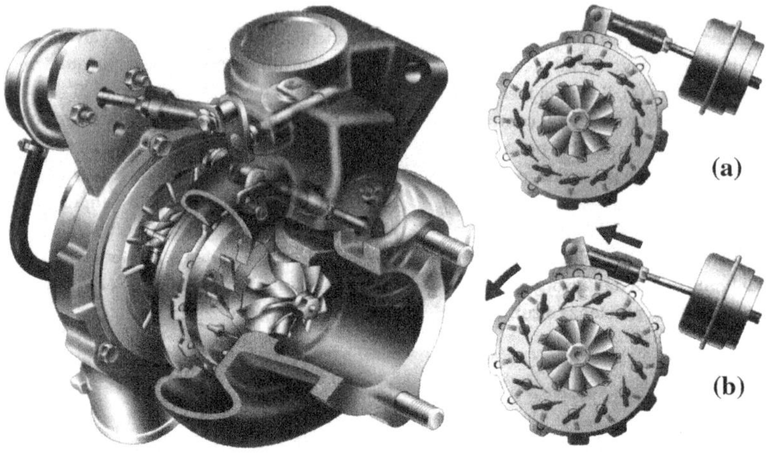

Fig. 10.36 Variable Geometry Turbocharger (VGT). **a** part load position: torque optimization at low load and speed; **b** full load position: maximum performances (courtesy of FIAT)

was improved featuring 4 lobes with helical shape, and also mechanical friction was significantly reduced. When operating in the transient conditions of the NEDC emission cycle, this new configuration allows one to obtain a 14 % fuel consumption reduction with respect to the basic port-fuel injection.

10.4.5 Variable Geometry Turbochargers

Traditionally, the advantages of turbocharger technology are much larger in compression ignition engines, with respect to spark ignition ones. When considering the large air excess accepted in compression ignition engines, it is clear that turbochargers allow an increase in the thermodynamic efficiency by increasing the feed pressure without a subsequent increase in fuel delivery. In spite of this, the turbo-lag effect is still an issue and, as already mentioned, turbine geometry defines, in a pre-determined way, the relationship between engine operating conditions and energy delivered by the turbine to the compressor.

Compression ignition engines overcome this critical point considering their lower exhaust gas temperature, which allows use of a variable geometry device, known as a Variable Geometry Turbocharger (VGT). This solution, featuring variable turbine vanes, is also known as Variable Nozzle Turbine (VNT) technology (Fig. 10.36). The higher exhaust gas temperature, even up to 950 °C, prevents the use of similar devices on spark ignition engines. This solution features a set of adjustable deflector blades, acting as diffuser, which maintain a quite constant gas velocity entering the turbine. As shown in the same figure, the blades rotate around an axis parallel to the turbine axis. At low engine speed (Fig. 10.36a) the gap between two consequent blades is reduced to a minimum so that, in spite of the limited flow of gases, the inlet local speed increases and, in addition, the gas flow is directed at the turbine blade periphery, thus determining a better leverage effect to drive the compressor. In the full load position (Fig. 10.36b) the blades allow a larger quantity of gases to enter the turbine. The final torque increases, in spite of the lower leverage effect. If the two conflicting effects are properly managed, and the maximum torque value is limited, it is possible to avoid using a wastegate valve. The blades are controlled by a pneumatic actuator similar to the one used to drive the wastegate, and the pneumatic signal, controlled via the engine ECU, defines the mapping of blade angles versus engine loads.

Finally this turbocharging solution features the minimal turbo-lag achievable when using small turbine sizes and can provide the full boost at a speed as low as 1,500 rpm, yet remaining quite efficient at higher engine speeds.

When, in premium class engines, the goal is achieving the shortest turbo-lag and the best values for both low-end torque and power, the more costly Variable Twin Turbocharger (VTT) or Two-Stage Turbo solution can be applied. It features the assembly of two turbine-compressor groups having different diameter. For instance, below 1,500 rpm only the smaller turbine operates, between 1,500 and 3,000 rpm, the first turbine operates and the larger one is kept at low speed. Above 3,000 rpm only the latter is operating. The engine ECU activates the pneumatic valves managing the exhaust flow following the engine speed and load signals. The average boost pressure and the specific power of the VGT are respectively 2.5 bar ad 55 kW/l, versus 3.0 bar and 70 kW/l typical of the latter approach. Following this trend, in 2012 BMW applied the tri-turbo version on a 6 cylinder-in-line engine, reaching values close to 4.0 bar and 94 kW/l.

10.4.6 Air to Fuel Ratio Control

Spark ignition engines use fuels that are volatile enough to be easily premixed with air, thus allowing one to obtain a homogeneous mixture before starting combustion. The first task of the induction system is delivery of the correct mass of air and fuel to obtain the required air to fuel ratio (AFR, for which symbol α is commonly used).

A second, but not less important, task is promoting a good mixing between air and fuel to obtain a good combustion speed and a correct combustion process from its beginning to its end. The correct final result is achievable by optimizing the two following mixture qualities:

- Atomization, which improves by reducing the fuel droplet diameter, and
- mixing, which improves with a quick fuel evaporation in the air stream and, of course, is strongly promoted by atomization.

Carburetors, widely used up to the 1990s, proved to have limited capability in promoting the above mentioned processes.

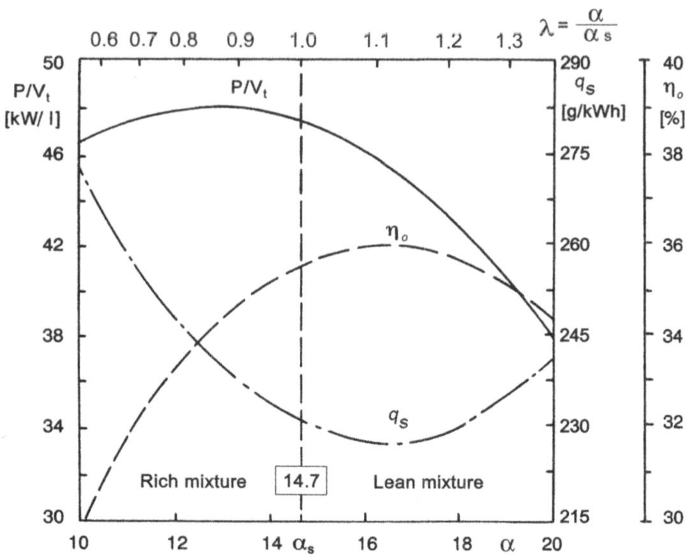

Fig. 10.37 Overall efficiency η_o, specific power P/V_t and specific fuel consumption q_s versus air to fuel ratio α or excess-air factor λ (from Giacosa 2000)

The multipoint, port-fuel injection, typically featuring a 3.5 bar injection pressure and a fuel jet having a correct shape and direction toward the intake port, is far better in achieving these goals. The pintle-type injector provides a fuel jet having the shape of a hollow cone and the fuel spray is directed toward the intake valve during its opening stroke. This geometry directs the fuel through the open section of the valve, thus avoiding that a significant fuel quantity impinges against metal walls of the ducts into the cylinder head; in so doing the charge quality is definitely improved. When two intake valves are used, the injector can be a multi-hole type (e.g. eight holes, with a 0.1 mm diameter each).

Air to fuel ratio is defined as the ratio of the air mass and the fuel mass participating to the combustion. Considering the hydrogen/carbon fuel composition, the complete combustion takes place for $\alpha \approx 14.7$, which defines the stoichiometric mixture ratio α_s. Considering the gasoline density, it means that to burn 1 l of gasoline, 9,500 l of air are needed. But it is possible to marginally deviate from the stoichiometric value using a rich or a lean mixture, i.e. using values of α lower or higher than 14.7, respectively. The optimal value of the air/fuel mixture for gasoline engine varies depending on its running conditions.

The value of α at which the engine delivers the maximum power at a given throttle opening lies on the rich side, as shown in Fig. 10.37. This condition promotes the maximum speed of the combustion front, provided air and fuel are duly mixed. The high combustion speed allows a reduction of the spark advance, and this combination makes the effective cycle closer to the ideal one. The air to fuel ratio enrichment is useful for another reason: the fuel excess, with its evaporation, reduces the temperature in the combustion chamber thus preventing detonation, particularly dangerous at high engine loads.

On the contrary, to obtain a fuel consumption reduction the mixture must be lean. A certain thermodynamic efficiency reduction takes place, due to slower combustion speed, but this is compensated by the fuel consumption reduction and, at the end, the overall efficiency improves.

It is well known that engine performance is significantly affected by the so-called mixture strength, which is also identified by the excess-air factor λ, defined as the ratio between the actual air to fuel ratio and the stoichiometric one ($\lambda = \alpha/\alpha_s$). Its value is 1 when the mixture is stoichiometric. Gasoline engines thus reach maximum power with a 10 % of air deficiency ($\lambda \approx 0.9$) and minimum fuel consumption with a 10 % of air excess ($\lambda \approx 1.1$).

The mixture strength at which traditional carburetors work lies between these two conditions: rich mixtures for full load and accelerations, lean mixtures when running at steady low-medium speed. To meet Euro 1 emission standards, since 1993, it was mandatory to introduce multipoint port-fuel injection, lambda sensors and three-way catalysts. Maintaining a certain fuel mixture enrichment at high loads but, in most part-load conditions, only the stoichiometric mixture can satisfy emission standards, allowing optimization of the catalyst efficiency.

Based on the above, it is clear that the range of air-to-fuel ratio providing the correct engine running is quite narrow and this is the reason why the only effective way to modify the power output is to control the air flow quantity. All the efforts performed for many decades to achieve acceptable spark ignition engine performance using an air to fuel ratio well above 14.7 to further reduce fuel consumption, did not reach satisfactory results. Only today the gasoline direct injection system is able to modify this situation and this result is obtained with the so called stratified mixture.

Compression ignition engines allow using a strong air excess and therefore an extremely lean mixture, $\lambda = 2.5 - 3.0$ in part load conditions, while a ratio approaching the stoichiometric value is obtained when the engine load increases up to the maximum performances.

10.4.7 Crankcase Ventilation

Gases filling the combustion chamber and the lubricating oil in the crankcase must be kept accurately separated. This separation is the goal of the so called crankcase ventilation which is essential both to allow a correct mechanical operation and to control emissions.

The components above the piston are periodically flooded by gases at high pressure or at slight negative pressures; on the contrary the components below the piston, within the crankcase, need an effective oil lubrication in steady pressure conditions. As already mentioned, the elastic rings, sliding along the cylinder surface, offer an efficient and continuous separation between the gases and the engine oil. Should only a minimal oil quantity reach the combustion chamber, unavoidably, the oil consumption increases and this is the reason why this consumption is a primary functional parameter and the driver is requested to periodically check the oil level. Through the last two decades, the oil consumption decreased, as an average, from 80 to 15 g/h, running under the most severe conditions, thanks to material technological improvements, improved crankcase design and constraints able to keep the cylinder shape as much as possible cylindrical during crankcase thermal expansion. Today oil consumption is almost negligible, so low that oil level checks are performed with increasingly lower frequency.

Elastic rings must also prevent the bleed of gases from the combustion chamber to the crankcase volume (blow-by gases) that can pressurize the crankcase causing oil leakage; in this event, abnormal mechanical stresses, with consequent damages, take place on the seals between crankshaft and crankcase.

The minimal oil quantity that is burned during combustion, is exhausted as heavy hydrocarbon and therefore it is mandatory to reduce this quantity as much as possible to meet exhaust hydrocarbon emission standards. In addition, combustion of heavy oil hydrocarbons produce Polynuclear Aromatics Hydrocarbons (PAH), known to be extremely noxious. Finally, oil consumption is responsible for lambda sensor and catalyst deterioration.

In the opposite direction, the blow-by gases, blowing from the combustion chamber to the crankcase, mix with oil vapor and produce a continuous plume of smoke, that is again quite harmful. Up to the 1960s, this mixture was directly vented to the atmosphere to avoid both crankcase abnormal pressure and damages to the sealing rings.

A circuit capable of maintaining a slight vacuum in the crankcase connecting it to the intake manifold thus providing the closed positive crankcase ventilation solves this problems (Fig. 10.38). It has been used since the 1970s and is still a valid solution for all engine configurations to avoid atmospheric pollution due to blow-by gases. Historically, it represents the first emission control system.

The engine air intake (1), the air filter (2) and the low pressure zone under the throttle (3) are shown in the figure. This vacuum source must be controlled by a valve to provide a source of limited negative pressure thus determining a small intake flow of blow-by gases and oil vapors (4) which, after reaching the intake manifold, are burned. It is important to avoid the oil vapor flow, due to the temperature increase in the oil sump, taking droplets of liquid oil. These droplets must be separated by using a liquid-vapor separator (5). At the end, the condensed vapor and oil droplets return into the sump through to the duct (6) and the system lets only gases escape.

This gas circulation may promote the formation of some deposits in the ducts from the air filter to the inlet valves and even on the valves themselves modifying the air to fuel ratio and determining a driveability worsening. To mitigate this situation specific detergents to be added to the fuel were developed. At the end, they proved effective.

Compression ignition engines generate only a limited amount of blow-by gases since only air is compressed during the compression stroke, therefore minimizing this kind of gases. They reach the crankcase only during the power stroke and blow-by gas pollution is reduced to no more than 10 % when compared with gasoline

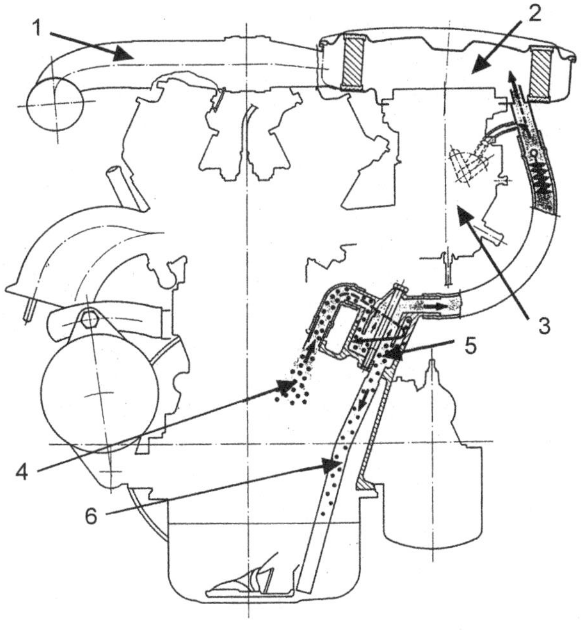

Fig. 10.38 Positive crankcase ventilation system. This system prevents oil vapors and combustion gases from blowing through the crankshaft seals toward the atmosphere

engines. The closed crankcase ventilation system became mandatory only recently for compression ignition engines.

Gasoline direct injection behaves in an intermediate way between port fuel injection and compression ignition engines.

10.5 Exhaust Emissions

The time allowed to complete the combustion and charge replacement is very short. Assuming a speed range between 600 and 6000 rpm, the time required for each revolution is between 100 and 10 ms. Since, as an average, the crankshaft rotates by 50° during the combustion phase, the time allowed for all chemical reactions to take place is between 15 and 1.5 ms. In such conditions a certain number of reactions cannot take place completely and this causes formation of a certain quantity of exhaust pollutants.

10.5.1 Exhaust Emission Formation

Engine exhaust gases are formed during combustion, in which chemical energy, stored in the fuel hydrocarbons, is released as heat of oxidation. The chemical reaction in the conversion process is the oxidation of hydrocarbons by oxygen supplied by the ambient air; the combustion reaction of heptane is:

$$C_7H_{16} + 11O_2 + 43N_2 \rightarrow 7CO_2 + 8H_2O + 43N_2 + \text{heat}\,.$$

Considering the air composition (21 % O_2, 78 % N_2, 1 % other gases), the reaction should require a sufficient oxygen quantity to oxidize carbon and hydrogen and to reach the final products: carbon dioxide and water. Unfortunately a high activation energy is required to complete these reactions and, therefore, they develop not as quickly as expected. Some intermediate steps, requiring a lower activation energy, take place first, causing the formation of partially oxidized products with the following simplified sequence: peroxides → aldehydes → CO, the latter being the most well-known partially oxidized product.

This sequence is responsible for CO emission and, finally, the combustion of 1 kg of gasoline at $\lambda = 1$, typically delivers 3.1 kg CO_2 and 1.3 kg H_2O, the remaining being N_2 which does not participate in the combustion process.

For compression ignition engines, where $\lambda = 2.5$–3.0, the CO_2 and H_2O quantities are similar with a significant increase of N_2 and O_2 as a consequence of the air excess.

During the combustion process, when the homogeneous mixture is delivered to the combustion chamber, a certain number of by-products or noxious pollutants are released, mainly CO, hydrocarbons and NO_x with a percentage that, in stoichiometric conditions, is about 1 % of the total exhaust gas quantity. For compression ignition engines, and to a lesser extent for gasoline direct injection engines, when the fuel direct injection delivers a non-homogeneous mixture, also particulate matter formation takes place. The above mentioned gaseous emissions started to be evaluated at the end of the 1960s. Their formation during the four cycle phases is outlined in Fig. 10.39.

During the compression stroke the valves are closed and therefore no exhaust takes place but fuel and oil start accumulating in the piston crevices around the compression rings and the piston crown. In addition, the lubricant thin film, which adheres to the cylinder walls, absorbs some hydrocarbons from the compressed fresh charge.

As already stated CO is formed in the reaction area during combustion. In the expansion stroke the temperature decreases so quickly that the oxidation process is frozen, preventing it from reaching the final chemical equilibrium and a certain amount of CO, depending on the air to fuel ratio, is always present.

The hydrocarbons formation can also be determined by an incomplete combustion, but has mostly a different origin. During combustion, the flame front is unavoidably quenched in the proximity of the walls where the temperature is too low to sustain the combustion thus determining the so called quenching effect. As a result, a hydro-

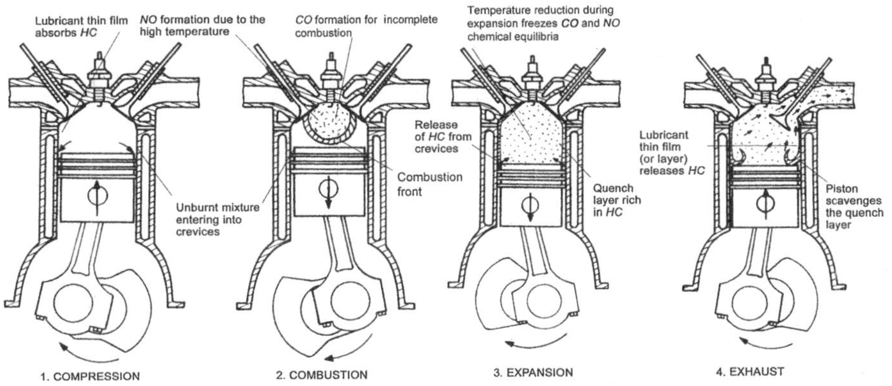

Fig. 10.39 CO, HC and NO_x formation during the various phases of a spark ignition engine

Table 10.4 Correlation between temperature peak at the end of combustion and NO_x concentration

Engine condition	Idle	Low-load	Medium-load	High-load
T (K)	1,500	2,000	2,250	2,500
NO_x (ppm)	200	1,000	3,000	6,000

carbon layer with an average thickness of 0.1 mm, containing most of the unburned hydrocarbons from both the lubricating oil and the fuel, remains.

When spark ignition engines run too lean, the flame front becomes particularly slow and, in this situation, the flame travel can be interrupted before reaching all the combustion chamber volume, thus justifying why it is systematically difficult to assure a proper engine running in the lean area. When this takes place, the quantity of unburned hydrocarbons rises substantially and is further increased by increased thickness of the oil/gasoline layer.

The formation of oxides of nitrogen (NO_x) in spark and compression ignition engines is due to the direct combination between nitrogen and oxygen. Their total amount raises proportionally to the temperature peak reached at the end of the combustion and a typical correlation between this latter parameter and NO_x concentration is shown in Table 10.4.

Medium-load conditions are seldom reached during the emission measurement, according to the NEDC cycle. The NEDC cycle will remain valid in Europe up to the Euro 6 level; most likely the cycle will be updated for the next emission step. The combustion temperature does not exceed 2,300–2,500 K considering the mixture enrichment which almost always takes place to reach maximum performance and to limit the temperatures of the mechanical components.

A similar correlation applies also to diesel engines as shown qualitatively in Fig. 10.57a, in which a maximum local temperature of 3,000 K is reported, considering that, in this case, no extra fuel is needed.

The oxides of nitrogen considered here are: Nitric oxide (NO), the prevailing one, nitrogen dioxide (NO_2) and traces of nitrous oxide (N_2O). The latter can marginally increase its concentration during some reducing processes applied to decrease the concentration of the previous two. Reasonably the NO_2 formation is promoted by the temperature and, more specifically, when the air-excess and the consequent oxygen concentration increases during the combustion process.

For spark ignition engines the typical oxygen concentrations in rich and in stoichiometric conditions are respectively 0.2 and 0.7 %. As a consequence, the NO_2 to NO ratio remains quite low spanning from zero to 1 %, following the oxygen rise.

Again, during the expansion stroke the temperature reduction freezes the chemical equilibria leaving, in addition to CO, also a certain concentration of NO_x.

As already stated, two steps can be recognized in the exhaust phase:

- The blowdown, during which hydrocarbons are violently exhausted, and
- the exhaust stroke, during which the piston rings scavenge the hydrocarbon layer deposited along the cylinder wall; hydrocarbons are therefore pushed out and mixed with the other exhaust gases.

Vehicles powered by a spark ignition engine release also evaporative emissions due to gasoline evaporating in the fuel tank. A certain amount of hydrocarbons are therefore released as vapor and are frequently identified as Volatile Organic Compounds (VOC).

In compression ignition engines, the air excess determines a lower CO and hydrocarbons formation and the lower exhaust gas temperature impairs any partial oxidation of these pollutants along the exhaust pipe.

The oxygen concentration varies in the range from 6 to 8 % at full load and from 10 to 15 % at low load, thus causing a NO_2 rise and an increase of the NO_2 to NO ratio from a few units to 25 %.

In addition to the above, when the fuel is injected directly into the combustion chamber, the internal part of the fuel spray is exposed to a rapid temperature rise determined by the initial combustion of the external part. This high temperature, together with a lack of oxygen due to insufficient mixing, favours a thermochemical decomposition of the spray droplets. Within the droplet, hydrogen burns quickly, while carbon leaves a solid residue characterized by an extremely extended specific surface. This thermochemical decomposition of organic materials at high temperatures without the participation of oxygen is referred to as pyrolysis, and is typical of compression ignition engines, but takes place also in gasoline direct injection engines, although to a significantly lesser extent. If the carbon does not have sufficient time to burn, it is exhausted as soot or dry particulate. In the past the strong soot concentration was known as black smoke, that now is no longer present thanks to a far more efficient atomization and mixing. During expansion, with the resulting temperature reduction, the carbon fraction contained in the soot, being extremely porous, promotes the condensation and aggregation of partly oxidized hydrocarbons, aldehydes, sulfates and water molecules finally determining an aggregation of inorganic and organic materials identified as particulate matter (PM).

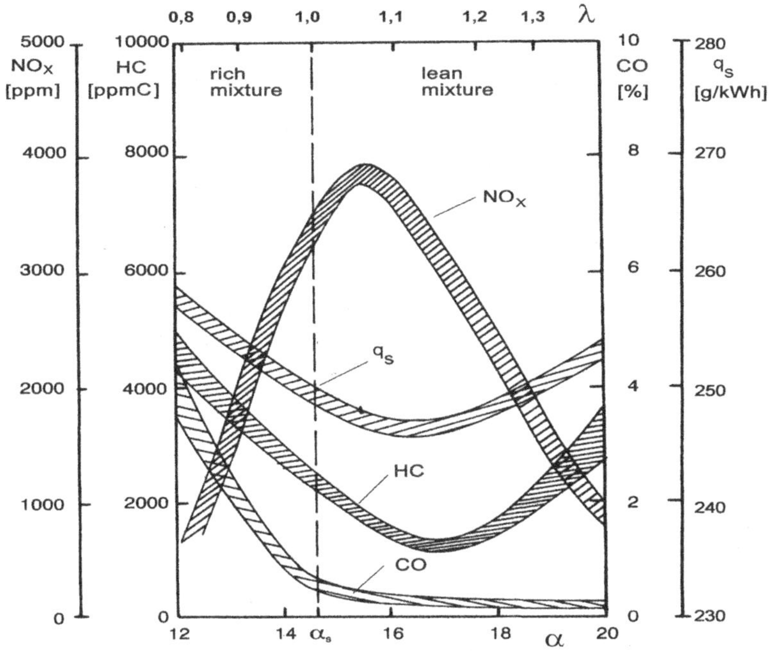

Fig. 10.40 Otto-cycle engines: CO, HC, NO_x concentrations and specific fuel consumption q_s versus air/fuel ratio α or excess-air factor λ (courtesy of FIAT)

Within particulate matter, it is possible to distinguish between the non-soluble fraction, or soot, mainly containing carbon, and Soluble Organic Fraction (SOF), containing the aggregated molecules. The soluble fraction is promoted also by blow-by gases and lubricant oil molecules due to deficiencies in the crankcase ventilation, more frequent during engine cranking and warm-up. During this transient period, the chemical structure of particulates is different and the exhaust is rich in water, aldehydes and aerosols, thus promoting the so-called 'white smoke'.

As it can be easily understood, the concentrations of pollutants in the exhaust gases of spark ignition engines strongly depend on the air to fuel ratio as shown in Fig. 10.40. Pollutant concentrations are expressed in percentage of volume or in parts per million (ppm; 1 % corresponds to 10,000 ppm); all data refer to steady state conditions. In the zone of rich mixtures the combustion is incomplete, due to oxygen deficiency; large hydrocarbons and CO quantities are therefore present. Around the stoichiometric value both reach their minimum values. As shown, for the stoichiometric value $\alpha = 14.7$, the CO content is equal to 0.7 % and also the residual oxygen content is close to the same value.

Entering in the lean mixture area, with a rising air excess, hydrocarbons content starts increasing again due to slow flame propagation and to the increased wall layer.

In the same field, the CO content remains low and its residual quantity is mainly due to the not complete mixture homogenization and distribution among the cylinders.

On the contrary, the NO_x content is low both in the rich and lean areas and peaks for a slightly lean mixture, approximately for the same value leading to maximum temperature and minimum fuel consumption. The NO_x content is linked to the temperature, which is limited by evaporation of the fuel excess in a rich mixture while it is limited by the excess of air if the mixture is lean. The diagram holds for multipoint, port-fuel injection systems; the air to fuel ratio values, extended up to $\alpha \approx 20$ ($\lambda \approx 1.36$), do not apply to the compression ignition combustion.

The simplest way to reduce CO and hydrocarbons is mixture leaning and this was applied from the 1960s to the 1970s and 1980s. Traditionally the average air to fuel ratio experienced during the NEDC cycle correlates quite well with the idle CO concentration, therefore it was possible to reduce the CO content from 35 g/km, base value, to the 1984 step, 12 g/km by reducing the idle CO content from 4.5 to 1.5 %, although using carburetors. The limited possibility of improving the charge quality did not allow further CO reductions. Only applying the multipoint injection has it been possible to achieve the stoichiometric value, further reducing the idle CO content down to 0.7 %, while obtaining also a reduction of the hydrocarbons content.

In spark ignition engines, thanks to the exhaust gas temperature and to the presence of residual oxygen, a further partial oxidation of CO and hydrocarbons can take place immediately downstream from the exhaust valve and in the first section of the exhaust ducts, strongly depending on the temperature, which may reach at least 600–700 °C in low-medium load conditions.

This oxidation process is favored by injecting additional air into the exhaust manifold with a specific air pump, thus obtaining a homogeneous post-combustion. This system was widely used in the 1970s, before catalysts were applied, and is still sometimes used today, using an electrically driven air pump. The pump is operated during cold start-up when the rich mixture, needed for starting, causes an increased concentration of CO.

When the air to fuel ratio approaches the stoichiometric value, it promotes an NO_x increase, therefore, during the same decades, to control NO_x emissions it was mandatory to implement suitable values of the compression ratio, of the valve timing and ignition retardation and dedicated systems such as Exhaust Gas Recirculation (EGR).

This mix of technologies is the baseline for engine emissions control; they may have a positive or negative effect on fuel consumption, and this can be evaluated only on a case by case basis.

When, in 1993, Euro 1 stricter levels became mandatory and the previous engine modifications were no longer sufficient, the multipoint, port-fuel injection and, in addition, the after-treatment system including lambda sensors and three-way catalysts became mandatory to reduce at the same time CO, hydrocarbons and NO_x. Engine modifications and after-treatment, with their further evolution up to Euro 6, are the only approaches able to reach emission levels close to zero or zero equivalent for the three pollutants. Exhaust emissions measured downstream from the catalysts, whichever it is, are defined as tailpipe emissions.

The above mentioned technologies are extremely important because their combination allows us to:

- Reach emissions level close to zero,
- implement fuel consumption reduction devices, while
- containing costs at an affordable level.

During the 1990s, with the goal of further reducing fuel consumption, it became interesting to examine how pollutants behave in the lean area, around $\lambda \approx 2.5$, in which compression ignition engines operate, thus introducing the multipoint gasoline direct injection with the related additional engine modifications. They allow extension of the correct engine operation area well into the lean area, obtaining a fuel consumption reduction close to 10 %. Within this area, the residual oxygen content increases linearly from 0.7 % onward and emissions are already low but still far from being considered as close to zero.

Multipoint gasoline direct injection engines, being positive-ignition engines, must comply with the same emission standards that apply to multipoint port fuel injection engines and, up to now, a catalyst technology able to reduce simultaneously the three pollutants down to levels close to zero, in the presence of significant oxygen concentrations, is still under development. At the beginning of the 2000s, this difficulty slowed the growth of gasoline direct injection configurations, because oxygen excess limits the three-way catalyst efficiency only for CO and hydrocarbons, therefore calling for innovative and costly solutions like the de-NO_x catalysts, which will be described later.

This limitation affects also compression ignition engines. They are, however, subjected to specific emission standards which, up to Euro 2, were defined to be achievable applying engine modifications, namely injection retardation and exhaust gas recirculation and adding the first type of after-treatment using oxidation catalysts able to further reduce CO and hydrocarbons. A remarkable improvement of the injection type, up to Common Rail, was needed to comply with Euro 3 and 4 regulations. The requested particulate matter reduction was obtained through the continuous evolution of injection systems and, partially, through oxidation catalysts. Euro 5 calls for a remarkable particulate matter reduction achieved by adding a dedicated after-treatment technology: The Diesel Particulate Filter (DPF).

Finally, the Euro 6 standard implements emission levels close to fuel neutral emissions: it means that compression ignition engine emissions must be close to spark ignition engine ones. It is easy to predict that this will be a tremendous technical challenge, mainly because a specific technology able to reduce NO_x will be required.

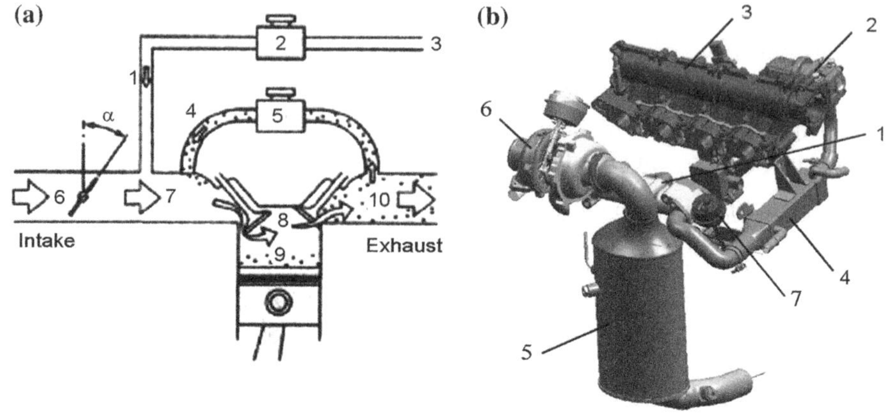

Fig. 10.41 **a** Otto-cycle engine charge formation: intake air, exhaust gas recirculation and fuel vapors from the carbon canister. **b** Diesel-cycle engines: exhaust gas recirculation layout and after-treatment system. (**a** Courtesy of Bosch, **b** Courtesy of FIAT)

10.5.2 Exhaust Gas Recirculation (EGR)

Spark Ignition Engines

In real-life the charge of spark ignition engines is not only the combination of air and gasoline, but something else is always added.

Quite often, to control the NO_x emissions, the Exhaust Gas Recirculation (EGR) system, shown in Fig. 10.41a, is added. It features a connection (4) inducing an exhaust gas flow in a percentage close to 10–15 % of the complete exhaust stream (10) into the intake manifold. The correct exhaust gas recirculation percentage is determined by a specific valve (5) able to deliver a suitable exhaust gas amount, based upon the requested NO_x reduction. The exhaust gases, mixed with the incoming air flow (7), reduce the temperature peak at the end of combustion. Actually, being inert, they cannot produce heat during combustion and in addition:

- The temperature reduction is also a consequence of their specific heat, significantly higher than the one of the air they replace;
- they reduce the oxygen concentration in the inlet air. With a 15 % of gas recirculating the incoming 21 % of oxygen is reduced by the same quantity, namely to 18 %.

An excess of recirculated flow leads to an hydrocarbon emission increase due the worsening of the combustion quality. On the contrary the exhaust flow leads to an improvement of the volumetric efficiency due to a wider opening of the throttle valve (6) required to compensate for the volume occupied by exhaust gases downstream of the throttle; the consequent de-throttling reduces fuel consumption by 2–3 %.

The latter is the main reason for applying this system. The mixture control achievable with the latest types of sensors and electronic control is so precise, that three-way catalysts can work with their best efficiency, thus reducing NO_x at the required levels to meet the standards.

Increasing the valve overlap, as made possible by the updated valve timing control devices, it is also possible to increase the residual gas quantity (9), from a typical 10 % in part load conditions, to a somewhat larger value. The mixture of residual gases with the fresh homogeneous charge (8) determines the so called internal exhaust gas recirculation and provides the desired temperature limitation.

In addition to the above, from Euro 1 onwards, all gasoline vehicles have been provided with the evaporative emission control system. The hose connection (1) which, through the control valve (2), introduces the gasoline vapors (3) released by the active carbon canister, to the engine intake to regenerate the carbon is shown in Fig. 10.41. The control valve is driven by a duty-cycle signal from the engine electronic control unit to meter the quantity of vapors to keep the air to fuel ratio within the requested value.

Compression Ignition Engines

Unlike gasoline engines, the exhaust gas recirculation application, with a flow up to 35 % in low load conditions, is mandatory on direct injection diesel engines, always provided with sophisticated recirculation systems to meet the NO_x standards. The recirculated gas percentage lowers with increasing load, considering that the engine needs a suitable charge. The layout applied to these engines since Euro 2 is shown in Fig. 10.41b, together with a Diesel Particulate Filter. In the early applications, the exhaust gas recirculation valve was located on the exhaust manifold. Today, considering that the valve actuation is always electric, with electronic drive system, the valve is located close to the intake manifold. Exhaust gases from the exhaust manifold (1) are directed to the valve (2) and finally to the intake manifold (3). To increase the recirculation flow, the exhaust gases must be cooled with engine cooling fluid using a suitable heat exchanger (4) which can be by-passed when the engine is cold by activating the heat exchanger by-pass valve (7). The cold recirculation flow must be prevented from causing a sharp increase of CO and hydrocarbons. To this end the after treatment device efficiency (5) must be increased to meet the final emission standards.

The variable geometry turbocharger (6) promotes a higher recirculation flow, where in spark ignition engines the exhaust gas recirculation quantity is activated by the differential pressure available between the exhaust and the intake manifold. The negative pressure, available in the diesel intake manifold, is limited because the throttle valve is not present. Therefore the backpressure increase due to the turbocharger is useful and is obtained when the blades are in their closed position when the NO_x reduction is required.

The layout now described is identified as high pressure recirculation system to distinguish it from the low pressure one, that will be described when dealing with Euro 6 standards.

10.6 Indirect Injection Spark Ignition Engines

During the 1970s and the 1980s many types of mechanical or electric/electronic indirect fuel injection system started to be fitted on some passenger cars of the medium-high segment. When compared to the prevailing systems based on carburetors, the new systems were offering a better fuel distribution among the cylinders, particularly when their number was higher than four. The flexibility offered by installing the fuel injector on each intake duct offered the possibility of a better fluid-dynamic shape of the intake manifold leading to improved engine performance. But the factor that forced to completely replace carburetors with port fuel injection was the need of reducing the emissions at the low level requested by standards, thus allowing three-way catalysts to be introduced.

An efficient electronic engine management system, able to strictly control the air to fuel ratio, is required for the application of catalysts. A closed loop control based on a lambda sensor was thus required and all the efforts done to accomplish this goal using traditional carburetors proved not sufficiently reliable.

The first car implementing this new system was launched on the USA market by Volvo in 1973 to meet their emission standards. The new system, with its subsequent updating, provides a completely new ignition and fuel control systems as described in the following sections.

10.6.1 Ignition System

The primary task of the ignition system is starting the combustion with a precise timing. In present day engines this is performed by a spark taking place between two electrodes.

The compression helps to completely vaporize the gasoline whose final boiling point is close to 210 °C. For compression ratios in the range of 8–12, the temperature at the end of compression reaches 400–500 °C. These values are still below the gasoline auto-ignition temperatures, considering the physical mixture conditions, and therefore the spark between the electrodes causes a localized temperature rise whose magnitude must provide the fuel ignition, i.e. a temperature at which the fuel has sufficient internal energy to break the carbon-hydrogen bond of the hydrocarbon thus starting its oxidation to carbon dioxide and water.

These temperature values explain why, in cold weather, several compression events are needed to increase the temperature at a value at which combustion can start. During this transient the mixture enrichment favours to reach the desired conditions.

The gasoline vapor interested by the spark starts its combustion with an exothermic reaction, thus rising the temperature and causing the front of the combustion to move with a maximum speed close to 25 m/s, as already discussed in the previous sections. With a bore in the range of 80–85 mm and a speed of 3,000 rpm, a crankshaft rotation angle of roughly 30° is required for the combustion to interest the whole charge. After the time required to perform this rotation, the combustion process involves the whole combustion chamber volume. At stoichiometric air to fuel ratio, the speed is typically reduced to 20 m/s and down to 15 m/s for a lean mixture. The speed at which combustion takes place during an explosion is roughly 2,000 m/s, so that the combustion process in internal combustion engine has nothing to do with an explosion.

Up to the 1990s, ignition and spark advance were provided with the traditional devices: ignition coils, mechanical distributors, vacuum capsules and contact points.

Progressively they were replaced by the fully-electronic type, able to generate a dedicated control signal for each cylinder, each equipped with its own ignition coil. The present-day inductive ignition system generates the high ignition voltage required to produce the spark between the plug electrodes and features a stationary voltage distribution. The engine electronic control unit uses suitable look-up tables to calculate the ignition timing based upon load, speed and temperature inputs.

Within each coil, the magnetic field is generated as current flows through its primary winding. When firing is required, the current flow is interrupted and the magnetic field collapses. This rapid field change induces the high voltage within the coil's secondary circuit and when this reaches the ignition voltage the spark is generated.

Present day systems are thus less noisy, reduce the number of high voltage connections and reduce electromagnetic interference. Finally they have a far better reliability and precision and allow increase of spark voltage, with respect to the mechanical systems, from 10 to 30 kV. The higher voltage and the faster spark rise are highly beneficial to:

- Meeting the more demanding emission standards under stoichiometric mixture,
- providing the spark even if the spark-plug electrodes are covered with liquid gasoline, likely to occur during cold starting and
- providing the spark in presence of the liquid fuel spray used in the stratified charging solution peculiar of gasoline direct injection.

These factors lead to a minimum demand for ignition energy of 30–50 mJ in multipoint port fuel injection systems both naturally aspirated and turbocharged. Higher values, up to 100 mJ, are however needed in gasoline direct injection configurations.

The main components of the ignition system are shown in Fig. 10.42. Each spark plug (1) is provided with its own ignition coil (2), shown in the figure in the typical location used in four-valve pent roof combustion chamber. The throttle switch provides the idle position (3). The engine control unit (4) transmits the driving signals to the ignition coil, in the correct instant, thus providing a spark advance which is a function of many parameters, like engine speed, intake manifold absolute pressure, engine temperature, throttle position and conditions leading to knocking. The engine temperature is controlled by sensor (5) and the knocking intensity by sensor (6). The engine speed sensor (7) detects both the speed and the reference mark, shown as a tooth gap on the engine flywheel (8), thus triggering the ignition process. The voltage of the battery (9) is monitored and the electronic control unit performs the voltage correction by extending or reducing the coil charging time. The ignition key

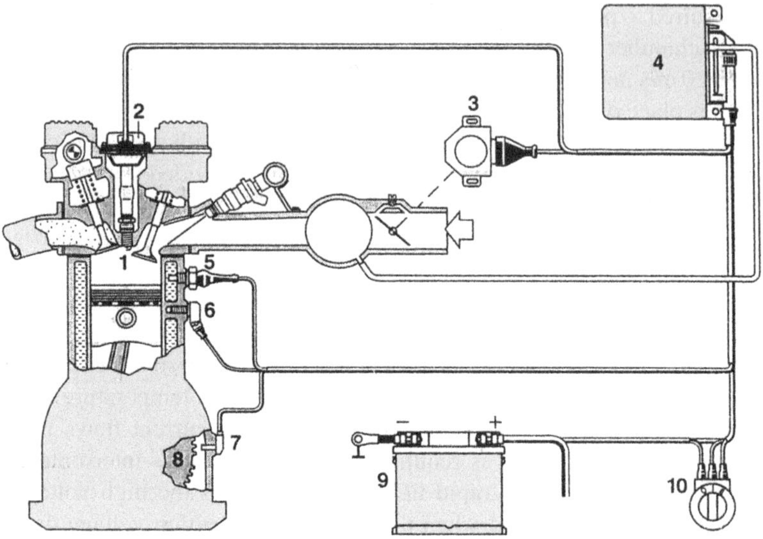

Fig. 10.42 Integrated ignition system with single ignition coil for each cylinder (courtesy of Bosch)

(10) completes the system. The ignition system is now integrated with the injection system, both being driven by the engine control unit.

Typically, the time lag between the spark and the completion of the combustion remains almost constant as long as the air to fuel ratio remains constant and therefore the firing signal must be progressively anticipated as the engine speed increases. As a general rule, the maximum power, for almost every engine speed, is reached setting a spark advance which assures that 50 % of the mixture has been burned when the TDC has been reached. In these conditions the peak of the combustion pressure takes place between 15 and 20° after the TDC.

The so called *Minimum spark advance for the Best Torque* (MBT) is the value of the advance allowing the best performance with minimum fuel consumption, because it optimizes the balance between areas II and III of the thermodynamic cycle, with reference to Fig. 10.3c.

The preliminary spark advance map is obtained by running the engine on its test bench in steady state conditions for a number of engine speed and load combinations. One of the best signals to monitor the load is provided by the absolute pressure sensor in the intake manifold. Subsequently the prototype vehicles must be driven on the road to obtain the best possible driveability during transient conditions, in cold and hot weather and using the different fuel qualities expected during the vehicle lifetime. Finally the vehicle must be tested on the dynamometer for verifying its compliance with emission standards.

A certain ignition retard lowers the hydrocarbons and NO_x content because, in so doing, the combustion takes place mostly during the expansion, thus lowering the temperature peak responsible for NO_x formation, but increasing the cylinder

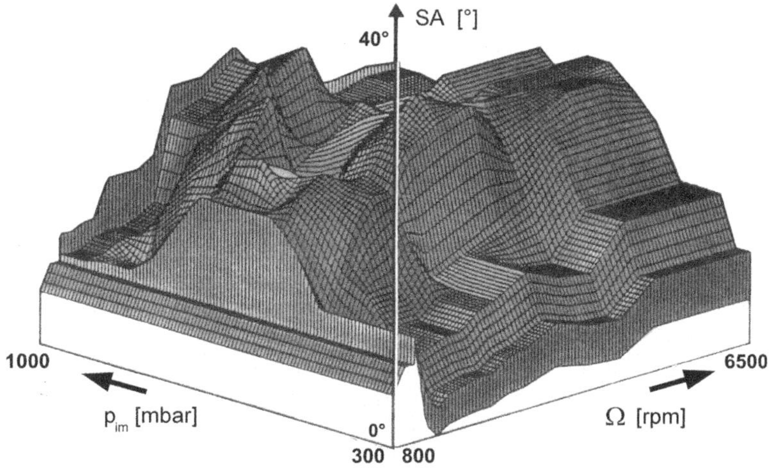

Fig. 10.43 Spark advance map. For each combination of engine load and speed the spark advance value is selected to optimize performance, driveability, fuel consumption and emissions (courtesy of Bosch)

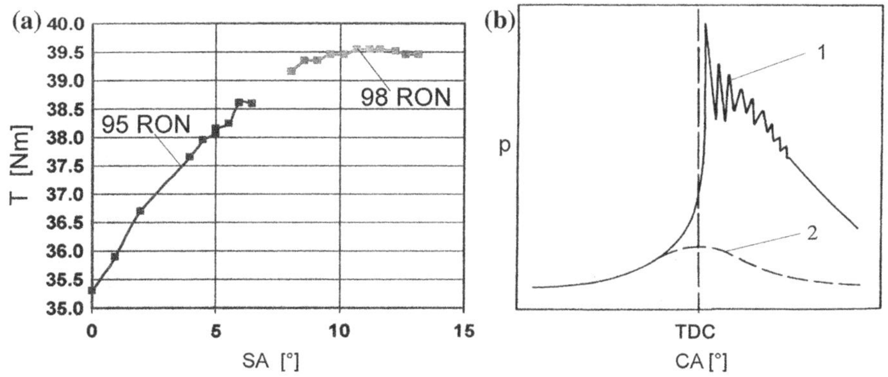

Fig. 10.44 Spark advance and detonation. Engine torque as function of spark advance and gasoline octane number (**a**). Peaks of pressure determined by detonation approaching TDC (**b**)

temperature at the end of the expansion. This temperature rise promotes hydrocarbons oxidation both in the cylinder and in the exhaust manifold.

At the end the preliminary map must be updated to achieve the best compromise among requirements, often conflicting with each other. A modified map is obtained and, hopefully, it will be the best for the application under development. This final map is stored and embedded in the memory of the control unit, as shown in Fig. 10.43 that clearly shows the retarded spark advance selected in the low load area, the most important for emission control.

Figure 10.44a shows how the spark advance, up to a certain extent, can provide a better torque, with an increase that, in this case, is of 12 %, from 35.3 to 39.6 Nm, increasing the spark advance from 0 to $-12°$. It must be noted that the original advance, from 0 to $-7°$, provides already a 9 % torque increase, from 35.3 to 38.6 Nm. This is the correct spark advance setting selected when running the engine under steady state conditions and using the European unleaded gasoline with 95 Research Octane Number (RON). Refuelling with 98 RON gasoline a further spark advance up to $-12°$ provides an appreciable 2.6 % further increase that cannot be achieved when using 95 RON, because detonation progressively disturbs the normal combustion.

Detonation is a type of abnormal combustion, due to auto-ignition or spontaneous ignition, typical of spark ignition engines. As shown, it limits the increase of the compression ratio when looking for increased performances. To avoid detonation, the fuel formulation must assure the correct octane number to allow using the required compression ratio value.

During the regular flame propagation, the flame front compresses and heats the fresh mixture contained in the volume of the so-called end-gases, not yet interested by the combustion. The flame front, almost spherical, is shown schematically in Fig. 10.39(2) while compressing the still unburned end-gases whose temperature increases. This temperature rise is likely to cause the auto-ignition within this volume not yet ignited which contains a mixture that is quite homogeneous and ready to burn. Therefore, differently from the normal combustion propagation, the combustion in the mentioned volume propagates almost instantly, thus determining an explosion. The consequent pressure peaks move with a speed exceeding the speed of sound and, immediately, all the combustion chamber is interested, and heavily affected, by an abnormal pressure situation as shown in Fig. 10.44b.

When running at low-medium loads, detonation can be detected from a typical metallic noise. When the engine load increases the detonation noise is covered by the mechanical and combustion noise and the driver cannot react reducing the load or retarding the advance and, when these conditions occur, irreversible mechanical damages are likely to take place due to high pressure shocks associated with the high load.

A typical example of these mechanical damages is the piston top perforation; it may take place when the pressure peaks approach or, even more, anticipates the TDC.

Detonation is monitored by a control system which includes a vibration sensor acting together with the electronically controlled ignition of the engine management system. When the knock detection takes place, the ignition timing of the interested cylinder is retarded, until the knock disappears; in so doing the fuel consumption increases but mechanical damages are avoided. This system has the advantage of allowing calibration of the spark advance mapping for a good quality gasoline, 95 or 98 RON, but, in case this quality is not available, it also allows running with a lower quality e.g. 91 RON, even if with some loss in performance and an increase in fuel consumption.

The knock sensors are mounted on the cylinder block between the pairs of cylinders, where vibration is best amplified by the thin walls of the coolant jacket. Four cylinder in-line engines are usually provided with one knock sensor positioned between each two pairs of cylinders. The sensor number increases proportionally with the number of cylinders to recognize the knocking cylinder.

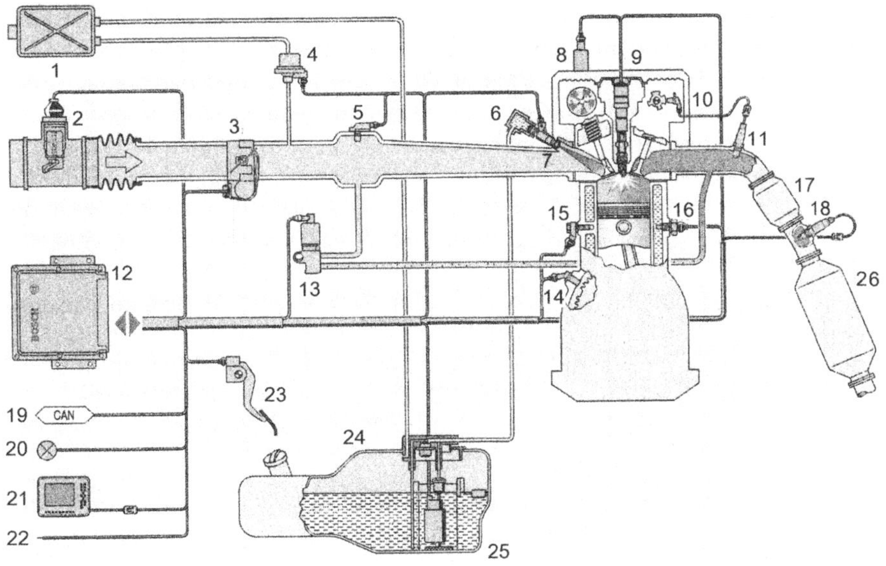

Fig. 10.45 Fuel injection, ignition and emission control system. Injection in the intake manifold or Port Fuel Injection (PFI) (courtesy of Bosch)

The auto-ignition process, now described for spark ignition engines, is a totally different process from the compression-ignition peculiar to diesel engines in which only the start of combustion is determined by compression, and the further combustion proceeds with the usual combustion speed.

10.6.2 Fuel Control System

The complete layout of the fuel injection and ignition systems used on present day engines is shown in Fig. 10.45. Many components have been already described and the figure allows us to comment on their interaction and recognize some functions, as follows:

- Evaporative emission control system with the activated carbon canister (1) and its duty-cycle valve (4) for canister regeneration ;
- fuel injection system, delivering a suitable air to fuel ratio already quite close to the final one. When exhaust gas recirculation is used the air-mass meter (2) senses the mass of air entering into the engine and the absolute pressure sensor (5) is used to evaluate the amount of recirculated gas. When recirculation is not needed, this latter sensor is sufficient to determine the engine load;
- throttle device (3) with the electronic throttle control;

- engine electronic control unit (12). This unit receives the load signal and, sensing the engine speed (14) and coolant temperature (16), defines the opening time of the fuel injectors (7). A first value of the air to fuel ratio is thus obtained, through a pre-determined, open-loop fuel map. To reach and maintain the stoichiometric value, an additional signal from the lambda sensor (11), placed upstream the pre-catalyst (17), is needed. A closed-loop control is thus activated so that the required three-way catalyst efficiency can be reached. The open-loop calibration is active during start-up and higher load running periods when the air to fuel ratio enrichment is requested;
- ignition system with the ignition coil—spark plug assembly (9) and knock sensor (15);
- three-way catalyst, made of a pre-catalyst (17) and a main catalyst (26), which constitutes the after-treatment device;
- exhaust gas recirculation system, when used, with its valve (13) to supply the correct exhaust gas percentage;
- fuel delivery system with the fuel tank (24) and the in-tank fuel pump (25), which delivers the injection pressure;
- fuel rail (6), usually made with reinforced polyamide, to distribute pressure to injectors;
- actuators and sensors (8) for variable valve timing and camshaft phase sensor (10);
- on Board Diagnostic (OBD) system with the lambda sensor downstream the pre-catalyst (18), the speed sensor (14), and the malfunction indicator lamp (20) located on the driver's instrument panel;
- interface with the Controlled Area Network (CAN) (19) and diagnostic tool (21), used off-board in the repair shop, able to identify the malfunctioning component, or function, causing the lamp to switch-on. When a diagnosis is needed, this tool is connected with the car using the standardized connector;
- interface with the immobilizer (22) linked with the ignition key;
- cut-off function to reduce the fuel consumption. It stops the fuel delivery when the accelerator pedal is fully released and the engine speed is higher than a pre-determined value, usually 1,300 rpm. Decreasing the speed, the fuel delivery is resumed to avoid a likely engine stall or torque fluctuations during the following acceleration or pick-up;
- intake duct, shaped in its final part close to the port, so that the mixture flow entering the combustion chamber provides the required turbulence level. A tumble motion (Fig. 10.46) proved effective, since it contributes to improving the mixture homogenization and the combustion stability, even if some losses in the flow capacity of the duct has to be accepted.

The system shown is the Sequential Multi Point Injection (SMPI) in which the timing of the fuel injectors is controlled sequentially with the intake valve opening thanks to the camshaft phase sensor (10). In so doing the injector minimizes fuel accumulation along the intake duct contributing to improve the air to fuel ratio quality and combustion stability.

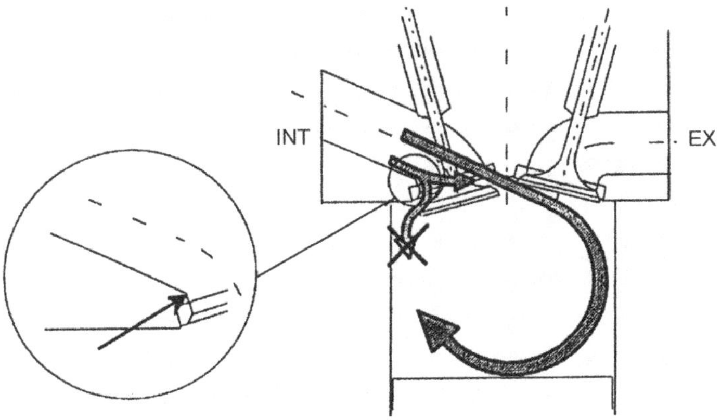

Fig. 10.46 Tumble air motion determined by the intake duct configuration (courtesy of FIAT)

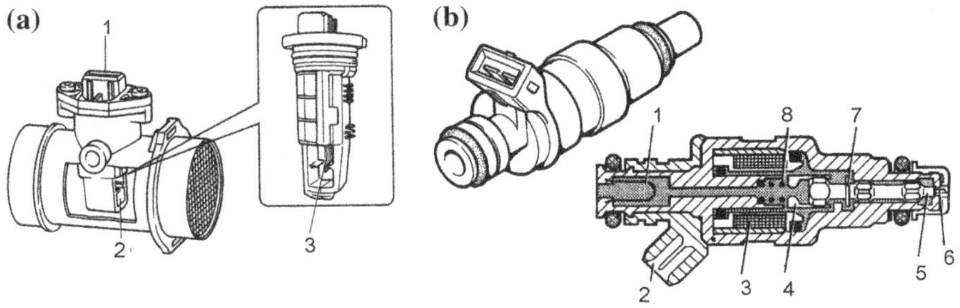

Fig. 10.47 Main injection components. **a** Air mass meter and **b** pintle shaped injector (courtesy of Bosch)

As already mentioned, during cold starting the mixture must be enriched, and the engine coolant temperature sensor modifies the injector opening time, according with the engine temperature to provide a richer mixture, $\lambda \leqslant 1$.

With reference to Fig. 10.45, the multipoint port fuel injection configuration features a drive by wire system in which there is no longer a mechanical link between the accelerator pedal (23) and the throttle valve (3). The link is electrically actuated: A potentiometer detects the pedal position and sends the signal to the engine control unit as an analog voltage signal. Based mainly on the load and speed sensor inputs, the control unit recognizes the load request and allows opening of the throttle, which is actuated by a servomotor integrated in the throttle body. Among the other capabilities, this system allows the idle speed to be maintained at its required value, for instance, whichever accessory is active (air conditioning, power steering, etc.).

Some details of the air mass meter (a) and of the fuel injector (b) are shown in Fig. 10.47. In the air mass sensor, the air stream, entering the engine, flows through the tubular meter housing in which a bracket (2) supports an element, the hot-wire (3), electrically heated at a temperature of about 120 °C. The incoming air flow cools

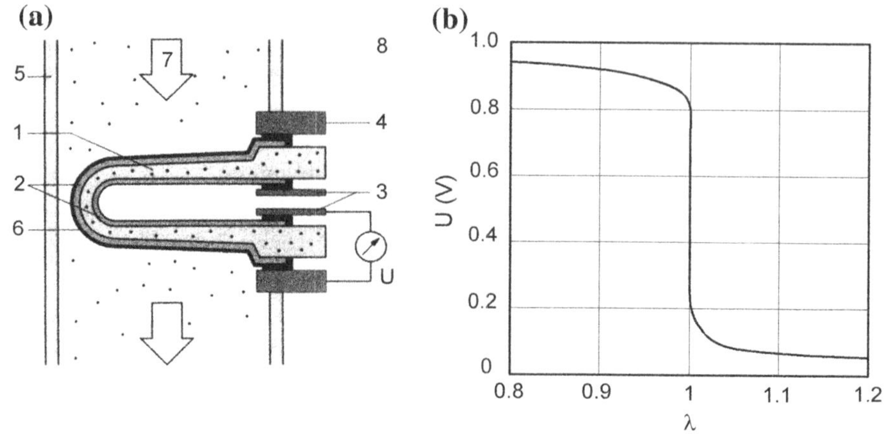

Fig. 10.48 Lambda sensor (**a**) and lambda sensor signal (**b**) to activate the closed-loop control (courtesy of Bosch)

the hot wire, so that by measuring the current needed to keep the temperature of the wire constant, a measurement of the air mass introduced into the engine is obtained. This electric signal is transmitted to the electronic control unit through connector (1).

The task of the injector is supplying the correct fuel quantity to the engine and atomizing it. The fuel is filtered when entering the injector (1). The fuel injection takes place thanks to the motion of the needle (5) mechanically connected with the magnetic nucleus (4), which slides back driven by the solenoid winding (3). The latter receives the current pulses through the connector (2). The needle (5) is kept in closed position by the spring (8) which pushes it in contact with the needle seat and its stroke can be controlled by the calibrated disc (7). The pintle shape (6) produces the final configuration of a tapered spray typically used in engines with one intake valve per cylinder. The tapered spray is directed into the opening between the intake valve head and the valve seat. Dual-spray formation is used in engines with two intake valves per cylinder.

The lambda sensor measures the content of residual oxygen concentration in the exhaust gases, comparing this variable value with the constant oxygen concentration in the external air, 21 %. The difference between the ideal value (O_2 content is 0.7 % for $\alpha = 14.7$) and the actual value is immediately monitored as a two-valued signal typically switching from 200 mV, lean side, to 800 mV, rich side as shown in Fig. 10.48b. When it recognizes that the combustion is not perfectly stoichiometric, a correction is applied starting from the next fuel injection and lasting up to the moment in which stoichiometric mixture is again restored. This continuous correction provides the closed-loop control, made possible by the fast lambda sensor response that oscillates between the rich and lean electric outputs with a frequency close to 3 Hz.

The lambda sensor configuration, introduced in the 1970s, is shown in Fig. 10.48a. The sensing element (1) is a solid state electrolyte which features a ceramic assembly, containing zirconium and yttrium, shaped as a tube closed at one end. The inner and outer surfaces of this element are covered with a porous platinum coating acting as an electrode (2), able to collecting the voltage differential between the contacts (3) and (4) that takes place when the actual value of the residual oxygen in the exhaust (7) changes from the ideal value. As mentioned, the reference oxygen value is that of the external air (8) entering in the tube. The sensor is assembled in a housing screwed into the exhaust pipe (5) and protected against contamination by a specific ceramic layer (6). The ceramic assembly is further protected against mechanical impacts and thermal shocks by a slotted metal tube.

This sensor is also known as the Exhaust Gas Oxygen (EGO) sensor. It only works when heated at approximately 300 °C so that a heating element is encased into the ceramic to bring its tip quickly to the required temperature level. Beginning in year 2000, the planar type sensor entered the market featuring a reduced size and incorporating the heater within the ceramic structure.

Lambda sensor deterioration is determined by its ceramic structure modification, induced by high temperatures and by chemical poisoning. Deterioration is monitored by the European On Board Diagnostic (EOBD) system which primarily detects the slower frequency of the rich/lean oscillation.

10.6.3 Three-Way Catalysts

The exhaust emission standards mandatory in Europe as Euro 1, that became operational in January 1993, required the application of the catalytic converter on all cars powered by spark ignition engines. The after-treatment of exhaust gases, or catalytic post-combustion, became thus compulsory. The goal of the three-way catalyst is to activate suitable chemical reactions able to reduce simultaneously the concentration of CO, hydrocarbons and NO_x. Three-way catalysts, so called because of the three regulated pollutants, are also efficient on a number of unregulated chemicals, like polynuclear hydrocarbons, aldehydes and benzene. As already mentioned, the multipoint, port fuel injection system, operating with homogeneous and stoichiometric mixture, is totally free from particulate matter emissions, and therefore no device is need to control this pollutant.

The catalyst volume is proportional to the engine displacement. A parameter defining the size of the catalyst is the so called *spatial velocity*, i.e. the ratio between the exhaust gas flow (in m^3/h) and the catalyst volume (in m^3). It is measured in h^{-1} and is usually in the range 100,000–150,000 h^{-1}. As an average, the catalyst volume is approximately 1.2 times the engine displacement.

The catalyst shown in Fig. 10.49, is assembled using an external stainless steel container with an internal substrate made by thousands of parallel channels, determining a cell density usually spanning from 400 to 600 cells/sq. inch (62–93 cells/cm^2), for a total of about 4,000 cells with an overall area of 3 m^2 in a catalyst of 1.5 l.

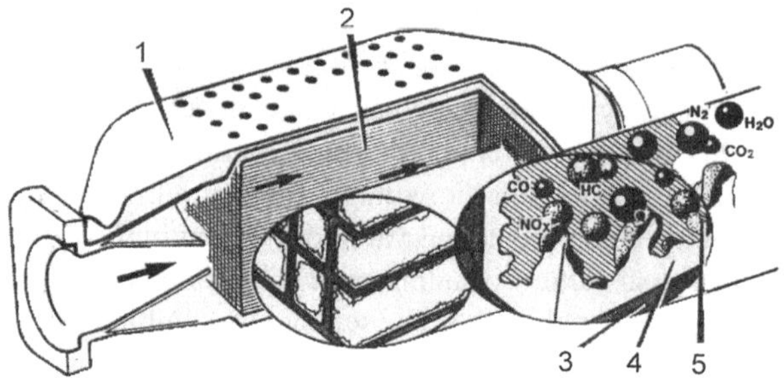

Fig. 10.49 Three-way catalytic converter to reduce CO, hydrocarbons and NO_x. External metal housing (*1*) and ceramic substrate (*2*) (courtesy of FIAT)

The wall thickness between contiguous cells is close to 0.13 mm. The number of cells can rise to 1,200 cells/sq. inch with a wall thickness reduced to 0.05 mm. The internal substrate is generally based on a ceramic material, Cordierite, thus obtaining a monolithic ceramic body. For special applications it is possible to use a metallic monolith, featuring a corrugated metal foil 0.05 mm thick, with the advantage of a lower resistance to the exhaust flow, thus reducing the exhaust backpressure. The external metal housing and the substrate are separated by a mineral swell mating having the goal of protecting the substrate from shocks and vibrations and, when ceramic is used, to compensate for the different thermal expansion.

The surface of both type of substrates (3) is impregnated with a so-called *wash-coat* (4), a ceramic material whose basic component is a very porous γ-alumina (Al_2O_3), on which crystals of noble metals (5) (platinum, palladium and rhodium) are finely dispersed. The specific surface area of the wash-coat, crystallized as γ-alumina, is 20 m^2/g, which produces an impressive increase of the total surface area up to 20,000 m^2, therefore bringing the exhaust gases in close contact with the noble metals. The loading of the latter is measured in grams per cubic feet, and lies in a range of 50–150 gcf. A 1.5 l catalyst contains from 2.5 to 8 g of noble metals with widely different proportions among the three of them. The cost of platinum and rhodium varies widely, but is not far from the cost of gold, while palladium is cheaper. Noble metals act as catalysts and therefore they do not participate in the following reactions:

- Oxidation of CO and hydrocarbons, with formation of CO_2 and H_2O, promoted by platinum and palladium

$$2CO + O_2 \rightarrow 2CO_2, \qquad 2C_2H_6 + 7O_2 \rightarrow 4CO_2 + 6H_2O.$$

- Reduction of NO_x, with formation of N_2 and O_2, promoted by rhodium:

$$2NO + 2CO \rightarrow N_2 + 2CO_2, \qquad 2NO_2 + 2CO \rightarrow N_2 + 2CO_2 + O_2.$$

The catalyst action starts taking place when its temperature exceeds the so-called light-off temperature. When the catalyst temperature exceeds 300 °C, the conversion efficiency reaches 99 % as shown in Fig. 10.50a. Hydrocarbons and CO conversion

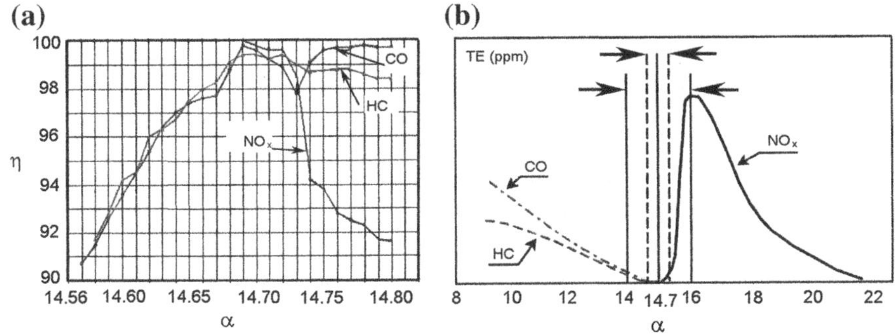

Fig. 10.50 Three-way catalyst conversion efficiency (η) around the stoichiometric mixture (**a**) and its effect on tailpipe emissions (*TE*) with a first and second generation of air/fuel control (**b**) (courtesy of FIAT)

efficiency is higher with higher oxygen concentration in the exhaust gases, therefore with $\lambda \geqslant 1$. On the contrary, NO_x are reduced to N_2 and O_2 in presence of oxygen deficiency, therefore when $\lambda \leqslant 1$. The same figure shows how important is to maintain the air to fuel ratio as close as possible to the stoichiometric value, reducing the width of the so-called stoichiometric window, determined by the air to fuel ratio control (Fig. 10.50b). The residual oxygen is used to complete the oxidizing reactions and, when the oxygen has been depleted, it is possible to start the NO_x reduction. If and when this condition is satisfied, the concentration of the three pollutants is reduced to zero as shown in Fig. 10.50b. This evidence explain why some engine modifications, primarily ignition retard, become no longer necessary by improving the electronic control of the air to fuel ratio, thus favouring a better fuel economy.

Starting from Euro 3, using on-board diagnostic systems, the stoichiometric window has been widened adding cerium in the catalyst formulation. This metal can store and release a certain quantity of oxygen so that:

- When the mixture drifts on the rich side, some oxygen is released to maintain the oxidizing reactions, and
- when the mixture drifts on the lean side, some oxygen is retained and stored thus still allowing the reducing reactions.

Catalyst technology is based on continuous developments and improvements of the ceramic substrate and of efficient washcoats. The former must better withstand thermal and mechanical stresses without mechanical failures, while the latter must promote lower light-off temperatures and higher conversion efficiencies.

Catalysts deterioration is due to the following reasons:

- High temperature exposition. The normal running temperature is in the range of 400–800 °C. At higher temperatures, γ-alumina crystals increase their size, causing reduction of the area exposed to exhaust gases thus reducing the catalyst efficiency (thermal aging). When the temperature exceeds 950 °C, the γ-alumina structure is heavily modified (γ-alumina transforms into α-alumina) with a strong reduction of the specific surface area. With a further temperature increase, at about 1,400 °C, the melting point of the ceramic substrate is reached.

- Accumulation of heavy metals from gasoline. A reduction of the gasoline lead content to minimal values was required to introduce three-way catalysts. The first Euro 1 level of 13 mg/l was further reduced, and now stabilized, to 5 mg/l with Euro 3. Similar reduction for sulphur from 500 to 50 mg/l and then to 10 mg/l was needed to meet the more strict requirements of new deNO$_x$ catalyst types. Their oxides are easily trapped within the washcoat porosity, resulting in a surface area reduction, thus preventing the effective contact between noble metals and pollutants.
- Accumulation of phosphorus and zinc contained in the lubricant oil: Oil consumption must be reduced to avoid a similar deterioration of the washcoat due to the same above mentioned reason.

Within the catalyst volume, temperatures exceeding 950 °C can be reached only during misfiring in one cylinder, when the catalyst is reached by an unusual quantity of unburned fuel and air which, instead of burning in the cylinder, is converted by the catalyst with an impressive temperature rise. This consideration stresses the importance of providing reliable ignition systems, built using high quality components (spark plugs, ignition coil, cables and relevant connections) and to detect even low misfiring percentages with the on-board diagnostic system.

Comparing the engine-out emissions E_e (system without catalyst) with the tailpipe emissions E_t (system with catalyst), it is possible to determine the catalyst efficiency measured during the complete type-approval cycle, which includes cold starting. Typically, a Euro 5 multipoint, port fuel injection engine produces a CO engine emission close to 6 g/km with a tailpipe emission close to 0.3 g/km; the overall efficiency of the catalyst is thus

$$\eta = \frac{E_e - E_t}{E_e} = 95\,\%. \tag{10.13}$$

The charge quality obtained with the port fuel injection is strongly beneficial to halve the CO engine-out emissions with respect to the best level achievable using carburetors: 12 g/km. This is a pre-requisite to obtain a long-lasting catalyst efficiency, able to meet the values for which the car has been initially type-approved, since the oxidizing reactions, responsible for the temperature increase in the catalyst, are thus limited. This is the result of the technology-forcing legal requirements which call both for reduced emissions in 'new' conditions and also for a long durability: Respectively 80,000, 100,000 and 160,000 km for Euro 1/2, 3/4 and 5/6. The in-use compliance and roadworthiness programs, activated by each Government, force the Customer to submit the catalyst to periodic efficiency checks, usually every two years.

An Euro 1/2 catalyst, suitable to be installed under the floor, is shown in Fig. 10.49. To satisfy the Euro 3 and Euro 4 standards, the catalyst must be installed immediately downstream from the exhaust manifold, implementing the so called close coupled (or cascade) configuration. When the volume available in the engine compartment is not sufficient for the catalyst, the more costly solution of a pre-catalyst and a main catalyst (Fig. 10.45) is used: it includes a small pre-catalyst on the exhaust manifold and the remaining catalyst volume located in the under floor position (main catalyst). Euro 5/6 configurations require exceeding the 90 % efficiency after only 50–150 s when starting with an ambient temperature of 20 °C; this result is achieved with suitable light-off temperature and engine calibrations during cranking and start up. After this time span a temperature of 300 °C is exceeded and the required efficiency is reached.

The above mentioned exhaust configuration has been used since Euro 3, January 2001, when also the European on-board diagnostic system was initially implemented. Its aim is to alert the driver of the reduced catalyst efficiency before reaching the date in which the roadworthiness control detects that the catalyst has worn out. The monitoring system requires adding a lambda sensor downstream from the catalyst, to compare the electric signals of the two sensors, monitoring the efficiency of the catalyst and warning the driver of the need to replace the worn-out catalyst.

An efficient catalyst uses all the residual oxygen to complete the oxidation reactions, so that the downstream sensor detects no-oxygen. When its efficiency starts lowering, the oxygen is no longer completely used, and the downstream sensor signal becomes progressively closer to the upstream signal.

10.6.4 Fuel Supply and Evaporative Emission Control System

Since Euro 1, the fuel feed system has been integrated with the evaporative emissions control system.

Figure 10.51 shows the fuel tank (1) usually made in polyamide material, with high mechanical strength and impermeable to fuel vapors. The electric pump (2), delivering the fuel at an average pressure of 3.5 bar, is included in an in-tank integrated assembly which also contains the fuel pressure regulator, the fuel-gauge sensor and the suction strainer for a first filtering stage.

A single feed line (3) connects the fuel tank to the injector fuel rail with the injectors (4). The main feature of this layout, representing the multipoint, port fuel injection system used since the second half of the 1990s, is the absence of the fuel return line previously connecting the rail to the fuel tank, still used on the first generation of multipoint injection systems.

When the temperature of the fuel rises, due to the day-night temperature variations or other causes, a certain quantity of hydrocarbons evaporates from the liquid gasoline surface. To prevent the pressure from increasing too much, thus controlling the tank pressure, the ventilation bleed valves (6) collect the hydrocarbon vapors and send them into the canister (5) containing the active carbon. The latter is characterized by an extremely high porosity and is able to absorb the light hydrocarbons, defined as Volatile Organic Compounds (VOC). The canister contains, as an average, 1 l of active carbon for a 50 l fuel tank. As already mentioned, the canister, trapping

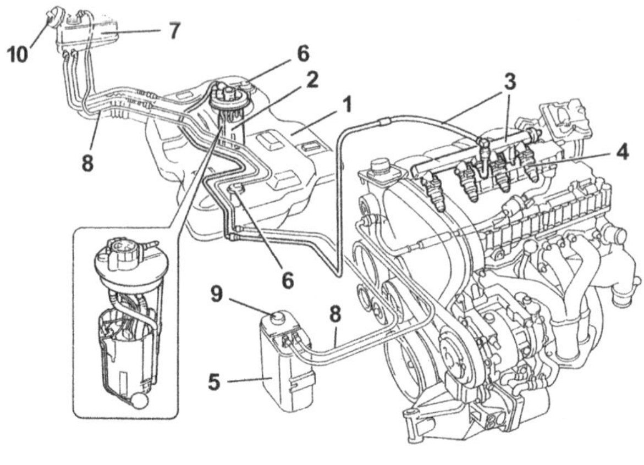

Fig. 10.51 Fuel feed system: the in-tank fuel pump delivers the fuel to injectors. Evaporative emission control system stores the fuel vapors, released by the fuel tank, into the activated carbon canister (courtesy of FIAT)

the hydrocarbons, avoids discharging them to the atmosphere with the ensuing air pollution.

The fuel vapors reach the active carbon canister through one or more bleed valves installed in elevated points of the fuel tank, from which the vapor, substantially free of liquid, can be collected. The liquid-vapor separator (7), positioned close to the filling cap (10), provides the return of the small amount of liquid fuel into the tank and allows only the vapors to reach the canister through the line (8). The liquid gasoline that might reach the active carbon can destroy its capability of absorbing the fuel vapors.

During engine running, part of the intake air is drawn through the canister inlet and this fresh air flow (9) desorbs the accumulated vapors purging the canister, thus regenerating the active carbon that can store 50–70 g of fuel vapors for each liter of carbon. The system efficiency is comparable with that of three-way catalysts, reducing the volatile organic compounds by at least 95 %.

The system fluidic resistances and valve calibrations assure that the fuel tank works with limited positive pressure. Only in unusual conditions (abnormal temperature variations or shocks) does a safety valve, located on the fuel feed hose or on the liquid/vapors separator open, avoiding high positive or negative pressure that may jeopardize the fuel tank's mechanical integrity (explosions or implosions).

10.6.5 On Board Diagnostics

As already mentioned, since January 2001, Euro 3 regulations required the European On Board Diagnostic (EOBD) system to be installed with the aim of monitoring the emission-related components and systems that, in case of failure, allow the emission level to exceed the stated thresholds when driving according to conditions close to the NEDC cycle. The emissions of CO, hydrocarbons and NO_x are monitored, while no particulate matter control is needed considering that multipoint, port fuel injection systems are completely free from this pollutant.

The system must alert the driver by activating the dashboard malfunction indicator lamp.

The key monitored functions or components, deemed as responsible for abnormal emissions are:

- Low catalyst efficiency,
- low lambda sensor efficiency,
- misfiring, i.e. loss of firing events, and
- emission related powertrain components, which are connected to a computer.

The monitoring of the evaporative emission control system only verifies the correct fuel filling cap installation.

10.7 Direct Injection Spark Ignition Engines

Historically the application of the Gasoline Direct Injection (GDI) configuration was determined by the need to achieve the following goals:

- Fuel consumption reduction, and
- performance improvement in aero and racing engines

During the first decades of the last century these solutions were developed on a number of prototype engines. In both cases remarkable results were obtained at the cost, from case to case, of huge increases of price or poor reliability. At the end, in spite of many attempts, the above solutions remained marginal in the automotive sector up to the point in which reliable and affordable electronic engine management systems became available only at the beginning of the 2000s.

The first application to achieve the first goal was obtained by Mitsubishi that, in 1997, launched on the market a medium-segment sedan powered by a 1.8 l GDI engine.

The first application to achieve the second goal dates back to 1954, with a Mercedes Benz top level sport car, the 300SL, with a 3 l engine, sporting outstanding performances. At that time the injection was provided by a mechanical pump following the mechanical configuration used on the best performing World War II aero engines. In the following decades no other engine was put in production with this system.

Today, when seeking for the above mentioned goals, the so called "stratified charge mode" or the "homogeneous charge mode" are implemented as shown in the following sections.

10.7.1 System Layout and Performances

This section outlines the major differences with respect to port fuel injection, briefly describing the mechanical components which allow obtaining specific engine performances able to accomplish the previously mentioned goals.

The gasoline direct injection system allows:

- A better cooling of the combustion chamber, thanks to the direct evaporation of the gasoline spray which, in turn, reduces the risk of detonation, allowing an increase in the compression ratio. In fact, at equal fuel octane numbers, gasoline direct injection systems operate with higher compression ratio than port fuel injection ones. A compression ratio increase from 10.5 to 11.7 was possible for naturally aspirated gasoline engines available in Europe in 2010, tuned for the 95 RON unleaded gasoline. Similarly, an increase from 10.1 to 11.1 was recorded when using 91 RON gasoline and from 11.0 to 12.1 when using 98 RON gasoline. With reference to Fig. 10.4, the gain in theoretical efficiency is in the range 2.5–3 %.
- To achieve a better fuel distribution between the cylinders, since the air-fuel mixture is no longer affected by the liquid fuel condensation along the intake ducts during the acceleration phases, when the intake pressure reaches its absolute minimum level.
- To increase the volumetric efficiency, since the air flow to the cylinders is no longer reduced by the volume of gasoline evaporating before entering the combustion chamber to provide the homogeneous charge. The estimated gain in volumetric efficiency is close to 5 %. With the same mixture strength, the above mentioned efficiency gains contribute to the overall efficiency improvement.
- To lean of the mixture up to $\lambda \approx 2.5$, thanks to the stratified charge. This mixture setting determines, under steady-state conditions, a fuel consumption reduction close to 15 %. This value is far different from the fuel consumption reduction achievable during an homogeneous combustion in which the reduction is limited to a few percent when moving from the stoichiometric air to fuel ratio to the lean side as shown in Fig. 10.40. The new value is more similar to that recorded for compression-ignition engines. When running the transient conditions of the NEDC emission cycle for a naturally aspirated engine in stratified mode, the fuel consumption reduction is limited to an average value close to 10 %. This final value includes all the efficiency benefits previously mentioned.

To obtain the above mentioned improvements many mechanical components must be modified with significant cost increases that are further boosted by the need to apply the so called, $deNO_x$ catalyst, in addition to the three-way one, when looking for a fuel consumption reduction. At the end both the fuel consumption reduction and the cost increase are about half those recorded when switching from a naturally aspirated port fuel injection to a common rail diesel.

In spite of the mentioned advantages, and based on the already mentioned survey of 450 engine configurations available in 2010, the higher costs, linked to innovative injection and ignition components, keep the percentage of gasoline direct injection

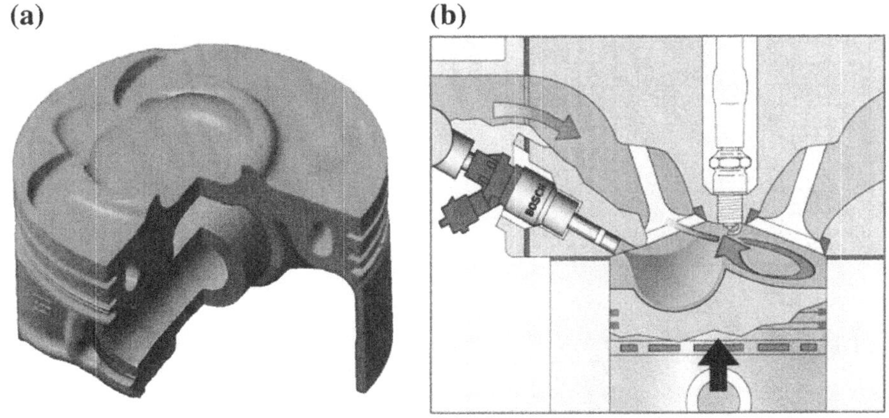

Fig. 10.52 Gasoline Direct Injection (GDI): piston shape (**a**), fuel spray and tumble formation (**b**). A suitable air to fuel ratio to the spark plug must be delivered for a correct combustion (**a**: courtesy of FIAT, **b**: courtesy of Bosch)

with respect to port fuel injection engines as low as 19 % for naturally aspirated engines (320 engines considered) and of 50 % for turbocharged engines (130 engines considered).

For the more powerful turbocharged versions, the overall cost increase is more acceptable and the turbocharger allows a significant engine downsizing with an improved fuel economy that can be twice the mentioned value up to a remarkable 30 %. This latter value is achieved when, in addition to the lower displacement, there is also a reduction in the number of cylinders: e.g. from 6 to 4, from 8 to 6, from 10 to 8. These reductions are attractive but applied to powerful and costly configuration whose market share is only a few percent. The emission standards are met running the engine with a homogeneous charge, therefore using the three-way catalyst technology. For example, the above mentioned configuration is applied to engine n° 5 in Table 10.2 and, under these conditions, the engine power density, i.e. the power to weight ratio, increases from a typical value of 0.6 kW/kg, for a naturally aspirated engine, to a remarkable 1.1 kW/kg.

The main differences in the components of the injection and ignition systems between port fuel injection and gasoline direct injection engines are here listed:

- The intake duct profile and piston design are shaped to produce an air motion which can widely vary according to the combustion process, stratified or homogeneous. An example of the incoming air flow and fuel spray in which the latter, deflected by the piston bowl, moves upward in the direction of the spark plug and meets the air flow which, during induction, already started a tumble motion, is shown in Fig. 10.52. This motion must be maintained up to the point at which both streams reach the spark plug.
- The two-steps lambda sensor is no longer sufficient and the new, more complex, broad-band lambda sensor is required as shown in Fig. 10.53a when looking for the

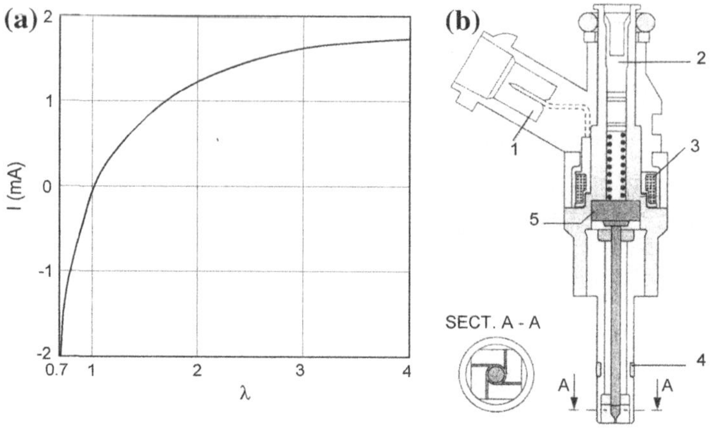

Fig. 10.53 Gasoline direct injection broad-band oxygen sensor to control the air to fuel ratio in the lean area (**a**). Fuel injector with swirl type geometry (**b**) (courtesy of Bosch)

stratified charge operation The broad-band sensor, introduced in the mid 1990s, is also called Universal Heated Exhaust Gas Oxygen (UHEGO) sensor, and differently from the traditional one, this sensor provides a continuous current signal, directly proportional to the oxygen content. In fact it is negative on the rich side ($\lambda = 0.7 - 1.0$), and positive on the lean side ($\lambda = 2.0 - 3.0$), Fig. 10.53a.

- The injector shape must reach the combustion chamber, as shown in Fig. 10.53b. Cross section A-A outlines the swirl type injector geometry in which the orifices, tangential to the vertical stream, split the jet and promote its quick mixing with the air stream. The other components are the usual ones: electric connector (1), fuel inlet (2), solenoid winding (3), magnetic nucleus (4) and gasket (5).
- The fuel pump which must provide a pressure in the order of 200 bar. It is driven by the camshaft and has a sensor to control the pressure. It discharges the excess of fuel through the return line to the tank. The mechanical configuration of this pump is often similar to that shown when describing diesel engine technologies.
- The fuel rail must be made of stainless-steel and be designed to allow the above mentioned maximum pressure values with the requested safety margins. The engine control unit, managing the pressure regulator, modulates the rail pressure according to the engine running conditions, within a pressure range 80–200 bar.
- The ignition system must be able to supply a higher energy to provide the spark even when liquid gasoline reaches the electrodes or when a second spark is needed due to the breakdown of the first determined by strong air motion. As already mentioned the energy typically increases from 50 to 100 mJ.
- The deNO$_x$ catalyst is mandatory when using stratified charge.
- The electronic control unit must have an increased software capability.

The evaporative emission control components are basically similar to the ones described for port fuel injection systems, with only marginal cost increases; the in-tank electric pump supplies the fuel to the high pressure one.

Also the on-board diagnostic strategy is similar to port fuel injection systems with the specific modifications requested depending on whether the charge is stratified or homogeneous.

10.7.2 Stratified Charge Mode

As already mentioned, many efforts were addressed to widen significantly the overall air to fuel ratio field in which a spark ignition engine can correctly operate. The final aim is to control the engine load similarly to compression ignition engines thus avoiding air throttling with consequent losses. To achieve this goal, the combustion sequence must offer a first phase in which the combustion takes place within a cloud of rich mixture and then propagates within the leaner area. Traditionally, to achieve this sequence a divided combustion chamber was often used. The first phase takes place in a small-volume chamber and then the lean combustion propagates within the main chamber thus following the scheme used in the pre-chamber diesel engines, Fig. 10.58a. This shape unavoidably introduces significant efficiency losses due to the gas lamination between the divided combustion chamber. Therefore this solution was progressively abandoned in favour of the open chamber solution which offers a better overall efficiency but requires a sophisticated control to obtain a stable combustion.

To outline the intrinsic difficulties of this new solution, it may be worth mentioning that in the port-injection solution the fuel is injected during the intake stroke, and therefore almost 300° of crankshaft angle are available to reach the complete stoichiometric and homogeneous fuel distribution: In practice almost one complete crankshaft revolution, something between 30 and 10 ms at 2,000 and 6,000 engine rpm, respectively.

In the direct injection stratified operating mode, applicable only to moderate engine load and speed, the injection takes place much later, namely during the compression stroke, whereby typically only 60° of crank angle are left for mixture preparation. At the engine speed of 2,000 rpm this corresponds to a time slot of only 5 ms and during this short time the fuel should vaporize and mix with air to provide a rich mixture cloud that can be ignited to start the combustion process. This cloud has to be directed over the spark plug.

When the combustion has been started, the fuel mixture becomes very lean, thus promoting the expected lower fuel consumption. This latter goal is also favoured by the reduction of the wall heat losses because the combustion takes place at the center of the combustion chamber surrounded by an insulating air-only atmosphere.

When compared with the homogeneous charge, after the ignition the stratified mode offers a faster process due to the higher flame propagation velocity arising from the rich cloud formation around the spark plug, thus providing a steeper mass fraction burned but, during the subsequent combustion phase, the process becomes much slower. In the above conditions the load can be adjusted by enlarging and shrinking the fuel cloud thus avoiding or reducing throttling operation with the associated pumping losses finally achieving, as already mentioned, in the most favorable steady-state conditions, a fuel consumption reduction close to 15 %.

Detonation cannot take place, since the more ignitable mixture is present only in the area close to spark plug electrodes and the remainder of the volume is filled with such a lean mixture that it prevents auto-ignition. The local combustion temperature is, however, high and an exhaust gas recirculation system is required to provide a first NO_x reduction since the three-way catalyst is inefficient in reducing NO_x, mostly owing to the high concentration of residual oxygen. The exhaust gas recirculation alone cannot limit NO_x emissions at the requested legal level, therefore a new deNO_x catalyst, able to reduce nitrogen oxides in spite of the high oxygen concentration, is needed. This catalyst is rather complex as it will be shown in the relevant section.

However, as the engine load increases, the stratified charge mode no longer complies with the engine performance requirements. At this point, switching to homogeneous operating mode is required. Finally, the maximum performance is achieved activating the homogeneous and than the rich mixture setting. During the switch-over from lean to rich, the drive by wire system suitably modifies the throttle opening to avoid that the driver has to compensate for unavoidable torque increase due to the mixture change.

10.7.3 Homogeneous Charge Mode

Starting from year 2000, a costly but interesting solution, open to further positive evolution, is integrating the latest turbo geometries, allowing us to achieve the already mentioned impressive power densities, with electronic gasoline direct injection. As a result, the achievable downsizing makes it possible to reach good performances together with a significant fuel consumption reduction.

The engine operates in the homogeneous mode delivering the injection during the induction stroke to allow sufficient time to complete an even atomization and mixing in the cylinder volume thus behaving like the port fuel injection configuration. Stoichiometric mixture and three-way catalyst are used.

This configuration, almost always includes a variable valve timing system which, as already mentioned, allows lowering the turbo-lag effect. The induction pressure in the intake manifold is higher than the exhaust backpressure and therefore the cam adjuster provides the wide overlap required to induce a suitable air flow rate which keeps the turbine at high speed thus obtaining a scavenging effect. The management of the latter option is a complex issue because there is a large difference between the air passing through the engine and the air participating to the combustion. The difference is given by the scavenging air, which depends on the double continuous cam phasers and the turbocharger delivery. During combustion, the air quantity must be only that needed to obtain a stoichiometric mixture, but the air flow through the

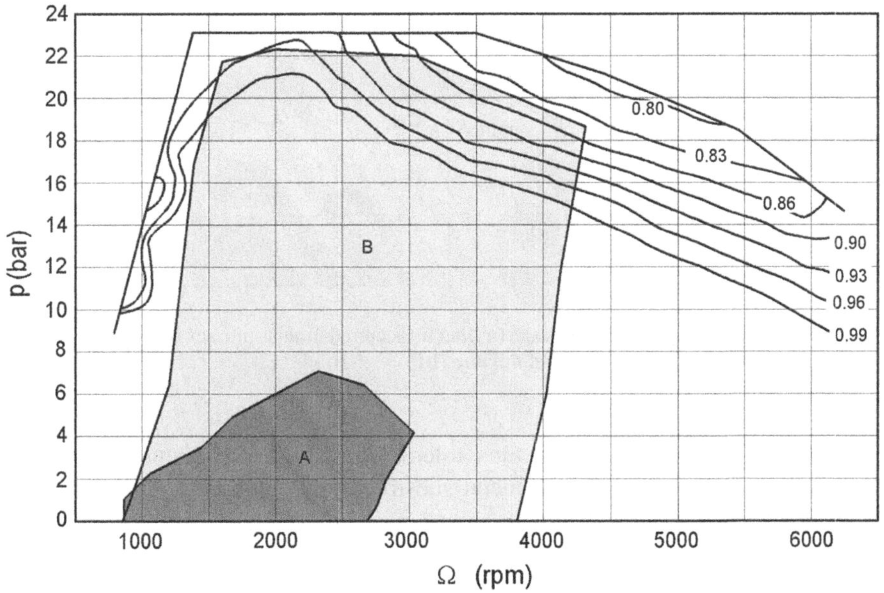

Fig. 10.54 Alfa Romeo 1.8 l gasoline direct injection mixture mapping: stoichiometric mixture in low and medium load area, rich mixture in high load conditions (courtesy of FIAT)

combustion chamber is definitely higher. The correct management is achieved with additional sensors and with model-based strategies.

The mean effective pressure map in Fig. 10.54 refers to the 2009 Alfa Romeo 1.8 l turbo direct injection engine and shows:

- A low load area (A) interested by the NEDC cycle for emission control characterized by the speed-load combination with $\Omega = 2{,}000\,\text{rpm}$ and $p_{me} = 2\,\text{bar}$, as already mentioned;
- a wide area (B) in which the engine still runs stoichiometric, and
- a high load area in which the mixture is enriched.

The mixture is enriched only to obtain the maximum performance, according to Fig. 10.37; this is quite different from the practice common in the 1980s for turbo engines, in which the exhaust temperature was limited by the extra fuel evaporation thus increasing fuel consumption.

To quantify how effective is this downsizing, the maximum value of the mean effective pressure, 23 bar, shown in the figure can be compared with the previously reported values for naturally aspirated engines, namely 10–12 bar. In addition, as shown, the maximum torque delivery is obtained from 1,400 rpm. The increase of the power density requires an improved mechanical strength. The circumferential gallery below the piston crown to protect the ring assembly is shown in Fig. 10.52. Moreover, the exhaust temperature can exceed 1,000 °C, thus requiring cast steel with an addition of 10 %, or even 20 %, of nickel for the exhaust manifold and the

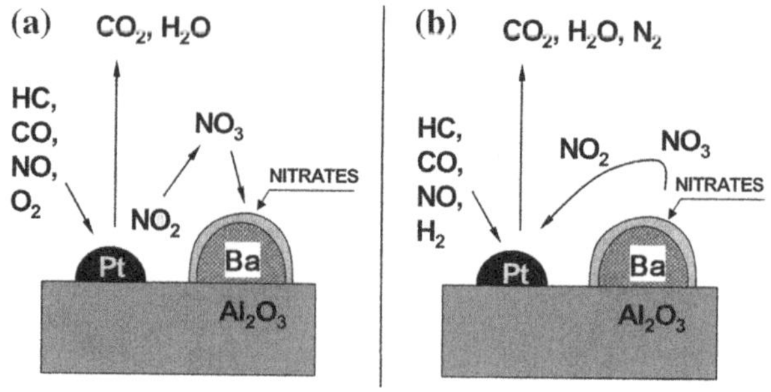

Fig. 10.55 DeNO_x catalyst NO_X storage (**a**) and its accumulation as nitrates on a barium surface. Regeneration phase with rich air to fuel mixture (**b**)

turbine housing. Also the catalyst must tolerate these high temperatures, so that the best materials, often including a metal substrate, must be used to meet the legal durability requirements.

10.7.4 Emission Control

The emission standards applied to gasoline direct injection engines are those applicable to spark ignition engines, adding the particulate standard defined for compression ignition engines. The spray injected to activate the stratified charge unavoidably faces the pyrolysis risk, causing a certain amount of particulate matter. Experience gained so far demonstrates that, differently from diesel engines, a specific after-treatment device to limit particulate matter may be not needed. As a consequence, an excellent mixing between the fuel and air during the injection process and properly designed injectors, of swirl or multihole type, are required.

The injector shape and a correct mixing are a prerequisite to meet the hydrocarbon standards avoiding wall wetting, responsible for the formation of an increased hydrocarbon layer along the cylinder surface with the additional risk of oil dilution.

As already mentioned, when running with stratified lean solution, an NO_x after-treatment system, traditionally referred to as deNO_x catalyst, is mandatory. It behaves like a NO_x adsorber, showing a geometrical shape similar to that of the three-way catalyst but they behave similarly to a trap or to a storage device able to accumulate NO_x in a net-oxidizing atmosphere and being, later on, regenerated. They are now identified as Lean NO_x Trap (LNT) or NO_x Storage Catalyst (NSC). The usual oxidizing noble metals, basically platinum, provide a further oxidation of NO to NO_2, which could be adsorbed by an alkaline earth element based on barium (Fig. 10.55). This rare earth element, by virtue of its basic properties, is able to form sufficiently stable nitrates starting from NO_2. Barium is bonded to a catalyst like barium carbonate and, due to the oxygen excess, can also be present as barium oxide, and therefore the NO_x, emitted during lean operation, accumulates as nitrates. The storage reactions can be simplified according to the following scheme:

$$BaCO_3 + 2NO_2 + \frac{1}{2}O_2 \rightarrow Ba(NO_3)_2 + CO_2$$

$$2BaO + 4NO_2 + O_2 \rightarrow 2Ba(NO_3)_2.$$

Accumulation takes place up to the point in which the barium is still able to store it. When the storage capability is depleted, to avoid an uncontrolled NO_x release, the alert can be raised using a model-based software or a specific NO_x sensor downstream from the catalyst which activates the regeneration strategy. This signal causes a switch-over of the engine mixture from stratified to homogeneous rich, e.g. $\lambda \leq 0.8$, to desorb the accumulated nitrates and reduce NO thanks to the reducing components. They are CO, hydrocarbons and H_2; the concentration of the latter being, as an average, half of that of CO. As an example, the following reactions show the removal of the nitrates accumulation and NO reduction performed by CO:

$$Ba(NO_3)_2 + 3CO \rightarrow 3CO_2 + BaO + 2NO, \qquad 2NO + 2CO \rightarrow N_2 + 2CO_2.$$

During this process, unwanted nitrous oxide (N_2O) and ammonia (NH_3) can be produced; their formation is avoided with a suitable process management and catalyst formulation. The need for periodical switching to rich or stoichiometric conditions for catalyst regeneration requires a complex engine control strategy in which the broad-band lambda sensor must cover a lambda range from 0.7 to 3.0.

As a consequence, the accumulation and regeneration periods strongly depend on the engine load, and therefore from the baseline engine emissions. Their likely values are respectively 60 and 2 s. The NO_x accumulation capability depends on the temperature level and reaches its maximum in the range 300–400 °C; it is thus much lower than the optimal values of the three-way catalysts. It means that the lean NO_x trap must be separated from the three-way catalyst and, usually, positioned downstream the first.

The presence of sulphur contained in the gasoline produces an additional difficulty in lean NO_x trap application, due to the link between barium and SO_3, which is stronger than the one with nitrogen dioxide (NO_2). The accumulation of sulphur compounds, resistant to high temperatures, causes a fast decrease of the NO_x trapping efficiency. For the above reason, the gasoline sulphur content was reduced from the previous 2005 level, 50 mg/l, to 10 mg/l starting from January 2009, to allow introducing this innovative technology. To limit the quantity of stored sulfates, which rises continuously with mileage accumulation, specific de-sulfating strategies must be developed and periodically performed. If the fuel contains 10 mg/l of sulphur, it can be estimated that the regeneration must take place every 5,000 km. During the de-sulfating process, the catalyst must be exposed to a temperature higher than 650 °C for at least 10–15 min with a rich mixture; in such condition, oxygen being totally absent, barium sulfate converts back to barium carbonate.

In spite of the above difficulties, and thanks to the improved gasoline quality, the first, already mentioned, applications by Mitsubishi were performed with a similar system, based on an iridium catalyst.

In the following years, Volkswagen applied a catalyst formulation already based on barium to the so called Fuel Stratified Injection (FSI) and other car makers implemented this technology to already existing engine configurations in the first decade of the 2000 s.

10.7.5 Development of Gasoline Direct Injection System

Direct injection offers a significant potential in the most significant areas of engine development, either fuel consumption reduction or performance increase.

Each automaker decides how to apply this technology in the most profitable way within the low, medium and high segment vehicles, both using the stratified or homogeneous mixture management based on fuel economy, driveability or performance goals. The suitable technological mix is of paramount importance to satisfy the weighted average CO_2 value and the requested reductions to comply with the 2015 and 2020 targets. The technical difficulties and the related costs define the, already mentioned, different share in the introduction of direct injection through the years. Its wider application is in progress, as demonstrated by the significant number of recently designed three-cylinder engines able to exploit the synergy between the gains in thermodynamic efficiency and the ones in mechanical efficiency linked to the reduced number of mechanical components, and new materials introduced to reduce engine friction which can contribute to fuel consumption reduction with an additional 6–8 %.

In some applications, regarding more sophisticated vehicles, engines featuring both direct and indirect injection systems with two set of injectors are used. For example, indirect injection is active at low-medium loads, while direct injection is the preferred solution at idle and medium-high loads.

10.8 Compression Ignition Engines

The main advantage of compression ignition engines is their higher combustion efficiency, due to the following reasons:

- High volumetric efficiency, since the power modulation is obtained modifying the fuel quantity following the power request,
- use of lean mixture,
- high thermodynamic efficiency, thanks to the high compression ratio, and
- higher fuel density, when compared with gasoline.

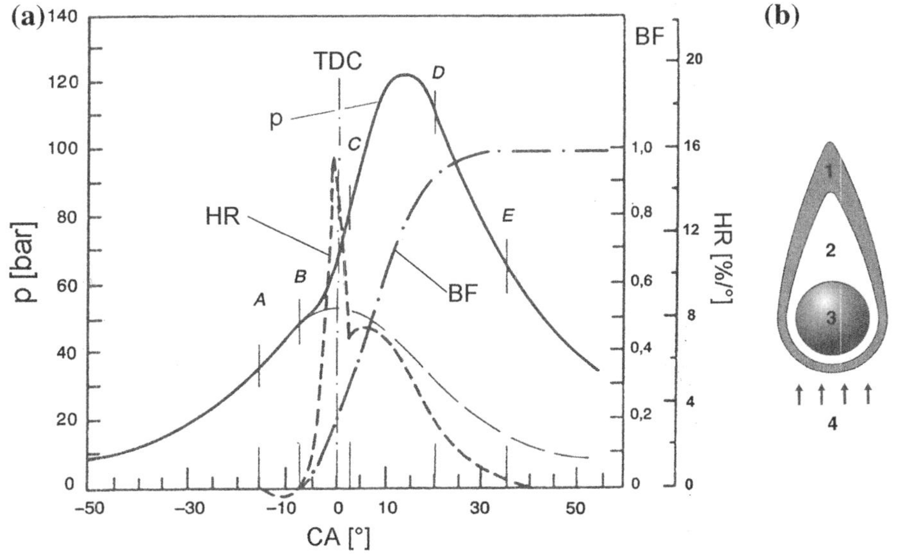

Fig. 10.56 Diesel combustion parameters. **a** Pressure *p*, Heat Release (*HR*) and Burnt Fraction (*BF*) as functions of the crank angle (*CA*). **b** Fuel droplet determining the combustion phases. Ignition delay (*A*–*B*), pre-mixed combustion (*B*–*C*), diffusive combustion (*C*–*D*), combustion end (*D*–*E*)

However, when trying to supply additional fuel with the goal of increasing the power at unchanged engine configuration, a strong limitation is the sharp increase of black smoke. This limitation is, in some respect, similar to the detonation limit that is encountered when increasing the compression ratio of a gasoline engine to improve the performance again keeping unchanged the engine configuration.

Compression Ignition-Cycle Combustion

As already mentioned, the fuel spray is injected into the cylinder slightly before the piston reaches its TDC. The plot in Fig. 10.56 shows an injection starting 15° before the TDC. During these few degrees, the external part of the spray vaporizes and mixes with air thus achieving, in its external part, a stoichiometric mixture so that ignition occurs after a delay of a few milliseconds. Figure 10.56b shows a fuel droplet (3) meeting the air stream (4). The flame can take place in zone (1), while the transition phase (2) is already vaporized but still too rich to start burning.

Also in compression ignition engines the injection advance must be increased with increasing engine speed. In this respect, Fig. 10.56a shows a medium-high load situation, in which the maximum pressure reaches 125 bar and injection advance is set to 15°. It is likely that, at full load, these values reach respectively 150 bar and 20°, while, at partial loads, both maximum pressure and injection advance decrease.

During this period the spray undertakes the following processes: atomization, heating, vaporization, mixing and diffusion. This process produces the final combustion shape shown in the figure, in which the following plots are reported:

- Pressure p inside the cylinder,
- heat release (HR line),
- burnt fraction (BF line), and
- pressure in case no ignition occurs (light dashed line).

During the injection-combustion process it is possible to distinguish four phases. To provide the required power while controlling the emissions, a key issue is to recognize their individual contributions.

Ignition delay

Fuel is injected into the combustion chamber (point A), about 15° before the TDC, when the air temperature, at the end of the compression phase, reaches 750–800 °C. Ignition (B) is caused by the increase of temperature due to compression, which definitely exceeds the diesel fuel auto-ignition temperature. Ignition is conventionally placed where the pressure curve exceeds the simple compression curve by 1 %. The ignition delay lasts thus from point A to point B. During this phase the heat release takes negative values due to fuel evaporation. Without ignition, at the end of compression the temperature typically reaches 900 °C.

The duration of the ignition delay strongly depends on the fuel quality; the cetane number is the indicator of the ignition quality: increasing its value, within the range 50–55, lowers the delay thus promoting a better ignition.

Pre-mixed combustion

Immediately after starting (point B), the pre-mixed combustion proceeds very quickly and ends in point C. The fuel quantity already accumulated increases the combustion velocity; as a result, this combustion takes place almost at constant volume, improving the thermodynamic efficiency but producing noise. The heat release takes its maximum value, causing a fast pressure rise associated with high temperature.

During the first part of the pre-mixed combustion its blue flame produces temperatures close to 2,000 °K, but, immediately after, the flame becomes red and 2,500 °K are reached. The high local temperatures and the excess of oxygen promotes the NO_x formation, higher than that typical of spark ignition engines (Fig. 10.57a), while being a fuel offering good ignition qualities can limit the NO_x formation. The local rich air to fuel mixture promotes the soot formation even if, at the end of injection, the excess-air factor λ varies within 1.3–1.5 at full load and approaches 10 at idle.

This phase produces the typical diesel combustion noise, and the related vibrations, mechanical stresses and temperature rise.

Diffusive combustion

The diffusive or controlled combustion phase starts from point C and ends in point D; it features the combustion of the main part of the jet which continues entering the combustion chamber. It is the typical diesel combustion, in which the overall air to fuel mixture is lean but strongly non-homogeneous. Its duration is defined by the quantity of the fuel injected to obtain the required pressure peak (125 bar in the

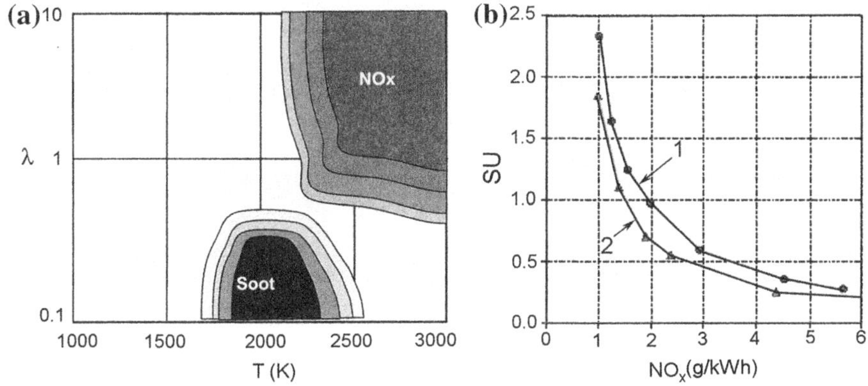

Fig. 10.57 Diesel combustion. **a** Soot and NO_x formation as functions of the temperature and of the excess-air factor. **b** Soot (Smoke Units SU) and NO_x trade-off: basic situation (*1*) and partially pre-mixed combustion (*2*) (courtesy of FIAT)

figure), needed to deliver the requested power. The heat release is controlled by the gradual delivery of the fuel, so that a more continuous and gradual heat release occurs as a consequence of the controlled fuel flow.

During this phase, the temperature lowers, causing particulate matter formation. A correctly working fuel injection system limits the hydrocarbons formation, which is always lower than in spark ignition engines because, due to the lean mixture, the fuel does not reach the cylinder walls. Similarly, the CO formation is lower, due to an air to fuel ratio always lower than stoichiometric.

Combustion end and engine emission control

From point D to point E the fuel already injected burns. Additional particulate matter formation takes place with an increase in particle size due both to the condensation processes and to the temperature decrease associated with expansion.

Examining the mentioned phases and recognizing that NO_x and particulate matter are the typical emissions of compression ignition engines, it is possible to state that their formation can be reduced with the following measures during the injection process:

- Reducing the ignition delay, and the following temperature peak, to lower the NO_x content,
- increasing the temperature during the diffusive combustion, to lower the particulate matter content, and
- reducing the combustion end duration, to lower the particulate matter content.

Finally, a temperature reduction and a better diffusion process promote a quick reduction of the baseline engine emissions. Unfortunately this combination is hardly achievable in all engine running conditions and the simultaneous reduction of NO_x and particulate matter is always difficult. To evaluate whether an engine modification is effective in producing a positive trend, the so called NO_x/soot trade-off shown in

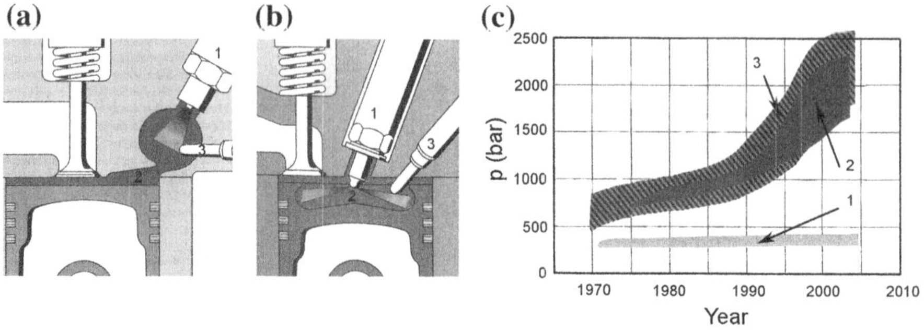

Fig. 10.58 Pre-chamber (**a**) and direct (**b**) injection configurations. **c** Evolution in time of injection pressure for pre-chamber (*1*) and direct injection engines, for heavy duty vehicles (*3*) and passenger cars (*2*) (courtesy of Bosch)

Fig. 10.57b can be plotted. This plot is widely used because soot correlates quite well with particulate matter emissions and, in addition, its concentration is more easily measured. The specific trade-off shown will be further commented when considering the multiple injection capabilities.

Diesel Direct Injection

Up to the 1990s diesel direct injection (Fig. 10.58b) was used only in engines for heavy duty vehicles while pre-chamber injection (Fig. 10.58a) was used in passenger cars engines. In both solutions, the key elements are the usual ones: the fuel injector (1), the combustion chamber (2) and the glow-plug (3). This situation was dictated by the need to maintain the noise at an acceptable level when using compression ignition engines on passenger cars. The second advantage of the pre-chamber solution (Fig. 10.58a) was its capability of providing a strong and effective air motion offering a good air to fuel mixing, thus improving combustion. In addition, cheaper and simple spring-loaded, pintle-shaped, needle injectors, providing a single fuel spray, could be used. On the contrary, it was already well known that the more compact combustion chamber, typical of the direct injection solution (Fig. 10.58b), is far more efficient. To make a step further in its improvement, it was necessary to compensate for its intrinsic less effective air motion with a higher injection pressure, able to promote atomization and mixing.

To summarize the outcome of the above mentioned trend, the evolution in time of the injection pressure is shown in Fig. 10.58c. At the end of the 1980s, the difference in injection pressure between pre-chamber and direct injection was respectively 300 and 800 bar, and, since that time, started increasing sharply. By applying electronically controlled fuel injection pumps, direct injection could be introduced in two-valve cylinder heads for the early passenger car applications as shown in Fig. 10.59.

The combustion chamber shape looks similar to a solid of revolution with an ω shaped generatrix and is completely machined into the piston head. Atomization

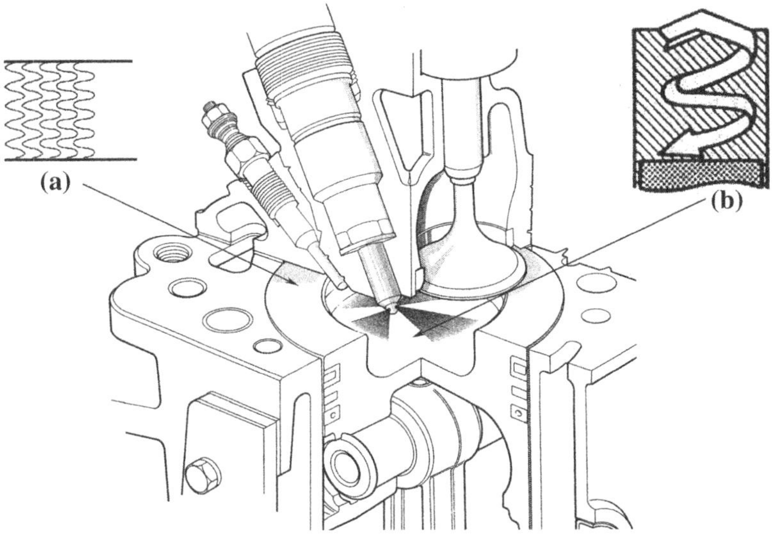

Fig. 10.59 First generation of Diesel direct injection for car engines, with squish area (**a**) and swirl motion (**b**) (courtesy of FIAT)

and mixing are promoted by the use of multi-orifices nozzles and by the extended turbulence of the air charge, obtained with a suitable configuration of the intake duct. The latter produces a swirl, i.e. a rotational motion of the air stream around the cylinder axis. In addition, the flat upper crown area of the piston mates closely with the flat cylinder head at the TDC, thus causing a squish of the charge and promoting a beneficial local micro-turbolence. The injector orifices, five or more, inject the fuel radially into the combustion chamber and the lean air to fuel ratio and swirl motion avoid that the fuel jets impinge against the metal wall of the chamber producing an unsuitable mixing and the increase of hydrocarbons emissions, up to smoke formation and loss of power.

The combustion taking place at constant volume is faster and improves the thermodynamic efficiency, even if the combustion noise increases due to the higher speed of flame propagation and higher peak pressure. A combustion efficiency higher than in pre-chamber injection can thus be obtained, and an overall efficiency close to 40 % can be reached. This achievement is also promoted by reducing the compression ratio which, in direct injection engines, can be as low as 16:1–18:1, versus 20:1–24:1, typical of pre-chamber engines.

The glow-plug needed to assure cold starting is also shown in Fig. 10.59; to reach an acceptable combustion stability after cold start, even in conditions with a low compression ratio, it must be able to reach 1,200 °C in a few seconds. During cold start the injection advance is increased by 5°–10°. After starting, the reduced thermal release toward the cooling system causes a longer warm-up time.

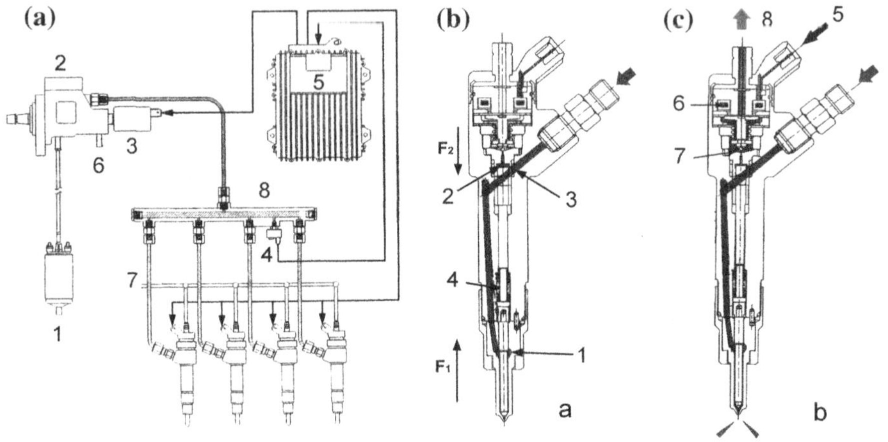

Fig. 10.60 Common Rail solution with solenoid valve injectors driven by the engine electronic control unit (courtesy of FIAT)

By applying these engine modifications it was possible to meet Euro 1 emission standards (January 1993). Injection retardation lowers NO_x while its advance lowers hydrocarbons and soot, so that a suitable trade-off could be obtained, together with the application of simple vacuum-actuated exhaust gas recirculation valves. To meet Euro 2 standards (January 1997) the above setting and components were refined by introducing Diesel Oxidation Catalysts (DOC). Their application was further extended to meet Euro 3 standards (January 2001).

Common Rail System

Rudolph Diesel predicted the advantages of this solution since 1912, but only in 1991 did Fiat demonstrate its industrial feasibility by applying it on small displacement experimental engines. The system reached full industrial production and its first application on an Alfa Romeo engine, in 1997. The essential Common Rail System (CRS) innovation consists in providing a constant and high fuel injection pressure to a single rail installed immediately upstream and very close to the injectors.

The scheme of the hydraulic layout is shown in Fig. 10.60a: the system features a low-pressure electric pre-supply pump (1), included in an in-tank unit as an integrated assembly, which incorporates a fuel filter and a water separator. This pump feeds the high pressure pump (2) which, via a high pressure fuel line, delivers the fuel to a common pressure accumulator, referred to as a 'rail'. From the rail, the fuel is distributed to the individual injectors, hence the 'common rail' name. A pressure regulator (3), provided with a pressure sensor (4), modulates the pressure to the injectors with some functional and geometrical similarities to the gasoline direct injection system.

The control signal provided by the electronic control unit (5) defines, with high precision, both the injection opening time and the injection timing according to the engine load and speed. Following the injector opening, the rail pressure causes the fuel delivery along the short ducts (7) to the injectors. The excess fuel (6) returns back to the fuel tank.

In the first generation common rail systems the injector pressure reached 1300–1400 bar at maximum engine speed and load. The system offered the key advantage of managing the pressure modulation depending on the final goal, performance and/or emission control, thus being independent from the basic engine parameters, such as speed and load, like in previous systems. The first generation of common rail systems could lower the engine baseline emissions below the Euro 3 standard, with limited specific emission devices such as a refined exhaust gas recirculation system, as shown in Fig. 10.41b, and the diesel oxidation catalyst, able to provide the usual high efficiency in oxidizing CO and hydrocarbons. This catalyst provides also a 15–20 % of efficiency for particulate matter, being capable of oxidizing the soluble organic fraction also responsible for unpleasant smell being rich, during some transients, in aldehydes; unfortunately it is inefficient in reducing soot.

The pilot injection was added to the main injection to reduce the combustion noise, a key feature for passenger car application. The pilot injection is delivered before the main injection and the noise reduction is determined by a preliminary release of partially oxidized products which promote the subsequent oxidation producing a less noisy combustion, allowing modification of the ignition delay of the main injection. The pilot injection is used only at low-medium loads and its advance, or dwell time, with respect to the main injection, is usually in the range of 2 ms with a typical fuel delivery of 1.5–3 mm^3/stroke compared with the main injection lasting, as an average, 1.2–1.5 ms and delivering 5 mm^3/stroke at idle and 10–20 mm^3/stroke at low-medium loads. At full load these values become 2 ms and 50 mm^3/stroke.

When, at the end of the 1990s, this system was launched together with turbocharging, it became a strong competitor of gasoline engines, offering a new and unique fun-to-drive feeling and, in addition, fuel consumption reductions by 25 % with respect to naturally aspirated gasoline port fuel injection engines but with significant cost increases linked to the sophisticated fuel injection system. This solution boosted the share of compression ignition engines close to 50 % of new car registration in West European Countries, starting from year 2000 and was the key factor able to reduce the CO_2 weighted average after the 1995 reference year.

A common rail solenoid injector is shown in Fig. 10.60b in its closed position. The high pressure reaches the chamber placed in the lower part of the needle (1), but cannot open the needle even if the force F_1 is applied. The same pressure reaches also the calibrated hole and the upper chamber (2) through the high pressure channel (3), so that the pressure and the spring (4) produce a force $F_2 > F_1$ which keeps the needle closed.

When the electric injection signal (5) triggers the electromagnet (6), the solenoid valve (7) opens the connection toward the low pressure pipe (8). In this way the pressure in the chamber exceeds the spring action, opens the needle, and causes the fuel to flow into the combustion chamber, as shown in Fig. 10.60c.

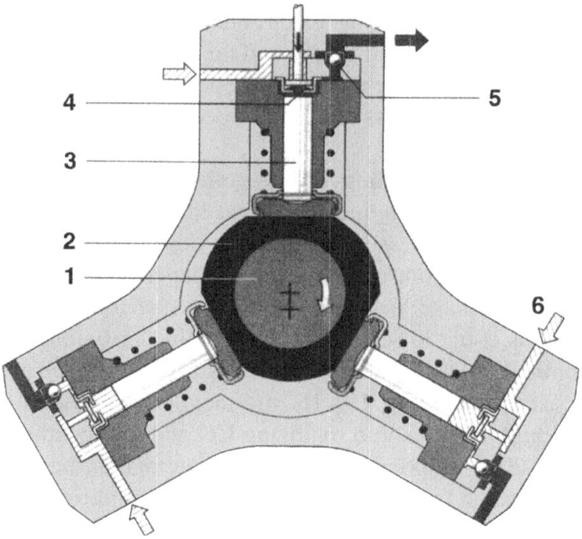

Fig. 10.61 Common Rail high pressure pump. The pressure delivered is 1600–1800 bar (courtesy of Bosch)

The injection orifices, facing the combustion chamber, have a typical diameter of 0.16 mm and are, usually six in number. The injection orifices are created by the electro-erosion process, considering that it is mandatory that their hydraulic diameters are, as much as possible, close to each other. If this condition is not met, the combustion propagation is badly affected with a negative effect on performance and emissions.

The high pressure pump, shown in Fig. 10.61, is placed between the low and the high pressure sections of the hydraulic circuit with the goal of supplying a constant pressure to feed the common rail. The pump shaft (1) is driven by the engine, frequently through a timing belt, with a fixed gear ratio which value depends on the need to supply the fuel quantity required for full load operation. The shaft drives the cam (2) which moves the plungers (3) positioned at 120° from each other. During every shaft revolution each plunger performs three useful strokes thus providing the required pressure. As mentioned, the pre-supply pump delivers the fuel to the high pressure pump inlet (6). The operating plunger, the upper one in the figure, causes the fuel to go through the inlet valve (4), then the rising pressure opens the outlet valve (5).

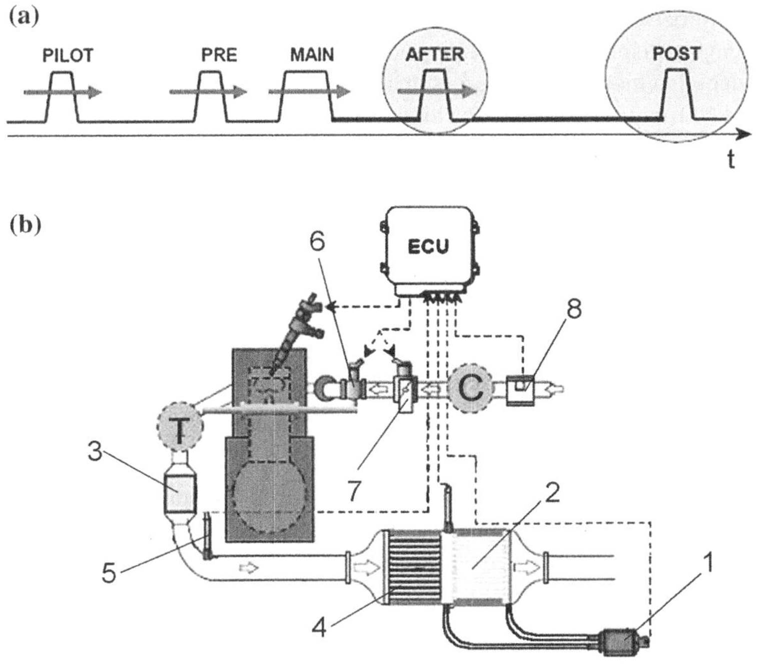

Fig. 10.62 Multiple injection and Euro 5 after-treatment layout (Cooled exhaust gas recirculation system, pre-catalyst, front catalyst and diesel particulate filter) (courtesy of FIAT)

10.8.1 Euro 4 Standards: Injection Evolution

Euro 4 standards (January 2006) required a significant evolution of the first generation of common rail system into its second generation featuring a refinement of existing solenoid valves to split the main injection into three phases, namely pre, main and after-injection, as shown in Fig. 10.62a, thus obtaining a multiple injection system. The minimum hydraulic dwell time (in the range of 0.25 ms) between the pre and the main injection is a key feature. The injection pressure increases to 1,600–1,800 bar and the injection orifices tend both to increase in their number and to reduce in bore size, down to 0.12 mm, to obtain an even finer dispersal of fuel.

The above mentioned engine modifications are able to obtain a more gradual combustion, thus reducing soot formation and allowing an increased EGR quantity to satisfy the NO_x standard. The after-injection is useful to promote the soot oxidation in the final part of the combustion.

This second generation allows the introduction of the post-injection, mandatory in view of the application of a diesel particulate filter for Euro 5, finally reaching five injection events.

When running the emission measurement, the NO_x emitted by a certain vehicle are strongly affected by its weight and its road load, in other words by the load factor, i.e. the ratio between the energy needed to operate the vehicle on the driving cycle and the engine displacement. For instance, on a given vehicle, a smaller engine will operate at higher mean effective pressure and, as consequence, it will be more difficult to control NO_x emission, due to the higher combustion temperatures and lower EGR rate. On the contrary, a larger engine will operate at a lower mean effective pressure, thus at lower temperatures and at higher EGR rate, but this choice entails higher fuel consumption, CO and hydrocarbons emissions. Unavoidably the most suitable sizing is fundamental to achieving both low NO_x emissions and good fuel economy.

This consideration explains why, to satisfy the subsequent standards, the selected technological mix strongly depends on the vehicle/engine combination.

10.8.2 Euro 5 Standards: Particulate Reduction

Euro 5 standards (January 2011) call for a further NO_x reduction. In this respect, a lower compression ratio, down to 16–16.5, proves its effectiveness as well as promoting an efficiency increase due to lower friction losses. When reducing so much the geometrical compression ratio some drawbacks must be considered. Startability is not the first issue, because the glow-plug technology improves very rapidly. More severe is the effect on CO and hydrocarbon emissions during engine warm-up, when even the hot EGR flow delivered by the EGR heat exchanger by-pass valve in open position (Fig. 10.41b), is not sufficient to keep the in-cylinder temperatures at a suitable level for completing the oxidation reactions. An improved oxidizing catalyst package, unavoidably more costly, can solve this issue.

But the air-mixing control is as important as fuel control and, in this regard, an additional swirl control proved effective by applying a specific swirl flap, along each intake duct to provide a more appropriate swirl index in the combustion chamber, depending on engine load and speed. In many applications the closed-loop control, obtained with the application of a broad-band lambda oxygen sensor, is needed to control air and fuel actuator drifts during the vehicle lifetime. This system accepts a further increase of the EGR flow, thus achieving the requested NO_x reduction.

The multiple injection system is mandatory to introduce the Diesel Particulate Filter (DPF) needed to satisfy the new particulate standard and, more and more frequently, the system reaches up to eight injection events.

The particulate filter behaves like a trap or a storage device able to accumulate particulate matter that will be regenerated later. With respect to the traditional catalysts substrate, it is assembled using an innovative kind of ceramic honeycomb structure made in Silica Carbide (SiC), able to efficiently trap particulates independently from their diameter, from 0.02 μm, upward.

The silica carbide structure tolerates, better than Cordierite, high temperatures and thermal shocks. The assembly differs from the traditional monolith, showing a segmented structure in which many prismatic elements are cemented together to reach the final circular section. The individual channels of each element are alternately plugged, at their end, by ceramic plugs (Fig. 10.63a) and, therefore the soot

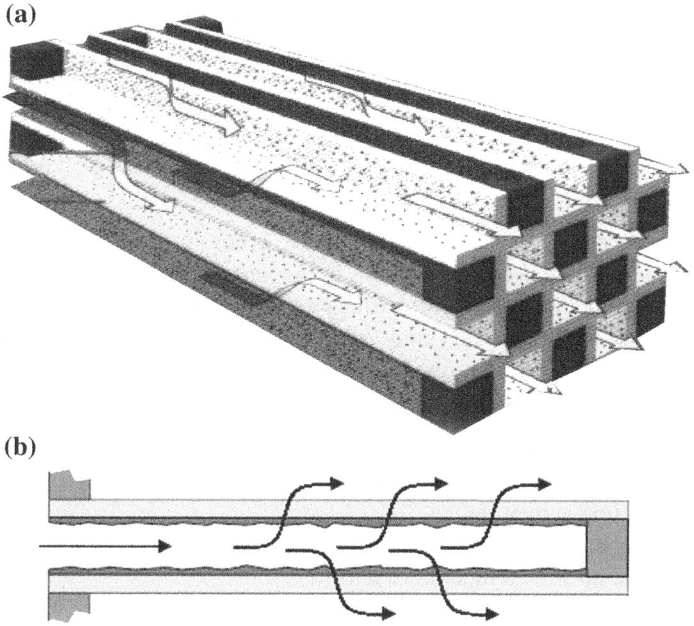

Fig. 10.63 Diesel wall-flow particulate filter structure

deposits along the wall of the closed channels (Fig. 10.63b), while the gases flow through the contiguous ceramic walls provided with a suitable and controlled porosity. The wall mean pore size is comparable with the mean size of particulate matter aggregates and this mechanism identifies this particulate filter as wall-flow type. The channel structure differs from the traditional catalyst substrate to provide a higher mechanical resistance. In fact the cell density is as low as 200–300 cells/sq. inch, the wall thickness increases to 0.30 mm or more with a typical filter volume being 1.5–2.0 times the engine displacement.

Depending on the filter material, this filter can store approximately 6–10 g/l of particulate before requiring regeneration or, in other words, before requiring the burning-off of the carbon soot accumulated. Therefore, depending on vehicle mission profile (city or highway), the engine can run from few to several hundreds kilometers before requiring filter regeneration which lasts about 10–15 min. After reaching filter saturation, the carbon is burned; if this process does not take place, the carbon built-up prevents the regular engine exhaust gas flow, causing the engine to stop due to the exhaust backpressure. During regeneration, gas emission increases and thus an additional emission test during this transient period is required and the average emissions, weighted with the mileage, must be computed; this average value is compared with the applicable emission standards.

The regeneration event is activated by the differential pressure sensor (1), integrated by a model-based load estimation, which monitor the particulate accumulation on the filter (2): Fig. 10.62b shows an early-type layout of a particulate filter system available in the mid 2000s. When the measured/estimated pressure drops above a threshold value, indicating that a significant carbon built-up is present, Carbon particles are oxidized to CO_2, using the oxygen present in the exhaust, when at least 600 °C are reached in front of the particulate filter according to the following simple reaction:

$$C + O_2 \rightarrow CO_2.$$

Reaching the above mentioned temperature is not straightforward, considering that at low load the exhaust temperature is in the range of 250 °C. The engine control unit must deeply modify the engine settings to provide the required temperature rise without affecting torque delivery and vehicle driveability. The main engine setting modifications and hardware components are the following:

- Retarding the main injection and deliver the after-injection at least 30°–50° after the TDC,
- modifying the multiple injection sequence, delivering the required extra-fuel with the post-injection activation,
- reducing the charge air pressure and throttle the intake air during cut-off conditions to increase the air flow temperature and
- reducing the exhaust gas recirculation to increase the oxygen quantity, at least, up to 5 %.

The post-injection, thanks to the pre and front-catalyst, (3) and (4), causes an exothermal reaction during the fuel oxidation with a subsequent temperature rise. For a 2 l engine, the catalyst volumes and noble metal loadings are respectively 0.6 l and 140 gcf, 1 l and 75 gcf in front of the filter with a 4 l volume and 30 gcf of platinum content. The platinum coating within the filter catalyzes the soot combustion, minimizes the residual soot and reduces CO and hydrocarbon emissions during regeneration. The total noble metal content is close to 10 g. The temperature sensor (5) placed upstream of the oxidation catalyst helps in determining the correct light-off.

The above mentioned oxidation capability leaves unchanged the overall NO_x, but promotes the NO_2 formation thus increasing the NO_2 to NO ratio.

The regeneration phase is critical, considering that, during carbon combustion, the local temperature can increase up to a critical range (1,000 °C) in a few seconds and local thermal gradients take place with such a severity that micro-cracks or even cracks of the ceramic substrate may occur. To this aim, the silica carbide structure must provide a high thermal resistance with no damages up to 1200 °C. The after and post-injections deliver the required extra-fuel with an average of 1–2 % consumption increase and a likely significant oil dilution, which must be controlled and monitored.

As already mentioned, the volumes of the particulate filter and of the oxidizing catalyst are quite large, when compared to the engine displacement; these components were therefore initially placed in the underfloor position, where the problems determined by their bulk and by the heat release, during regeneration, can more easily be controlled. Unfortunately, in this position, it is more difficult to reach the regeneration start-up temperature and the overall assembly is quite expensive. To solve these difficulties, the particulate filter and the oxidizing catalyst are now assembled in the same container, immediately downstream from the turbine, in a close-coupled position as shown in Fig. 10.41b. This layout allows a volume reduction, together with an average temperature increase of 50 °C and a significant cost saving. It will be almost mandatory when, to meet Euro 6 standards, the NO_x aftertreatment will be required.

The exhaust gas recirculation circuit is shown in Fig. 10.62b. In addition to the EGR valve (6), the additional throttle valve (7) and the air-flow meter (8) are used to, respectively, increase the intake vacuum, when necessary, and to meter the air flow, thus delivering a more precise exhaust gas flow, in particular during transient conditions.

A limited number of car maker, namely PSA group, to reduce the regeneration temperature from 600 to 450 °C, use a specific additive, cerium oxide, although other additives, based on iron compounds, were also tested. The additive is contained in a separate tank and is injected in the fuel main stream by a separate injector. This tank must be refilled after a specified, and quite high, mileage. The additive cannot be mixed with the fuel because it would be directly emitted in the atmosphere by vehicles not equipped with the diesel particulate filter instead of being trapped by the latter. The use of this additive use is considered with a certain suspicion by the Environmental Agencies and, therefore, it is allowed only on cars equipped with filters that can trap it. The additives, based on metal molecules, cannot be burned and are deposited as ash which accumulate and gradually clog the filter, which therefore must be periodically cleaned.

The application of particulate filters allows achievement of more than 90 % of particulate matter reduction, therefore more than the 80 % as requested by Euro 5 standard.

The systems described, although being the most common, are not the only ones. The following alternatives are worth mentioning:

- Continuous Regeneration Trap (CRT) introduced by a catalyst supplier, Johnson Matthey, and suitable for commercial vehicles considering their prevailing highway driving conditions. It features a particulate filter and an oxidizing catalyst able to promote the NO_2 formation and thus the soot combustion according the following reaction:

$$NO_2 + C \rightarrow \frac{1}{2}N_2 + CO_2.$$

- Diesel NO_x Particulate Reduction (DNPR), applied by Toyota on their passenger cars. It maintains the diesel particulate geometry but, with a specific catalyst composition, promoting the Lean NO_x Trap effect described ahead, performs other reactions within the filter volume.

A more detailed description is out of the scope of the present book and is available in the technical literature.

10.8.3 Euro 6 Standards: NO_x Reduction

Euro 6 standards (September 2015) require a 55 % reduction of NO_x with respect to Euro 5 and represent a target close to fuel neutral emission.

To achieve this goal a promising solution is using partially pre-mixed combustion for the simultaneous reduction of NO_x and soot as shown by an improved trade-off shape (Fig. 10.57b). This combustion mode is based both on a long ignition delay for enhancing the mixing process and on high EGR rates, to reduce the oxygen content and the local flame temperature. To reduce the value of the latter, a lower compression ratio is effective also because it increases the ignition delay which, in turn, helps the mixing process before the beginning of combustion. The high EGR flow is available only in low load conditions, where this type of combustion can shift the combustion temperatures toward lower values and thus within an area at which both soot and NO_x production start decreasing (Fig. 10.57a).

When the diffusive combustion takes place, the *rate of injection* directly affects the *rate of combustion* to provide their best combination able to keep the combustion process into the above mentioned area. A suitable rate of combustion improves the control of the heat released during combustion with the aim of reducing the maximum temperature. In the above conditions, it is likely to reach the combustion stability limit, that means high cycle-to-cycle and cylinder-to-cylinder combustion fluctuations and misfiring. In such a situation it can be useful to activate a closed loop combustion control, based on cylinder pressure sensors integrated in the glow plug and on dedicated control strategies, developed with the aim of managing such sensors and acting on the engine for combustion control.

Applying this concept to an Euro 5 engine on a medium segment vehicle, a significant NO_x reduction can be achieved, even if with a CO and hydrocarbons increase.

A third generation of solenoid injectors, able to perform consecutive injections without any limitation of hydraulic dwell time, was developed to achieve the above mentioned goal. This feature means that pre and main-injection became a nearly continuous ignition law with a smoother starting phase followed by a stronger main phase.

The Injection Rate Shaping (IRS) so obtained, keeps under control the combustion noise and combustion stability without increasing soot formation and achieving a fuel consumption reduction in the range of 2 %. When it is deemed useful, the injection system can provide a further rise of injection pressure up to 2,000 bar and, in the future, even more. To achieve this goal, piezoelectric injectors can be useful, in spite of their higher costs, due to their higher injection pressure. Finally, both solutions, solenoid valve and piezo injectors must meet contradictory requirements: high needle speeds and extremely small injection quantities.

The relative periods of ignition delay and diffusive combustion must be suitably selected according to engine speed and load.

To comply with the NO_x Euro 6 standard, the updated close-coupled position of oxidation catalyst and particulate filter (Fig. 10.64b) paves the way to introducing

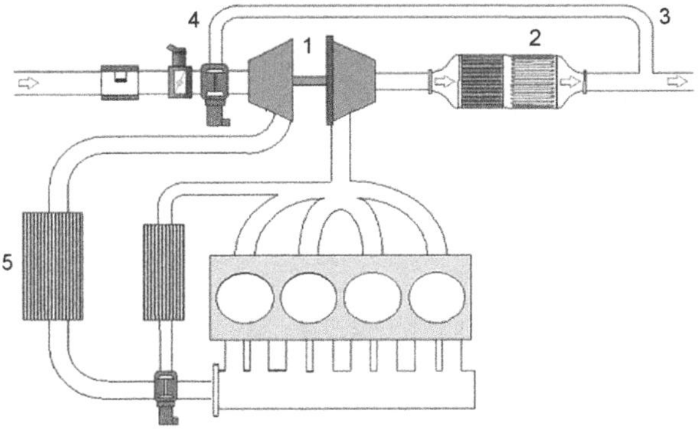

Fig. 10.64 Long route exhaust gas recirculation system (courtesy of FIAT)

the so-called low pressure exhaust gas recirculation, shown in Fig. 10.64, which is an effective way to obtain high exhaust gas recirculation rates at acceptable air to fuel ratio values. With this system, a clean and cool exhaust gas (3) is taken downstream of the turbocharger (1) and particulate filter (2), in a close-coupled position, and is inserted upstream of the turbocharger compressor (4). In this way, the mixture of fresh air and exhaust gas recirculation flows through the charge air cooler (5), getting a very efficient cooling. Compared to a standard Euro 4/5 high pressure exhaust gas recirculation system, the low pressure system can guarantee an overall higher cooling efficiency reaching a 30 % NO_x reduction. Moreover, low pressure recirculation could immediately solve two major problems resulting from the application of high exhaust gas recirculation rates together with standard configuration: fouling of the intake system and cylinder-to-cylinder exhaust gas recirculation distribution.

On the contrary, it is clear that the response time of a low pressure recirculation is higher than that of an high pressure system, due to the longer route to be covered by the exhaust gases before reaching the intake manifold. The combination of a low and high pressure exhaust gas recirculation combines a high cooling efficiency with a fast response time and gives the possibility of controlling the exhaust gas recirculation temperature during the engine warm-up phase, when hot exhaust gas recirculation is preferred.

Looking at these results, the first low pressure recirculation systems have been introduced since 2012 solving the following main drawbacks due to a strong exhaust gas recirculation:

- Possible damages to the compressor wheel, due to the impact of small particles coming from the exhaust, and
- water condensation and chemically aggressive condensates, owing to the presence of nitrogen and sulphur, that could produce intercooler corrosion, also due to air mixture cooling.

The above mentioned solutions are not sufficient to reach the NO_x standard and, therefore, depending on the load factor, additional and specific NO_x after-treatment

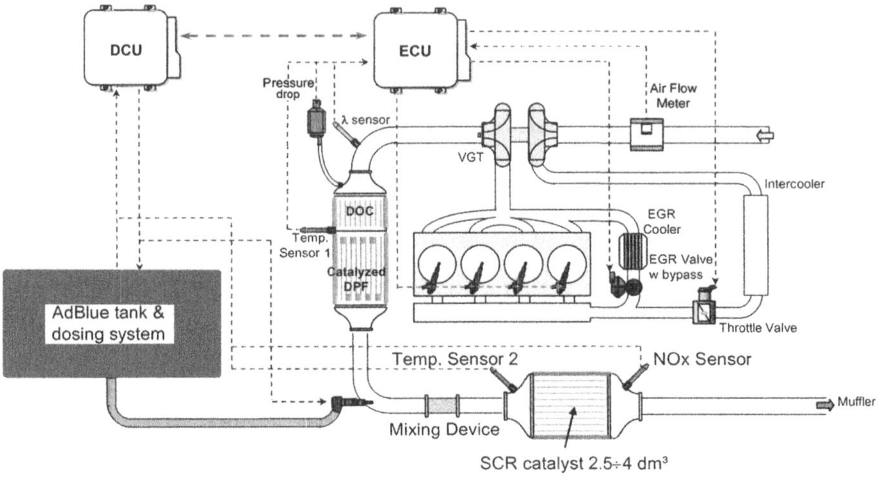

Fig. 10.65 Selective catalytic reduction system (courtesy of FIAT)

technologies must be introduced, to a lower or higher extend. At present the following solutions can be applied:

- NO_x adsorbers, identified as Lean NO_x Trap (LNT) or NO_x Storage Catalysts (NSC), and
- urea-based Selective Catalytic Reduction (SCR).

As already mentioned, the scheme of the lean NO_x trap does not substantially differ from the one already introduced in the gasoline direct injection section as shown in Fig. 10.55. The system is based on the concept of trapping NO_x during engine normal operation and then reducing it during engine operation in stoichiometric or rich mixture and high temperature with a frequency based on engine baseline emissions.

A conversion efficiency higher than 70 % can be reached with fresh components, but, also in this case, the effects of aging and sulphur contamination reduces the system efficiency which, in any case, remains satisfactory. In the case of diesel engines, the temperature increase required to perform the de-sulfating process is obtained similarly to the temperature rise provided for the particulate filter regeneration. Clearly it is not so easy to manage these different types of regeneration events which must take place with both different frequency and different mixture setting. One additional difficulty, with respect to gasoline direct injection, is the intrinsic difficulty in running the diesel engine under rich conditions due to smoke formation.

In spite of what is said above and thanks to the improved diesel fuel quality, this system has started to be applied on passenger cars mainly for European emission standards.

Urea-based Selective Catalytic Reduction (SCR) is the most mature NO_x control technology, already developed for industrial vehicle applications and now already applied on high displacement passenger cars and SUVs since 2010 (Fig. 10.65). In this systems, urea, $CO(NH_2)_2$, well known as a fertilizer, is hydrolyzed to ammonia NH_3, thus providing a strong NO_x reductant. Urea is quite soluble in water, and can be contained in a specific tank, as urea/water solution, and injected within the exhaust gas flow upstream of the selective catalytic reduction catalyst, typically based on vanadium or zeolite, where NO_x is then reduced by reacting with NH_3. The urea solution is distributed under the brand name of AdBlue, a stable, non-flammable and colorless fluid, which contains a 32.5 % of urea in water with a freezing point of −11 °C, to refill vehicles equipped with this system. To insure the system operability in very cold weather, specific heating provisions must be applied both on dispensing facilities and vehicles. AdBlue consumption, for commercial vehicles, is typically from 3 to 5 % of the diesel fuel and the size of its tank is such as to require a refill every few normal tank refillings. Its consumption is lower in passenger cars, and depends on the engine baseline emissions and vehicle load factor.

By correctly dosing and mixing the injected urea, a potential NO_x reduction up to 90 % can be achieved. Moreover, this very high efficiency can be reached in every engine operating point almost without any effect on fuel consumption, particulate matter, CO and hydrocarbon engine emissions. The only relevant constraint is linked to maintaining the system within the correct temperature window, that is typically between 200 and 550 °C. Above this range a fast catalyst ageing can occur. The impact on layout definition is clear and a precise gas temperature control must be developed to hasten the catalyst activation and to prevent high temperature peaks at full load, during particulate filter regeneration, and in every other critical condition.

A likely installation in which the selective catalytic reduction catalyst is located downstream the closed coupled package of oxidation catalyst and particulate filter is shown in Fig. 10.65. The mixing device is located with the following goal in mind:

- Complete mixing of AdBlue droplets and exhaust gas,
- avoiding urea deposits on the pipe walls and
- quick conversion of urea into ammonia.

The AdBlue injector must provide a droplet diameter below 70 μm, avoiding large droplets above 120 μm that may impinge on the pipe wall, with subsequent urea deposition.

The main processes within the system take place according to the following sequence:

- Oxidation catalyst promotes an additional NO_2 formation

$$2NO + O_2 \rightarrow 2NO_2 \ ;$$

- hydrolysis reaction converts ammonia into urea:

$$CO(NH_2)_2 + H_2O \rightarrow 2NH_3 + CO_2 \ ;$$

- urea reduces NO and NO_2:

$$4NH_3 + 4NO + O_2 \rightarrow 4N_2 + 6H_2O$$

$$4NH_3 + 2NO + 2NO_2 \rightarrow 4N_2 + 6H_2O\,.$$

- A second oxidizing section downstream from the reducing catalyst, not shown, provides the so called anti-slip function which is mandatory to avoid the ammonia slipping which takes place when an excess of reducing agent is delivered to the catalyst, finally resulting in ammonia release from vehicle exhaust:

$$4NH_3 + 3O_2 \rightarrow 2N_2 + 6H_2O\,.$$

Ammonia is listed as a toxic chemical product, having a very low odor threshold at 15 ppm. An NO_x sensor or a new ammonia sensor, with a ± 5 ppm ammonia detection accuracy, located downstream from the system, continuously monitors the overall efficiency.

The great potential of the SCR system is counterbalanced by a system complexity that includes various components like the urea-tank, sensors, pump and dosing system (DCU, in the figure), injector, mixer, SCR catalyst and electronic control. The impact on vehicle layout and cost is clear and it is a roadblock that limits a possible SCR application to high segment vehicles only.

10.8.4 On Board Diagnostic

During the Euro 3 standard implementation, starting from January 2004, the application of the European on-board diagnostic system became mandatory also for diesel engines. Similarly to spark ignition engines, this system must monitor the emission-related components and systems that in case of failure make the emission level to exceed the system thresholds when driving according to conditions close to the NEDC cycle.

Differently from spark ignition engines, diesel engines do not include emission control components having a physical output directly linked to the emission behavior. The system is thus rather more complex and this is also the reason why a certain delay was allowed for its first implementation. The interaction of relevant components must be monitored using model-based strategies able to recognize that a drift, exceeding certain ranges during standardized conditions, is responsible for an emission non-compliance. This is true for all the main emission-related components such as exhaust gas recirculation, oxidation catalyst and particulate filter. Their monitoring is based on the correlation between temperature, pressure and air to fuel ratio measurements and engine load and speed. When using additives, the level in their tank can be easily checked, alerting the driver and verifying that the tank has been refilled with the correct additive. If the system detects a response different from the expected one, it has the authority to reduce the engine performances.

10.8.5 New Types of Combustion

Considering how much the after-treatment systems impact on the final diesel engine configuration, it becomes a priority to investigate how a different type of combustion can be beneficial in reducing engine baseline emissions.

When dealing with the pre-mixed combustion, the combustion sensitivity to dispersions and combustion noise, during transients, has still to be improved. A key role is played by engine control strategies and particularly by the air and exhaust gas recirculation flow controls that need to be fast and precise to manage the engine relevant parameters: effective exhaust gas flow re-circulation and the excess-air factor λ.

In present diesel engines, the ignition delay strongly decreases as the speed and the load increase; premixed combustion is thus applicable only to low speed and low load conditions. In particular, the use of premixed combustion is not profitable above a threshold of mean effective pressure of about 5 bar, since the advantages in NO_x reduction are overcompensated by a strong increase of CO and hydrocarbons emissions and fuel consumption. The reason for this is that injection and combustion must be delayed, thus worsening fuel consumption, to obtain a sufficient ignition delay. As a consequence, it is easy to understand that the benefits of premixed combustion are limited to conditions characterized by a reduced load factor.

To extend the application of pre-mixed combustion to higher loads is one of the main future activities and, in this perspective, the integration of low pressure exhaust gas recirculation, compression ratio reduction, boost improvement, and variable valve actuation looks promising when applied together with premixed combustion.

From a theoretical point of view, the Homogeneous Compression Combustion Ignition (HCCI) engine is the engine closest to the ideal Otto cycle, where the combustion occurs at a constant volume. This concept aims to produce a combustion evolving in a very short time, smaller than spark ignition and diesel engines. To this end a pre-mixed homogeneous charge is desirable. Depending on the fuel type, different systems can be used for the fuel supply, but the homogeneous mixture formation is possible only if a sufficient time is available for the mixing process.

As it is well known, combustion starts when the ignition conditions, determined by compression, are reached. These conditions are functions of the in-cylinder temperature, pressure and air to fuel ratio. In the case of a pre-mixed homogeneous charge, the ignition conditions should occur in the whole combustion chamber, with the formation of multiple points of near-simultaneous ignition, which do not contain localized high temperature regions of flame front. As a consequence, there is no clear distinction between the burned and unburned mass fraction, defined by the flame front propagation. In practice, the auto-ignition already described for spark ignition engines which determines knocking, share much in common with homogeneous compression combustion ignition although the temperature and pressure time histories the fuel undergoes are quite different.

It is thus useful to point out that the main problem with homogeneous compression combustion is the auto-ignition control. A suitable management of the ignition delay, based on engine speed and load, is the pre-requisite to obtaining the expected performances with the required emission control. Within the likely working range, the main potential advantage of homogeneous compression combustion is represented by its capability to extend the diesel high efficiency, close to 40 %, in the medium load area (e.g.: $p_{ime} = 6.7$ bar; peak combustion temperature = 1720 K; exhaust temperature = 561 K = 288 °C; indicated efficiency 40.9 %). A key difficulty is extending these results to transient conditions taking place in the actual use of the vehicle during the NEDC cycle.

Compared to spark ignition engines, the homogeneous compression combustion higher efficiency is determined by the use of a higher compression ratio, a load control performed by varying the air to fuel ratio without using the throttle valve and a faster combustion around the top dead center.

With reference to Fig. 10.57a, the average combustion temperature is shifted from the 2,250 K area to the left, around 1,700 K, with a strong charge homogenization and with an air to fuel ratio not far from stoichiometric. This low temperature combustion is characterized by a cool flame able to provide fuel oxidation at low temperature. Finally, low NO_x emissions results from the lower combustion temperature, while low particulate emissions results from a relatively high air to fuel ratio of an homogeneous premixed charge, thus avoiding local over-rich zones.

The practical application of homogeneous compression combustion has however showed, up to now, many difficulties to keep the combustion conditions sufficiently close to the optimal ones during engine transient operation as it occurs during the NEDC emission cycle. Nevertheless, intermediate steps in this direction can be implemented, to achieve a significantly reduced in-cylinder NO_x generation. In this perspective, the parameter playing a fundamental role is the charge dilution through exhaust gas recirculation, thus achieving a totally premixed combustion, i.e. the homogeneous compression combustion ignition.

The increased exhaust gas recirculation rate calls for a tight recirculation rate control not only in steady-state operation, but also in transient conditions, to avoid peaks of NO_x or soot emission, to avoid unacceptable particulate filter loading. This objective can be pursued by following different strategies. One is to improve the exhaust gas recirculation tolerance with a reduced air to fuel ratio, lean but not far from stoichiometric. Another way is to increase the recirculation rate at the same mixture ratio; in this case the density of either the re-circulated gases or the air mass must be increased, particularly at part load. In this direction, the air boosting provided by the Two-Stage Turbo can be appreciable.

The disadvantage of homogeneous compression combustion is the CO and hydrocarbon rise determined by the wall quenching effect, like in spark ignition engines. In addition, the extremely quick combustion causes a high pressure rise, with a consequent distinctive noise.

The homogeneous compression combustion ignition is a matter of theoretical studies and experiments for both gasoline and diesel engines.

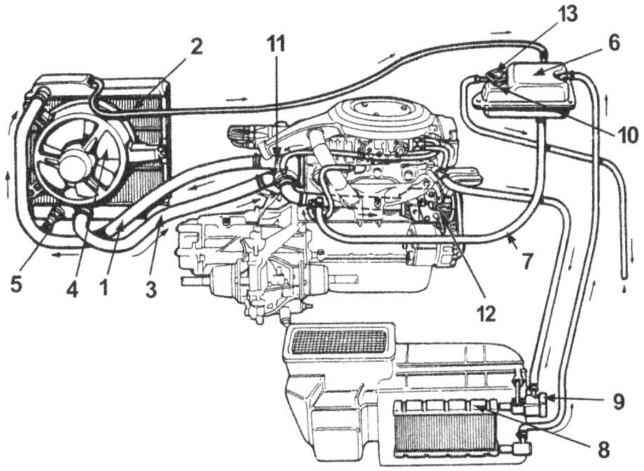

Fig. 10.66 Cooling system (courtesy of FIAT)

10.9 Auxiliary Engine Systems

10.9.1 Cooling System

A significant amount (about 15 %) of the fuel energy is dissipated as heat by the cooling system, whose task is avoiding that the engine reaches dangerous temperatures. In most vehicular applications liquid cooling is used, with a closed circuit provided with a circulation pump located between the radiator and the engine. The system is pressurized and the operating pressure depends on the pressure control valve calibration, while the radiator is connected to an auxiliary tank, which allows the fluid to expand.

In sealed circuits the cooling fluid, a mixture (normally 50/50 %) of water and ethyl glycol having a freezing point at $-50\,^{\circ}$C, is of the permanent type. The ethyl glycol contains specific additives having anti-oxidizing, anti-corrosion, anti-foam and anti-deposits properties. This package of additives has the goal of maintaining the initial specifications for the different circuit components, mainly for rubber hoses, therefore assuring good performance for a long mileage and avoiding the need for fluid replacement.

The key components of the system are shown in Fig. 10.66. The cooling liquid leaves the engine through pipe (1), reaches the radiator (2) and exit from it through pipe (3). The water circulation is activated by the water pump (12). The radiator fan (4) is driven by an electric motor. Its speed is determined by the temperature of the cooling fluid and the effective cooling needs. The on-off switch actuation is supplied by a sensor (5) located on the lower tank of the radiator. The cooling fluid temperature can thus be maintained at its optimal level, avoiding an excess of cooling when the ambient temperature or the requested power are low. The possibility of switching off the fan reduces the fuel consumption and maintains the engine components at their correct running temperature. A quick warm-up of the engine is possible.

The installation described is typical in the transversal front engine configuration. The system includes the auxiliary tank (6) connected to the remaining circuit through pipe (7). The tank is provided with a safety valve (10), included in the filling cap (13), to protect the circuit in case of overpressure. The figure shows also the passenger compartment heating radiator (8) with its regulation valve (9).

The thermostat (11) allows the cooling fluid to reach a stabilized temperature in the shortest possible time and to maintain the set temperature value independently from engine operating conditions and from ambient temperature. To reach this goal, the thermostat, a valve operated by a capsule containing a suitable type of wax, which increases its volume depending on the temperature, reduces the cooling fluid flow from the engine to the radiator, thus preventing an excessive cooling.

The cooling system can contribute to improve the engine mechanical efficiency splitting the circulation between cylinder head and crankcase, thus allowing the mechanical components to reach their optimum running temperature. More recently, the ECU is given the task of coordinating the so-called engine thermo-management, which includes an electrically driven water pump, a map-controlled thermostat and an electric fan. This system allows further optimization of the temperatures of the different engine components in any operating condition. The electric water pump can remain operational to remove waste heat from the turbocharger, for instance, when the engine is switched off after full-load operation and can be completely shut down to achieve a faster warm-up.

10.9.2 Engine Lubrication Circuit

The lubrication circuit, sketched in Fig. 10.67, provides the required oil quantity to the engine components where metal surfaces in relative motion are in contact.

Oil is stored in the oil sump which provides an effective oil cooling since it is exposed to the air stream. The engine-driven gear pump (1) sucks the oil from the sump (3) and delivers it, at the correct pressure, in the range 3.5–5 bar, at 100 °C, to the engine components through a system of ducts and pipes. The filter (8) and the calibrated relief valve (2) are located between the pump and the longitudinal main duct (5). An electric pressure transducer allows the continuous monitoring of the oil pressure in the main duct.

The system of ducts and galleries is machined, partly in the crankcase and partly in the components which need lubrication. The purpose of the main duct is distributing the oil to the crankshaft main bearings. Within the crankshaft (4), a suitable duct system provides the oil transfer between the main bearings and the connecting rod bearings. The additional secondary duct system (6) starts from the main one and allows the oil to reach the cylinder head, the camshafts, valves, camshaft driving mechanism and turbocharger shaft (9). The return lines (7) and (10), recover the hot oil into the oil sump.

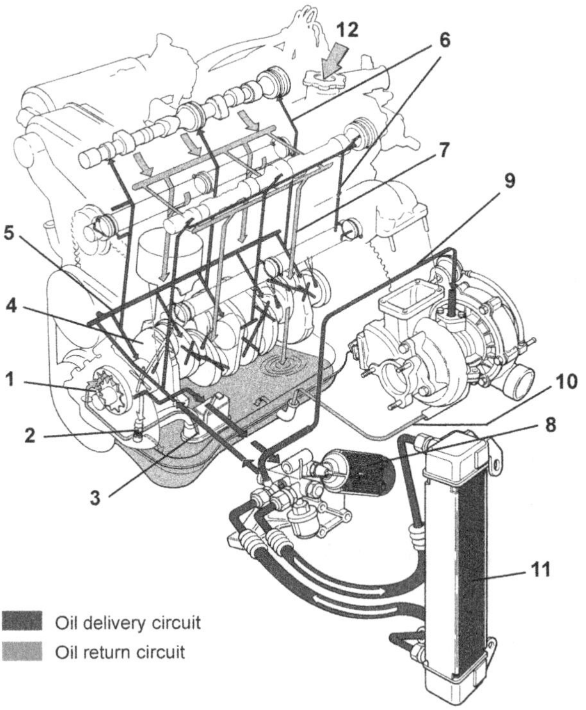

Fig. 10.67 Lubrication circuit (courtesy of FIAT)

To reduce the parasitic power loss, a new type of capacity-controlled oil pump, delivering only the amount of oil demanded by the engine, is increasingly used. Its capacity is maximum at idle and is then reduced with increasing engine speed.

The circuit must provide a suitable cooling capability to assure that the temperature remains in the range 100–120 °C; a 140 °C temperature value is considered as a limit. When necessary, a cooling-fluid/oil or air/oil heat exchanger can be added (11). The filling cap (12) allows the oil refilling.

Almost always, a specific oil jet operates through a pinhole drilled through the bearing and the connecting rod big end to direct an oil jet along the thrust side of the piston, to improve the lubrication of the latter. In both compression ignition and gasoline turbocharged engines the piston features a circumferential gallery (Fig. 10.52a), to more effectively cooling the piston ring assembly.

The graduated dipstick is always present to check the oil level in the sump, but oil level sensors are increasingly used to alert the driver if the level drops too low.

(a)

	180°	360°	540°	720°
1	Intake	Compr.	Comb. Expans.	Exhaust
2	Compr.	Comb. Expans.	Exhaust	Intake
3	Exhaust	Intake	Compr.	Comb. Expans.
4	Comb. Expans.	Exhaust	Intake	Compr.

(b)

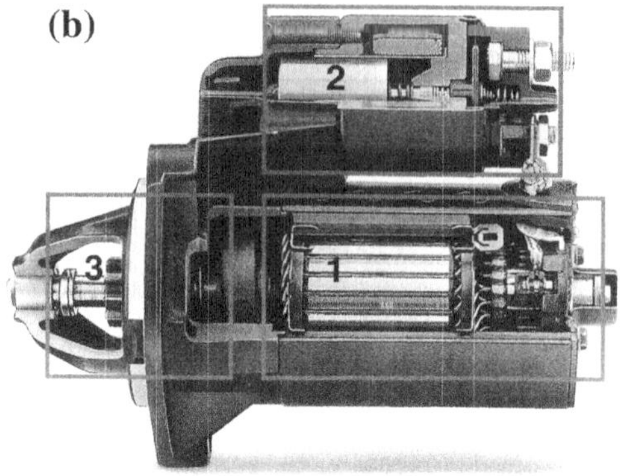

Fig. 10.68 **a** Phases in the various cylinders of an in-line, four-cylinders engines as functions of the crank angle. **b** Sketch of a starter motor (from P. Scolari, *Motor Vehicles and their evolution*, 10th edition, 2010)

10.9.3 Starter Motor and Alternator

The starting system of internal combustion engines includes a battery-driven DC starter motor, able to drive the engine at a speed at which engine starting is possible. The table in Fig. 10.68a, referred to an in-line, four-cylinders gasoline engine, shows that one cylinder is always in its intake phase, in whichever position the crankshaft stops. Theoretically, one cylinder is always in the condition to ignite within a single revolution. In the real word, if the engine speed is below a certain threshold, the energy available during the first expansion phase is not sufficient to complete the mixture compression phase in the subsequent cylinder, for the following main reasons:

- The first cold cycle cannot be completed,
- friction, in cold conditions, reach its maximum level, and
- the engine flywheel does not accumulate sufficient energy.

A power of 800–3000 W is required for the started motor, and thus in a 12 V system, a current of a few hundred Amperes is required.

The main components of the starter motor, namely the rotor (1), the engagement solenoid (2) and the pinion (3) meshing with the engine flywheel with a reduction ratio of about 10 or 15, are shown in Fig. 10.68b. An average of 2,000 starting operations per year can be computed from a mileage of 15,000 and 7.5 km of average travelled distance.

Compression ignition engines, of course, require a more powerful starter motor, mainly considering the higher compression ratio and the high torque required to drive the high pressure pump.

The alternator, shown in Fig. 10.69, must supply the vehicle's electrical system with a sufficient quantity of current under all operating conditions to ensure that the battery is consistently maintained at an adequate charge level. The battery feeds all the accessories, which must be in the condition to run, for a certain time, even with

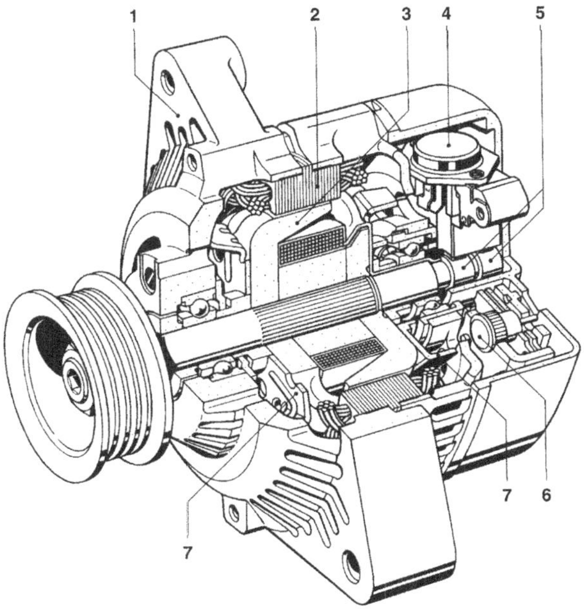

Fig. 10.69 Alternator (courtesy of Bosch)

Table 10.5 Typical values of power for some electric loads

Item	Power (W)
Starter motor	800–3,000
ECU	175–200
Cooling fan	200
Diesel glow-plugs (4 cyl.)	400
Fuel pump	50–70
Windscreen wiper	60–90
Headlights (each)	50–60
Passenger compartment lamp	5

the engine stopped. The electric loads for typical accessories and engine components are listed in Table 10.5.

The cross section of a typical automotive alternator is shown in Fig. 10.69. It is entrained with a multiplying gear ratio of 1:2 or 1:3. Its internal components are cooled by two small internal fans (7). The stator (2) is located between the two external semi-parts (1) constituting a supporting structure. The DC excitation current, required by the excitation winding, is supplied to the rotor (3) through collector rings (5) and sliding contacts (carbon brushes, 4). As direct current is required for charging the battery, a set of diode rectifiers, normally six (6), must convert the alternator's three phase AC into DC. The diodes also prevent the battery from discharging when the vehicle is stationary. The diodes are located in a suitable structure, adequately cooled and the electronic regulator is integrated in the structure supporting the carbon brushes.

In its latest configurations, the alternator is able to convert into electric energy part of the braking energy during vehicle slowing down.

Chapter 11
Transmission

11.1 Transmission Layout

Most motor vehicles need a transmission, since the speed requirements of the wheels are different from those of the engine. This is particularly true when the engine is a reciprocating internal combustion engine: in this case the transmission includes a start-up device (either clutch or torque converter), a variable ratio gearbox and the final drive and differential, which operates the wheel by means of shafts and joints. The transmission thus includes the entire kinematic chain connecting the wheels to the engine.

The function of the transmission is thus to match the available torque from the engine to the needs of the vehicle, which are imposed by the nature of the road, the will of the driver and environmental requirements. The transmission is particularly relevant in determining a number of vehicle functions, such as dynamic performance, fuel consumption, emissions and driveability.

Transmission technology can be assumed to be mature, but this does not imply that there will be no further evolution, particularly in automatic transmissions. As a matter of fact, significant developments in the field of automatic transmissions are expected, particularly in the European market. The developments in the mechatronic field, both in terms of performance and cost, are making it possible to develop transmissions that essentially consist of conventional manual transmissions, automated by actuators and controlled by microprocessors. They are meant not to affect negatively vehicle performance while leaving the driver the pleasure of driving manually, if he chooses so.

At present, the most common type of automatic transmission is the power-shift transmission, with four or more gear ratios, where speed shifts are made without engine disengagement. Nevertheless some manufacturers use for their cars a continuously variable transmission, with a steel belt and variable diameter pulleys. In spite of some disadvantages related to their mechanical efficiency, they show better vehicle performance because of the availability of an infinity of transmission ratios.

G. Genta et al., *The Motor Car*, Mechanical Engineering Series,
DOI: 10.1007/978-94-007-8552-6_11,

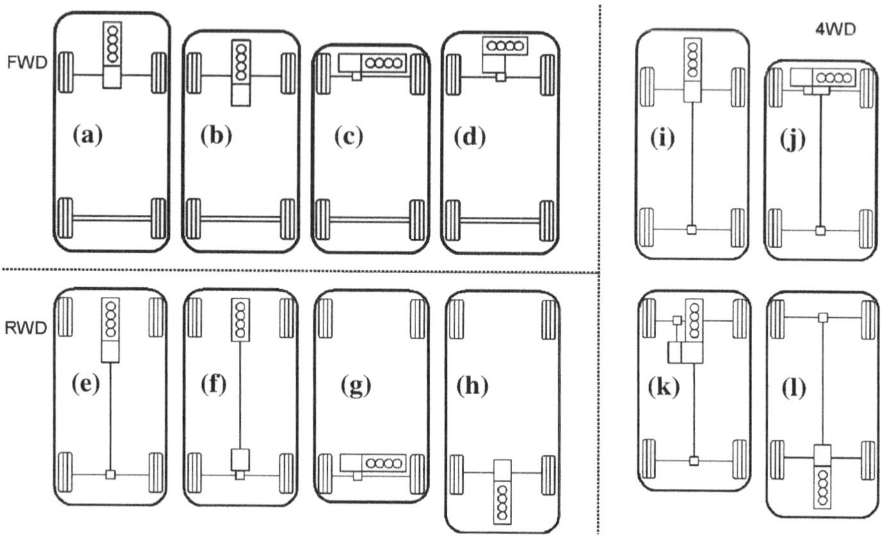

Fig. 11.1 Different types of transmission architectures, used in for front-wheels-drive cars (**a**–**d**), rear-wheels-drive cars (**e**–**h**) and 4WD cars (**i**–**l**). Solution **k** is particularly suitable for off-road vehicles, because an additional range-change unit can be easily integrated into the transfer box. The impact on external dimensions is deliberately exaggerated

Concerning manual transmissions, six gear transmissions are no longer an exception, while the five gear ratio configuration is standard on all applications; many of these transmissions sport an electro-hydraulic actuation system, with the aim of offering a similar or superior performance when compared to conventional automatic transmissions, without affecting the traditional advantages of lower cost, high efficiency, and driving pleasure.

The functions that can be obtained from this kind of transmission include clutch automation, servo assistance in selecting and shifting gears, with positive effects on comfort and actuation speed, availability of more ergonomic sequential shifting commands, which can be included on the steering wheel, and a fully automatic choice of transmission ratio and shifting sequence.

Transmission architectures are determined by the decision about which wheels must perform the driving task. This choice has a strong impact on vehicle performance, in terms of handling, ride comfort, safety and interior space organization.

The engine can be installed either in the front or in the back of the vehicle, even if in some cars it is located between the axles (central engine). In all cases the driving axle may be the front or the rear one, with the possibility that all wheels are driving (4WD), however, the solution with rear engine and front wheels drive is extremely rare. The solution with rear wheels drive, with longitudinal engine in the front, is usually considered the conventional or traditional layout.

Figure 11.1 shows the architectures used on front-wheels-drive cars (a–d), rear-wheels-drive cars (e–h) and 4WD cars (i–l). Sketch (a) shows the longitudinal layout, with the powertrain on the driving axle; it also shows how some advantage can be obtained for the maximum steering angle of the wheel, with some disadvantage for the front overhang. In this case every transmission ratio can be obtained with a single stage of gears; the rotation axis of the gearbox output shaft must be turned 90° with a pair of bevel gears.

Sketch (b) shows the same layout, where the front overhang is shortened by installing the engine over the front axle; in this case an additional shaft and some modification of the oil sump are needed.

Sketch (c) shows the most common solution: transversal powertrain with engine and transmission in line. Here a single stage transmission is necessary; the output shaft is connected to the differential through a pair of spur gears. Vehicles can be more compact because the front overhang can be shorter; on the other hand the minimum turning radius is negatively affected. This solution is particularly suitable when the engine is not too long.

Finally, sketch (d) shows a not widely applied solution, where engine and gearbox are transversal and parallel; the gearbox can be in the front, behind or beneath the engine. Advantages and disadvantages are mixed, as compared with the previous case; an additional double stage transmission or a chain transmission is needed to connect the engine shaft with the gearbox input shaft.

In all these cases the subsystem including engine and transmission is called the *powertrain* or *transaxle*.

Sketches (e) and (f) show two possible architectures applied to conventional vehicles; type (e) is the most widely used, while type (f) is applied primarily in sports cars, where a more uniform weight distribution is required. In these cases the gearbox must be of the double stage type (also known as the countershaft type), where input and output shafts are in line. A bevel gear final drive is also necessary. The front overhang can be reduced; the interior space is negatively affected by the transmission shaft under the floor.

Two possible versions of the rear engine lay-out are also shown: the transverse powertrain arrangement (g) and the longitudinal arrangement (h). This kind of architecture is today only applied in sports cars, where weight distribution and yaw inertia are more important than interior space; the same considerations made for solution (a) and (c) apply.

The architectures that are known for four wheel drive vehicles are shown in sketches (i), (j), (k) and (l). Configurations (i), (j) and (l) can be derived from the corresponding layouts for single driving axle vehicles. These are primarily applied to road vehicles or sport utility vehicles (SUV) with permanent four wheel drive; solutions (i) and (l) are easily designed, because the transfer box can be integrated in the powertrain. Architecture (k) is applied to vehicles specialized for off-road missions; the transfer gearbox required to move the other axle can easily include a range-change unit and differential lock.

Independent of transmission configuration, vehicle transmissions include the following components, which are listed starting from the engine.

Start-up device

Usually a friction clutch, operated by the driver through a pedal, is used with manual gearboxes; sometimes, the driver is assisted by an external source of energy (electric, pneumatic or hydraulic). In most automatic gearboxes, the start-up device is based on a hydraulic torque converter with a lock-up clutch.

Gearbox

The gearbox is the transmission component that allows the driver to change the transmission ratio; it can be based on ordinary or planetary gears. Gear shifts can be performed manually (with only the force of the driver), semi-automatically (with the help of auxiliary energy) or automatically. In the gearbox there are also *gear shifting mechanisms* and *synchronizers*, whose aim is engaging and disengaging the pairs of gear wheels needed to obtain the desired transmission ratio.

Final drive and differential

The final drive and differential can be integrated in the gearbox or located in a separate box. The former further reduces the rotational speed of the wheels, while the latter theoretically operates the two wheels of the same axle with the same torque even if the speeds are different. It may be free, self-locking or manually-locking.

Transfer boxes

Transfer boxes are gearboxes that, like a differential subdivides the torque between two wheels, subdivide the torque among the axles of a vehicle with more than one driving axle.

Shafts and joints

Shafts and joints connect the several rotating sections of the transmission; in conventional drives they connect the gearbox with the differential. In independent suspensions, they connect the wheels with the differential.

11.2 Manual Gearbox

Gearboxes are normally classified according to the number of gear wheel pairs (*stages*) involved in the transmission of motion at a given speed. In the case of manual vehicle transmissions, the number to be taken into account is that of the forward speeds only, without considering the final gear, even if included in the gearbox. The usual types are:

- *Single stage* gearboxes,
- *dual stage* or *countershaft* gearboxes, and
- *multi-stage* gearboxes.

Figure 11.2 shows the three configurations for a four speed gearbox. Each wheel is represented by a line whose length is proportional to the pitch diameter of the wheel; the line ends with two horizontal short lines, representing the width of the teeth. If the line is interrupted where crossing the shaft, the gear wheel is idle, while if the line crosses the line representing the shaft without interruption the wheel is rigidly attached to the shaft and rotates with it. Hubs are represented according to the

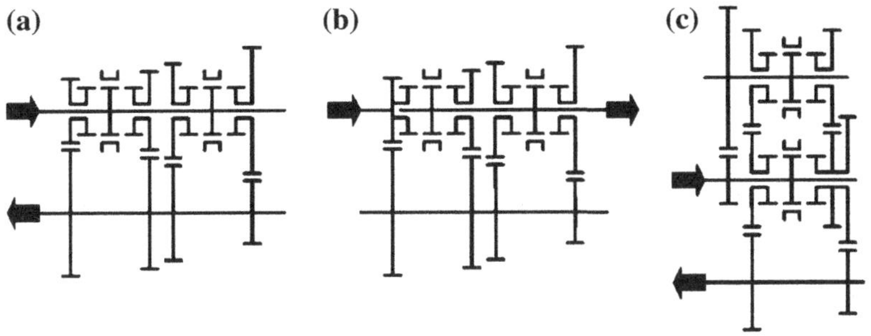

Fig. 11.2 Schemes for a four speed gearbox shown in three different configurations. **a** Single stage, **b** double stage and **c** triple stage

same rules, while sleeves are represented by a pair of horizontal short lines. Arrows show the input and output shafts.

Single stage gearboxes are primarily applied to front-wheels drive vehicles, because in this case it is useful that the input and the output shaft are offset; in conventional vehicles, on the other hand, it is better that input and output shafts are aligned: This is why rear-wheels drive vehicles usually are provided with a double stage gearbox.

In triple stage gearboxes, shown in Fig. 11.2c, the three shafts do not lie in the same plane, as the scheme seems to show. In a front view, the outline of the three shafts should be represented as the vertices of a triangle; this lay-out reduces the transversal dimension of the gearbox. In this case and others, as will be shown later, the drawing is represented by rotating the plane of the input shaft and of the counter shaft on the plane of the counter shaft and of the output shaft.

Gear trains used in reverse speed are listed separately. The inversion of speed is achieved by using an additional gear. In a train of three gear wheels, the output shaft has the same direction as the input shaft, while in trains of two gears the output shaft has an opposite direction; the intermediate gear is usually referred to as an *idler*.

The main configurations are reported in Fig. 11.3. In scheme (a), an added countershaft is provided with a sliding idler, which can match two close gears that are not in contact, as, for example, the input gear of the first speed and the output gear of the second speed. In this scheme, the drawing does not preserve the actual dimension of the parts.

Scheme (b) has two sliding idlers, rotating together; this arrangement offers additional freedom in obtaining a given transmission ratio. The countershaft is offset from the drawing plane; arrows show the gear wheels that match when the reverse speed is engaged.

Scheme (c) is similar to (a) in relation to the idler; it couples an added specific wheel on the output shaft with a gear wheel cut on the shifting sleeve of the first and second speed, when it is in idle position.

Configuration (d) shows a dedicated pair of gears, with a fixed idler and a shifting sleeve.

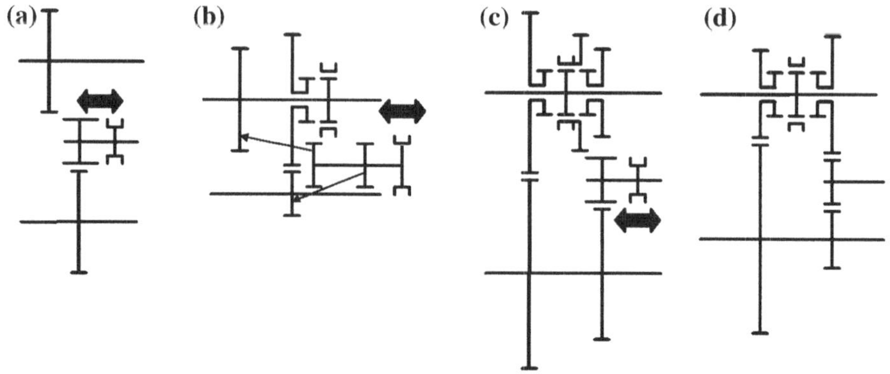

Fig. 11.3 Schemes used for reverse speed; such schemes fit every type of gearbox lay-out

The advantages and disadvantages of the configurations shown in the figure are:

- Schemes (a), (b) and (c) are simpler, but preclude the application of synchronizers (because the gears are not always engaged), and do not allow the use of helical gears (because wheels must be shifted by sliding).
- Scheme (d) is more complex but can include a synchronizer and can use helical gears.
- Schemes (a), (b) and (c) do not increase the gearbox length.

In manual gearboxes, changing speed and engaging and disengaging the clutch are performed by the driver force only. This kind of gearbox is provided with helical gears and each speed has a synchronizer; some gearboxes do not use the synchronizer for the reverse speed, particularly those in economy cars.

Single stage gearboxes are used in trans-axles; they are applied, with some exceptions, to front-wheels drive cars with front engine and rear-wheels drive cars with rear engine; this is true with both longitudinal and transversal engines.

Countershaft double stage gearboxes are used in conventionally driven cars, where the engine is mounted longitudinally in the front and the driving axle is the rear axle.

Four speed gearboxes represented the most common solution in Europe until the 1970s, with some economy cars having only three speeds. With the increase of engine power, the improvement in aerodynamic performance and increasing attention to fuel economy, the transmission ratio of the last speed had to be increased, with that of the first remaining at the same values; since car weight continued to increase and engine minimum speed did not change significantly. To achieve satisfactory performance all manufacturers developed five speed gearboxes, and this solution is now standard, while many examples of six speed gearboxes are available on the market, not only on sports cars.

An example of a six speeds, double stage transmission with the fifth gear in direct drive is shown in Fig. 11.4; in this case the first and second pair of wheels are close to the bearing. This rule is not generally accepted: On one hand having the most stressed pairs of wheels close to the bearing allows to reduce the weight of the shaft

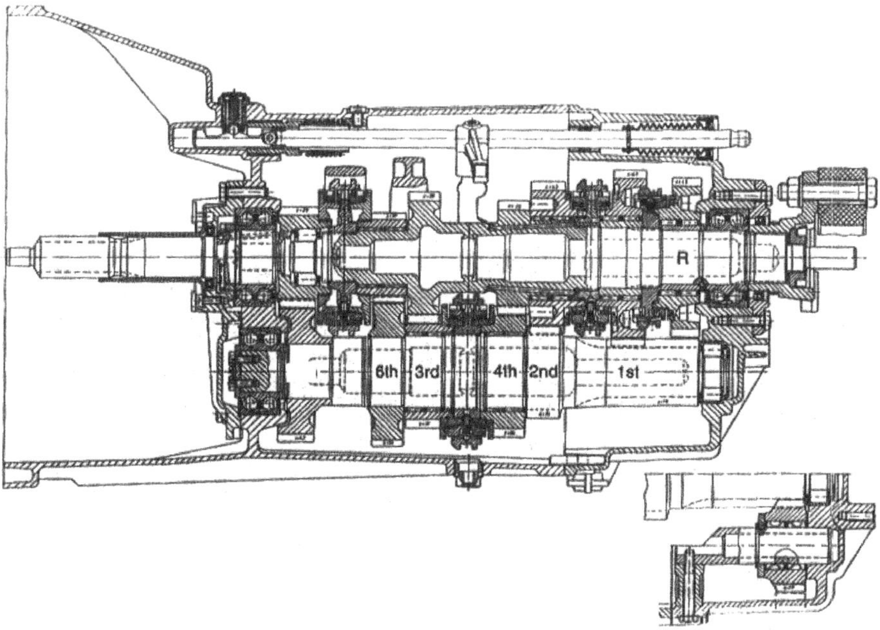

Fig. 11.4 Double stage six speeds gearbox (courtesy of GETRAG)

but, on the other hand, having the most frequently used pairs of wheels close to the bearing reduces the noise due to shaft deflection.

The synchronizers of the fourth and third speed are mounted on the countershaft; this layout reduces the work of synchronization, improving shifting quality by an amount proportional to the size of the synchronizing rings. The synchronizers of the first and second gear are on the output shaft, because of their diameter, larger than those of the corresponding gear; the increase of the synchronization work is compensated by the use of a double ring synchronizer. Synchronizers on the countershaft offer a further advantage: In idle position the gears are stopped and produce no *rattle* because of engine speed variations.

An example of a single stage gearbox for a front longitudinal engine is shown in Fig. 11.5. The input upper shaft must cross over the differential and the increased length of the shafts suggested adopting a hollow section. Because of this length, the box is divided into two sections; additional bearings are located on the flange connecting the two sections to reduce shaft deflections. The input shaft features a ball bearing close to the engine and three needle bearings that support radial loads only. The output shaft has two conical roller bearings on the differential side and a cylindrical roller bearing on the opposite side. This choice is justified by the relevant axial thrust due to the bevel gears.

The first and second speed synchronizers are on the output shaft and feature a double ring. The reverse speed gears are placed immediately after the joint (the idler

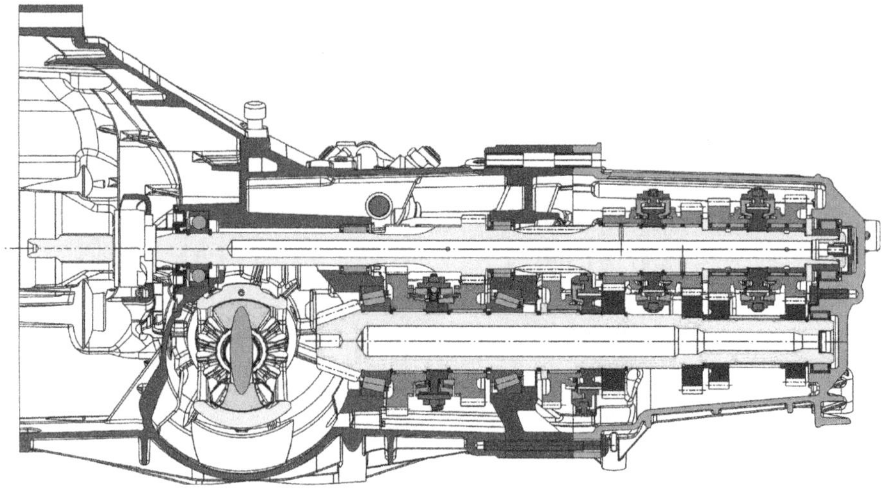

Fig. 11.5 Single stage six speeds gearbox for longitudinal front engine (courtesy of Audi)

gear is not visible) and have a synchronized shift. The remaining synchronizers are set in the second section of the box on the input shaft. The output shaft ends with the bevel pinion, a part of the final drive. The gears of the first, second and reverse speeds are directly cut on the input shaft, to reduce the overall dimensions.

Since most contemporary cars have front-wheels drive with transversal engine; in most gearboxes the final drive has cylindrical helical gears and is located in the gearbox. The pairs of gears are mounted from the first to the last speed, starting from the engine side, as shown in the example in Fig. 11.6.

Like many other transmissions originally designed with only four speeds, the fifth speed is located outside the aluminium box and is enclosed by a thin steel sheet cover; this is done to limit the transversal size of the powertrain, in the area where there is potential interference with the left wheel in the completely steered position. This solution is questionable as far as the total length is concerned but shows some advantages in reducing the span between the bearings. All bearings are of the ball type; on the side opposite to the engine the outer ring of the bearing can move axially, to compensate for differential thermal displacements. One of the gear wheels of the reverse speed is cut on the first and second shifting sleeve.

The casing is open on both sides and one of the bearings of the final drive is located on the end of the casing. A large cover closes the casing on the engine side and, in the meantime, provides installation for the second bearing of the final drive and the space for the clutch mechanism; it is also used to join the gearbox to the engine. In this gearbox, synchronizers are placed partly on the input shaft and partly on the output shaft.

A drawing of a more modern six speeds gearbox, in which it was possible to install all the gears in a conventional single stage arrangement, thanks to the moderate value of the rated torque, is shown in Fig. 11.7. Gears are arranged from the first to the

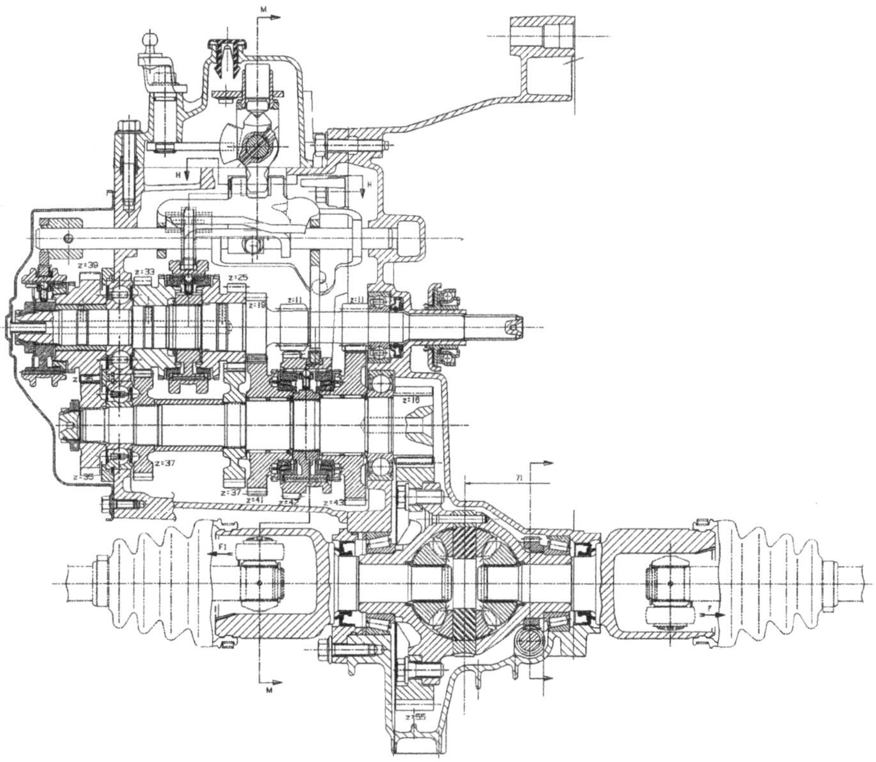

Fig. 11.6 Five speeds transmission for a transversal front engine (courtesy of FIAT)

sixth, starting from the engine side; as was already said this arrangement is required to minimize shaft deflection. Only the synchronizers of the first and second speed found no place on the input shaft; they are of the double ring type, as for the first speed. The reverse speed is synchronized and is located on a countershaft (not shown in this drawing).

11.2.1 Shifting Mechanisms

Mechanical devices that enable the driver to engage gears to obtain the desired transmission ratio are called *shifting mechanisms*; they are *internal mechanisms* if they are contained within the gearbox casing, or *external mechanisms* when they are mounted partly on the outside of the gearbox casing and partly on the vehicle body. In the latter case they connect the gearshift lever with the internal shifting mechanisms.

Shifting mechanisms perform an important function, because they are responsible for the response the driver receives from the shift stick, and therefore for the ease

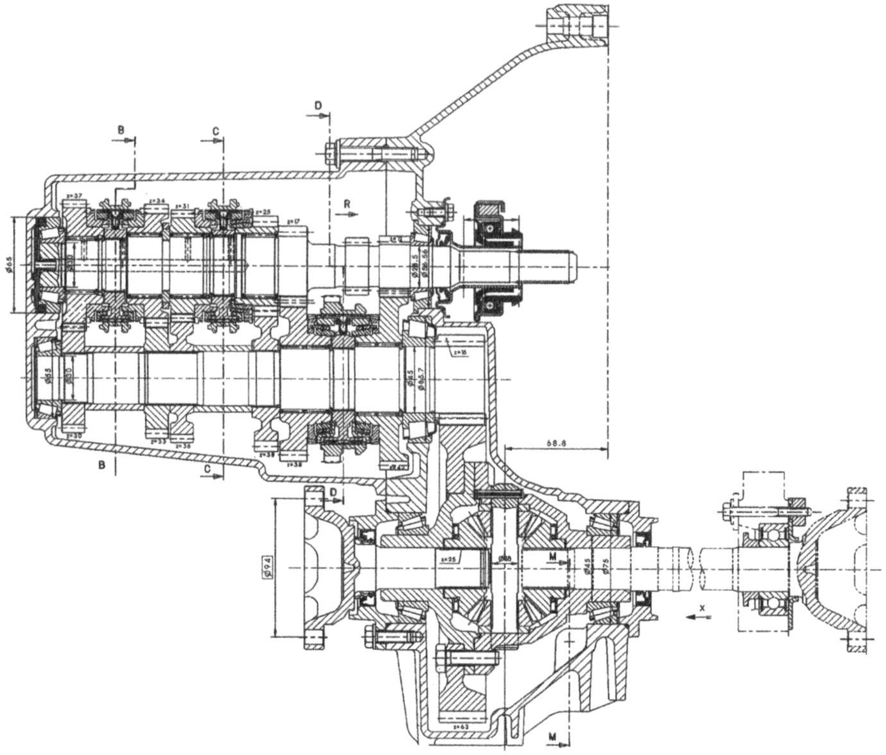

Fig. 11.7 Six speed transmission for a transversal front engine (courtesy of FIAT)

with which the gearbox can be operated. An ideal response should be positive and precise. From an ergonomic viewpoint, friction and clearance between moving parts of the shifting mechanisms can increase shifting time and make the identification of the engagement positions of the shift stick more difficult. These positions, in fact, are not shown by any device but are learned by the driver through the 'feel' of the shift stick.

Internal mechanisms include selector bars and selector forks; the latter move the shifting sleeves and synchronizers. To understand how the internal mechanisms works, reference will be made to one of the gearboxes seen in the previous section neglecting, initially, the synchronizers, since they do not change the nature of the shifting manoeuvre.

The shifting sleeves engage, through an internal spline, with both a hub on the gearbox shaft and another hub on the gear wheel. Wheels are kept in the correct axial position by spacers and the diameter transitions on the shafts and are free to rotate; a gear wheel is locked on its shaft by shifting the sleeve on its hub.

Each sleeve has three positions; with reference to the sleeve on the lower right of Fig. 11.4 for the first and second speed:

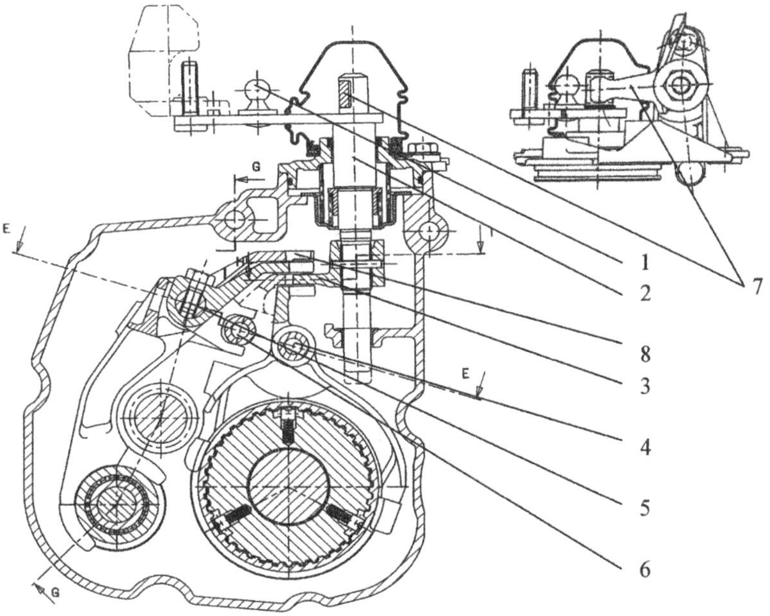

Fig. 11.8 Cross section of a transversal single stage gearbox on the internal shifting mechanism; *4, 5* and *6* are the shifting bars and forks moving the sleeves; a section of the sleeve on the output shaft, a section of the input shaft (*upper left*) and a section of the idling shaft of the reverse speed can also be seen (courtesy of FIAT)

- It is in neutral position when engaged with no wheel hubs, as shown in the figure.
- If shifted to the left, it locks the second speed gear to the output shaft and applies the second gear ratio.
- If shifted to the right, it locks the first speed gear to the output shaft and applies the first gear ratio.

The gearbox has four different *sleeves*: one for the first and second speeds, one for the third and fourth speeds, one for the fifth speed and the last for the reverse speed. In this gearbox the reverse speed is obtained by shifting an idler, not shown in the figure; when the reverse speed is engaged the idler meshes with a gear wheel obtained by machining the teeth on the sleeve of the second and first gear (in neutral) and with a specific wheel on the input shaft.

A typical mechanism suitable for this task is shown in Fig. 11.8. It is made by three selector bars (4, 5, 6), each one with a fork that engages with a groove on the shifting sleeve; the desired speed can be engaged by moving one bar at a time. Each bar is moved by a gate (8), existing on each *fork*.

A *finger* lever (3) on a shaft (2) can either rotate around the shaft center line or shift along this center line. When shifted it selects the correct selector bar; when rotated, in one of the three gates, it shifts the desired speed. The finger matches, with

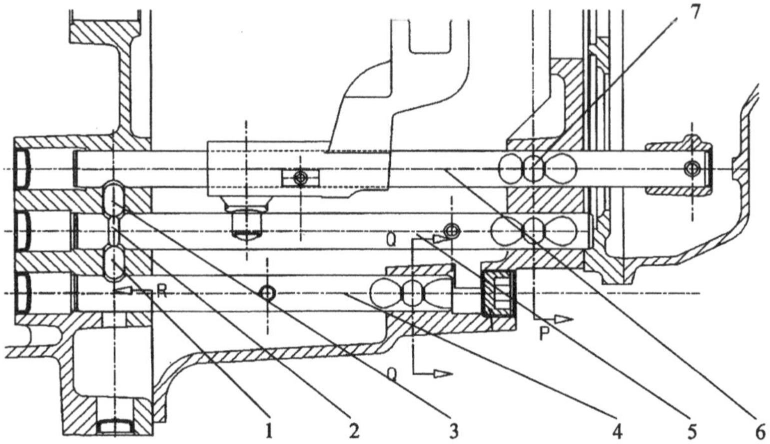

Fig. 11.9 Detail of the plunger and pin interlocking device, applied on the selector bars, to avoid the simultaneous shift of two bars; if one bar is moved the plungers *1* and *3*, through the pin *2*, keep the remaining bars in idle position (courtesy of FIAT)

a suitable low clearance, with the gates on the forks. In this gearbox the reverse fork slides on the selector bar of the fifth speed but features its own gate. Acting on levers (1) and (7), shifting and selection motions can be accomplished. When the gearbox is in neutral, all gates are aligned and the finger is set in the gate of the third and fourth fork.

To achieve a correct operation, there should be auxiliary devices suitable for:

- Maintaining the mechanisms in the selected position, even if the driver abandons the shift stick, and
- ensuring that more than one bar cannot be moved in the same time, with fatal consequence for the gearbox integrity.

The first function is performed by plungers with spherical ends, matching spherical grooves on the selector bars. The plungers are pushed into their grooves by springs that keep the bar in one of the three positions. The detail of such grooves (detail 7) are shown in Fig. 11.9; the central groove is for the idle position, the other two for engaging the required speed.

The second function is performed by *plungers* that are similar to those previously discussed; they have their center line on the same plane as the bars' center line (1 and 3 in Fig. 11.9); these plungers are set in two holes in the box, between the bar of the first and second speed, and that of the fifth and reverse speed. The bar of the third and fourth speed contains a hole, aligned with the others in the neutral position, with needle (2) inside; the length of the plungers and needle is designed to allow the motion of one bar at a time. A spring system must keep the finger in the idle position in the third and fourth speed gate; stroke limiters must stop the finger at both ends of the selection stroke, to avoid sticking.

The external shifting mechanisms are used to move the internal mechanism when the driver operates the gear shift lever. This is easy when the engine is located in front of the car in longitudinal position and the driving wheels are the rear ones, because the internal mechanism is close to the natural position of the driver's hand. Selection and engagement motions are consistent with those of the shifting bars and forks. The engagement positions of the gear shifting lever were historically defined by the simplest solution of the most common architecture.

The solution is much more complicated when engine and gearbox are mounted transversely or when the lever is not mounted on the tunnel, but on the dashboard or on the steering wheel shaft. The tasks to be performed by the shifting mechanisms are two:

- To convey the motion of the lever in a different direction; for example, in a front-wheels drive car with transversal engine, the engagement motion in the gearbox is transversal, while the corresponding motion of the shift stick is longitudinal, when the lever is on the tunnel or on the dashboard. It is almost vertical when the lever is on the steering wheel shaft.
- To maintain the engagement position of the shift stick unchanged, with reference to the neutral, even if the powertrain is moving because of vehicle vertical acceleration or torque variation; this problem is particularly important, because on front-wheels drive cars the powertrain also reacts to the wheel tractive forces.

A simple mechanism that can perform this function is the bar mechanism. It is a mechanical system made of bars articulated by means of spherical heads. Some bars in the longitudinal direction replicate the motion of the lever in a position close to the engine; other bars in the transversal direction connect the end of the former bars with the internal shifting mechanism.

The motion of the powertrain (rotations caused by the reaction torque and vertical oscillations due to uneven road surface) are mostly contained in a vertical plane; these motions will not interfere with the transversal motion of the mentioned bars.

An example of such a mechanism in shown in Fig. 11.10. The shift stick is connected to the bar 1 for the selection motion, with a pivot with transversal axis; the selection motion will rotate the bar around its axis and will leave the lever free to perform the engagement motion. Bar 1 is mounted longitudinally. A second bar 2 is linked to the shift stick at a different point, again with a pivot with transversal axis; the engagement motion of the lever shifts this bar longitudinally. The two motions are decoupled; the end of the bar 1 carries a finger that moves a second bar 3 transversely. A rocker arm 4 converts the longitudinal motion into transversal motion of the bar 5. Bars 3 and 5 move the internal shifting mechanism in the gearbox.

All articulated heads of bars 1 and 2 and of the rocker arm 4 are mounted on the plates A and B, fixed to the car body. If bars 3 and 5 have a sufficient length, all powertrain motions contained in the vertical plane will move the levers negligibly. The powertrain motion in a transversal plane is usually small, because the engine suspension in this direction is quite stiff since there are no vibrations to be filtered. The behavior of this mechanism can be adequate if the clearance of the pivots is limited and the lubrication of the bearings is acceptable.

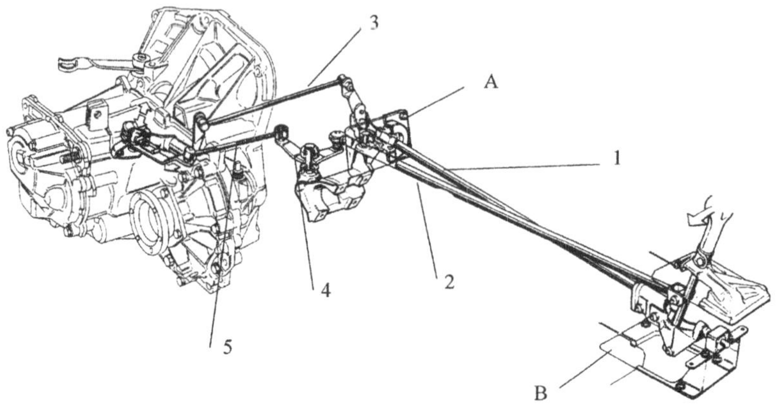

Fig. 11.10 View of a bar external shifting mechanism; bars *3* and *5* are not affected by the suspension motions of the powertrain; therefore the suspension motion does not affect bars *1* and *2* and the engagement position of the shift stick (courtesy of FIAT)

11.2.2 Synchronizers

The function of a synchronizer is to enable meshing gears to be changed on a moving vehicle without noise or damage. Nevertheless, during synchronization the clutch must be disengaged. The most widely applied synchronizer is the Borg Warner, named after the manufacturer who developed it. Figure 11.11 shows a cross section of this synchronizer, in its single cone version; there are, as usual, two synchronizers, for managing two neighboring gears on the same shaft.

The gears are idle and are mounted either on the input or output shaft on needle bearings and are constantly meshing together. Each gear carries a synchronizer *hub* 3 that can be either welded (by laser welding, as in the figure) or mounted with a spline shaft. Teeth can also be directly cut or stamped on the side of the gear. In the last case, there must be a sufficient difference in radial dimensions between the gearing teeth of the wheel and the selector teeth of the synchronizer hub. The different fabrication technologies of the joint between the hub and the wheel have a direct consequence for the length of the synchronizer assembly and the gearbox itself.

The hub 3 has a row of teeth on its outer circumference; the *synchronizer ring* is free to move axially and matches with the hub through a tapered friction surface. It can also match with the selector teeth of sleeve 4, in such a way as to lock the rotation, leaving the axial motion free. The synchronizer ring 2, between the hub and the sleeve, acts as a friction clutch that can synchronize the elements before the engagement of the teeth; in a following section it will be shown how this function can be achieved.

Single cone synchronizers were widely used until a few years ago; recently the target of shortening shift time without increasing the force applied by the driver forced to introduce multiple cone synchronizers, particularly on low gears where the synchronization work is more demanding.

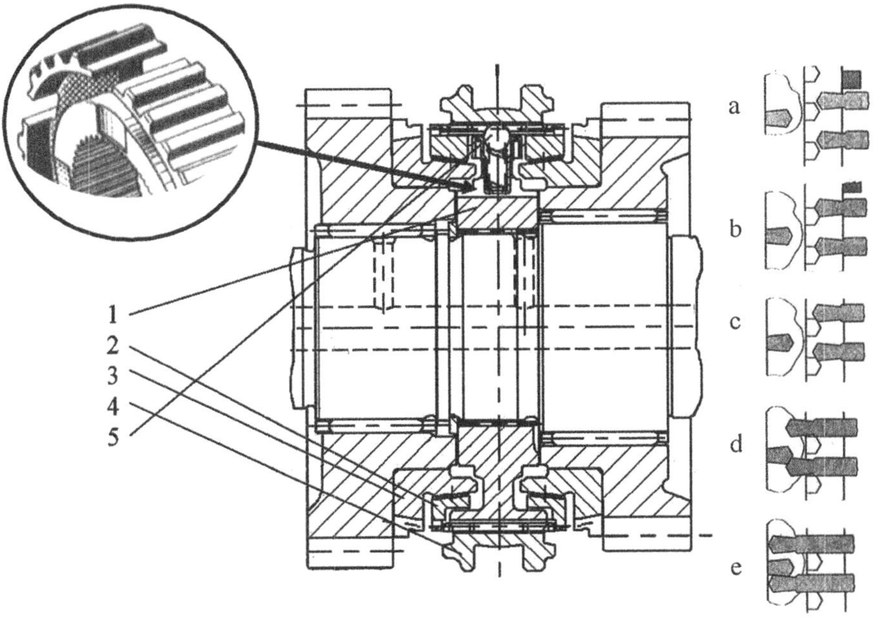

Fig. 11.11 Cross section of a single cone Borg Warner synchronizer. A detailed view of the synchronizer body is shown in the circle. On the right the positions of the teeth of the hub *3*, of the ring *2* and the sleeve *4* are sketched, in the different phases of the synchronization process (redrawn from Genta and Morello 2009)

A double cone synchronizer is shown in Fig. 11.12; this arrangement can be appropriate to the mid-gears of a large gearbox or the lowest gears of a small one. The hub is reduced to a row of teeth but there is a tapered friction ring connected to the hub and rotating with it, with some freedom to move in other directions.

Friction surfaces are machined on the inside and outside of this ring. The matching tapered surfaces are made with two different elements fixed to the selector sleeve in a particular way. The outside surface has the same shape as a single cone synchronizer; the inside surface is matched to the outside through a spline shaft that does not rotate but leaves a certain freedom along the radial and axial directions. This freedom is needed to compensate for the wear and alignment errors of the friction surfaces.

With reference to Fig. 11.11, a magnification of the synchronizer body is shown at the upper left of the figure. The body 5 has a spline surface with six interruptions at 60°; these interruptions are replicated on the synchronizer ring 2.

The ring interruptions match, with a defined angular clearance, with three large teeth on the inside diameter of the sleeve. The remaining three interruptions at 120° match with three push elements 5 that are free to slide inside the three mentioned elements, which are the sleeve, the body and the ring; the push elements has a radial hole in which a plunger with spherical head is located and is pushed by a coil spring against a seat on the inside of the sleeve.

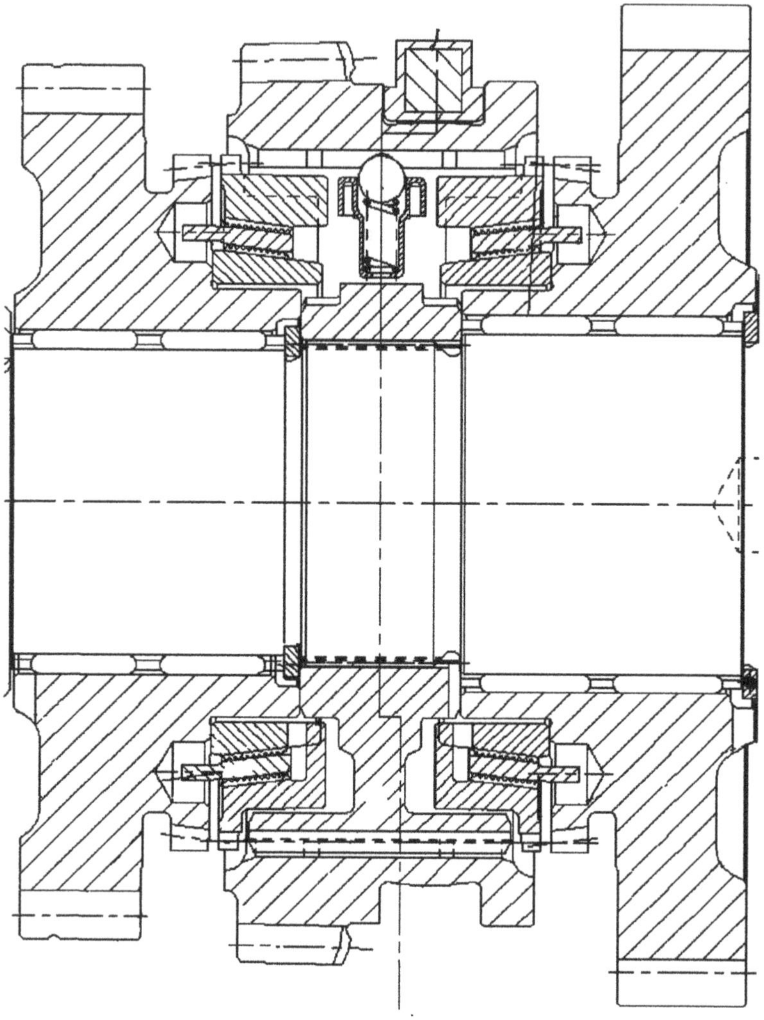

Fig. 11.12 Cross section of a double cone synchronizer (from Genta and Morello 2009)

Assume that the gearbox is in neutral, with both engine and vehicle in motion, and that the left wheel must be engaged. Five different phases of the shifting manoeuvre of the sleeve, actuated by the shift stick, can be identified; for each of these phases a schematic cross section is drawn, showing the teeth of sleeve, ring and hub developed on a plane.

(a) The sleeve is pushed to the left against the reaction force of coil springs that try to keep the push elements 5 in neutral position. Elements 5 and the large teeth push the ring 2 and bring into contact the two matching tapered friction surfaces of ring and hub. Because the two parts have different speeds, a friction torque will arise, opposing the ring and hub relative speed. The circumferential clearance between the large teeth and ring will allow a small rotation of the latter; the clearance is designed in such a way that the ring teeth stop the axial motion of the sleeve. This phase is called *pre-synchronization* because it does not imply a real energy exchange between moving parts.

(b) Friction torque between ring and hub transfers energy from the heaviest part of the system (the vehicle) to the lightest, to slow or to accelerate the rotating parts fixed to the wheel to be shifted. Friction torque *measures* the relative angular speed because it will go to zero together with the relative speed. The same torque allows sleeve motion to stop until synchronization is complete. This phase is called *synchronization*.

(c) Before describing the next phase, consider the shape of the dog teeth on the hub and sleeve. They have a tapered end and a counter-tapered surface. When ring and hub move at the same speed, the friction torque is zero and the tapered ends of the dog teeth rotate the hub of the angular thickness of half a tooth, under the pressure exerted by the shift stick.

(d) The sleeve is now free to move across the ring and to match the hub; again the tapered end of the teeth will rotate the hub if necessary. In this phase the push element of the plungers are completely retracted and no longer withstand the sleeve motion.

(e) Now a *positive engagement* is made; at the end of this phase the driver stops pushing the shift stick. The counter-tapered shape of the dog teeth will retain the sleeve in the engaged position under the action of the engine torque. The same shape will exert a certain reaction on the shift stick, when the driver begins to disengage the gear for the next speed shift.

11.3 Automatic Gearboxes

Automatic gearboxes were initially designed for comfortable family cars with large displacement engines, with the primary goal of freeing the driver from the physical effort of operating the clutch pedal during start-up and gears shifts. Driving was made easier also by eliminating the need to identify the most suitable moment for shifting and to coordinate operations, particularly in an uphill start. For passengers, the objective was to limit the *jerk* (the change in the vehicle acceleration) during gear shifts and starts; the human body is quite sensitive to this parameter, making it relevant to travel comfort. They are now increasing their market share in small and sports cars, a market where customer's expectations are more addressed to performance, economy and driveability than to comfort; in addition, much more attention is now devoted to emissions and energy conservation. Technical solutions to these problems have proliferated, but a prevailing solution has not yet been identified.

The following types of automatic gearboxes can be identified:

- *Full automatic* gearboxes; gear shift and start-up functions are both carried out automatically, with mechanisms developed for this purpose. Among different available working modes, there is usually a semi-automatic mode where the choice of the gear ratio is left to the driver, the shifting sequence being entirely automated.
- *Semi-automatic* gearboxes; one of the above automatic functions is partly or totally missing. For example, start-up device operation is automatic, but not the gearshift; or this second function is automatic, but decisions about the most suitable shifting moment are left to the driver. In this second version, the accelerator pedal must be operated by the driver in coordination with clutch operation, to avoid engine overspeeding or stalling. This kind of gearbox is no longer of interest to car designers, because it does not have a favorable cost/benefit trade-off. It is, however, used on some industrial vehicles, because of the effort required to operate the commands manually; these gearboxes sometimes share most components with their manual counterparts.
- *Robotized* or *automated* gearboxes; mechanisms are deliberately derived from manual gearboxes, including the friction clutch. These cannot be defined as semi-automatic gearboxes, like in the previous example, because their control systems can handle all functions automatically. The decision to use a modified manual gearbox can be justified by other goals, such as reducing production costs, since the existing production facilities can be used. Operating costs can be also reduced, due to the reduced mechanical losses in comparison with automatic gearboxes. There is also a positive impact on sports car performance, because shifting and start-up times can be reduced in a way not readily achievable even by professional drivers.

Gearboxes of the latter type are increasing their market share and can have further developments in the close future.

An important aspect conditioning automatic gearbox configuration is the gearshift mode, being possible to distinguish between gear shifts with and without power interruption or *powershifts*. In the first case gears are shifted as in a manual gearbox using synchronized or non-synchronized dog clutches; during gear shift the power flow going through the gearbox must be interrupted to allow the disengagement of the existing speed and the engagement of the next. The power available to move the vehicle goes to zero during this manoeuvre and the vehicle slows down; the jerk can thus be significant.

Qualitative diagrams of the traction force and vehicle speed as functions of time during a gear shift with and without power interruption are shown in Fig. 11.13. If the manoeuvre occurs without interruption of power, it is possible to reduce the jerk to a minimum and to continue to accelerate the vehicle. This result can be achieved on stepped gearboxes if each synchronizer is replaced by a clutch and disengagements and engagements overlap; it can be easily understood that in this case a gear shift sequence implies a larger power dissipation than in a conventional synchronizer. In this procedure shifting time is not as important and can be compromised in favour of a better driving comfort.

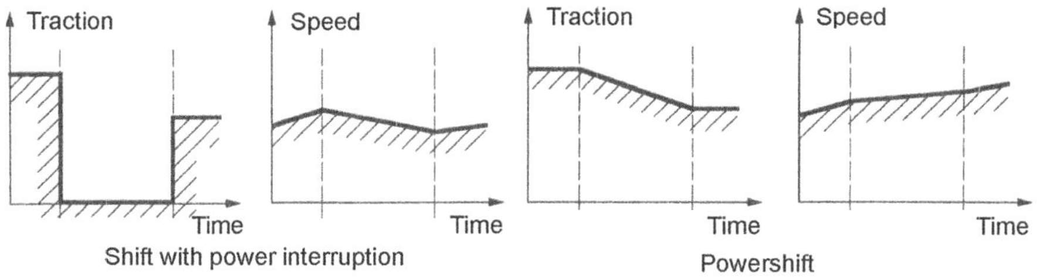

Fig. 11.13 Qualitative diagrams of traction force and vehicle speed as functions of time for gear shift with and without interruption of power (redrawn from Genta and Morello 2009)

Automatic gearboxes are said to be *stepped* if a limited number of transmission ratios (*steps*) is available, like in manual gearboxes. The total number of speeds has increased recently, in spite of the increased mechanical complexity, to reduce fuel consumption; many examples exist with eight speeds and a few with nine. When deciding the number of speeds of an automatic gearbox with torque converter, it is useful to remember that this start-up device is itself a continuously variable transmission. In this case it is usual to avoid the first speed of the corresponding manual gearbox, since its function is performed by the torque converter from stall to lock-up and therefore the number of available speeds is reduced by one unit without affecting vehicle performance: For instance, a four speed torque converter automatic gearbox can be compared to a five gear manual gearbox with clutch.

Stepless or *continuously variable transmissions*, CVT for short, feature an unlimited number of transmission ratios between a lower and an upper limit. On CVTs, shifting manoeuvres are, by definition, of the powershift type: Considering the small difference between the previous and the next transmission ratios, the energy wasted during gear shift is minimal.

Gear trains used on stepped automatic gearboxes can be based either on *ordinary gears* (gear wheels with fixed rotation axis), similar to those used in a manual gearbox, or *planetary* (epicycloidal) *gears* of different configuration that, together with band brakes or multi disc clutches, can achieve different ratios including reverse speed, always featuring coaxial input and output shafts. The first solution is simple but seldom used on powershift gearboxes because of the space taken by the clutches. Some examples of *synchronizer gearboxes* (automated gearboxes sometimes with powershift capability), *epicycloidal gearboxes* (automatic powershift gearboxes) and of *CVTs* (automatic powershift gearboxes) are described in the following sections.

11.3.1 Synchronizer Gearboxes

Automatic gearboxes with synchronizers are often derived from manual gearboxes. External shifting mechanisms are usually electrohydraulically servo actuated. On hydraulic systems, the external shifting mechanism includes two types of

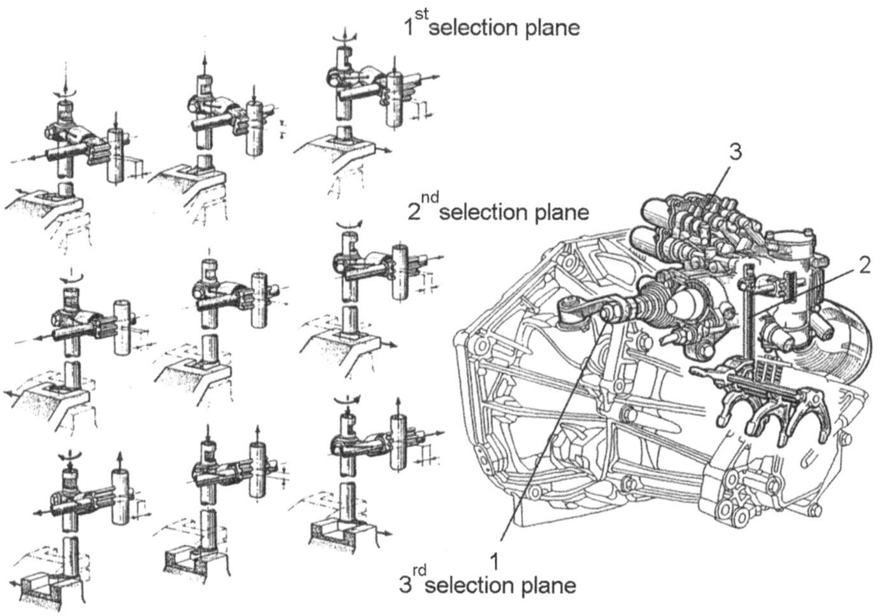

Fig. 11.14 Scheme of an electro-hydraulically actuated internal shifting mechanism. A Hydraulic actuator *1* disengages and engages the clutch, while a second combined actuator provides for rotation of the shaft *2*; a hydraulic group *3* includes an oil pump, a pressure accumulator and the electric actuation valves (courtesy of Marelli)

components: Those providing for hydraulic energy generation, regulation and distribution, and a set of actuators. Actuators are as many as the controlled movements to be performed: One for the clutch and two, at least, for the gearbox (one for selection and one for engagement).

A system of this kind is shown in Fig. 11.14. A hydraulic actuator 1 engages and disengages the clutch, while a second double actuator provides for shifting and turning the shaft 2 and for engagement and selection motions; the hydraulic group 3 includes an oil pump, a pressure accumulator and electric actuation valves. The selection mechanism is handled by a rack that can raise the shaft 2 matching the finger with the three selection gates (first and second speed, third and fourth speed, fifth and reverse speed). The second actuator can rotate the same shaft to obtain engagement and disengagement. It must feature three different positions: The engagement of the two neighboring speeds and the intermediate idle position.

A peculiar configuration scheme is shown in Fig. 11.15; it is called *dual-clutch* because two different clutches provide for start-up and power-shift. The scheme is suitable for a front-wheels drive car with transversal engine, but could be easily adapted to different drive configurations. Clutches F_1 and F_2 can entrain two coaxial shafts, where one is fixed to the odd speeds input wheels and the other to the even speeds input wheels. In first speed, for example, clutch F_2 is engaged, while clutch F_1

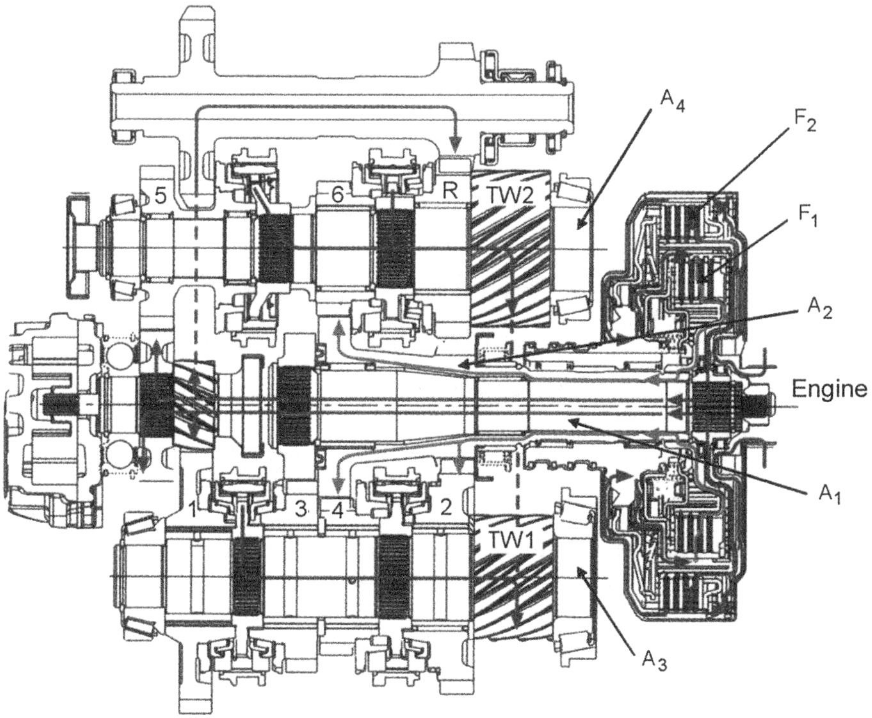

Fig. 11.15 Working scheme of a powershift dual-clutch six speed gearbox (courtesy of Audi)

is disengaged; the second gear shaft is idle and receives no torque. If the sleeve of the second speed is engaged, the internal shaft begins to move but remains at idle; at any time clutch F_1 can be engaged and clutch F_2 disengaged, according to a powershift procedure. In the same way, the first gear can be disengaged, to engage the third, in preparation for the next shift. Speeds must be used sequentially and double up or down shifts are impossible.

The clutch F_2 is also used for starting up the vehicle. The peculiar architecture of this gearbox features a reduced number of idle turning clutches and no torque converter. The effect on mechanical efficiency is surely positive, but this advantage is offset by an increased design complexity.

The scheme is drawn on a single plane, as usual; on the engine shaft the two clutches F_1 and F_2 (start-up device) that move shafts A_1 and A_2 are installed; shaft A_1 operates the input wheels of the first, third and fifth speed, while shaft A_2 operates the input wheels of the second, fourth and sixth (a single wheel) speed. Two countershafts, A_3 and A_4, match the final drive wheel and are respectively fixed to the driven wheels of the fifth and sixth speeds and with the driven wheels of the first, second third and fourth speeds.

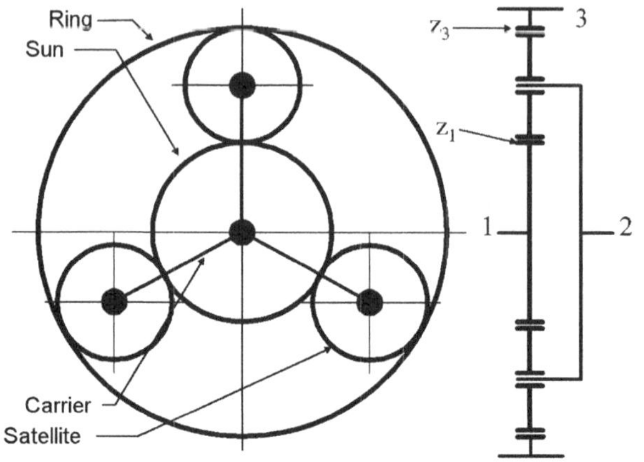

Fig. 11.16 Scheme of a simple epicycloidal gear train (from Genta and Morello 2009)

The reverse speed has a third idle countershaft, not visible in the diagram. This countershaft meshes with the driving wheel of the first speed on shaft A_2 and with the R wheel on the shaft A_3; its synchronizer is close to that of the sixth speed.

11.3.2 Planetary Gearboxes

Most automatic gearboxes are based on planetary gears with several different layouts. These trains are always made with cylindrical spur gears, even if in the past some examples of planetary gears made with bevel gears were available.

The simplest configuration for a planetary gear train is shown in Fig. 11.16. The relevant wheels are an internal gear (*ring*, or *annulus*) located outside, a co-axial external gear (*sun*) located at the center and a number of wheels (*satellites*) whose hubs are supported by shafts mounted on a rotating structure co-axial to the sun, called the *carrier*. There are normally three equi-spaced satellites so that the radial components of the gearing forces are balanced.

The system has three degrees of freedom and, to transmit a torque from an input to an output element, the third element must be fixed to the gearbox casing (stator). Three transmission ratios can be obtained with a train of this kind (as many as the elements that can be fixed to the stator: sun, carrier, ring), multiplied by two (the number of possibilities for a given element to be the input or the output shaft). To these six possible transmission ratios, one more should be added as a direct drive, by fixing the input and the output shafts together.

The ordinary transmission ratio that can be obtained by fixing the carrier is

$$\tau_0 = -\frac{z_3}{z_1} .$$

Table 11.1 Possible values of the transmission ratio of a simple planetary gear, in terms of the constraints and of ordinary transmission ratio τ_0, with reference to Fig. 11.16

Input	Output	Stator	Transmission ratio τ
1	3	2	$\tau = \Omega_1/\Omega_3 = \tau_o$
3	1	2	$\tau = \Omega_3/\Omega_1 = 1/\tau_o$
1	2	3	$\tau = \Omega_1/\Omega_2 = 1 - \tau_o$
2	1	3	$\tau = \Omega_2/\Omega_1 = 1/(1 - \tau_o)$
2	3	1	$\tau = \Omega_2/\Omega_3 = 1/(1 - 1/\tau_o)$
3	2	1	$\tau = \Omega_3/\Omega_2 = 1 - 1/\tau_o$

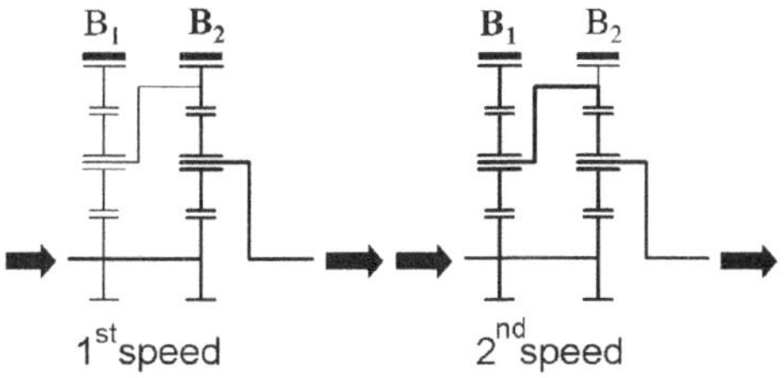

Fig. 11.17 Scheme of a Wilson compound ring gear train; two speeds can be obtained by stopping one of the two ring wheels with B_1 and B_2 band brakes (from Genta and Morello 2009)

The six different values of the transmission ratio τ that can be obtained are reported in Table 11.1.

Ratio τ_0 is negative, because when the carrier is locked, the output shaft is counter rotating with respect to the input shaft; not all ratios can be used on the same gearbox, because of the obvious difficulty to have the same element working as input and output in the same mechanism. With this mechanism two ratios (one reduced and one direct drive) can be obtained in practice. To obtain a higher number of transmission ratios, more planetary gear trains must be combined together; an obvious solution is to put more simple gears in series, where each one of them yields one of the required ratios, alone or in combination with the remaining gears.

A second method is to use compound gear trains, where different elements are integrated together. This second way, preferred on somewhat older configurations, allows design simplification but with limited flexibility in determining the transmission ratios.

The Wilson planetary gear train, shown in Fig. 11.17 , is a first example of a compound train. In this mechanism two simple trains are matched in such a way that the carrier of the first train is fixed to the annulus of the second; by stopping one of the ring wheels at a time, two different, not inverted, transmission ratios can be obtained. By fixing the two carriers together, a direct drive can be obtained.

A different compound gear train configuration, widely used in automatic gearboxes, is that of the so-called Ravigneaux train, shown in Fig. 11.18. In this train a single ring wheel is provided with two satellites s_1 and s_2 on the same carrier P; the satellites mesh with two different sun wheels S_1 and S_2. Satellites s_2 mesh with the smaller sun S_2; they do not mesh with the annulus but with the satellites s_1 of

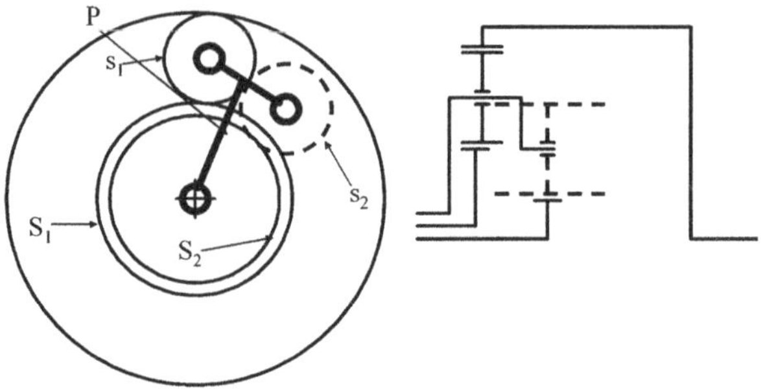

Fig. 11.18 Scheme of a compound Ravigneaux planetary gear train; the satellite s_1 is represented conventionally with *dotted lines*, because its drawing plane is not on the sketch plane (from Genta and Morello 2009)

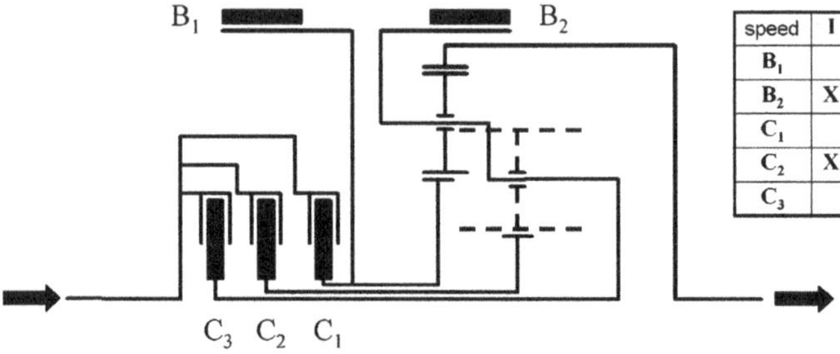

speed	I	II	III	IV	R
B_1		X		X	
B_2	X				X
C_1			X		X
C_2	X	X			
C_3			X	X	

Fig. 11.19 Scheme of an automatic gearbox with a Ravigneaux planetary gear train; the table shows the status of clutches and brakes at different speeds (from Genta and Morello 2009)

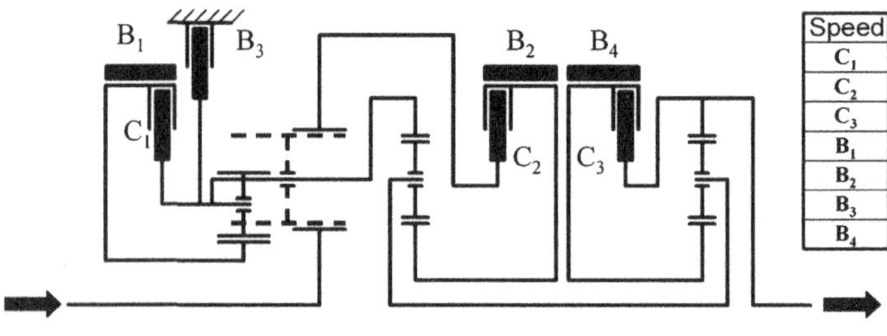

Speed	I	II	III	IV	V	RM
C_1			X	X	X	
C_2	X			X	X	X
C_3	X	X	X	X		X
B_1		X				
B_2	X	X	X			
B_3						X
B_4					X	

Fig. 11.20 Scheme of an automatic powershift gearbox featuring three planetary gear trains; five different speeds are obtained (from Genta and Morello 2009)

the larger sun S_1. Because satellites s_2 have their center line on a different drawing plane, they are conventionally sketched with dotted lines.

The status of clutches and brakes is shown in Fig. 11.19, using the established procedure. Using this scheme, with three multi-disc clutches and two band brakes, four different forward speeds and one reverse speed can be obtained.

An example of a Ravigneaux compound gear train joined to two simple planetary gear trains is shown in Fig. 11.20; the number of theoretically available speeds is well

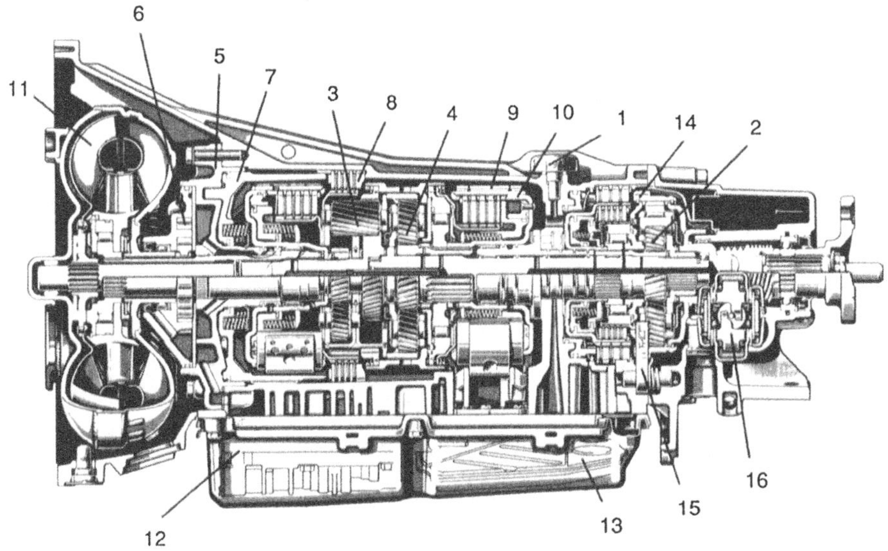

Fig. 11.21 Scheme for a five speed automatic powershift gearbox (courtesy of Mercedes)

over the five practically implemented. Because many possible combinations were discarded, the implemented transmission ratios come close to assuming the desired values and therefore allow a better optimization of performance, fuel consumption and emission.

The example in Fig. 11.21 corresponds to the scheme in Fig. 11.20 and is based on a Ravigneaux gear train 3 and two planetary speed reducers 2 and 4, downstream of the first; it features five forward speeds and one reverse speed.

On manual gearboxes it is possible to keep the vehicle braked while parking, by engaging only the lowest speed, due to the engine compression and friction torque. In the case of automatic gearboxes the torque converter, inserted on the transmission line, acts as a free wheel joint at low speed and the lock-up clutch, if any, cannot keep the car braked, because the actuation pressure is not indefinitely available when the engine is stopped. To keep the vehicle stationary, a manually actuated park pawl is added, matching a sprocket wheel on the output shaft; in the figure the pawl is the element 15 and the corresponding sprocket wheel is the element 14. The torque converter (without lock-up clutch), the oil pump 6 and the different multi-disc wet clutches used for gear engagement are also shown in the figure.

Automatic gearboxes for front-wheels drive cars, with longitudinal or transversal engine, are complicated by the need of having an output shaft at a convenient distance from the input shaft instead of being co-axial as in planetary gearboxes; in addition, space limitations due to the front steering wheels are particularly critical on transversal engines. Nevertheless, automatic gearboxes have gear trains that are substantially similar to those of conventional drives. Additional single stage trains are added to locate the output shaft at the desired final drive pinion position.

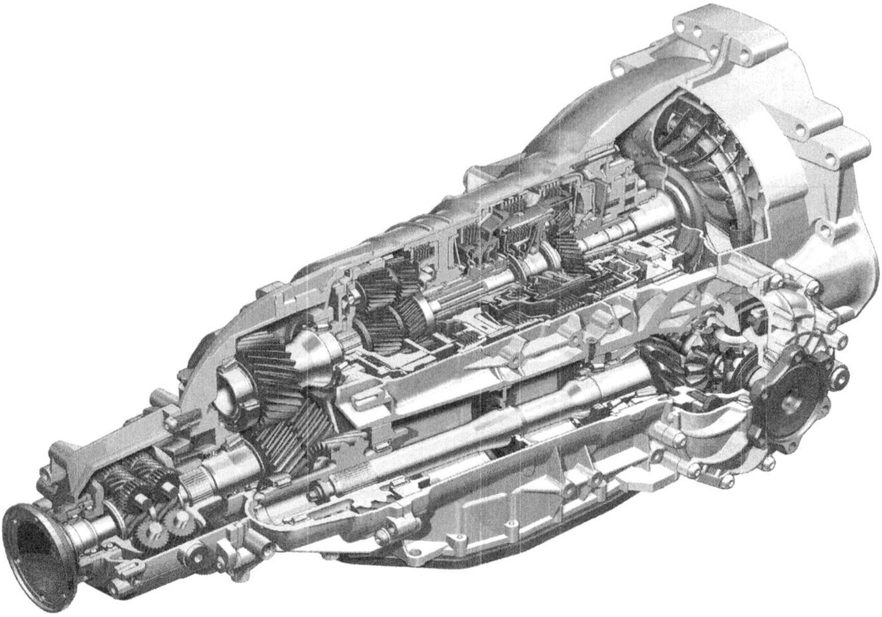

Fig. 11.22 Six speed automatic gearbox with epicycloidal gear trains; the lateral shaft connects the gearbox output shaft with the differential pinion, behind the engine (courtesy of Audi)

An example of this architecture is reported in Fig. 11.22 , for a front-wheels drive car with longitudinal engine. The scheme includes both a simple and a Ravigneaux gear train that, in conjunction with three clutches and two brakes, make six speeds available. The Ravigneaux gear train output shaft moves the differential final drive through an auxiliary shaft. The presence of the all-wheels-drive does not influence the layout.

Two solutions are available for front-wheels drive cars with transversal engine. A first solution is based on a torque converter and oil pump in-line with the engine; the torque converter output shaft carries a sprocket wheel that, through a silent chain, moves the input shaft of the gearbox in an offset position. Every geometrical problem linked with the transversal engine is thus solved, but, on the other hand, a large front overhang of the car must be accepted. A second solution divides the planetary gearbox into two sections. The first section may include a Ravigneaux gear train co-axial to the engine; its output shaft moves, through a single stage train, a simple planetary gear train co-axial to the differential pinion. It is thus possible to have five speeds within the space available on a car with front-wheels drive with transversal engine.

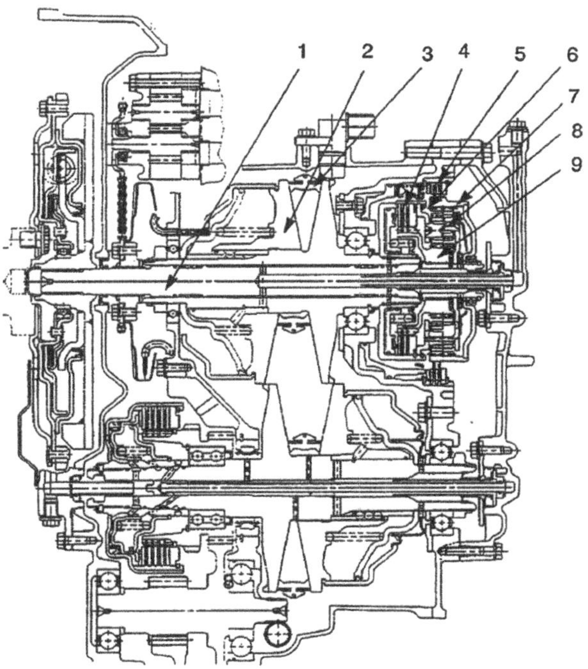

Fig. 11.23 Van Doorne steel belt CVT; the gearbox is characterized by the start-up multi-disc clutch on the output shaft (courtesy of Honda)

11.3.3 CVTs

CVTs are mostly found on front-wheels drive vehicles. They are particularly appreciated since they allow for an unlimited number of transmission ratios, with benefits for comfort, performance and fuel economy. Possible systems suitable for a CVT include in principle electric transmissions, hydrodynamic transmissions (among them the already mentioned torque converter), hydrostatic transmissions and variable geometry mechanical transmissions. The first three categories have been frequently considered in the past, but have been abandoned on conventional vehicles because of poor transmission efficiency, a problem that jeopardizes the theoretical advantages of CVTs. Steel belt transmissions, mostly of the Van Doorne type, are the most common technical solution.

The cross section of a CVT with a Van Doorne steel belt is shown in Fig. 11.23. The gearbox in the figure does not include a torque converter, because start-up functions are performed by an electronically controlled multi-disc clutch on the output shaft. The engine is connected to the carrier 8 of a planetary gear train with twin satellites that are used for forward and reverse speed; the gearbox input shaft 1 is instead fixed to the sun gear 9 of the same train. By using the clutch 4 it is possible to set the train in direct drive; by engaging instead the brake 5, the annulus is locked and the

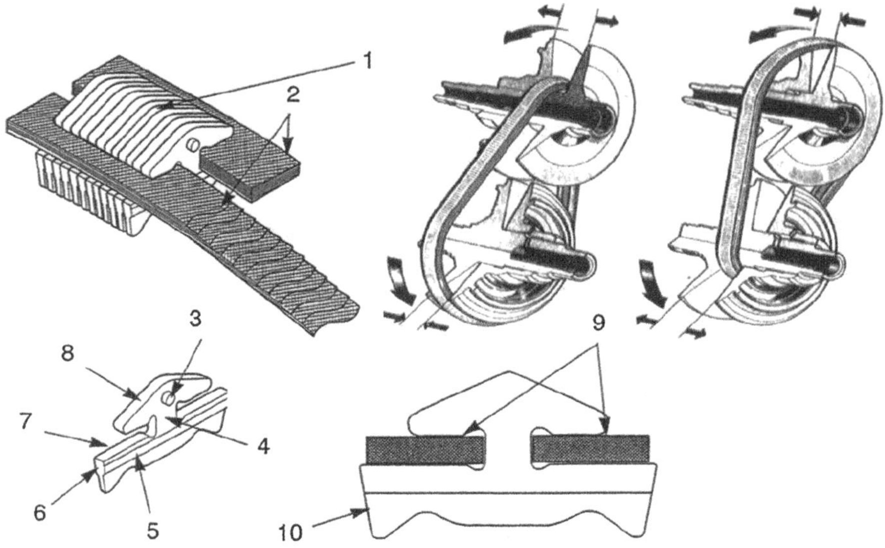

Fig. 11.24 Detail of the pulley motion in a Van Doorne steel belt CVT (*upper part*); detail of the belt (*lower side*). The torque is transmitted by belt compression (from Genta and Morello 2009)

sun gear rotation is opposite to that of the carrier, obtaining the reverse speed. With this design it would be possible, if necessary, to change transmission ratio in both forward and reverse drives.

The transmission ratio variation is obtained through two variable gauge V pulleys that match the belt (see detail in Fig. 11.24). Each pulley consists of two steel cones that can slide axially but are fixed together when rotating. The left input semi-pulley and the right output semi-pulley are also fixed in the axial direction; the other two semi-pulleys are instead free to slide.

The belt is quite stiff, so that its length cannot change. By closing together the two driving semi-pulleys, the driving pulley primitive diameter increases and the two driven semi-pulleys must accordingly separate, to reduce their primitive diameter. The belt shifts to the right and the transmission ratio decreases. With the opposite motion the transmission ratio increases.

A given transmission ratio is obtained between a minimum overdrive ratio and a maximum reduced ratio. The two limit ratios are reciprocal numbers; this situation, where the 1:1 value is in the middle of the ratio range, must be adapted to the vehicle requirements by adding a final drive that brings the higher ratio close to 1. For this reason there is an idler gear final drive between the output pulley and the differential pinion; in any case the system must be designed to fit in the engine compartment lay-out. This final drive is also used to adapt the same gearbox to different cars.

The active surfaces of the semi-pulleys are moved by hydraulic pistons, so that the oil pressure determines the contact force of the semi-pulley with the belt. Controlling this parameter is vitally important, because the contact force determines torque transmission through friction between the pulleys and the belt; if this force is insufficient, slip occurs, causing loss of efficiency and surface damage. If this force is too high, mechanical losses would also be higher than necessary, because the contact area between belt and pulley exceeds the kinematically correct primitive circumference.

Van Doorne belts use many thrust elements 1, such as those on the lower side of Fig. 11.24. Each of these elements contacts the next and is set along the belt center line; the elements are kept aligned by a number of flexible steel ring bands 2, one inserted inside the other with no clearance. Thrust elements have a trapezoidal shape with an appendix at their center; the two inclined sides 10 match the sides of the pulleys and exchange the friction force needed for torque transmission.

The central appendix 4 of each thrust element matches the two sets of concentric rings 9; the contact surfaces of the rings are inclined in such a way that the rings do not come in contact with the sides of the pulley. The rings must be very thin to keep the curvature stresses within reasonable values; the required tensile force is withstood by an adequate number of rings, whose length must allow to assemble them without radial clearance to distribute evenly the tensile force.

Thrust elements can only exchange compression forces with neighboring elements, so that the Van Doorne belt cannot work with positive tension and transmits the torque through the compression of the driven part of the belt. The sole purpose of the rings is to keep the elements at the correct radial position in the pulleys, and to keep them aligned when outside of the pulleys.

Friction forces between the belt elements and pulleys would waste no energy if there were no relative motions. These forces are limited because there is no macroscopic slip between parts; nevertheless, local microscopic relative motions are present because:

- The contact area of the thrust elements must have a certain radial extension to limit the contact pressure; at each contact area a small slip at points outside the primitive radius occurs, and
- Thrust elements entering and leaving the pulleys must slide because they change their path from straight to circular and viceversa.

The power wasted by these small relative motions is, at any rate, determined by the pressure acting between the belt sides and the pulleys; this pressure must be limited to the minimum value needed to avoid macroscopic slip. The contact pressure must therefore be carefully adjusted as a function of the transmitted torque.

There are also belt CVTs with a torque converter between the engine and the speed variation unit.

11.4 Start-Up Devices

11.4.1 Clutch

Friction is used in clutches to transmit a torque between an input and an output shaft. A clutch of the type most commonly used in cars is based on three discs; two, the *driving plates*, are connected to the engine crankshaft and one, the *driven plate*, located between the other two, is connected to the gearbox input shaft. Friction forces between these discs transmit and modulate torque. The driving plate can be displaced axially while a spring system provides the pressure force between the driving plates and the driven plate, so that friction forces can be exerted between the discs; the value of this pressure controls the transmitted torque.

The clutch has the following tasks:

- To transmit the torque from the engine crankshaft to the gearbox input shaft, when there is a difference of speed between the two parts, particularly when the vehicle is not moving or moving backwards;
- to connect the two shafts positively, once they are synchronized, in such a way as to transmit all the engine torque, and
- to separate the engine from the transmission, when the gearbox speed must be changed or the car stopped, without stopping the engine.

In addition to these tasks, the following functions have been added in recent times:

- To absorb torque pulses caused by the engine polar inertia, in case the clutch is not correctly operated (main *torsion damper*), and
- to control the driveline torsional stiffness, in such a way as to avoid vibration and noise when one of the harmonics of the engine torque match one of the transmission vibration modes (secondary *torsion damper*).

The springs of the clutch, formerly of the coil (helical) type, are now of the *diaphragm* type, because the latter allows reducing the powertrain length, for the same value of torque, the mass imbalance and the disengagement load on the clutch pedal.

The cut-out view of a diaphragm spring clutch is shown in Fig. 11.25; the engine shaft 1 is flanged to the flywheel 2, which carries one of the active surfaces of the driving plate. Cover 3 is also flanged to the flywheel and carries a second driving plate 2', the *pressure plate*, that is pushed axially by the diaphragm spring. The friction linings 4 are mounted on the driven plate, which has a spline engaging with the gearbox input shaft 5. The thrust bearing 6 can be axially moved to reduce the pressure of the spring, until the driven plate is disengaged. The bearing can slide on the tube 7 and is moved by the fork 8. The assembly of the pressure plate, the cover and the related elements used for the disengagement form what is called the *disengagement mechanism*.

A drawing of a driven plate is shown in Fig. 11.26. Its design is quite complex, particularly near the hub where the torsional dampers are located. The plate is built with several steel discs. The first, marked with number 1, is the hub, with a spline fitting the gearbox input shaft. Two other elements marked 2 and 3 are made with two steel discs riveted with spacers; the element 4 that supports the friction linings is also riveted to disc 3.

Disc 4 has a particular shape (see also the detail at the lower right), and shows a number of separate sectors on which the friction linings are riveted. The sectors are bent off the main disc plane and the rivets are set on parallel planes. With this arrangement the friction linings are spaced and the two friction surfaces can displace elastically along the axial direction. The result is that the pressure force is limited by the disc elasticity and the clutch engagement is more progressive. In addition, because the thermal gradient on the pressure plate can cause conical deformations, linings can more easily match this shape, maintaining a uniform contact on the surface of the driving discs. The assembly of discs 2, 3, 4 is free to rotate by a certain angle, with reference to the hub 1.

The three elements 1, 2 and 3 carry four rectangular windows, where four coil springs are press fit. If the torque acting on the disc is less than the total pre-compression moment of the springs, the disc acts as a rigid body; if the torque is higher, disc and hub can rotate with respect to each other. Between the rotating elements, there are also two friction discs 5 and 6, pressed through a diaphragm spring; when the hub is moving it is possible to dissipate part of the elastic deformation work of the springs and damp torsional vibrations. This assembly is the main *torsion damper*.

11.4.2 Start-Up Devices for Automatic Gearboxes

A devices that can start up the vehicle smoothly, without help from the driver, is required in automatic gearboxes; for this reason, devices different from the clutch in which an intermediate non-mechanical device, electric or hydraulic, is used, were developed. Today this issue is not as important, because it has proven possible to control automatically a conventional friction clutch, designed for pedal actuation, with satisfactory results by using an electronically controlled electrohydraulic actuator.

The torque converter is a particular hydraulic machine that allows two shafts to be connected with a continuously variable transmission ratio. Unlike the friction clutches, in torque converters the input and the output torque are not bound to be equal, but they are determined by more complex relationships depending on the transmission ratio. To understand how the converter operates, it can be thought as a transmission made by a hydraulic pump and a hydraulic turbine connected to each other through suitable piping. The pump rotates together with the engine and moves a certain flow of oil from a suction pipe to an outlet pipe; the pipes are also connected to a turbine that is mechanically linked to the gearbox input shaft. A transmission like this has been used on ships to connect engines with propellers.

Unlike mechanical transmissions, there is no positive link between the two shafts. The transmission ratio is determined by the inertia of the oil flow in the piping and by

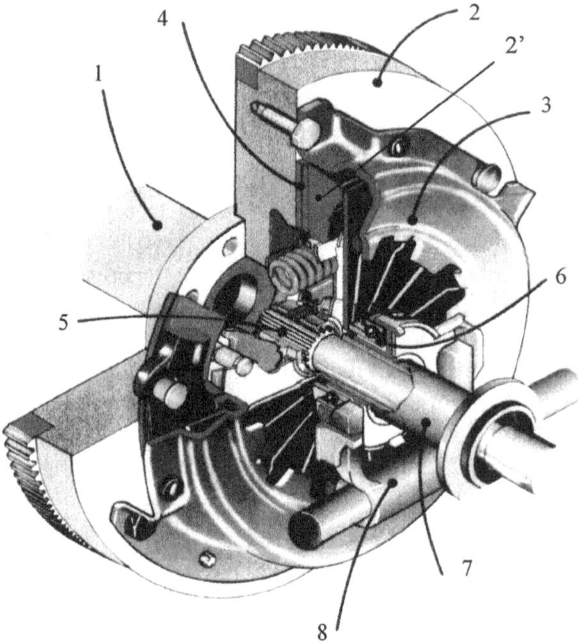

Fig. 11.25 Cut-out view of a diaphragm spring clutch. *1* Engine shaft. *2* Engine flywheel and driving plate. *2'* Pressure plate. *3* Cover and internal engagement mechanism. *4* Driven plate. *5* Gearbox input shaft. *6* Thrust bearing. *7* Thrust bearing guide. *8* Disengagement fork (from Genta and Morello 2009)

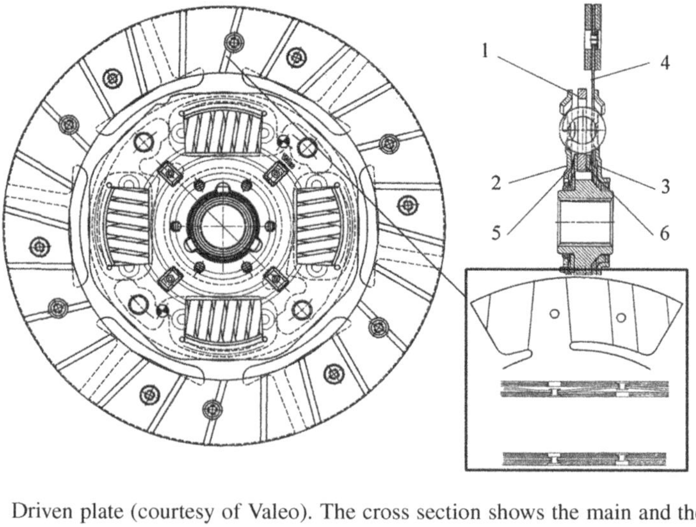

Fig. 11.26 Driven plate (courtesy of Valeo). The cross section shows the main and the auxiliary torsion dampers. The detail on the *right* shows how friction liners are mounted on the disc, in order to make the engagement progressive

the hydraulic machines. It is therefore possible to have the engine running when the transmission shaft stalls for any reason; as a matter of fact the turbine blades cannot stop the oil and can, in the meantime, receive a force from the flow.

Such transmission shows the following advantages:

- The transmission ratio can be continuously changed as a function of the torque ratio,
- there are no parts in contact and no wear,
- there is a high damping capacity for torsional vibrations and
- there is no danger of stalling the engine as a consequence of too high a torque being applied to the gearbox input shaft.

On the other hand there are some disadvantages:

- The transmission efficiency is low in all conditions, and
- there is an increased transmission complexity.

A converter cannot be easily disconnected from the driveline and has a significant polar inertia; this makes the application of conventional synchronizers impossible. It is therefore necessary to use powershift gearboxes, where shifting gears can occur in longer time frames, because there is no torque interruption.

The kind of layout described above is too bulky for vehicle application; Öttinger had the idea of integrating the pump and the turbine in a single compact machine, avoiding connecting pipes. This result was obtained by using turbine and pump wheels having the same diameter, so that the wheels can face each other, creating the hydraulic circuit without additional elements.

This lay-out is shown in Fig. 11.27; the wheel P, connected to the engine crankshaft, acts as a radial centrifugal pump; the disc supporting the blades has a cavity where the centripetal turbine T, connected with the output shaft, can be installed. The volume where the oil flows is limited by the rotating discs, the blades and the closure surface C, which can also be avoided. The turbine and the pump are almost identical; the two wheels must be co-axial and a rotating seal must be provided, to keep the oil inside. The proposed scheme shows no stationary wheel suitable for receiving a reaction torque. Input and output torque must in this case be equal, and this particular machine is called a *hydraulic clutch,* because it acts like a friction clutch. This machine cannot transmit any torque if there is no speed difference between pump and turbine; the oil flow rate controls the torque that reduces to zero when the speeds are equal.

The scheme of a *torque converter* is shown in Fig. 11.28. Unlike the previous hydraulic clutch, that includes only the pump P and turbine T, here there is a third bladed wheel S, called the *reactor* wheel, connected to a stationary element, such as the gearbox housing. In this case the torque on the pump and on the turbine might be different, because the reacting element can exert a torque equal to the algebraic difference of other two torques. If, as it usually occurs, the output torque is greater than the input torque, the reaction torque must have the same direction as the input torque. If the output torque is less than the input torque, the reaction torque has a direction opposite to the input torque.

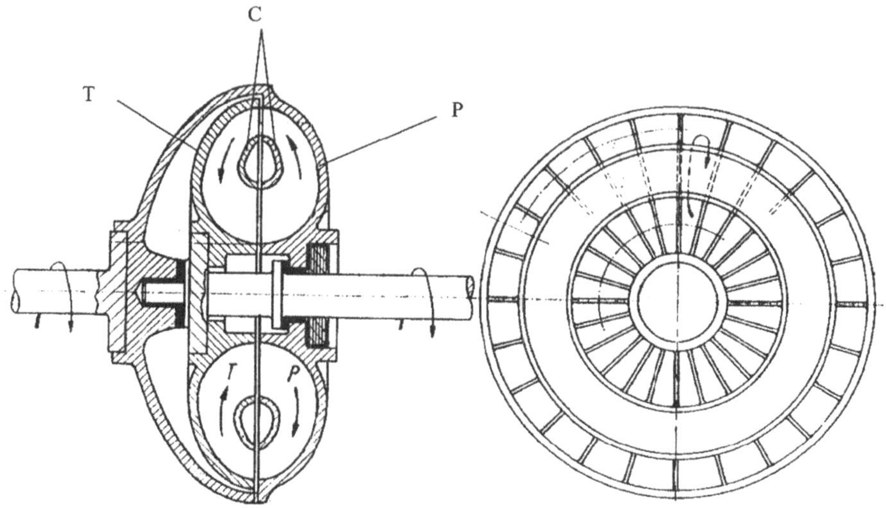

Fig. 11.27 Hydrodynamic clutch scheme; *P* is the centrifugal pump and *T* the turbine. The *arrows* show the direction of flow when the pump speed is greater than the turbine speed (from Genta and Morello 2009)

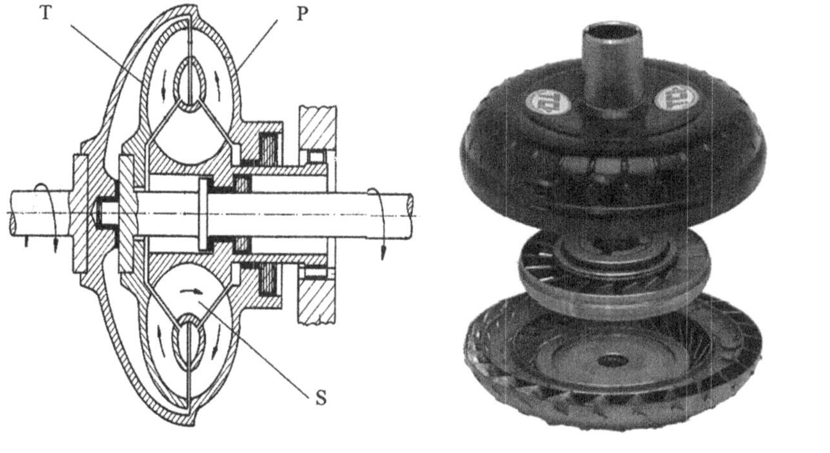

Fig. 11.28 A torque converter scheme is represented on the *left*. A picture of the elements of a torque converter is shown on the *right* (from Genta and Morello 2009)

If the stator is connected to the gearbox housing with a freewheel that can only withstand a torque with the same direction as the input torque, the machine can multiply the input torque or behave like a hydraulic clutch, when the output torque equals or is greater than the input torque. The picture in the figure shows a disassembled torque converter made of stamped steel sheet parts.

11.5 Differentials and Final Drives

The terms differential, transfer box and final drive are sometimes used imprecisely; in addition, these mechanisms are often integrated in the same subsystem in a wide variety of combinations. In this section some definitions that should help to clarify this issue are introduced.

The *differential* is a mechanism that allows the torque from an input shaft to be subdivided into two predetermined parts, flowing through two output shafts; torque ratios are independent of the speed ratios of the same shafts. This mechanism can be used either to divide the torque coming out of the final drive into equal parts acting on the driving wheels of the same axle, or to subdivide the torque coming out of the gearbox into two predetermined parts acting on different axles of the same vehicle. This second application is sometimes called the *transfer box differential* or *central differential*.

The *final drive* is a gear train that further reduces the speed of the gearbox output shaft to adapt it to the driving wheels; this gear train is usually integrated with a differential mechanism.

The *transfer box* is a mechanism that provides for the movement of two or more drivelines through the single output shaft of the gearbox; it is used on vehicles with multiple driving axles. When multiple axle drive is permanent, a differential is also needed to allow different rotation speeds on the axles; in such cases the differential train is usually integrated in the transfer box.

On rear-wheels drive cars, final drives carry out the task of rotating the driveline axis by 90°, from the longitudinal gearbox center line, to the transversal axle center line, as shown in Fig. 11.29. Center line rotation and speed reduction are achieved through a pair of bevel gears with spiral teeth; the wheel is bolted to a hub in order to allow the easy adaptation of the same production line to different final transmission ratios. The shafts are supported by conical roller bearings, because the axial thrust is relevant. The differential train has straight teeth bevel gears, because their rotation is not continuous and their relative speed is low. The *planetary wheels* are fixed to the wheel axles through splines, while *satellites* are idle on a short shaft fixed to a carrier through a pin; the bevel gear wheel of the final drive is bolted onto this carrier.

If the rear wheels have independent suspensions, the differential assembly is suspended to the body and the wheels are operated by half axles through constant speed joints, as in a front-wheels drive car. A suspended differential casing must react to the resultant of the wheel drive torques, acting along the wheel axis, and of the gearbox output torque, acting along the propeller shaft axis; for this reason the casing is attached to the car body through a strong fixation interface, usually through an auxiliary frame. A small elastic tube located on the bevel pinion shaft is shown in the same figure; it allows the axial pre-load to be controlled on bevel roller bearings, compensating for axial dimensional tolerance.

In this application the final drive bevel gears feature *hypoid* teeth, so that their center lines can lie on different planes; this configuration allows the transmission line to be placed in a lower position with advantages for interior space. The thrust

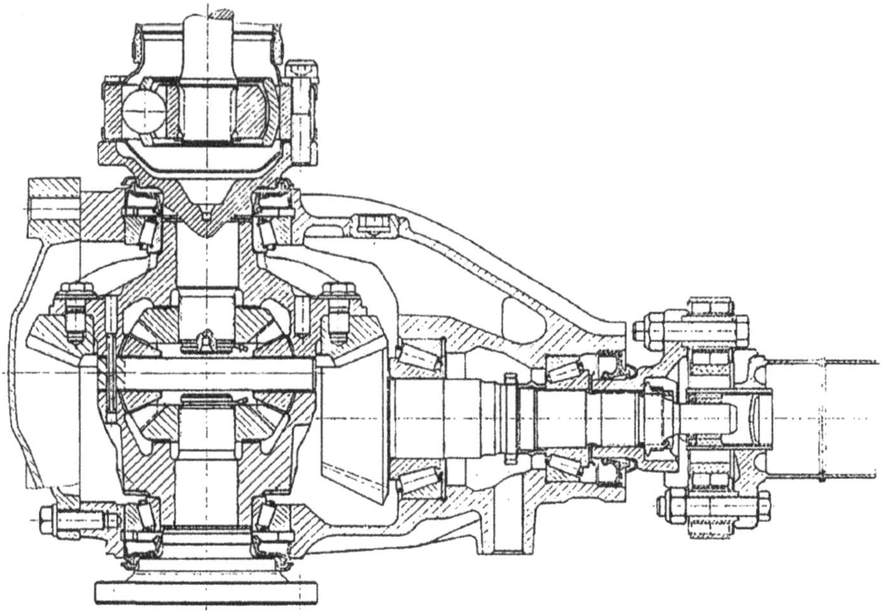

Fig. 11.29 Cross section of a differential assembly for a rear-wheels drive car with rear independent suspensions (from Genta and Morello 2009)

bearings for the planetary wheels and the satellites are simple wear-resistant washers, because of the low relative speed.

On front-wheels drive cars, the differential and the final drive are integrated in the gearbox; the reaction torque therefore acts on the powertrain suspension. Two cases, where the engine is transversal or longitudinal, can be identified. While in the second case the differential is similar to that of conventional drivelines, in the first case, shown in Fig. 11.30, the gearbox output shaft and wheels center lines are parallel and a pair of cylindrical gears with helical teeth can be used. The pinion is cut on the output shaft and meshes with a wheel, which in this case is also flanged to the differential carrier. The differential mechanism is similar to the previous one. In this case the thrust bearings of the planet wheels are made by needle cages to improve the mechanical efficiency.

11.5.1 All Wheel Drive Transfer Boxes

On all-wheels drive vehicles, the transfer box architecture is conditioned by two factors :

- The configuration of the single axle drive vehicle from which the all-wheels drive version is derived, with reference to the architectures shown in Fig. 11.1, and

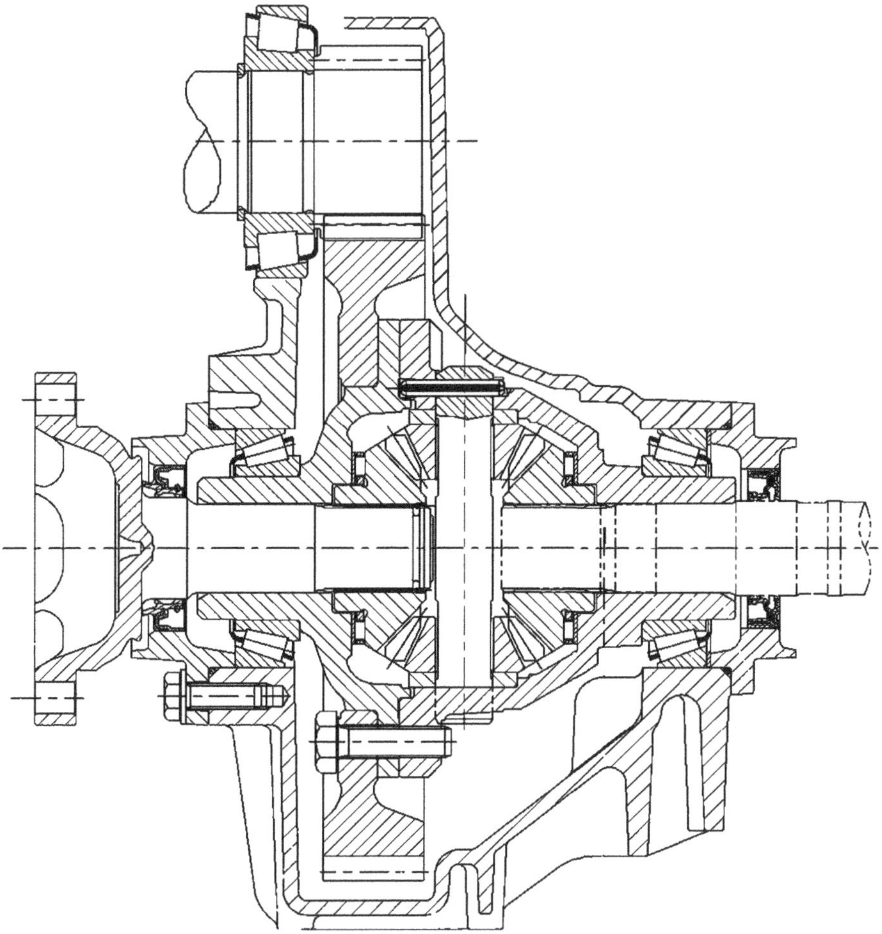

Fig. 11.30 Differential and final drive for a front-wheels drive car with transversal engine (from Genta and Morello 2009)

- whether the drive is *permanent* (can be used at any vehicle speed) or *non-permanent* (can be used occasionally at low speed on bad roads). In the first case a central transfer differential is also needed, to avoid unnecessary tire wear and additional rolling resistance when wheel speeds are not equal; in the latter case the differential can be avoided.

In modified conventional rear-wheels drive cars many architectures are used and a possible classification can be derived from the vehicle use. It is possible to identify off-road vehicles, where the main objective is to obtain good mobility on dirty and slippery roads, and road vehicles, where the objective is to achieve better stability and handling on paved roads at high speed. In the former case (scheme k, Fig. 11.1), the transmission line is offset in comparison to the engine crankshaft to increase ground

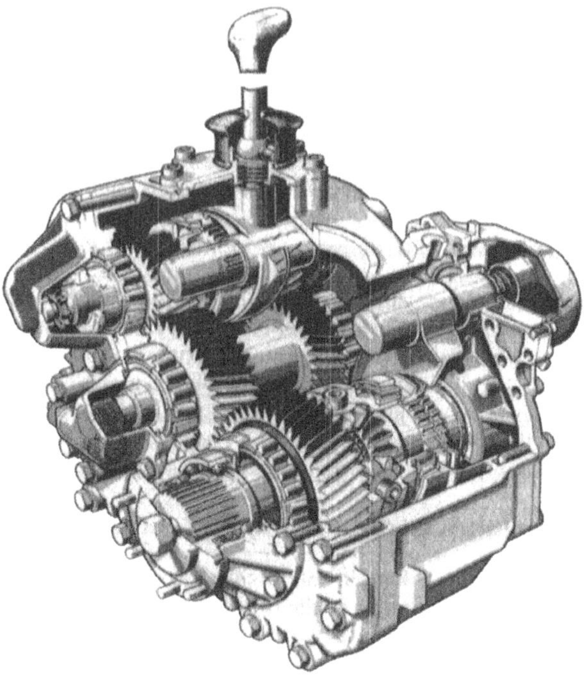

Fig. 11.31 Transfer box for a professional off-road vehicle (courtesy of Mercedes). The transfer box integrates the two axle drive through a bevel gear lockable differential and a range speed reducer with two synchronized speeds

clearance and the transfer box drives the rear axle only. The transfer box lay-out is visible in Fig. 11.31.

Often other functions are present: In the figure, the gearbox output shaft drives a synchronized two speed range reducer, that operates a third shaft featuring a bevel gear differential, allowing the torque to be split between the two axles. In the same figure a dog clutch is also shown; it can lock the differential carrier with the front planet gear, in this way locking the differential for slippery roads at low speed.

The offset between crankshaft and transmission line is useful because it allows rigid axle suspensions to be used without interference with the oil sump. This type of transfer box can have a simple dog clutch instead of the differential to insert the front wheel drive on low speed slippery road.

On on-road vehicles, the original rear wheel drive line position is maintained and a transfer box is added to operate the front axle, using a central differential at all times. An example is shown in Fig. 11.32, where the front axle is driven through an idler. The differential is of the planetary spur gears type, where the carrier is fixed to the rear axle transmission shaft, the sun gear moves the front drive transmission shaft, and the ring gear is driven by the gearbox output shaft.

A multi-disc wet clutch controls the torque applied to the two axles automatically. A second clutch, at the right of the figure, can put the front axle at idle. Three oper-

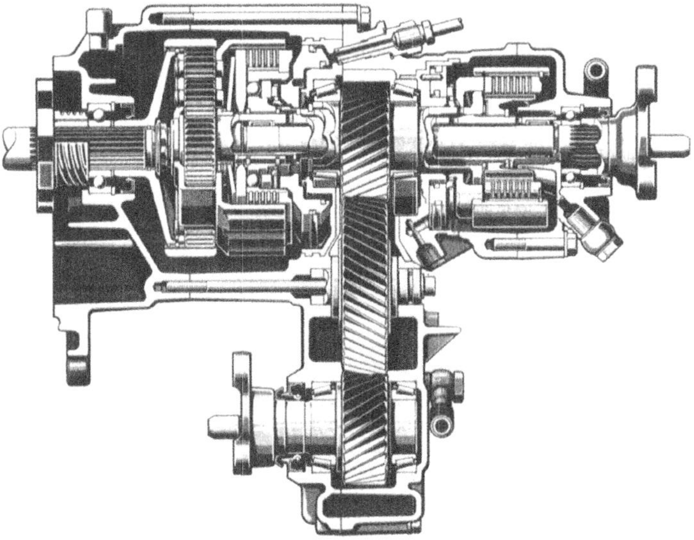

Fig. 11.32 Transfer box for an all-wheels drive on-road vehicle with longitudinal front engine (courtesy of Mercedes). The box integrates the axle drive with a planetary differential with slip control

ating modes are available: Rear-wheels drive, constant rate all-wheels drive, locked differential all-wheels drive. The carrier shows twin satellites, to allow the correct rotation direction for the front axle. The use of a spur gear planetary differential allows any torque breakdown rate other than 50/50; this is useful when axle loads are different and, in any case, correctly controls vehicle stability.

In the case of modified front-wheels drive cars, a distinction can be made between transversal and longitudinal engines. In the most common types of the former, as shown in Fig. 11.33, the front axle differential is coupled with a simple bevel gear drive featuring a transmission ratio close to 1. The driving bevel gear is fixed to the front axle differential carrier and the rear axle transmission shaft is operated by the driven bevel gear. There may be a simple dog clutch to move the rear axle, in case of a non-permanent all wheel drive or an electronically controlled friction clutch in case of a permanent all wheel drive; both clutches are not shown in the figure.

The case of a vehicle with a longitudinal engine is shown in Fig. 11.34. The all-wheels drive looks simpler: The gearbox output shaft is hollow and the shaft moving the front differential bevel pinion is left free to rotate in its cavity. The gearbox output shaft drives a Torsen type differential, not described here. The planet wheels of this differential operate the front and the rear axle. This differential transfers and controls the torque to the two axles.

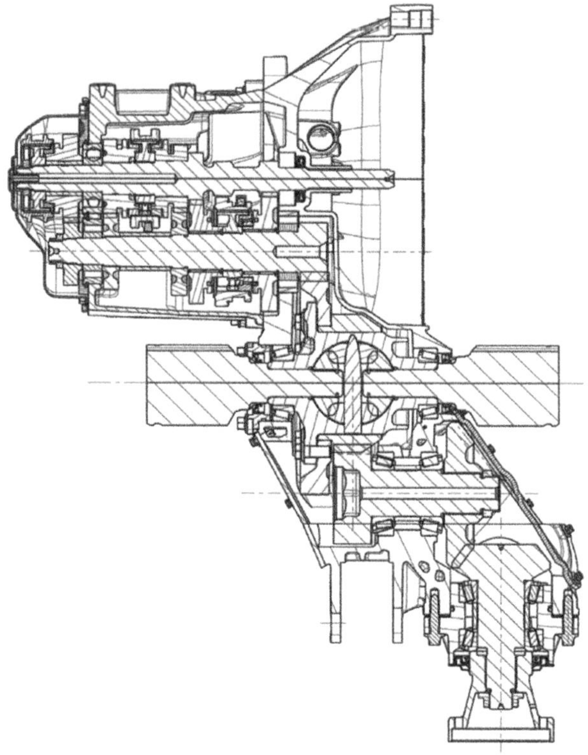

Fig. 11.33 Gearbox, differential and transfer box of an all-wheels drive car with transversal front engine (courtesy of FIAT)

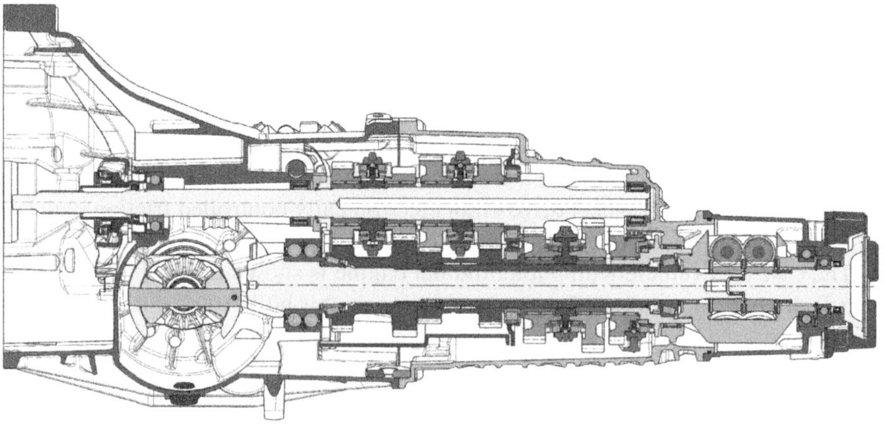

Fig. 11.34 Gearbox, differential and transfer box of a permanent all-wheels drive car with longitudinal front engine (courtesy of Audi). The gearbox output shaft is hollow, to install the front wheel axle drive shaft

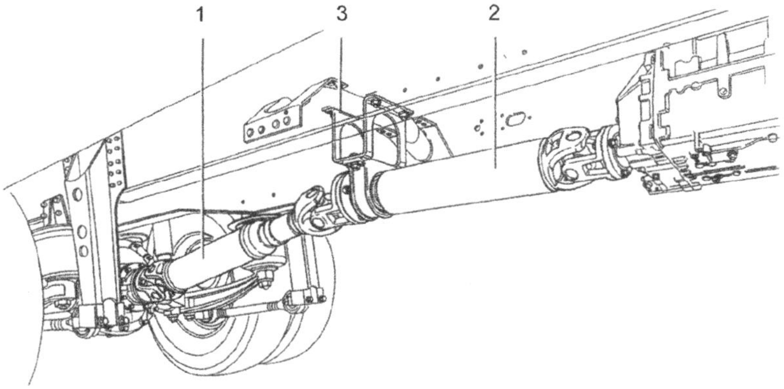

Fig. 11.35 Two-pieces propeller shaft for the rigid rear axle of a heavy duty industrial vehicle (courtesy of Iveco)

11.5.2 Shafts and Joints

Transmission shafts are used to connect shafts that cannot be perfectly aligned, like:

- Gearbox output shaft or reducer output shaft with final drive input shaft, for conventional drive vehicles using either rigid axle or independent suspensions; this transmission shaft is also called the *propeller shaft;*
- differential output shaft with driving wheels in all independent suspensions; these are also called *half shafts*.

Transmission shafts must be able of working with non-negligible offset and angles, because of the suspension stroke and, for front wheels, steering angles. Transmission components directly connected to the chassis structure like, for example, gearbox and differential box in conventional drive vehicles with independent suspensions, cannot be joined by rigid flanges, because of the flexibility of the chassis and of the presence of elastic mountings, whose goal is damping vibration.

Shafts fall, therefore, under two different categories: Propeller shaft and half shafts. A typical propeller shaft is that shown in Fig. 11.35, related to an industrial vehicle. The final drive on the rigid axle is connected to the gearbox output shaft through a two spans propeller shaft: The swinging shaft 1 and the fixed shaft 2. An intermediate bearing 3 holds the fixed shaft aligned with the gearbox output shaft and supports the swinging shaft. The latter is subject to changes in inclination with reference to the side beams of the chassis because of rear suspension bounce; to evaluate the offset to be covered, lateral displacement due to suspension and chassis flexibility should be accounted for.

Because the reaction to the driving or braking torque applied to the axle is also reacted to by the elastic suspension elements, some axle rotation about the y axis can be expected due to the S-deformation of leaf springs. The two piece design is required due to the relevant distance between the flanges to be connected, considering

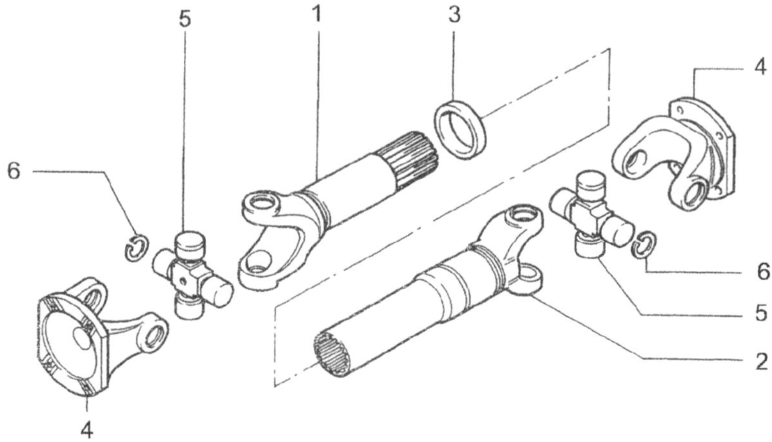

Fig. 11.36 Exploded view of the moving part of a propeller shaft with universal Hooke joints (from Genta and Morello 2009)

the natural bending frequencies of the shaft; moreover, it is advisable that these frequencies are quite different from the natural bending frequencies of the chassis to avoid annoying booms. For this purpose, an elastic suspension is also provided for the intermediate bearing.

For shorter vehicles, such as cars, semi-trailer tractors or light duty industrial vehicles, a single-piece propeller shaft can be used. High strength steel or aluminum are used for the shafts, but sometimes, on sports cars, filament wound reinforced plastic materials, like kevlar-epoxy composites, are also used.

A universal Hooke joint is normally used, as shown in Fig. 11.36. The propeller shaft consists of two tubes 1 and 2, connected by a spline that allows length variations due to the motion of the axle suspension. The spline is lubricated by grease sealed by a rubber ring 3. The two ends of the moving piece carry the *yokes* of the universal joints. Two other yokes 4 are flanged to the gearbox output shaft and to the differential input shaft or fixed transmission piece, if any.

Two *trunnion crosses* 5 connect the yokes; the connection between crosses and yokes is usually made by sealed needle bearings, kept in place by retaining rings 6.

The half shaft layout of an independent-wheels suspension for a front-wheels drive car is shown in Fig. 11.37. As the driving axle is also a steering axle, joints must be designed for large working angles. The transversely mounted powertrain always implies half shafts of different length. The different lengths can be a problem, particularly in wide cars with powerful engines, if, as a result, the half shafts feature a different torsional stiffness. Under high values of driving torque, self-steering moments can be applied to the steering mechanism, because the stiffer half shaft transmits a higher torque. The problem can be solved by using larger cross sections on the longer half shaft. The different geometry of the two half shafts could cause the natural bending frequency of the longer half shaft to be too low; to avoid this

Fig. 11.37 Complete transmission for a mid-size front-wheels drive car with transversal engine; the transmission is seen from the back. (courtesy of FIAT)

problem, the longer half shaft is divided into two pieces as with propeller shafts, the first one being fixed to the powertrain assembly.

Because of the limited space and wide angles, constant velocity joints based upon the Rzeppa principle are applied to half shafts rather than universal joints.

The solutions used on independent rear suspensions are similar to those seen for front driving suspensions. The double universal joint cannot be used in the transmission of an independent suspension to the steering wheels. This solution has been applied in the past for non-steering rear wheels but does not appear to be acceptable today because of the limitations on camber recovery, to keep the first and third shafts of the transmission parallel. In independent wheels suspensions only constant velocity joints of the Rzeppa type are used, such as those shown in Fig. 11.38a; the sketch shows a section of a complete half shaft for a front-wheels drive car. The wheel, not represented, is on the right.

The joint on the wheel side must be able to rotate because of two components: The steering angle, in an almost horizontal plane, and the angle imposed by the suspension motion (stroke and camber) in an almost vertical plane; the total working angle allowed by these joints is about 45°.

The other joint on the gearbox side should be subject only to the angle caused by the suspension stroke, but since the joint on the wheel side cannot be set on the king-pin axis, the wheel steering motion causes the joint center to move on an almost horizontal plane causing an angle in the same plane to the other joint, whose total working angle is about 20°. The wheel steering and the suspension stroke impose on the joint on the wheel side a trajectory not coincident with a circle centered on

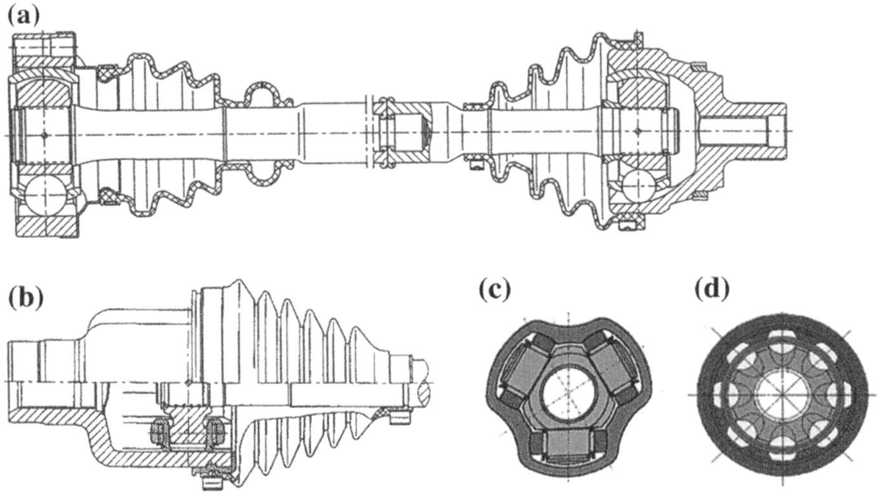

Fig. 11.38 Complete half shaft for a front-wheels drive car **a** with a fixed joint on the right and a sliding joint on the left. Scheme **b** shows a tripod joint that can be used in place of a sliding Rzeppa joint. Details **c** and **d** show the cross sections of these joints (from Genta and Morello 2009)

the joint on the gearbox side; for this reason a sliding joint must be used, that must allow a displacement of up to 50 mm.

Rzeppa sliding and fixed joints are made by a row of balls engaged in corresponding grooves on the inner nut and on the outer cup, where the wheel or the differential output shafts are connected. The balls are kept on the same plane by a cage, made by a ring with holes. Scheme d in the figure shows a section of the fixed joint, not dissimilar in this view from the sliding joint section. The shape of the ball grooves on the nut and cup is such as to determine four different contact points where the surface in contact is almost perpendicular to the force to be exchanged.

The intersection of the plane containing the ball centers with the joint rotation axis defines the joint center or the half shaft articulation point; the cup of the joint on the wheel side has a groove curvature that avoids ball displacements along the rotation axis, while the cup on the gearbox side has straight grooves to allow the half shaft to slide. No suspension position (usually the most dangerous one is at rebound with full steering angle) must result in a ball getting out of its groove. Cups are sealed with rubber bellows, rotating together with the half shaft, that keep the balls lubricated for life.

The constant speed behavior of these joints can be explained by Fig. 11.39. In this scheme only one ball (3) is represented, engaged with two grooves at the same time; these grooves are fixed respectively to the two parts of the joint 1 and 2; only the part of these grooves represented with a full line actually exists on the nut and cup. The shape of the groove is such as to cause all balls to have their centers on the bisector plane of the angle made by the two rotation axes of the shafts of the joint; in this

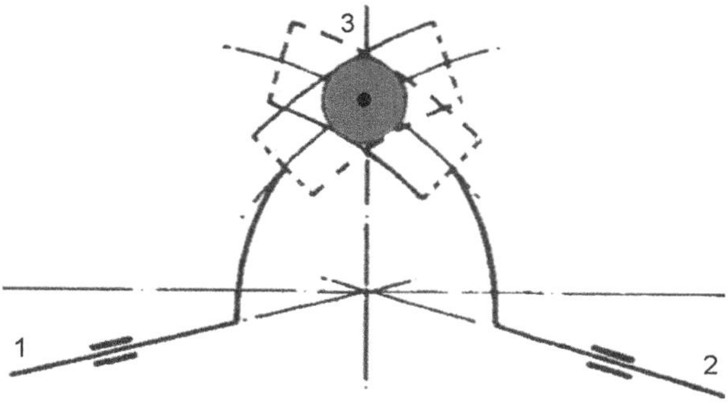

Fig. 11.39 Scheme of a constant speed Rzeppa joint; only one ball *3* is represented, engaged with the two grooves on shafts *1* and *2* (from Genta and Morello 2009)

condition the distance between each ball and the two rotation axes will be constant, determining the identical rotation speed of the two shafts. During joint rotation the balls will be subject to displacements along their grooves proportional to the joint working angle.

From the description of their operation, it is clear that friction in ball joints must be carefully considered; in a complete revolution of the joint each ball is displaced along its groove proportionally to the working angle and this displacement cannot be reduced to pure rolling of the balls in their grooves. Friction causes a small energy loss in the transmission and an exchange of forces in the transversal direction, because of suspension and powertrain motions; these forces can cause vibrations and discomfort that is not efficiently filtered by the powertrain suspension. For this reason a limitation to the force needed to have the joint sliding under torque must be carefully specified and taken under control; the tripod joint is an attempt to solve this problem with a different architecture. A scheme of this joint is shown in Fig. 11.38b: The balls are replaced by three rollers, mounted on pins with needle bearings, rolling in the straight grooves of the cup. This design is suitable to contain friction caused by joint sliding motion under torque.

Chapter 12
Alternative Powertrains

At present, the large majority of road vehicles are powered by fossil fuels derived from oil, even if a growing number of electric vehicles and of vehicles powered by alternative fuels can be seen on the road. Since these alternative energy sources and vehicle schemes are more a perspective for the future than a reality of today, a thorough discussion on these topics will be reported in Part III. In the following sections, however, the existing propulsion systems that are alternative or complementary to engines burning fossil fuels will be outlined, including electric and hybrid vehicles and vehicles propelled by fuel cells and gaseous fuels.

12.1 Battery Electric Vehicles

Electric vehicles are once again arousing the interest of car manufacturers worldwide for the role they may have in satisfying the demand for mobility and the need for virtually zero impact on air quality in urban areas. The advantages of electric vehicles are linked primarily to the possibility of moving the pollution from where the vehicle is used to where the power is generated, taking advantage of the better pollution control of power stations versus small engines. Another advantage is the possibility of regenerative braking.

The disadvantages are also well known: The performance of electric drives is affected by losses in both the engine and the batteries, and above all by the difficulties batteries have in delivering high power and, even more, accepting the power bursts that occur in braking. The quantity of energy that can be actually recovered by regenerative braking is thus only a fraction of that theoretically available. The main limitation of electric vehicles is still a number of characteristics that caused their use to be practically discontinued at the beginning of the twentieth century; their reduced range, the limited duration and high cost of the batteries and their high mass. However the performance of electric vehicles is sufficient for urban use.

G. Genta et al., *The Motor Car*, Mechanical Engineering Series,

DOI: 10.1007/978-94-007-8552-6_12,

Fig. 12.1 The LRV, designed and built by Boeing at the end of the 1960's, was used on several *Apollo* lunar missions during the 1970's (courtesy of NASA)

From the point of view of energy the advantages of battery powered electric vehicles (BEV) are still in doubt: When the primary source is a fossil fuel, in spite of the greater efficiency of the primary conversion and regenerative braking, the overall consumption is comparable to, and usually higher than, that of internal combustion engines. This point will be dealt with in detail in Sect. 14.1.

A technical review of modern vehicular electric drive systems can start with the description of an absolutely special, emblematic electric vehicle: the Lunar Roving Vehicle (LRV) used on the Moon by the Astronauts of the last *Apollo* missions (Fig. 12.1). It was first used during the *Apollo 15* mission, which lifted off on July 26, 1971 launched by a *Saturn V* rocket. The mission landed on the Moon 4 days later, on July 30, near the Hadley Range and returned to Earth on August 7, after approximately 12 days in space and 3 days on the Moon. They orbited twice around Earth and 74 times around the Moon. Mission Commander David Scott and astronaut James Irwin travelled a total distance of approximately 30 km on the surface of the Moon using the LRV (astronaut Alfred Worden remained in orbit around the Moon on the command module) collecting approximately 80 kg of lunar soil samples along the way.

The LRV was designed and built by Boeing at the end of the 1960's. Its main specifications were:

- Empty mass: 240 kg (weight: 2350 N on Earth, 360 N on the Moon); payload mass (two astronauts and equipment): 490 kg;
- Length: 3.1 m; width 1.8 m; height 1.1 m; wheelbase 2.3 m;
- Multilayer metallic mesh wheels, reinforced with titanium alloy chevrons for traction;
- Four independently suspended and steering wheels, steer-by-wire system;
- Four electric motors, one for each wheel, 18 kW each;
- Traction batteries: two 36 V silver-zinc type;
- Range: 65 km;
- Maximum speed: 17 km/h;

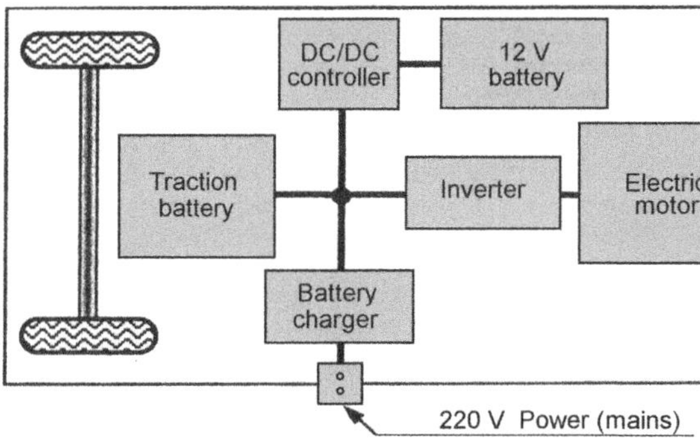

Fig. 12.2 Scheme of the propulsion system of an electric car

Owing to the short planned duration and range, the vehicle had primary (non rechargeable) batteries: the high energy density of this kind of batteries made it possible to keep the mass of the vehicle to a minimum. The LRV was a strange mix of very advanced features, anticipating some trends that may in the future become common in the automotive field, and simple solutions that at that time already belonged to the past. For instance, it had four-wheels driving and steering, but the steering control was based on kinematic steering; it had a steer-by-wire control, but the brakes were operated by cables.

Apart from this, very peculiar, vehicle, electric cars always had their small niche in the automotive market, having lost the large share of the market they had at the beginning of the twentieth century (see Part I). The biggest problem electric vehicles always had to face is the low performance of lead-acid electric batteries, the only rechargeable batteries that, until not many years ago, could be realistically used in the automotive field. This situation is now changing, owing to new battery types that are now available (see Chap. 13).

The most important problem to be solved in designing the layout of Battery Electric Vehicles (BEV) is battery installation, because of their significant volume and weight. The best results can be obtained only if the vehicle is specifically designed as a BEV from the beginning or, at least, if, in the design of the electric versions of a conventional car, provisions are taken to find space for the batteries at the very beginning of the development process.

The scheme of a very simple electric drive system is shown in Fig. 12.2; it essentially consists of a number of traction batteries, an inverter, a three-phase asynchronous electric motor and a DC/DC electric converter. The inverter is an electronic device converting direct voltage from the traction battery into three-phase, alternating, regulated voltage for powering the asynchronous motor; the DC/DC converter converts traction battery voltage into a lower voltage (12 V) for recharging the service battery.

Three-phase asynchronous motors are preferred to the direct current motors used in vehicular applications in the past because they:

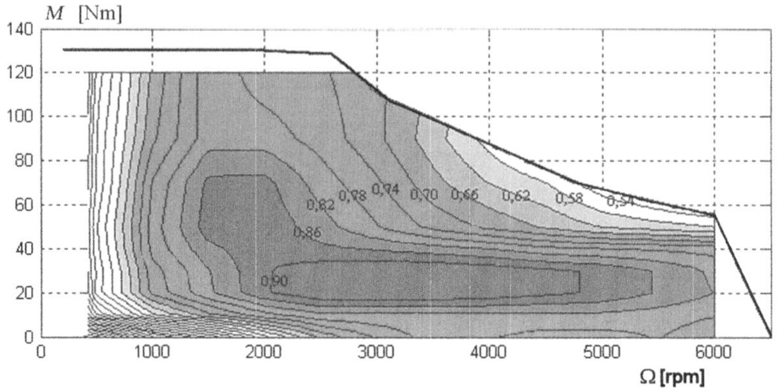

Fig. 12.3 Operating map of an induction AC electric motor. Plot of the torque M as a function of the speed Ω, with constant efficiency lines (from Genta and Morello 2009)

- are more compact, lightweight and have a high specific power;
- have a wide operating range at constant power;
- have a better energy efficiency and
- are highly reliable and, due to the absence of brushes, are virtually maintenance-free.

These advantages are even more marked if a brushless DC motor is used, like in the more advanced applications.

Since the electric motor can start under a load, there is no need for a clutch and usually no need for a gearbox with more than one transmission ratio. The asynchronous motor can thus be conveniently arranged on the rear end of the vehicle, if the rear wheels are the driving ones. It drives the rear wheels via a mechanical one-speed gearbox and a differential unit.

The map of the efficiency of an induction traction AC motor with a nominal power of 35 kW is shown in Fig. 12.3. The efficiency is much less sensitive to operating conditions (speed Ω and torque M) than in conventional internal combustion engines. However, also in this case a variable speed gearbox could help in achieving the best fuel efficiency.

As an alternative to this traditional architecture, with the motor operating the wheels through a mechanical transmission, it is possible to put two or more motors directly in the wheels. As an alternative, instead of locating directly the motors in the wheels, where they increase the unsprung masses, there can be one electric motor per wheel located under the vehicle body, driving the relevant wheel through a half-shaft. This configuration was suggested and tried several times in the past with limited success except for special vehicles, but it seems to be ready for large-scale application today.

At any rate, traditional or brushless DC or AC motors require a mechanical transmission, since they supply an insufficient torque and operate at a speed that is higher than that of the wheels. High torque motors (torque motors, both with internal and external rotor) that can be connected directly to the wheels without an interposed reduction gear are at present available. Apart from the advantage, which may be important in some applications, of allowing an arbitrarily large steering angle, even up to 360°, putting the motor in the wheels without using a reduction gear leads to high efficiency, low noise and a large degree of freedom in placing the various subsystems on the vehicle. Torque motors are, however, still heavier than high speed motors plus the required reduction gear and imply an increase of the unsprung masses, which is detrimental to both handling and comfort: direct drive, torque motors seem to be more a perspective for a future, nobody knows how far, than a possibility for today. When two electric motors operate the two wheels of the same axle, either located on the wheels or attached to the body and operating the wheels through semi-shafts, the motor control system can perform the electronic differential function, distributing the torque to the wheels of the axle, possibly simulating all the functions of a limited slip (or in general controlled) differential.

Apart from the traction batteries, that supply a voltage in the range 96 ÷ 300 V, on board electric vehicles there is usually a 12 V battery powering all services. A DC-DC converter transforms high voltage direct current into 12 V direct current, for keeping the low-voltage battery charged. This redundancy is due to the need to ensure emergency operation with flat traction batteries and also for safety reasons: high voltage must be confined to the few points where it is absolutely needed, and even so the dangers of high DC voltage in electric cars must not be understated. The choice of 12 V for low voltage accessories is also motivated by the economic advantage of using components that are standard in existing vehicles.

Traction batteries may be recharged either by the on-board battery charger, from the AC mains at 220 V, or by a dedicated fixed system which powers the DC batteries directly. Eighty percent of the maximum battery charge is reached in approximately 4 h while the time needed for a complete recharge cycle is about 8 h. Batteries must be fully recharged frequently during normal vehicle use to ensure good battery performance; the batteries may be conveniently recharged at night. Modern (non lead-acid) batteries require careful charging and sophisticated chargers are used both for improving the whole charge-discharge process and for safety.

Each major car manufacturer has electric cars on the market or, at least, under development: Renault, taken as an example, presents a very complete range, now on sale on selected European markets. The models of this range, whose pictures are shown in Fig. 12.4, are many and feature technical specifications not so different from their internal combustion engine counterpart, as shown in Table 12.1.

When electric vehicles are compared with conventional vehicles on the grounds of performance, flexibility, comfort and safety levels their traditional shortcomings, namely: heavy weight, low range, long battery recharging time and high costs, are still there and it is predictable that this situation will improve slowly, as it will be shown in Part III.

Fig. 12.4 The range of electric vehicles produced by Renault are not so different from they conventional counterparts (courtesy of Renault)

Table 12.1 Specifications of the Renault electric vehicles available for sale in selected European countries

Name	Fluence	Zoe	Twizy	Kangoo
Type	Large sedan	Compact sedan	Quadricycle	Van
Length [mm]	4,748	4,086	2,337	4,213
Width [mm]	1,813	1,788	1,237	1,828
Power [kW]	70	60	13	44
Torque [Nm]	226	222	57	226
Top speed [km/h]	135	135	80	130
Range [km]	185	160	96	170
Seat number	5	5	2	2
GVW [kg]	1,543	1,392	450	1,410
Cargo	–	–	–	650 kg

12.2 Traction Batteries

Electrochemical batteries are devices in which an electric current is generated as a consequence of a chemical reaction occurring between two reactants. An example of the electrochemical reaction is that taking place in lead acid batteries in which, when fully charged, the anode is made of lead, the cathode is made of lead oxide and the electrolyte is diluted sulfuric acid (in the form $H_2SO_4 \rightarrow H^+ + HSO_4^-$). During discharge, both plates are turned into lead sulfate, releasing electrons at the anode and getting electrons at the cathode. The reactions are:

$$\begin{array}{ll} Pb + HSO_4 \rightarrow PbSO_4 + H^+ + 2e^- & \text{Anode,} \\ PbO_2 + HSO_4 + 3H^+ + 2e^- \rightarrow PbSO_4 + 2H_2O & \text{Cathode,} \end{array}$$

i.e., as an overall reaction,

$$Pb + PbO_2 + 2H_2SO_4 \rightarrow 2PbSO_4 + 2H_2O\,.$$

Like in all secondary (rechargeable) batteries, this chemical reaction is reversible and can be run backwards by passing a current through the cell. This decomposes the lead sulfate into lead and lead oxide, through the reactions

$$\begin{array}{ll} PbSO_4 + H^+ + 2e^- \rightarrow Pb + HSO_4 & \text{Anode,} \\ PbSO_4 + 2H_2O \rightarrow PbO_2 + HSO_4 + 3H^+ + 2e^- & \text{Cathode.} \end{array}$$

This reversibility is never complete, like in all rechargeable batteries, and the battery cannot be recharged an infinite number of times: at every recharge the performance of the energy conversion somewhat deteriorates until the cell cannot be recharged any more. The performance of any battery depends on many factors and, above all, its capacity is affected by how fast the charge and discharge process is performed.

At present, the types of secondary batteries that are available or are predictable for the near future are:

- Lead-acid cells. They were used for almost all 'old' electric vehicles and also the attempts to revive them were based mostly on batteries of this kind. While in origin they were based on an open container full of electrolyte (sulfuric acid), more modern types are sealed and provided with a valve to prevent pressure build up. The electrolyte can be semi-solid (gel) or can be absorbed in a special fiberglass matting. Their success is mainly due to their relatively low cost, low maintenance, safety and easy handling, their manufacturing simplicity and their interesting electrochemical features.
- Nickel-based cells. A wide family of rechargeable batteries are based on nickel chemistry, like nickel-cadmium (NiCd),[1] nickel-iron (NiFe), nickel-zinc (NiZn), nickel metal hydride (NiMH) cells. NiCd batteries are no longer legal, because of the ban on cadmium, while nickel-metal hydride batteries are becoming increasingly popular: For the same weight, these batteries double the vehicle range, with respect to lead-acid batteries at the disadvantage of a higher cost.
- Rechargeable alkaline cells. They are seldom considered for applications in which a large quantity of energy has to be stored.
- Lithium-based cells. They include lithium-ion (Li-ion), lithium ion-polymer (LiPo), lithium-iron-phosphate (LiPh) and lithium-sulfur (LiS) cells. In general the electrodes are made of lithium cobalt (or manganese, or nickel or iron) oxide and of carbon, in different forms. The electrolyte is made by lithium salts in a

[1] Actually NiCd is a proprietory name and should not be used to indicate NiCd cells in general.

Table 12.2 Main characteristics of some types of secondary batteries (e/m: mass energy density, e/v: volume energy density, P/m: power density, V: cell voltage, η: charge/discharge efficiency, d: self-discharge, c: number of cycles)

Type	e/m (Wh/kg)	e/v (Wh/dm^3)	P/m (W/kg)	V (V)	η (%)	d (%/month)	c
Lead-acid	30–40	60–70	180	2.0	70–92	3–4	500–800
NiCd	40–60	50–150	150	1.2	70–90	20	1,500
NiFe	50	–	100	1.2	65	20–40	–
NiZn	60	170	900	1.2	–	–	100–500
NiMh	30–80	140–300	250–1,000	1.2	66	30	500–1,000
Alkaline	85	250	50	1.5	99	<0.3	100–1,000
Li-ion	150–250	250–360	1,800	3.6	80–90	5–10	1,200
LiPo	130–200	300	3,000	3.7	–	2.8–5	500–1,000
LiPh	80–12	170	1,400	3.25	–	0.7–3	2,000
LiS	400	350	–	–	–	–	100

liquid organic solvent or in a solid polymer. The basic reactions, referred to 1 mol (which explains the coefficient x), are:

$$\begin{array}{ll} LiCoO_2 \rightarrow Li_{1-x}CoO_2 + xLi^+ + xe^- & \text{Anode,} \\ 6C + xLi + xe^- \rightarrow Li_xC_6 & \text{Cathode.} \end{array}$$

The use of Li-ion batteries is expected to provide a threefold increase in vehicle range for the same weight

As a general consideration, nickel based batteries have a larger self-discharge, while lithium batteries are better from the viewpoint of quick charging and discharging. The highest energy density, i.e., the ratio between stored energy and mass, is obtained from lithium-sulphur cells that, however, have a short life in terms of cycles.

The main characteristics of some types of secondary batteries are reported in Table 12.2; a plot of the mass energy density versus the volume energy density is reported in Fig. 12.5.

The energy density, and even more the power density of batteries is much lower than those of gasoline or diesel fuel: for the energy density the figures are $30 \div 200$ Wh/kg against approximately 13,000 Wh/kg. In addition, liquid fuel tanks can be filled in a matter of minutes, in comparison with the 4 or 8 h needed for electric batteries. If a range of $\sim$120 km is deemed necessary to accomplish the daily mission of a compact car, and this corresponds to an energy pack of 20 kWh, the mass of the required battery is 670 kg if a lead–acid battery is used, or 330 kg with a Ni–MH battery and to 170 kg with a Li–ion battery. This can be compared with about 10 kg of fuel, computed assuming its use with an efficiency of just 15 % (and assuming a 100 % efficiency for the electric motor).

The charge of any battery is critical, since the efficiency and the life of a battery depends on how accurately the energy is introduced into the system. Some batteries

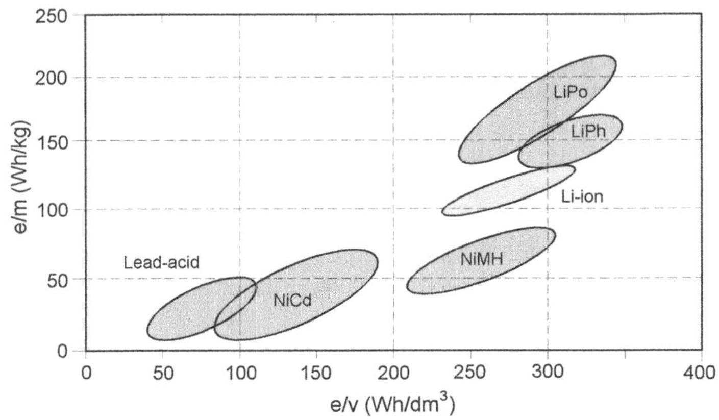

Fig. 12.5 Mass and volume energy density for the main types of secondary batteries

are less critical from this viewpoint, like lead-acid and NiCd cells, although the latter display what is usually referred to as a memory effect, consisting in the tendency to lose some of its capacity when recharged repeatedly after being only partially discharged.

Advanced batteries are more critical, and may even become dangerous if not properly charged, with the risk of fire and explosions. This is solved by using accurately controlled, microprocessor-based, chargers. Battery packs are increasingly provided with on-board electronics that monitor continuously the charge conditions and keeps the current flowing through the various cells under control.

Generally speaking, batteries cannot be used at the same time at high power density and at high energy density: As already stated, when required to supply high power (discharge with high current) the efficiency, and consequently the capacity, decreases. They show also a reduction of their useful life when used in these conditions. Lead-acid batteries are particularly sensitive to this, while some kinds of NiCd and other more advanced batteries can operate with high currents, both during charging (quick charge) and during discharge (high power output).

An ideal battery should be characterized by

- High energy density,
- Almost constant voltage during discharge (a flat discharge characteristics),
- Low internal resistance,
- High discharge current,
- Possibility of operating at both high and low temperatures,
- Long operating life and high number of charge-discharge cycles,
- High efficiency in recharge,
- Low cost.

In several of these points, no actual battery has particularly good performance.

The voltage of a secondary battery decreases during discharge and the plot of the voltage as a function of time is referred to as the discharge characteristics of the cell. Initially there is a sharp drop from the maximum voltage, typical of the fully charged state, to a lower value that is maintained, with a slight decrease, for most of the discharge time. When the discharged conditions are approached there is a sharp drop again. This third phase of the discharge curve must not be used: discharges that are too deep are detrimental to the possibility of fully recharging the battery and can, in the long run, deteriorate it. The discharge curve is much influenced by how fast the discharge is: if the current is large the voltage decrease in the intermediate phase may be larger, depending on the battery type.

At present lead-acid batteries are largely used in the automotive industry, mostly as starting batteries: according to a 2003 report,[2] the batteries of vehicles on the road contained an estimated 2,600,000 t of lead, for a yearly use of over 1,000,000 t. This is made possible only by recycling the old lead from used batteries: in the United States 97 % of all battery lead was recycled between 1997 and 2001.[3] A first difficulty regarding the diffusion of BEV, if done by using conventional lead-acid batteries, is the huge quantities of lead that are required. In 2010 a total of 4.14 MT of lead were mined, and the total production of lead, including recycling, was 9.602 MT, out of which 71 % was used for batteries for various applications. This factor alone would rule out the possibility of converting to battery-electric vehicles a substantial percentage of the vehicles at present on the road: the huge increase of lead to be mined would push the price of this metal beyond what is acceptable for the automotive use.

The situation is not much better with lithium batteries. The annual world production of lithium and the estimated reserves are respectively 18,000 t and 9.9 MT (2009 data[4]): to equip 50 % of the 72 million cars produced in 2007 with a lithium battery containing just 1 kg of lithium each, a total of 36,000 t, twice the annual world production, is needed.

12.3 Hybrid Vehicles

All vehicles considered up to this point work on a single source of energy stored on board, be it the chemical energy contained in a fuel or the electric energy stored in an electrochemical battery. A vehicle that has on board energy stored in two or more forms is said to be a *hybrid* vehicle. A hybrid vehicle must not be confused with a *bimodal* vehicle, that may work both on energy stored on board and energy supplied from the outside during motion, like a trolleybus having a number of electric batteries to obtain a certain off-line range.

A solution to the limited range of electric vehicles consists in associating a second source of power, like a more or less conventional internal combustion engine, to the electric power system. The combined thermal engine-electric batteries hybrid drive

---

[2] *Getting the Lead Out*, Environmental Defense and the Ecology Center of Ann Arbor, Mich.

[3] http://www.batterycouncil.org/LeadAcidBatteries/BatteryRecycling/tabid/71/Default.aspx

[4] U.S. Geological Survey, 2009, commodity summaries 2009: U.S. Geological Survey.

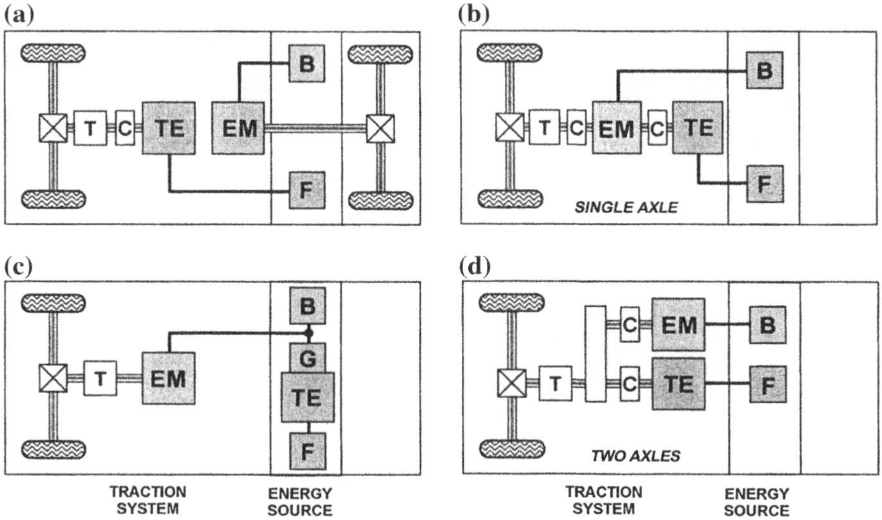

Fig. 12.6 Schemes for hybrid propulsion systems. **a** dual mode propulsion system; **b** series propulsion system; **c** parallel single-axis propulsion system; **d** parallel double-axis propulsion system. *TE* thermal engine, *EM* electric motor, *G* generator, *B* batteries, *F* fuel, *T* transmission, *C* clutch

system lends itself to different interesting applications. However, it is not the only type of hybrid system studied: prototypes of vehicles in which one of the storage system works on elastic energy (hydraulic or pneumatic accumulator), kinetic energy (flywheel) or energy of an electric field (supercapacitors) were studied and even built at a prototype level, and a solution with an internal combustion engine, batteries and flywheel seemed to be particularly interesting. These more complex concepts seem to be interesting for large vehicles, in particular city buses, while only internal combustion engine-battery hybrid systems was seriously considered for cars.

Hybrid drive systems can be used to:

- Drive in highly polluted urban areas in zero emission conditions as an electric vehicle with much less stringent limitations of range,
- improve the energy efficiency of the internal combustion engine since its efficiency decreases together with the power demand; this effect may be obtained by using the internal combustion engine at a higher power than needed to recharge the batteries with the power in excess, and
- recover the kinetic energy during vehicle deceleration for recharging the batteries, while in conventional vehicles brakes would dissipate totally the kinetic energy.

There are essentially three possible hybrid system configurations: *dual-mode*, *series* and *parallel*, as shown in Fig. 12.6. In the dual-mode configuration (Fig. 12.6a), the two drive systems are entirely separate like when using a conventional internal combustion engine to drive the front axle and an electric motor to drive the rear axle. The resulting functions are comparable to those of the parallel hybrid configuration illustrated below.

In the series configuration (Fig. 12.6c), the electric motor drives the wheels and propels the vehicle. The energy is mostly stored in the fuel and transformed into electric energy by the generator driven by the internal combustion engine to be partially stored in a battery and partially directly used to drive the wheels. Series hybrids are designed essentially to minimize emissions, because the internal combustion engine can run in stationary, low emissions, conditions. Moreover, the full potential of the catalytic converter can be used to further minimize emissions.

Parallel hybrids (Fig. 12.6b and d) allow integration of the mechanical energy coming from the fuel through the internal combustion engine and that coming from the batteries through the electric motor. Both can transmit their torque to the wheels, at the same time, working in a strictly integrated way, so that the fuel energy does not undergo a double mechanical/electrical conversion.

With parallel hybrids it is possible to select the size of the electric motor according to the urban missions for which the vehicle is intended (small size), to convert fuel energy into mechanical energy directly and to use the fuel energy if electric power is suddenly unavailable. The electric motor may also be used either to back up the internal combustion engine, by integrating traction power demands, or as a generator for energy recovery during vehicle deceleration.

The parallel hybrid system can be implemented following different configurations. One particularly compact arrangement is the single-axis design (Fig. 12.6b) in which the electric motor is placed between engine and transmission with interposed clutches allowing various operating modes. The two-axle design (Fig. 12.6d) becomes a possibility if the mass of the electric motor can be reduced.

Optimum configuration (series or parallel), size of components and hybrid vehicle management logic depend on the class of vehicle and the missions for which it is intended. Ultimately, a series hybrid configuration is best suited for vehicles intended mainly for urban use, where zero emissions could become mandatory, whereas a parallel hybrid configuration will be expedient for out-of-town use, requiring comparable or better performance than that of thermal engine vehicles.

A particular way of applying hybrid propulsion systems are the so-called *plug-in* hybrids. With these systems the electric energy stored in the battery comes not only from the thermal engine or from the recovered kinetic energy, but also from an external source of energy, like the electric mains; in this case the vehicle is intended to be used as a pure electric vehicle for most of the time, mainly in urban environment, while the thermal engine has the role of a range extender to allow the use of the vehicle when the batteries are discharged or in long journeys.

Before dealing with contemporary hybrid vehicles, a few words must be said about what is usually considered as the oldest hybrid vehicle, the Lohner-Porsche Elektromobil (Fig. 12.7). It was built by Ferdinand Porsche as a BEV and shown at the Paris Exibition in 1900. It had a 61 km range and was peculiar, even in a time when BEVs were common, because of its electric motors located directly in the front wheels. In 1901 it was modified by adding a gasoline internal combustion engine driving a generator to recharge the batteries, becoming a series hybrid vehicle.

The Toyota Prius was the first hybrid car in the world to be mass-produced; it is a parallel single-axis hybrid in which the internal combustion engine and an

Fig. 12.7 The Lohner-Porsche Elektromobil was built in 1900 as a BEV, but was then converted into a hybrid vehicle

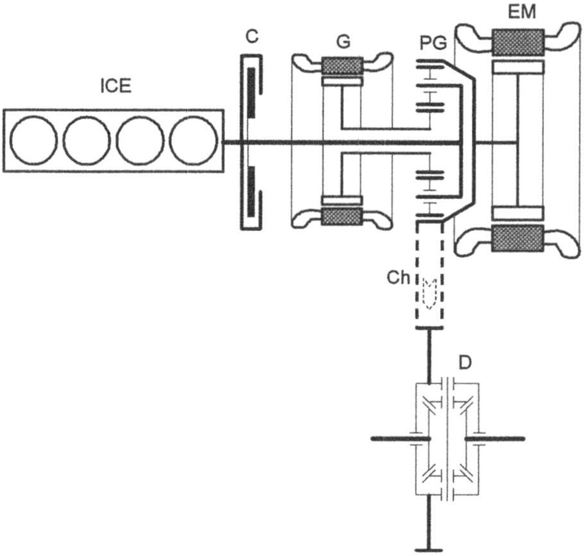

Fig. 12.8 Layout of the hybrid power system of the Toyota Prius; *ICE* thermal engine, *C* automatic clutch, *G* generator, *EM* driving electric motor, *PG* planetary gear, *Ch* chain driving the final gear at the differential D (from Genta and Morello 2009)

electric motor, arranged on the same axle, drive the wheels either separately or at the same time. As shown in the sketch of Fig. 12.8, the system includes a planetary gear for splitting the engine torque between the electric generator and the driveline. The generator, in turn, may either charge the battery or directly power the electric traction motor through the inverter. At start-up and at very slow speed, the electric motor works alone. The heat engine cuts in automatically at medium speed. The planetary torque splitter (central differential) replaces the gearbox because the electric traction motor seamlessly regulates speed.[5]

This system works both as a parallel and a series hybrid. The angular velocity of the thermal engine adapts to that of the vehicle by changing the speed of the generator, something that can occur only by subtracting a torque, through the generator. By doing this, some power from the thermal engine charges the battery, as in the series layout. At low speed, a part of the power needed for motion is supplied by the electric motor, which takes it from the battery. Finally, at a very low speed only the electric motor operates, as in parallel layouts. This also occurs when the speed of the thermal engine can adapt itself to that of the vehicle, without the generator subtracting any power. When the vehicle slows down, the available kinetic energy is recovered.

This method allows the working range of the thermal engine to be restricted to that where minimum fuel consumption is obtained, for a given power requirement. It is also possible to stop the engine when the vehicle stops and to restart it easily at a speed greater than those at which conventional starter motors operate, owing to the generator that is now used as a motor. The batteries are never recharged from outside the vehicle.

The internal combustion engine operates with gasoline, according to an Atkinson cycle, to reduce pumping losses at partial load, by reducing air intake by a later closing of the intake valve instead of using a throttle valve.

The main specifications given by the manufacturer are:

- Internal combustion engine displacement and maximum power: 1,798 cc and 73 kW at 5,200 rpm;
- Total engine and electric motor power output: 99 kW at 5,200 rpm;
- Permanent magnets electric motor;
- Ni-MH, 200 V batteries;
- Empty vehicle mass; 1,395 kg, and
- Fuel consumption (NEDC): 4 l/100 km.

The Peugeot 3,008 Hybrid 4, shown in Fig. 12.9, is the first hybrid car available with a diesel engine; it is a two-axis parallel hybrid system, in which the internal combustion engine works on the front wheels and the electric motor on the rear axle. In this way an all-wheels drive vehicle is easily obtained from the conventional vehicle with front-wheels drive.

The functions of this system include a start-stop device, with a cranking motor of suitable size; the same auxiliary motor is used for operating an air conditioning compressor, the oil pump for power steering and a vacuum pump for power brakes. The internal combustion engine drives the front axle through a robotized synchromesh gearbox; the electric driveline can simulate powershifs with a suitable control of the traction force.

[5] M. Duoba et al. *In-situ mapping and Analysis of the Toyota Prius HEV Engine*, SAE Paper 2000-01-3096.

Fig. 12.9 The Peugeot 3,008 Hybrid 4 is the first hybrid car available with a diesel engine; it is a two-axis parallel hybrid system, in which the internal combustion engine works on the front wheels and the electric motor on the rear axle (courtesy of Peugeot)

The fuel consumption reduction in comparison with the conventional diesel powertrain with the same power is in the range of 40 % on the New European Driving Cycle. Different control strategies are available:

- For minimum fuel consumption, the electric motor recharges the battery during braking or when the ICE would operate with poor efficiency.
- For maximum performance, the electric motor adds its power to that of the electric motor and the latter improves the vehicle dynamic control with the rear axle torque.
- For Zero Emissions, only the electric motor works, with a reduced range.
- For maximum traction, the electric motor is controlled simulating a limited slip central differential.

The main specifications stated by the Manufacturer are:

- Internal combustion turbocharged Diesel engine displacement: 1,997 cc;
- Maximum thermal power output: 120 kW at 3,750 rpm;
- Maximum thermal and electric power output: 147 kW at 3,750 rpm;
- Permanent magnets electric motor;
- Ni-MH, 200 V batteries;
- Vehicle mass; 1,660 kg;
- Fuel consumption (NEDC): 3,8 l/100 km.

The Opel Ampera, a particular kind of hybrid car, which could also be considered as an extended-range electric car or a plug-in hybrid, provides a final example. It is designed to travel up to 80 km on battery power alone, with a 16 kWh, Li-ion battery pack. When the battery runs low, a compact gasoline powered generator is activated automatically and supplies energy with the sole purpose of allowing driving beyond

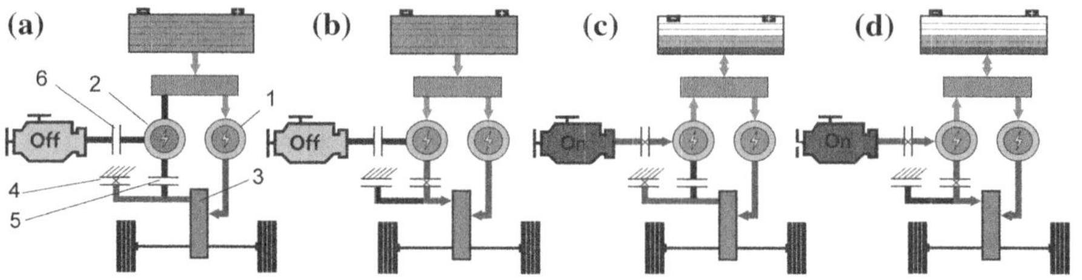

Fig. 12.10 Schemes of the Opel Ampera propulsion system, with two electric motors *1* and *2*, a planetary gear *3* and two electric clutches *4* and *5*. **a** single motor electric mode; **b** dual motor electric mode; **c** single motor extended range mode; **d** dual motor extended range mode (courtesy of Opel)

the range of the battery. It runs at selected speed and load conditions, which have been chosen to supply the requested power with maximum fuel efficiency.

When the engine is running, the system is working as a pure series hybrid one. Obviously, also in these conditions, additional energy recovered during braking is converted into electricity and stored in the battery pack. More precisely, the propulsion system is equipped by two different electric motors 1 and 2, as shown in the scheme in Fig. 12.10a. These units are connected via a planetary gear 3 and two electric clutches 4 and 5, to provide power output for propulsion in four operating modes:

- Single motor electric mode (Fig. 12.10a): The primary motor 1 runs on battery power; its maximum propulsion power is 111 kW; clutch 4 is engaged, while clutch 5 is disengaged; the planetary gear 3 works as a single speed gearbox.
- Dual motor electric mode (Fig. 12.10b): At vehicle speeds higher than 110 km/h, the secondary motor 2 engages over clutch 5; the planetary is free because clutch 4 is disengaged, so that it adds the speed and power of the secondary motor to increase the car speed without further increase of the speed of the primary motor that is running more efficiently. This allows a lower energy consumption to be achieved, without increasing the maximum available power.
- Single motor extended mode (Fig. 12.10c): When the battery reaches its minimum charge, it triggers the combustion engine on. Clutch 4 is again engaged, while clutch 5 is disengaged. The engine drives the secondary motor 2 through the clutch 6, which is now engaged; the motor 2 is used as a generator, to keep the minimum battery charging level. The primary motor can still provide its 111 kW for a short time.
- Dual motor extended mode (Fig. 12.10d): The electric motors are used again in dual configuration with increased efficiency at higher speeds. Additionally, the combustion engine contributes propulsion power via the planetary gear. The power drained from the battery is less than in dual mode electric for the same propulsion power, thus extending the range.

At the end of the mission the battery can be fully recharged in about 6 h by plugging into any standard household 230 V outlet. The battery itself is T-shaped, and is located

Fig. 12.11 The battery of the Opel Ampera is T-shaped; one branch of the T is along the floor tunnel while the other is across the car, under the rear seat (courtesy of Opel)

centrally in the chassis, as shown in Fig. 12.11; one branch of the T is along the floor tunnel, the other across the car, under the rear seat. This keeps the center of mass low and does not affect the interior space and the baggage compartment significantly.

Main specifications stated by the manufacturer are:

- Displacement of the internal combustion gasoline engine: 1,398 cc.
- Maximum thermal power output: 63 kW at 5,600 rpm.
- Maximum electric power output (primary motor): 111 kW at 3,750 rpm.
- Maximum electric power output (secondary motor): 54 kW at 3,750 rpm.
- Permanent magnets electric motors.
- Vehicle mass: 1,715 kg.
- Fuel consumption (NEDC): 1,2 l/100 km.

According to UN ECE R101 European regulation, fuel consumption in plug-in hybrid cars is calculated with the following formula:

$$C = \frac{D_e C_1 + D_{av} C_2}{D_e + D_{av}},$$

where:

C is the fuel consumption in l/100 km;

C_1 is the fuel consumption in l/100 km, with a fully charged electrical energy storage device;

C_2 is the fuel consumption in l/100 km with the electrical energy storage device in minimum state of charge;

D_{av} is the vehicle's electric range;

D_e is the distance between two battery recharges, assumed to be conventionally 25 km.

Since, in this case, the pure electric range is about 80 km, the electric energy, needed to recharge the battery, should be added to the figure of 1,2 l, needed to drive the car for 100 km, to obtain the total energy cost.

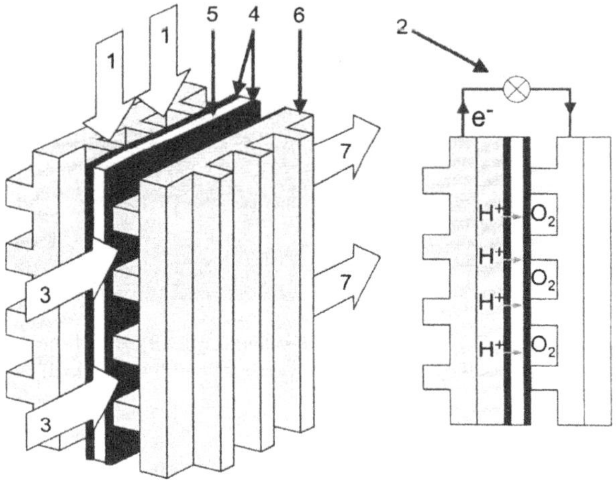

Fig. 12.12 Fuel cells consist of two electrodes *4* separated by an electrolyte *5*. Hydrogen is introduced in the anode area through grooves *1*, while air is introduced through grooves *3*. Nitrogen and water steam are exhausted through the grooves *7*.

12.4 Fuel Cells

The objective of building zero emission drive systems has spurred an increasing interest in developing fuel cells for automotive applications. They are devices that generate pollution-free electric energy by electrochemical reaction of hydrogen with oxygen, producing water as a by-product.

Fuel cells were invented by Sir William Grove in 1839 but were developed in the 1960s for space applications, namely the *Gemini* and the *Apollo* missions and then were widely used on the *Space Shuttle*.

Structurally, fuel cells are similar to batteries (see Fig. 12.12), i.e., they consist of two electrodes (4) separated by an electrolyte (5). However, unlike batteries, they do not use previously accumulated energy. Powered by a fuel (hydrogen) and an oxidant (oxygen), fuel cells generate electrical energy, and can run for as long as the hydrogen fuel supply lasts, but are not rechargeable by electric energy. In this way, the reaction between the fuel and the oxidizer is not a combustion process producing heat that is later converted into mechanical or electric energy, but an electrochemical reaction, producing directly electric energy. For this reason, the efficiency of fuel cells can be higher than that of devices based on thermal engines.

The electrodes of a fuel cell are normally made of graphite with the addition of a catalyst (platinum, palladium, or organic-metallic compounds). Various electrolytes may be used. The most recent developments employ a solid polymeric electrolyte, such as Nafion (brand name of perfluorosulphonic acid), permeable to hydrogen and oxygen ions.

Two conductive plates (6) with built-in grooves are used to organize the flow of the reaction gases. Hydrogen is introduced in the anode area through grooves (1), while air is introduced through grooves (3). In the anode area electrons are released leaving positively charged hydrogen ions. The oxygen, contained in the air and introduced around the cathode, combines with hydrogen ions, acquiring electrons and releasing water steam that is exhausted through grooves (7).

The chemical reaction is:

$$2H^+ + 2e^- + \frac{1}{2}O_2 \rightarrow H_2O \ .$$

An electric load 2 can close the circuit utilizing the generated energy, that depends on the quantity of hydrogen and oxygen absorbed by the cell. A single fuel cell generates ~ 0.8 V. Cells may be connected in series, forming batteries able of powering an electric motor; in this case, the size of the electric power package is comparable to that of the corresponding conventional internal combustion engine, obviously not including the fuel (hydrogen) tank.

Theoretically, other fuels may be used, but most fuel cells are much sensitive to poisoning by carbon monoxide and impurities, so that operating with fuels other than hydrogen and with oxidizers other than pure oxygen, or in some cases by air (carefully purified) is problematic. This is one of the main reasons that makes the transfer of fuel cell technology from the aerospace field (where pure hydrogen and oxygen are easily available and costs are much less a problem) to the automotive field so difficult.

The catalyst, the electrolyte and the membrane separating the electrodes may be of different types, and consequently different types of fuel calls, each one with its peculiar advantages and drawbacks for the different applications, exist.

- Alkaline fuel cells (AFC) use a liquid, corrosive, electrolyte and must be fuelled by pure hydrogen and oxygen, since impurities in the fuel poison the cell. Their efficiency is about 50 %, or somewhat higher. They are used in space applications since when they were developed for the *Gemini* missions; their building and operating cost is fairly low and they do not require complex ancillary equipment, but are somewhat bulky. Hydrogen-oxygen AFC for space use are a mature technology and need no specific research.
- Proton exchange membrane fuel cells (PEMFC) use a polymer electrolyte and require pure hydrogen as fuel. Contaminants like sulfur compounds and carbon monoxide poison the cell. Owing to their compact design and high energy density they are suited for automotive use, but require complex and costly equipment, like compressors and pumps, that use about 30 % of the energy produced. That notwithstanding, their efficiency is around 30 %. They operate at low temperature, about 80 °C.
- Molten carbonate fuel cells (MCFC) use an electrolyte composed of a molten carbonate salt mixture suspended in a porous, chemically inert ceramic matrix. They are tolerant of the impurities in the fuel and can run on carbon monoxide. Thus they accept different hydrocarbons like natural gas, that can be converted

to hydrogen and carbon oxides or gases made from coal. They operate at high temperatures (650 °C), which reduces their useful life. The efficiency is about 60 %, but can be increased up to 85 % if the waste heat is reused.
- Phosphoric acid fuel cells (PAFC) use liquid phosphoric acid as electrolyte. They are not affected by carbon monoxide impurities in the fuel. Their operating temperature is 150–200 °C. Their efficiency is low (37–42 %), but can be increased if the waste heat is reused. They have a limited service life and use a costly catalyst.
- Solid oxide fuel cells (SOFC) use a solid oxide material as electrolyte. They are not affected by poisoning from carbon monoxide and do not need high-cost, platinum based, catalyst, but are affected by poisoning due to sulfur impurities. The operating temperature is quite high, from 500 to 1,000 °C. Owing to the high temperature, they can use methane, or butane or even liquid fuels that are externally reformed. Their efficiency can reach 60 %, and can be used for cogeneration of electric power and heat. They can operate on light hydrocarbons such as propane and methane without a reformer, or can run on higher hydrocarbons with only partial reforming, but the high temperature and slow start-up time of these fuel cells are problematic for automotive applications.
- Direct methanol fuel cells (DMFC). They are similar to PEMFC, but use directly methanol as a fuel. Their operating temperature is in the range of 50–120 °C, but their efficiency is low, about 20 %. They do not require a reformer, but provide a lower energy density compared to conventional fuel cells, although this could be counterbalanced with the much better volume energy densities of ethanol and methanol over hydrogen. Bio-alcohol fuel is a renewable resource.

If oxygen-hydrogen fuel cells are used, the reaction product is water, that can be stored and again converted into oxygen and hydrogen by an electrolyzer. This combination of fuel cell and electrolyzer is usually referred to as a regenerative fuel cell, and in practice works as a rechargeable battery. No material is consumed (except for some losses) and the system needs only energy.

Much research is at present devoted to fuel cells for vehicular application, both for reducing their cost and for using different types of fuel. The choice of the fuel is quite limited: an interesting alternative to hydrogen is methane, that is much easier to store. If the lower energy density is not a problem, methanol or formic acid can be used as liquid fuel. Recent developments have been aimed at experimenting with direct methanol fuel cells where alcohol instead of pure hydrogen is injected into the cell. Alcohol cells may solve the currently almost insurmountable problems arising from large-scale use of hydrogen, at a cost of a low efficiency. The oxidizer is usually at any rate oxygen.

Current objectives of fuel cell development programs for automotive applications are reducing weight and volume while providing equal levels of available energy and cutting costs.

Systems for storing the fuel (hydrogen) on board and in the supply network—or for producing the fuel on board employing hydrocarbons—must be developed. On-vehicle fuel storage is possibly the biggest problem when designing fuel cell-powered vehicles. Several solutions are being considered, like storing hydrogen gas in high pressure cylinders (~300 bar) or in medium-pressure, very low temperature cryogenic tanks. Alternatively, organic fuels (natural gas, methanol, gasoline, diesel) could be suitably treated using on-vehicle systems, to extract the hydrogen needed to fuel the cell pack. The latter approach has the difficulty of meeting the sudden, instant demand for power during acceleration. A solution may involve the use of auxiliary batteries to obtain the instant response needed, although this increases the weight and size of the package. Since the problem of storing hydrogen is a key point in the hypothetical future *hydrogen society*, this point will be dealt with in detail in Part III.

The average tank-to-wheel efficiency of a fuel cell vehicle on a driving cycle like the NEDC is about 36 %, a value to be compared with 22 % of a diesel vehicle, with a maximum of the order of 45 % at low loads. However, if the losses due to hydrogen production, transportation, and storage are taken into account the 36 % efficiency reduces to a power-plant-to-wheel efficiency of 22 % if the hydrogen is stored as high-pressure gas, and 17 % if it is stored as liquid hydrogen. At any rate, fuel cells are efficient relative to combustion engines, but are not as efficient as batteries.

All major manufacturers have produced demonstration prototypes with fuel cell powered drives. System architectures are very different and reflect the different points of view of the various manufacturers and their research establishments. Essentially, architectures fall within two large categories, namely electric-powered vehicles and electric-heat engine powered hybrid vehicles. Within each category, some manufacturers have opted for a pure hydrogen fuel cell system (fuelled with hydrogen in gas or liquid form) while others are developing organic liquid fuel systems (methanol, gasoline, etc.) using reformers to extract hydrogen from the fuel. In this case, batteries are used to cope with power peaks during accelerations.

The variety of solutions is proof of the current uncertainty surrounding developments in product technology and the feasibility of producing new fuels on an industrial scale and distributing them through extensive networks. Fuel cell powered vehicles will become a reality only if international agreements will cause basic technology to converge toward the development of standardized components and a common distribution network for the new fuel (pure hydrogen or other). For obvious reasons, such developments must be seen as long-term projects (~30 years ahead or more).

12.5 Gaseous Fuels

Gaseous fuels, such as Liquefied Petroleum Gas (LPG) or Compressed Natural Gas (CNG), have always been considered as possible alternatives to gasoline or diesel fuel in motor vehicle engines. Their use only necessitates non-structural changes to the standard gasoline engine and the installation of a special fuel tank. In ordinary vehicles, the tank is nearly always placed in the trunk.

The resulting loss of luggage space in addition to the limited scope of the fuel distribution network, restricted large-scale use to certain historical periods or situations

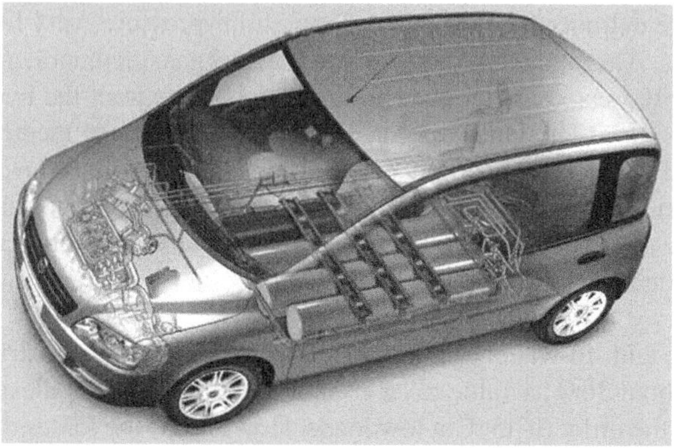

Fig. 12.13 Natural gas vehicle developed by FIAT; the MPV architecture with high floor surface allows an easy installation of gas bottles without negative impact on interior space (courtesy of FIAT)

of fuel shortage. This was the case in Italy in response to pre-war sanctions and immediately after World War II. Today, high taxation on conventional fuels makes the use of natural gas and LPG economically appealing at least on particular vehicles than can easily accept the added fuel tank. This explains why there are about 1,200,000 LPG vehicles and (unknown elsewhere in Europe) some 700,000 natural gas powered vehicles on Italian roads (year 2005).

The need for technological solutions able of lessening the impact of vehicle traffic on the environment has recently given impetus to the development of vehicles operating on gaseous fuels. They do not contain aromatic hydrocarbons or olefins and therefore the unburned hydrocarbon emissions deriving from their use are less reactive and less toxic than the emissions of unburned hydrocarbons from gasoline and diesel engines.

The carbon content of butane and propane molecules (the basic components of LPG) or natural gas is lower than that of hydrocarbon molecules in gasoline and diesel fuel. Consequently, for the same fuel consumption by weight, CO_2 emissions are also lower. With these considerations in mind, the ideal gaseous fuel would be hydrogen because, if burned in an engine of a conventional power plant, its emissions would be very low, namely no CO_2, CO or particulate, low NO_x (which is inevitable on account of the nitrogen content of air), and minimum hydrocarbon (due only to the burnt oil).

An example of natural gas vehicle (actually, a gas-gasoline bi-fuel vehicle) developed by FIAT is shown in Fig. 12.13; the MPV architecture with high floor allows an easy installation of gas bottles without negative impact on interior space . The bi-fuel system is needed since the fuel supply network of natural gas for automotive purposes is still scanty and non-existent in some areas.

Compressed natural gas is stored in cylinders at 200 bar and therefore in gaseous form; this normally limits the vehicle range because of the limited space available, but this kind of vehicle can accommodate large volume cylinders that allow a range of about 500 km, comparable to that of conventional vehicles.

A possible evolution is the storage of natural gas in liquid form employing very low temperature cryogenic tanks. This solution has already been implemented on trucks and is being tested in the USA. Thus, natural gas systems may offer an opportunity to test and refine a technology, which will be required if hydrogen is going to be the fuel of the future. All LPG vehicles marketed today feature also a dual fuel gas-gasoline engine. Again, this is done to offset the limitations of the fuel supply network.

LPG is easily stored as a liquid in tanks at a pressure of 7 bar. The presence of two tanks, one for LPG and one for gasoline, causes problems similar but with lower impact on space than those of natural gas vehicles. However, on account of the higher density of the fuel, the range with a full tank of LPG is 400–450 km.

Natural gas is perhaps the gaseous fuel offering the largest advantages, since it is:

- An excellent fuel with high octane rating. In natural gas-only engines, this property allows partial recovery of the loss of volumetric efficiency due to the low density of natural gas;
- available in nature in large volumes in various areas of the planet, not necessarily in oil fields;
- readily available in Europe thanks to a capillary distribution network;
- obtainable by decomposing organic waste and, finally,
- the fuel closest to hydrogen because its molecules have a highest hydrogen/carbon atomic ratio.

Therefore, it is a true alternative fuel, capable of relieving the current dependency on petroleum as a source of energy for motor vehicles. Various programs are currently underway in many European centers in an effort to refine natural gas technology for application to a wide range of vehicles, including buses, trucks, door-to-door delivery vans, taxis, private cars and utility vehicles.

Part III
Future

Chapter 13
Introduction to Part III

The evolutionary lines along which the future development of motor cars will proceed are well traced from some points of view and quite fuzzy from others. The problems the designer will face in the future have been known for decades and have not changed much in the past years, and the basic lines along which to look for adequate solutions are also known. There is little hope of a breakthrough that can make things suddenly easier and sudden dramatic changes that will jeopardize the very future of the industry are likewise unlikely.

However, most of the statements that are at the basis of prediction of future changes are controversial and there is no agreement on some of the basic figures. Some of these statements and figures have a strong political impact and imply many economical, industrial, strategical and national interests and thus it is difficult to find unbiased statements and opinions. The emotional and irrational issues are strong and it is often difficult to distinguish between simple opinions, often based on fears, hopes or outright ignorance but forwarded as if they were fact and considerations based on evidence and scientifically proven facts (for what the last definition can mean).

For instance, a basic point that will influence the future of automotive industry is the availability of the required energy resources and, in particular, of the fossil fuels that at present power the large majority of road vehicles (and air vehicles too). But this very issue is controversial, and the statements that can be found (and are presented by those who forward them as matters of fact) span from the alarming assertion that the world oil reserves are almost depleted (or at least the peak production has already been reached, and the peak of the price of oil that occurred in the recent past was by some interpreted in this way) to the tranquilizing statement that there will be no problem in the supply of oil in the twenty-first century.

According to a scenario developed by the US Department of Energy, the worlds' oil reserves will be depleted by the end of this century. The prospect is better for natural gas, and slightly more so for coal. Figure 13.1 offers an overall picture of the energy consumption forecasts in the world; the consumption is measured in billion barrels of oil per year, regardless the actual nature of each of the energy sources

G. Genta et al., *The Motor Car*, Mechanical Engineering Series,
DOI: 10.1007/978-94-007-8552-6_13,

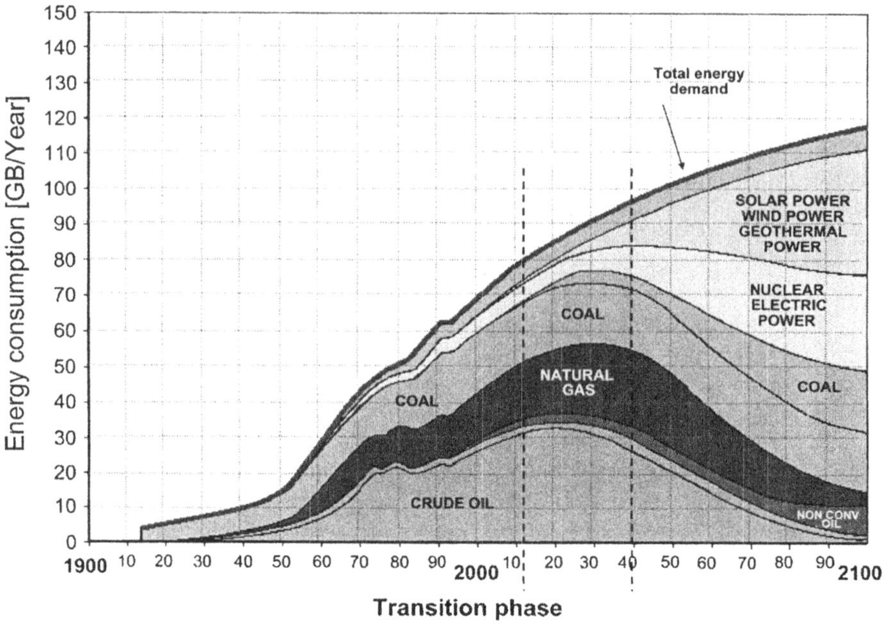

Fig. 13.1 Overall picture of the energy consumption forecasts in the world, following a scenario developed by the U.S. Department of Energy. The consumption is measured in billion barrels of oil per year, regardless the actual nature of each of the energy sources taken into consideration

taken into consideration. Other predictions yield much less pessimistic prediction for oil, but above all for natural gas that at present seems to be much more abundant than it is.

Another critical point of this scenario, and of many other similar exercises, is the assumption that research on fusion energy generation will yield no practical result within the present century: if energy from nuclear fusion becomes available in good quantity within some decades, a completely different scenario may result and this will have a deep effect on the future of the car industry. Apart from fusion, in all these scenarios no allowance is given for technological advances that may reduce the energy consumption or make new sources available: no doubt it is wise not to take into account possible future advances that may not materialize, but in this way the resulting scenarios are necessarily overly pessimistic.

At any rate, being immaterial which scenario will actually materialize, alternative energy sources need to be developed, and in particular nuclear power and solar power (photovoltaic, solar-thermal, wind energy and hydroelectric power all unquestionably come from the Sun). During a so-called transition phase (from now to 2040), energy from fossil fuels will likely be employed to produce alternative fuels with lower environmental impact, for fuelling the internal combustion engines of tomorrow and possibly, in the longer term, fuel cells.

The idea that, before the middle of the century, hydrogen will become the fuel of choice in the automotive field is widespread. The concept of a *hydrogen civilization* first emerged in the 1970s when the production of hydrogen from nuclear energy was prospected. Hydrogen is not an energy source, but an energy carrier and producing hydrogen requires more energy than that hydrogen can produce when used. The hydrogen civilization is thus possible only if abundant and cheap energy is available, and in practice requires a widespread use of nuclear power, from fission and, even more, from fusion. No alternative technologies capable of producing large amounts of hydrogen, with low CO_2 emissions into the atmosphere, at reasonable cost are available or predictable. The issue of how to fuel the automobiles of the future is today entirely unresolved. It is however likely that, before a viable solution is found to the problem of hydrogen production and transportation, the use of other fuels, such as natural gas, synthetic fuels and alcohol-based fuels, will prevail.

The spread of opinions on environmental problems is even wider, if possible. In particular, the controversial points on the global warming (now renamed climate change) issue are such that in the face of the statements of the IPCC (Intergovernmental Panel on Climate Change) endorsing the increase of the temperature of the planet due to human activities, and in particular to the burning of fossil fuels, there are opposite statements from some scientists that either deny the fact that these changes are induced by human activities or even deny the effect as a whole. Actually, most of the research in this field is based on mathematical models that have never been fully validated and thus are open to criticism. The point is that the regulatory bodies that prepare the standards on emission and on fuel consumption base their work on these statements and thus the directions the automotive industry will take in the future depend on them.

This part of the book will follow a pattern different from that characterizing the previous ones: instead of being subdivided following the various subsystems of the vehicle (body, chassis, drivetrain, etc.), topics will be subdivided following the problems that will induce the changes or the new technologies that will produce them. A final chapter dealing with deeper architectural changes that in a more distant future may change the very concept of what a motor vehicle is will follow.

Chapter 14
Energy and Environmental Issues

As stated in Part II, two of the problems the automotive industry had to solve starting from the 1960s were related to the environmental impact due to the widespread use of road vehicles, in particular in highly populated areas like cities, and the use of energy, and in particular fossil fuels. Since then strict regulations have been imposed regarding emissions, while it was left to the market forces to take the task of convincing manufacturers to invest in reduction of fuel consumption. This trend was reinforced in the following years and the standards on emissions and then also on consumption, became increasingly strict.

At present, environmental regulations do not deal only with substances that can be defined as pollutants (unburned hydrocarbons, NO_x, CO, particulates, etc.), but also with greenhouse gases like carbon dioxide (CO_2). The latter cannot be considered as a pollutant, since it is in itself not dangerous for life-forms (it is essential to plant life), but above a certain global concentration may cause an increase of the temperature of Earth. Note that some greenhouse effect is required, since without it Earth would be too cold.

The two problems are here dealt with together, since some solutions that would reduce pollution could cause an increase of fuel consumption and vice-versa. For a correct approach, they should be studied in a global way: it is not just a problem of how much pollutants are produced or energy is used by a certain vehicle when it operates, but what is the global pollution and energy consumption of a given vehicle through its whole life-cycle, i.e. the pollution and energy requirements in a vehicle's production, maintenance and disposal must be also accounted for. Operating in this way the total distance travelled by a vehicle in its useful life becomes important: If a vehicle is little used what most matters is the pollution and energy consumption in construction and disposal, while the importance of these two phases of the vehicle lifecycle is reduced with increasing distance travelled by the vehicle during its life.

Any change in a vehicle that is introduced to reduce its emissions or to improve its fuel efficiency, but affects negatively the pollution or energy use in its construction (for instance owing to the greater complexity or the use of particular materials) has a positive effect starting only from a given actual utilization of the vehicle. For instance,

G. Genta et al., *The Motor Car*, Mechanical Engineering Series,
DOI: 10.1007/978-94-007-8552-6_14,

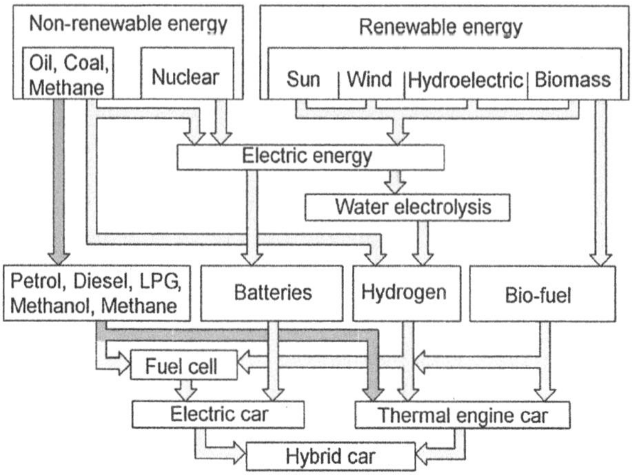

Fig. 14.1 This chart shows how primary energy sources, classified as renewable and non-renewable, can be conveyed to their final use. The darker path is the traditional one

a complex hybrid vehicle may be expedient if used as a taxi, but not as a private, seldom used, city car.

The ways proposed for improving future vehicles from these viewpoints are many, but the main ones are related with innovative drivelines: electric vehicles, hybrid vehicles, vehicles using 'alternative' fuels. This does not cover all the possible innovations in this sector, since any improvement that can be made in energy efficiency in general, like reducing aerodynamic drag, rolling resistance, weight, etc. causes a reduction of energy consumption and, as a consequence, reduces pollution owing to decrease of the fuel that has to be burned.

The energy needed for motor vehicle propulsion may be generated and stored in various ways. Traditionally, non-renewable sources of energy, mostly oil and derivate products (gasoline and diesel fuel) have assumed a dominating, although not exclusive, role. Technological developments today allow us to design different scenarios in which primary renewable sources of energy may come into play, contributing to both improving environmental conditions and safeguarding the reserves of natural resources which may no longer be available in the future.

The chart in Fig. 14.1 shows the primary energy sources that can be used by each kind of propulsion system and how they are conveyed to their final use. The intermediate forms of energy can also be called *energy vectors*. The chart is far from being complete, but it covers the traditional approach (darker path) and many of those that have been suggested for the future. In particular, hydrogen has been assumed to be produced through water electrolysis, while at present only 3 % of the total hydrogen production is treated in this way.

A vehicle fleet including vehicles propelled by internal combustion engines (gasoline, diesel, LPG, natural gas and hydrogen), electric batteries, hybrid systems and fuel cells can have its energy requirements met by diversifying the sources of energy

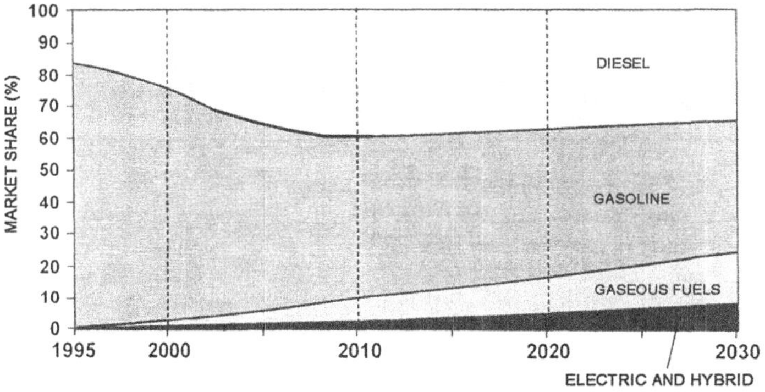

Fig. 14.2 Possible scenario for market shares of present and future fuels/power plants

including renewable, and therefore inexhaustible, ones. Even internal combustion engine fuels may be derived from oil, coal, natural gas (methane), biomass decomposition or water hydrolysis (hydrogen). On the other hand, the electricity needed to fuel electric and hybrid vehicles and to produce hydrogen through water electrolysis may be generated by thermal, nuclear, hydroelectric, wind or solar power stations.

In conclusion, today's exclusive dependency on oil and derivate products could be mitigated by diversifying the type of vehicles in the fleet and therefore the sources of energy needed to meet demand. In this light, to ensure sustainable mobility whilst protecting the environment and saving energy, the motor vehicle fleet will gradually evolve by adopting a diversity of power plant solutions. A prediction of the possible future development of car power plants in Europe in the next 20 years is shown in Fig. 14.2.

In this prediction, gasoline and diesel engines will continue to dominate the market in the foreseeable future. The introduction of direct injection and current progress in engine management strategies and exhaust gas post-treatment systems will ensure that the ambitious emission and fuel consumption targets stated by new emission and consumption standards are met without sacrificing the flexibility, convenience and driving pleasure we are used to enjoy. Concurrently, the market share of gaseous fuels powered engines—LPG and especially natural gas—is expected to continue to increase as in general, motorists consider their performance to be comparable to that of conventional engines, are cheaper to run (provided the current level of fuel taxation is maintained) and are more environmentally friendly.

Electric cars, despite their limited range and high costs (bound to remain so for many years to come) may offer the right solution for urban mobility, if made available at city entry points where motorists can park their own vehicle and hire an electric car for use in town. Initiatives of this kind, tried in European cities, are proof of the feasibility of such solutions. Fleets of vehicles owned by public authorities and private companies may also be converted to battery electric vehicles without losing in efficiency and with significant advantages for the urban environment.

Table 14.1 From electrochemical energy density ED to battery energy density, in batteries considered for electic vehicals

Battery type	Electrochemical ED (Wh/kg)	Cell ED (Wh/kg)	Battery ED (Wh/kg)
Lead–Acid	170	40	30
Ni–MH	180	70	60
Li–ion	710	150	120

Finally, whilst hybrid or fuel cell powered cars can offer near-zero emissions and solve the problem of limited range of electric cars, their impact is unlikely to radically alter the future scenario, because of their cost; nevertheless they represent the only available technology to reduce consumption and emissions to a minimum, without changes in our everyday life. A prudential estimate suggests that the total of electric and hybrid vehicles is expected to reach a 10 % share of the market.

14.1 Battery Electric Vehicles

At the end of Part II it was said that the main problems that today hamper the diffusion of battery electric vehicles are still the same problems that caused them to be abandoned in the past. The situation will be mitigated in the future, thanks to the introduction of new and better batteries, but the improvements in this area can only be pushed up to a point. The available space for improvement can be estimated from Table 14.1, in which the maximum theoretical energy density of some battery systems is compared with the energy density actually achieved at a cell and a battery level: It is clear that the theoretical energy density, that takes into account only the mass of the electrochemical reactants, is a maximum that cannot be closely approached. Some space for improvement is shown also by the comparison of the cell energy density with the actual battery energy density; some savings in the mass of the battery structure integrating cells, installation interfaces and cooling system that are usually needed can improve the energy density of all battery types.

However, the problems related with battery electric vehicles are more linked with their mass production in large quantities than with designing and building vehicles of this type that can satisfy the customers' needs. The first point, as already discussed, is the large scale availability of the materials needed to build the batteries, like lead or lithium.

Even worse problems are related to the energy requirements. The energy consumption of a battery electric vehicle is not better than that of a conventional vehicle, but it is likely worse. Electric energy has to be produced in power stations from the primary source, then transported along the electric lines, used to charge the vehicle battery and then transformed into mechanical energy in the electric motor. Each step has its own efficiency, in particular the charging and discharging of the battery has an efficiency that depends on many factors, but cannot be very high. The average efficiency with which electric energy is produced at the power station is much higher than the efficiency of the internal combustion engines of cars (typically 39 % against and average of 20–25 %) but this is usually not enough to compensate for the larger number of energy transformations from the primary source to the wheels.

Battery electric vehicles have the energy advantage of allowing regenerative braking, a thing that improves their energy efficiency, but also the disadvantage of not allowing the use of waste heat produced on board for heating. Particularly in cold weather, the heating of the passenger compartment must be made by tapping energy from the batteries or burning a small quantity of fuel. While in the second case the efficiency is quite good, but the implementation is tricky and an on-board tank for the heating fuel is required, in the first case the efficiency is low, since the efficiencies of the chain starting from the power station and ending at the battery must be accounted for. Slightly better is the alternative of storing thermal energy on board, since at least the losses linked with the charge and discharge of the batteries are avoided, but an additional thermal energy storage system must be introduced, and the efficiency of the thermal insulation of the passenger compartment must be improved over present standards. For air conditioning there is little to say: energy must come from the battery, decreasing the range and increasing energy consumption.

Similar considerations hold for pollution and greenhouse gas emissions. Electric vehicles are considered to be Zero Emission Vehicles (ZEV) , but actually, as already said, they have the advantage of moving the pollution from the place of use of the vehicle to the place where the electric energy is generated. This may be a good advantage in avoiding pollution in congested urban areas, but has no global effect, except for the better pollution control that can be achieved at the power station.

The result for both energy and environmental issues depends strictly on the mix of primary sources used in any particular country. If the electric power is generated mostly using oil-based fuels, the energy advantages are at best irrelevant and usually counter productive, as is the impact on greenhouse gas production. The well-to wheel energy consumption is bound to increase with a large use of electric vehicles. If large use is made of coal, the greenhouse gas production is worsened. Only if the primary source is nuclear, or for the small percentage they can contribute to, renewable sources, there is an advantage in a large use of electric vehicles, that are a solution suited only to an energy-rich society, possibly based on nuclear power.

Note that all the above considerations are in a way incorrect, because they do not take into account the whole life-cycle balance, and specifically do not take into account the energy required to build the batteries (that have a limited life), and the energy and costs related with the required increase of generation capacity and with the increase of the capacity of electric power lines, even if the last two can be mitigated by charging the batteries in off-peak hours.

Other improvements, that in the last years made the construction of effective electric cars possible, are linked with advancements in electric motors and power electronics technology. Brushless, rare earths, permanent magnet motors, can reduce the bulk and weight of the powerplant as well as increasing their efficiency, even if the average efficiency of the small electric motors used on cars is quite lower than that often quoted figure of 90 %.

Here another limitation can be predicted: the availability of neodymium, essential in the production of powerful neodymium–iron–boron magnets for electric motors and generators, is limited and its production is at present less than 20,000 tons per year. A severe shortage of neodymium, whose production is concentrated (about 97 %) in a single country (China) and the subsequent sharp price increase, is predicted for the foreseeable future. For sure, the neodymium reserves are insufficient to build a number of electric motors corresponding to the conversion of a large portion of vehicles to electric power.

As a conclusion, battery electric vehicles are improving their performance, and there is no doubt that prototypes with performance similar to those of conventional vehicles can be built already at present and, even more, in the near future. To build a good number of such vehicles at a cost and with an ease of use matching what the customers at present expect from a car is a more difficult task, although a task manageable in the future. What at present seems to be unfeasible, and even inadvisable, is an attempt to substitute altogether present internal combustion vehicles with electric vehicles.

Even worse are the perspectives for electric vehicles powered through fuel cells; some experts believe that fuel cell cars will never become economically competitive with other technologies or that it will take decades for them to become profitable. In July 2011, the Chairman and CEO of General Motors, Daniel Akerson, stated that while the cost of hydrogen fuel cell cars is decreasing, "The car is still too expensive and probably won't be practical until the 2020-plus period, I don't know."[1] Similarly, Steven Chu, the US Secretary of Energy of the Obama administration, stated that hydrogen vehicles "will not be practical over the next 10–20 years."[2] Furthermore, he stated that since hydrogen is primarily obtained by reforming natural gas, some of the energy content of natural gas is lost and there does not yet exist a good storage mechanism for transportation.

Fuel cell technology thus seems to be a hypothetical alternative for the far future, in a scenario of abundant energy produced by a primary source different from fossil fuels and after the problems related with production, storage and distribution of hydrogen (see below) have been solved.

14.2 Hybrid Vehicles

Hybrid vehicles provided with an internal combustion engine, one or more electric motors and an electrochemical battery are already on the road and were described when dealing with the present state of the art.

---

[1] Shepardson, David. "*GM CEO: Fuel cell vehicles not yet practical*". The Detroit News, July 30, 2011.

[2] Chu, Steven. *Winning the Future with a Responsible Budget*. U.S. Dept. of Energy, February 11, 2011.

The question whether they represent the future of automotive vehicles is often asked. There is little doubt that:

- They require smaller batteries and (in some configurations) smaller electric motors than battery electric vehicles. The difficulties linked with large scale production and to the limited supply of battery materials (lead or lithium) and motor material (neodymium) are thus less severe;
- they are more energy efficient than both conventional internal combustion engine vehicles and battery electric vehicles;
- their emissions are less harmful that those of conventional vehicles. Although they pollute more than electric vehicles in the place of utilization, the comparison in terms of global pollution and the production of greenhouse gases depends on the mix of primary energy sources with which the batteries of the latter are charged.
- they are free of some of the typical drawback of battery electric vehicles, like limited range, recharge time, increase of electric energy consumption and increase of the electric energy transportation capability.

On the other side, they are surely more complex than both conventional and battery electric vehicles and the energy required for their construction is likely to be larger than that required for at least conventional vehicles.

An assessment on this issue is made more difficult by the wide variety of hybrid vehicles that are presently built and are proposed for the future: They go from what are practically battery electric vehicles with a small on board charging facility, to almost conventional vehicles with a device to recover braking energy and to improve acceleration. In this sense, racing cars with a Kinetic Energy Recovery Storage (KERS) could qualify as hybrid vehicles, even if usually this term is restricted to vehicles in which the amount of energy stored is larger.

The present and, even more, the future improvements in the battery field answered to the objection that batteries do not have the power density for applications of this kind, or at least that they work in inadequate condition on hybrid vehicles, causing a short life. To avoid this problem, mostly related to lead acid batteries, hybrid vehicles with supercapacitors, flywheels or compressed air storage were proposed in the past, either as simple hybrids (with, for instance, internal combustion engine and flywheel) or ternary hybrid (internal combustion engine, batteries and flywheel). The latter solution seemed to be optimal, because both energy storage devices work in optimal conditions, but their complexity is larger, perhaps impractically so.

Most of these problems were solved with advanced batteries, and lithium batteries lend themselves optimally to the use on hybrid vehicles, although with some safety problems.

The efficiency of the internal combustion engine is important, and thus a definite improvement is obtained by using a small diesel engine instead of a spark ignition engine. Attempts to use small gas turbines , on the basis of their higher efficiency, have been done (see below). Specifically designed engine-generator units are under study for series hybrid, in which the unit has no mechanical energy output. Configurations can be used in which the pistons operate linear generators directly, without needing to transform reciprocating into rotational motion.

In conclusion, while it is certain that hybrid vehicles are the best solution for many applications, such as vehicles used mostly in an urban environment and that travel a long distance during their life (taxis being a typical example), it is questionable whether vehicles which are used less, or operate mostly outside cities, could really benefit from hybrid technology. It is possible to predict that the improvements in battery, motor and power converter technologies will cause the number of hybrid vehicles to increase in the future but a complete substitution of conventional vehicles can be considered as unlikely.

14.3 Non-conventional Fuels

14.3.1 Hydrocarbons and Oxygenated Fuels

The large majority of motor vehicles at present use two types of fuel, both derived from oil: gasoline and diesel fuel. They are made by hydrocarbon molecules that are fairly big and complex. For instance, a typical molecule found in gasoline is isooctane, C_8H_{18}, and a typical molecule of diesel fuel is cetane, or n-hexadecane, $C_{16}H_{34}$. Since they contain a fairly large quantity of carbon, they produce a large quantity of carbon dioxide when burned.

The simplest hydrocarbon, and the one that produces the least carbon dioxide, is methane, CH_4, the primary constituent of liquefied or compressed natural gas. A slightly more complex one is propane, C_3H_8, the primary constituent of liquefied petroleum gas (LPG).

Non conventional fuels contain also constituents that are not strictly hydrocarbons, but contain also a small quantity of oxygen, like alcohol fuels such as methanol, CH_3OH and ethanol C_2H_5OH, and biodiesel. They are often referred to as oxygenated fuels.

The pollution produced by burning all these fuels depends mostly on how they are burned and how the exhaust gases are dealt with; it is thus more a matter of the engine than of the fuel: in ideal conditions the combustion products of all these fuels would be only carbon dioxide and water. On the contrary, the amount of carbon dioxide produced depends on the ratio between the ratios of the carbon and the hydrogen atoms the fuel contains. The specific CO_2 emission referred to the energy contained in the fuel is 0.25 kg/kWh for gasoline, 0.27 kg/kWh for diesel fuel, 0.23 kg/kWh for liquid petroleum gas and 0.20 kg/kWh for natural gas. These values depend however on the exact composition of the fuel and on the assumption made on how well the combustion is carried out. At any rate they refer to the energy contained in the fuel; to refer them to the energy produced by the engine the efficiency of the latter must be kept into account. This is important above all when comparing gasoline with diesel fuel, since diesel engines have a better efficiency than spark ignition ones.

The use of LPG or natural gas and above all methane (the hydrocarbon with least carbon) instead of gasoline or diesel would reduce CO_2 emission, although not by a large proportion. They are also considered beneficial for pollutant emissions in general, since it is easier to keep low the contents of most pollutants, in particular unburned hydrocarbons and particulate.

A completely different issue is the use of biofuels, including alcohol produced from plants and biodiesel: They are usually considered to have a specific CO_2 emission equal to zero. The rationale behind this statement is that by producing the biological material from which the fuel is produced the plants subtract from the atmosphere the same quantity of carbon dioxide that will be reintroduced when burning the fuel, with a zero greenhouse gases balance. The problem with biofuels is not technological, but an economical and ethical one: if crops are specifically cultivated for producing fuels in any globally consistent amount, this will result in a reduction of the soil dedicated to growing edible plants and thus in a decrease of the food production and an increase of food price. A different matter is whether biofuels are produced from crop residues and waste (straw, cellulosic biomass, etc.). These considerations are controversial, with some asserting that the food price increase of 2007–2008 was due to the increase of biofuel production, and defining the latter as a crime against humanity, and others who stated that the impact of biofuel production on the price of food was small. At any rate the idea of using land to cultivate plants for the primary reason of producing fuel does not seem to be good.

As a general conclusion it can be stated that the use of alternative fuels does not require substantial changes as far as vehicles are concerned, since most engines can be easily adapted at the factory to burn different fuels. This is the reason why so many prototype vehicles burning a wide variety of alternative fuels were presented and demonstrated at shows. The problems mostly regard the distribution networks to make these fuels available to the general public and thus are of organizational and not of technical nature. In some countries, like Brazil, there is already a good experience in using biofuels (e.g. bioethanol) or various mixtures of traditional and alternative fuels (like the so called gasohol, 10 % ethanol and 90 % gasoline, or E85, 85 % ethanol and 15 % gasoline). In any case, the use of these fuels can improve the situation as far as greenhouse gases themselves are concerned, and can decrease the dependence of a country on imported energy products. From a global viewpoint, however, in countries where a good fraction of the electric energy is produced by fossil fuels, it is not of much importance whether biofuels are burned in cars or in power stations.

14.3.2 Hydrogen

14.3.2.1 Generation

Hydrogen was discovered by Cavendish in 1766 (Fig. 14.3). Lavoisier realized that it was an element, and named it hydrogen; immediately it was realized that it was very flammable and difficult to contain since it escaped from even the smallest porosities of the container. Nonetheless, just 17 years after it had been discovered, large

Fig. 14.3 Hydrogen: from its discovery by Cavendish to its possible future use as an energy vector for cars

quantities of the gas were used by Jacques Charles to inflate an aerostat which flew on December 1st, 1783. This was almost the only use of hydrogen for 200 years (apart from applications in chemical laboratories[3] or industry for the production of methanol, ammonia, etc.), until the Hindemburg disaster (1937) ended its use in lighter-than-air machines, which have been since then inflated by helium. Since the 1960s it became a widespread fuel for rocket engines, owing to the high specific impulse of the hydrogen-oxygen propellant combination.

When burning hydrogen no carbon dioxide is produced and so no contribution to the greenhouse effect can result. The exhaust is also much cleaner, with no unburned hydrocarbons, particulate and carbon monoxide. The only emissions can be related to NO_x.

In the automotive field hydrogen can be used in two different ways: burning it in a more or less conventional engine or using it in a fuel cell to produce directly electricity.

At any rate hydrogen is not an energy source: although being by far the most common substance in the universe, our planet does not contain any free hydrogen. We can say that all the hydrogen available on Earth is either 'already burnt' (water) or combined with a variety of other elements in many compounds. Energy is required to extract hydrogen, and this energy is then given back when it is recombined. Hydrogen is thus an energy accumulator, a sort of chemical battery.

The (conceptually) simplest way to produce hydrogen is by electrolyzing water, but this requires huge quantities of electric energy. An electrolytic cell plus a fuel cell forms what is usually called a reversible fuel cell, that is a sort of rechargeable battery, with a disadvantage that it requires intermediate storage of hydrogen, which is something not easy at all, as it will be seen later. Water can be decomposed also by heat, but the temperature at which it spontaneously dissociates is around 2,500 °C, too high for industrial applications of thermolysis. Catalysts can be used to reduce

[3] It should not be forgotten that also the very first internal combustion engines, such as the prototypes developed by Barsanti and Matteucci or by De Rivaz, ran on Hydrogen.

this temperature. The sulphur-iodine cycle is promising for producing hydrogen from water in high temperature nuclear reactors.

Actually just about 3 % of hydrogen is at present produced by electrolysis, while the remainder is obtained from chemical reactions involving a fossil fuel, like natural gas, carbon or oil. A common approach is steam reforming, in which steam reacts at high temperature (700–1100 °C) with methane:

$$CH_4 + H_2O \rightarrow CO + 3H_2 \ .$$

The carbon monoxide reacts again with water at about 130 °C, yielding:

$$CO + H_2O \rightarrow CO_2 + H_2 \ .$$

This way of producing hydrogen has an energy efficiency of 80 %, so that when hydrogen is burned to produce energy, less energy is obtained than it would by burning methane directly. These reactions produce large quantities of carbon dioxide, and overall the quantity of greenhouse gases is larger when methane is transformed into hydrogen and burned than when it is burned directly (at equal thermal energy produced by the combustion). If the hydrogen is used in a fuel cell, and if the efficiency of the system made up of the latter plus the electric motor is higher than that of the thermal engine, then an improvement from this viewpoint may be obtained.

Other reactions of the same type can be used for different hydrocarbons, although all of them produce huge quantities of carbon dioxide. There are however reactions whose end product is not hydrogen and carbon dioxide, but hydrogen and carbon black: operating in this way however, less than half of the energy contained in the original hydrocarbons is at the end contained in hydrogen. A good quantity of energy ends up in the carbon, and cannot be extracted without producing CO_2.

Quite promising is the production of biohydrogen, i.e. the conversion of biomass and biological waste into biohydrogen through biomass gasification, steam reforming or biological conversion like biocatalysed electrolysis or fermentative hydrogen production. The latter involves the use of bacteria, through biolysis, photofermentation, dark fermentation, enzymatic hydrogen generation or biocatalysed electrolysis.

As a result, a widespread use of hydrogen in vehicles makes sense only if the primary source is different from fossil fuels: the true advantage of hydrogen is that it transforms into a portable energy source those sources that are not portable, such as nuclear energy, but also hydroelectric, solar, wind and other renewable sources. In particular, the use of processes that do not pass through electric energy allow production of hydrogen directly from nuclear energy, so allowing us to power vehicles from nuclear energy in an indirect way.

14.3.2.2 Storage

The same feature that makes hydrogen so well suited to aerostatic applications, namely its low density, is its main drawback in its applications as a fuel. When used as a rocket propellant hydrogen is stored as a cryogenic liquid: its boiling temperature at atmospheric pressure is about 20 K, i.e. −253 °C. In these conditions its density is just 71 kg/m^3, i.e. 1/14 of the density of water. The size of the huge external tank of the space shuttle is due to this reason; if it contained a hydrocarbon like kerosene it would be about ten times smaller. Moreover, the liquid hydrogen tank must be filled a short time before launch, and any time spent between filling and utilization causes a severe boil-off of the fuel, with the ensuing waste and above all the risk of fire, owing to the high flammability of the mixture of hydrogen and air.

In automotive applications it is possible to use liquid hydrogen, but the tank must be a true insulated cryogenic tank, able to keep its content at a temperature of −253 °C. It is practically impossible to insulate the tank well enough by passive means to prevent boil-off and a slow evaporation must be taken into account, unless active cooling is provided, a process that has its costs in terms of energy. Also the difficulties of filling safely the tank with such a liquid are clear. Anyway, BMW chose this way for the BMW Hydrogen 7 research car.

An alternative is storing hydrogen as a gas in pressurized tanks. Bottles able to keep a pressure of 350 or 700 bar are usually employed and even higher pressures are under study. The density of hydrogen at 800 bar is 36.6 kg/m^3, about half of its density in liquid form at atmospheric pressure. The energy needed to compress the hydrogen is not too large, but also not negligible, amounting to something like 2 % of the total energy content.

The safety problems linked with storing several kilograms of a highly flammable gas at such high pressures on board a vehicle are clear. Filling operations are difficult too, and time consuming, although faster than recharging the batteries of a battery electric vehicle. Further increases of storage pressure would produce an increase of the weight of the tank: an optimum between the decrease of the volume and the increase of the wall thickness with increasing pressure must be sought.

To increase safety, the pressurized hydrogen tanks are made in three layers: a polymeric, gas tight, inner one, a carbon fiber intermediate one able to withstand the internal pressure and an outer layer able to protect the inner ones against mechanical and corrosion damages. Several manufacturers such as GM, Honda or Nissan built experimental vehicles in which hydrogen is stored in this way.

The two techniques can be combined in cryo-compressed storage devices: hydrogen is stored at cryogenic temperatures, but when it boils off it is initially not vented and the pressure is allowed to rise up to about 350 bar, so that much less hydrogen is lost, in particular if in the meantime the vehicle is used and the fuel is consumed.

Other approaches have been followed: they can be subdivided into chemical and physical systems. In the most common chemical systems hydrogen is stored in metal hydrides, such as MgH_2, $NaAlH_4$, $LiAlH_4$, LiH, $LaNi_5H_6$, and $TiFeH_2$, with varying degrees of efficiency. Some of these hydrides are liquid at ambient temperature and pressure, while others are solid. Their energy density by volume is usually good, while their energy density by weight is usually worse than that of hydrocarbon fuels. The problem with this kind of storage is that energy is required both to store and to release the hydrogen from the hydrides. The weaker the bond of the hydrogen atoms within the compound the more energy is required in the first phase and the less in the second, so that also here a compromise is required. The goal is a temperature lower than 100 °C for release and a pressure lower than 700 bar for recharge. Catalysts can help in these issues, but a satisfactory solution has not yet been found.

Hydrogen can be bound to form carbohydrates that can be stored in liquid or solid form. Glucose, cellulose, starch, have been tested. Also synthetic hydrocarbons can be used as hydrogen carriers, but the disadvantage is the need of a reformer on board to extract the hydrogen from the compound, that makes the powertrain complex and costly.

Ammonia is a hydrogen carrier that is produced in large quantities and can be burned in slightly modified gasoline automobile engines but at standard temperature and pressure is a toxic gas with quite a bad smell. Other chemicals that can be used as hydrogen carriers are amine borane complexes, formic acid, imidazolium ionic liquids, phosphonium borate carbonite substances, etc.

Apart from cryogenic and compressed gas storage, which can be dubbed as storage by physical means, other forms of physical storage of hydrogen exist. For instance, hydrogen can be stored in nanostructured carbon, like carbon nanotubes, although how much hydrogen can be stored in this way is still controversial. Other porous materials, like metal-organic frameworks, glass capillary arrays, glass microspheres, can store hydrogen, in particular at low temperatures.

The research is proceeding and new chemical and physical storage devices are constantly found, but they all are a long way from being practical, safe and cost effective energy storage devices. As it was said in the preface, the very wide variety of the proposed methods show that these technologies are still quite young and a long time may be required to establish standard methods and techniques.

The various devices are characterized by a performance index, namely the ratio between the mass of the contained hydrogen and the mass of the full tank (hydrogen plus container). The American Department Of Energy (DOE) has stated a goal of 0.065 for this performance index, i.e. for every kilogram of hydrogen a tank of 14 kg is required, but lower values are now considered as satisfactory, like the value 0.04 obtained with storage pressures of 450 bar. If all ancillary equipment, like pressure and temperature control devices, etc. is considered, even lower values can be achieved.

The ratio between the mass of the hydrogen and the volume of the tank is about 70,6 kg/m^3, roughly equal to the density of liquid hydrogen.

14.3.2.3 Utilization

As already stated, hydrogen is the cleanest fuel and can be used by burning it in a more or less conventional engine or by having it react with oxygen in a fuel cell. The first alternative is for sure the simplest one, since a conventional internal combustion engine can be easily converted to hydrogen usage, but is characterized by a low efficiency, since the chemical energy of the fuel must be first converted into thermal energy and then into mechanical energy and the efficiency of the latter transformation has the limitations stated by the second principle of thermodynamics.

In a fuel cell the same reaction goes on without liberating thermal energy (or at least too much thermal energy, since some fuel cells operate at high temperature, although lower than that characterizing combustion) and chemical energy is converted directly into electric energy. However, although there is no theoretical strict limitation to the efficiency of this process, the efficiency of a fuel cell is much lower than the theoretical 100 % maximum. Moreover, the electric energy must be furthermore converted into mechanical energy in an electric motor, and the efficiency of this latter transformation must be accounted for. The overall efficiency is between 20 and 50 %, depending on the type of fuel cell, motor and controller used.

As a conclusion, the problems with hydrogen vehicles provided with internal combustion engines are mostly related with the storage of the fuel and the refuelling techniques and facilities. Prototypes with satisfactory performance can be built and have actually been produced and even a small series production is within the possibilities for the near future, particularly if the vehicle has dual fuel capabilities, so that it can work on gasoline when no hydrogen supply is available. The problems with large scale production and cost containment are however still to be solved and above all it makes little sense to produce hydrogen as an energy medium to use in internal combustion vehicles with the present mix of primary energy sources, even in countries where the share of electric power generated by nuclear energy is large, like in France (75.2 %): It is much better to use fossil fuels for vehicles while using different sources for producing electricity.

14.4 Reduction of the Resistance to Motion

The most obvious way to reduce the energy consumption of a vehicle is to reduce the so-called road load, i.e. the total resistance to motion. Any improvement in this area is not only effective in reducing fuel consumption, but also in decreasing emissions, since the amount of pollutant and greenhouse gas emitted is proportional, everything else remaining equal, to the amount of fuel burned, or more in general, to the amount of energy spent.

What has been said above applies in general, but there are cases where this effect is particularly important: in case of battery electric vehicles, for instance, the reduction of the road load leads to an increase of range, which is the most critical issue for BEV or, if the range is kept constant, in a reduction of the mass of the batteries. The mass of the electric motors and, consequently, the total mass of strategic materials needed for building batteries and electric motors, are reduced too.

In most of the vehicles seen in the previous parts of this chapter (hydrogen vehicles, electric vehicles of all types, including those powered by fuel cells, etc.) the total amount of energy carried on board is quite critical, and so anything that can reduce the total road load increases their feasibility.

14.4.1 Aerodynamic Drag

Much has been done in recent years to reduce aerodynamic drag of vehicles, in particular as a result of the energy crisis of the 1970s. Unfortunately, further improvements are at the same time more difficult and less effective: more difficult because it is difficult to further improve the already streamlined shape of present cars, and less effective because while aerodynamic drag was a substantial fraction of the total road load when the average C_x of cars was 0.45–0.5, now that it approaches 0.25–0.3 it is much less important.

Moreover, aerodynamic drag is important only at constant high speed, in practice in motorway driving, and the mileage driven by vehicles in these condition is a small fraction of the total, in many cases averaging something like 7 or 8 %. In the most frequent driving conditions, namely urban and suburban driving, the importance of aerodynamic drag goes from nil to marginal. This is even more so because of the speed limits imposed by the law in most countries: in North America, for instance, where motorway driving is more important than in Europe, the low speed limits make aerodynamic drag of little consequence. Germany is the only European country in which there are motorways with no speed limitations, and in these driving conditions the reduction of aerodynamic drag can be an interesting energy conservation practice.

This does not mean that a good aerodynamic design will not be important in the future: car aerodynamics does not deal only with drag reduction. Aerodynamic design deals with comfort (mostly noise reduction), safety (improving handling and stability) and other issues, like avoiding the deposition of dirt on head lamps and windows. The point is that in most cases these requirements are conflicting, and what is done to improve handling may for instance cause an increase of drag or an improvement in noise may cause a decrease of handling characteristics. It is well known that aerodynamic negative lift, essential to improve the performance of sports and racing cars by increasing the forces the tires can transfer to the ground, causes a strong increase of drag.

Aerodynamics is also strictly related with style, and with the habitability of a vehicle. As a general conclusion, we can expect that in the future motor vehicles will be carefully studied from the aerodynamic viewpoint but, nevertheless, the reduction of aerodynamic drag will contribute only marginally to the issues studied in the present chapter.

14.4.2 Rolling Resistance

The reduction of rolling resistance is on the contrary one of the most important ways to achieve a reduction of fuel consumption. Rolling resistance is the most important form of resistance to motion at low speed, and thus its reduction is important in urban driving, the condition in which cars are mostly used. It has two aspects:

reduction of the weight of the vehicle (see next section) and improvement of the rolling characteristics of the tires.

The first substantial improvement in this sense took place in the 1960s with the introduction of radial tires, whose rolling resistance is about 20 % lower than that of cross ply tires. In Europe radial tires became eventually compulsory, mostly for this reason.

Rolling resistance is mainly due to energy dissipation in the tire, and thus can be lowered by decreasing the damping of the tire material; in the usual formulation of elastomers used in tires, natural rubber has a lower damping than synthetic rubber and damping increases with increasing amount of carbon black in the compound. But while lowering damping lowers the rolling resistance at low speed, it makes the tire more prone to vibrate, and vibration causes a sharp increase of energy dissipation at high speed. At a certain speed (the critical speed) this increase causes overheating of the tire and this must be avoided: it is thus impossible to decrease too much the damping of the rubber material, at least in tires meant to be used on fast vehicles.

In the 1970s low speed, low resistance tires were built for electric vehicles, with the goal of increasing their range. At the end of the 1990s a new approach was introduced: by substituting silica for carbon black it was possible to introduce a dependence of the damping on the frequency, yielding tires with superior performances, accompanied by a lower rolling resistance. These tires were first introduced by Michelin and dubbed as *green tires* , with a play on words: they were *green* in the sense they used less energy, so lowering their environmental impact, but owing to the lack of carbon black they could be built in any color and, in the initial intention of their manufacturer, they would have also been green in color. Market studies showed that customers would not like tires of colors making them look like toys, and so they were colored in black, and the term 'green' applied only in a figurative sense. Notice that the term 'green tire' is also used in a wider sense, for any tire allegedly more environmentally friendly than traditional ones, and these claims are sometimes little substantiated.

The tendency toward a reduction of rolling resistance will continue in the future, with the obvious caveat that a tire is a complex object and that any change in its structure, material, production technology, etc. will affect its performance as a whole, from the ability to produce forces to safety considerations, from rolling resistance to cost and duration. A tire must be a compromise among all these features, and it makes no sense to try to optimize a single feature neglecting the other ones. It seems that at least a partial substitution of silica for carbon black is effective and that future cars will have a lower rolling resistance than present ones.

14.4.3 Vehicle Mass

Reduction of the mass of a vehicle has a twofold positive effect on the reduction of fuel consumption: on one side it reduces rolling resistance, and on the other it reduces the energy needed to accelerate, together with the energy to be dissipated in the brakes. The energy needed to accelerate can be reduced also by reducing the moment of inertia of the rotating parts, at equal mass. These effects are all particularly important in urban driving, and so reducing the vehicle mass plays an important role in the conditions that are statistically the most frequent.

In spite of these considerations the mass of cars has increased in the recent past and it is difficult to predict a decrease in the near future. This is a result of several mechanisms, partly linked with customer's requirements and partly with the regulations imposed on the car industry.

From experience in the aeronautic industry, it is well known that weight reduction is not a matter of cutting much weight in a few selected spots of the vehicle, a thing that is usually impossible, but to shave a few grams in a large number of places. In a similar way, an increase of weight is not usually the result of adding a few large masses, a thing that is easily controlled, but of adding in many places masses that, each one considered in itself, seem to be negligible.

The first of these requirements are those related to passive safety. Modern cars must pass a number of safety tests that require not only an accurate structural design, but also the introduction of safety related elements. Safety has costs, and the increase of the vehicle mass is one of them. Structurally, the requirements of absorbing crash energy and at same time of providing a rigid survival cell, providing rigid attachment points for safety belts, introducing elements like air bags, with all the related electronics and deployment systems, anti-intrusion beams and plates, fireproof elements, etc. cause an increase of the vehicle mass.

Three wheeled vehicles, that in most countries are not required to comply with the safety regulations that apply to cars but to the more relaxed rules that apply to motorcycles, can be much lighter. The same is true for those microcars or quadricycles that, owing to their mass and power lower than given limits stated by the law, are not required to comply with the automotive safety rules. No doubt they are much lighter, but they are also much less safe, by design.

Active safety has a lower impact on the vehicle mass, but all the devices like ABS, antispin, VDC, EPS and so on have their electronic control unit, actuators and other devices that add to the mass of the vehicle. Large windows to improve visibility are another weight increase factor.

The ever-increasing search for comfort implies the introduction of phono-absorbent materials, vibration insulators and vibration absorbers and improved internal finishing materials. Even electronic devices, sometimes with an utilitarian scope, others just entertainment gadgets, have their mass.

Many customers still like large cars, and in many countries, particularly in Europe, the average size of cars has not decreased in the past years. In North America, after a decrease in the 1970s, as a consequence of the energy crisis, the size of cars started to increase again mostly due to the success of SUVs and the continuing popularity of pick-ups. This trend is not likely to be reversed, except for unexpected events like a generalized increase of the cost of energy or the introduction of regulations penalizing large cars.

From what has been said, it is predictable that the trend toward larger vehicle masses is going to continue. The only factors that may fight against it are the improved stress analysis techniques, that can contain the structural mass through a more optimized design, and the introduction of new materials. These two factors, and in particular the first one, are not new, and in the past allowed reduction of the increase of the mass of vehicles below what could have been if the traditional approach was used. Both will be dealt with in specific sections.

14.5 Innovative Engines

Many proposal have been put forward in the past to radically change the type of engine used on motor vehicles and in particular on cars. The proposed alternatives came from many sources, but most of them could be summarized in the following trends:

- Internal combustion engines:

  - gas turbines
  - rotary piston engines

- External combustion engines

  - steam engines
  - hot gas engines.

None of these proposals had any practical success for a number of reasons. For sure one of them is the fact that standard reciprocating engines benefitted from a long history of development and a huge cumulative investment in research. The issue of whether any of the solutions above (or one of the others that may be conceived) could be the most successful engine type if it had the same century long history of development and improvements is questionable, and after all it is also of little interest. In the actual world things went as they did, and the engines we now have are the result of this development. The only things we can document are the reasons that caused the failure of these attempts.

14.5.1 Internal Combustion Engines

Gas turbines were regarded in the 1950s as a natural candidate to substitute reciprocating engines in vehicles, like they did in aircraft and helicopters, owing to their higher specific power and absence of reciprocating parts, which results in a smoother running and, at least in part, higher reliability and longer lasting. Turbines, particularly small ones, are much faster than reciprocating engines (the turbine of the 1963 Chrysler Gas Turbine Car, Fig. 14.4a, had a top speed of 44,500 rpm), implying a more complicated transmission. Moreover, they were much more difficult to control, at least when these attempts were made (presently, owing to simple microprocessor-based

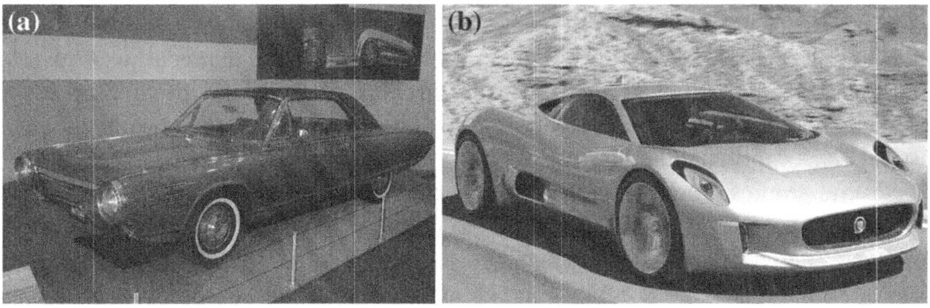

Fig. 14.4 **a** Chrysler turbine car of 1963. Five prototypes, plus 50 'production' vehicles were built. **b** Jaguar C-X75, hybrid car with 2 small gas turbines and 4 electric motors, built in 2011

controls, miniaturized gas turbines were successfully built), their fuel consumption is higher and their acceleration performance is poor. When the oil crisis broke out, their high fuel consumption stopped all attempts in this area and the many prototypes most manufacturers built in the 1960s were scrapped or ended in museums.

Now it is possible that gas turbines will make their comeback in the automotive industry for hybrid vehicles. In fact, modern ceramic materials and control technology allow miniaturization of this kind of engine, the possibility of coupling it directly with a very high speed generators allows us to dispense with a transmission (small turbojet and turboprop units used in model airplanes turn at 160,000 rpm or more, are low cost and fairly simple to use, although their life is too short for automotive application), there is no acceleration problem and above all their power density is high and their efficiency, up to 40 %, is better than that of diesel engines, and not much lower than that of fuel cells, while accepting a variety of fuels. Jaguar has built an interesting prototype of a turbine-powered hybrid car (Fig. 14.4b).

Rotary piston engines, like the Wankel engine, were introduced in the 1950s in the automotive field and had initially some success. Their main advantage is their compactness and smooth running, owing to the lack of reciprocating parts, but they have always been plagued by several problems, such as the wear of the rotor seals. Although from time to time some non automotive application resurfaces, after the oil crisis and the environmental laws their higher fuel consumption and the difficulty in complying with emissions regulations caused them to disappear from cars. In this case a comeback is unlikely.

14.5.2 External Combustion Engines

As described in Part I, steam engines were used to power road vehicles in the nineteenth century and at the beginning of the 20th, when their use was discontinued. The idea of returning to steam engines was revived during the oil crisis of the 1970s: the rationale was that while an external combustion engine can burn any kind of fuel in its boiler, the control of emissions is easier in a continuous combustion than in an intermittent one. The usual advantage of not needing a clutch and gearbox with several transmission ratios and the drawback of requiring some time to build up steam pressure were not considered very important at that time, if compared with the issues of fuel consumption and emissions.

Several studies were done, but little reached the prototype stage. Sophisticated steam engines, with superheaters and other complex devices reached high values of efficiency, and as recently as 2006 an Automotive Engineering International Technical Award was given at the SAE World Congress in Detroit to Cyclone Technologies LLP, developer of steam engines for automotive applications. Their claim is that the engine has a 30 % efficiency.

With respect to better controlling emissions and using a variety of fuels, hot air engines have the same advantages as steam engines. They can be based on different thermodynamic cycles; the ones usually considered are closed cycles like the Stirling cycle (an engine that incorporated an innovative regenerator, then dubbed Economizer, patented in 1816 and built in 1818) or the Brayton cycle. Their advantage is high efficiency, and for this reason they were several times proposed for automotive applications but their low power density, both referred to the engine mass and volume, discouraged practical applications. Their use in extending the range of battery electric vehicles has however been attempted.

14.5.3 Improving the Standard Reciprocating Engine

While these innovative engines did not enter into production, and likely will not find applications at least in the near future, conventional reciprocating internal combustion engines, both of the spark ignition and diesel types, are continuously improving and future improvements are predictable. These improvements regard both performance, in terms of specific fuel consumption, specific power, emissions, and overall characteristics, like ease of maintenance and construction, and durability. In many cases these improvements were forced by the increasingly restrictive regulations regarding emissions, in other cases by market requirements and customers' expectations.

A powerful factor allowing these advancements was the introduction of microprocessor based controls for most of the engine functions, so that all working parameters could be kept under strict control and optimized. Other factors were the improvements of design techniques, made possible by Computer Aided Engineering, constructional techniques and new materials.

All these trends are still at work, and we can expect improvements in all these areas, so that even more strict requirements regarding emissions and fuel consumption will be met by the evolution of standard engines, without having to resort to more complex and costly solutions.

Chapter 15
The Impact of New Technologies

Technology evolves mostly by small incremental steps and, as stated in Part I, true revolution are quite rare. However, there are cases where a strong technology transfer occurs between different fields of technology, and this can cause sudden changes that are felt as true revolutions.

In the last part of the twentieth century electronics and computer science caused a revolutionary change in how we process and exchange information that had deep consequences in the whole of our society and at large in our way of living. The term ICT (Information and Communication Technologies) Revolution has been used to summarize this accumulation of new technologies. It had a deep effect also on automotive technology, setting new standards to define how we use ground vehicles and what we expect from them, but also changing the ways vehicles are designed and built.

Another field in which technology is evolving fast is that of materials. The materials used in motor vehicle construction are, with some exceptions, conventional ones, with steel being the main structural material. It has been almost 50 years since a prediction was made that a radical change in the area was imminent, sometimes advocating the use of light alloys and others the use of reinforced plastics, but these changes remain mostly on paper. However, recent advances in the field of new materials have increased their pace and radically new alternatives are becoming available, or at least are thought to be due within a reasonable time. Are we at the onset of another technological revolution, this one pushed by the introduction of new materials?

15.1 ICTs: Process Innovations

Traditional design used mathematical analysis only for a few selected critical components and for the remaining parts of the machine or the structure the design was based on empirical consideration or on past experience. Prototypes were then built and extensive experimentation on them allowed introduction of modifications until

G. Genta et al., *The Motor Car*, Mechanical Engineering Series,

DOI: 10.1007/978-94-007-8552-6_15,

satisfactory performances were produced. Only at that point could mass production be undertaken. Clearly this procedure depended on many factors, like the number of specimens that had to be produced, the cost and the possibility of the experiments, the dangers linked to a failure and even on the sheer size of the object and the difficulties linked with the construction of a mathematical model that could yield significant results.

Civil engineering structures, for instance, required accurate structural analysis, and most of the methods for stress analysis were developed in the nineteenth century in connection with railway bridges. However, it was also common to build reduced scale models to experiment, particularly with dynamic loading like seismic loading. Little functional and structural analysis was performed for the early motor cars, that were mostly developed through trial and error procedures.

In the twentieth century much research work was devoted to developing mathematical methods, particularly for aviation, and then aerospace and nuclear engineering. When computers became available, in the 1950s, they were first used to automatically perform those computational procedures that required long and tedious work, for which electromechanical calculators had been widely used for decades. At the end of the 1950s computations that nobody could even think to perform without using computers became routine work and programs of increasing complexity were made available to designers. In the 1960s the situation evolved further, and the first commercial codes based on the Finite Element Method (FEM) appeared on the market. One of the first general purpose FEM codes was NASTRAN, funded by NASA as a part of that technological effort that lead to the *Apollo* programme. In the 1970s general-purpose codes that can tackle a wide variety of different problems were commonly used in almost all field of technology, including the automotive industry, that soon became one of the largest users of computers and supercomputers, as at that time the largest computing machines were called.

The size of the mathematical models computers could handle increased with their increasing power, that roughly doubled every 2 years (Moore's law), enabling designers to model, at first components, then subassemblies and finally vehicles as whole machines. Such complex models, useful for simulating many characteristics of the machine, may be considered as true *virtual prototypes*. Virtual reality techniques allow these models to yield a large quantity of information, not only on performance and the dynamic behavior of the vehicle, but also on the space taken by the various components, the adequacy of details and their esthetic qualities, that is comparable to what was once obtainable only from physical prototypes.

Mathematical models are useful not only to designers, but also to test engineers in interpreting the results of testing and performing all adjustments. Numerical experimentation is increasingly substituting physical experimentation and, while it is clear that it will never substitute for the latter completely, the number of prototypes built and of tests required is thus reduced, reducing the time and the cost needed to build a new vehicle.

A warning must however be stated: mathematical models must be validated experimentally and to rely on not yet validated models is a danger that must be avoided: this is the basic objection that many scientists raise about most of the studies on climate change. Oden and Bathe[1] defined a sort of disease, they called *number crunching syndrome*, that may affect those who deal with computational mechanics. It consists in a "blatant overconfidence, indeed the arrogance, of many working in the field [of computational mechanics] ... that is becoming a disease of epidemic proportions in the computational mechanics community. Acute symptoms are the naive viewpoint that because gargantuan computers are now available, one can code all the complicated equations of physics, grind out some numbers, and thereby describe every physical phenomena of interest to mankind." These words were written in 1978, but they are still an important warning against overconfidence in numerical modelling.

This danger must, and can, be avoided. If so, the constant progress in the field of computational hardware and software is a unique opportunity to improve the performance of vehicles well beyond the advances of the recent decades, while reducing the cost and allowing development of new models with a shorter time-to-market, which is essential in the present highly competitive market. The examples of details that once were created by trial and error and now are accurately modelled and simulated are many, from the circulation of air within head-lamps, dealt with computational fluid-dynamics (CFD) codes (with the advantage of improving the cooling of the bulbs, extending their life) to the study of the vibration dynamics of parts that once were considered negligible, like air conditioning pipes (improving the noise and vibrational comfort). The fact that car production deals always with large product numbers makes everything easier: the cost of modelling and accurately studying a small detail may be high, but if the study applies to a large number of units produced, the unitary cost may be reasonably low.

Computer aided engineering is increasingly covering the whole production cycle of a car: instead of drawings on paper there is a single data base, in which the three-dimensional properties of all components are fully represented. In this sense, we are back to the early concepts of the Renaissance, where the designer conceived a machine in its 3-D material essence, as can be seen from the 3-D sketches of the theatres of machines. Then, starting from the end of the eighteenth century, the need of precise representations to allow factory workers to build the various components, forced a shift from 3-D representations to 2-D technical drawings, thanks to projective geometry. Now computers have closed the cycle: today it is possible to define directly a 3-D geometry with the precision, but without the limitations, of drawings on paper. This is a conceptual revolution, allowing machines to be conceived in their whole 3-D material essence and not through projections. One of the best examples in the automotive field is linked to suspensions: before computers their geometry was usually conditioned by the need to project their kinematics on a plane to allow kinematic study. Now the freedom of the designer is much wider, and true 3-D geometries are possible.

From the data base containing the geometry of all components it is possible to obtain directly mathematical models for kinematic and dynamic study, stress analysis, performance evaluation, up to the virtual prototypes. The same mathematical models

---

[1] Oden T.J., Bathe K.J., *A Commentary on Computational Mechanics*, Applied Mechanics Review, 31, p.1053, 1978.

supply information to the computer controlled machines that work on the parts that are machined, the molds and punches for sheet metal parts, the models for castings, etc. The term Product Lifecycle Management (PLM) is used to define a strategic approach to the management of all information processes and resources regarding the whole lifecycle of a product, from its initial conception, to its design, construction, placement on the market, maintenance and finally scrapping.

This revolution is still not complete, and will deeply influence the future of automotive vehicles.

15.2 ICTs: Product Innovations

The new technologies linked with electronics and computer science are not only affecting the way automotive vehicles are designed and built, but may change in the future the way they operate, to the point of introducing deep changes in the product itself. The first of these changes will affect the interactions between the vehicle and the human who is in charge of controlling it.

15.2.1 Classical Manual Control

In all classical vehicles produced since the 1990s, the driver had to perform all control and monitoring tasks. The only assistance came from devices like power steering or power brakes that amplified the force the driver exerted on the controls. In this situation, the human controller is fully inserted in the control loop or, as usually said, the system includes a *human in the loop*.

Actually the driver must control high-level functions (choice of the trajectory, decisions about speed and driving style, manoeuvres like overtaking, etc.) and intermediate-level functions (reacting to perturbations coming from the air and the road, following the chosen trajectory, etc.). Only stability at the lowest level, involving the sideslip angle and the yaw velocity, is provided by the dynamic behavior of the vehicle. In two-wheeled vehicles like motorbikes, the driver must also act as a stabilizer against roll over.

In particular, in the twentieth century the following controls were standardized in all cars:

- Direction control is implemented by applying a torque to the steering wheel that is then transmitted through a mechanical system (steering box, steering arms, various linkages) to the steering wheels, which are always the front wheels. The torque exerted by the driver may be increased by a hydro-pneumatic or electromechanical system (power steering) that nonetheless never replaces the driver by exerting the whole moment. The required sensitivity is provided by the torque the steering system exerts on the driver through the aligning torque and the contact forces at the wheel-road interface. These, in turn, depend upon the geometry of the steering system (caster angle, toe-in, offsets, etc.).

- The control of the power supplied by the engine is managed through the accelerator pedal, operating directly through a mechanical leverage. Sensitivity is supplied by the elastic reaction of a spring that reacts to the motion of the pedal. The driver must control the power accurately enough so that the maximum force the wheel can exert on the ground is not exceeded.
- Engine control is accompanied by control of the gearbox and the clutch, which operate through the clutch pedal and the gear stick. These controls are often automatic.
- Braking control is performed by applying a force on the brake pedal that is then transmitted through a system (usually hydraulic, but pneumatic in industrial vehicles) to the brakes located in all wheels. Here the force exerted by the driver can also be augmented by a hydro-pneumatic device (power braking). In all cases, sensitivity is granted by the fact that the force exerted by the driver is proportional (or at least depends in an almost linear way) to the braking torque and then to the braking force. The driver must control the braking force so that the wheels do not lock.

These basic controls are accompanied by many secondary controls, such as those of the lighting systems, window cleaning and defrosting, parking brake, etc. Although not directly used to control the motion of the vehicle, these are extremely important for driving safety. The basic controls are standardized on all vehicles, with some difference in special vehicles, and are subjected to detailed standards. In the case of particular arrangements, to be used by persons with disabilities of various kinds that do not allow them to operate conventional controls directly, a non-conventional *user interface* is provided, designed as needed for each particular installation. The transmission of commands, however, remains the same: for instance, the accelerator control may be brought to the steering wheel with a ring coaxial to the wheel that can be moved axially. This, in turn, operates the conventional accelerator control through levers.

The situation with two-wheeled vehicles is essentially the same, the only difference being that the driver can change the inertial and geometrical characteristics of the vehicle by moving his own body, using these changes as control inputs: for instance, he can move the center of mass sideways or change the aerodynamic characteristics. The controls are obviously different with the front and rear brakes often operating independently.

15.2.2 Automatic Controls

ICTs promised to change all this in a more or less remote future through the possibility of introducing automatic control devices in road vehicles so that an increasing number of these functions would be performed automatically. A goal for the future can be transforming the driver into a passenger, but this goal is still distant and, as with predictions of the so-called strong artificial intelligence, there are doubts as to its achievability, at least with today's technologies and those likely to be developed in the foreseeable future. However, while the goal of building a fully automatic road vehicle may be a long way off, many partial applications have already been implemented or are about to be realized.

At present, the fields in which control devices are more common or are at least being actively studied are:

- Engine control systems. All modern automotive internal combustion engines are provided with one or more Electronic Control Units (ECU) that control its main functions. The motor control may be conventional or *by wire*, but in the latter case there is no problem in supplying the driver with adequate sensory inputs.
- Longitudinal slip control in traction (ASR, Anti Spin Regulator[2]). These are systems that detect the beginning of driving wheel skid and reduce the power supplied by the engine or operate the brakes of some wheels. Theoretically, they should measure the longitudinal slip of the tires, but in practice they measure the acceleration of the driving wheels.
- Longitudinal slip control in braking (ABS, Antilock Braking System). They are systems that detect the beginning of wheel skid and reduce the braking torques. They, too, should measure the longitudinal slip of the tires, but actually measure the deceleration of the wheels.
- Vehicle dynamics control systems (VDC, Vehicle Dynamic Control, ESP, Enhanced Stability Program, DSC, Dynamics Stability Control). The goal of these systems is to improve the dynamic response of the vehicle. They often act by differentially braking (and sometimes differentially driving) the wheels of the same axle to produce a yaw torque. The driver controls the trajectory normally through the steering wheel, while the control device tries to counteract the difference between the behavior required and that actually obtained by applying yaw torques.
- Suspension control systems. Many different types of controlled, semi-active and active suspensions have been and are being developed. These can simply adapt the suspension characteristics to the type and conditions of the road or, in the most advanced cases, completely substitute an active system for the conventional suspension.
- Electric Power Steering (EPS). Strictly speaking, EPS should not be considered a control system any more than conventional power steering, but electric actuation allows steering control functions to be added. EPS, then, may be considered as a first step towards *steer by wire*.
- Electric braking. A wide span of functions are available through electric braking, from simple electric power braking with an electric actuator on the master cylinder of a conventional hydraulic system to a true *brake by wire* system, with the electric actuators at the wheels.

---

[2] The acronyms here mentioned are often trade marks of a particular manufacturer, even if many of them have entered technical jargon to designate a variety of similar devices.

- Servo controlled gearbox and clutch. These systems provide automatic gearbox functions by controlling a more or less conventional manual transmission using suitable actuators, with all the advantages of classic automatic transmissions but with a much more efficient mechanical transmission without a torque converter.
- Finally, the parking brake must be counted among the secondary controls that may be made automatic. The advantages are that it is possible to ensure that the brake is applied every time the driver leaves the vehicle, without the possibility of forgetting it, while reducing the effort needed to engage and disengage the brake. An electric parking brake yields a larger freedom to the designer of the interior of the vehicle.

All the mentioned systems allow the tasks of the driver to be simplified and safety increased, assuming that they meet reliability standards. The driver is still in the control loop, but his/her work is made simpler by avoiding low-level control tasks so that he/she can concentrate on high-level decisions.

15.2.3 By Wire *Systems*

The controlled active systems used on motor vehicles are usually based on hydraulic components or, particularly on industrial vehicles, on pneumatic devices. Control of the actuators is performed by electrovalves and the actuation power is supplied by pumps, which are usually powered by the engine or by electric motors. Traditionally, the same holds for systems helping the driver, such as power steering or power brakes.

The development of electromechanical components and systems led to the consideration of alternative solutions in which electric actuators control directly the various functions without the need for electrovalves, pumps and other hydraulic or pneumatic devices. The tendency to replace hydraulic and pneumatic devices with electric systems is widespread and in many fields *more-electric* or even *all-electric* systems are considered. The advantages are many, among them the following:

- Interfacing electric devices to control systems is easier than hydraulic devices;
- the transmission of power and command signals requires electric cables, which allow much greater freedom of layout than hydraulic pipes or mechanical transmissions as used with mechanical controls;
- electric devices are less affected by environmental conditions, above all temperature, than hydraulic systems. In particular, in the latter the viscosity of hydraulic fluids is strongly dependent on temperature;
- the fluids used in hydraulic devices require anti-pollution measures during construction, maintenance and ultimately disposal of the vehicle that are not required in electric devices.

The drawbacks of electric devices that have up to now hampered their diffusion are:

- Electric actuators are in general heavier and often more bulky than the corresponding hydraulic actuators;

- thc cost of high performance magnetic materials (in particular rare earth magnets) is still high for automotive applications; in the future a shortage of neodymium may cause a new increase of their price;
- electric systems are in general stiffer than hydraulic systems requiring a more precise control. They may cause more noise and vibration.

In a way what has been done in the field of aeronautics can be a model. Since World War II, devices able to keep an aircraft at a given attitude and on a prescribed course, allowing the pilot to leave the controls for a more or less prolonged time, have entered common use. Such devices do not need to sense external conditions and adapt to them; they are simple regulators, that only need to maintain the predetermined motion conditions. Devices of this kind have only a limited use in road vehicles (for instance, *cruise control* devices) because road vehicles must continuously adapt their motion to the road and traffic conditions.

Military aircraft are increasingly built using configurations that reduce intrinsic stability or are even unstable, with the goal of improving manoeuvrability; the task of stabilizing the aircraft is given to suitable control devices. Moreover, the senses of the pilot have been enhanced by supplying additional information through the control devices, such as devices that shake the control stick when stall conditions approach. Artificial stability may prove interesting in the vehicular field as well, not so much for improving manoeuvrability as for allowing the use of configurations that are advantageous but reduce stability.

Devices providing an artificial sensibility, often referred to as *haptic*, are those that provide a reaction force through *by wire* controls that is similar to the reaction that conventional mechanical controls would supply. They may also add further information, like the devices that cause the accelerator or the brake pedal to shake when getting close to slip conditions in traction or braking. Such devices are intrinsically necessary when controls are made automatic. They are at present under study and in same cases already on the market.

Nowadays in the aeronautical field commands are no longer transmitted by mechanical (rods, cables, etc.) or hydraulic devices but by electric systems (*fly by wire*), the only exception being small and low cost aircraft. There are two main advantages: first, freedom in architecture and layout is greatly increased (it is much easier to route electric cables than mechanical controls), resulting in mass reduction. Second, it is much easier to integrate control systems, which are mostly electronic, in *by wire* than in conventional architectures. In the most modern aircraft, the pilot interacts with a computer that in turn actuates the control surfaces through *by wire* devices.

A similar evolution is also underway in the automotive industry. Here the term *steer by wire* is used for the steering control, *brake by wire* for the braking function and *drive by wire* for the accelerator control. The generic term for these systems is *X by wire*, where the generic *X* stands for the various controls. The advantages are similar to those in the aeronautical field, with the added bonus of allowing the use of different user interfaces that can, for example, be designed specifically for disabled persons and even adapted for individual cases.

However, the transfer from the aeronautic *fly by wire* to the automotive *X by wire* is not simple. A first difference between the two fields is linked to cost, or better, to the ratio reliability/cost. The total cost of an aircraft is greater than the cost of a motor vehicle by orders of magnitude, allowing the use of control systems and components much more expensive than those that may be used in vehicles. Something similar can be said for the low cost segment of the aeronautical market: *fly by wire* systems are still not used in light and ultralight aviation.

The scale of production may mitigate this problem: development costs are subdivided, in the automotive market, into a much greater (even by orders of magnitude) number of machines than in the aeronautical market. Reliability is strictly linked to costs: when dealing with functions that are vital for safety, like steering or braking, the need for extremely high reliability leads to high costs, because the required safety is obtained through redundancy of sensors, actuators, control units and communication lines as well as high quality components. Electronic and computer based devices have been available for motor vehicles for several years in non-vital functions and, more often, in gadgets performing tasks that are sometimes of little practical use.

However, it is not just a matter of cost: motor vehicles are designed for general use; their mission analysis is less determinate than that of aircraft, and they must be able to work in conditions far from those for which they have been designed, with a less stringent respect for maintenance schedules. This makes technology transfer from aeronautics to automotive industry even more difficult, particularly where complex and even critical technologies are concerned.

One field where technology transfer may be facilitated is that of racing cars, and in particular Formula 1 racers, because these vehicles must be optimized with a limited number of parameters in mind, accrue higher costs and are used in controlled conditions. However their design specifications are strictly linked to racing regulations, which at present do not allow the use of automatic control devices.

In the last decade of the twentieth century it was a common opinion that the *X-by wire* technology was about to enter massively automotive productions: this is another prediction that failed to materialize. The few prototypes that were built didn't show the large advantages that were expected and the enthusiasm about this new way of controlling road vehicles subsided. If we can risk advancing predictions, we can say that the introduction of this technology was still premature and that we can predict that its development will be much slower. At present there are several practical problems that must still be understood and solved.

An apparently marginal problem that is hampering the introduction of *by wire* devices is the voltage of the on-board electric system. The increase in the power of the electric devices on motor vehicles makes it convenient to increase the voltage from the traditional (on cars) 12 V to at least 24 V, as on industrial vehicles, or even to 36 or 48 V. In this way, the current needed by the various devices would be reduced, with advantages in cost and weight. This is not a marginal change, as it might seem, because it would require the simultaneous redesign, production and marketing of a large number of new electric components (batteries, bulbs, switches, electric motors, etc.).

Another problem under intense study is electromagnetic compatibility. The electromagnetic environment in which automotive electromechanical and electronic devices must operate is very dirty, which may induce malfunctioning of different kinds. These must be by all means prevented when electric systems are entrusted functions that are vital for the safety of the vehicle.

By wire systems do not comply with many safety regulations which, for instance, require a mechanical link between the steering wheel and the steering box. These regulations can be changed, but to do so the safety of the proposed solutions must be unequivocally demonstrated. Thus, the problems to be solved before vehicles completely controlled *by wire* may be marketed are not only technical (design, production, marketing, etc.) but also legal, regulatory and standards-based.

Finally, it is not certain that the customer would appreciate this new technology. Most of the users of cars would not benefit from the advantages and will not even realize that there is something new in the control transmission chain. A typical example are the 4 Wheel Steering (4WS) systems, which may be made easier in connection with drive by wire: the average driver will not benefit from its introduction (the performance of conventional 2 wheel steering is enough for him), but he/she may be disappointed by the different and unnatural feel of the vehicle, that requires the learning of new reflexes and skills. Only if the new technology will produce obvious advantages (lower cost, lower fuel consumption, increased reliability, lower maintenance etc.) are there perspectives that it will have an appeal outside the niche market of the technology or sports car enthusiasts.

The *by wire* systems which predictably will enter first into mass production may be the following ones:

- Steer by wire. Electric steering systems like Electric Power Steering (EPS) are in common use, primarily in cars in the low or medium market segment. Their application does not imply substitution for the mechanical steering system, but simply the presence of an electric actuator in parallel with the manual steering system. The electric motor may act directly, exerting a torque on the steering wheel shaft, or it may exert a force on the rack. The torque exerted by the power steering is proportional to that applied by the driver, following strategies that may depend on many parameters. The steering control may be made 'harder' with increasing speed, to compensate for the decrease of steering torque typical of many cars and to induce the driver to act on the wheel with more care. The steering actuator may act independently of the command given by the driver, as when using a differential gear having as inputs the steering wheel shaft and an actuator controlled independently. In this case, direct control remains, but the authority of the control system is greater. In a true *by wire* system, there is no direct control link and the steering wheel is connected solely to a rotation sensor (potentiometer, encoder…) or a torque sensor supplying the value of the angle or the moment to the system that controls the steering actuator. A system of this kind is much more flexible, allowing different control strategies to be used, such as a variable ratio between the rotation of the steering wheel and the steering rotation of the wheels. This allows the command from the driver to interact in a more complex way with the command from the control system. Because there is no direct link between the wheels and the steering wheel, there must be an actuator exerting a torque on the steering wheel to supply the driver with information on the working conditions of the wheels (haptic controls).

- Brake by wire. Electric power brakes may be simple devices in which an electric motor actuates a pump amplifying the command given by the driver through the master cylinder connected with the brake pedal. In more complicated systems no pump is connected to the brake pedal, with the latter simply supplying a position or force signal that, through a control system, acts on the actuators at the brakes. The actuator may be a single pump supplying high pressure fluid to a more or less conventional braking system, or an electric actuator located in each wheel. In the latter case, the actuator may pressurize a fluid acting on the pistons of the caliper or may directly actuate the caliper through a ball screw or a mechanical system of other kinds. The choice among the various solutions must take into account the mass, the cost and the reliability of the system, the need for maintenance and adjustments, and the possibility of self-adjusting. Clearly, an actuator in each wheel allows functions like ABS, Traction Control System (TCS), VDC, etc. to be performed without the need of valves discharging the pressure from the branch of the system in each wheel or of pumps that recover the fluid. The brake in each wheel is controlled independently following the commands from the driver and the various control devices.
- Electromechanical suspensions. Semi-active suspensions already use systems that are at least partially electromechanical, such as the shock absorbers based on electro-rheological or magneto-rheological fluids. Electric actuators may replace hydraulic actuators in active suspensions (for instance in Active Roll Control (ARC) systems) and above all, electromechanical eddy current dampers may replace classical shock absorbers. In particular, eddy current dampers may work as passive, uncontrolled components (in this case they may behave as almost perfect viscous dampers, without the drawback of the presence of the fluid and with a greater stability in changing environmental conditions), they may be inserted in an electric circuit containing controlled elements or they may work in a fully active way. At present, their mass is comparable to that with similar hydraulic devices and their cost, although still higher, is decreasing.
- Robotized gearbox. The term robotized gearbox is used for a driveline whose mechanical layout is essentially equal to that of a manual transmission, but in which the control of gear switching and of the clutch is performed by actuators controlled by an automatic device. They may be *by wire* systems, but strictly speaking they may also fail to classify as such, since they may be operated by hydraulic actuators. As already stated, with respect to standard automatic transmissions they have the advantage of not requiring a torque converter and consequently their efficiency is higher and the overall system may be simpler. With respect to manual transmissions they have the advantage of avoiding mechanical linkages between the gearbox and the passenger compartment, which may transmit vibration. An added advantage is that it is possible to operate the system manually, by excluding the controller and operating directly the actuators through a human/machine interface. Whether or not the system is less costly than a traditional automatic transmission depends

on the cost of the controller and actuating devices. A difficulty of this concept, that delayed its entering into the market, is that in many cases simply adding the actuators to an existing manual transmission doesn't work. The precision of the mechanical device must be much higher, even after wear has occurred, since the gear selection actuators and controller do not have the sensitivity and the smartness of a human operator.

In general, all functions of the vehicle may be controlled by electromechanical actuators. Because the control of the engine is the simplest, conversions to *by wire* accelerators are now common. Secondary controls, such as the parking brake, may also be replaced by electric devices, with the advantages of automation and of allowing a greater freedom in command placement and design of the user interface, as well as a much simpler design of the control transmission.

15.2.4 Higher Level Control

As already stated, the goal of installing on board control devices able to take high level decisions and taking control of the vehicle up to the point of transforming the driver into a passenger is still far in the future. An autonomous car that can drive itself to its destination, obviously stated through a voice system by its owner, still belongs to science fiction.

Some steps in this direction have already been successfully done. Parking systems, in which the vehicles parks by itself, controlling the engine, the steering system and the brakes, are already on the market. Although a spectacular achievement, it could be obtained by integrating the parking sensors, the by-wire engine control, the automatic transmission and the braking system, putting them under the control of a computer-based controller that has to deal with a highly standardized situation.

Another device that is on the market is a cruise control system able to maintain a given distance when following another vehicle. This is essentially an adaptive cruise control and again is not too difficult to implement.

Many experiments were performed to build a vehicle that can follow a road by itself. In this case different technologies were attempted, some involving only the vehicle, like an optical system that 'sees' the road and follows it, and others involving also the infrastructure, like embedding an antenna into the road surface and providing the vehicle with a receiver that feels the road and follows it. This second alternative is, at least apparently, easier since it is based on well-known technologies like the automatic industrial transportation systems working on the same principle, but the difficulties of allowing higher speeds and of operating in a less structured environment makes this technology transfer more difficult. The high cost of modifying the existing roads for this kind of operation lead to privileging self-contained on-board optical systems.

An intermediate possibility is that of assembling trains of vehicles, linked together by a non-material connection, following a leading vehicle that is operated by a

Fig. 15.1 A 'road train' made of vehicles that follow automatically a lead vehicle driven by a human driver. Picture from the SATRE project by Volvo

human driver. In 2009 Volvo launched the Safe Road Trains for the Environment (SATRE) project (Fig. 15.1), in which the vehicles of the train follow the leader using sensors and radio signals that control the accelerator, brake and steering controls. In the figure, a train of five vehicles, two trucks and three cars is shown. A speed up to 90 km/h can be reached in this configuration. The advantages are many, like general fuel consumption reduction, enhanced utilization of the infrastructure and increased safety, obtained without the need of modifying existing infrastructure and only limited changes of existing vehicles. In case vehicles have already steer by wire, vehicle to vehicle communications and sensors to monitor the distance from the preceding values, the changes are mostly in the software of the on-board computer. The system is very flexible, since the driver can enter the train driving his car in the standard way, and link the car to the train information network. After this he is free to do whatever he likes, being free from any driving task. When he wants to leave the train, he just logs out of the train network, and drives his car manually.

Much research work was conducted, both by universities and car manufacturers on all the aspects of autonomous road transportation. One of the first successful prototypes was a Mercedes-Benz van that in the 1980s was able to drive under its robotic vision system for 100 km on streets without traffic. In the period 1987–1995 the European Commission started the EUREKA Prometheus Project on autonomous vehicles that had a funding of 800 million euro. In the same years the United States, with the Defence Advanced Research Projects Agency (DARPA), started funding a large research project on Autonomous Land Vehicles (ALV) following different alternatives, like laser radar, computer vision and autonomous robotic control for both road following and off-road map based navigation.

An interesting University project was the ARGO project (by A. Broggi of the University of Parma) that in 1996 lead to a successful 2000 km trip of a modified Lancia Thema on the motorways of northern Italy, with an average speed of 90 km/h, in which the car was under automatic control (with human supervision, for safety reasons) for 94 % of the time. The control system was based on two black-and-white

Fig. 15.2 **a** Chevrolet EN-V electric autonomous vehicle. **b** Toyota Prius hybrid, modified by Google into an autonomous vehicle

low-cost video cameras, to demonstrate the feasibility of low cost automatic driving systems.

Since the American effort is led by DARPA, it concentrates more on off-road driving, while other projects are more oriented towards motorway travel. The first two competitions sponsored by DARPA (the 2004 and 2005 DARPA Grand Challenge) involved off-road driving, while the third (the 2007 DARPA Urban Challenge) took place in city traffic. Further challenges, inspired by those sponsored by DARPA, were held in the following years. The longest trip by self-driven vehicles was the VIAC Challenge, in which four vehicles drove from Italy to China on a 13,000 km trip with only limited human intervention in 2010.

At present automated road vehicles are not allowed on open roads, with the exception of Nevada where a law allowing the use of autonomous vehicles, defined as"a motor vehicle that uses artificial intelligence, sensors and global positioning system coordinated to drive itself without the active intervention of a human operator", was passed in June 2011.

At present many prototypes of autonomous vehicles are under experimentation by automotive companies, like General Motors, Volkswagen, Audi, BMW, Volvo or even by companies like Google whose interests are more in information technologies. Some of them disclosed plans to market these vehicles in a future ranging from 2018 to much later dates. The Chevrolet EN-V (Fig. 15.2a) is a prototype that can be driven manually or autonomously, through its GPS, vehicle-to-vehicle communications and distance-sensors. It carries two people and its lithium-ion batteries grant it a 40 km range (see next chapter). The car shown in Fig. 15.2b is a more conventional Toyota Prius hybrid that was modified by Google into an autonomous car. These cars are controlled by computers processing a combination of mapping data, radar, laser sensors and video feeds and are now road-legal in Nevada, although they need to carry a distinctive red registration plate so that they can be easily recognized by other road users.

15.2.5 Navigation and Communication Devices and on Board Computers

Many electronic devices are not directly aimed to take control, at low or high level, of the vehicle, but simply to help the driver, by supplying him/her information about the road, by monitoring his/her conditions or the surrounding environment, by allowing communication with other vehicles or with fixed stations and other similar tasks. Others are simply aimed to entertain the passengers to make the journey more pleasant. Most of them enter the category of infotainment (information-entertainment) devices.

Some of these devices could quickly enter mass production owing to their non-critical character or to their low cost. After all, an entertainment system for passengers or other gadgets of this kind need not have a high reliability and the worst that can happen in case of failure is a mild complaint from the user.

Perhaps the most common of these systems are Global Positioning System (GPS) navigation systems, offered as an optional on most cars and even as an add-on for any type of vehicles. GPS technology in itself is a very advanced and complex technology, that was developed for military applications during the cold war. Once developed, simplified versions for civilian use could be mass produced at low cost and it is likely that in the future all vehicles will be provided with them.

More specific systems, that have a direct impact on safety and that are increasingly installed on vehicles, are radar-based collision avoidance devices, Lane Departure Warning System (LDWS), devices that warn the driver of impending condition of his own loss of attention, devices that enhance vision in blind spots (like rear view cameras) or in poor visibility conditions (night vision devices, fog vision devices), devices that help in parking maneuvers, etc. Some of these systems are now reliable enough to be directly connected with the controls of the vehicles, like a device that monitors the driver conditions and that, in case the driver falls asleep, slows the car down, checks the lanes on the right and, when they are free, moves the vehicle in the rightmost lane and stops it without creating dangers to other vehicles. This simple maneuver is at any rate quite a complex one to be carried on by an automatic system, but it is within the possibility of present technology.

The term infomobility is used to designate a situation in which vehicular communication systems are integrated into an Intelligent Transportation Systems (ITS) in which the vehicles exchange informations about road conditions, weather, etc., integrating private and public transportation so that an optimal use of the existing infrastructure can be made.

One of the basic features of Intelligent Transportation Systems are the Vehicular Communication Systems, communication networks in which both vehicles and roadside units are the communicating nodes, using the so-called Dedicated Short Range Communications (DSRC) technology. These devices work in the 5.9 GHz band with a bandwidth of 75 MHz and an approximate range of 1000 m. Usually a distinction is made between vehicle-to-vehicle (V2V) communications and vehicle-to-infrastructure (V2I) communications.

The main aim of these communication system is to provide the driver with information such as safety warnings and traffic information that can improve road safety and traffic fluidity. The rationale is that a cooperative approach can be more effective in avoiding accidents and traffic congestions than if each road user tries to solve these problems individually. However there is a much wider scope: people are now accustomed to living in a connected environment, and these systems will allow road users, drivers and passengers alike, to remain connected during the time they spend in their cars.

As an example of the potential importance of vehicle to vehicle communications to enhance safety, it can be remembered that the US Department of Transportation estimates that 21,000 of the annual 43,000 deaths due to road accidents in the United States are caused by vehicles leaving the roadway or approaching road intersections. These accidents can be, at least in theory, drastically reduced by vehicle-to-vehicle communications if the former can inform other vehicles of their intention of departing the highway and the latter can warn other cars traversing that intersection. Vehicle to infrastructure communication systems can allow drivers to be aware of the presence of pedestrians, animals and other obstacles on the road, preventing accidents to both pedestrian and motorists.

In the same way, communication systems can allow saving at least a part of the huge amount of money that is lost due to traffic congestion and poor road management. Highways provided with communication systems may in the future allow higher speeds, since the automatic devices are not limited by human response time. The goal, in a longer perspective, is that of the so-called automated highways, in which traffic can be faster and vehicles can drive closer to each other, increasing the capacity of a given highway without increasing the number of lanes.

Among the many possible vehicle to infrastructure applications, one that has been operating for many years, is the automatic toll collection.

15.3 New Materials

At present the structure of cars is based on sheet metal components and the basic material is low-carbon steel, chosen more for the ease of stamping and spot welding than for its mechanical properties. This is something that is unchanged since more than half a century, and it is unlikely that radical changes will enter the mass production market in the near future. The mechanical parts of the driveline are made from a wider choice of materials, like high strength steel, aluminium alloys, cast iron, etc.

It is since the 1960s that predictions of revolutionary changes in the materials used in the automotive industry, in particular for the structural parts of the body and the chassis, are heard, but these changes, although based on prototypes and small scale production, failed to enter the market of mass produced cars. This is manly due to the fact that any change in materials requires a change in technologies, which has a deep effect on the final cost of the vehicle. All the materials proposed would improve the performance of the final product, in particular through a weight reduction, but unless this is possible at a reasonable cost, the application of the innovative material will be restricted to a niche and will be at best marginal.

The use of more advanced materials in the driveline is a different matter. Here the construction technologies are more differentiated and the possibilities of using advanced materials (as an example, ceramics for the hottest parts of the engine, metal matrix composites, etc.) are more open. These applications will not be dealt with any further.

15.3.1 High Strength Steel

The use of high strength steel is the logical alternative to low carbon steel for the most loaded parts of the chassis and the body, and could result in a non negligible weight reduction. This consideration is mitigated by the fact that high performance steel is stronger but not stiffer (its Young's modulus is roughly the same as for low carbon steel) and thus it leads to no advantages in all those components that are designed for stiffness and not for strength. An example is all the thin wall parts subject to compression, where the danger comes from elastic instability, and those parts that give the required torsional stiffness to the body, an essential requirement for good handling performance. Moreover, in the parts of the body and the chassis that must absorb the vehicle's energy during a crash, an essential requirement is the ability of the material to absorb large quantities of deformation energy with large deformations, a feature that may not be better in high strength steels.

All this not withstanding, the use of high strength steel in selected sheet metal parts could lead to lighter and better vehicles.

The reason why this has not yet become common practice is not so much the cost of the material, that is moderately higher, but the larger difficulties in obtaining the required shapes through the conventional stamping technologies. The larger welding difficulties and the greater wear of the dies are other reasons that caused these materials not to be widely used in car production.

15.3.2 Light Alloys

Light alloys (aluminium alloys) are commonly used in the construction of the engine and of the driveline (light alloy wheels are at present common), and for some selected chassis components like some parts of the suspensions, but are seldom considered for vehicle bodies, except in the case of sports cars or off-road vehicles produced in limited numbers.

Here again the problem is not so much linked to the higher cost of the material (that is anyway higher by not a small amount): The difficulties come mostly from construction technologies. Initially, assembly technologies derived from aeronautics were used and the body was made of parts riveted to each other: this is too costly for the automotive industry and leads to difficulties in obtaining a finish like that customers are used to. Later, welding techniques were used, but welding aluminium is difficult and is not a practice suitable to mass production.

15.3.3 Reinforced Plastics

The extensive use of reinforced plastics in car bodies is in a way an obvious development, and was and was for almost half a century expected to appear momentarily, but did not materialize. Why? After all, similar developments occurred in industrial sectors, which are not apparently far from the automotive one. A typical example is that of small boats and yachts, where wood, and some times metal hulls have now completely yielded to fiberglass hulls. This caused several advantages, like reduced cost, weight and maintenance and increased designer freedom. In aeronautics composite materials are also increasingly used, in particular high performance composites like those based on kevlar or carbon (both high modulus and high strength) fibers.

In the automotive sector, racing cars are now built using carbon fiber-epoxy composites and many special cars produced in small quantities have a composite body, mostly glass-epoxy or even glass-polyester. This choice is linked to performance, like in the case of racing cars where only the use of advanced composites, mostly based on Carbon Reinforced Plastics (CRP) with thermoset resin, could allow reaching the required high stiffness, low weight characteristics, or to avoiding the high tooling costs that may be unbearable in case of low production volumes.

Two of the reasons sometimes mentioned for the failure of composite materials to enter massively the automotive market are the difficulty of obtaining a satisfactory finish on fiberglass components and the difficulty of repairing fiberglass bodies. These reasons are not really convincing: while it is true that it is difficult to obtain the typical automotive finish in fiberglass bodies, it is not impossible, and customers can be convinced to accept this 'new look'. About repairs, it must be said that car bodies are increasingly not repaired in the classical sense. It is impossible to insure the crashworthiness of a heavily repaired steel body, and so repair shops now substitute the damaged parts instead of fixing them, a thing that is possible also with fiberglass elements.

The point is that structural composite material components lend themselves to low scale production but, when used for mass production, are much more expensive than traditional steel constructions. High performance matrices, like epoxy resins, have a long cure time and require high temperature curing: a large number of molds and huge autoclaves are required to produce a large number of reinforced plastic car bodies, leading to high costs. The energy required for a cure operation is more than the energy required for a conventional steel stamping process, and the safety and environmental measures needed when working with toxic resins add much to the cost of production. And, finally, steel car bodies are easily recycled at the end of the life of the vehicle, while recycling a fiberglass body is much more difficult, and produces a material of much worse characteristics that, in large quantities, would prove useless.

In case of small production numbers, the lower cost of the molds with respect to traditional dies may overcome these problems, contrary to what occurs in mass production, typical of the automotive industry.

Matters are completely different when non-structural parts are concerned. In this case it is possible to use thermoplastic resins and often short or chopped fibers reinforcement, and thus lower cost, mass production technologies are available. This is the reason why hatches, hoods, bumpers and other parts that do not carry much load are often built using reinforced plastics.

There are possibilities to see an increase of composite materials in car bodies in the future, since the need of reducing weight is felt more and more strongly, in particular with the aim of reducing fuel consumption, but this is linked to the introduction of new materials and technologies able to reduce the cost of mass production.

15.3.4 Smart and Advanced Materials

Much research work is being performed in the material field, and great expectations lie on new and revolutionary materials. Nanotechnologies promise to produce new materials which will allow 'bottom up' manufacturing, i.e. building parts by assembling single atoms or molecules in the way living organisms build themselves, in contrast with the usual 'top down' manufacturing, in which parts are machined from solid pieces of material.

Bottom up manufacturing could allow the construction of complex parts using materials whose performance will be much higher than those of today's materials, at a much lower cost. Materials having the possibility of self-repairing, of changing their structure as a response to the stresses, of responding to external stimuli in a way similar to biological materials, are predicted. Some of these innovative technologies are at present at the experimental stage, and predictions on the possibility that they will be ready for practical application within a short time have been put forward. However these predictions have not materialized, and the way leading to applications of this kind seems to be quite long, even in fields like aerospace, robotics and bioengineering characterized by a lower scale of production and lower importance of containing costs.

An easy prediction is that, short of an unexpected breakthrough, their diffusion in the automotive field is still unpredictable and confined to the far future. It is highly unlikely that we will, in a predictable time, see cars that self-repair, adapt themselves to the conditions in which they operate, and do not need a human operator.

Chapter 16
Ground Vehicles in a Longer Perspective

The ideas that our way of providing mobility for most people—by producing cars and other vehicles suitable for private operation—are basically wrong, are not uncommon; these views are voiced strongly by a motivated and determined minority. These ideas are motivated by a variety of ideologies and political, economical or ethical beliefs, from sheer antitechnological feelings to the fear that the present way of providing mobility damages the environment, from the idea that our mobility system is based on the exploitation of a part of humankind to the belief that it dehumanizes our lives.

While the first point (is the way we provide mobility wrong?) is something well beyond the scope of this book, since it deals with political choices and has much more to do with the values we believe in than with the technical choices designers have to make in their everyday life, the second point (is the way we design vehicles wrong?) is of more technical nature and must be at least touched upon when speaking of the future of automotive vehicles.

On the issue of how cars are designed almost all people who use cars, at least occasionally, has his or her own opinion. We are all self-appointed experts in car technology, like all sports enthusiasts are self-appointed coaches for their favorite team. The authors remember a bumper sticker popular in the United States in the 1960s, saying 'made in Detroit by idiots'. Those who went around with this sticker were trying to show their strong dissent from the way motor cars were designed and built. This may seem just a joke, but these feelings are more widespread than what motor vehicles experts think and may influence as well professionals of disciplines that affect the automotive industry, like architects, city planners, etc.

As stated in the introduction, car design (in its wider meaning, including also the design of the technologies, tools and equipment to build them) is a complex business. Not only is it difficult for the user or in general the layman to understand why a certain design choice has been done, but often, owing to specialization, it is also difficult for experts in the automotive industry to have clear ideas about the decisions taken by other experts in different branches of the same industry. Nobody can be deep in all specializations involved, from mechanical engineering to electronic engineering,

G. Genta et al., *The Motor Car*, Mechanical Engineering Series,
DOI: 10.1007/978-94-007-8552-6_16,

from fluid dynamics to thermodynamics, from material science to stress analysis, and so on. If also economics and business, public relations and management are considered, the spectrum of know-how needed to build a successful car is so wide that only well organized multidisciplinary teams may have the knowledge that can cope with the task.

So, if the way we design cars is wrong, or at least if we can do much better, hopefully the cars of the future can be more intelligent, and thus more or less different from what we now have. In the following sections some of the most common criticisms of the way we build cars are briefly analyzed, accompanied by discussions of the type of vehicles we expect our ways to lead to. Some changes to road vehicles, more radical than have been predicted in the past, are dealt with, to see whether they were just fantasies of the past or may become a reality in a more or less distant future.

16.1 Radically Smaller, Lighter, Cheaper and Greener Vehicles

A common criticism, that has not only technical but also political implications, is that the cars we build are too big and heavy for the needs of people using them. A consequence of this statement is that much smaller, lighter, cheaper and consequently environmentally friendly, cars can be built. These vehicles are mostly thought for urban use, for a number of reasons:

- Cars are presently used mostly within the city boundaries, and thus improving the situation in these conditions has the largest effect on the problems we want to solve (fuel consumption, pollution, etc.);
- the same problems are more critical in urban areas, where the size of present cars make congestion, and the related decrease of the average speed, worse;
- present vehicles seem to be designed with long range travel in mind, and they appear to be unnecessarily fast, powerful, big and even unnecessarily comfortable when used on short distances.

The radical response to this challenge is to introduce innovative vehicles, usually referred to as Urban Personal Mobility vehicles, that are much smaller and lighter, are designed to carry one or two persons and possibly are powered by electric motors. Some examples are shown in Fig. 16.1.

The first vehicle is the well known Segway personal transporter, a two wheel vehicle that is stabilized by an active control system. At present, GM and Segway are developing a similar vehicle with a closed body able to carry two people, called PUMA, an acronym for Personal Urban Mobility and Accessibility. The Chevrolet EN-V shown in Fig. 15.2a is a development of the same concept. The second image refers to the Toyota i-Real, a three wheeled vehicle, while the third one is the Nissan Pivo 2, an electric microcar able to carry two people.

Other, less extreme, solutions are based on three-wheels vehicles, possibly tilting, like the ones shown in Fig. 16.2. These solutions may be more similar to motorcycles, possibly with an added wheel, or to cars, possibly with a wheel less, but what all of

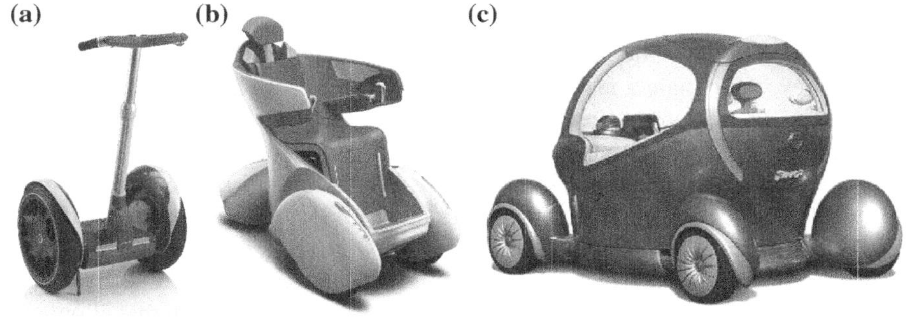

Fig. 16.1 Urban Personal Mobility vehicles. **a** Segway PT, **b** Toyota i-Real, **c** Nissan Pivo 2

Fig. 16.2 Prototypes of tilting vehicles. **a** BMW Clever; **b** Mercedes F 300

them have in common is to be included in the motorcycle or microcar classes for what the regulations are concerned. As a consequence, their safety is much lower than that typical of cars, and is more or less equivalent to that typical of motorcycles.

In fact, Urban Personal Mobility vehicles are not as innovative as they seem: we have already something similar, i.e., small motor bikes and mopeds, possibly in their modern electric version, and microcars, that can also be electric. And these new 'alternative' vehicles have the same drawbacks as the old ones: they are much less comfortable, in particular in case of bad weather, they carry only a small quantity of luggage and their cost is comparable to, and in case of electric vehicles, often much higher than, that of small city cars, using this term to designate vehicles that fully qualify as cars.

However, the biggest problem of these vehicles is their low safety. In spite of their limited speed, that often is such only theoretically, they offer only a limited protection to the driver and possible passengers. And this is not something that can be corrected by a better design: In a way they are dangerous by design, since they do not comply with the safety regulations of cars. If they did, they would need to incorporate crashworthy structures, safety devices as airbags etc. and their weight would be higher. At this point they would need larger engines, better braking systems and so on, and their mass, power, fuel consumption and environmental impact would be those of city cars. It is true that they could be battery electric vehicles but then they would have the same problems of all other electric vehicles. Owing to these factors, they will remain a niche product and will cover the same niche that now is occupied by small motorcycles and mopeds.

Small city cars, that satisfy all regulations to be considered cars, are obviously another matter, and it is predictable that they will become more common in the future. However, they will be not much lighter and smaller and will not use much less energy (fuel or electric energy) than present city cars, of which they represent an evolution.

A possible exception are three wheeled tilting city cars, like the ones shown in Fig. 16.2. Their essential characteristic is that of having a mechanism that allows the body to tilt when managing a bend, so that the local vertical (the direction of the resultant of the gravitational and the centrifugal acceleration) remains always in the direction of the head-feet axis of the passengers. This is not so much a matter of comfort, which at any rate is improved, but a matter of stability. A tilting vehicle can be narrower and taller, with the passengers seating in a more upright position; a feature that allows the vehicle to be also shorter. If this is applied to a city car, it results in a much smaller area on the ground and a reduction of traffic congestion.

Tilting vehicles that give the driver the feeling of driving a car and not of riding a motorcycle need to have a sophisticated position control for the body and this control must be absolutely reliable, since if it fails the vehicle rolls over. The result is that problems like those posed by X-by-wire systems are encountered. Another point is that the safety regulations of cars must be applied, otherwise we are back to a vehicle that has the low safety of all motorcycles, with all the consequences on the weight, required power, etc. They would be at any rate an example of progress with respect to present city cars in reducing traffic congestion in cities.

16.2 Radically Safer Cars

Often we feel motor cars as are unsafe means of transportation. According to World Health Organizations (WHO), road accidents annually cause approximately 1.2 million deaths worldwide; one fourth of all deaths caused by injury. Also about 50 million persons are injured in traffic accidents. A study from the American Automobile Association (AAA) concluded that car crashes cost the United States $300 billion per year.

As a consequence, we often claim that we need radically safer vehicles: Here we are speaking about human lives, and even a single dead person on the road is a tragedy that must be avoided. Also the cost of road accidents is huge and, particularly in times of economic restrictions, a cut of these costs would be a welcome achievement. On the other side any human activity involves some risk, and the idea of transportation without accidents is just an utopian goal. The risks linked with any activity may be dealt with only statistically, trying to reduce them but always remembering that safety has a cost and that this cost must always be taken into account. If we forget it, we can easily take counter-productive measures.

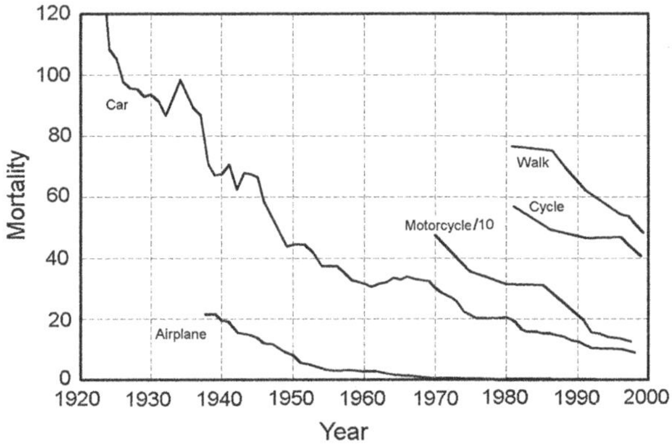

Fig. 16.3 Mortality, in deaths per billion passenger km, due to use of road vehicles and, for comparison, to air travel. In the case of cars the data are referred to billion vehicles km. Note that the curve related to motorcycles has been divided by a factor of 10. The data refer to the United States, except for those related to cycling and walking that are referred to Great Britain (data from Bjørn Lomborg, The Skeptical Environmentalist, Cambridge University Press, Cambidge, 2005)

Consider for instance that in a country a law is passed imposing very high safety standards (something like the Experimental Safety Vehicles of the 1970s, but this time actually working ones). As a consequence the cost of cars would have a sharp increase, and a number of people, who cannot afford these vehicles or simply want to spend their money in another way, would shift to other means of transportation like bicycles or motorcycles, that are much more dangerous. As a result of that law the number of casualties on the road would have a sharp increase instead of the expected decrease.

Another point is that the way we perceive safety is quite subjective and changes from time to time and from country to country. We think that the problem of car accidents and casualties due to transportation is linked with our way of moving around, based on cars. But before cars were used the casualties due to accidents with horses and carts were much more than those now due to the use of motor cars. And not surprisingly, since there was no safety rule (no rule of any type) for carts, the road were much worse, and so on. But at that time the public concern about safety was much less and nobody bothered about it.

The mortality, in deaths per billion passenger km, due to the use of road vehicles and, for comparison, to air travel is reported in Fig. 16.3 (data referred to the United States, except for those about cycling and walking that are referred to Great Britain). In the case of cars the data are referred to billion vehicles km: Since the average occupancy of cars is 1.6 passengers, to obtain comparable results the curve should be divided by 1.6. Note that the curve related to motorcycles has been divided by a factor of 10.

Traveling by car is thus seven times safer than cycling, eight times safer than walking and 22 times safer than riding a motorcycle. From this plot we can draw the conclusion that cars are the safest way to move, except from aircraft (and trains, not included in the plot). This in a way is obvious: aircraft and trains are operated by well trained professionals, whose physical conditions are frequently checked, and are subjected to a maintenance routine managed by *ad hoc* organizations. In case of personal mobility devices the situation can be improved by having vehicles that help the driver in his/her task and can correct some of his/her errors (active safety), that require less maintenance, retaining their safety for a longer time and finally that minimize the effects of a crash on the people on board and on people hit by the vehicle (passive safety).

A further point is that the largest decrease of mortality occurred starting from the 1930s and not from the introductions of the safety regulations (from the 1970s) whose effects do not seem to have been so large; perhaps, however, they were effective in stopping the small increase that took place in the 1960s. The decrease of mortality seems to be more an effect of the evolution of cars, and of the infrastructures (better design, more care to maintenance, better infrastructures, etc.) than of specific regulations.

A final point that can be seen in the plot is that the decrease of mortality accompanied the increase of the average speed of cars: the common opinion that speed is dangerous in itself is not confirmed at all and the introduction of speed limits, that initially were not justified by safety reasons but by the need of saving fuel during the oil crisis of the 1970, had little effect in increasing the trend toward safer road transportation.

The figure shows that also the rate due to other ways of travelling on roads, walking, cycling, riding motor bikes, decreased. In part this can be ascribed to the safety regulations for cars aimed to reduce the damages that pedestrian and cyclists suffer when colliding with a car.

In any case the trend toward a lower accident rate continued after 1998, the year at which the curves of Fig. 16.3 stop. This decrease must continue also for the future, and the evolutionary trend leading to improvements in vehicle safety must go on. Also here, however, it seems that there is no need to resort to revolutionary changes in vehicle design, but only to the application of those provisions that now are available, or will be available in the future, to help the driver in his/her task and to protect the occupants in case of a crash. Non technical measures, like those designed to prevent people from driving in conditions leading to dangers or to improve infrastructures, will have an important effect.

As already stated, future diffusion of vehicular communication systems can improve safety. This concept has been pushed to its ultimate consequences, stating that once each car is connected to the web, and is aware of the position of all other vehicles, fixed obstacles and other road users (pedestrians, bicyclists, but also animals, etc.) traffic accidents would be totally eliminated. As a consequence, passive safety precautions would no more be needed, and the Urban Personal Mobility vehicles described in the previous section could become fully feasible.[1] While it

[1] W.J. Mitchell, C.H. Borroni-Bird, L.D. Burns, *Reinventing the Automobile*, the MIT Press, Cambridge, 2010.

Fig. 16.4 The active wheel built by Michelin: a complete corner unit, provided with drive motor, brake, suspension and steering

is clear that communication systems could decrease the number of accidents, it is questionable whether we could rely only on them to protect people travelling by car and other road users, and the possibility of reducing drastically the mass, and consequently the power, fuel consumption and environmental impact of cars in this way is questionable.

The increase of safety due to autonomous vehicles may lead to the possibility of increasing the average speed on motorways, by raising the speed limitation now enforced in the majority of countries.

16.3 Modular Vehicles

The possibility of locating electric motors in the wheels has already been discussed. It is possible to go further in this direction, by concentrating all the functions regarding a wheel, namely power, suspension, braking and steering, in a single unit, usually defined a *corner*. Each corner can thus be assembled on the car using a mechanical, power and control interface.

An example of this approach is the Active Wheel built by Michelin, shown in Fig. 16.4. In the figure it is easy to see the active suspension, with its spring and electric actuator, the traction motor, the steering system with its actuator, the disc brake and the mechanical interface, consisting in a flange used to bolt the unit to the vehicle frame. The active wheel needs also a power interface, to supply power to the traction and steering motors but also to the active suspension unit, a command interface, that carries all the x-by-wire controls for the steering and brake actuators, and a sensor interface to send back the outputs of all the sensors the wheel is provided with to the electronic control unit of the vehicle. The sensors must include a velocity sensor, accelerometers, pressure sensors for the tire, and many other sensing units that allow us to keep the operation of the corner under close control.

In this case the drive motor is provided by a fixed ratio speed reducer, and this allows us to keep the unsprung mass to a low value to improve both comfort and handling.

Fig. 16.5 The 'electric skateboard' concept by GM. The frame is here shown complete with its four active corners

Four active corners of this type (but also three, in case of a 3-wheeled tilting vehicle or six for a larger van) can be directly mounted on a frame, that can be a more or less classical unibody or may be just a chassis like in the *electric skateboard* concept: a unit of this kind, built by GM is shown in Fig. 16.5. In the latter case, the chassis contains the power pack, made of batteries or of a fuel cell, and in turn carries a body that has little or no structural functions.

The modularity is thus pushed at its limits, returning to the approach that was common at the beginning of the automotive evolution, when the frame and the body were separate units. New materials, design methods and power plants might make this concept competitive again with the more traditional one. The ultimate goal is that the units can be designed and built separately, being possible to assemble them with a 'plug-and-play' philosophy like at present is done with computers. The customer could thus buy a frame, four active corners (all with driving motors or just two motorized and two free wheeling), a body plus all the accessories needed to obtain a fully customized vehicle. This approach will easily be technically feasible, but it is still to be demonstrated that it is also economically feasible, above all if the scale of production and the safety requirements are also taken into account.

16.4 Possibilities of Breakthroughs

From the previous sections a conclusion can be drawn: it is unlikely that the car of the future will be radically different from that of today, due to revolutionary changes, while it is likely that it will accumulate a large number of small improvements that can be seen as a continuation of the evolution that has occurred in the past. The accumulation of these improvements will eventually lead to vehicles that will be much better than the present ones in a scenario of continuity.

However, we cannot exclude that sooner or later this evolution will lead to a radically new vehicle that could be considered as a motor car only because it will

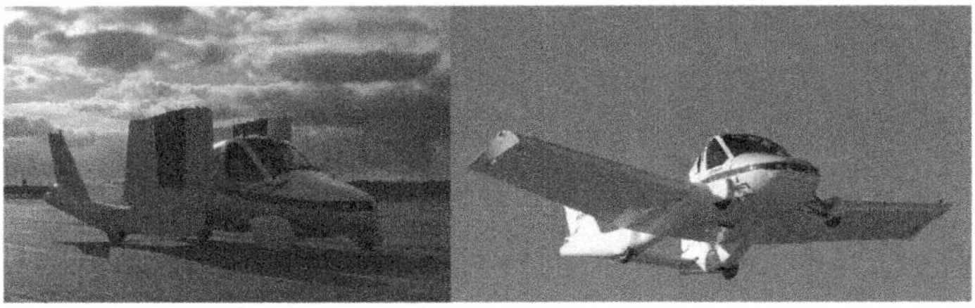

Fig. 16.6 The Transition®, a roadable aircraft presented by Terrefugia at the 2012 New York Auto Show

fulfil the task that now cars are performing in our society. The point is the usual one: In the past several times prototypes of vehicles intended to perform the same tasks of cars in different ways were built, some of them performed in a more or less successful way, but they were never produced in significant quantities and they never substituted for conventional motor vehicles.

16.4.1 Flying Cars

The enthusiasm for light aviation that was so widespread in the 1950s and 1960s, in particular in the United States, lead to the idea that it was possible to build vehicles that could travel on normal roads and, when required, fly like aircraft. They were usually called *flying cars*, but the term *roadable aircraft* was also used. As usual with hybrid concepts it is possible to conceive a variety of solutions from a true car that could (sometime) fly to a true aircraft that could taxi and take off from roads open to the public.

Since the beginning it was clear that the two concepts of car and aircraft are in a way alternative and that it was difficult to strike a compromise. A car must cover a small area to share the road with other vehicles, and in particular cannot be wider than a given size, imposed by the width of a road lane, that is in turn established by laws. An aircraft needs a lifting surface to collect aerodynamic forces, a surface that cannot be too small owing to the low pressure air can exert on it and the width of this surface, the wingspan, is critical. Moreover, the smaller is this area, the higher is the take-off speed and the higher the power needed to fly, not to mention the difficulties to control the aircraft. Owing to these contradictory requirements, in some attempts, mostly the earliest (the first prototype that actually flew was built by Waldo Waterman in 1937), the vehicle had two configurations and had to be manually reconfigured to pass from one to the other. More modern prototypes are able to reconfigure themselves using a number of actuators that unfold the wings or the rotor blades. A recent prototype of the last type is the flying car presented by Terrefugia at the 2012 New York Auto Show, named The Transition® (Fig. 16.6). More than a flying car, it is a roadable aircraft.

Fig. 16.7 The Skycar built by Moller International

Both fixed wings and rotary wings prototypes were built, the latter both in the form of helicopters and autogyros. After a period of stagnation, a new golden age of these attempts started about year 2000, and the new prototypes could take advantage of the advancements in automatic control systems to make flight control easier and safer. Both NASA and DARPA took interest in developing the concept. In 2007 NASA funded a 2-million dollars contest to build flying cars, dubbed for the occasion Personal Air Vehicles (PAVs) and DARPA started a 65-million dollars program to develop a four-places roadable aircraft by 2015, able to take off vertically and having a 450 km range. Also the Society of Automotive Engineers (SAE) promoted conventions on the subject of flying cars in various cities.

One of the most interesting concepts is the quadcopter/convertiplane Skycar M400 built by Moller International (Fig. 16.7).

The interesting point of this concept is that it does not need to be re-configured for flying and that, owing to its strongly automated control and GPS-based navigation systems, can be flown safely by anyone, or at least this is the claim of the builder. On the other side it seems that the designers were more concerned with the behavior of the vehicle as a flying machine than as a car: the wheels are apparently more the undercarriage of a helicopter than the running gear of a road vehicle, but this can be improved later.

The problems with flying cars are however not only technical, even if just to have them flying successfully and safely is a great achievement: the bigger concerns are how to control the flight of many of these flying vehicles in a small, possibly urban, area. In fact this was the point raised by the FAA (Federal Aviation Administration) when Henry Ford proposed to mass produce flying cars, and this is still the point today, in spite of the improvements of the air traffic control systems.

Perhaps the only way is to have automatic route planning and collision avoidance devices based on networking of the computers installed in each car, but this is a difficult task, much harder than the similar systems proposed for dealing with traffic control on motorways and not yet implemented.

Issues about energy consumption and environmental sustainability will also be raised. As a conclusion, flying cars will be built in the future and a niche market may

Fig. 16.8 Volkswagen Aqua, a hovercraft powered by hydrogen fuel cells that won a Volkswagen-sponsored design competition. Like for other futuristic vehicles, only a design or a mockup exists

develop, but it seems extremely unlikely they will become a popular transportation means, with a diffusion comparable with that of present motor cars.

16.4.2 Hovering Vehicles

Starting from the late 1950s, particularly in the United Kingdom, a new kind of vehicle was developed and raised great hopes: the hovercraft. This device was based on an air cushion created by vertical axis fans, and could travel on land and water, not needing a prepared surface, although the power required for hovering grows fast with increasing ground irregularities. On very smooth ground the power required for motion is comparable with that of wheeled vehicles. In spite of the predictions about hovering cars, the development of hovercraft was toward large commercial or military crafts, and even those started declining. Small personal hovercraft are now limited to amateur usage and there are several clubs of people who build their own vehicles often from drawings or kits. A hovering car based on the hovercraft principle is the Volkswagen Aqua, an All-Terrain Vehicle (ATV) powered by hydrogen fuel cells (Fig. 16.8). Its designer, Yuhan Zhang, won a Volkswagen-sponsored design competition and the vehicle bears the VW name for this reason.

At any rate, hovercraft are much more suited to moving over water than over solid ground and, in the latter application, there is no reason for which they should be considered better than conventional wheeled vehicles: they use more power, are more noisy, more complicated and much more difficult to drive since driving, braking and cornering forces are not supplied by the contact between the vehicle and the ground, but by aerodynamic forces. It is easily predictable that there are no hovering cars in the future of ground transportation.

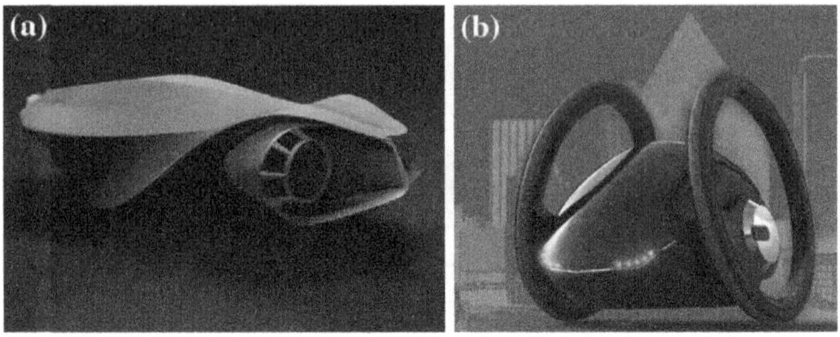

Fig. 16.9 **a** Design for a maglev car which won the Design Competition organized by the Royal College of Art and sponsored by Opel in 2009. **b** The ZiV (Zero impact Vehicle) designed by Chris Latta

16.4.3 Magnetic Levitation Vehicles

Magnetic levitation vehicles have been extensively experimented and there are commercial short lines served by vehicles of this kind. All these vehicles are however guided vehicles and are comparable with trains and not with cars. After many contrasting decisions, at present it seems that magnetic levitation (Maglev) trains will not be an option for the near future.

Even dimmer is the perspective of building magnetic levitation cars. In the enthusiasm raised by the discovery of high temperature superconductors in 1986 many predictions about the possibility of levitating objects as big as vehicles were forwarded. These ideas were based on the prediction that higher and higher temperature (possibly up to room temperature) superconductors could become possible and that their performance could be improved. These predictions didn't materialize, and non-guided magnetic levitation vehicles remained just dreams, but patents have been filed and designs have been forwarded, like the design which won the Design Competition organized by the Royal College of Art and sponsored by Opel in 2009 (Fig. 16.9a). Magnetic repulsion vehicles are however common in science fiction, like in the Star Wars saga. Moreover, like in the case of hovercraft, the problem of generating driving and cornering forces without a contact with the ground remain.

Another more realistic idea has also been forwarded: a wheeled vehicle whose wheels are supported by magnetic bearings instead of rolling element bearings, or in a more fantastic way, vehicles whose tires are suspended to the wheels through magnetic bearings. An example is the ZiV (Zero impact Vehicle) designed by Chris Latta (Fig. 16.9b). The vehicle, that in a way is similar in architecture to the Segway, is powered by fuel cells that provide the energy needed for both the levitation magnets and the propulsion magnets. The tire floats millimeters from the stationary rim.

To suspend the wheels on magnetic bearings is fully possible (less to suspend the tires, a thing that anyway has no advantage) but the advantage in doing so is not clear. With respect to a conventional wheeled vehicle the only energy saving is due to the reduced bearing drag, that is at any rate much less than the rolling resistance of the tires. In trains the advantage of magnetic levitation is linked with the possibilities of reaching speeds well above the maximum speed allowed by the wheels, and even in this case the progress of conventional railways that allowed reaching higher and higher speeds makes questionable the actual large scale use of maglev trains. In cars the possibility of actually travelling that fast is ruled out, so that there is little space for dreams of maglev cars.

As a conclusion, short of radical advances leading to 'science fiction' technologies allowing levitation of a vehicle over a road surface, it seems that we can safely exclude this possibility in a foreseeable future.

16.4.4 Other Very Advanced Breakthroughs

The only one possibility for radically new types of vehicles that remains is antigravity vehicles. Although the possibility of controlling gravity is accepted as a theoretical possibility by a number of scientists and a number of experiments have been conducted by respectable scientific institutions, sometimes even with the sponsorship of space agencies like NASA, there is no proof that it is actually possible, and no serious hint about how to do it. If it is possible at all, we still need a deep understanding of the underlying physics, and then the development of new technologies before thinking of applications. And these applications would be likely in fields like space flight where the advantages would be enormous. Using antigravity for cars, or for lifts, like in some Isaac Asimov novels, seems to be really a waste! In particular because the advantage would be only to avoid rolling resistance, that is responsible for a not too large part of the energy required for motion while all other problems would remain there, and some would become much worse, like producing tractive, braking and cornering forces.

Appendix A
Body Vehicle Design Process

The purpose of this appendix is to provide the reader with an overview of product development technology and, in particular, of the role that computers play in it. To keep the size of this appendix within reasonable limits, our attention is concentrated on the car body only.

The reason for this choice is that the car body has some peculiar aspects and features that make it different from that of the other car subsystems, since the body is designed keeping in mind not only its technical characteristics and production technologies, but also the aesthetic characteristics of its '*style*', which plays a fundamental role in determining the commercial success of a car.

From a technical standpoint, a conventional steel body includes the body shell and trimming which are primarily made by thin-wall elements, whose external surface performs usually aesthetic functions; this fact alone justifies peculiar engineering techniques. These thin sheet elements of complex shape required the development of particular representation rules, that are different from those used for mechanical components (engine, transmission, suspensions, etc.), and that are limited to a few views and sections, aimed to keep the representation as simple as possible.

A further specific characteristic is related to the high capital investments needed to mass produce the vehicle body, with respect to its relatively short production life, usually limited to a few years. The body shell and trimming are totally revised at each new model launch, i.e. every $5 \div 7$ years on average, and this contrasts with what happens with many other components, invisible to the customer, that may be reused with only incremental improvements on next generation models.

This short life may be justified by the accepted tradition of transmitting the image of new product contents through a new shape, which is supposed to correspond better to the aesthetic tastes and style trends of the moment. The simultaneous needs for careful management of costs and cash flows for new investments on one hand, and for developing aesthetic forms consistent with the customer demand at the time of the commercial launch on the other, result in a strong pressure to shorten the time allocated to development. Although this requirement is shared also by other car components, it is a critical issue when dealing with the car body.

G. Genta et al., *The Motor Car*, Mechanical Engineering Series,
DOI: 10.1007/978-94-007-8552-6,

When considering body drawings, it should be remembered that they have different applications; they provide not only the description of parts to be manufactured but they are also the starting point for the development of other drawings that are just as important. With respect to a body shell stamped panel drawing, for example, the following items must be mutually consistent:

- The stamp set for production;
- The positioning and fixing tools for welding with the neighboring parts;
- The characteristics of the working robots for painting and assembling (usually applied for many model generations);
- The spare parts catalogues;
- The disassembling and assembling schemes for servicing and repairing.

The development of any of these items also includes activities which rely on body drawings, often including tests on actual or virtual prototypes. The latter are based on mathematical models that describe completely the geometry of the relevant components, to allow Finite Elements (FE) analysis or the construction of a Digital Mock-up (DMU) used to verify the geometrical compatibility between parts or to verify their kinematic behavior (e.g.: doors opening, wiper motion, side glasses opening, etc.).

A.1 Typical Activity Planning

Body drawings should be available early enough with respect to the start of production to enable all related activities to be performed. The project has thus to be organized is such a way that many activities can be performed in parallel and to accomplish this a minimum of information needed for performing each of these activities must be determined.

This so-called *simultaneous engineering* or *concurrent engineering* provides a means for organizing each elementary operation so that the results needed to initiate the successive operations with a minimum delay are produced. For instance, to start the design of a stamp or a positioning tool, it may be sufficient to determine the overall dimensions of the steel sheet rather than the details of its aesthetic shape; therefore tool drafting can be started before the body shell panel it will produce has been defined in detail. This way of organizing activities increases considerably the number of modifications to be introduced, either for completing drawings with additional detail, or for correcting mistakes that arise due to the lack of information available at the moment when the design was performed.

Correspondingly, the rate of modification becomes just as important as the drafting speed; in addition, the knowledge and know-how of those involved in the development process, and of those who perform subsequent activities, become critical factors for success. Consequently each of those involved in the process must aim for *final* customer satisfaction while effectively working also for an *internal* client who will apply the results to other equally important activities from the perspective of production or commercialisation.

In the case of the body, organization of the development work is further complicated by the fact that each design engineer is not only the supplier to an internal client but is, at the same time, the customer of activities performed by designers developing the aesthetic shapes; actually, body engineering must be performed in parallel with style development.

Today an overall development time of about 24 months, from the style model choice to the start of production, represents the best performance achieved by major car manufacturers, although this is only possible thanks to the wide-ranging application of computers, regularly used for:

- Computer Aided Styling (CAS) applied to develop visible shapes,
- Computer Aided Drafting (CAD) applied to engineering activities related to final product or production tools,
- Computer Aided Manufacturing (CAM) applied to the design of some aspects of the production process,
- Digital Mock-Up (DMU) applied to represent complex assemblies for virtual testing or to develop production and assembly plant layout.

Computers not only reduced the time needed for drafting and for the implementation of modifications, but also made it possible to set up a unique data base that enables all the operations needed to keep product development aligned and updated.

The outside surface of a body panel, the result of the body style development, a part of the component drawings, stamp drawings, etc. may be in common, avoiding replication for different working environments and the risk of errors and delays.

The development process may be outlined as in Fig. A.1. The scheme demonstrates that a high project development speed (here engineering activities are divided in planning and engineering) can be justified by the existence of a phase, here termed

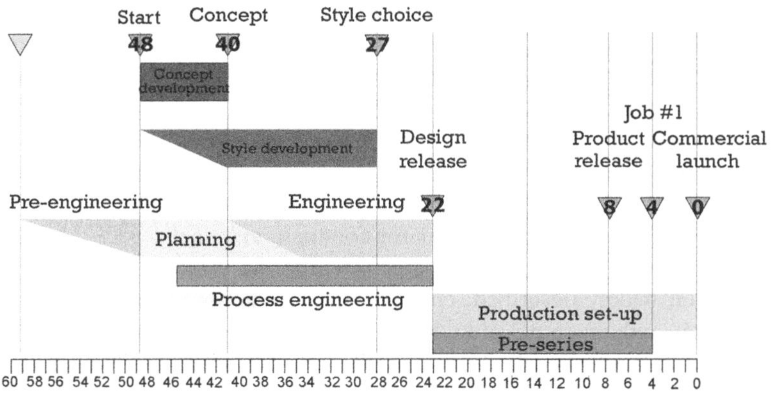

Fig. A.1 The logical scheme of the main phases of the development process, with specific reference to the body. The time scale, at the *bottom*, is measured in months from the commercial launch and the elapsed time is highlighted with respect to the major milestones (from Morello et al. 2011)

pre-engineering (or *shelf-engineering*), where alternative solutions are studied and validated experimentally, without a real finalization; these activities contribute to filling an ideal shelf where good ideas are stored, ready for future application.

The *concept* development phase, or feasibility study, is performed with the cooperation of engineers, marketing experts and stylists, who define a product specification (concept) by examining alternative solutions in parallel; this concept specification must simultaneously satisfy customer's aesthetic and functional expectations and achieve technical and economical feasibility.

Manufacturing of prototypes is started at almost the same time as engineering activities; consequently, prototypes relate to partial aspects only, allowing design features to be validated one at a time. Only at the end will prototypes similar to the final product be available to be used for design approval (or design release), to certify that each specification defined by the concept has been satisfied.

The so-called *pre-series* are other prototypes that, as opposed to the first ones, are built using mass production tools and plants; the positive conclusion of their tests is used for product approval (product release), i.e. providing certification that the production means and organization are adequate to obtain a product complying with specifications.

The *Job #1* vehicle is the first that can be delivered to final customers.

During the time elapsed between concept definition to product definition (from 40 to 22 months before the commercial launch) many activities are performed in parallel such as planning, engineering, style definition, planning and development of stamping, welding and assembling processes, economical evaluation and optimization, in addition to the manufacturing of prototypes that are increasingly complete and closer to the final product.

Many activities that, logically, would be executed in sequence are instead performed in parallel and, as a consequence, are based upon provisional and continually changing information. This organization of activities, apparently nonlogical, can provide advantages in term of cost and time assuming that design tools, in this case CAS, CAD and CAM, enable rapid modification and reutilizing of models for different purposes.

This procedure enables a product solution consistent with production processes, that will not require major changes during the following industrialization phase, to be developed 22 months before the commercial launch; it is therefore possible to start building production tools in time for validation eight months before commercial launch at the supplier's premises. Subsequently the production tools are transferred to the plant of the car manufacturer in order to initiate producing pre-series vehicles and accumulate the necessary numbers for commercial launch.

The advantages of CAD in terms of shortening the time for engineering activities have been widely described; correspondingly the significant compression of timescales have only been possible through the pervasive application of computers to all development activities. Fig. A.2 illustrates the existing links between engineering activities and corresponding relevant computer-based tools.

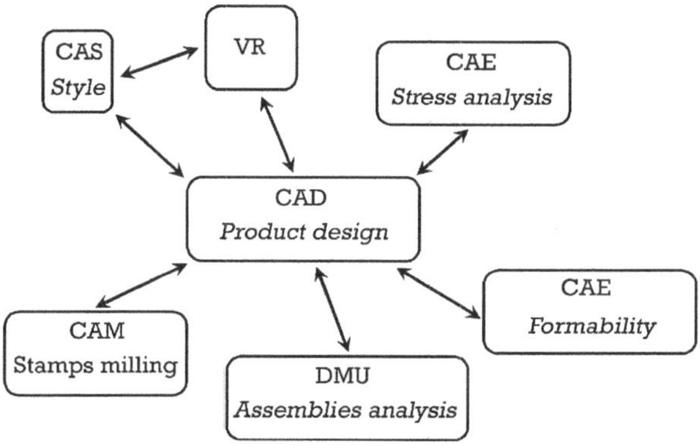

Fig. A.2 Links between product development activities and dedicated informatics tools (from Morello et al. 2011)

- Style development applies CAS tools with two different outcomes: Improving style architect productivity and generating an output (mathematical style model) that can be reused as it is by body engineers.
- Virtual reality is a computer based tool that enables representation of exterior visible surfaces with enhanced realism in context; this makes it possible to evaluate the forms and light reflections they will have in a real environment and enable their visualization from any point of view or in motion; this technique enables decision makers not familiar with drawings or information technology to be involved effectively in the decision process.
- Structural analysis is one of the major tools included in CAE that enables the evaluation, not only of the body structural integrity, but also many other aspects, from weight to drag coefficient. The body mesh, the starting point of any FE analysis, is performed by computers almost automatically, starting from CAD mathematical models.
- A particular FE analysis, performed by considering the elasto-plastic material behavior, can be applied to the design of stamps, again using CAE tools; the same body panel mathematical model, defined by product engineers, can be reused by process engineers to verify that sheets can be stamped to their shape, without risk of rips and wrinkles; similar tools are also used to assess plastic parts formability.
- CAM enables definition of the cutting tool path for milling machines that will manufacture the stamps; again the same body panel mathematical model that has been used for stamp design can be used also for stamp manufacturing.
- The computing power of modern computers enables the preparation of complete assemblies renderings with simplified surfaces, including also related production tools, applying DMU techniques, that also take advantage of virtual reality.

Such renderings offer many potential applications by:

- Style architects: Evaluating aesthetic results obtained by joining different panels taking into account gaps or profile errors.

- Product engineers: Evaluating kinematic mechanisms or interference between neighboring parts.
- Production engineers: Assessing the motion of parts along the production line while checking for collisions.
- Service engineers: Evaluating assembling and disassembling operation for repairs.

A.2 CAS, Computer Aided Styling

When the development of a new model is started, a set of general vehicle specifications is defined, essentially consisting in the following information:

- The type of car, market segment and expected production volumes,
- The relevant exterior and interior dimensions,
- The engines, gearboxes and tires to be used,
- The parts to be carried-over from previously developed models,
- The manufacturing and assembling technologies to be used in connection with the production plant selected,
- The performance and cost targets.

A preliminary layout sketch of the car can be drafted from the first five items; from this sketch, body style and structure developments are initiated. An example of a preliminary car layout sketch is shown in Fig. A.3.

Usually, each new car is born as a replacement or as a part of an existing car family; this fact generates constraints regarding the utilization of an existing platform, including car floor, chassis components and powertrain. The crucial objective is to generate a form that can excite positive emotions in future customers, which are consistent with the commercial target of the model to be developed and with the existing boundary conditions (i.e. production plant, suppliers, cost, etc.).

In this context two distinct phases in the style development can be identified:

- Form generation, and
- Mathematical model generation.

A.2.1 Form Generation

The aim of this section is to identify a number of relevant facts without trying to analyze or rationalize the creative process behind them. In developing cars, or industrial products of similar complexity, designers and engineers must closely cooperate in order to define the form of the product. The designer conveys insight through graphical elements, while the engineer brings this insight to completion, transforming these graphical elements into a technologically feasible object. To ensure continuity, it is

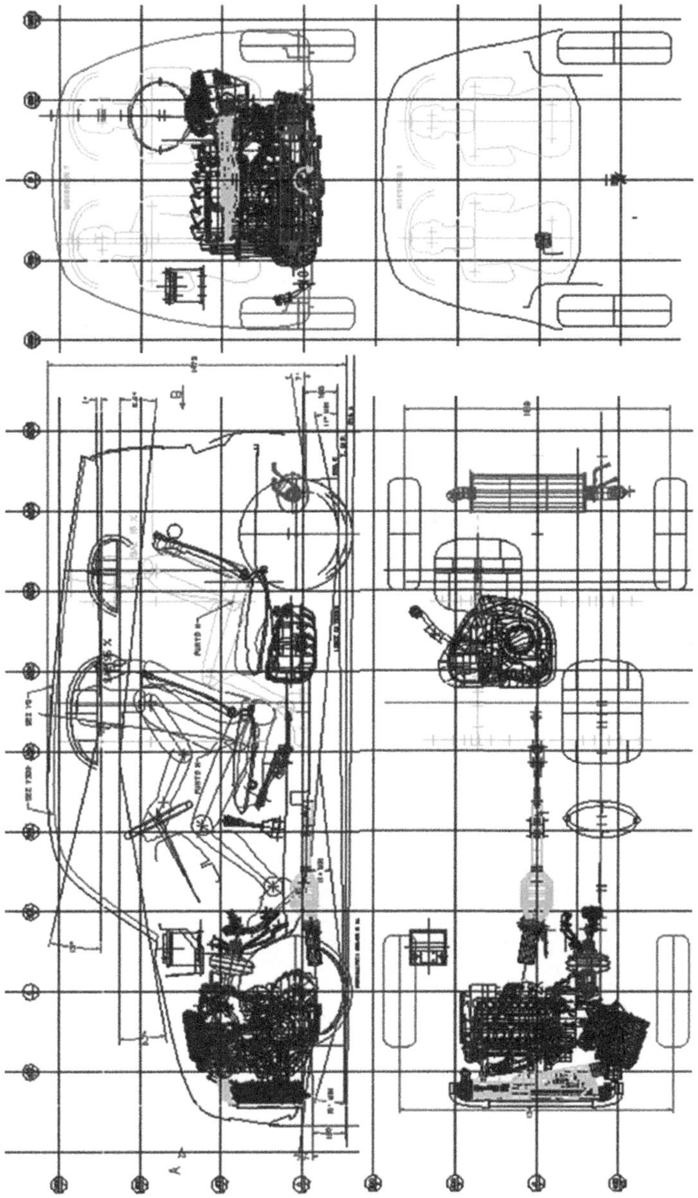

Fig. A.3 Example of preliminary car layout sketch, suitable to verify packaging, roominess and visibility; this is the starting point of body style and structure development (redrawn from Morello et al. 2011)

necessary that part of the creative background of designers is available to engineers and vice versa.

In principle there could be two different approaches to product design: develop an idea, a solution to a problem, with the assistance of scientific analyses, or alternatively use scientific analyses to define an idea. This second approach, even if desirable from a certain perspective, yields limited results and is in general only useful to solve relatively simple problems. The first approach leads usually to the best results.

Style development is a balanced mix of intuition and scientific approach, where the success of the development process requires an appropriate equilibrium between creative and structured activities. What is important is thus full understanding between two different thought processes: while it is true that the best idea without engineering development cannot become a viable product, it is also clear that the best engineering method without insight will not translate into an interesting product. During form generation, the product's appearance will summarize the designer's understanding of model targets; this understanding will materialize through partial sketches, like those shown in Fig. A.4, that only later will be integrated into a consistent form. These sketches give form to the ideas generated while capturing product targets to be integrated successively. They do not have all the characteristics required for an objective representation, but rather represent the idea as interpreted by the designer.

Still only very immediate tools such as a sheet of paper and a pencil for a freehand sketch can materialize insights and make them available for further processing. Indeed no software tool is yet available for creative activity which is quite as effective as freehand sketching, and in general few designers are able to materialize their ideas with draft and paint software tools available; only when key ideas are conceived can they be integrated into a detailed digital model. This is already a rationalization phase that seldom receives further creative contributions.

Fig. A.4 Hand drafted sketches representing the designer's perception of product objectives (from Morello et al. 2011)

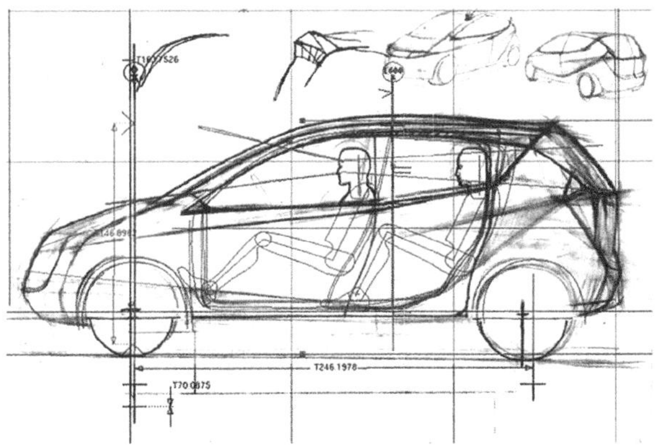

Fig. A.5 Transforming sketches into coordinate lines (on the xz and yz planes) allows obtaining a style sketch that will start discussions between designers and engineers (from L. Morello et al. 2011)

The digitalization of the initial sketches, according to the coordinate lines in the xz and yz planes, is shown in Fig. A.5. It enables the creation of a complete sketch of the car body, that can be observed from any point of view; it will initiate discussions between designers and engineers. In the past designers and engineers could discuss only using scaled clay models, available later in the style development process; nowadays rapid sketching and paint software tools enable an early interaction between these two different cultures.

A.2.2 Mathematical Model Generation

It looks unlikely, at least for the time being, that a CAD system designed to provide a complete and very detailed description of the body surface can be also used for a creation process that concentrates only on more specific aspects. Nevertheless CAD systems are useful immediately after form generation to investigate the compatibility of what has been conceived with the desired performance and the technologies involved with its production; product and process engineering require a very complete and sharp representation.

The few lines, initially defined, are now converted into spatial coordinate surfaces with the help of software tools, as shown in Fig. A.6. Missing details are added; this operation is performed almost routinely since it does not require conceiving new forms but is just a consistent development of the previous insight.

Image processing and rendering systems allow results to be verified regularly and the needed corrections to be introduced; Fig. A.7 shows one such representation. At this stage, these surfaces are used as an input for a CAD system to enable further details to be added. At this point, many shape details are added to complete

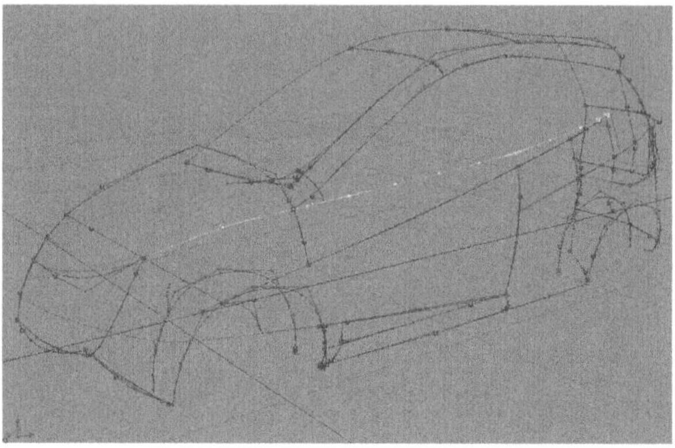

Fig. A.6 The few lines, initially defined, are digitalized and converted into spatial surfaces by means of dedicated software tools (from Morello et al. 2011)

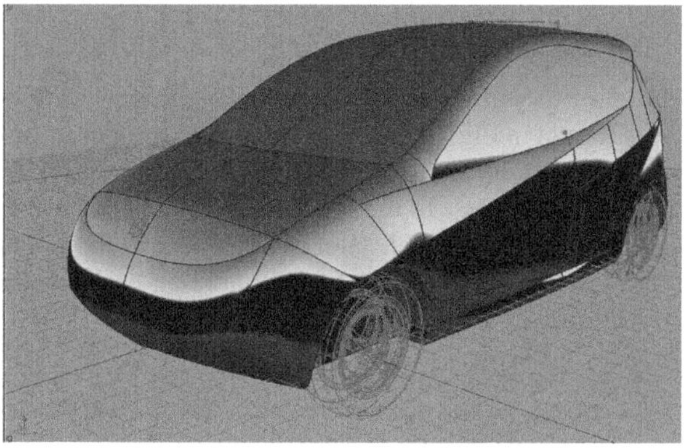

Fig. A.7 Three-dimensional rendering of initial sketches allows correction of added elements, according to their consistence with the initial insight (from Morello et al. 2011)

the surfaces; the transformation into mathematical lines of the detail of door handles is shown in Fig. A.8.

During the engineering development of the style shape many problems may arise that require a new reinterpretation or redefinition of the initial form. These problems do not regard the body taken as a whole, but more often relate to single details or the transitions between different surface elements.

A CAD mathematical model usually enables the effective evaluation of the car exterior shape, using for instance virtual reality. Nevertheless, improvements in these systems are still needed to allow each person involved in the decisions regarding the appearance of the car to be familiar with them; this is why physical models are still in use to confirm decisions taken on virtual models. Sometimes direct viewing, and the sense of touch on a full scale object, are useful to perceive the correct impression.

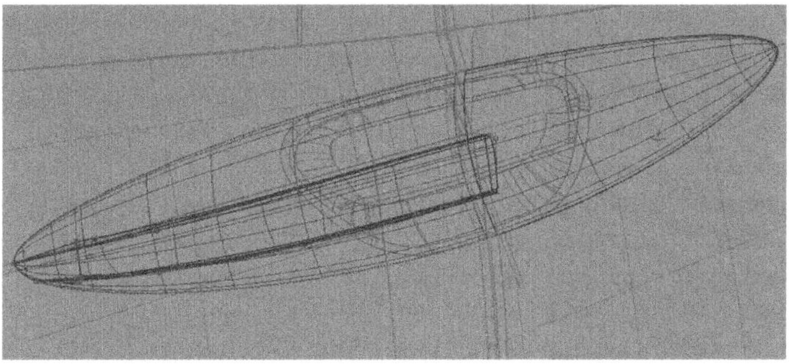

Fig. A.8 Transformation into mathematical lines of details, like door handles (from Morello et al. 2011)

Fig. A.9 Milled surface cut from a sawdust and epoxy resin block, treated with plaster and paint, to simulate the final appearance (from Morello et al. 2011)

Figure A.9 shows one of these models. They are built automatically by means of computer aided milling of a synthetic material block (a mixture of wood sawdust and epoxy resin, for instance Epowood®), starting from the CAD mathematical model directly. These milled blocks can achieve a highly realistic aspect by finishing, painting and polishing by hand.

Sometimes, corrections are introduced by designers on the model directly, to improve the appearance or to visualize and discuss possible modifications. In this case the mathematical model will be updated by digitalizing the new surface directly.

A.3 CAD, Computer Aided Design

Modern computer aided instruments enormously simplify the detailed and exhaustive representation of the surface geometry that is now represented by continuous mathematical equations; nevertheless it introduces new issues depending on the type of

CAD system in use. Those currently available on the market have comparable modeling capacity, but not identical techniques for representing surfaces mathematically; therefore the use of different CAD systems implies the use of so called universal translators such as STEP or IGES, which are two of the most common, to convert the data generated in one system so that they can be used in a different one. In certain cases dedicated translators have to be developed. Additional time must anyway be spent to convert, repair and complete the data, with consequent costs that could have been avoided by choosing a single CAD system, at least for the development of a given model.

In addition significant training time is required to use any CAD system, that, like any computer code, undergoes continuous improvements and modifications. Different systems imply different procedures to create the same geometric entities and the skills acquired by an engineer on a certain system can be applied to a different one only with some difficulty.

A decision about adopting a given CAD system is a strategic one, because of the capital investments involved for software acquisition and training, and the cost of reusing drawings developed in the past. This has an amplified effect on partners and suppliers. In spite of these inconveniences, it is relatively easy to find instances, in the recent history of car design, of major car manufacturers using different CAD systems (e.g. body engineering and powertrain engineering) at the same time.

The transfer from CAS to CAD system of these mathematical models usually does not require reworking because surfaces are simple (only the visible parts are represented) without comments, dimensions and tolerance information; the effective communication between these two systems is a major advantage of these computer tools: the same mathematical model used to represent the aesthetic surface is also used to represent part of the component and to program the milling machines cutting stamps.

Aesthetic surfaces contain not only information about their shape, but must define also their contour, including the gaps between different panels of the body skin or play between fixed parts and parts that can be opened or replaced, or between body panels and glasses. The definition of these details is a consequence of the compromise between aesthetic concepts and engineering requirements, required by part function and manufacturability.

In the next section the development of a car body is described, with particular reference to the body shell. The example described refers to a new body development, starting from an existing platform, which represents the most frequent case. In fact, the development of any major part of the platform (engine, gearbox, suspensions, etc.) requires a significant development effort; for this reason, such parts continue to be used for more than 10 years, also on very different cars.

A.3.1 Body Modelling

The most important steps of body modelling are typically the following:

- Structure architecture definition, as a function of expected production volumes, technologies and materials;
- Definition of the main sections of the skeleton, partly independent of the style model;
- First approximation modelling of body exterior panels;
- Detailed analysis of junctions between side members, cross members and pillars;
- Trade-off between aesthetic, structural and manufacturing requirements.

Structure architecture

The first step is the definition of technological and design solutions that will be used for the body under development. To shorten development time, including modelling, virtual and physical validation, solutions are preferred that have already been applied on other cars or validated in previous non-finalized development activities (so-called pre-engineering): these solutions are usually conceived as *archetypes*, meaning a generalized model, useful for subsequent application to an assigned exterior style or shape .

Archetypes are not only the consequence of a structure architecture, but also of the assembling process or materials selected. For instance, on a large production car, the most common body archetype is the unitized one made of stamped steel, while for a small production volume a partly not unitized solution could prove more convenient.

The term archetype is all-embracing because it includes many different design aspects such as, for instance, the impact of aluminium use, with the possibility of using extruded or cast parts, or laser welding, with the elimination of welding flanges and electrodes passage holes.

Main sections

According to the selected archetype, including the structure architecture, material, applicable forming and assembling technologies, characteristic cross sections of the skeleton are defined (see, for example, Fig. A.10).

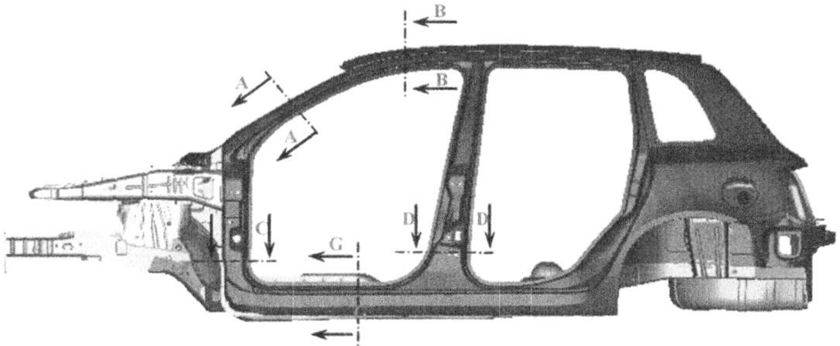

Fig. A.10 Some typical body cross sections that can be defined starting from an archetype (from Morello et al. 2011)

Starting from an archetype, even if not yet applied to an assigned aesthetic surface, many functional parameters are conditioned such as:

- The dimensions of resistant cross sections,
- The thickness and material for each of its parts,
- The aesthetic matching with neighboring parts, for instance parts that can be opened or made of glass,
- The assembling solutions for small components such as weather strips, hinges, locks, etc.

To understand how all this information can be focused in a few lines in practice, reference can be made to Fig. A.11, where alternative configurations for section B–B of Fig. A.10 are shown, corresponding to the part of the roof above the front door.

Four different archetypes can be observed:

- Archetype 1, characterized by a longitudinal aesthetic strip covering the welded joint between the roof and the body side, by a sliding glass without visible outside frame and by a weather strip integrated with the guide liner of the glass.
- Archetype 2, characterized by an enveloping roof with no covering strip, framed glasses and weather strip mounted on the welding flange between the roof and the body side.
- Archetype 3, again characterized by a longitudinal cover strip, integrating also the function of weather strip.
- Archetype 4, similar to archetype 2, but with frameless glasses and integrated weather strips and glass guide liner.

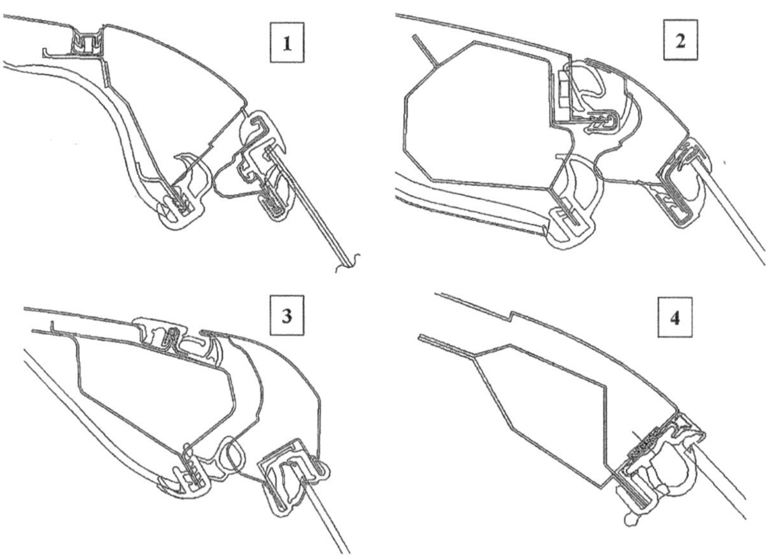

Fig. A.11 Some different archetypes for joining the roof with body sides, depending on the door archetypes that can be matched (from Morello et al. 2011)

Different archetypes have different aesthetic contents, also corresponding to different capital investments, and procurement and production costs, as a function of the different structure architecture and materials. It is important to remember that these sections can be applied to the specific body only if they are adapted to the mission to be accomplished and to the specific shape of the car style.

This archetype selection and adaptation process can be performed not only for major components, but also for details such as weather strips and glasses. To speed-up this process and avoid mistakes, it is important to access a wide database of already developed in-house or competitors' solutions.

This operation consists of repositioning on the new style and resizing already existing elements, corresponding to production cars or pre-engineering results; other components may be used as-is, i.e. without adaptation, e.g. the weather strip section, locks, hinges, etc. This geometry will be subsequently refined following analysis and virtual or physical prototype testing.

Modelling

Once a number of sections have been positioned and sized in sufficient quantity to define the body shell critical areas, engineers proceed modelling the remaining body shell parts, doors, hood, etc. Critical areas usually include the structural skeleton and the doors hinges.

To clarify this process, a body shell conceived as a space frame may be considered, as in the example of Figs. A.12 and A.13, since it is easier to identify beams and junctions on the body side. For instance, in section B–B (Figs. A.11 and Fig. A.10), the archetype that better matches the example is 1. Based upon this cross section (and clearly a number of others along the door contour) the upper part of the body side and its upper beam can be modelled in detail. Existing CAD systems enable this operation to be performed almost automatically, thanks to their parametric and associative features.

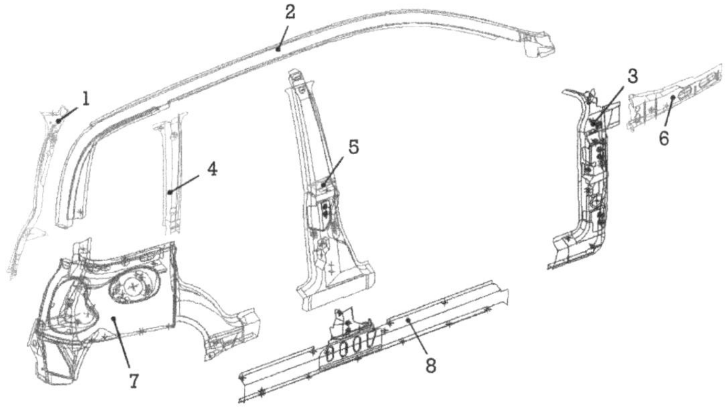

Fig. A.12 Basic components of the body side of a space frame type body (from Morello et al. 2011)

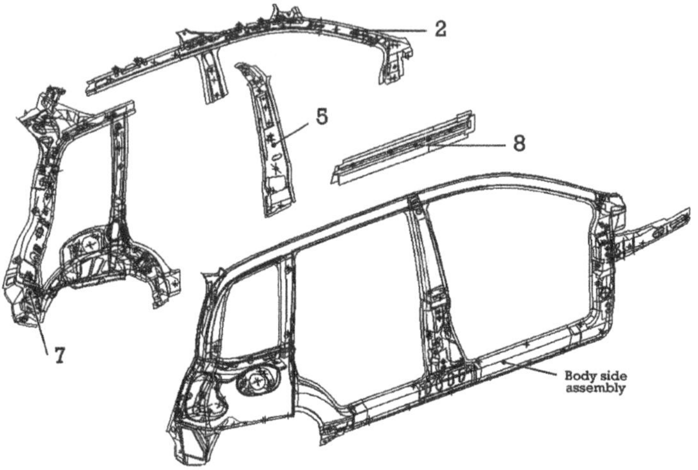

Fig. A.13 Body side assembly of the previous space frame body (from Morello et al. 2011)

The upper beam to be modelled is logically connected to the aesthetic surface (it is associated with the surface) and its size is defined by the measurements as a function of a limited number of fundamental dimensions (parameters). These parametric and associative features also allow implementation of the many modifications to be expected in a short time; for instance, CAD systems with these features allow an entire model to be regenerated, as a consequence of a style modification, maintaining the archetype logical relationships automatically.

The power of the associative features can be exploited to investigate the effect of a modification also to other geometric entities, such as stamps, milling paths, finite elements analysis models, etc. This kind of approach can be repeated for all body elements whose geometry can be assimilated to a cross section extruded along a curved path.

Stamped steel or aluminium body panels may be represented as a portion of solid space included between the aesthetic surface and a similar surface obtained by shifting the previous one by the panel thickness. A line obtained on a drawing by cutting the aesthetic surface is called *style profile*; the other line defining the panel section is obtained with a pure shift. This process may be applied with some caution to thicker elements, such as plastic components. Further elements, such as ribs, roundings, pins, etc. may also be defined using archetypes.

Some car manufacturers, on the contrary, opted to model also surface-like parts with solid bodies, thanks to CAD systems designed for this feature: this procedure involves greater modelling time, but enhances the subsequent development steps, such as designing stamps for high thickness plastic components.

Junctions modelling

When structural beam elements of the body are modelled, the junctions at beam intersections can be defined in detail. To designate these junctions, a standard nomenclature has been defined using letters: it is interesting to note that the order adopted puts in the foreground the areas more bound to style than to structural behavior. Consider, for instance, junction A, at the intersection of body side with windshield and roof: The components of the body shell involved (see Fig. A.14) are the roof, upper windshield cross member, exterior body side, front or A pillar cross section and reinforcement member. At this joint also the windshield and its structural adhesive bond contribute, and the geometry of any liner may be involved. The front door and their weather strips are also affected by this geometry, as is the space available for driver and passenger and their visibility.

Together with the space requirements for the installation of these elements, the structural requirements relevant to torsional stiffness, front impact, roll-over should also be considered; the assembly cycle feasibility should also be considered. If classical spot welding is chosen, also access for the electrodes should be taken into account. For example (see Fig. A.14 , at right), the eyelet surrounded by a dotted line allows access to the electrode of the welding tool from the inside of the body and therefore makes the junction between roof and body side possible.

Also the task of designing different junctions in detail is simplified by CAD systems, enabling 3 D modelling, with a clearer perception of geometrical issues, together with quicker modification or adaptation of existing pre-validated solutions, therefore shortening the development time.

Test validation

Once the body has been defined, including doors, hood, hatch and removable parts, together with all detail interfacing body components (such as weather strips, locks, hinges, etc.), a complete 3 D model is available that can be easily tested with respect to different perspectives. Each virtual test will result in modifications for surfaces or sections; with a parametric associative CAD system, each will be implemented in a very short time, sometimes automatically, and transferred also to associated entities such as stamps, assembly fixtures or other dedicated production tools.

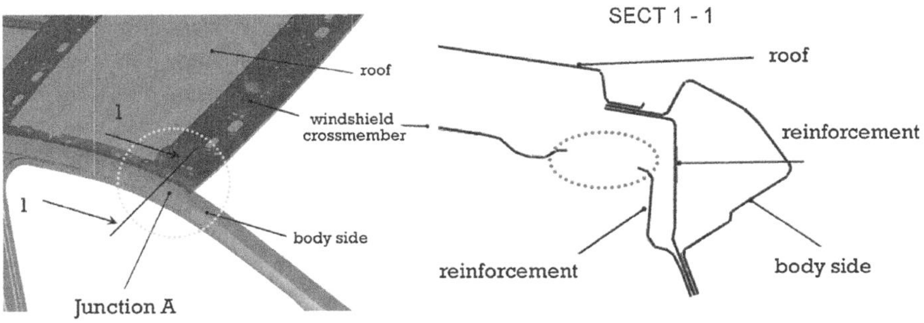

Fig. A.14 Junction A modelling, at the intersection of body side with windshield and roof (redrawn from Morello et al. 2011)

CAD system speed is particularly expedient when introducing a large number of modifications resulting from potential malfunctions detected during the virtual validation tests.

As far as aesthetics is concerned, mathematical models obtained with this procedure enable the refinement of the exterior surfaces with style architects and certify consistency between shape, performance and manufacturability: for instance, the shape required for the windshield by the style concept is verified in terms of visibility and installation of wipers, or for side glasses again in terms of visibility and sliding motion into the door.

The body mathematical model can be transferred for evaluation in terms of structure analysis, regarding dimensions, local thickness, material choice and welded junctions. Also in this case, computer tools contribute to shorten development times by generating the calculation mesh almost automatically. In this way, the mathematical models become an effective communication tool among engineers, stress analysts and, again, style architects, to negotiate modifications and define trade-offs. A similar situation enables process engineers to evaluate and analyze the stamping and assembling process.

The mathematical model is built in the ideal hypothesis of nominal dimensions, without taking into account the effect of fabrication tolerances. However, using dedicated software tools, mathematical modelling enables investigating the effect of dimension variations on the assembling process and compatibility with product performance.

A.3.2 Rules and Common-Practice in CAD Modelling

CAD models are the transposition into mathematical parameters of a set of numbers that regard not only the geometric nature of a part, but also its technical and organizational features. The information regarding geometry includes the coordinates of points, surface and lines equations, etc., needed to define the shape of a part completely. Part of this geometric information has also some organizational content; for example, the associative features describe, in computer language, the operations accomplished by the design engineer while modelling a part (design intent) and their relationship with neighboring models (welding and assembling interfaces); this information allows the regeneration of the mathematical model automatically when some of its geometrical parameter are modified.

Technical features cannot directly infer geometric information, containing for example dimensional tolerances, positioning reference points, functional specifications, etc. that are needed for part operation, manufacturing, or production fixtures. Nominal dimensions are instead usually inferred from the mathematical model.

Organizational features consist in data needed to manage the part within the company information system according to usual company practice. Among the organizational information, the description of the product configuration is the most relevant; product configuration describes the logical relationship between any of the parts of the vehicle. Product configuration enables the correct identification of any subassembly or elementary part and the preparation of a bill of materials.

Product configuration description allows also the creation of any assembly drawing and the management of material flow along the production line, orders to suppliers and the delivery and storage of spare parts.

A further organizational feature to be recorded on mathematical models is a note concerning their development status; it is needed to inform any of the potential users of a mathematical model about the level of risk of processing information that has not been fully validated since many activities are performed in parallel by people with real-time access to a common data base.

Finally, CAD models are classified with reference to their content and to their preparation technique. The following classification is usually applied with regard to the content:

- Outline drawings, essentially useful to the design engineer to develop ideas, define critical matching between parts and solve problems of geometric compatibility.
- Assembly drawings, useful for documenting how parts are assembled together.
- Part drawings, useful for exhaustive documentation of dimensions, materials, specifications of a single part.

The following classification refers, instead, to preparation technique influencing the kind of model used, the main ones being:

- Wire frame models, i.e. 3 D models where solid bodies are represented by spatial curves describing their edges; between one curve and the next, the surface shape is unknown and must be interpolated.
- Surface models, i.e. 3 D models where a part is described through its contour surfaces; the coordinates of each points are fully determined.
- Solid models, i.e. 3 D models including a complete description of the space included between the boundaries of the model and representing every detail of a part, including its physical properties such as mass, moments of inertia, etc.
- Conventional drawings. They are paper print-outs, similar to a pencil drawing, used for communication only and unsuitable for further development.

Reference frame

The reference frame traditionally used by body engineers is different from that suggested by the SAE J670 standard for vehicle dynamics mathematical models; here an xyz axis system is defined, with its origin in the body symmetry plane, at the intersection with the front wheels axis of rotation and with the axes defined as follows:

- the x axis is contained in the symmetry plane and set horizontally, pointing rearwards,
- the y axis is perpendicular to the symmetry plane, pointing to the right with respect to the driving direction,
- the z axis is consequently vertical, pointing upwards.

If the vehicle body were not symmetric, the xz plane would be assumed to be coincident with the vertical plane.

The definition of the origin depends upon the position of the suspension, which determines the centre of the front wheels and the body pitch angle; therefore, this definition must be applied assuming a reference load condition, usually corresponding to a full tank of fuel and four passengers of 70 kg of mass each, with 40 kg of luggage in the trunk for cars with four or five seats, or two passengers and their luggage in two-seater cars. For different kinds of vehicles (vans, minibuses, trucks, etc.) a load condition close to full load is usually selected.

Considering the body size, and the fact that drawings should be printed at least at full scale, a complete body assembly is seldom represented on the same drawing for sake of clarity.

To enable the correct location of partial assembly drawings or part drawings, a quoted grid is represented on each drawing, according to the local applicable coordinates.

Surface representation

Body part drawings usually contain surface elements that are fully defined by a mathematical model; large surfaces are represented by joining a sufficient number of patches, whose size depends on the local curvature: a higher curvature requires a larger number of small patches.

Visible parts of the body surface are called aesthetic surfaces and are coincident with the style model.[1] In the case that the aesthetic surface is not available as a mathematical model (this was commonplace until just a few years ago), style architects must provide a physical full scale 3 D support representing this surface; this support will be digitalized point by point, for example at the intersections of the surface with the body reference reticule; the lines obtained are joined with interpolated mathematical patches as in the other case.

Each section of the drawing representing a part contoured by the aesthetic surface will contain also the style profile, the intersection of the aesthetic surface with the section surface. The concept of style profile applies also to solid models, for instance high thickness plastic parts.

Skeleton surfaces do not usually contain parts of the aesthetic surface; nevertheless they are represented with the same system, developing a surface mathematical model for the side of the skeleton matching with visible parts, taking care to define a single style profile for each of its parts. Where the aesthetic surface is interrupted, for example at a door contour or near a removable part, the mathematical surface will be cut by a surface determined by an envelope of constant diameter circles with their centre on the aesthetic surface; this operation enables a constant gap between different parts to be generated.

---

[1] The following indications are not standardized but refer to the praxis of a large car manufacturer and his suppliers. There are no internationally applicable standard rules for the time being, even if similar rules are widespread.

Subsequently the two parts of the surface are rounded to reproduce the effective shape of the part that can be obtained by stamping or bending. Flat surfaces might be added, to join the outside body panel to the skeleton. These are called completion surfaces, for comparison with the aesthetic surface.

Drawing surfaces are usually classified in classes (A, B and C class) with reference to the accuracy of their mathematical model. The accuracy is measured by comparing the mathematical surface with the aesthetic surface and measuring the continuity between patches. Classes are useful to speed-up the activity of style architects and body engineers, improving time and manpower efficiency with regard to applying the finishing touches when the style freeze event is still far away and many major modifications could still be needed.

Class type represents, therefore, not only the accuracy of the mathematical model, but also the maturity of the project: from class C, used for feasibility studies and the first style proposals, drawings are refined up until class A, at the end of the project, to represent the part to be put in production.

Common practice is that:

- At the initial stage, when more alternative style proposals are under evaluation, class C is allowed in each mathematical model;
- when a style alternative is selected, but uncertainties and variants are still present, class B is allowed;
- final drawings, ready for production, must contain class A surfaces only.

Specifications for each class are the following:

- Position continuity. The distance between each point of the edges of two neighboring patches must be no more than 0.01 mm, for Class A; no more than 0.02 mm, for Class B and no more than 0.05 mm, for Class C.
- Tangents continuity. The angle between the tangents to the surface on the edges of two neighboring patches must be no more than 0.1° , for Class A; no more than 0.2°, for Class B and no more than 0.5° , for Class C.
- Shape tolerance. The shape tolerances are measured by comparison of a surface patch with the counterpart on the aesthetic surface. they must be class no more than ±0.5 mm, for Class A, no more than ±1.0 mm, for Class B, while for Class C surfaces shape tolerance is not applied.

However class A surfaces could show other unacceptable aesthetic defects; to correct these before stamps milling and finishing, aesthetic surfaces are represented by means of special rendering systems, sometimes available in CAD systems, suitable to represent light reflection on the actual painted body surface.

The simulated lighting system is, traditionally, a linear source lamp (neon tube lighting) that can highlight each of the defects to be corrected; Fig. A.15 allows the difference between inaccurate surfaces or class A, well refined, surfaces to be perceived.

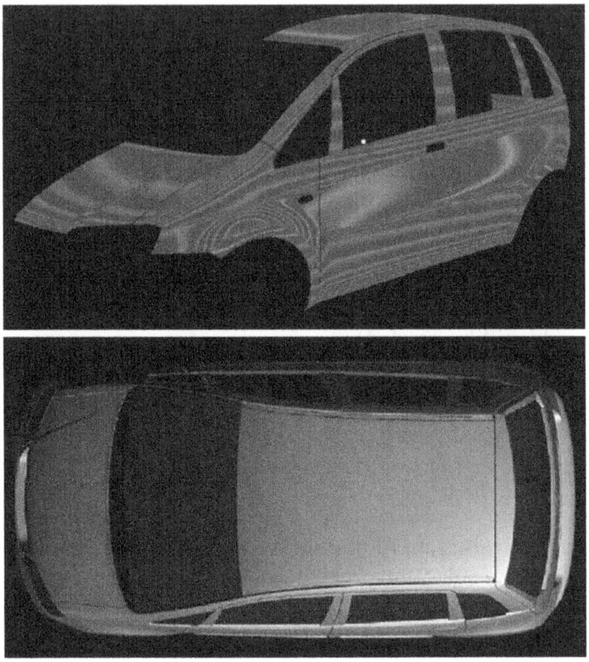

Fig. A.15 Light reflection representation for a complete body with good (*below*) and poor (*above*) quality surfaces (from Morello et al. 2011)

A.4 DMU, Digital Mock-Up

A major portion of the development costs of a new car corresponds to prototype building and testing; newest development projects take advantage of DMU to lighten part of this burden with the so-called virtual prototypes and virtual tests, providing an enormous advantage through the increasingly detailed graphical representation of real parts.

An example of two complete under-hood environments is shown in Fig. A.16. The high graphical complexity makes this kind of representation not viable using a traditional CAD system. A DMU system enables the assembly, in a suitable application environment, of all parts that should be examined; these parts must be available as independent CAD mathematical models.

A DMU system allows studying the interaction between each component with the assembly, in order to search for potential interference in their final position or during assembly, disassembly or motion according to their function. It is therefore a virtual representation of objects supporting the development process of both product and production tools.

Virtual reality, although based on a different technology, can be considered to be an implementation of DMU featuring a higher representation realism. DMU takes

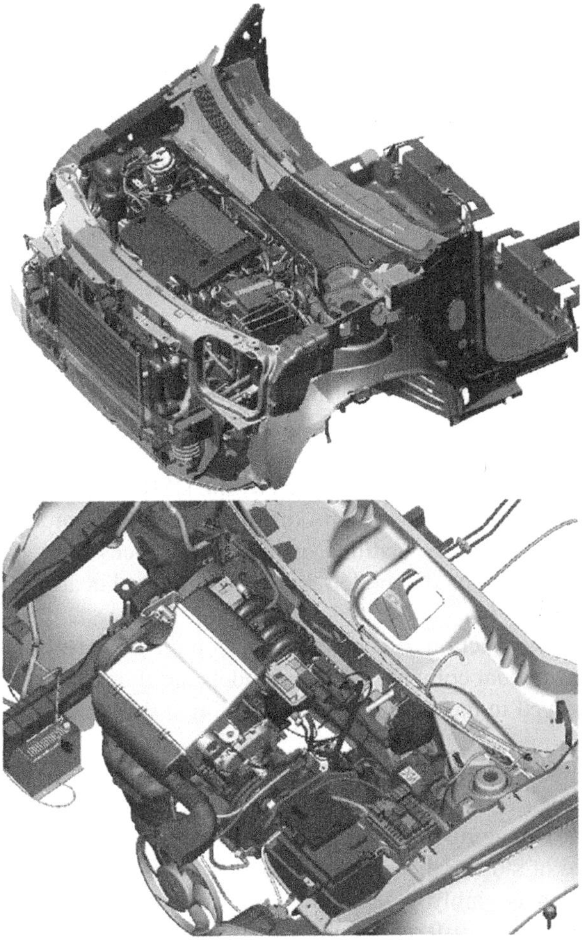

Fig. A.16 Two examples of a DMU of the complete under-hood environment; the high graphical complexity makes this representation not viable with a traditional CAD system (from Morello et al. 2011)

advantage of visualization systems that are able to manage very large assemblies, such as an entire car body, with interior trimming, an engine compartment or a platform, complete with all mechanical components. These systems approximate the boundary surface of each component with simpler elements, e.g. triangles, obtaining a simplification in terms of the information describing the objects to be represented. This technique is called tessellation or tiling.

To provide a general picture, the file representing the assembly can be reduced by ten times in size. This reduction is usually performed off-line with suitable translators, interfacing original CAD models with the visualization system; the representation error introduced by this operation can be controlled by the user.

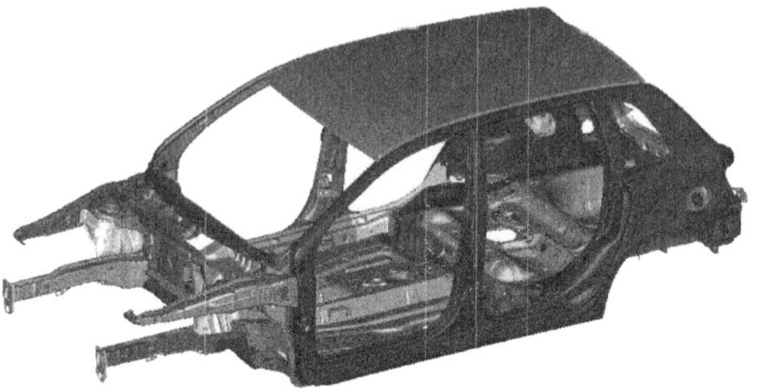

Fig. A.17 DMU of a complete body shell. This figure reproduces what displayed by a computer monitor, showing the product breakdown structure on the left, with the names and part numbers of each component or subassembly included in the model (from Morello et al. 2011).

An important further advantage provided by the DMU system is the possibility to manage, in the same representation environment, mathematical models from different CAD systems; their diversity, justified by previous choices or by particular design aspects, is overridden when the models are to be translated in any case to a different common format.

In addition, the reduction in mathematical complexity of these models makes them usable on personal computers; these models are thus available also to people not directly involved in the design process, making a wider sharing of information possible.

The virtual prototype (the entire vehicle or a large part of it) represented by a DMU system can offer each working group a single and reliable reference, offering the advantage of managing each component in its real environment. Finally DMU systems are able of storing the product breakdown structure, identifying each part included in a given assembly. A last important implication of DMU is that working in different locations or different companies is facilitated by a common assembly model of reasonable size than can also be shared through the internet.

An example of a complete body shell DMU is shown in Fig. A.17. It shows what is displayed by the computer monitor; on the left the work breakdown structure shows part number and name of each component or subassembly in the assembly.

The product breakdown structure is usually imported from the company information system before initiating the implementation of the digital mock-up, helping to build up a complete assembly from its parts without omitting any. The same structure is used as a navigation tool to reach easily each of the parts in the assembly or to build and check partial subassemblies, available along the production line, before completion.

The digital mock-up not only simulates the final shapes of the body but stores its parts in hierarchical systems able to simulate the body assembly process, in the same sequence as developed along the production line. The first application of the digital mock-up is composing assemblies, to look for errors in their components, since the high complexity of components geometry makes shape errors quite likely.

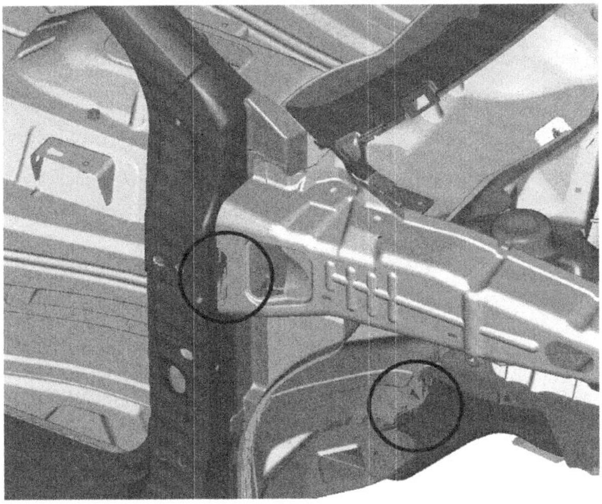

Fig. A.18 The *circled areas* in this picture show color changes that indicate interpenetrations of contacting parts due to modeling errors (redrawn from Morello et al. 2011)

Assemblies visualizations are usually made by assigning to each component a different color that is altered when the relevant part is selected (although the grey tones prints of this book are unable to illustrate this). By examining Fig. A.18 it can be realized how flanges or spot welds disappear when assembling a large number of components in a single assembly and an interpenetration between different elements can be discovered. This occurrence is more likely during the initial phase of the mathematical model development of body panels or when assembling a trimming element on the body for the first time.

To correct these errors, the following checks can be performed directly within the digital mock-up environment:

- Geometry compatibility between visible elements of different parts.
- Building local cross sections and measurements of distance between parts not immediately visible. This operation can be made quickly, even on two remote personal computers connected on line, for example during a teleconference.
- Potential interference during assembly.
- Geometry compatibility in dynamic conditions, when some element is deformed by external forces.
- Feasibility of kinematic displacements.

Figure A.19 shows how it is possible to represent the complete body shell with a multi color shadowed model in the digital mock-up environment and to open, in the meantime, a window where a cross section along an assigned plane is quickly made. This plane can be translated in any direction to investigate the entire interface between two parts. This procedure allows one to easily study a new assembly, as it may occur when an assembly is developed by a supplier or is carried-over from an

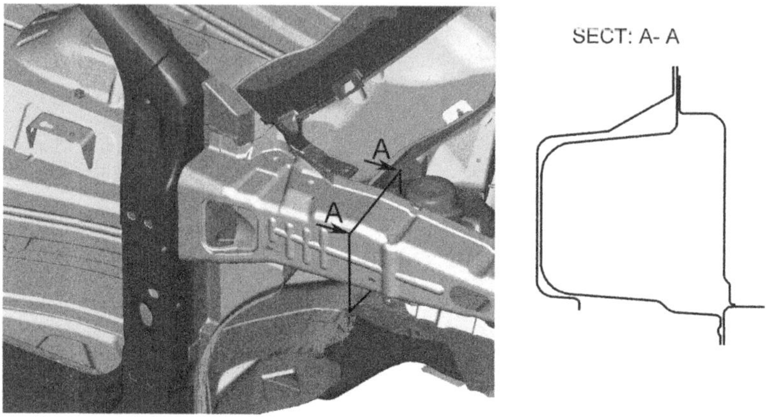

Fig. A.19 Automatic creation of a cross section to check for body interpenetration (redrawn from Morello et al. 2011)

existing model. Apart from instant sectioning, the temporary elimination of some part is also possible, to concentrate the attention on single elements of the assembly.

By looking at the enlargement at the right in Fig. A.19, the lines correspond to the style surface of different parts. When actual parts are in contact, section lines of the style surface are separated by the sheet thickness. This feature must be kept in mind to avoid misinterpretations while checking models for gaps and interpenetrations. Measurements can be taken also on cross sections (see Fig. A.20): This function enables a quick check of part dimensions or gaps in the welding flanges areas. As mentioned previously, these operations can be performed by people not totally familiar with CAD modeling.

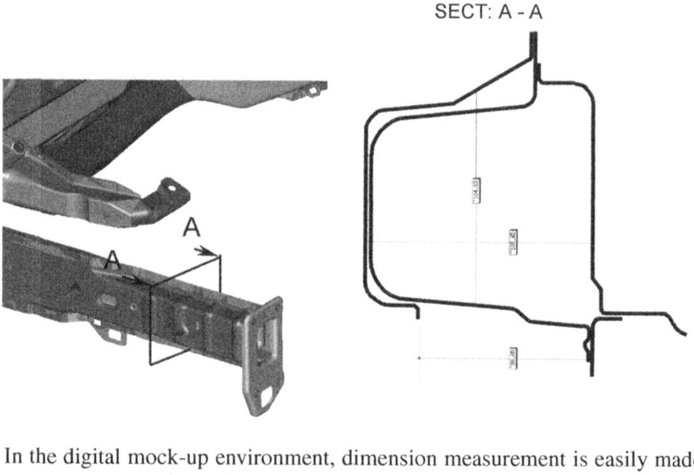

Fig. A.20 In the digital mock-up environment, dimension measurement is easily made, to check for part geometry or gaps (redrawn from Morello et al. 2011)

The only virtual tests not performed with digital mock-up systems are those related to dynamic and structural performance, that are investigated by different simulation models. the main examples of digital mock-up applications are:

- Drawings check,
- kinematic analysis, and
- Evaluation of assembling and disassembling feasibility.

A.4.1 Drawings Check

Drawings are checked to verify their accuracy and compliance with company's design rules. Digital mock-up assemblies are examined according to dedicated check-lists that enumerate rules to be observed and certify the compliance of related drawings. The person in charge of this check is not always the person in charge of the drawing and its related modification, and could also be an internal customer of this drawing. These check-lists become thus a tool to enhance a structured discussion between members of the different working groups.

Check-lists of an assigned assembly could consist in, for example, a table reporting, in the rows, the functions to be performed and, in the columns, the components of the assembly involved in these functions. Each of the verifications to be made should refer to a written rule documenting this function objectively. At each intersection between rows and columns the table should report 'yes', 'no' or 'not applicable' with respect to the question of compliance with relevant design rules.

If during prototype testing a malfunction is detected in an area covered by an item in a check-list, some decision must be taken by all those involved in the design process of the related component or subassembly. Not only must this error be corrected, but the rules that allowed this error to occur must be modified as well. Thus company rules are modified to avoid this error occurring again in the future or new control items are added to the check list.

A typical list of functions to be verified in a part or subassembly of the body should include the following categories of checks:

- Drawing completeness (dimensions, tolerances, surface finishing, etc.).
- Visibility of stamp split lines (which should never be visible from viewpoints corresponding to the normal positions of driver or passengers or should be concealed by dedicated trimming).
- Assembling and disassembling (these conditions are not equivalent, because assembly shop cycles cannot be reproduced in service shops; for obvious reasons, in principle, each part should be replaced without requiring disassembly of other parts).
- Interference between parts or contacts (if not correctly treated, contact points are a potential source of squeaks and rattles).
- Utilization of parts with dimensions at the limits of the allowed tolerance field.

- Dust and water tightness of passenger compartment and trunk (treatment of junctions between parts separating trunk and passenger compartment from the outside).
- Separation between electric harness or hot parts and flammable materials (prevention of fires after vehicle crash).
- Separation between power and signal electric harness (prevention of malfunctions due to electromagnetic interference).

As we have seen, the digital mock-up easily allows the measurement of distances between parts; this can be done by drawing cross sections in the critical areas of the assembly. This procedure looks to be insufficiently productive when verifying large body assemblies, such as the engine compartment, the underbody or the very congested space between dashboard an firewall. Figure A.16, illustrating two assemblies of engine compartments, highlights the difficulty in identifying all potential interferences or dangerous proximity points between parts to be checked. In addition, such verifications should be performed not only in static conditions, but also in any position a component may take during vehicle operation, considering for instance the powertrain motion on its suspension due to road bumps and torque variation.

All digital mock-up systems offer the possibility of automatically looking for points of neighboring parts closer than an assigned set point and sorting out components in the assembly according to their function in the product breakdown structure.

Movable parts (for instance wheels, during their steering motion) can be introduced in the assembly as an envelope body encompassing all potential positions during vehicle operation.

A.4.2 Kinematic Analysis

Kinematic analysis finds important applications in body engineering while designing mechanisms of moving parts, such as doors, hood, trunk lid and other mechanisms as wipers, sliding glasses, locks, outside mirrors, etc. The objectives of this analysis are:

- Validation of the working of the mechanism,
- interference of moving parts that become closer during their motion,
- positions of stops limiting the mechanism motion, and
- evaluation of speeds, accelerations and forces during mechanism operation.

Consider, for example, Fig. A.21 showing the extendable hinge of a lateral sliding rear door of a minivan: The digital mock-up of this subassembly includes the mechanism itself (the extendable hinge), the fixed related components, such as the body side, the sliding rail and the front door, and the movable part, such as the sliding door and its interior covering panel.

One highly useful feature of a digital mock-up is that each part in the assembly is loaded into the model from the company data base at its present status of development; once the model is set up, it is possible to repeat any verification as soon as a new

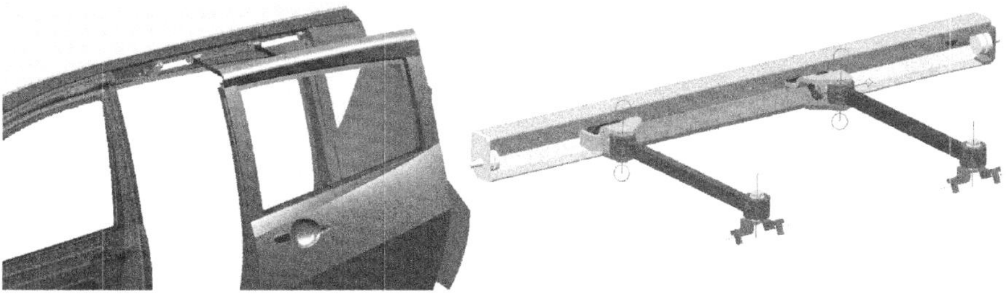

Fig. A.21 Digital mock-up of an extendable hinge of a sliding door (from Morello et al. 2011)

element in any of its parts is added or a modification introduced. If new modifications are needed to allow the operation of mechanisms, they can be introduced in the same working section and checked consequently.

The mechanism is represented by defining the following elements:

- Kinematic couplings between mechanism parts, such as cylindrical, spherical or sliding bearings,
- external shape of moving and fixed parts, and
- extension of the rotary or sliding stroke of the elements of the mechanism.

The digital mock-up allows only kinematic motion without friction and can only verify geometric compliance; nevertheless it can be used to calculate the input functions for a multibody model of the system dynamics, suitable for providing a full evaluation of system performance, including dynamic forces.

A.4.3 Assembling and Disassembling Evaluations

A further design requirement of complex systems (such as the car body) is the compliance of shapes and dimensions of each component with an existing assembling shop and cycle; this compliance is not only related to parts dimensions, but also to the compatibility of these parts with the welding, assembling and transportation fixtures of the production lines. This problem has already been mentioned when referring to check-lists; this section deals with the digital mock-up functions involved in this verification.

Considering, for example, the front part of the engine compartment lower rail: The complete assembly obtained as a result of a number of operations defined by the assembling cycle is shown in Fig. A.22. In the same figure an exploded view of the parts involved in the assembling is also represented: Without entering into the details involved in this set of operations, it is possible to imagine identifying, through a number of trials, possible operations sequences achieving the required result. If different sequences obtain the same result, the most expedient should be selected by taking into account:

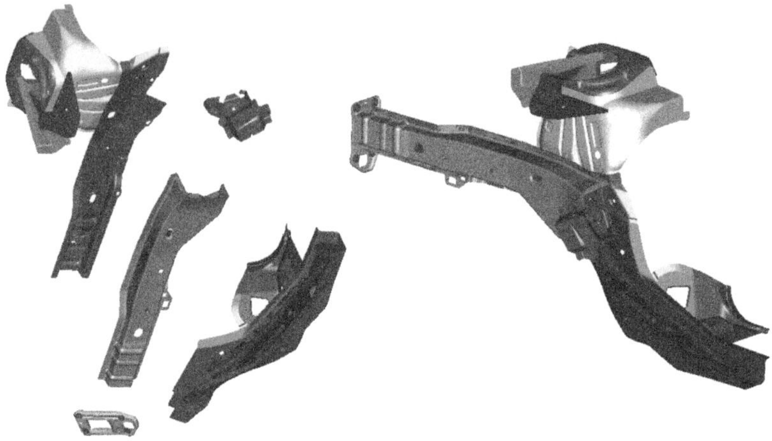

Fig. A.22 Automatic design of a subassembly made by a digital mock-up system, starting from the CAD mathematical model of each part (from Morello et al. 2011)

- The number of loading and unloading operations of parts on the welding and assembling fixture,
- the final dimensional tolerance that can be obtained, and
- the possible deformations on the assembly due to thermal expansion caused by the spot welding sequence.

The digital mock-up proves to be a very useful tool to initiate working team discussions or topical verifications. Many significant mistakes can be avoided during a free preliminary multidisciplinary discussion. The examples introduced should enable the usefulness of the digital mock-up in its main applications to be appreciated.

These virtual analyses become increasingly accurate as the digital tools improve; nevertheless actual tests on physical prototypes must be performed, at least for final product validation; their cost is however lower due to the increasing number of errors corrected at the virtual stage.

A.4.4 Virtual Reality

This final section highlights the role of Virtual Reality (VR) in enhancing engineering design, describing this tool in conjunction with digital mock-up since, in fact, virtual reality may be thought as the most sophisticated tool to represent and navigate a large detailed assembly. This technique allows one or more user to interact simultaneously in a simulation environment controlled by computers; this simulation can achieve greater realism than that available while interacting with a model by using a conventional monitor and a mouse, making operations more intuitive, effective and quick.

Due to the fact that these systems are still evolving, some information regarding the hardware is also included, while realizing that new developments will make virtual reality simpler, less expensive and, therefore, more widely applied. Many interface devices exist that provide the illusion of seeing, touching and manipulating virtual objects; their common characteristic is granting people, working with this simulation, the possibility of viewing from any direction, providing high levels of visual perception and sense of involvement; visual feeling can be enhanced by hearing and touch.

Virtual reality is a technology that can be used to enhance natural human capabilities or exploit those usual capabilities that appear only in front of a physical prototype of the object under examination. The ultimate target is, again, to reduce development time and costs, bearing in mind that traditional interfaces such as keyboard, mouse and monitor effectively force engineers to work within a limited and unnatural bi-dimensional space, requiring a high capacity for abstraction and imagination which not everyone has.

An ideal virtual reality system should possess at least the following features:

- High resolution stereoscopic visualization; human eyes can distinguish punctiform objects seen under an angle of $1'$ (i.e. two dots with mutual distance of 0.58 mm, at 2 m from the eyes), and give a full representation of any object in the semi-space containing the optical axis, and
- update of objects in motion with respect to the point of view, in a time lower than 10 ms and with a time lag of less than 20 ms. When images are obtained by using multiple projectors, outputs synchronization must be perfect.

In addition, systems possessing also non-optical outputs should provide:

- Full stereo sound output of both environment and object noise, with a frequency response in the range $16 \div 25{,}000$ Hz, dynamic range of at least 90 dB and background noise lower than 30 dB, and
- tactile and force feed-back, as functions of object displacement suitable for human motion capability.

The first virtual reality experiments were performed with flight simulators, developed by the Massachusetts Institute of Technology in the 1960s; a crucial contribution to this technology came in the 1980s from Silicon Graphics, a company leading the production of powerful and relatively inexpensive computers. The first applications mostly addressed the training of aircraft pilots. Virtual reality was thus developed to better visualize objects simulated by computers. Further developments followed, allowing to simulate the physical properties of a vehicle to verify different functions, such as internal and external visibility, access to the passenger compartment, space availability while interacting with the vehicle, etc. In the case of style development, a more complete representation of an object, in motion or seen by a moving observer, can be obtained.

Figure A.23 can offer an idea of the performance currently available for representing the outside of a car, starting from a CAD surface mathematical model,

Fig. A.23 Virtual reality model of the exterior shape of a car obtained by CAD surface mathematical models (from Morello et al. 2011)

in spite of the relatively poor quality of the black and white picture and the fact that the results of VR cannot be represented on paper.

Not only a high resolution representation, but also the interaction with the model are required, allowing the operator to manipulate the objects in the virtual environment. The interaction of the user with the model is possible with a full-immersive environment or by improvements of the current practice. A simulation environment is considered to be full-immersive if the operator is completely insulated from the real world, independently of the technology applied to the simulation.

An immersive simulation can be, on the other hand, more or less intrusive depending on the extent the simulation tools available to the user feel natural. For example, a 3 D visualization helmet could be immersive and intrusive at the same time. Another issue is the so-called presence level of the user in the virtual environment. The presence level is high if the user can also see his body during his experience in the virtual environment; on the contrary it is low if his body and those of other operators can only be shown by virtual representation.

The features of an immersive simulation include the following:

- The virtual environment can be observed in a natural way, offering realistic views when the head is rotated;
- The view is stereoscopic and allows perception of distances between virtual objects;
- Virtual objects are perceived in their natural size;
- There are devices addressed to the manipulation and control of virtual objects.

The illusion of living in the virtual environment may be enhanced by sound.

Virtual environment

The virtual environment usually consists of a realistic representation of virtual objects (a vehicle and its components) in a related outside scenario (i.e. show room, production shop, repair shop, etc.). Geometrical accuracy is very important, as is the accurate definition of colors, textures, light spots and their correct reflection on surfaces, according to their gloss.

Texture is, in this case, the repetition of an ornamental drawing on a surface, that can be deformed, e.g. representing a decorated fabric upholstery covering seat stuffing or the grain of a dashboard surface, simulating leather. This can be better explained by Fig. A.24: The simulation system is supplied with an elementary drawing of the cloth decorative pattern. The simulation software repeats this drawing indefinitely, covering an assigned surface (in this case, the interior side panel of the door), taking into account the deformation of the cloth on stuffing, that is simulated by a full deformable, non-extendable surface.

Virtual objects can be created by a dedicated 3 D modeling tool or can be imported from an existing CAS or CAD model. In any case, the description of these models not only includes the mathematical equations describing their surface, but also information about the aesthetic nature of their visible surface (influencing how light is reflected or diffused) and, sometimes, about forces (mass forces, resistance, elasticity) related to their displacement.

The virtual environment controls object motion with procedures considering dynamic rendering and dynamic simulation; while parts are displaced, potential collisions are controlled by dedicated software.

The environment status is influenced by inputs that are usually the positions of head and hands of operators, while outputs are video, audio and feed-back force. Some dynamic objects can be defined without any space constriction, while others are physically constrained to move within a set of boundary surfaces. For instance an observer's head evaluating the external visibility of the car, from the driver's seat, is constrained to stay within the passenger compartments. When the head is moving

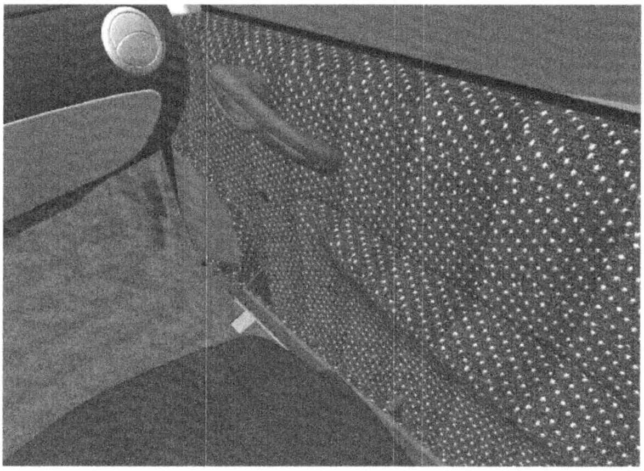

Fig. A.24 Visualization software multiplies indefinitely an elementary ornamental drawing, taking into account the actual deformation consequent to the adaptation of the cloth on the stuffing (from Morello et al. 2011)

outside the allowed space, at least an error signal should be sent to the operator or, for a perfect simulation, a reaction force applied to his head.

Generally speaking, any object is constrained for displacements and rotations with respect to a local reference frame. These constraints are chosen as process parameters and they should be controlled to stay within allowable limits.

3 D graphics requires an intensive usage of computer memory storage, considering object size, number and real-time refresh. The numerical complexity of the model plays a primary role in determining the refresh speed. To keep this speed at high values, different models of the same part, with different number of details, are generated and contemporarily stored in the computer memory. The operating system selects the most suitable model for the current simulation: detailed models are suitable for close-ups, when few parts are present at the same time in the visible environment, while coarse details are sufficient for viewing from a larger distance, when many parts are to be described. Models change, therefore, with operators motion. In addition, for each exterior surface the quantity of reflected and diffused light must be calculated, as a function of the surface gloss.

Input signals

Input signals regard position and orientation of head and hands of users inside the virtual environment. The user's head position defines the point of view for the virtual environment. This input must be detected at a frequency of 100 Hz at least.

There are four kinds of sensors in use to detect position and orientation of an object in a 3 D space, based upon electromechanical, electromagnetic, acoustic and optic techniques respectively. Electromechanical sensors were the first to be considered; they consist of an articulated arm connected to a helmet worn by the user, leaving adequate freedom to him: The angular displacements of the articulated arm joints allow to calculate the head position and orientation.

Electromagnetic monitoring systems are now more frequently used. A fixed device generates electromagnetic signals that are measured by a set of receivers on the user's head (helmet or stereoscopic glasses). The time delay in receiving the signals is used to calculate the head position and orientation. Acoustic sensors operate in a similar way.

Optical systems are under consideration owing to the performance improvement of processors. Many kinds of sensors are available, from video cameras to optical diodes. Infrared light is particularly suitable because does not interfere with users eyesight.

Position and orientation of the user's hand are used in the virtual environment to input commands to the computer in a more efficient way than with a keyboard or mouse; in particular cases (i.e.: assembling and disassembling simulations, ergonomic evaluation of a vehicle control) these signals are used to move parts of the mathematical model. Particular articulated gloves are used to detect fingers' motion and to apply the correct force feed-back to them.

Interaction with the virtual environment

The correct feeling, received by touching or moving objects, may be reproduced by suitable devices as joysticks with six degrees of freedom (three displacement and three rotations) or interactive gloves. This feeling can be enhanced by including the user's hands in the representation of the virtual environment. The system can detect whether the user's hand can reach and touch any of the virtual objects in the virtual environment. If there is a contact, the system can be programmed to move this part in a consistent direction or to make it integral with the user's hand. If the glove integrates suitable actuators, it is possible to give the user the tactile feeling by the corresponding forces that this object could apply in the real world.

The so called *haptic* technology allows interfacing with the user via the sense of touch, by applying forces, vibrations, and motions to his/her fingers. The word haptic, from the ancient Greek *haptikos*, means pertaining to the sense of touch. This technology contributes to our understanding of how touch and its underlying brain functions work.

An example is shown in Fig. A.25; the user observes the virtual world (in this case the interior of a passenger compartment) through a helmet with binocular display. The seat is a real physical prototype, but the controls are simulated by means of a haptic glove that restores the same feeling the user will have when operating an actual control device. The scene is showing the user while operating a virtual switch on the dashboard, to evaluate ergonomics of this control.

In the future haptic technologies will study a wider range of touch sensations, including the nature of the surface to be touched (smoothness or roughness, temperature, humidity, viscosity, etc.); for the time being, only reaction and inertia forces can be reasonably considered.

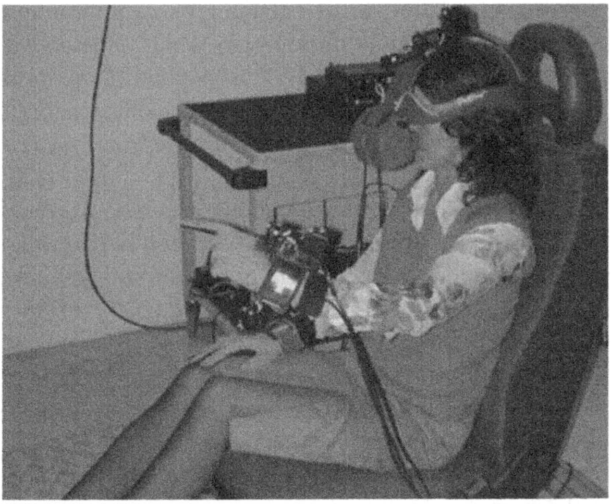

Fig. A.25 The user observes the virtual world (in this case the interior of a passenger compartment) through a helmet with a binocular display. The seat is a real physical prototype; controls are simulated by means of a haptic glove (from Morello et al. 2011)

All haptic devices should obviously be connected to a hand tracking system, to avoid that hands are penetrating into virtual objects, by increasing contact forces significantly. These devices are particularly suitable to investigate the characteristics that are expected of an ideal control or are needed in drive simulators to restore a realistic feeling on the steering wheel.

Visualization devices

Visualization devices may be subdivided into two groups, depending on the technology applied to display the virtual environment: direct vision or projection. The first group includes:

- Head Mounted Displays (HMD), and
- Binocular Omni-Orientation Monitors (BOOM).

With these two systems, the user observes the virtual scenario directly through the internal monitors of the device. An example of HMD is shown in Fig. A.25.

The second group includes the CAVE[2] and the projection backlit walls (Power walls) or tables.

Head mounted displays were the first devices able to allow an immersive experience to the user. They include two miniature displays, like the lenses of a pair of glasses, that simulate the stereoscopic view. A moving tracer on the HMD measures the head position and orientation continuously allowing the computer to calculate consistent views. This device allows more than one user to walk and look around in the virtual environment. The computer processing speed must be high to prevent image slow down or jumps; even if processing speeds dramatically increased, virtual scenes must be simplified to allow the computer to update the images in a realistic way (at least ten times in a second).

BOOM devices contain an optical and display system similar to HMD, but this is contained in a box suspended to an articulated arm, equilibrating its weight. The user looks at the virtual environment through two holes and can explore the space by moving or orienting the device. Suitable sensors control any motion. Devices of both types may be integrated in test rigs, containing also physical parts, such as seats, steering wheel, pedals and shift stick, to study posture comfort, reach of controls, realistic visibility of a car.

Projection based devices consist of special stereoscopic glasses used to observe a semi-transparent screen. The 3 D effect is based upon the fact that each eye sees slightly different images under the perspective they should have in the real world. The human brain interprets these different views to recognize distances and volumes of objects in the environment (stereoscopic view). Stereoscopic glasses separate images perceived by the right and left eye using liquid crystals shutters, synchronized with the actual projection. With suitable infrared pulses sent to the glasses, different projections can be triggered to different eyes.

A typical immersive, non-intrusive projector for virtual reality is the CAVE (Cave Automatic Virtual Environment), a cubic room build up with backlit projection

---

[2] Power Wall is a registered trade mark of BARCO; CAVE is a registered trade mark of Fakespace.

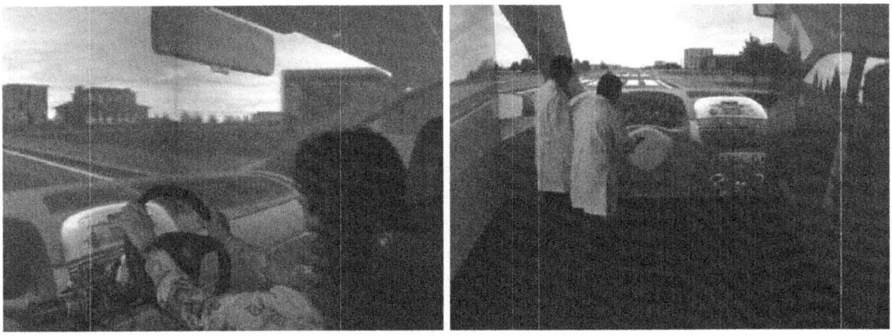

Fig. A.26 Two pictures of a CAVE including also real elements interfacing the user in a driving simulator (from Morello et al. 2011)

screens, where users can move and interact with the virtual environment. Inside this room, many other devices can be used, such as haptic gloves or joysticks, enhancing user's interaction. Images are projected on the walls, on the ceiling and sometimes also on the floor.

Two pictures of a CAVE including also real elements (haptic pedals and steering wheel) interfacing the user in a drive simulator are shown in Fig. A.26.

If several people are present in a CAVE, or other immersive projection systems, part of them will see the virtual environment from the same point of view as a reference user, irrespective of their actual position. The other problem is that they can cast their shadow on the screen obstructing the view of other users. Also in this case haptic gloves or joysticks can be used.

The main advantages of this device are:

- An ample field of view and a high resolution and
- a work environment accessible to more than one user, with some limitation.

Obvious disadvantages are the space and cost investment required. Power walls offer a cheaper compromise by limiting the projection to a single wall of large dimensions with multiple projectors. Backlit projection tables are suitable to observe spatial phenomena on an object of limited dimensions as a component or a scale model of a vehicle. An example of a virtual wind tunnel, where a computer, solving a mathematical model of the air flow around a vehicle model, can display the shape of the aerodynamic wake, is shown in Fig. A.27. The projection table can be rotated to simulate yaw or pitch to observe their effect on the wake.

Applications to product development

The diffusion of virtual reality has launched a process where many engineering applications of mathematical models convert their outputs, from graphs or printouts, to 3 D models that can be explored and discovered by interacting with them more directly. The success of this approach is due to the higher reproduction fidelity of 3 D outputs that can be observed from any point of view, discovering critical issues that

Fig. A.27 Virtual wind tunnel for scale model of a vehicle (from Morello et al. 2011)

sometimes may be hidden in a conventional printout, representing a plane section made in the wrong place.

For the time being and in the short term, existing mathematical models of virtual reality must be adapted, by developing suitable hardware architectures and visualization software, to enable the simulation and visualization environments to interact correctly at the appropriate synchronization speed. For this reason the applications more often considered in product development regard the interactive visualization of the car exterior and interior for style evaluation, and the digital mock-up exploration for the purposes described previously.

Successful applications can be also mentioned regarding exploring a digital mock-up of virtually crashed or virtually stressed cars; the advantages of virtual reality enable all critical local issues to be grasped quickly. The advantages are demonstrated in practice; in style development, for example, virtual reality applications allow a drastic reduction of the number of physical models required: On occasions they have been reduced from 10 to 2, needed for the final evaluation of two alternatives.

The recent diffusion of virtual reality in product development and research is due to the fact that the best way to evaluate and improve an industrial product is to visualize and use it. This principle also applies to the analysis of performance made by engineers and to perception, acceptance and usability evaluations made by the potential customer.

Virtual reality applications are also starting to spread in multi-sensorial analyses, such as acoustic comfort and human reaction to vehicle controls and displays.

Appendix B
Basics on Vehicle Dynamics

B.1 Mathematical Modelling

The automotive market is increasingly competitive and subjected to an increasing number of standards and rules, primarily regarding safety and environmental impact, issued by regulating bodies and governments and offers its products to more and more demanding customers. While the tasks designers must face become increasingly complex, the time between the conception of a vehicle and its entering the market, the so called *time to market*, is an essential factor for its commercial success. The traditional approach, based on the construction of prototypes, followed by long experimentation and modification phases, is no longer adequate.

The availability of increasingly powerful computational hardware and software gives designers the possibility of predicting the behavior of motor vehicles with increasing precision by using more and more complex and realistic mathematical models with which they can perform detailed numerical experiments. The goal, at present still far from being achieved, is to reduce the importance of physical testing to a simple verification of what virtual testing has ascertained, so that the whole development process can be faster and less expensive, while yielding vehicles of better and better characteristics.

The *mathematical models* on which numerical experimentation is based consist of a number of equations, whose behavior is similar to that of the physical system it replaces. In the case of dynamic models, such as those used to predict the performance of motor vehicles, the model is usually built from a number of ordinary differential equations[3] (ODE).

The complexity of the model depends on many factors, that represent the first choice the analyst has to make. The model must be complex enough to allow a realistic simulation of the system's characteristics of interest, but no more. The more complex the model, the more data it requires, and the more complicated are the solution and interpretation of results. Today it is possible to built very complex

[3] A dynamic model, or a dynamic system, is a model expressed by one or more differential equations containing derivatives with respect to time.

G. Genta et al., *The Motor Car*, Mechanical Engineering Series,
DOI: 10.1007/978-94-007-8552-6,

models, but overly complex models yield results from which it is difficult to extract useful insights into the behavior of the system.

Before building the model, the analyst must consider accurately what he wants to obtain from it. If the goal is a good physical understanding of the underlying phenomena, without the need for numerically precise results, simple models are best. If, on the contrary, the aim is precise quantitative results, even at the price of more difficult interpretation, the use of complex models becomes unavoidable.

Finally, it is important to take into account the available data at the stage reached by the project: Early in the definition phase, when most data are still not available, it can be useless to use complex models, into which more or less arbitrary estimates of the numerical values must be introduced. Simplified, or synthetic, models are the most suitable for a preliminary analysis. As the design is gradually defined, new features may be introduced into the model, reaching comprehensive and complex models for the final simulations.

The models of a given vehicle often evolve initially toward a greater complexity, from synthetic models to virtual prototypes, to return later to simpler models. Models are useful not only to the designers in defining the vehicle and its components, but also to test engineers in interpreting the results of testing and performing all adjustments. Simplified models can be used on the test track to allow the test engineers to understand the effect of adjustments and reduce the number of tests required, provided they are simple enough to give an immediate idea of the effect of the relevant parameters. Here the final goal is to adjust the virtual prototype on the computer, transferring the results to the physical vehicle and hoping that at the end of this process only a few validation tests are required.

Simplified models that can be integrated in real time on relatively low power hardware are also useful in control systems. A mathematical model of the controlled system (*plant*, in control jargon) may constitute an *observer* (in the sense of the term used in control theory) and be a part of the control architecture.

The aim of the present appendix is to summarize the basics of mathematical modelling of the dynamic behavior of motor vehicles, showing in particular the synthetic models that are useful in the early stages of vehicle design.

It is possible to demonstrate that under a number of simplifying assumptions the dynamic behavior of any vehicle can be studied by three uncoupled models, dealing with its

- longitudinal behavior,
- lateral behavior (often referred to as handling) and
- vertical behavior (often referred to as comfort).

The basic assumptions that allow this are:

- full linearity of the model, and
- existence of a symmetry plane.

The first assumption implies that the vehicle moves with its longitudinal axis not too removed from the tangent to the trajectory and the steering angles are not too large (i.e. the radius of the trajectory is much larger than the wheelbase), conditions that are usually verified in the normal use of the vehicle, but not in conditions approaching the limits. Furthermore, the suspension springs and shock absorbers must be linear, and the tires must have a linear behavior in vertical direction. No such assumption is required for the lateral behavior of the tires.

The second assumption is usually verified, at least approximately: except for a few exceptions (motorcycles with a sidecar, some special industrial vehicles) motor vehicles have a vertical plane of symmetry with respect to their shape. The distribution of their internal elements is usually not exactly symmetrical, but this causes a deviation from symmetry that is not too large.

B.2 Reference Frame

The study of the motion of motor vehicles is usually performed with reference to some reference frames that are more or less standardized. They are (Fig. B.1):

- *Earth-fixed axis system* XYZ. This is a right-hand reference frame fixed on the road. It will be regarded as an inertial frame, even if strictly speaking it is not such as it moves along with the Earth, but the inertial effects due to this motion are so small that they can be neglected in all phenomena studied in motor vehicle dynamics. Axes X and Y lie in a horizontal plane while axis Z is vertical, pointing upwards.[4]
- *Vehicle axis system* xyz. This is a right-hand reference frame fixed to the vehicle's center of mass and moving with it. As already stated, if the vehicle has a symmetry plane, the center of mass is assumed to lie in it. The x-axis lies in the symmetry

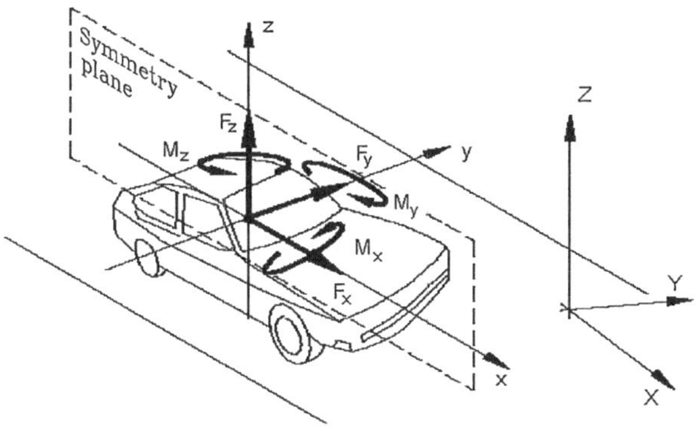

Fig. B.1 Reference frames, forces and moments used for the dynamic study of motor vehicles

[4] Recommendation SAE J670 and ISO/TC 22/SC9 standard state that the Z-axis is vertical and points downwards. In the present text the direction of the Y and Z axes is opposite to that suggested in the mentioned standard.

plane of the vehicle in an almost horizontal direction.[5] The z -axis lies in the symmetry plane, is perpendicular to the x-axis, and points upwards. The y-axis is perpendicular to the other two.[6] The y-axis points to the left of the driver. In the case of vehicles without a symmetry plane, the xz plane is identified by the direction of motion when the wheels are not steered and by a direction perpendicular to the road in the reference position of the vehicle.

B.3 Longitudinal Dynamics

When computing the performance of a vehicle in longitudinal motion (maximum speed, gradeability, fuel consumption, braking, etc.), the vehicle is modelled as a rigid body, or in an even simpler way, as a point mass.

B.3.1 Vehicles as a Point Mass: Resistance to Motion

The presence of suspensions and the compliance of tires are then neglected and the motion is described by a single equation, the equilibrium equation in the longitudinal direction. Assuming that the vehicle moves in the direction of the inertial X-axis, the longitudinal equilibrium equation reduces to

$$m\ddot{X} = \sum_{\forall i} F_{x_i}, \tag{B.1}$$

where F_{x_i} are the various forces acting on the vehicle in the longitudinal direction (aerodynamic drag, rolling resistance, traction, braking forces, etc.).

The rolling resistance of the i-th wheel is assumed, as a first approximation, to be equal to the force pressing the ith wheel against the ground F_{z_i} multiplied by a rolling coefficient f, that depends on many parameters. It is also a function of the speed, and this dependence can be approximated by a quadratic law, so that the rolling resistance can be expressed as

$$R_r = \sum_{\forall i} F_{z_i} \left(f_0 + K V^2 \right). \tag{B.2}$$

[5] The mentioned standard states that the x-axis is contained in the plane of symmetry of the vehicle, is "substantially" horizontal and points forward.

[6] Here again there is a deviation from the mentioned standards.

Assuming that the rolling coefficient is the same for all wheels,[7] it can be brought out from the sum. The sum of the forces perpendicular to the ground on a road with a longitudinal grade angle α, taking into account aerodynamic lift $F_{z_{aer}}$, is

$$\sum_{\forall i} F_{z_i} = \left[mg\cos(\alpha) - F_{z_{aer}}\right] = \left[mg\cos(\alpha) - \frac{1}{2}\rho V_r^2 S C_z\right], \tag{B.3}$$

where the usual expression for the aerodynamic forces has been used. In this expression, ρ is the air density, V_r is the velocity with respect to the air (it is equal to the vehicle speed V when there is no wind, S is a reference surface (usually the area of the cross section of the vehicle) and C_z is a coefficient that depends on the Reynold's and Mach numbers, but is considered as approximately constant. Thus it follows that

$$R_r = \left[mg\cos(\alpha) - \frac{1}{2}\rho V_r^2 S C_z\right](f_0 + K V^2). \tag{B.4}$$

Aerodynamic drag (or, better, the aerodynamic force in the x direction) is expressed by a formula similar to that of force in z direction:

$$R_a = \frac{1}{2}\rho {V_r}^2 S C_x \tag{B.5}$$

where also C_x is a coefficient that depends on the Reynold's and Mach numbers, but is considered as approximately constant

With increasing speed, the importance of the aerodynamic drag grows and at a given value of the speed it becomes more important than rolling resistance. This speed is lower for small cars while for larger vehicles, particularly for trucks at full load, the rolling resistance is the primary form of drag. Another factor is that usually the mass of the vehicle grows with its size more rapidly than the area of its cross section.

If the road is not level, the component of weight acting in a direction parallel to the velocity V, i.e. the grade force

$$R_g = mg\sin(\alpha) \tag{B.6}$$

must be added to the resistance to motion.

The grade force becomes far more important than all other forms of drag even for moderate values of grade. The total resistance to motion, or road load, as it is commonly referred to, can be written in the form

$$R = \left[mg\cos(\alpha) - \frac{1}{2}\rho V^2 S C_z\right](f_0 + K V^2) + \frac{1}{2}\rho V^2 S C_x + mg\sin(\alpha), \tag{B.7}$$

[7] This assumption holds only as a first approximation, since it does not take into account the dependence of f on the driving or braking conditions or other variables.

where, assuming that the air is still, the velocity with respect to air V_r becomes conflated with velocity V.

To highlight its dependence on speed, the road load can be written as

$$R = A + BV^2 + CV^4, \tag{B.8}$$

where

$$A = mg\,[f_0 \cos(\alpha) + \sin(\alpha)],$$

$$B = mgK \cos(\alpha) + \frac{1}{2}\rho S[C_x - C_z f_0],$$

$$C = -\frac{1}{2}\rho SKC_z.$$

The last term in Eq. (B.8) becomes important only at very high speed in the case of vehicles with strong (negative) lift: It is usually neglected except for racing cars.

Since the grade angle of roads open to vehicular traffic is usually not large, it is possible to assume that

$$\cos(\alpha) \approx 1, \quad \sin(\alpha) \approx \tan(\alpha) \approx i,$$

where i is the grade of the road. In this case coefficient B is independent of the grade of the road and

$$A \approx mg(f_0 + i)$$

depends linearly on it. C never depends on the grade.

B.3.2 Power Needed For Motion

The power needed to move at constant speed V is obtained simply by multiplying the road load given by Eq. (B.8) by the value of the velocity

$$P_n = VR = AV + BV^3 + CV^5. \tag{B.9}$$

Motion at constant speed is possible only if the power available at the wheels at least equals the required power. This means that the engine must supply sufficient power, taking into account losses in the transmission as well, and that the road-wheel contact must be able to transmit this power.

The engine drives the wheels through a mechanical transmission whose task is essentially that of reducing the angular velocity of the engine to that required at the wheels. If a reciprocating internal combustion engine is used, the transmission also has the task of uncoupling the engine from the wheels at a stop or at low speed, for which reason the driveline includes a clutch or a torque converter as well.

It is possible, at least as a first approximation, to state a value of the efficiency η_t for any type of driveline. The power available at the wheels is thus

$$P_a = P_e \eta_t . \tag{B.10}$$

Depending on the type of transmission, the efficiency can be considered as a constant (obviously only as a first approximation), or may be computed as a function of several parameters, but only after assessing the working conditions of the driveline and above all of the torque converter, if present.

The equation linking the speed of the engine Ω_e to that of the wheels is simply

$$V = \Omega_e R_e \tau_t , \tag{B.11}$$

where τ_t is the overall gear ratio , defined as the ratio between the speed of the wheels and that of the engine shaft, and is usually smaller than 1. Once the gear ratios of all parts of the transmission are known, the power available at the wheels can be plotted as a function of the speed of the vehicle on the same plot where the power needed for motion at constant speed is reported.

If the curves of the required and available power are plotted on a logarithmic plot, any change of the gear ratio causes a translation of the curves along the V-axis, while a change in the efficiency of the driveline causes a translation of the curve related to the available power along the P-axis (Fig. B.2a. If a continuous transmission (CVT) is present, the position of the latter curve is a function of the gear ratio, and then it is possible to define a zone on the VP-plane where all possible working conditions are included.

B.3.3 Maximum Speed

The maximum speed that can be reached on level road with a given transmission ratio can be found by intersecting the curve of the available power at the wheels with that of the required power on level road. The transmission ratio causing this intersection to occur at the maximum available power allows the vehicle to reach the highest speed that can be attained by a given vehicle-engine combination (curve 1 in Fig. B.2b).

The computation of the maximum speed and of the overall gear ratio τ_t required to reach it is straightforward. By intersecting the required power curve with the horizontal straight line

$$P = P_{a_{max}} = P_{e_{max}} \eta_t ,$$

a fifth degree equation is obtained

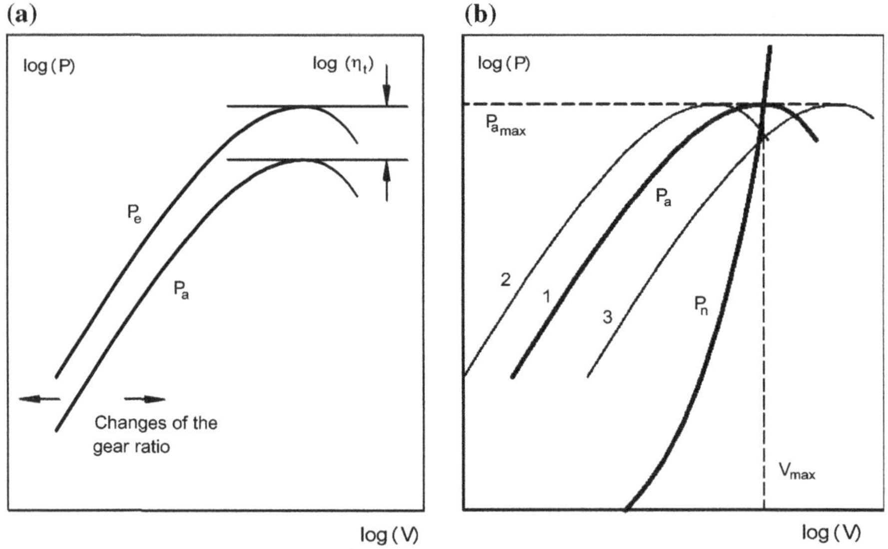

Fig. B.2 (**a**): Curves of maximum engine power and power available at the wheels plotted with logarithmic scales. Changing the efficiency of the transmission and the overall transmission ratio. (**b**): Maximum speed for a vehicle with internal combustion engine

$$AV + BV^3 + CV^5 = P_{e_{max}} \eta_t, \tag{B.12}$$

whose solution directly yields the maximum value of the speed.

If aerodynamic lift is neglected (actually it is sufficient to neglect the contribution to rolling resistance proportional to the square of the speed due to lift), C vanishes and the equation is cubic. Its solution can be obtained in closed form

$$V_{max} = A^* \left(\sqrt[3]{B^* + 1} - \sqrt[3]{B^* - 1} \right), \tag{B.13}$$

where

$$A^* = \sqrt[3]{\frac{P_{e_{max}} \eta_t}{2mgK + \rho S C_X}} = \sqrt[3]{\frac{P_{e_{max}} \eta_t}{2B}},$$

$$B^* = \sqrt{1 + \frac{8m^3 g^3 f_0^3}{27 P_{m_{max}}^2 \eta_t^2 (2mgK + \rho S C_X)}} = \sqrt{1 + \frac{4A^3}{27 P_{m_{max}}^2 \eta_t^2 B}}.$$

Once the maximum speed has been obtained, the gear ratio allowing the vehicle to reach it is

$$\tau_t = \frac{V_{max}}{R_e (\Omega_e)_{P_{max}}}, \tag{B.14}$$

where $(\Omega_e)_{P_{max}}$ is the engine speed at which the peak power is obtained.

If the transmission is of the mechanical type, the overall gear ratio is the product of the gear ratio at the gearbox τ_g (in the relevant gear) and that of the final drive τ_f

$$\tau_t = \tau_g \tau_f .$$

The transmission ratio of the gearbox, which in top gear is usually close to 1, can be stated and consequently the gear ratio τ_f at the final drive can be computed.

Note that this procedure is based on the assumption that the intersection in Fig. B.2 occurs at the peak of the engine power curve. This can, however, occur in only one given condition, since not only the load, but also the rolling resistance coefficient and even the air density, affect the road load curve. Air density also affects the engine power curve. If the intersection occurs in the descending branch of the curve (situation 2 in Fig. B.2) the vehicle is said to be "undergeared", i.e., the overall transmission ratio is "too short". Conversely, if the intersection occurs in the ascending branch of the curve (situation 3 in Fig. B.2), the vehicle is "overgeared" and the overall transmission ratio is "too long".

There are thus two ways of choosing the top gear ratio: One has already been stated, namely a "fast" gear ratio, chosen in order to reach the maximum speed. A different approach is to use a longer overgeared ratio, with the goal of reducing fuel consumption (see below). In practical terms, this trade-off is typical of five or six speed transmissions: Either the maximum speed is reached in fifth (sixth) gear or in fourth (fifth) gear, the longest one being an overdrive gear.

This strategy works only in the case of vehicles with high power/weight ratio: In low powered vehicles, this "economy" gear would be difficult to use since any increase of the required power, e.g. due to a slight grade, headwind, etc., would compel a shift to a shorter gear. Undergearing may be a necessity in this case.

B.3.4 Gradeability

The maximum grade that can be managed with a given gear ratio may be obtained by plotting the curves of the required power at various values of the slope and looking for the curve that is tangent to the curve of the available power (Fig. B.3). The slope so obtained is, however, only a theoretical result, since it can be managed only at a single value of the speed: If the vehicle travels at a higher speed, it slows down because the power is not sufficient, but if its speed is reduced the power is insufficient and the vehicle slows down further: The condition is therefore unstable and the vehicle stops.

To be able to manage a specified slope safely, the curve of the available power must be above that of the required power in a whole range of speeds, starting from a value low enough to assure that starting on that slope is possible. To choose a value of the gear ratio of the bottom gear allowing the vehicle to start on a given grade,

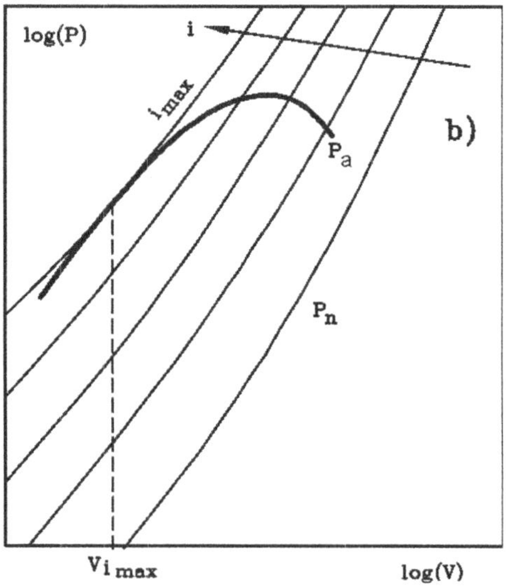

Fig. B.3 Maximum slope for a vehicle with internal combustion engine

it is possible to state a reference speed and to compute the gear ratio in such a way that at that speed the P_a and P_n curves intersect.

As the vehicle is moving at low speed, only the first term of the required power curve needs to be accounted for. As the power developed by the engine can be written in the form

$$P_e = M_e \Omega_e = M_e \frac{V}{R_e \tau_g \tau_f}, \tag{B.15}$$

where M_e is the engine torque, the equilibrium condition allows the overall gear ratio to be computed as

$$\tau_t = \frac{M_e \eta_t}{R_e m g [f_0 \cos(\alpha) + \sin(\alpha)]}. \tag{B.16}$$

The value of the engine torque to be introduced into Eq. (B.16) can be the maximum torque available at the minimum engine speed, possibly multiplied by a number smaller than 1 for safety. The mass of the vehicle must be that at full load, including the maximum trailer mass the vehicle is allowed to tow. For the grade, values of 25 % or even 33 % for road vehicles can be considered, but it must be kept in mind that in some cases, as in ferry ramps or private garage ramps, very steep grades may be encountered.

Another consideration in the choice of the gear ratio for the bottom gear is to assure a regular working of the engine at a speed chosen so as to avoid a prolonged use of the clutch in low speed driving. A reference value may be 6 or 8 km/h. Both criteria must be satisfied. Once the ratios of the bottom and top gears have been chosen, the intermediate ones can be stated using different criteria. The simplest is to set them in geometric sequence, i.e., stating that the ratios between two subsequent gear ratios are all equal. Operating in this way, the available power curves on the $P(V)$ logarithmic plot are all equispaced.

B.3.5 Fuel Consumption at Constant Speed

The energy needed to travel at constant speed can be immediately computed by multiplying the power required for constant speed driving by the time

$$e = P_n t = \frac{P_n d}{V}, \tag{B.17}$$

where d is the distance travelled. Note that Eq. (B.17) gives the energy required at the wheels: To obtain the energy actually required, it must be divided by the various efficiencies (transmission, engine, etc.).

If the efficiency of the engine η_e and the thermal value H of the fuel are known, the fuel consumption can be computed. Introducing the expression derived from Eq. (B.7) for the total road load into the expression for the power, the fuel consumption per unit distance Q is

$$Q = \frac{A + BV^2 + CV^4}{\eta_t \eta_e H \rho_f}, \tag{B.18}$$

where ρ_f is the density of the fuel, introduced to obtain the consumption in terms of volume of fuel per unit of distance. In S.I. units it is measured in m^3/m, while liters per 100 km is a more practical, although not consistent, unit. Often the reciprocal of Q, expressed in km per liter or miles per gallon, is used.

To compute the consumption Q, the simplest procedure is to obtain the power required at the wheels as a function of the speed and hence to compute the power the engine must supply to travel at constant speed

$$P_e = \frac{P_n}{\eta_t}.$$

Once the transmission ratio has been stated, the rotational speed of the engine is known and thus the working point on the map of the engine is located. From it the efficiency η_e or, which is the same, the specific fuel consumption

$$q = \frac{H}{\eta_e}$$

is obtained and the fuel consumption can be computed as

$$Q = \frac{q P_n}{\eta_t V \rho_f}. \tag{B.19}$$

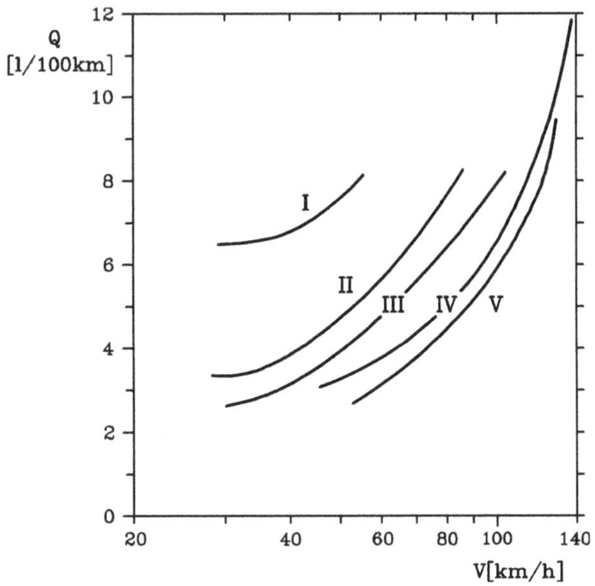

Fig. B.4 Fuel consumption with different gear ratios at constant speed on level road. Passenger vehicle with a five-speed gearbox

The curves $Q(V)$ are of the type shown in Fig. B.4. They usually have a minimum at low speed, obtained in conditions in which the engine works at low power with low efficiency. Since the conditions in which the engine works depend on the overall transmission ratio, the fuel consumption is also largely influenced by the value of the gear ratio. Usually the longer the ratio, the lower the consumption, as a "long" ratio allows the engine to be used at low speed in conditions which are close to the maximum power, where the specific fuel consumption is low.

B.3.6 Acceleration

If the curve of the required power lies, at a certain speed, below that of the power available at the wheels, the difference $P_a - P_n$ between the two is the power available to accelerate the vehicle .

Consider a vehicle with a mechanical transmission with a number of different gear ratios. During acceleration a number of rotating elements (wheels, transmission, the engine itself) must increase their angular velocity, and it is expedient to write an equation linking the engine power with the kinetic energy $\mathcal{T}$ of the vehicle

$$\eta_t P_e - P_n = \frac{d\mathcal{T}}{dt} \,. \tag{B.20}$$

The transmission efficiency should not be included for the part or the engine power needed to accelerate the engine, but the error so introduced is usually negligible.

Once the transmission ratio has been chosen, Eq. (B.11) gives the relationship between the speed of the vehicle and the rotational speed of the engine. Similar relationships may be used for the other rotating elements that must be accelerated when the vehicle speeds up.

The kinetic energy of the vehicle can thus be expressed as

$$\mathcal{T} = \frac{1}{2}mV^2 + \frac{1}{2}\sum_{\forall i} J_i\Omega_i^2 = \frac{1}{2}m_eV^2, \tag{B.21}$$

where the sum extends to all rotating elements which must be accelerated when the vehicle speeds up. The term m_e is the equivalent or apparent mass of the vehicle, i.e., the mass of an object that, when moving at the same speed as the vehicle, has the same total kinetic energy. Usually it is written in the form

$$m_e = m + \frac{J_w}{R_e^2} + \frac{J_t}{R_e^2\tau_f^2} + \frac{J_e}{R_e^2\tau_f^2\tau_g^2}, \tag{B.22}$$

where J_w is the total moment of inertia of the wheels, which are assumed to have the same radius and hence to rotate at the same speed, and of all elements rotating at their speed, J_t is the moment of inertia of the propeller shaft and of all elements of the transmission, and J_e is the moment of inertia of the engine, the clutch and all the elements rotating at speed Ω_e.

To account for the fact that the engine is accelerated directly, at least in an approximate way, the last term is sometimes multiplied by η_t. The modifications to Eq. (B.22) to take the presence of different wheels on different axles into account are obvious.

Of the three latter terms, the first is usually small, the second negligible, while the third may become important, particularly in low gear. As only the last term depends on the transmission ratio at the gearbox, the equivalent mass can be written in the form

$$m_e = F + \frac{G}{\tau_g^2}, \tag{B.23}$$

where

$$F = m + \frac{J_w}{R_e^2} + \frac{J_t}{R_e^2\tau_f^2}, \quad G = \frac{J_e}{R_e^2\tau_f^2}$$

or, possibly,

$$G = \frac{J_e\eta_t}{R_e^2\tau_f^2}.$$

As the equivalent mass is a constant, once the gear ratio has been chosen, Eq. (B.20) yields

$$\eta_t P_e - P_n = m_e V \frac{dV}{dt}. \tag{B.24}$$

From Eq. (B.24), the maximum acceleration the vehicle is capable of at various speeds is immediately obtained

$$\left(\frac{dV}{dt}\right)_{max} = \frac{\eta_t P_e - P_n}{m_e V}, \tag{B.25}$$

where the engine power P_e is the maximum power the engine can deliver at the speed Ω_e, corresponding to speed V.

The minimum time needed to accelerate from speed V_1 to speed V_2 can be computed by separating the variables in Eq. (B.25)

$$dt = \frac{m_e V dV}{\eta_t P_e - P_n} \tag{B.26}$$

and integrating

$$T_{V_1 \to V_2} = \int_{V_1}^{V_2} \frac{m_e}{\eta_t P_e - P_n} V dV. \tag{B.27}$$

The integral must be performed separately for each velocity range in which the equivalent mass is constant, i.e. the gearbox works with a fixed transmission ratio. Although it is possible to integrate Eq. (B.27) analytically if the maximum power curve is a polynomial, numerical integration is usually performed.

A graphical interpretation of the integration is shown in Fig. B.5 : The area under the curve

$$\frac{V m_e}{\eta_t P_e - P_n} = \frac{1}{a}$$

versus V is the time required for the acceleration.

The speeds at which gear shifting must occur to minimize the acceleration time are readily identified on the plot $1/a(V)$. Since the area under the curve is the acceleration time or the time to speed, the area must be minimized and gears must be shifted at the intersection of the various curves. If they do not intersect, the shorter gear must be used up to the maximum engine speed.

B.3.7 Braking

The study of braking on straight road is performed using Eq. (B.1). Apart from cases in which the vehicle is slowed by the braking effect of the engine, which can dissipate a non-negligible power, and by regenerative braking in electric and hybrid vehicles, braking is performed in all modern vehicles on all wheels. Subscript i thus extends to

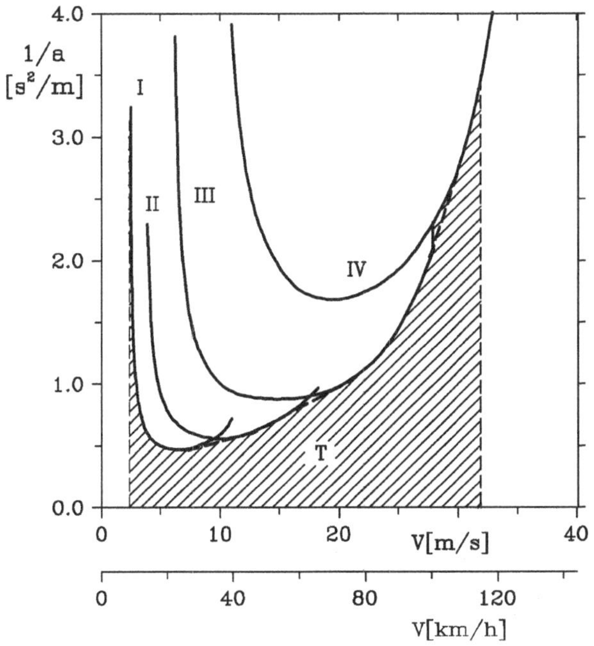

Fig. B.5 Function $1/a(V)$ showing the optimum speeds for gear shifting. The hatched area is the time to speed

all wheels or, when thinking in terms of axles, as is usual for motion in symmetrical conditions, on all axles.

Ideal braking can be defined as the condition in which all wheels brake with the same longitudinal force coefficient μ_x, defined as the ratio between the longitudinal force a wheel is exerting and the forcing loading it on the road.

The study of braking forces the vehicle can exert follows the same scheme seen in the previous section, the only obvious difference being that braking forces, like the corresponding longitudinal force coefficients and the longitudinal slip, are negative. The acceleration can thus be computed as

$$\ddot{X} = \frac{dV}{dt} = \frac{1}{m} \sum_{\forall i} \mu_{x_i} F_{z_i}, \tag{B.28}$$

remembering that when braking both the longitudinal force coefficient and the acceleration are negative.

The longitudinal equation of motion of the vehicle is thus

$$\frac{dV}{dt} = \frac{\sum_{\forall i} \mu_{x_i} F_{z_i} - \frac{1}{2}\rho V^2 S C_X - f \sum_{\forall i} F_{z_i} - mg \sin(\alpha)}{m}, \tag{B.29}$$

where m is the actual mass of the vehicle and not the equivalent mass, and α is positive for uphill grades. The rotating parts of the vehicle are slowed down directly by the brakes, and hence do not enter into the evaluation of the forces exchanged between vehicle and road. They must be accounted for when assessing the required braking power of the brakes and the energy that must be dissipated.

Aerodynamic drag and rolling resistance can be neglected in a simplified study of braking, since they are usually far smaller than braking forces. Also, rolling resistance can be considered as causing a braking moment on the wheel more than a direct braking force on the ground.

Since in ideal braking all force coefficients μ_{x_i} are assumed to be equal, the deceleration is

$$\frac{dV}{dt} = \mu_x \left[g\cos(\alpha) - \frac{1}{2m}\rho V^2 S C_Z \right] - g\sin(\alpha). \tag{B.30}$$

On level road, for a vehicle with no aerodynamic lift, Eq. (B.30) reduces to

$$\frac{dV}{dt} = \mu_x g. \tag{B.31}$$

The maximum deceleration in ideal conditions can be obtained by introducing the maximum negative value of μ_x into Eq. (B.30) or (B.31). The assumption of ideal braking implies that the braking torques applied on the various wheels are proportional to the forces F_z, if the radii of the wheels are all equal. This may occur in only one condition, unless some sophisticated control device is implemented to allow braking in ideal conditions.

In general, it is possible to define an efficiency of braking η_b as the ratio between the acceleration obtained in actual conditions and that occurring in ideal conditions, so that the actual deceleration on level road is

$$\frac{dV}{dt} = \eta_b \mu_x g. \tag{B.32}$$

The efficiency of braking η_b is obviously smaller than 1 and depends on many factors, like the road conditions and the force the driver exerts on the brakes.

The brakes cannot dissipate the kinetic energy of the vehicle directly; they usually work as a heat sink, storing some of the energy in the form of thermal energy and dissipating it in due time. Care must obviously be exerted to design the brakes in such a way that they can store the required energy without reaching excessively high temperatures and so that adequate ventilation for cooling is ensured. The average value of the braking power must, at any rate, be lower than the thermal power the brakes can dissipate.

B.4 Lateral Dynamics (Handling)

To steer a wheeled vehicle the driver operates the steering wheel causing some wheels to work with a sideslip and to generate lateral forces. These forces cause a change of attitude of the vehicle (the attitude or sideslip angle β is defined as the angle between the longitudinal direction of the vehicle and the velocity vector, see Fig. B.6) and then a sideslip of all wheels: The resulting forces bend the trajectory. However, the linearity of the behavior of the tires, at least for small sideslip angles, the high value of the cornering stiffness and the short time delay with which the wheels respond to changes in the sideslip and camber angles, give the driver the impression of a kinematic steering, i.e. that the wheels are in pure rolling, without any sideslip, and the trajectory seems to be determined by the directions of the midplanes of the wheels.

This impression has influenced the study of the handling of wheeled vehicles for a long time, originating the very concept of kinematic steering and in a sense hiding the true meaning of the relevant phenomena. Actually, the concept of kinematic steering was applied first to horse drawn wagons: Ackermann patented in 1818 the idea that the lines perpendicular to the midplanes of the wheels passing through the contact points on the ground should converge in the instant center of rotation of the vehicle.[8] This concept worked well in connection with carriage wheels provided with steel tires and later was applied to motor vehicles. Only in the 1930s did some experiments show the importance of the sideslip angle in generating lateral forces and this concept was finally formalized by Olley in 1937.

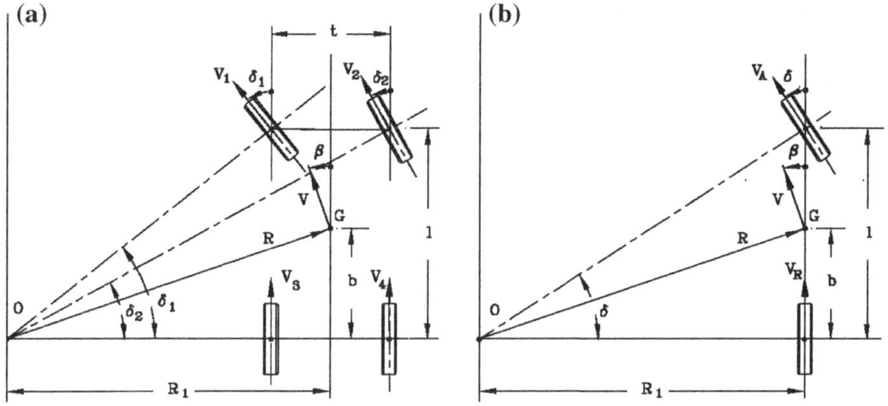

Fig. B.6 Kinematic steering of a four-wheeled and a two-wheeled vehicle

[8] Erasmus Darwin had actually this idea in 1758 but the one who applied it first was a German carriage builder who, in 1818, had his agent in England, Rudolph Ackermann, patent what is now called the Ackermann steering geometry.

The impressions received by the driver are in good accordance with the kinematic approach, at least for all the linear part of the behavior of the tire. When high values of the sideslip angles are reached, the average driver has the impression of losing control of the vehicle, much more so if this occurs abruptly. This impression is confirmed by the fact that in normal road conditions, particularly if radial tires are used, the sideslip angles become large only when approaching the limit lateral forces.

B.4.1 Kinematic Steering

Low speed or kinematic steering is, as already stated, defined as the motion of a wheeled vehicle determined by pure rolling of the wheels. The velocities of the centres of all the wheels lie in their midplane or, in other words, the sideslip angles α_i are vanishingly small. In this condition, the wheels cannot exert any cornering force to balance the centrifugal force due to the curvature of the path. Kinematic steering is thus possible only if the velocity is vanishingly small.

Consider a vehicle with four wheels, two of which can steer (Fig. B.6). The relationship that must be verified to allow kinematic steering is easily found, by imposing that the perpendiculars to the midplanes of the front wheels meet the ones of the rear wheels in the same point:

$$\tan(\delta_1) = \frac{l}{R_1 - \frac{t}{2}}, \qquad \tan(\delta_2) = \frac{l}{R_1 + \frac{t}{2}}. \tag{B.33}$$

By eliminating R_1 between the two equations, a direct relationship between δ_1 and δ_2 is readily found:

$$\cot(\delta_1) - \cot(\delta_2) = \frac{t}{l}. \tag{B.34}$$

A device allowing steering of wheels in exact compliance with Eq. (B.34), is usually referred to as *Ackerman steering* or *Ackerman geometry*. No actual steering mechanism allows one to follow such a law exactly and a *steering error* $\Delta\delta_2$, defined as the difference between the actual value of δ_2 and that obtained from Eq. (B.34) can be obtained as a function of δ_1.

The radius of the trajectory of the centre of mass of the vehicle is:

$$R = \sqrt{b^2 + R_1^2} = \sqrt{b^2 + l^2 \cot^2(\delta)}, \tag{B.35}$$

where δ is the steering angle of the equivalent two-wheeled vehicle (Fig. B.6b), also called *monotrace* model. Although it should be computed by averaging the cotangents of the angles of the two wheels, it is very close to the direct average of the angles.

In case the radius of the trajectory is large if compared to the wheelbase of the vehicle, Eq. (B.35) reduces to:

$$R \approx l \cot(\delta) \approx \frac{l}{\delta}. \tag{B.36}$$

Equation (B.36) can be rewritten in the form:

$$\frac{1}{R\delta} \approx \frac{1}{l}. \tag{B.37}$$

The expression $1/R\delta$ has an important physical meaning: It is the ratio between the response of the vehicle, in terms of curvature $1/R$ of the trajectory, and the input which causes it. It is therefore a sort of transfer function for the directional control and can be referred to as *trajectory curvature gain* . In kinematic steering conditions it is equal to the reciprocal of the wheelbase.

B.4.2 Ideal Steering

If the speed is not vanishingly small, the wheels must move with suitable sideslip angles to generate cornering forces. A simple evaluation of the steady-state steering of a vehicle in high-speed or dynamic steering conditions may be performed as follows. Consider a rigid vehicle moving on level road with transversal slope angle α_t and neglect the aerodynamic side force. Define a η-axis parallel to the road surface, passing through the centre of mass of the vehicle and intersecting the vertical for the centre of the path, which in steady-state condition is circular (Fig. B.7). Axis η does not coincide with the y axis, except at one particular speed.

Assume that the road is flat and neglect aerodynamic forces. The equilibrium equation in η direction can be written by equating the centrifugal force mV^2/R to the sum of the forces P_η due to the tires

$$\frac{mV^2}{R} = \sum_{\forall i} P_{\eta_i}. \tag{B.38}$$

For a first approximation study, forces P_η may be conflated with the cornering forces F_y of the tires and all wheels may be assumed to work with the same side force coefficient μ_y. Since the last assumption is similar to that seen for braking in ideal conditions, this approach will be referred to as *ideal steering*. These two assumptions lead to substituting the expression $\sum_{\forall i} P_{\eta_i}$ with $\mu_y F_z$.

Force F_z exerted by the vehicle on the road is

$$F_z = \sum F_{z_i} = mg. \tag{B.39}$$

By introducing Eq. (B.39) into Eq. (B.38) the ratio between the lateral acceleration and the gravitational acceleration g is

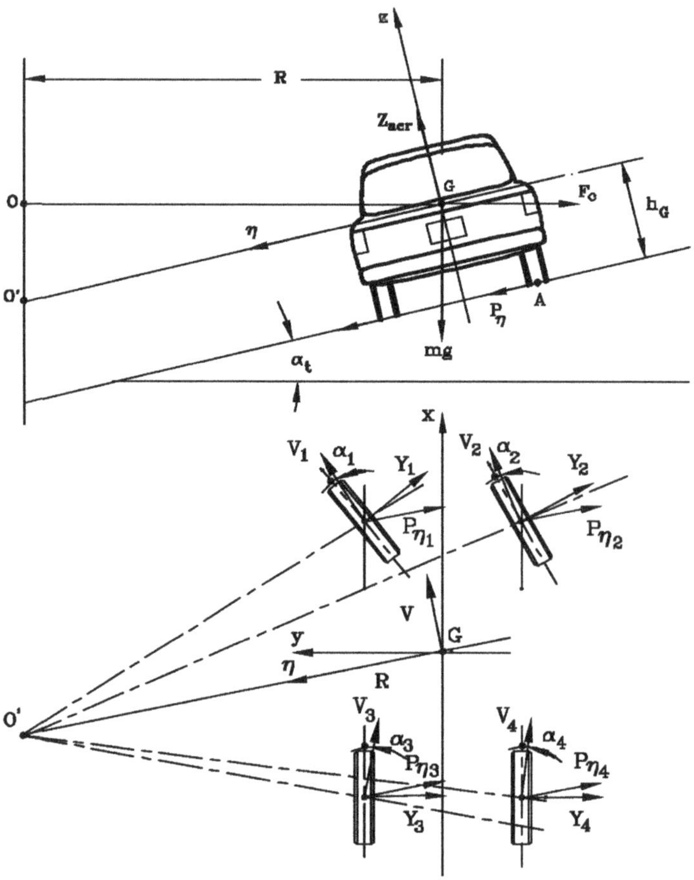

Fig. B.7 Simplified model for dynamic steering

$$\frac{V^2}{Rg} = \mu_y. \tag{B.40}$$

By introducing the maximum value of the side force coefficient μ_{y_p} into Eq. (B.40), it is possible to obtain the maximum value of the lateral acceleration

$$\left(\frac{V^2}{R}\right)_{max} = g\mu_{y_p}. \tag{B.41}$$

The maximum speed at which a bend with radius R can be negotiated is

$$V_{max} = \sqrt{Rg}\sqrt{\mu_{y_p}}. \tag{B.42}$$

The limitation to the maximum lateral acceleration due to the cornering force that the tires can exert is, however, not the only limitation, at least theoretically. Another can come from the danger of rollover occurring if the resultant of forces in the yz plane crosses the road surface outside point A (Fig. B.7).

The moment of the forces applied to the vehicle in the ηz-plane about point A is

$$M_A = -\frac{t}{2}mg + h_G\frac{mV^2}{R}. \tag{B.43}$$

The limit condition for rollover can then be computed by equating moment M_A to zero,

$$\left(\frac{V^2}{R}\right)_{max} = g\frac{t}{2h_G}. \tag{B.44}$$

The rollover condition is identical to the sliding conditions, once ratio $\frac{t}{2h_G}$ has been substituted for μ_{y_p}.

The maximum lateral acceleration is thus

$$\left(\frac{V^2}{R}\right)_{max} = g \min\left\{\mu_{y_p}, \frac{t}{2h_G}\right\}. \tag{B.45}$$

Whether the limit condition first reached is that related to sliding, with subsequent spin out of the vehicle, or related to rolling over depends on the relative magnitude of μ_{y_p} and $\frac{t}{2h_G}$. If the former is smaller than the latter, as often occurs, the vehicle spins out. This condition can be written in the form

$$\mu_{y_p} < \frac{t}{2h_G}.$$

If a negative aerodynamic lift is present, like in racing cars, at high speed strong lateral accelerations are possible. Also a transversal slope of the road may have a large effect on the lateral acceleration.

The value of μ_{y_p} at which rollover may occur is as high as $1.2 \div 1.7$ for sports cars, $1.1 \div 1.6$ for saloon cars, $0.8 \div 1.1$ for pickup and vans and $0.4 \div 0.8$ for heavy and medium trucks. Only in the latter case does rollover seem to be a possibility, at least if the lateral forces acting on the vehicle are restricted to the cornering forces of the tires.

This model is only a rough approximation of the actual situation, as it is based on the assumption that the side force coefficients μ_y of all wheels are equal, implying that all wheels work with the same sideslip angle α. It also ignores the effect of the different directions of the cornering forces of the various wheels, which should be considered as perpendicular to the midplanes of the wheels and not directed along the η axis. The load transfer between the wheels of the same axle and the presence of the suspensions have also been neglected, two other assumptions contributing to the lack of precision of this model.

If the maximum speed at which a circular path can be negotiated is measured in a steering pad test and the value of the lateral force coefficient is computed through Eq. (B.42), a value of μ_{y_p} well below that obtained from tests on the tires is obtained.

The cornering force coefficient obtained in this way is that of the vehicle as a whole, and the difference between its value and that related to the tires gives a measure of how well the vehicle is able to exploit the cornering characteristics of its wheels.

The side force coefficient measured on the whole vehicle also depends on the radius of the path, with a decrease on narrow bends. The majority of industrial and passenger vehicles are able to use only a fraction, from 50 to 80 %, of the potential cornering force of the tires, with higher values found only in sports cars. This reduction of the lateral forces makes the danger of rollover even more remote.

Actually, rolling over in a quasi-static condition is impossible for most vehicles, notwithstanding the fact that rollover actually occurs in many road accidents. Rollover can usually be ascribed to dynamic phenomena in nonstationary conditions or to lateral forces caused by side contacts, e.g. of the wheels with the curb, that rule out the possibility of side slipping while causing far stronger lateral forces to be exerted on the wheels. Moreover, the presence of the suspensions makes rollover a likely outcome of many accidents.

B.4.3 High-Speed Cornering: Simplified Approach

To go beyond the extremely simplified model of ideal steering, the distribution of cornering forces between the axles, the sideslip angle of the vehicle on its path and the sideslip angles of the wheels must be taken into account .

Assume that the vehicle is moving at constant speed on a circular path and that the road is level. Moreover, assume that the radius of the path R is much larger than the wheelbase l and, as a consequence, all sideslip angles are small. The small size of all angles allows the "monotrack" or "bicycle" model to be used . Neglecting aerodynamic forces and aligning torques of the wheels, the forces acting in the xy plane at the tire-road interface are shown in Fig. B.8.

The equilibrium equation in the direction of the y axis is similar to Eq. (B.38), except for the presence of the sideslip and steering angles

$$\frac{mV^2}{R}\cos(\beta) = \sum_{\forall i} F_{x_i}\sin(\delta_i) + \sum_{\forall i} F_{y_i}\cos(\delta_i). \tag{B.46}$$

The equilibrium to rotations about point G can be expressed as

$$\sum_{\forall i} F_{x_i}\sin(\delta_i)x_i + \sum_{\forall i} F_{y_i}\cos(\delta_i)x_i = 0. \tag{B.47}$$

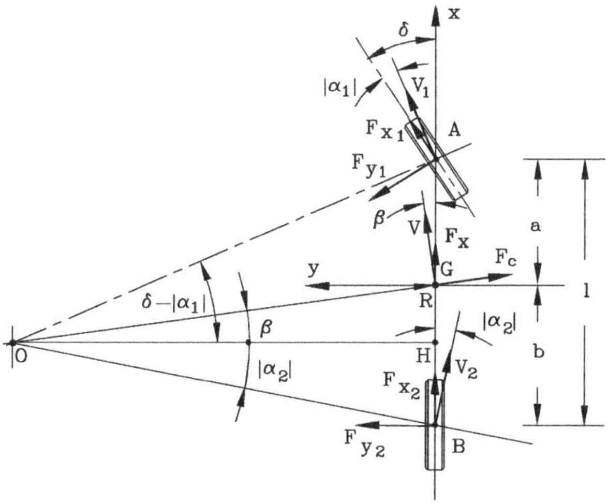

Fig. B.8 Simplified model (monotrack vehicle) for studying the handling of a two-axle vehicle

Since angles β and δ_i are assumed to be small, the terms containing the longitudinal forces of the tires can be neglected and the equilibrium equations reduce to

$$\begin{cases} \sum_{\forall i} F_{y_i} = \frac{mV^2}{R} \\ \sum_{\forall i} F_{y_i} x_i = 0. \end{cases} \tag{B.48}$$

For a two-axle vehicle, they can be immediately solved, yielding

$$F_{y_1} = \frac{mV^2}{R}\frac{b}{l}, \quad F_{y_2} = \frac{mV^2}{R}\frac{a}{l}. \tag{B.49}$$

When the sideslip angles of the wheels are small, the lateral forces (cornering forces) of the axles can be assumed to be proportional to the sideslip angles α_i through their cornering stiffness C_i. As a consequence, it follows that

$$\alpha_i = -\frac{F_{y_i}}{C_1}, \tag{B.50}$$

i.e.

$$\alpha_1 = -\frac{mV^2}{R}\frac{b}{lC_1}, \quad \alpha_2 = -\frac{mV^2}{R}\frac{a}{lC_2}. \tag{B.51}$$

The cornering stiffness of the axles is equal to the cornering stiffness of the wheels multiplied by the number of wheels of the axle.

A relationship between the sideslip and steering angles can be found with simple geometrical considerations from Fig. B.8,

$$\delta - \alpha_1 + \alpha_2 = \frac{l}{R}. \tag{B.52}$$

Introducing the expressions of the sideslip angles into Eq. (B.52), it follows that

$$\delta = \frac{l}{R} + \frac{mV^2}{Rl}\left(\frac{b}{C_1} - \frac{a}{C_2}\right), \tag{B.53}$$

or, in terms of path curvature gain ,

$$\frac{1}{R\delta} = \frac{1}{l}\frac{1}{1 + K_{us}\frac{V^2}{gl}}, \tag{B.54}$$

where

$$K_{us} = \frac{mg}{l^2}\left(\frac{b}{C_1} - \frac{a}{C_2}\right) \tag{B.55}$$

is the so-called *understeer coefficient* or *understeer gradient* of the vehicle. The understeer coefficient is a non-dimensional quantity, and is often expressed in radians.

As already stated, in kinematic conditions

$$\left(\frac{1}{R\delta}\right)_{\text{kin}} = \frac{1}{l}. \tag{B.56}$$

The expression $1 + K_{us}V^2/gl$ can be considered as a correction factor giving the response of the vehicle in dynamic conditions as opposed to kinematic conditions.

From Eq. (B.53) it follows that

$$\delta - \delta_{\text{kin}} = \frac{V^2}{Rg}K_{us}, \tag{B.57}$$

i.e.,

$$K_{us} = \frac{g}{a_y}\left(\delta - \delta_{\text{kin}}\right). \tag{B.58}$$

The understeer factor can thus be interpreted as the difference between the steering angles in kinematic and dynamic conditions divided by the centrifugal acceleration expressed as a multiple of the gravitational acceleration.

Sometimes, instead of the understeer coefficient, a *stability factor*

$$K = \frac{m}{l^2}\left(\frac{b}{C_1} - \frac{a}{C_2}\right) \tag{B.59}$$

is defined.

As a first approximation, K and K_{us} may be considered as constants for a given vehicle and load condition. However, their dependence on speed cannot be neglected for more precise assessments.

It is possible to define a *lateral acceleration* gain as the ratio between the lateral acceleration and the steering input:

$$\frac{V^2}{R\delta} = \frac{V^2}{l} \frac{1}{1 + K_{us}\frac{V^2}{gl}}. \tag{B.60}$$

The sideslip angle can be obtained through simple geometrical considerations, yielding

$$\beta = \frac{b}{R} - \alpha_2. \tag{B.61}$$

A *sideslip angle gain* , expressing the ratio between the sideslip angle and the steering angle can be defined as well. Its value is

$$\frac{\beta}{\delta} = \frac{b}{l}\left(1 - \frac{maV^2}{blC_2}\right)\frac{1}{1 + K_{us}\frac{V^2}{gl}}, \tag{B.62}$$

B.4.4 Definition of Understeer and Oversteer

If $K_{us} = 0$ the value of $1/R\delta$ is constant and equal to the value characterizing kinematic steering; i.e. the response of the vehicle to a steering input is, at any speed, equal to that in kinematic conditions. This does not mean, however, that the vehicle is in kinematic conditions, since the value of the sideslip angle β is not equal to its kinematic value and the values of the sideslip angles of the wheels are not equal to zero.

A vehicle behaving in this way is said to be *neutral-steer* (Fig. B.9a).

If $K_{us} > 0$ the value of $1/R\delta$ decreases with increasing speed. The response of the vehicle is thus smaller than in kinematic conditions and, to maintain a constant radius of the path, the steering angle must be increased as the speed increases. A vehicle behaving in this way is said to be *understeer* .

A quantitative measure of the understeer of a vehicle is given by the *characteristic speed*, defined as the speed at which the steering angle needed to negotiate a turn is equal to twice the Ackerman angle, i.e. the path curvature gain is equal to $1/2l$.

Using the simplified approach outlined above, the characteristic speed is

$$V_{car} = \sqrt{\frac{gl}{K_{us}}} = \sqrt{\frac{1}{K}}. \tag{B.63}$$

If $K_{us} < 0$ the value of $1/R\delta$ increases with increasing speed until, for a speed

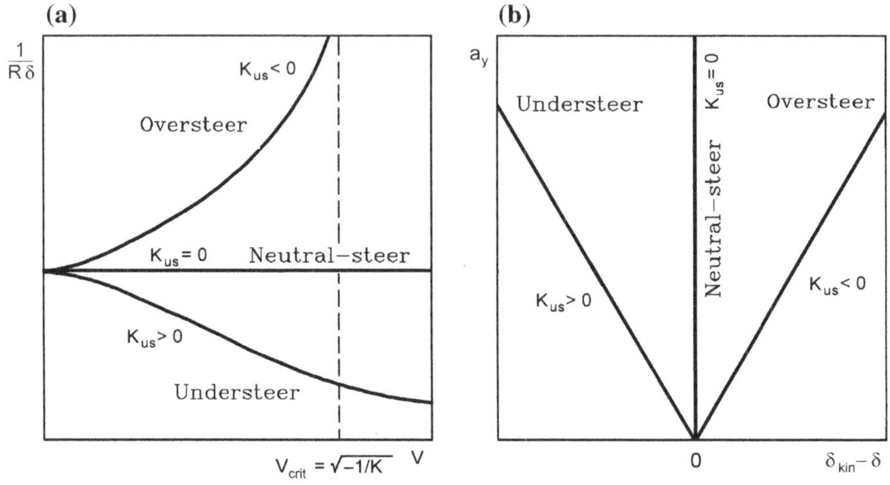

Fig. B.9 Steady state response to a steering input. Plot of the path curvature gain as a function of speed (**a**) and handling diagram (**b**) for an oversteer, an understeer and a neutral steer vehicle. The understeer factor is assumed to be independent of speed

$$V_{crit} = \sqrt{-\frac{gl}{K_{us}}} = \sqrt{-\frac{1}{K}} \tag{B.64}$$

the response tends to infinity, i.e., the system develops an unstable behavior.

A vehicle behaving in this way is said to be *oversteer* , and the speed given by Eq. (B.64) is called the *critical speed.* The critical speed of any oversteer vehicle must be well above the maximum speed it can reach, at least in normal road conditions.

Instead of plotting the path curvature gain as a function of the speed, it is possible to plot the *handling diagram*, i.e., the plot of the lateral acceleration a_y as a function of $\delta_{\text{kin}} - \delta$ (Fig. B.9b). If the vehicle is neutral steer, the plot is a vertical straight line, if it is oversteer it is a straight line sloping to the right, while in case of an understeer vehicle it slopes to the left.

The value of β, or better, of β/δ, decreases with the speed from the kinematic value up to the speed

$$(V)_{\beta=0} = \sqrt{\frac{blC_2}{am}} \tag{B.65}$$

at which it vanishes. At higher speed it becomes negative, tending to infinity when approaching the critical speed for oversteer vehicles and tending to

$$\frac{aC_1}{aC_1 - bC_2}$$

when the speed tends to infinity in the case of understeer vehicles.

B.4.5 High-Speed Cornering

The study of handling seen in the previous sections was based on the assumption of steady-state operation. Moreover, only the cornering forces acting on the tires were considered.

A simple mathematical model for the handling behavior of a rigid vehicle that overcomes the above limitations can, however, be built. The vehicle is considered as a rigid body moving on a surface, and thus a model with three degrees of freedom is needed for the study of its motion. If the road is considered as a flat surface, the motion is planar. By using the inertial reference frame[9] XY shown in Fig. B.10, it is possible to use the coordinates X and Y of the centre of mass G of the vehicle and the yaw angle ψ between the x and X axes as generalized coordinates.

To keep the model as simple as possible, the sideslip angle of the vehicle β and of the wheels α_i are small. The yaw angular velocity $\dot{\psi}$ is also considered as a small quantity.

The equations of motion of the vehicle are

$$\begin{cases} m\ddot{X} = F_X \\ m\ddot{Y} = F_Y \\ J_z\ddot{\psi} = M_z, \end{cases} \tag{B.66}$$

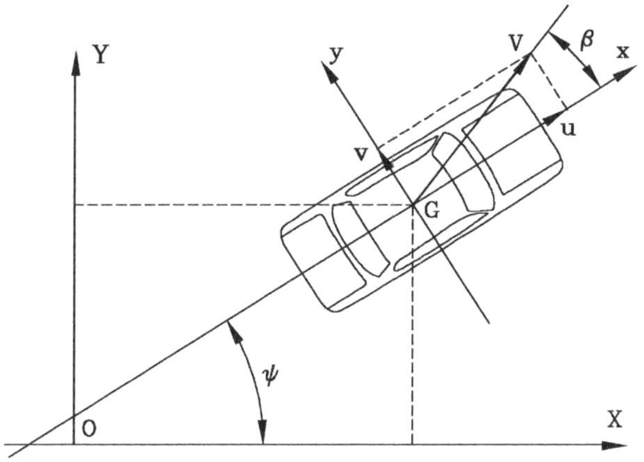

Fig. B.10 Reference frame for the study of the motion of a rigid vehicle. The vehicle has three degrees of freedom, and the coordinates X and Y of the centre of mass G and the yaw angle ψ can be used as generalized coordinates

[9] As already stated, such a reference frame is not, strictly speaking, inertial, since it is fixed to the road surface and hence follows the motion of Earth. It is, however, inertial "enough" for the problems here studied, and this issue will not be dealt with any further.

where F_X, F_Y and M_z are the total forces acting in the X and Y directions and the total yaw moment. Equation (B.66) are very simple but include the forces acting on the vehicle in the direction of the axes of the inertial frame. They are clearly linked with the forces acting in the directions of axes x and y of the vehicle by the obvious relationship

$$\begin{Bmatrix} F_X \\ F_Y \end{Bmatrix} = \begin{bmatrix} \cos(\psi) & -\sin(\psi) \\ \sin(\psi) & \cos(\psi) \end{bmatrix} \begin{Bmatrix} F_x \\ F_y \end{Bmatrix}. \tag{B.67}$$

If the model is used to perform a numerical integration in time, they can be used directly without any difficulty.

However, if the model has to be used to obtain linearized equations to gain a general insight into the behavior of the vehicle, it is better to write the equations of motion with reference to the non-inertial xy frame, to avoid dealing with the trigonometric functions of angle ψ, which in general is not a small angle, and would make linearizations impossible.

To write the equations of motion with reference to the body-fixed frame xyz , it is expedient to use the components u and v of the speed in the directions of the x and y axes and the yaw angular velocity

$$r = \dot{\psi}.$$

There are many ways to obtain the mathematical model, but perhaps the simplest is to remember that the derivative with respect to time of a generic vector $\vec{A}$, expressed in the body-fixed frame, but performed in the inertial frame

$$\left.\frac{d\vec{A}}{dt}\right|_i,$$

can be expressed starting from the derivative performed in the body fixed frame

$$\left.\frac{d\vec{A}}{dt}\right|_m$$

as

$$\left.\frac{d\vec{A}}{dt}\right|_i = \left.\frac{d\vec{A}}{dt}\right|_m + \vec{\Omega} \wedge \vec{A}, \tag{B.68}$$

where $\vec{\Omega}$ is the absolute angular velocity of the body-fixed frame.

Remembering that the velocity V is contained in the xy plane,

$$\left.\frac{d\vec{V}}{dt}\right|_m + \vec{\Omega} \wedge \vec{V} = \begin{Bmatrix} \dot{u} - rv \\ \dot{v} + ru \\ 0 \end{Bmatrix}, \tag{B.69}$$

the equations of motion of the vehicle, expressed with reference to the xyz frame, are thus

$$\begin{cases} m(\dot{u} - rv) = F_x \\ m(\dot{v} + ru) = F_y \\ J_z\dot{r} = M_z. \end{cases} \tag{B.70}$$

Equation (B.70) are nonlinear in the velocities u, v and r but, since the sideslip angle β is small and its trigonometric functions can be linearized, the linearization of the equations is possible. The components of velocity V can be written as

$$\begin{cases} u = V\cos(\beta) \approx V \\ v = V\sin(\beta) \approx V\beta. \end{cases} \tag{B.71}$$

Product $\dot{\psi}v$ can be considered the product of two small quantities and it is thus of the same order as the first term ignored in the series for the cosine. It is therefore cancelled.

The speed V can be considered as a known function of time, which amounts to studying the motion with a given law $V(t)$ (in many cases at constant speed) and assuming as an unknown the driving or the braking force needed to follow such a law. The unknown for the degree of freedom related to translation along the x-axis in this case is the force F_{x_d} exerted by the driving wheels. When braking, force F_x is the total braking force exerted by all wheels.

Equation (B.70) reduces to the linear form in F_x, v and r:

$$\begin{cases} m\dot{V} = F_x \\ m\left(\dot{v} + rV\right) = F_y\ . \\ J_z\dot{r} = M_z \end{cases} \tag{B.72}$$

If the interaction between longitudinal and transversal forces due to the tires is neglected or accounted for in an approximate way, the first equation of motion, which has already been studied in the section dealing with the longitudinal performance of the vehicle, uncouples from the other two.

This amounts to saying that the lateral behavior is uncoupled from the longitudinal behavior and can be studied using just two variables, either velocities v and r:

$$\begin{cases} m\left(\dot{v} + rV\right) = F_y \\ J_z\dot{r} = M_z \end{cases}, \tag{B.73}$$

or β and r if the equations are written in the equivalent form

$$\begin{cases} mV\left(\dot{\beta} + r\right) + m\beta\dot{V} = F_y \\ J_z\dot{r} = M_z \end{cases}. \tag{B.74}$$

The sideslip angles of the wheels may be expressed easily in terms of the generalized velocities. With simple computations it follows that

$$\alpha_i = \arctan\left(\frac{v + \dot{\psi} x_i}{u - \dot{\psi} y_i}\right) - \delta_i \tag{B.75}$$

where x_i and y_i are the coordinates of the center of the contact area of the ith wheel in the reference frame of the vehicle.

Equation (B.75) can be easily linearized. By noting that $y_i \dot{\psi}$ is far smaller than the speed V, it follows that

$$\alpha_i = \beta_i - \delta_i \approx \frac{v + r x_i}{V} - \delta_i = \beta + \frac{x_i}{V} r - \delta_i . \tag{B.76}$$

Coordinate y_i of the centre of the contact area of the wheel does not appear in the expression for the sideslip angle α_i. If the differences between the steering angles δ_i of the wheels of the same axle are neglected, the values of their sideslip angles are equal. This allows one to work in terms of axles instead of single wheels and to substitute a model of the type of Fig. B.6b for that of Fig. B.6a, i.e. to use a monotrack or bicycle model.

The explicit expressions of the sideslip angles of the front and rear axles of a vehicle with two axles are thus

$$\begin{cases} \alpha_1 = \beta + \dfrac{a}{V} r - \delta_1 \\ \\ \alpha_2 = \beta - \dfrac{b}{V} r - \delta_2 \end{cases} . \tag{B.77}$$

In the majority of cases only the front axle can steer and $\delta_2 = 0$.

The total lateral force is a function of the sideslip angle β and the yaw velocity r, which can be considered as the variables of the motion, and the steering angle δ, which can be considered as the control variable. By developing in Taylor series the force $F_y(\beta, r, \delta)$ about the condition $\beta = r = \delta = 0$, and truncating it after the linear term, it follows that

$$F_y = Y_\beta \beta + Y_r r + Y_\delta \delta + F_{y_e}, \tag{B.78}$$

where coefficients Y_β, Y_r and Y_δ are the derivatives of the force with respect to the three variables β, r and δ and may be obtained in any way, even experimentally, if possible. They are called the derivatives of stability of the vehicle.

In the same way, the linearized expression for the yawing moments is

$$M_z = N_\beta \beta + N_r r + N_\delta \delta + M_{z_e}, \tag{B.79}$$

where N_β, N_r and N_δ are other derivatives of stability of the vehicle. F_{y_e} and M_{z_e} are external forces and moments acting on the vehicle and can be considered as disturbances.

Equation (B.74) can be written directly as a state equation

$$\dot{\mathbf{z}} = \mathbf{A}\mathbf{z} + \mathbf{B}_c\mathbf{u}_c + \mathbf{B}_e\mathbf{u}_e, \tag{B.80}$$

where the state and input vectors $\mathbf{z}$, $\mathbf{u}_c$ and $\mathbf{u}_e$ are

$$\mathbf{z} = \begin{Bmatrix} \beta \\ r \end{Bmatrix}, \qquad \mathbf{u}_c = \delta, \qquad \mathbf{u}_e = \begin{Bmatrix} F_{y_e} \\ M_{z_e} \end{Bmatrix},$$

the dynamic matrix is

$$\mathbf{A} = \begin{bmatrix} \dfrac{Y_\beta}{mV} - \dfrac{\dot{V}}{V} & \dfrac{Y_r}{mV} - 1 \\ \dfrac{N_\beta}{J_z} & \dfrac{N_r}{J_z} \end{bmatrix}$$

and the input gain matrices are

$$\mathbf{B}_c = \begin{bmatrix} \dfrac{Y_\delta}{mV} \\ \dfrac{N_\delta}{J_z} \end{bmatrix}, \quad \mathbf{B}_e = \begin{bmatrix} \dfrac{1}{mV} & 0 \\ 0 & \dfrac{1}{J_z} \end{bmatrix}.$$

The study of the system is straightforward: The eigenvalues of the dynamic matrix allow one to see immediately whether the behavior is stable or not, and the study of the solution to given constant inputs yields the steady state response to a steering input or to external forces and moments.

B.5 Vertical Dynamics (Comfort)

In all motor vehicles, with the exception of a few slow industrial vehicles, the wheels are connected to the chassis through elements provided with elasticity and damping: the suspensions. Their aim is to:

- Provide that the forces exchanged by the wheels with the ground comply with design specifications in every load condition;
- determine the vehicle *trim* under the action of static and quasi static forces;
- absorb and smooth down shocks that are received by the wheels from road irregularities and transmitted to the body.

Owing to the presence of the suspensions, the vehicle body, that is usually assumed to be a rigid body, can move in a direction perpendicular to the ground (z-axis in Fig. B.1, heave motion) and can rotate about x -axis (roll motion) and about y-axis (pitch motion). The vehicle is thus assumed to be composed by a *sprung mass*, coinciding with the body, plus a number of *unsprung masses*, each containing a wheel and some mechanical components.

If the assumptions of symmetry and linearity mentioned in Sect. B.1 hold, the heave and pitch motions are coupled with each other, while roll motion uncouples or, better, roll motion couples with the handling behavior of the vehicle.

B.5.1 Heave and Pitch Motion

In the study of the low frequency motion of the sprung mass, the tires can be considered as rigid bodies, i.e. the unsprung masses can be assumed as rigidly connected with the ground. The simplest model for studying the heave-pitch coupling is shown in Fig. B.11a. Its equation of motion is that of a beam on two elastic and damped supports

$$\begin{bmatrix} m_S & 0 \\ 0 & J_y \end{bmatrix} \begin{Bmatrix} \ddot{Z}_s \\ \ddot{\theta} \end{Bmatrix} + \begin{bmatrix} c_1 + c_2 & -ac_1 + bc_2 \\ -ac_1 + bc_2 & a^2 c_1 + b^2 c_2 \end{bmatrix} \begin{Bmatrix} \dot{Z}_s \\ \dot{\theta} \end{Bmatrix} +$$
$$+ \begin{bmatrix} K_1 + K_2 & -aK_1 + bK_2 \\ -aK_1 + bK_2 & a^2 K_1 + b^2 K_2 \end{bmatrix} \begin{Bmatrix} Z_s \\ \theta \end{Bmatrix} = \tag{B.81}$$
$$= \begin{Bmatrix} c_1 \dot{h}_A + c_2 \dot{h}_B + K_1 h_A + K_2 h_B \\ -ac_1 \dot{h}_A + bc_2 \dot{h}_B - aK_1 h_A + bK_2 h_B \end{Bmatrix}$$

where h_A and h_B are the z coordinate of the centers of the wheels, that under the assumption of rigid tires coincide, except for a constant, with the z coordinate of the wheel-ground contact points. They thus define the ground profile, or better, the average between the ground profiles under the right and left wheels.

The overturning moment due to weight (term $-m_S gh$ to be added in position 22 in the stiffness matrix) is not included in Eq. (B.81), because no assumption has been made on the height of the pitch center over the road plane, nor is any aerodynamic term introduced into the equation of motion. The longitudinal position of the springs and the shock absorbers has been assumed to be the same.

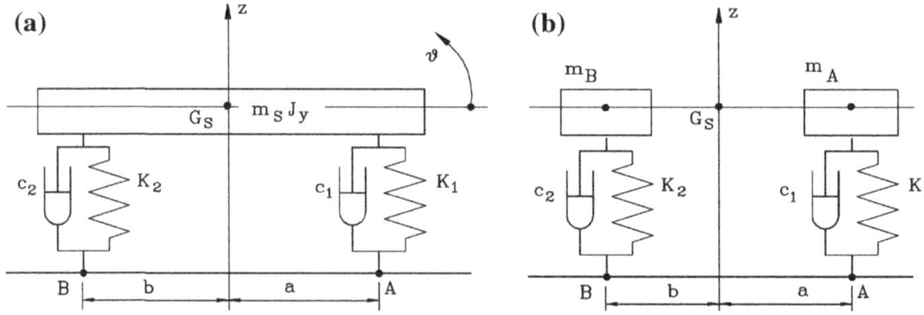

Fig. B.11 Beam models for heave and pitch motions

The forcing functions were written in a form that considers only the vertical motion of points A and B, neglecting horizontal forces at the ground-wheels interface and the possible coupling between vertical and horizontal motions due to suspensions.

If mass m_s and moment of inertia J_y are those of the whole sprung mass, the stiffnesses K_i and the damping coefficients c_i are those of a whole axle and are thus twice those of a single spring or shock absorber.

A *dynamic index* I_d of the sprung mass can be defined as

$$I_d = \frac{J_y}{m_s ab} = \frac{r_y^2}{ab}, \tag{B.82}$$

where r_y is the radius of gyration of the sprung mass. If I_d is equal to unity, i.e. if

$$J_y = m_S ab, \quad r_y^2 = ab,$$

the model of Fig. B.11a reduces to that of Fig. B.11b, i.e. the motion of the sprung mass reduce to the motion of two masses

$$m_A = m_S \frac{b}{l} \text{ and } m_B = m_S \frac{a}{l},$$

suspended on the front and rear suspensions.

This condition is usually not verified in practice. The tendency to increase the wheelbase for stability reasons leads to values of the dynamic index usually smaller than 1, even smaller than 0.8.

The natural frequencies of the undamped system may be computed using the homogeneous equation associated with Eq. (B.81), after cancelling the damping term. They are thus the roots of the characteristic equation

$$\omega^4 - \omega^2 \frac{K_1(r_y^2 + a^2) + K_2(r_y^2 + b^2)}{m_S r_y^2} + K_1 K_2 \frac{l^2}{m_S^2 r_y^2} = 0, \tag{B.83}$$

that yields

$$\omega_i = \frac{\sqrt{\left(b^2 + r_y^2\right) K_2 + \left(a^2 + r_y^2\right) K_1 \pm \Delta}}{r_y \sqrt{2m_S}}, \tag{B.84}$$

where

$$\Delta = \sqrt{\left(b^2 + r_y^2\right)^2 K_2{}^2 + 2K_1 K_2 \left[\left(ab - r_y^2\right)^2 - r_y^2 l^2\right] + \left(a^2 + r_y^2\right)^2 K_1{}^2}.$$

The solution with the (+) sign yields two positive values, that generally are not equal to each other. The motion of the beam is neither a rotation about its center of

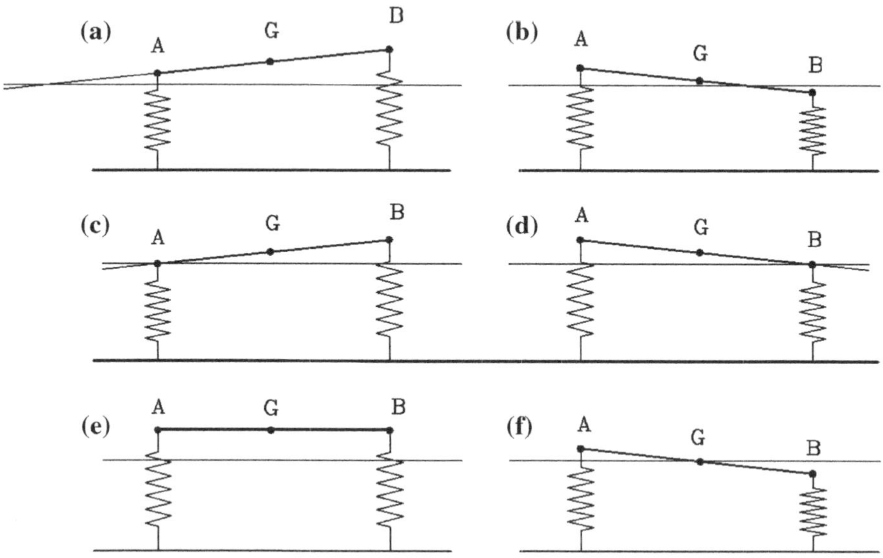

Fig. B.12 Heave and pitching motions. **a** and **b**: General case; primarily heave (**a**) and primarily pitching (**b**) motions. **c** and **d**: Case with $I_d = 1$. **e** and **f**: Case with $aK_1 = bK_2$.

mass (pitch) nor a translational motion in the z direction (heave), but the very fact that the displacements at the front and rear axles have the same sign means that the node (the point with zero displacement) lies outside of the wheelbase, and thus the motion is primarily translational (heave, Fig. B.12a). The solution with the $(-)$ sign (Fig. B.12b) yields a positive and a negative value: the displacements of the front and rear axles are one positive and one negative and the node is within the wheelbase. The motion is primarily rotational, even if not about the center of mass, and it is primarily a pitching motion.

If the dynamic index I_d has a unit value (Fig. B.12c, d), it follows that

$$\omega_1 = \sqrt{\frac{lK_2}{am_s}}, \quad \omega_2 = \sqrt{\frac{lK_1}{bm_s}}. \tag{B.85}$$

As previously stated, the natural frequencies in this case are those of the two separate masses of Fig. B.11b and the free oscillations of the sprung mass are rotations about the points where the suspensions are connected to the body. It is impossible to identify a heave and a pitch mode; it is more accurate to speak about a front-axle and a rear-axle mode.

The other limiting case (Fig. B.12e, f) is when

$$aK_1 = bK_2. \tag{B.86}$$

The heave motion uncouples from the pitch motion. The first is translational, while the second is rotational and occurs about the center of mass. The natural frequencies are thus

$$\begin{cases} \omega_1 = \sqrt{\dfrac{lK_1}{bm_S}} & \text{heave,} \\ \omega_2 = \sqrt{\dfrac{laK_1}{r_y^2 m_S}} = \omega_1 \sqrt{\dfrac{ab}{r_y^2}} & \text{pitch.} \end{cases} \tag{B.87}$$

The two limiting cases may also occur simultaneously. From Eq. (B.87) it follows that when this is the case the two natural frequencies have the same value. This solves an apparent inconsistency; if

$$aK_1 = bK_2,$$

the centers of rotation are one in the centre of mass (pitch mode) and one at infinity (heave mode), while when the dynamic index has a unit value they are at the ends of the wheelbase. When both conditions occur simultaneously, the two natural frequencies coincide; in this case, any linear combination of the eigenvectors is itself an eigenvector. Thus in the case of a rigid beam, any point of the beam (or better, of the straight line constituting the beam axis) may be considered as a center of rotation.

In the study of high frequency motions, the compliance of tires cannot be neglected, and the model must contain also the unsprung masses. The minimum number of degrees of freedom needed to study heave and pitch motions is thus four.

B.5.2 Quarter-Car Models

The simplest model for studying the suspension motions of a vehicle is the so-called *quarter-car model* , including a single wheel with the related suspension and the part of the body whose weight is imposed on it. Often, in four-wheeled vehicles, the quarter vehicle including the suspension and the wheels is called a *corner* of the vehicle. The quarter-car model may be more or less complex, including not only the compliance of the suspension, but also the compliance of the tire and that of the rubber mounts connecting the frame carrying the suspension to the body. It may even include the inertia of the tire.

Three models based on the quarter-car approach are shown in Fig. B.13.

The first model has a single degree of freedom. The tires are considered rigid bodies and the only mass considered is the sprung mass. This model holds well for motions taking place at low frequency, in the range of the natural frequency of the sprung mass (in most cases, up to 3–5 Hz, in the range defined as *ride* by SAE).

The second model has two degrees of freedom. The tire is considered as a massless spring, and both the unsprung and the sprung masses are considered. This model holds well for frequencies up to the natural frequency of the unsprung mass and slightly

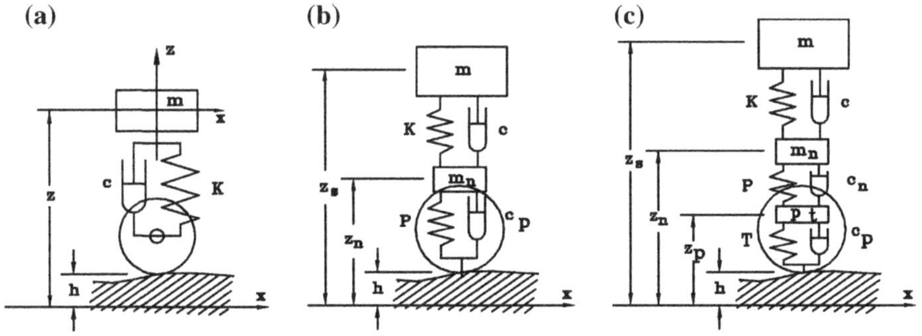

Fig. B.13 Quarter-car models with one (**a**), two (**b**) and three (**c**) degrees of freedom

over (in most cases, up to 30–50 Hz, including the ranges defined as *ride*, *shake* and partially *harshness*).

The third model has three degrees of freedom. The tire is modelled as a spring-mass-damper system, representing its dynamic characteristics in the lowest mode. This model allows us to study motions taking place at frequencies in excess of the first natural frequency of the tires (up to 120 – 150 Hz, including *harshness*).

If higher frequencies must be accounted for, it is possible to introduce a higher number of tire modes by inserting other masses. These models, essentially based on the modal analysis of the suspension-tire system, are clearly approximated because a tire can be considered as a damped system and one that is usually nonlinear.

Consider the simplest quarter-car model shown in Fig. B.13a. It is a simple mass-spring-damper system that, among other things, has been used in the past to demonstrate that the shock absorber must be a linear, symmetrical viscous damper.[10]

The equation of motion of the system is simple. Using the symbols shown in the figure it is

$$m\ddot{z} + c\dot{z} + Kz = c\dot{h} + Kh, \tag{B.88}$$

where $z(t)$ is the displacement from the static equilibrium position, referred to an inertial frame, and $h(t)$ is the vertical displacement of the supporting point due to road irregularities.[11]

The frequency response of the quarter-car can be obtained simply by stating a harmonic input of the type

$$h = h_0 e^{i\omega t}.$$

The output is itself harmonic and can be expressed as

---

[10] Bourcier De Carbon C.: *Théorie mathématiques et ré alisation pratique de la suspension amortie des vehicules terrestres*, Proceedings SIA Conference, Paris, 1950.

[11] The z coordinate must be considered as the displacement from the static equilibrium position. By doing this, the static problem of finding the equilibrium position is separated from the dynamic problem here dealt with. This can only be done because of the linearity of the model.

$$z = z_0 e^{i\omega t},$$

where both amplitudes h_0 and z_0 are complex numbers to account for the different phasing of response and excitation due to damping.

By introducing the time histories of the forcing function and of the response into the equation of motion, an algebraic equation is obtained:

$$\left(-\omega^2 m + i\omega c + K\right) z_0 = (i\omega c + K)\, h_0. \tag{B.89}$$

It links the amplitude of the response to that of the excitation, and yields

$$z_0 = h_0 \frac{i\omega c + K}{-\omega^2 m + i\omega c + K}. \tag{B.90}$$

If h_0 is real (that is, if the equation is written in phase with the excitation), the real part of the response (the in-phase component of the response) can be separated easily from its imaginary part (in-quadrature component)

$$\begin{cases} \dfrac{Re\,(z_0)}{h_0} = \dfrac{K\left(K - m\omega^2\right) + c^2\omega^2}{\left(K - m\omega^2\right)^2 + c^2\omega^2} \\ \dfrac{Im\,(z_0)}{h_0} = \dfrac{-cm\omega^3}{\left(K - m\omega^2\right)^2 + c^2\omega^2}. \end{cases} \tag{B.91}$$

The amplification factor, i.e., the ratio between the absolute values of the amplitudes of the response and the excitation and the phase of the first with respect to the second, can be easily shown to be (Figs. B.14a, c)

$$\begin{cases} \dfrac{|z_0|}{|h_0|} = \sqrt{\dfrac{K^2 + c^2\omega^2}{\left(K - m\omega^2\right)^2 + c^2\omega^2}} \\ \Phi = \arctan\left(\dfrac{-cm\omega^3}{K\left(K - m\omega^2\right) + c^2\omega^2}\right). \end{cases} \tag{B.92}$$

More than the frequency response expressing the ratio between the amplitudes of response and excitation, what matters in motor vehicle suspensions is the ratio between the amplitudes of the acceleration of the sprung mass and that of the displacement of the supporting point. Because in harmonic motion the amplitude of the acceleration is equal to the amplitude of the displacement multiplied by the square of the frequency, it follows that:

$$\frac{|(\ddot{z})_0|}{|h_0|} = \omega^2 \frac{|z_0|}{|h_0|}.$$

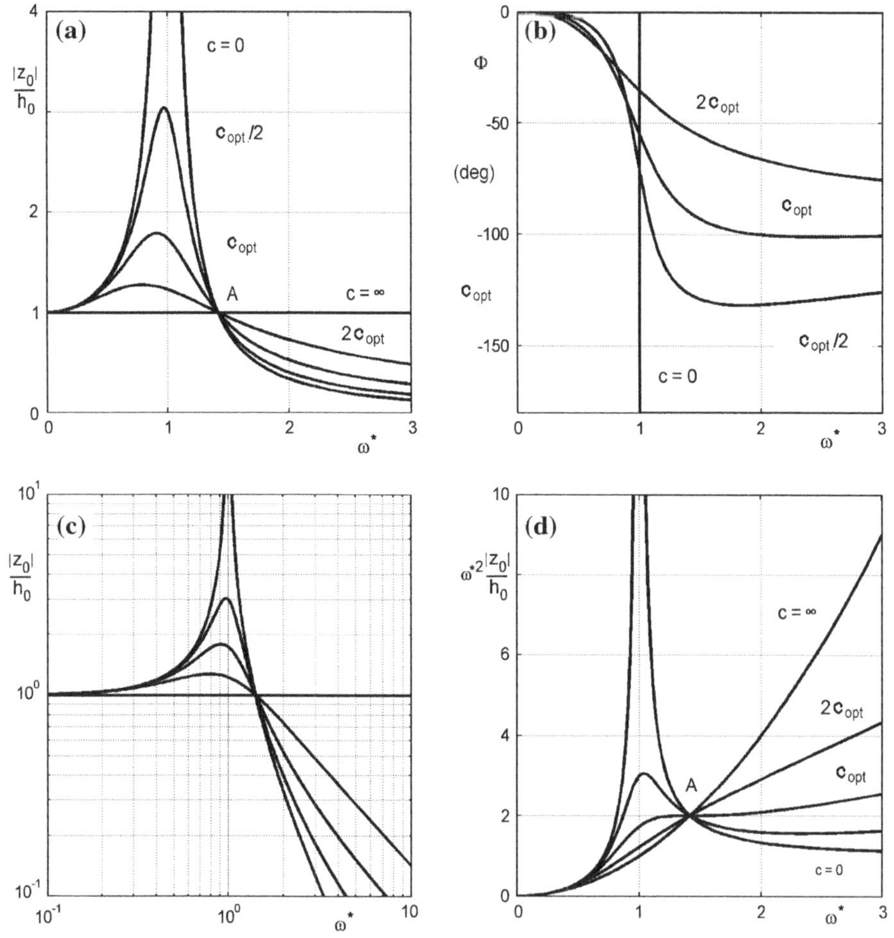

Fig. B.14 Quarter-car with a single degree of freedom, response to harmonic excitation. **a** and **b** Ratio between the amplitude of the response (in terms of displacement, linear scale) and the amplitude of the ground displacement, and phase as functions of the non-dimensional frequency $\omega^* = \omega\sqrt{m/K}$. **c** Same as **a**, but plotted in a logarithmic scale. **d** Same as **a**, but for the acceleration of the sprung mass

Both frequency responses are plotted in Fig. B.14 for different values of the damping of the shock absorber, together with the phase Φ . The responses are plotted as functions of the nondimensional frequency

$$\omega^* = \omega\sqrt{\frac{m}{K}}$$

and damping is expressed as a function of the optimum damping defined below.

All curves pass through point A, located at a frequency equal to $\sqrt{2K/m}$. Because the acceleration of the sprung mass must be kept to a minimum to produce a comfortable ride, a reasonable way to optimize the suspension is to choose a value of shock absorber damping that leads to a relative maximum, or at least a stationary point, at point A on the curve related to acceleration. By differentiating the expression of

$$\omega^2 \frac{|z_0|}{|h_0|}$$

with respect to ω and equating the derivative to zero at point A, the following value of the optimum damping is obtained:

$$c_{opt} = \sqrt{\frac{Km}{2}} = \frac{c_{cr}}{2\sqrt{2}}, \tag{B.93}$$

where

$$c_{cr} = 2\sqrt{Km}$$

is the critical damping of the suspension .

Although this method for optimizing the suspension can be readily criticized, because the comfort of a suspension is far more complex than simple reduction of the vertical acceleration (the so-called "jerk", i.e., the derivative of the acceleration with respect to time d^3z/dt^3 also plays an important role), it nonetheless gives important indications.

The dynamic component of the force the tire exerts on the ground is

$$F_z = c\left(\dot{z} - \dot{h}\right) + K(z - h) = -m\ddot{z}. \tag{B.94}$$

Minimizing the vertical acceleration leads to minimizing the dynamic component of the vertical load on the tire, which has a negative influence on the ability of the tire to exert cornering forces. The condition leading to optimum comfort seems, thus, to coincide with that leading to optimum handling performance. Equation (B.93) allows one to choose the value of the damping coefficient c. For the value of the stiffness K there is no such optimization: To minimize both the acceleration and the dynamic component of the force, K should be kept as low as possible. The only limit to the softness of the springs is the space available.

This reasoning has, however, the following limitation: The softer the springs, the larger the oscillations of the sprung mass. Large displacements must be avoided because they may cause large errors in the working angles of the tire, causing the tires to work in conditions that may be far from optimal.

An empirical rule states that soft suspensions improve comfort while hard suspensions improve handling. This is even more true if aerodynamic devices are used to produce negative lift: suspensions allowing a large degree of travel cause major changes of vehicle position with respect to the airflow, producing changes of the aerodynamic lift that are detrimental.

Moreover, at a fixed value of the sprung mass, the lower the stiffness of the spring, the lower the natural frequency. Very soft suspensions easily lead to natural frequencies of about 1 Hz or even less, which may cause motion sickness and a reduction in comfort that varies from person to person. Cars with soft suspensions with large travel, typical of some manufacturers, are popular with some customers but considered uncomfortable by others.

The optimum value of the damping expressed by Eq. (B.93) is lower than the critical damping. The quarter-car is then underdamped and may undergo free oscillations, even if these generally damp out quickly because the damping ratio $\zeta = c/c_{cr}$ is not very low:

$$\frac{c_{opt}}{c_{cr}} = \frac{1}{2\sqrt{2}} \approx 0.354. \tag{B.95}$$

As for the previous case, when studying higher frequency motions, the compliance of the tires must be accounted for and more complex quarter car models must be used.

B.5.3 Roll Motion

As already stated, roll is coupled with handling and not with ride comfort. However it is also true that rolling can affect strongly the subjective feeling of riding comfort.

The simplest model for studying roll motion is made by a rigid body, simulating the sprung mass, free to rotate about the roll axis, constrained to the ground by a set of springs and dampers with a stiffness and a damping coefficient equal to those of the suspensions (Fig. B.15). If J_x, m_s, χ_i and Γ_i are respectively the moment of inertia about the roll axis, the sprung mass, the torsional stiffness and the damping coefficient of the ith suspension, the equation of motion is

$$J_x\ddot{\phi} + (\Gamma_1 + \Gamma_2)\,\dot{\phi} + (\chi_1 + \chi_2)\,\phi - m_s g h_G \sin(\phi) = \tag{B.96}$$

$$= \Gamma_1\dot{\alpha}_{t_1} + \Gamma_2\dot{\alpha}_{t_2} + \chi_1\alpha_{t_1} + \chi_2\alpha_{t_2},$$

where the forcing functions are those due to the transversal inclination of the road α_{t_i} at the ith suspension.

The inertia of the unsprung masses and the compliance of the tires are not included in such a simple model, which is formally identical to the quarter-car with a single degree of freedom.

For small values of the roll angle, the model may be linearized, stating $\sin(\phi) \approx \phi$. The roll natural frequency is thus

$$\omega_{roll} = \sqrt{\frac{\chi_1 + \chi_2 - m_s g h_G}{J_x}}. \tag{B.97}$$

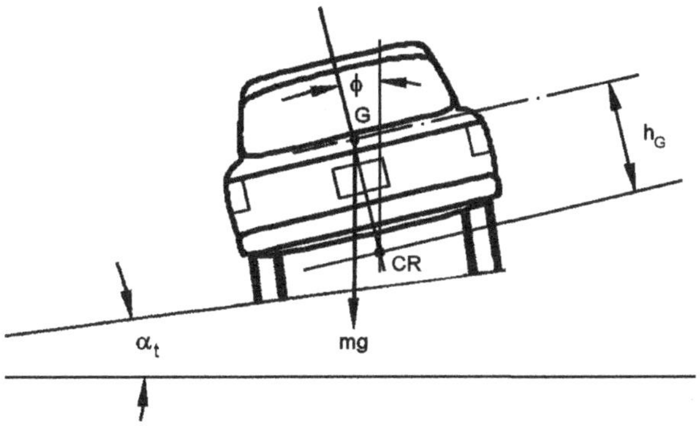

Fig. B.15 Model with a single degree of freedom for the study of roll motion. Cross section in a plane containing the center of mass G of the sprung mass. The roll axis goes through point CR

The optimum damping value may be obtained from Eq. (B.93):

$$\Gamma_{opt} = \sqrt{\frac{J_x\left(\chi_1 + \chi_2 - m_s g h_G\right)}{2}}. \tag{B.98}$$

This condition is generally not satisfied, particularly if the vehicle has anti-roll bars. The torsional damping of the suspensions is supplied by the same shock absorbers normally designed to optimize vertical motion; the roll damping they supply is usually lower than needed. The increase in stiffness due to anti-roll bars is not accompanied by an increase in damping, which causes a decrease of the damping ratio, together with an increase of the natural frequency.

The stiffer the suspension in torsion, the more underdamped the roll behavior, if the increase in stiffness is due to anti-roll bars. Although reducing rolling in stationary conditions, they may increase it in dynamic conditions. An overelongation in the step response, as when roll is due to a moment abruptly applied (steering step input, wind gusts or other similar instances), may then result. A large roll in dynamic conditions may cause rollover.

The stationary value of the roll angle on a road with transversal slope $\alpha_t = \alpha_{t_1} = \alpha_{t_2}$ is

$$\phi = \alpha_t \frac{\chi_1 + \chi_2}{\chi_1 + \chi_2 - m_s g h_G}. \tag{B.99}$$

The importance of a center of mass not too high on the roll axis and stiff suspensions (in roll) is thus clear. The last condition may contradict the need for a small roll in dynamic conditions.

B.6 Coupled Dynamics

When either the above mentioned assumptions do not hold, or the analyst wants to reach a higher precision, the whole vehicle must be modelled as a single system. A vehicle on elastic suspensions is usually modelled as a system made by a certain number of rigid bodies connected with each other by mechanisms of various kinds and by a set of massless springs and dampers simulating the suspensions. A vehicle with four wheels can be modelled as a system with 10 degrees of freedom, six for the body and one for each wheel. This holds for any type of suspension, if the motion of the wheels due to the compliance of the system constraining the motion of the suspensions (longitudinal and transversal compliance of the suspensions) is neglected. Additional degrees of freedom, such as the rotation of the wheels about their axis or about the kingpin, can be inserted into the model to allow the longitudinal slip or the compliance of the steering system to be taken into account.

The multibody approach can be pushed much further , by modelling, for instance, each of the links of the suspensions as a rigid body. To model a double wishbone suspension it is possible to resort to three rigid bodies, simulating the lower and upper triangles and the strut, plus a further rigid body simulating the steering bar. While modelling the system in greater detail, the number of rigid bodies included in the model increases. However, if the compliance of the various elements is neglected (i.e. if these bodies are rigid bodies), the number of degrees of freedom does not increase along with the number of bodies: a double wishbone suspension always has a single degree of freedom, even if it is made up of a number of rigid bodies simulating its various elements.

The mathematical model of a multibody system is thus made up of the equations of motion of the various elements, which in tridimensional space are $6n$, if n is the number of the rigid bodies, plus a suitable number of constraint equations.

Consider for instance an articulated truck with 6 solid axles (3 for the tractor plus 3 for the trailer). The rigid bodies are 8 (tractor, trailer and 6 rigid axles) and thus the equations of motion are 48 second-order differential equations. The constraint equations are 27: 3 equations for the constraint between tractor and trailer (these state that the coordinates of the center of the hitch, assumed to be a spherical hinge, are the same whether this point is seen as belonging to the tractor or to the trailer) and 4 equations for each axle, leaving to each one of them just two, out of its 6 degrees of freedom. The 27 constraint equations are algebraic equations containing only the generalized coordinates but not the velocities (holonomic constraints).

By using the 27 constraint equations to eliminate 27 of the generalized coordinates, a set of 21 equations in the 21 independent generalized coordinates is obtained. It must be emphasized that in the 48 equations of motion originally written, the forces the various bodies exchange at the constraints are included; these forces are then eliminated when the constraint equations are introduced. The 27 equations so eliminated can be used to compute the constraint forces.

This approach is the most general, and is usually implemented in *general purpose* multibody computer codes.

It is possible to resort to a simpler approach in the case of the multibody models used in motor vehicle dynamics, because the internal constraints are holonomic and the system is branched. One of the bodies (the vehicle's body) may be chosen as *main body*, to which a number of secondary, *first level* bodies are attached. Other secondary bodies, considered as *second level* bodies, are then attached, and so on. Secondary bodies have only the degrees of freedom allowed by the constraints. In this way, the minimum number of equations needed for the study are directly obtained. Such equations are all differential equations, usually of the second order. The forces exchanged at the constraints between the bodies do not appear explicitly in the equations and need not be computed in the dynamic study of the system as a whole. A model of the mentioned articulated truck with 6 solid axles can be obtained in this way: the tractor is considered as the main body, the axles of the tractor and the trailer are first level secondary bodies, and the axles of the trailer are second level secondary bodies.

The first approach is that followed by many commercial codes, like ADAMS, whose specialized version ADAMS-CAR is specifically adapted to automotive simulation. Some specialized commercial codes designed for motor vehicle simulation, like CarSim, are based on the second approach.

B.7 Symbols

a	Acceleration distance between centre of mass and front axle
b	Distance between centre of mass and rear axle
c	Viscous damping coefficient
c_{cr}	Critical damping
c_{opt}	Optimal damping
e	Energy
f	Rolling coefficient
f_0	Rolling coefficient at zero speed
g	Gravitational acceleration
i	Grade of the road
l	Wheelbase
m	Mass
m_e	Equivalent mass
q	Specific fuel consumption
r	Yaw velocity
t	Time; track
u, v, w	Velocities in the xyz frame
xyz	Body-fixed reference frame
$\mathbf{z}$	State vector
$\mathbf{A}$	Dynamic matrix in the state space
$\mathbf{B}$	Input gain matrix
C	Cornering stiffness
C_x, C_z	Aerodynamic coefficients
F	force
H	Thermal value of fuel
I_d	Dynamic index
J	Moment of inertia
K	Stiffness, stability factor, coefficient for the rolling resistance
P	Power
R	Radius of the wheel (unloaded); radius of the trajectory; resistance to motion
R_e	Effective rolling radius
S	Reference surface
$\mathcal{T}$	Kinetic energy
V	Vehicle speed
XYZ	Inertial frame
α	Sideslip angle; grade angle of the road
α_t	Transversal grade angle of the road
β	Sideslip angle of the vehicle
γ	Camber angle
δ	Steering angle
η	Efficiency

θ Pitch angle
μ_x Longitudinal force coefficient
μ_y Cornering force coefficient
ρ Density
τ Transmission ratio
ϕ Roll angle
χ Torsional stiffness
ψ Yaw angle
ω Frequency
ω_n Natural frequency
Γ Rotational damping coefficient
Φ Phase
Ω Angular velocity

References

—, *Diesel Engine Management, Robert Bosch GmbH* (Professional Engineering Publishing, Stuttgart, 2005)
—, Dinglers Politechnischer Journal, Band **310** (1898)
—, Dinglers Politechnischer Journal, Band **314** (1899)
—, Dinglers Politechnischer Journal, Band **315** (1900)
—, *Gasoline engine management, robert bosch GmbH* (Professional Engineering Publishing, Stuttgart, 2004)
—, *L'industria automobilistica mondiale 2000–2010*, (Anfia, Torino, 2010)
—, Progetto archivio storico FIAT, I primi quindici anni della FIAT, Franco Angeli (1987)
—, *The automobile, a century of progress* (SAE, Warrendale, 1997)
—, *The global automotive industry value chain* (UNIDO, Vienna, 2003)
D. Baker, B. Hammerle, R. Backes, *Urea infrastrucuture for control ov vehicular* NO_x *emissions* (Aachener Kolloquium Fahrzeug und Motorentechnik, Aachen, 2000)
P.L. Bassignana, *Storia fotografica dell'industria automobilistica italiana* (Bollati Boringhieri, Torino, 1998)
L. Baudry de Saunier, *L'automobile theorique e pratique*, 2 volumes (Omnia, Paris, 1900)
L. Baudry de Saunier, *L'allumage dans les moteurs a explosion* (Dunod, Paris, 1905)
C. Beatrice, C. Bertoli, M. Migliaccio, *Basic concepts of the homogeneous charge compression engines*, ATA Ingegneria dell'Autoveicolo, vol 55 (Torino, Turin, 2002)
L. Bernard, A. Ferrari, M. Ferrera, D. Micelli, Air Control and Direct Injection on Turbocharged Gasoline Engines, in *ATA Ingegneria dell'Autoveicolo*, vol 59 (2006)
G.P. Blair, *Design and simulation of four-stroke engines* (SAE International, Warrendale, 1999)
M. Boisseaux, *L'automobile méthodes de calcul* (Dunod, Paris, 1948)
E. Borchi, R. Macii, G. Ricci, I motori della scuola fiorentina, comitato manifestazioni 150° anniversario dell'invenzione del motore a scoppio (2003)
E. Borchi, R. Macii, G. Ricci, I padri del motore a scoppio, comitato manifestazioni 150° anniversario dell'invenzione del motore a scoppio (2003)
E.B. Butler, *Transmission gears* (Griffin, London, 1917)
C. Campbell, R. Bentley, *The sports car*, (Cambridge, 1954)
G. Canestrini, *L'automobile, il contributo italiano all'avvento e all'evoluzione dell'autoveicolo* (R.A.C.I., Kokomo, 1938)
F. Caputo, M. Martorelli, *Disegno e progettazione per la gestione industriale* (ESI, Napoli, 2003)
W.H. Crouse, *Automotive transmissions and power trains* (Mc Graw-Hill, New York, 1955)
C.L. Cummins, *Internal fire* (SAE, Warrendale, 1989)

G. Genta et al., *The Motor Car*, Mechanical Engineering Series,
DOI: 10.1007/978-94-007-8552-6,

W.F. Decker, *The story of the engine* (Scribner, London, 1932)
R. Devillers, P. Merces, *Le moteur a explosion* (Dunod, Paris, 1935)
V. Doria, A. Ferrari, M. Lucatello, U. Nasi, *The fiat uniair variable valve actuation system* (Variable valvetrain Conference, Essen, 2002)
C. Dugald, *The gas, petrol and oil engine* (Longmans, London, 1916)
E. Eckermann, *World history of the automobile* (SAE, Warrendale, 2001)
A. Eiser, Innovative technology for future powertrains, in *ATA Ingegneria dell'Autoveicolo*, vol 60 (2007)
R. Flieri, M. Klüting, *The third genaration of valvetrains: new fully variable valvetrains for throttle-free load control* (SAE Paper, Warrendale, 2000-01-1227)
G. Genta, L. Morello, *The automotive chassis*, 2 volumes (Springer, New York, 2009)
D. Giacosa, *Motori endotermici* (Hoepli, Milan, 1941)
D. Giacosa, *Progetti alla FIAT* (Automobilia, Torino, 1988)
B. Gille, Storia delle tecniche, Editori Riuniti (1985)
K.H. Glück, U. Göbel, H. Hahn, J. Höhne, R. Krebs, T. Kreuzer, E. Pott, *Die Abgasreinigung de FSI-Motoren von Volkswagen*, MTZ Motortechnische Zeitschrift (2000)
J. Gunn, *The practical design of motor cars* (Arnold, London, 1910)
J.B. Heywood, *Internal combustion engine fundamentals* (McGraw-Hill Book Company, New York, 1988)
A. Jamieson, E.S. Andrews, *Heat and heat engines* (Griffin and Co, Washington D.C., 1925)
R. Kaiser, K. Rusch, *Designing of SCR systems for reducing nitrogen oxide in diesel engines*, MTZ, Heft 12 (2007)
S. Kalpakjian, *Manufactring engineering and technology*, (Addison Wesley, Boston, 1989)
H. Klingenberg, D. Schürmann, *Unregulated motor vehicle exhaust gas components* (Volkswagen AG, Wolfsburg, 1989)
A. Levi Cases, *Atti del primo congresso del motore a scoppio* (Padova, 1927)
G. Lief, *The firestone story* (McGraw-Hill, New York, 1951)
J.C. Maroselli, *L'automobile et ses grands problèmes* (Larousse, Paris, 1958)
A. Mentasti, *Carrozzerie d'automobile* (Hoepli, Milano, 1929)
L. Morello, L. Rosti, G. Pia, A. Tonoli, *The automotive body*, 2 volumes (Springer, New York, 2011)
W. Neubecker, *Antique automobile body construction and restoration* (Post Publication, Arcadia, 1912)
J.P. Norbye, *Sports car suspension* (Sport Car Presse, New York, 1965)
G. Oliver, *Cars and coach building* (Sotheby Parke Bernet, London, 1981)
V.W. Page, *The modern gasoline automobile* (Hodder and Stoughton, London, 1917)
M. Peter, *Der Kraftwagen* (RC Schmidt, Berlin, 1937)
C. Riedl, *Konstruktion und berechnung moderner automobil- und kraftradmotoren* (Schmidt, Berlin, 1930)
G. Rogliatti, C. Valier, L. Giovanetti, *La frizione nel tempo* (Valeo, Torino, 1980)
C. Schmidt, *Practical Treatise on Automobiles* (The American Text-book, Philadelphia, 1909)
M. Schwaderlapp, Friction reduction: the mechanical contribution to fuel economy, in *ATA Ingegneria dell'Autoveicolo*, vol 61 (2008)
A. Seniga, *Il meccanismo di trasmissione negli automobili* (Biblioteca di automobilismo e aviazione, Milano, 1912)
M. Serruys, *La suspension et la direction des véhicules routiers* (Dunod, Paris, 1947)
F. Steinparzer, W. Mattes, P. Nefischer, T. Steinmayr, *Der Neue Vierzylinder-Dieselmotor von BMW*, MTZ, Heft 11 (2007)
J. Szybist, B. Bunting, Chemistry Impacs in Gasoline HCCI Engines, in *Literature Review for CRC Project AVFL-13* (Fuel Oak Ridge National Laboratory, 2006)
E. Tompkins, *The history of the pneumatic tyre*, Dunlop Archive Project (Eastland Press, London, 1981)

Index

G. Genta et al., *The Motor Car*, Mechanical Engineering Series,
DOI: 10.1007/978-94-007-8552-6, © Springer Science+Business Media Dordrecht 2014

The manufacturer's authorised representative in the EU is Springer Nature Customer Service Centre GmbH, Europaplatz 3, 69115 Heidelberg, Germany. If you have any concerns regarding our products, please contact ProductSafety@springernature.com

Printed and bound by CPI Group (UK) Ltd, Croydon, CR0 4YY
30/09/2026
02225444-0017